# Plant Genetics and Genomics: Crops and Models

**Series Editor:** Richard A. Jorgensen

For further volumes:
http://www.springer.com/series/7397

Kevin M. Folta · Susan E. Gardiner
Editors

# Genetics and Genomics of Rosaceae

Springer

*Editors*
Kevin M. Folta
Department of Horticultural Sciences
University of Florida
1301 Fifield Hall P.O.Box 110690
Gainesville FL 32611-0690
USA
kfolta@ifas.ufl.edu

Susan E. Gardiner
The New Zealand Institute for Plant and Food Research Limited
Palmerston North Research Centre
Private Bag 11600
Palmerston North 4442
New Zealand
SGardiner@hortresearch.co.nz

ISBN 978-0-387-77490-9 e-ISBN 978-0-387-77491-6
DOI 10.1007/978-0-387-77491-6

Library of Congress Control Number: 2008942718



Printed on acid-free paper

springer.com

# Preface

The Rosaceae Family is represented by approximately 3,000 species of diverse plants, primarily confined to temperate climates. The family has a rich variety of architectural forms and contains herbaceous, tree and shrub species. Many family members are readily recognizable because of their edible seasonal fruits that are prized for their unique flavors, colors and nutritious properties (e.g. apple, strawberry, raspberry, pear, cherry, plum, apricot, pear), as well as familiar ornamentals (e.g. roses) and nuts (e.g. almonds).

Today's rosaceous cultivars have been derived from centuries of careful selection and breeding, using a palette of some of evolution's most curious creations. The careful sculpting that has transformed the germplasm was not trivial, as several of the most coveted fruit species maintain complicated genomes—in some cases among the most complex of cultivated plants (e.g. strawberry). Other species in the family are represented by large perennial tree crops that exhibit substantial juvenility phases, posing a barrier to standard breeding and genetic analyses. Yet, today's superior cultivars feature robust growth, substantial yields and resistance to common biotic and abiotic stresses; traits fostered by human intervention. When the hurdles to efficient cultivation, breeding and selection are considered, the quality and quantity of rosaceous plant products derived from traditional breeding techniques is little short of amazing.

Surprisingly, the breeding practices that have yielded today's finest cultivars have remained almost unchanged for the last 50 years. While selection and cultivation practices have moved forward with only minor revisions, the world has changed significantly, posing new challenges to sustained and profitable, yet environmentally compatible production. Changes in public policy have limited access to water, power, labor and land. Effective fumigants and fungicides have fallen from favor and are now restricted from traditional uses. The pest and pathogen landscape continually changes, remodeled by variations in climate and spreading human populations. Meanwhile, consumer demand increases annually for flawless fruit with superior flavor and health attributes, and for perfect flowers with a long vase life—all the time at lower cost and in a short time-frame.

This means that new tools and strategies are needed to assist efficient development of future generations of innovative cultivars across the Rosaceae. Recent advances in genome-level analyses have proven a valuable route to deriving an understanding of the molecular events that govern plant responses in model systems. Applying these proven techniques to questions within the Rosaceae will make it possible to provide new tools to aid selection and enhance production. However, the Rosaceae research community overall has experienced a dearth of plant and molecular resources in the public domain, a trend that has only recently been reversed. Today the important crops of the Rosaceae are beginning to enjoy the same benefits once confined to model plant systems. It is possible to quickly assess linkage relationships between genes, test the relevance of a gene or genes to traits of interest, and develop molecular markers to assist traditional selection. Although these concepts have come a long way in the last decade, genomics-level studies in the Rosaceae still are in their infancy.

Unlike other genomics texts that feature the products of a decade of high-throughput molecular and genome-level analyses, our work in Rosaceae is much more the end of the beginning, rather than the beginning of the end. In the last two years significant EST datasets have been developed for *Malus* and *Prunus* (~260,000 and 85,000 ESTs respectively in public databases.) Genomics-level investigations into the structure of genomes, the physical associations between genes, and specific roles of given genes in physiological functions of importance are now emerging. The 3rd International Rosaceae Genomics Conference in New Zealand in March 2006 initiated broad scale interactions among researchers in the Rosaceae internationally. These were further developed further at the 4th International Rosaceae Genomics Conference in Chile in March 2008, with new research investigations crossing international lines. This reality makes the timing of this volume even more appropriate, as it marks a starting point for an explosion of genomics investigations in the Rosaceae. Complete curated genome sequence is expected for apple, peach and strawberry within the realistic range of finite calendar pages. These sequences will accelerate the next wave of studies exploring and comparing the form and function of the many genomes that define the striking differences in form within the Rosaceae e.g., a herbaceous plant vs. a tree, or a fleshy rather than a dehiscent fruit. Among the many dividends of this research will be the development of superior products for consumers, a better understanding of the genetic elements that contribute to agronomic traits of interest, an enhanced vision of Rosaceae evolution, as well as answers to some of the fundamental questions of plant biology, particularly around the specification of plant architecture, that may be best answered by species within this family.

This book covers recent progress in genomic research among the Rosaceae family of crops, grounding recent findings firmly in a historical context of genetic studies. The current status of the application of genomics technologies for crop development is examined. A general introduction precedes summaries of genomics research and applications on a crop by crop basis, each authored by a panel of active researchers on that particular species. This volume will be of considerable value as a resource for workers in the Rosaceae operating at all levels, from research scientists in genetics, genomics and breeding, to graduate and undergraduate students.

Palmerston North, New Zealand — Susan E. Gardiner
Gainesville, Florida — Kevin M. Folta

# Contents

# Contributors

**Albert G. Abbott**
Department of Biochemistry and Genetics, Clemson University 29634, USA

**Herb Aldwinckle**
Department of Plant Pathology, Cornell University, Geneva, NY 14456, USA

**Andrew C. Allan**
The New Zealand Institute for Plant and Food Research Limited, Mt Albert Research Centre, Private Bag 92169, 1142 Auckland, New Zealand

**Pere Arús**
IRTA. Centre de Recerca en Agrigenòmica CSIC-IRTA-UAB. Carretera de Cabrils Km 2. 08348 Cabrils, Spain

**Maria L. Badenes**
Fruit Breeding Department, Instituto Valenciano de Investigaciones Agrarias, Apartado Oficial 46113, Moncada, Valencia, Spain

**Nahla Bassil**
USDA ARS, National Clonal Germplasm Repository, Corvallis, OR USA

**Douglas Bielenberg**
Department of Biological Sciences, Clemson University, Clemson, SC USA

**Emily Buck**
The New Zealand Institute for Plant and Food Research Limited, Palmerston North Research Centre, Private Bag 11600, Palmerston North 4442, New Zealand

**Roger Bumgarner**
Department of Microbiology, School of Medicine, University of Washington, Seattle, Washington 98109, USA

**Vincent G.M. Bus**
The New Zealand Institute for Plant and Food Research Limited, Hawke's Bay Research Centre, 265 Lawn Rd, Hastings 4172, New Zealand

**David H. Byrne**
Department of Horticultural Sciences, Texas A&M University, College Station, TX USA

**David Chagné**
The New Zealand Institute for Plant and Food Research Limited, Palmerston North Research Centre, Private Bag 11600, Palmerston North 4442, New Zealand

**Jose X. Chaparro**
Horticultural Sciences Department, University of Florida, Gainesville, FL USA

**Elisabeth Chevreau**
UMR- 1259 (GenHort), INRA-INH-Université d'Angers, INRA, 42 rue Georges Morel, BP 57, 49071 Beaucouzé Cedex, France

**Jacques Claverie**
Institut National de la Recherche Agronomique, Centre de Bordeaux, Unité de Recherches sur les Espèces Fruitières et la Vigne, BP 81, 33883, Villenave d'Ornon Cedex, France.

**Ross Crowhurst**
The New Zealand Institute for Plant and Food Research Limited, Mt Albert Research Centre, Private Bag 92169, Auckland 1142, New Zealand

**Chris Dardick**
Department of Genetic Improvement of Fruit Crops, USDA-ARS-Appalachian Fruit Research Station, 2217 Wiltshire Road, WV 25430, USA

**Thomas M. Davis**
Department of Biological Sciences, University of New Hampshire, Durham, NH, 03824, USA

**Thomas Debener**
Institute of Plant Genetics, Department of Molecular Breeding, Hannover, Germany

**Béatrice Denoyes-Rothan**
Institut National de la Recherche Agronomique, Centre de Bordeaux, Unité de Recherches sur les Espèces Fruitières, BP 81, 33883, Villenave d'Ornon Cedex. France

**Amit Dhingra**
Department of Horticulture and Landscape Architecture; Department of Electrical Engineering and Computer Sciences, Washington State University, Pullman, WA 99164, USA

**Elisabeth Dirlewanger**
Institut National de la Recherche Agronomique, Centre de Bordeaux, Unité de Recherches sur les Espèces Fruitières, BP 81, 33883, Villenave d'Ornon Cedex. France

**Daniel Esmenjaud**

Unité "Interactions Plantes-Microorganismes et Santé Végétale", Equipe de Nématologie, Institut National de la Recherche Agronomique, B.P. 2078, 06606 Antibes Cedex, France

**Kevin M. Folta**

Horticultural Sciences Department and the Graduate Program in Plant Molecular and Cellular Biology, University of Florida, Gainesville, FL 32611, USA

**Fabrice Foucher**

*Inra* Angers, UMR Génétique et Horticulture (GenHort), Angers, France

**Susan E. Gardiner**

The New Zealand Institute for Plant and Food Research Limited, Palmerston North Research Centre, Private Bag 11600, Palmerston North 4442, New Zealand

**Ksenja Gasic**

Horticulture Department, Clemson University, Clemson, SC USA

**Andrew Gleave**

The New Zealand Institute for Plant and Food Research Limited, Mt Albert Research Centre, Private Bag 92169, Auckland 1142, New Zealand

**Thomas Gradziel**

University of California, Department of Plant Science, Davis, CA 95616, USA

**Julie Graham**

SCRI, Invergowrie, Dundee, DD2 5DA

**James Hancock**

Department of Horticulture, Michigan State University, East Lansing, MI 48824, USA

**Laurence Hibrand-Saint Oyant**

UMR Genetique and Horticulture (GenHort) – INRA / INH / UA – BP60057 49071 Beaucouze CEDEX

**Xuming Huang**

College of Horticulture, South China Agricultural University, Guangzhou 510642, P. R. of China

**Kim E. Hummer**

USDA ARS National Clonal Germplasm Repository, 33447 Peoria Road, Corvallis, Oregon 97333-2521

**Amy F. Iezzoni**
Department of Horticulture, Michigan State University, East Lansing, Michigan. 48824-1325, USA

**Jules Janick**
Department of Horticulture and Landscape Architecture, 625 Agriculture Mall Drive, Purdue University, West Lafayette, IN 47907-2010, USA

**Ananth Kalyanaraman**
Department of Horticulture and Landscape Architecture; Department of Electrical Engineering and Computer Sciences, Washington State University, Pullman, WA 99164, USA

**Markus Kellerhals**
Research Station Agroscope Changins-Wädenswil, Wädenswil, Switzerland

**Schuyler S. Korban**
Department of Natural Resources and Environmental Sciences, University of Illinois, Urbana, Illinois USA 61801

**François Laurens**
Institut National de la Recherche Agronomique, Centre de Recherches d'Angers, BP57, 49071 Beaucouzé Cédex, France

**Kimberly Lewers**
USDA ARS, Beltsville Agricultural Research Center, Genetic Improvement Fruits & Vegetables Lab, Beltsville, MD USA

**Shunquan Lin**
College of Horticulture, South China Agricultural University, Guangzhou 510642, P. R. of China

**Chengming Liu**
College of Horticulture, South China Agricultural University, Guangzhou 510642, P. R. of China

**Mickael Malnoy**
Istituto Agrario San Michele all'Adige, Italy

**M. Margarida Oliveira**
Univ. Lisboa, Fac. Ciências, Department of Plant Biology, Campo Grande 1749-016, Lisboa, Portugal; ITQB, Quinta Marquês, P-2784-505 Oeiras, Portugal

**Bruno Mezzetti**
Dipartimento di Scienze Ambientali e delle Produzioni Vegetali – Marche Polytechnic University, Via Brecce Bianche, Ancona, 60100-IT

**Richard Newcomb**
The New Zealand Institute for Plant and Food Research Limited, Mt Albert Research Centre, Private Bag 92169, Auckland 1142, New Zealand

**John L. Norelli**
USDA-ARS-AFRS, 2217 Wiltshire Road, Kearneysville, WV 25430

**Hilde Nybom**
Swedish University of Agricultural Sciences, Kristianstad, Sweden

**Ariel Orellana**
The Millennium Nucleus in Plant Cell Biotechnology, Fundación Ciencia para la Vida, Center of Plant Biotechnology, Andres Bello University, Santiago, Chile

**Cameron Peace**
Department of Horticulture and Landscape Architecture, Washington State University, Pullman WA 99164

**Cesar Petri**
Department of Genetic Improvement of Fruit Crops, USDA-ARS-Appalachian Fruit Research Station, 2217 Wiltshire Road, WV 25430, USA

**Carlo Pozzi**
Parco Tecnologico Padano, via Einstein Loc. C.na Codazza, Lodi 26900

**Daniel Rokhsar**
Joint Genomics Institute, Walnut Creek, CA 94598, USA

**Carlo Rosati**
ENEA, Trisaia Research Centre; Department of Genetics and Genomics; S.S. 106, km 419+500; 75026 Rotondella (MT), Italy

**Daniel J. Sargent**
East Malling Research, East Malling Research, East Malling, Kent ME 19 6 BJ, UK

**Jan G. Schaart**
Plant Research International, Wageningen University and Research Centre, PO Box 16, 6700 AA Wageningen, The Netherlands

**Robert Schaffer**
The New Zealand Institute for Plant and Food Research Limited, Mt Albert Research Centre, Private Bag 92169, Auckland 1142, New Zealand

**Wilfried Schwab**
Biomolecular Food Technology, Technical University Munich, Lise-Meitner-Str. 34, 85354 Freising, Germany

**Ralph Scorza**
Department of Genetic Improvement of Fruit Crops, USDA-ARS-Appalachian Fruit Research Station, 2217 Wiltshire Road, Kearneysville, WV 25430, USA

**Vladimir Shulaev**
Virginia Bioinformatics Institute and Department of Horticulture, Virginia Polytechnic Institute and State University, Blacksburg, VA 24061, USA

**David W. Simpson**
East Malling Research, East Malling Research, East Malling, Kent ME 19 6 BJ, UK

**Bryon Sosinski**
Department of Horticultural Science, Centennial Campus, North Carolina State University, Raleigh NC 27695, USA

**Ryutaro Tao**
Kyoto University, Graduate School of Agriculture, Laboratory of Pomology, Sakyo, Kyoto 6068502 Japan

**Stefano Tartarini**
Department of Fruit and Woody Plant Sciences, Alma Mater Studiorum – Bologna University, 40127 Bologna, Italy

**Alberto Vecchietti**
Parco Tecnologico Padano, via Einstein Loc. C.na Codazza, Lodi 26900

**Riccardo Velasco**
Department of Genetics and Molecular Biology, IASMA Research Center, San Michele all'Adige TN 38010, Italy

**Ignazio Verde**
Istituto Sperimentale per la Frutticoltura, Rome Italy

**Mary Woodhead**
SCRI, Invergowrie, Dundee, DD2 5DA

**Ana Wünsch**
Unitad de Fructicultura, Centro de Investigación y Tecnología Agroalimentaria de Aragon, Avda. Montañana 930, 50059 Zaragoza, Spain

**Toshiya Yamamoto**
National Institute of Fruit Tree Science, Tsukuba, Ibaraki 305-8605, Japan

**Xianghui Yang**
College of Horticulture, South China Agricultural University, Guangzhou 510642, P. R. of China

# Introduction

The disciplines of plant genetics and plant genomics are revolutionizing plant biology research by fundamentally changing the way plant biologists perform research and the way they view and understand plants. Not only are tools and data sets expanding more rapidly than they can be analyzed and understood, but also plant biologists must assimilate, assess and interpret data from diverse sources and in many plant species. No longer can plant biologists focus on a single discipline, such as physiology, development, or biochemistry - all now depend on each other and all are deeply rooted in genetics and genomics. As a result, it is often said, "we are all *biologists* now!", meaning that it is no longer meaningful to think of ourselves as physiologists, biochemists or geneticists, but rather we must think of ourselves as multidisciplinary biologists, working across and breaking down traditional disciplinary boundaries. This creates a great need for basic information that is accessible to all plant biologists, which is the purpose of this series - to bring novel, fundamental information about genetics and genomics of plants to the entire plant biology community.

The book series *Plant Genetics and Genomics: Crops and Models* is designed to provide current overviews and summaries of the state of the art in genetics and genomics for all of the major crops or groups of crops, as well as for each major genetic model system for which a significant need exists. Most volumes will focus on a single crop, species, group of close relatives, or group of plants with similar biology (such as Tropical Crops). Other volumes will have a specific disciplinary or technological focus, such as cytogenetics, comparative genomics, translational genomics, and epigenomics, encompassing the plant kingdom. In this way, we hope that all the most important areas of interests to both basic and applied plant scientists will be covered.

Series Editor, Plant Genetics and Genomics — Richard A. Jorgensen

# 1. Rosaceae: Taxonomy, Economic Importance, Genomics

**Kim E. Hummer and Jules Janick**

*A rose by any other name would smell as sweet.*
Shakespeare

*A rose is a rose is a rose.*
Gertrude Stein

**The Rose Family**
*The rose is a rose*
*And was always a rose;*
*But the theory now goes*
*That the apple's a rose,*
*And the pear is, and so's*
*The plum, I suppose.*
*The dear only knows*
*What will next prove a rose.*
*You, of course, are a rose,*
*But were always a rose.*
Robert Frost

## 1 Nomenclature and Taxonomy

### *1.1 Origins*

The magnificent simplicity, or to some, the monotonous consistency, of the actiniomorphic flowers of the rose family has been recognized for millennia. The origin of the name rose is summarized in the American Heritage Dictionary (2000):

> The English word rose comes from Latin and Old French. Latin rosa may be an Etruscan form of Greek Rhodia, "Rhodian, originating from Rhodes." The Attic Greek word for rose

K.E. Hummer (✉)
U. S. Department of Agriculture, Agricultural Research Service, National Clonal Germplasm Repository, 33447 Peoria Road, Corvallis, Oregon, 97333, USA

K.M. Folta, S.E. Gardiner (eds.), *Genetics and Genomics of Rosaceae*, Plant Genetics and Genomics: Crops and Models 6, DOI 10.1007/978-0-387-77491-6_1,

> is rhodon, and in Sappho's Aeolic dialect of Greek it is wrodon. In Avestan, the language of the Persian prophet Zoroaster, "rose" is varda and in Armenian vard, words both related to the Aeolic form. The Modern Persian word for "rose" is gul.

Soon after Linnaeus published his *Systema Naturae* (1735), botanists worked to improve systematic classification. Michel Adanson (1763, 1963) was first to publish "Rosaceae" as the name for the rose family, although the International Code of Botanical Nomenclature (ICBN) (2006) now accepts Antoine Laurent de Jussieu (1789) as the author. The ICBN conserved Jussieu's names for 76 plant families, because he combined the Linnaean concept of binominal nomenclature with Adanson's methodology for defining groups based on multiple characteristics.

Recently, controversies and deficiencies in angiosperm classifications (Cronquist, 1981; Dahlgren, 1980; Thorne, 1992, 2000; Takhtajan, 1997) are being resolved by phylogenetic approaches based on analysis by the angiosperm phylogeny group (APG I, 1998; APG II, 2003). The APG system is based on the analysis of chloroplast and ribosomal coding genes in association with morphological charateristics.

Fossil record shows that Rosaceae is cogent with ancient dicotyledons (Heywood, 2007). Turonian fossils from 90 million years before the present (mybp) are attributed to this family (Crepet et al., 2004). Wikström et al. (2001) estimates that the stem rosaceous group dates to ca. 76 mybp, and crown group divergence (Rosoideae not included) from 47 to 46 mybp. Rosaceous physiological structure and anthecology also suggest that it is primitive. Hutchinson (1964) states that Rosaceae is an offshoot of the ancient woody magnolias, and on a common evolutionary line leading to orders such as Leguminales (Fabales), Araliales (Umbellales), Fagales, and Juglandales, that have more specialized inflorescences.

The APG phylogeny has separated plant orders and families on a linear time scale into basal angiosperms, eudicots, early diverging dicots, and core dicots. Research supporting APG (Soltis and et al., 2005) considers the rosids, of which rosales and its typical family, Rosaceae to be of the core eudicot group. Although the Rosaceae has great morphological diversity, to the point of being "indefinable" (Dickinson et al., 2002), the family is robust as along with morphological and chemical assessments (Challice, 1974), analyses of rbcL sequences strongly support the monophyly of Rosaceae (Fig. 1) (Morgan et al., 1994).

## *1.2 Morphology*

Judd et al. (1999) describe Rosaceae to include herbs, shrubs, or trees, which are sometimes rhizomatous, climbing, or thorny. Plant hairs are simple or stellate, and can be present with prickles. Leaves are usually alternate, and are simple to palmately or pinnately compound. Stipules are usually present and flowers are often showy, bisexual or infrequently unisexual. The hypanthium ranges from flat to cup-shaped or cylindrical and either free from or adnate to the carpels, often enlarging in fruit. Sepals are usually pentamerous, sometimes alternating with epicalyx lobes.

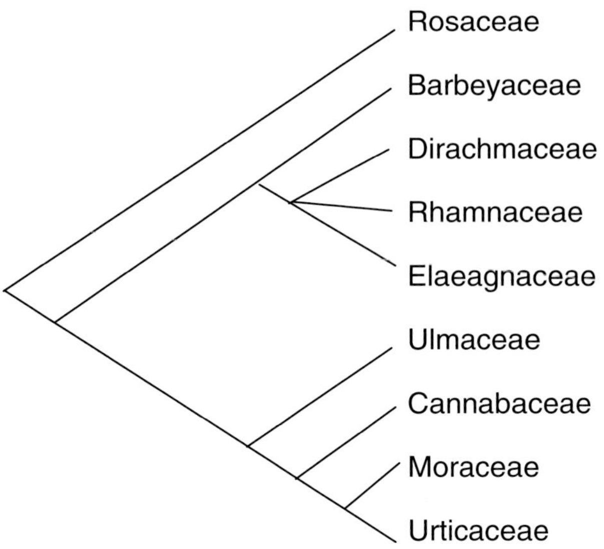


**Fig. 1** Cladogram based on the *rbcL* of the Rosaceae within Rosales (tree phylogeny from APG II, APW, 2007, based on Morgan et al., 1994)

Petals are usually pentamerous, while stamens are usually 15 or more, sometimes 10 or fewer. Filaments are distinct or basally fused to the nectar disk. Pollen grains are tricolpate. Carpels are 1 to many, and are distinct or connate, though sometimes adnate to the hypanthium. The ovary varies from superior to inferior depending on the genus. The styles are present in the same number as carpels. The fruit can be a follicle, achene, pome, drupe, aggregate or accessory with drupelets or achenes, or rarely a capsule. Endosperm is usually absent from the seed. The occurrence of numerous stamens and the absence of endosperm have been key structural apomorphies for systematic classification.

## *1.3 Distribution and Ecology*

Rose family distribution is cosmopolitan (Judd et al., 1999) to sub-cosmopolitan, but is diversified, particularly in the Northern hemisphere. The herbaceous species grow in temperate forests as understory plants, in salt or freshwater marshes, in arctic tundra, in old fields, and along roadsides. Woody members are pioneer species, and are prominent in the early stages of forest succession. Rosaceous trees may also be minor components of mature mixed deciduous forests.

The Rosaceae is the 19th largest family of plants (APW, 2007). It includes from 95 to more than 100 genera and 2830–3100 species (Judd et al., 1999; Mabberley, 1987). Familial synonyms include: Agrimoniaceae Gray, Alchemillaceae Martinov, Amygdalaceae (Juss.) D. Don, Cercocarpaceae J. Agardh, Cliffortiaceae Mart., nom. nud., Coleogynaceae J. Agardh, Dryadaceae Gray, Fragariaceae Rich. ex Nestl., Lindleyaceae J. Agardh, Malaceae Small ex Britton, Neilliaceae

Miq., Potentillaceae (Juss.) Wilbr., Prunaceae Bercht. & J. Presl, Quillajaceae D. Don, Rhodotypaceae J. Agardh, Sanguisorbaceae Durande, Spiraeaceae Bertuch, Ulmariaceae Gray. The ICBN accepted genera for the family are listed (Table 1).

**Table 1** International code of botanical nomenclature (ICBN) accepted genus names within Rosaceae

| | |
|---|---|
| *Acaena* Mutis ex L. | *Osteomeles* Lindl. |
| *Adenostoma* Hook. & Arn. | *Pentactina* Nakai |
| *Agrimonia* L. | *Peraphyllum* Nutt. 60 |
| *Alchemilla* L. | *Petrophytum* (Nutt. ex Torr. & A. Gray) Rydb. |
| *Amelanchier* Medik. | *Photinia* Lindl. |
| *Aphanes* L., | *Physocarpus* (Cambess.) Raf., nom. cons. |
| *Aremonia* Neck. Ex Nestl., nom. cons. | *Polylepis* Ruiz & Pav. |
| *Aria* (Pers.) Host, *Aronia* Medik., nom. cons. | *Potaninia* Maxim. |
| *Aruncus* L. | *Potentilla* L. |
| *Bencomia* Webb & Berthel. | *Prinsepia* Royle |
| *Brachycaulos* R. D. Dixit & Panigrahi | *Prunus* L. |
| *Cercocarpus* Kunth | *Pseudocydonia* (C. K. Schneid.) C. K. Schneid. |
| *Chaenomeles* Lindl., nom. cons. | *Purshia* DC. ex Poir. 70 |
| *Chamaebatia* Benth. | *Pyracantha* M. Roem. |
| *Chamaebatiaria* (Porter ex W. H. Brewer & S. Watson) Maxim. | *Pyrus* L. |
| *Chamaemeles* Lindl. | *Quillaja* Molina |
| *Chamaemespilus* Medik. | *Rhaphiolepis* Lindl., nom. cons. |
| *Chamaerhodos* Bunge | *Rhodotypos* Siebold & Zucc. |
| *Cliffortia* L. | *Rosa* L., nom. cons. prop. |
| *Coleogyne* Torr. | *Rubus* L., nom. cons. prop. |
| *Coluria* R. Br. | *Sanguisorba* L. |
| *Cormus* Spach | *Sarcopoterium* Spach |
| *Cotoneaster* Medik. | *Sibbaldia* L. |
| *Cowania* D. Don | *Petrophytum* (Nutt. ex Torr. & A. Gray) Rydb. |
| *Crataegus* L. | *Photinia* Lindl. |
| *Cydonia* Mill. | *Physocarpus* (Cambess.) Raf., nom. cons. |
| *Dalibarda* L. | *Polylepis* Ruiz & Pav. |
| *Dichotomanthes* Kurz | *Potaninia* Maxim. |
| *Docynia* Decne. | *Potentilla* L. |
| *Docyniopsis* (C. K. Schneid.) Koidz. | *Prinsepia* Royle |
| *Dryas* L. | *Prunus* L. |
| *Duchesnea* Sm. | *Pseudocydonia* (C. K. Schneid.) C. K. Schneid. |
| *Eriobotrya* Lindl. | *Purshia* DC. ex Poir. |
| *Eriolobus* (DC.) M. Roem. | *Pyracantha* M. Roem. |
| *Exochorda* Lindl. | *Pyrus* L. |
| *Fallugia* Endl. | *Quillaja* Molina |
| *Filipendula* Mill. | *Rhaphiolepis* Lindl., nom. cons. |
| *Fragaria* L. | *Rhodotypos* Siebold & Zucc. |
| *Geum* L. | *Rosa* L., nom. cons. prop. |
| *Gillenia* Moench | *Rubus* L., nom. Cons. prop. |
| *Guamatela* Donn. Sm. | *Sanguisorba* L. |

**Table 1** (continued)

| | |
|---|---|
| *Hagenia* J. F. Gmel. | *Sarcopoterium* Spach |
| *Hesperomeles* Lindl. | *Sibbaldia* L. |
| *Heteromeles* M. Roem. | *Sibiraea* Maxim. |
| *Holodiscus* (K. Koch) Maxim., nom. cons. | *Sieversia* Willd. |
| *Horkelia* Cham. & Schltdl. | *Sorbaria* (Ser. ex DC.) A. Braun, nom. cons. |
| *Horkeliella* (Rydb.) Rydb. | *Sorbus* L. |
| *Ivesia* Torr. & A. Gray | *Spenceria* Trimen |
| *Kageneckia* Ruiz & Pav. | *Spiraea* L. |
| *Kelseya* (S. Watson) Rydb. | *Spiraeanthus* (Fisch. & C. A. Mey.) Maxim. |
| *Kerria* DC. | *Stephanandra* Siebold & Zucc. |
| *Leucosidea* Eckl. & Zeyh. | *Taihangia* T. T. Yu & C. L. Li |
| *Lindleya* Kunth, nom. cons. | *Tetraglochin* Poepp. |
| *Luetkea* Bong. | *Torminalis* Medik., |
| *Neviusia* A. Gray | *Vauquelinia* Corrêa ex Bonpl. |
| *Oemleria* Rchb. | *Waldsteinia* Willd. |
| *Orthurus* Juz. | *Xerospiraea* Henr. |

As the gynoecium varies greatly among species of this family, this variation has been utilized for classification within the family to subfamilies Takhtajan (1997):

Rosoideae – many apocarpous pistils mature into achenes; Amygdaloideae (prunoideae) – a single monocarpellate pistil matures into a drupe; Spiraeoideae – the gynoecium consists of two or more apocarpous pistils that mature into follicles; Maloideae (Pomoideae) – the ovary is compound and inferior, and an epigynous zone may occur.

## *1.4 Conventional Taxonomy*

In addition to the subgenera, tribes have been described to accommodate outlying genera. The composition of genera within subfamilies and tribes has been problematic. Intergeneric hybridization occurs within these groups. Species definitions are extended to account for apomixis or hybridization. Some treatments have suggested that some of the subfamilies have familial status (Hutchinson, 1964); others overlook subfamilies and worked solely with tribes (GRIN, 2007).

Although molecular analysis agrees with traditional determinations that the Rosaceae family is robustly monophyletic, it has added to the debate on the subfamily and tribal groupings. Genetic analysis has been performed to examine phylogenies for the family, individual subfamilies, tribes, and the larger genera.

Dickinson et al. (2002) has produced a parsimonious tree derived from the cladistic analysis of 125 combined morphological and molecular characters from members of the Rosaceae (Fig. 2) and added several genera to Maloideae. Spiraeoideae, no longer monophyletic, is split by Amygdaloideae (Prunoideae). The Rosoideae have been subdivided into individual tribes. Thus, Dickinson et al. recircumscribes Maloideae and Rosoideae, the two largest subfamilies. They

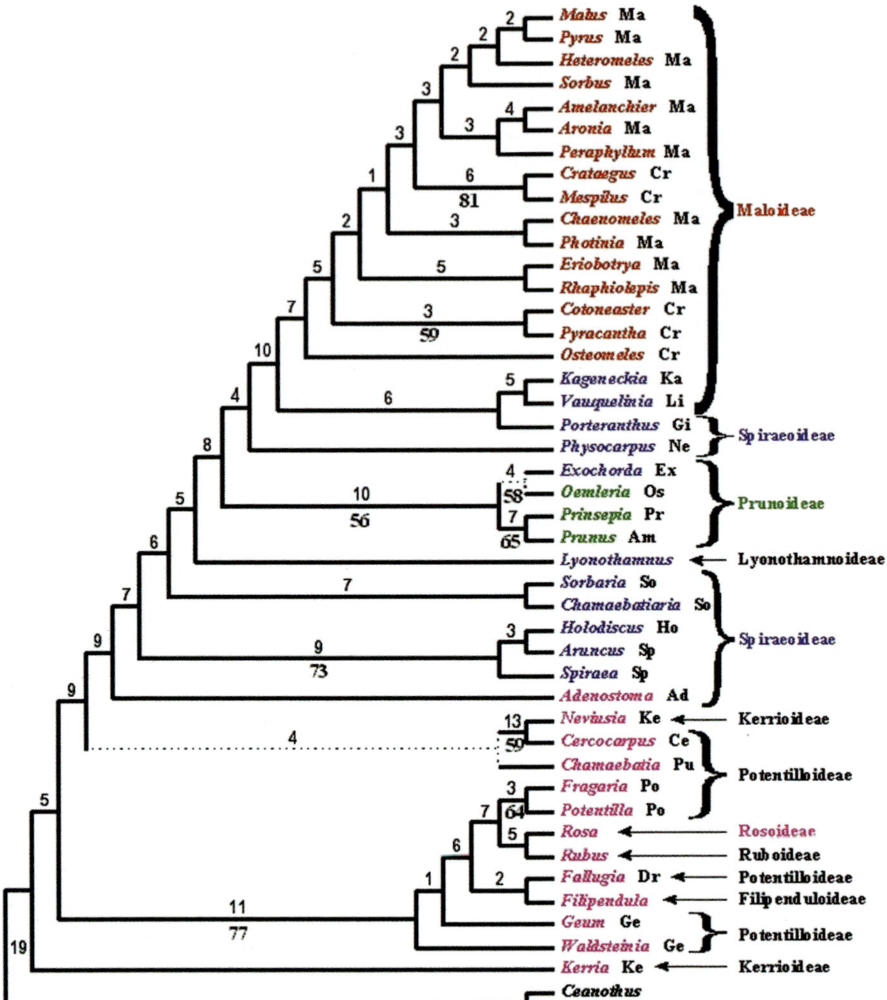


**Fig. 2** Parsimonious tree derived from the cladistic analysis of 125 combined characters (61 non-molecular + 64 from molecular tree branches (ndhF, rbcL, nrITS). Adapted from: Dickinson et al. (2002). The colors represent traditional subfamilies (red = maloideae; blue = spiraeoideae; green = amygdaloideae; pink = rosoideae). *Ceanothus* and *Rhamnus* are outgroup taxa. The classification to tribes are as proposed by Takhtajan (1997): (**Ad**enostomateae, **Am**ygdaleae, **Ce**rcocarpeae, **Dr**yadeae, **Ex**ochordeae, **Ge**eae, **Cr**ataegeae, **Gi**llenieae, **Ho**lodisceae, **Ka**geneckieae, **Ke**rrieae, **L**indleyieae, **Ma**leae, **N**eillieae, **Os**maronieae, **Po**tentilleae, **Pr**insepieae, **Pu**rshieae, **So**rbarieae, **Sp**iraeeae). Colored names outside of brackets correspond to traditional subfamilies, black names to subfamilies proposed by Takhtajan (1997)

rejected Amygdaloideae and Spiraeoideae, neither of which proved to be monophyletic. Lastly, they redefined old tribes and suggested new ones.

Not only does this study offer insights on monophyly and relationship between the groups, but it can be interpreted to provide commentary concerning the origin of subfamilies.

The traditional hypothesis suggests that the origin of the Maloideae (Pomoideae) occurred in an ancient polyploidization event, because the maloid base chromosome number is $x = 17$, whereas other Rosaceae are $x = 7$, 8, or 9. For the hybridization, Challice (1974; 1981) suggested that the Maloideae were created by an allotetraploidization event following an ancient hybridization between Amygdaloideae ($x = 8$) and "Spiraeoideae" ($x = 9$) ancestors. A second theory suggests an allo- or auto-polyploidization event occurred solely within the "Spiraeoideae." If the Maloideae originated within the "Spiraoideae," then the fleshy "pome" fruit, e.g., apple, must have been derived from the expansion of the hypanthium (floral cup). Incorporation of the ovaries by the enlarged hypanthium resulted in the inferior ovaries present in the majority of Maloideae genera (Dickinson, 2007). Molecular analysis is continuing and has not yet eliminated the possibility of an allotetraploidization event. The uniting of formerly "Spiraeoid" genera to the Maloideae, however, gives credence to the solely "Spiraeoid" event hypothesis.

Although definition of the four major rosaceous subfamilies may be collapsing from a taxonomic view, these grouping still have great utility from an economic and horticultural standpoint. The ultimate decision on subfamilies may depend on future development in genomics, the subject of this volume.

## 2 Economic Importance

The Rosaceae include many well known and beloved species of economic importance particularly edible temperate zone fruits (Janick, 2005) and ornamentals, but also some timber crops and medicinals or neutriceuticals. Some economically important taxa of the Rosaceae are summarized in (Table 2). The more important species are briefly reviewed below by subfamilies. The total world production of the edible rosaceous fruits in 2005 based on FAO statistics is about 113 million tonnes. At a very conservative farm gate value of US$400 per tonne this translates to $45 billion. Adding in the world value of almonds, cut roses, rose plants, and other products suggests that that rosaceous plants could be worth at least $60 billion annually at the farm gate, with a consumer value of triple this amount, $180 billion.

### *2.1 Amygdyloideae*

The stone fruits, i.e. species of *Prunus*, include delicious gifts of summer (peach, cherry, plum, and apricot), as well as almond, the most important nut worldwide.

**Table 2** Some economically important species of Rosaceae by subfamily

| Subfamily | Genus | Species | Common name | Uses |
|---|---|---|---|---|
| Amygyloideae | *Prunus* | *armeniaca* | Apricot | Fresh and processed fruit |
| | | *avium* | Sweet cherry | Fresh and processed fruit |
| | | *cerasus* | Tart (sour) cherry | Fresh and processed fruit |
| | | *domestica* | European plum | Fresh and processed fruit |
| | | *dulcis* | Almond | Fresh and processed nut |
| | | *mume* | Mume | Ornamental |
| | | *persica* | Peach, nectarine | Fresh and processed fruit |
| | | *serotina* | Black cherry | Timber species |
| *Maloideae* | *Amelanchier* | *alnifolia* | Saskatoon, serviceberry; shadbush | Landscape ornamental |
| | *Aronia* | *melanocarpa* | Black chokeberry | Processed fruit for juice, nutriceutical |
| | *Chaenomales* | *japonica* | Japanese quince | Landscape ornamental, processed fruit |
| | *Cotoneaster* | *spp.* | Cotoneaster | Landscape ornamental |
| | *Crataegus* | *spp.* | Hawthorn, thornapple | Landscape ornamental, craft uses for wood |
| | *Cydonia* | *oblonga* | European quince | Fresh and processed fruit, dwarfing rootstock for pear and loquat |
| | *Eriobotrya* | *mespilus* | Loquat | Fresh and processed fruit |
| | *Malus* | *×domestica (M. pumila)* | Apple | |
| | | *spp.* | Crabapples | Landscape ornamentals |
| | *Pyrus* | *calleryana* | Callery pear | Landscape ornamental |
| | | *communis* | European pear | Fresh and processed fruit |
| | | *serotina* | Japanese pear (nashi) | Fresh fruit |
| | | *ussurienses* | Chinese pear | Fresh fruit |
| | *Mespilus* | *germanica* | Medlar | Fresh fruit (bletted) |
| | *Photinia* | *spp.* | Photinia | Landscape ornamental |
| | *Pyracantha* | *spp.* | Firethorn | Landscape ornamental |
| | *Sorbus* | *spp.* | Mountain ah, rowan | Landscape ornamental |
| Rosoideae | *Fragaria* | *×ananassa* | Strawberry | Fresh and processed fruit |
| | *Geum* | *spp.* | Avens | Herbaceous perennial |
| | *Kerria* | *japonica* | Kerria | Landscape ornamental |
| | *Potentilla* | *spp.* | Cinquefoil | Landscape ornamental |
| | *Rosa* | *spp.* | Rose | Cut flowers, landscape ornamental, perfume oil, medicinal |
| | *Rubus* | *spp. and hybrids* | Blackberry, raspberry, hybrid berry | Fresh and processed fruit |
| Spiraeoideae | *Spirea* | *prunifolia* | Bridal wreath | Landscape ornamental |
| | *Exochorda* | *racemosa* | Exochorda | Landscape ornamental |
| | *Physocarpus* | *opulitolius* | Ninebark | Landscape ornamental |

The stone fruits are less hardy than the pome fruits; they tend to flower earlier, making them very susceptible to spring frosts. The fruits tend to be soft at maturity and have much poorer storage life than the pome fruits, however their exquisite flavors have made them much admired. Many are consumed dried, especially plum and apricot. Black cherry, *P. serotina*, is an important timber species. A number of interspecfic crosses within *Prunus,* such as plum with apricot, have led to new fruits such as plumcot, pluot®, and aprium®.

**Almond.** *Prunus dulcis* (synonym = *Amydalis*) originates in Asia and is an ancient central Asian crop cultivated in the Mideast since antiquity and apparently re-introduced to Spain during the incursions of Arabs into Europe in the $8^{th}$ century. The domesticated almond with "sweet," in contrast to "bitter" (high amygdalin) seeds have made almond a popular nut for both fresh and processed products for millennia. World production of this nut (2005) was 1.7 million tonnes (Mt). A favorite confection made from almond paste mixed with sugar, molded, and painted to resemble other fruits and products was introduced by Arabs to Sicily where it has become a unique culinary art form. Immature almonds are also eaten in the Mideast. Almond has been transformed into a large scale,industry in California, that now produces over 70% of the world crop, and ranks as seventh largest US food export.

**Apricots.** The delectable apricot, *Prunus armeniaca*, is consumed fresh and canned, but is principally known in the form of semi-dried fruit, usually sulfured to maintain color and longevity. World production exceeds 3 Mt. The major dried industry is located in southwest Turkey (Malatya). Some seeds are also consumed, similarly to almonds. The beautiful flowers of the Japanese apricot, *P. mume*, make them popular ornamentals in Asia.

**Cherries.** Cherry is the common name of several *Prunus* species that include sweet cherry (*P. avium)*, tart or sour cherry (*P. cerasus*), Duke cherries (hybrids between *P. avium* and *P. cerasus*). World cherry production is about 3 Mt. Black cherry (*P. serotina*) an important timber species and there are a number of flowering species, principally *P. serrulata* that are included in a group called ornamental cherries, however their taxonomy is confused.

Sweet cherry is native to the Caucuses and has become a much beloved fruit but is restricted to cooler climates. Cherry is consumed fresh as a gourmet fruit, and also consumed in soups, tarts, pies, and candied, often covered with chocolate and as an almost artificial product called *maraschino*. Cherry is used for liqueurs and wines. Cherry wood is used for furniture and veneers, as well as decorative paneling and is cultivated for this purpose in plantations in Europe.

Tart or sour cherry is hardier than sweet cherry and requires a longer season to mature. Its use is restricted to processing, mainly as a filling for pies and pastries, however is now becoming popular as a dried fruit and as a source of syrup that is promoted as having health properties. Tetraploid natural hybrids between the diploid sweet cherries and the tetraploid sour cherries, known as Duke cherries, are a distinct minor crop.

Black cherry, *P. serotina* is a prized timber species used for veneer, furniture, and paneling due to its beautiful red color. It is usually the highest or second highest valued fine hardwood. Black cherry is cultivated in plantations and breeding programs are underway in Indiana and Pennsylvania. The bark has been used medicinally.

Ornamental cherries are favorite landscape species of Asia. The flowering cherry known as *Sakura* were donated to the United States as a gift in 1912, promoted by the wife of President Taft. Flowering cherry has become a landmark and important symbol for Washington, D.C.

**Peach and Nectarine.** *Prunus persica*, despite its name, is native to warm areas of China. The peach and its smooth skin mutation, the nectarine, are the most important contributors to stone fruit production. In 2005 this production reached almost 16 Mt. Peach thrives in hot summer climates, however, many low chill cultivars have enabled expansion of peach production to the subtropics. The nectarine or fuzzless peach has been increasing in popularity. Peaches occur in a wide variety of fruit and flesh color from yellow to red and shape. There are basically two types of flesh texture, melting and rubbery. The rubbery flesh types are clingstone and are used for processing principally, however their use is now being promoted for fresh consumption because they can be shipped in a riper state. Peaches are processed into juice and sliced or diced products, while some are dried. Peaches have been beloved in Asia as a symbol of longevity. Major problems of peaches have been quality problems, especially cottony flesh texture, associated with long distance shipping and chilling injury. Peach and nectarine tend to be susceptible to summer diseases such as peach leaf curl and brown rot.

**Plums.** Plums represent a diverse group of fruits that include European, Asian, and American species. The cultivated European plum, *P. domestica* is a hexaploid ($2n = 6x = 42$) that probably originated as a hybrid between *P. cerasifera* (diploid) and *P. spinosa* (a tetraploid). Asian species include *P. salicina* (Japanese plum) and *P. simonii* (Simon or apricot plum). There are at least five American species: *P. americana* (common wild plum), *P. nigra* (Canada plum), *P. angustifolia* (chickasaw), *P. horulana* (hortulna plum) and P. *munsoniana* (wild goose plum). The principal economically important plum species are *P. domestica* and *P. salicina* with world production in 2005 exceeding 9 Mt.

While plums are mostly consumed as a fresh or processed fruit, the prune-type plums (French prune or "Agen") produce a well known product called prunes or dried plums. Principally produced in California, prunes have long been a health product promoting digestive regularity among seniors, although the industry has developed moist types as a snack food.

## *2.2 Maloideae*

This subfamily includes important edible temperate fruit species (known collectively as pome fruits) and a great number of landscape plants (Table 2). The most economically important members are apple and pear, in addition there are a number of minor fruits such as medlar, loquat, and quince.

**Apples.** *Malus* ×*domestica*= *M. pumila* is the most economically important rosaceous species with annual world fruit production (2005) in excess of 62 Mt, the fourth most important fruit after citrus, grapes, and banana. Apples are produced in all temperate and subtropical countries of the world, with minor production in high altitudes of tropical countries. The popularity of apple derives from the fact that the fruit has multiple uses. It can be consumed fresh and some cultivars can be stored for an entire year, while a considerable proportion of the crop is processed into sauces, juice, and slices which are a favored ingredient for pastries, cakes, pies, and tarts. Some of the juice is sold as a fermented product called cider (often special apples, many of which are may be hybrids with native species such as *M. sylvestris*) and a small portion is distilled as apple brandy (calvados). Much vinegar production derives from apple cider.

Various species of *Malus*, usually small fruited species referred to crabapples, are very popular as ornamentals for their spring flowers and fall fruit. Crabapples are particularly popular in the US nursery trade. Apple wood (*M. sylvestris*) has been used in furniture and has specialized uses for turning, mallet heads, croquet and skittle balls, umbrella handles, toys, cog wheels, wood screws, canes, pianos, tool handles, drawing instruments, and bookbinder screws. The wood is widely used for smoking meats and for barbecues and is especially valued for fireplace burning.

**Aronia.** *Aronia melonicarpa* (black chokeberry), native to North America, is grown in Europe for its juice which has high antioxidant properties. Fruit extract has been used as a component of nutrient supplements.

**Hawthorn.** There are about 1000 *Crataegus* species that are used in landscape plantings. A number have edible fruit and a few species such *as C. monygna* (English hawthorn) are used for timber. *Crataegus pinnatifida*, Chinese hawthorn, is processed for juice in China.

**Loquat.** *Eriobotrya japonica*, despite its name is native to China, where it is a favorite since it is the first fruit that appears in the summer. Production is now increasing rapidly in China and production is also advancing in Mediterranean countries. Spain is the principal exporter of loquat. Development of seedless triploids could transform this fruit into a very popular crop worldwide.

**Medlar.** *Mespilus germanica*, a monotypic genus, is an ancient fruit that must be consumed after it undergoes a fermentation called bletting, where the fruit softens and develops a spicy flavor. Medlar is still marketed in Northern Italy and Germany, but is now considered a minor fruit elsewhere.

**Mountain Ash.** There are about 80 *Sorbus* species in North America, Europe and Asia that are widely used as ornamentals or as windbreaks. Most species have bitter fruits, but some are sweet and have been suggested as a possible new fruit. There are intergeneric hybrids between *Sorbus* with *Amelanchier* and *Pyrus*.

**Pears.** Species of *Pyrus* are the second most important rosaceous fruit species, with world pear production of about 10 Mt (2005). There are three economically important species, *P. communis* (European pear), *P. pyrifolia*, (Japanese pear or

Nashi), and *P. usuriensis* (Chinese pear). Japanese and Korean cultivars are complex hybrids of *P. pyrifolia* and *P. ussuriensis*. Pear has similar uses to apple, although its popularity may be somewhat lower because the best quality is ephemeral in European pear. In European winter pears this eating quality is achieved by ripening after harvest. Pear cider is usually made from cultivars of *P. nivalis* and is called perry. The pear tree is also an important ornamental and is beloved in Asia where pear is considered a sign of good luck. In the United States, the most popular ornamental pear trees were selections of *Pyrus calleryana,* the Callery pear. These street trees can be found from Oregon to Ohio to New York, and south to Alabama and Georgia. *Pyrus koehnii*, an evergreen species native to Taiwan, is planted in California and Florida.

**Quinces.** Two closely related genera, *Cydonia* and *Chaenomeles* are referred to as quince. The genus *Cydonia* consists of a single species *Cydonia oblonga*, native to southwest Asia. Quince is now a minor crop used principally for processing into preserves, although there is considerable production in Argentina. Most cultivars are too astringent to consume fresh, however there are non-astringent types grown in Iran, India, Afghanistan, and Central Asia, suggesting this fruit has developmental possibilities. Quince is used as a dwarfing rootstock for pear and loquat. Intergeneric hybrids have been reported between quince and pear (Trabut, 1917).

Asian flowering quinces, *Chaenomeles,* are closely related to *Cydonia*, *Pyrus*, and *Malus*. The Japanese quince (*C. japonica*) is a shrubby plant with attractive red flowers and aromatic hard fruit that resemble loquat in appearance. It is a popular landscape ornamental. Japanese quince fruit are a minor crop in the Baltic countries, and efforts are underway to commercialize this species for processing

**Serviceberry.** *Amelanchier* species known by various local names including Serviceberry, Juneberry, or Saskatoons are hardy plants used as landscape plants. Various attempts have been made at domestication for uses as a new fruit crops, but it appears that these species crops will be best used as edible landscape species.

## *2.3 Rosoideae*

This subfamily is the home of the rose, the species that has provided the name for Rosaceae, and includes a number of other ornamentals such as *Potentilla* and small fruits including strawberry (*Fragaria)* and the brambles blackberry, raspberry, and various hybridberries (*Rubus*).

**Brambles.** Bramble is a collective term for various prickly shrubs sometimes classified horticulturally as small, bush, or berry fruits. All brambles are species of *Rubus*, a taxonomically complex group that includes blackberry (European and American species), raspberries (*R. ideaus*) and various hybrids. Hybrid combinations include upright and trailing types such as eastern dewberry (*R. trivialis*) and blackberry × raspberry crosses such as youngberry, marionberry, loganberry, and tayberry. Raspberries include red raspberry, black raspberry (*R. occidentalis* and *R. leucodermis*), and purple raspberry (black raspberry × black raspberry). World production of brambles is now increasing with the development of new cultivars and air transport and in 2005 was 0.6 Mt (154 thousand t of blackberry and 498 thousand t of raspberry.

**Roses:** *Rosa* is one of the major economically important genera of ornamental horticulture and the area under cultivation continues to expand. The rose, admired since antiquity for its beauty and fragrance has multiple uses: cut flowers, landscape plant, oils (attar of rose) for perfume as well as culinary use (rose water), and hips (fruits) as a source of Vitamin C. *Rosa* species are found throughout the colder and temperate regions of the Northern hemisphere form the Artic to the subtropics with more than 100 species but modern cultivars are mostly interspecific hybrids deriving from only 10 of these: *R. canina*, *R. chinensis*, *H. foetida*, *R. gallica*, *R. gigantea*, *R. moschata*, *R. multiflora*, *R. phoenicea* and *R. rugosa*, and *R. wichuraina*. The multiflora rose *(R. multiflora)* is considered a noxious weed. The cut flower industry is becoming globalized with production moving to South and Central America and Africa, however plant production for landscape use is usually local because of the problems of transporting soil.

**Strawberry.** The domesticated strawberry *F.* ×*ananassa* is a hybrid between *F. chiloensis* and *F. virginiana* that was first found in France in the 18th century. Subsequent breeding efforts have produced large size, high quality fruits produced by in field cultivation or in protected culture. Most of the crop is grown for fresh fruit, but a small portion of the crop is frozen or used to make preserves. Total 2005 world production has been estimated at 3.6 Mt.

## 3 Rosaceae Genomics

The Rosaceae is an ancient plant family containing many genera and a plethora of systematic challenges that are beyond the scope of morphological determination. A coordinated effort in developing crop genomics in the Rosaceae will unravel complex relationships within the cultivated genera, provide insights into basic physiological understanding of species within the family, and assist in genetic improvement of individual crop species. With these goals in view, members of the scientific community who are devoted to species of the Rosaceae have been studying the genomics of individual crops, and the time is ripe to coordinate this effort to resolve overarching issues within the family.

### *3.1 Systematic Challenges*

Each of the crops within Rosaceae has nomenclatural challenges. Traditional descriptions of certain genera, such as those in the Maloideae, seem unrestricted by the boundaries of botanical species or generic definitions. In this group cross- and graft-compatibilities are known to occur not only between species but between genera as well.

Widely disjunct populations of species and genera abound in the Rosaceae, a globally distributed family. New populations of some genera, such as *Mespilus canescens*, an Arkansan distribution of an otherwise European genus, have been found. In other cases, such as *Rubus*, species have been defined in unprecedented numbers and species aggregates are used as a taxonomic reference.

The breakdown of the present taxonomy of the Rosaceae has been an issue among taxonomists and the family is presently under revision. Crop specific genomics and studies of synteny within the family are needed to clarify these relationships.

## 3.2 Physiological Development

Model plants such as *Arabidopsis* have been extensively studied to provide a basis for understanding gene structure and function, in such important processes as flowering, morphology, and development. However, this species has limitations for extrapolation to the Rosaceae (Aŕus et al., 2006). The Rosaceae offers genera with small genomes, such as that of *Fragaria vesca* with 164 Mbp (Aŕus et al., 2006), which could become model systems with direct, economically important application to the many cultivated species within the family. The genomics of rosaceous species offers special opportunities to clarify problems in fruit development and ripening, incompatibility systems, hardiness, apomixis, and, given the unique system of gamete formation in *Rosa canina*, meiosis. Further, the complex ploidy systems and interspecific hybrids within Rosaceous genera also offer opportunity to look beyond simple diploids.

## 3.3 Genetic Improvement

Advances in Rosaceaous genomics and synteny will provide new ways to obtain genetic improvement within economically important rosaceous species. Genes that control fruit quality including flavor and texture are of immediate interest. The development of saturated maps with codominant and transferable markers within *Prunus*, *Malus*, and *Fragaria* will foster genetic research and should lead to immediate practical results through marker-assisted selection. In *Prunus*, 28 major genes affecting morphological and agronomic characters can be located on a reference map (Table 3).

Identification of genes producing desirable quality traits within the family could provide a means of crop improvement by shifting genes within the family. For example, the genes for hardiness in *Malus* might be transferred to *Prunus* while resistance genes for black spot caused by *Diplocarpon rosae* in apple could be transferred to rose. We suggest that the transfer of genes within the Rosaceae, if properly explained, will be less likely to engender consumer resistance to genetic transgenesis, than would the movement of genes from more phenologically disparate plants.

**Table 3** Description of 28 major genes affecting characters in different *Prunus* crops that have been placed on a reference map (After Dirlewanger et al., 2004)

| Linkage group | Character | Crop | Population |
|---|---|---|---|
| | Sharka resistance | Apricot | Lito × Lito |
| | Evergrowing | Peach | Empress op dwarf × P1442380 |
| | Flower color | Almond × peach | Garfi × Nemared |
| G2 | Root-knot nematode resistance | Peach | P.2175 × Felinem, Akame × Juseitou, Lowell × Nemared Garfi × Nemared; Padre × 54P455 |
| | Shell Hardness | Almond | Ferragnes × Tuono |
| | Broomy (pillar) growth habit | Peach | Various progenies |
| | Double flower | Peach | NC174RL × P1 |
| G3 | Flesh color around stone | Peach | Akame × Jusetou |
| | Anther color (yellow/anthocyanic) | Almond × peach | Texas × Earlygold |
| | Polycarpel | Peach | Padre × 54P455 |
| | Flower color | Peach | Akame ×Jusetou |
| G4 | Blooming time | Almond | D.3.5 × Bertina |
| | Flesh adhesion | Peach | (*P. ferganesis* × IF310828) BC1; Akeme × Jusetou |
| G5 | Non-acid fruit | Peach | Ferjalou Jalousia × Fantasia |
| | Kernel taste (bitter/sweet) | Almond | Padre × 54P455 |
| | Skin hairiness (nectarine/peach) | Peach | Ferjalou Jalousia × Fantasia |
| G6 | Leaf shape (narrow/wide) | Peach | Akame × Juseitou |
| | Plant height (normal/dwarf) | Peach | Akame × Juseitou |
| | Male sterility | Peach | Ferjalou jalousie × Fantasia |
| | Fruit shape (flat/round) | Peach | Ferjalou jalousie × Fantasia |
| | Self-incompatibility | Almond | Ferragnes × Tuono;D.3.5 × Bertina Padre ×54P455 |
| | | Apricot | Lito × Lito |
| G6–G8 | Fruit skin color | Peach | Akame × Juseitou |
| | Leaf color (red/green) | Peach | Garfi × Nemared; P.2175 × Felinem, Akeme × Juseitou |
| G7 | Root-knot nematode resistance | Myrobalan plum | P.2175 × Felinem |
| | Resistance to powdery mildew | Peach | (*P. ferganensis* IF310828)NC1 |
| | Leaf gland (reniform/globose) | Peach | (*P. ferganensis* × IF310828)NC1 |

## 4 The Future

Full genomic sequencing, as has been achieved in human and a number of agronomic crops, is now possible for specialty horticultural crops, such as those in the Rosaceae. Representative crops within the Rosaceae are indeed currently being sequenced. Genomic libraries and physical maps for genera within Rosaceae are also forthcoming. Expressed sequence tags (ESTs), unigene sets, microarrays, proteomic and metabolomic tools are on available for analysis of significant Rosaceous crop and will be presented in later chapters. Study of Rosaceae synteny is only now beginning (Arús et al., 2006). Gene function validation, including high throughput plant transformation systems with cross validation are also being developed.

Bioinformatics software and databases must be developed for Rosaceous crops. Ontology must be consistent within the global community for the broadest communication of data and information. Phenotype databases that include origin and pedigree information as well as observational descriptive data should be developed to promote linkages of data and maximum use. Inventories of genebanks of wild and commercial germplasm in current gene banks, which are fortunately available, and biological resource centers that store mutants, libraries and clones are available for research use through international consortia.

Challenges do lie ahead for the Rosaceous genomics community. Traditional breeding practices have produced a standard of excellence for yesterday and today's finest cultivars for fruit and ornamentals. Though recent advances in genome-level analyses have proven a valuable route to deriving an understanding of the molecular events that govern plant responses in model systems, these new technologies must prove that they can do as well in producing useable crop products for future generations.

## References

Adanson, M. 1763. *Familles des plantes.* Vincent, Paris.

Adanson, M. 1963. The Bicentennial of Michel Adanson's "Familles des plantes." George H. M. Lawrence (ed.). Part 1. xii, 392 pp.

American Heritage Dictionary. 2000. The American Heritage® Dictionary of the English Language, 4th ed. Houghton Mifflin Co.

Angiosperm Phylogeny Group. 1998. Angiosperm Phylogeny Group classification for the orders and families of flowering plants: APG I Bot. J. Linn. Soc. 85:531–553.

Angiosperm Phylogeny Group. 2003. An update of the Angiosperm Phylogeny Group classification for the orders and families of flowering plants: APG II. Bot. J. Linn. Soc. 141: 399–436.

Arús, P., T. Yamamoto, E. Dirlewanter, and A. Abbott. 2006. Synteny in the Rosaceae. Plant Breed. Rev. 27: 175–211.

AWP. 2007. Angiosperm Phylogeny Website. Version 7, May 2006. http://www.mobot.org/MOBOT/research/APweb/ managed by Stevens, P. F. Accessed 22 June 2007.

Challice J. S. 1974. Rosaceae chemotaxonomy and the origins of the Pomoideae. Bot. J. Linn. Soc. 69:239–259.

Challice J. S. 1981. Chemotaxonomic studies in the Rosaceae and the evolutionary origins of the subfamily Maloideae. Preslia 53:289–304.

Crepet, W. L., E. M. Friis, and M. A. Gandolfo. 2004. Fossil evidence and phylogeny: The age of major angiosperm clades based on mesofossil and macrofossil evidence from Cretaceous deposits. American J. Bot. 91: 1666–1682.

Cronquist. A. 1981. An integrated System of Classification of Flowering Plants. Columbia University Press, New York.

Dahlgren, R. M. T. 1980. A revised system of classification of the angiosperms. Bot. J. Linn. Soc. 80:91–124.

Dickinson, T. A. 2007. (http://www.botany.utoronto.ca/faculty/dickinson/Rosaceaeevolution/evolution.html.

Dickinson, T. A, R. C. Evans, and C. S. Campbell. 2002. Rosaceae classification and phylogeny: introduction and overview. http://www.botany2002.org./symposl3/abstracts/5.shtml accessed 28 Janauary 2009.

Dirlewanger, E., E. Graziano, T. Joobeur, F. Garriga-Calderé, P. Cosson, W. Howad, and P. Arús. 2004. Comparative mapping and marker-assisted selection in Rosaceae fruit crops. PNAS. www.pnas.org/cgi/doi/10.1073/pnas.0307937101

GRIN, 2007. USDA, ARS, National Genetic Resources Program. Germplasm Resources Information Network – (GRIN) [Online Database]. National Germplasm Resources Laboratory, Beltsville, Maryland. URL: http://www.ars-grin.gov/cgi-bin/npgs/html/family.pl?972 (30 April 2007).

Heywood, V. H. 2007 Flowering plants of the world. Mayflower Books. Elsavier Publisher, New York, NY. pp. 141–144.

Hutchinson, J. 1964. The genera of flowering plants. Vol 1. Dicotyledons. Clarendon, Oxford.

Janick, J. 2005. The origins of fruits, fruit growing, and fruit breeding. Plant Breed. Rev. 25: 255–320.

Judd, W. S., C. S. Campbell, E. A. Kellogg, and P. F. Stevens. 1999. Plant Systematics: A phylogenetic approach. Sinauer Associates, Inc. Sunderland, MA. 290–306.

Jussieu, Antoine Laurent de. 1789. Genera plantarum.

Linneaus, C. 1735. Systema Natural.

Mabberley, D. J. 1987. The Plant-Book. Cambridge University Press, Cambridge, U.K. pp. 506–507.

Morgan, D. R., D.E. Soltis, K. R. Robertson. 1994. Systematic and evolutionary implications of rbcL sequence variation in rosaceae. Am. J. Bot. 81: 890–903.

Soltis, D. E., P. S. Soltis, P. K. Endress, M.W. Chase. 2005. Phylogeny and Evolution of Angiosperms. Sinauer, Sunderland, Massachusetts. pp. 237–248.

Takhtajan, A. 1997. Diversity and Classification of Flowering Plants. Columbia University Press, New York.

Thorne, R. F. 1992. Classification and geography of flowering plants. Bot. Rev. 58: 225–348.

Thorne, R. F. 2000. The classification and geography of the flowering plants: dicotyledons of the class Angiospermae. Bot. Rev. 66: 441–647.

Trabut, L. (1917). *Pyronia*. J. Hered. 7: 416–417.

Wikström, N., V. Savolainen, and M. W. Chase. 2001. Evolution of the angiosperms: Calibrating the family tree. Proc. Roy. Soc. London B, 268: 2211–2220.

# 2. Genomics Approaches to Crop Improvement in the Rosaceae

**Cameron Peace and John L. Norelli**

## 1 Use of Genomics in Rosaceae

Genomic research in Rosaceae crops is commonly directed at understanding the genetic control of important agronomic traits with the aim of improving product quality and reducing production costs. Genomic knowledge can be used for genetic improvement of cultivars through breeding or genetic engineering. Genomic knowledge can also be used for the development of new cultural practices and the tailoring of existing production practices according to genetic categories of cultivars. The translation of genomic data and fundamental discoveries into practical results with real world applications is often termed "translational genomics". However, this term is also used to describe the transfer of genomic knowledge from model organisms, such as *Arabidopsis*, to crop species, with practical application sometimes only implied.

Many crop attributes are limited by the underlying genetics of the cultivars at hand. Breeders seek to raise the bar with each generation, and provide new genetic possibilities. New cultivars are designed to possess improved potential for horticultural performance, whether as incremental gains over previous cultivars, or with novel attributes that set them apart. Decisions regarding parent selection for crossing and progeny selection for advancing potential cultivars are based on knowledge, as well as educated guesses and hunches, of how controlling genes combine and are expressed in breeding populations. Genomics can shortcut or enhance the scope of genetic studies to elucidate the genetic architecture of traits by identifying, quantifying, and validating important genomic regions. It can also identify the genes that control trait variation and determine their strength of expression under varying production conditions. Armed with such knowledge, breeders can more efficiently manipulate germplasm over generations to produce optimum genetic combinations and novel genetic possibilities in the form of new cultivars that perform better for

C. Peace (✉)
Department of Horticulture and Landscape Architecture, Washington State University, Pullman WA 99164, USA

K.M. Folta, S.E. Gardiner (eds.), *Genetics and Genomics of Rosaceae*, Plant Genetics and Genomics: Crops and Models 6, DOI 10.1007/978-0-387-77491-6_2, 

growers and produce superior products for handlers, processors, marketers, and ultimately, consumers. Breeders are therefore genetic architects, designing new products from the manipulation of genetic components.

Genomics can also impact the production of established plantings. In horticultural production, many crop attributes are readily influenced by cultural practices. However, some cultivars may respond poorly to treatments or respond differently across different environments and seasons. Knowledge of the genetics underlying the performance of each cultivar could lead to genetic "diagnostics" that allow cultural practices to be tailored to a specific functional genetic group of cultivars. Another approach is genomic-based crop "therapeutics", or "chemical genomics", an emerging field of research that also allows improvement of plants already under cultivation. Where specific genes are known to influence important traits, compounds can be designed that enhance or interfere with their expression to improve a crop's performance or product quality.

New genomic technologies are also valuable for more fundamental studies. Basic biological research has traditionally avoided Rosaceae species, and numerous organismal systems are much more tractable to studying many basic biological mechanisms. However, certain biological phenomena in Rosaceae crops such as perenniality, dormancy, extended juvenility, scion-rootstock interaction, complex polyploidy, and diverse plant, flower, and fruit form are usually absent in model organisms. The genetic systems of one or more species in the Rosaceae family can offer useful platforms for uncovering the genomic networks underlying these attributes, mechanisms, and processes. In this approach, the rosaceous species is the model organism. Ultimately, fundamental genomic studies in this plant family can be turned to practical use by providing the valuable knowledge that aids in understanding and manipulating existing cultivars, and breeding the next generation.

## *1.1 Genetic Basis of Agronomically Significant traits*

A collection of attributes sets most rosaceous crops apart from field crops and model species. These attributes include perenniality, large plant size, extended juvenility, use of rootstocks, clonal propagation and highly perishable products (strawberry being an exception to the first four attributes). Product quality, rather than yield, is critical for profitability. From the perspective of Rosaceae crop industries, key needs are to (1) improve fresh and processed product quality, shelf life and safety, including the development of novel or improved flavors, textures, aromas, and colors, for a healthier and more satisfied consumer; (2) reduce chemical pesticide use and develop stress tolerant plants for greater environmental sustainability; and (3) decrease labor and energy costs of crop production (The U.S. Rosaceae Genomics, Genetics, and Breeding Initiative White Paper, 2006). The traits associated with these needs, including (1) fruit, nut, and flower postharvest quality, (2) pest, disease and abiotic stress resistance, and (3) plant architecture and phenology, are currently designated as the highest priority targets for improvement by the U.S. genomics, genetics, and breeding community. These priorities are mirrored in the international arena.

### 1.1.1 Genetics or Genomics?

Improvement of Rosaceae crops in the era before formal knowledge of genetics principles, prior to the widespread acceptance of Mendel's laws, was often based on selection and clonal propagation of superior individuals. The principles of Mendelian genetics, followed by their elaboration into quantitative genetics theory, provided a powerful framework for genetic improvement through dedicated plant breeding. The discipline of genetics holds that heritable traits are controlled by interacting alleles of individual genes, themselves interacting with a finite number of other genes in the background of the whole genome. Genomics takes a more holistic approach from the outset, considering the complexities of gene networks and gradually narrowing the focus to specific genetic elements, at which point, genetic approaches can be effectively engaged.

The success of genetics and genomics approaches to crop improvement is strongly dictated by the underlying genetic architecture of traits of interest. The main components of genetic architecture of a trait are heritability (degree of genetic as opposed to environmental control), the number of influencing loci, the genetic action and magnitude of effect of alleles at controlling loci, and genetic linkages with other traits. Important traits can be categorized as qualitative (also known as simple, Mendelian, or discrete traits) or quantitative (also known as complex or continuous traits). Qualitative traits are typically controlled by variation in one gene with high heritability, and are usually readily tackled by genetics. Quantitative traits may be influenced by many genes or by a few genes with low heritability, and can be approached by genetics or genomics, separately or together. Some quantitative horticultural traits have been previously addressed by genetics without significant success, and genomic technologies offer powerful new tools for their elucidation.

Current knowledge of the genetic architecture of important traits in Rosaceae crops can be exemplified by fruit texture attributes. Components of fruit texture, including firmness, softening rate and pattern, hardness, crispness, crunchiness, juiciness, mealiness, fibrousness, turgor, and others, cover the spectrum of genetic architecture. Various genetic and genomic approaches, integrated with physiology, molecular biology, and practical aspects of breeding, have been employed to study fruit texture, as described below for flesh softening and mealiness in peach and nectarine.

### 1.1.2 "Melting Flesh" in Peach and Nectarine

Fruit flesh softening is of considerable interest to the peach and nectarine industry. The market is divided into fresh, which are usually the quick-softening "melting flesh" (MF) types, and canning, which uses almost exclusively non-melting flesh types. Breeding programs are usually separated for fresh-market and canning cultivars, with crosses conducted within, but rarely between, these two categories and typically targeting greater firmness of fruit to facilitate harvest and transport. While the melting texture is most desired by consumers for fresh eating, breeders in some regions, such as Florida and Spain, have developed very firm non-melting flesh (NMF) peach cultivars that are suitable for the fresh market as they do not have a rubbery texture like canning peaches.

The MF/NMF attribute is qualitative, as each tree produces fruit that is either MF or NMF in most cases, and is easily determined by squeezing or biting into ripe fruit. The genetic control of this qualitative trait is stable across seasons and locations, and thus heritability is very high. The trait has long been described as under the control of a single locus, *Melting flesh*, and was part of the first linkage group described for peach (Bailey and French 1949). Basic genetic analysis of segregating populations easily identifies that MF is dominant over NMF (Bailey and French 1949; Peace et al. 2005b).

Various molecular genetic tools were used in several labs (Lester et al. 1994, 1996; Callahan et al. 2004; Peace et al. 2005b) to identify the controlling gene as that encoding endopolygalacturonase (endoPG), an enzyme that metabolizes pectin in the cell wall and is implicated in fruit softening and abscission processes of various crops (Hadfield and Bennett 1998). Our current understanding is that presence of the *Melting flesh* endoPG gene results in MF fruit, while absence of the gene results in NMF fruit. A simple PCR test is available for making this distinction (Peace et al. 2005b). With this knowledge of the gene behind an important horticultural trait, the genetic predisposition of currently grown cultivars can be better understood, and the genetic marker can be used in breeding. These applications are indeed occurring. However, this locus has other interesting aspects.

In loci where genetic polymorphism causes qualitative differences in phenotype, alleles with quantitative effects on phenotype can also be detected (Robertson 1989). Qualitative mutants are often the result of critical mutations in a gene producing a non-functional gene product or no product at all. Quantitative alleles can result from point mutations in the gene sequence, giving a less efficient product and a subtle difference in phenotype (Pflieger et al. 2001). This theory has been supported in studies of *Arabidopsis* (Koornneef et al. 1998), maize (Beavis et al. 1991), and *Drosophila* (Mackay 2004). Recent examination of allelic variation in the *Melting flesh* gene has uncovered evidence that this phenomenon may also occur for softening rate in peach and nectarine (C. Peace et al., manuscript in prep.).

Another revelation from probing the genetic basis of *Melting flesh* was that for some cultivars, another functional endoPG gene resides on the same locus, less than 50 kbp upstream of the *Melting flesh* gene. This second gene encodes an identical amino acid sequence to the *Melting flesh* gene, but differs slightly in the DNA sequence of its introns and promoter region (Peace et al. 2005b, 2007; A. Callahan and C. Peace, manuscript in preparation). These differences presumably alter its transcription and/or translation, and subsequent timing and location of enzymatic action, as this second endoPG gene controls the *Freestone* trait (Peace et al. 2007). Presence of the gene produces freestone fruit, where the flesh fibers are detached from the stone in ripe fruit and the stone comes away freely. Absence of the

*Freestone* endoPG gene is associated with clingstone fruit, where fibers remain attached to the stone. It is possible that minor allelic variants may underlie less extreme phenotypes of flesh fiber adhesion. Together, the two endoPG genes underlie the *Freestone-Melting flesh* (*F-M*) locus, and although the two genes are separate, their identical encoded amino acid sequence and same genomic location identifies the *F-M* locus as being pleiotropic. Combinations of presence/absence of the two genes exist in naturally occurring trees, representing the four known major functional alleles of *F-M* (Fig. 1).

Interactions within the *F-M* locus include additivity, dominance, pleiotropy, and perhaps epistasis. The *Freestone* and *Melting flesh* endoPG genes combine additively in the F allele (actually a haplotype as it encompasses two adjacent genes) to give the FMF phenotype (Fig. 1). In diploid combination in peach and nectarine trees, the F allele is dominant over the other three alleles (thus F- = FMF), the f allele is dominant over the f1 and f2 alleles (ff, ff1, and f2 = CMF), and the f2 allele is recessive to all (f1f1 and f1f2 = CNMF, f2f2 = CNSF) (Fig. 1). The f1 and n alleles appear to have arisen from deletions in the F allele of the *Freestone* gene and both genes, respectively. Flesh of fruit that are homozygous for the f2 allele lose very little firmness during ripening whereas the f1 allele (in homozygosity or as f1f2) results in gradual softening to a rubbery texture. Although it exhibits only the *Melting flesh* gene like the f allele, the f1 allele does not provide the expected clingstone melting flesh (CMF) phenotype – instead it is CNMF. With a different origin to the f allele, the f1 allele's *Melting flesh* gene is either functionally impaired through as yet undetected mutation or the melting flesh phenotype of the F allele requires the presence of both genes. A stable FNMF phenotype has not been reported although theoretically it could arise through deletion of the *Freestone* gene in an F allele.

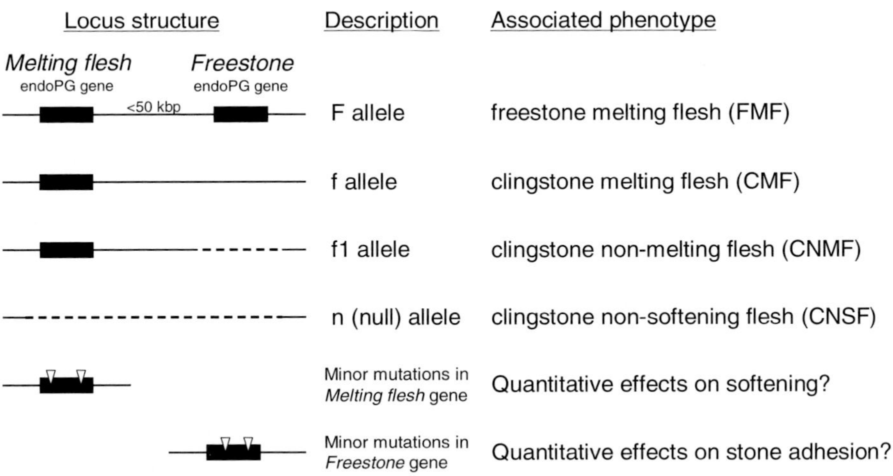


**Fig. 1** Structural organization, alleles, and associated functions of the *Freestone-Melting flesh* locus of peach and nectarine (*Prunus persica*) that contains multiple endoPG genes. The origin of the f1 and f2 allele haplotypes have apparently arisen through gene deletions of the F allele

In conclusion, while *Melting flesh* was originally considered a simple Mendelian trait, two copies of the controlling gene were discovered at the locus, with several major phenotypes resulting from their allelic variants caused by presence/absence of functional gene copies. The locus also appears to underlie quantitative variation in fruit softening. The *F-M* locus occurs at the distal end of *Prunus* linkage group 4, which genetically links it with nearby quantitative trait loci (QTLs) for other important traits such as bleeding, soluble solids concentration, titratable acidity, and flowering date (Peace et al. 2005a, 2006).

### 1.1.3 Mealiness in Peach and Nectarine

Mealiness is another texture-related trait that is of major concern to the peach and nectarine industry, due to its broad dislike by consumers. A large proportion of world cultivars of fresh market peach and nectarine produce fruit that often become mealy (dry and soft with a grainy mouth-feel) after cold storage of a few weeks. Such fruit have the outward appearance of good quality, and are often sold to unaware, and ultimately unsatisfied, consumers (Crisosto et al. 1999). Yet cold storage is required to halt softening and bruising while fruit are shipped to distant markets. Some cultivars appear more susceptible than others, suggesting a genetic component.

The endoPG enzyme has long been implicated in the development of mealiness in susceptible cultivars (Buescher and Furmanski 1978). Study of mealiness susceptibility in a peach population segregating at the *F-M* locus concluded that endoPG plays a qualitative role in the trait's expression, where a functional *Melting flesh* endoPG gene must be present for endoPG activity to occur during but not after cold storage (Peace et al. 2006). Although the melting phase does not occur in fruit that become mealy, partial endoPG activity in storage leads to gradual softening after storage, and appears to enable other genes involved in mealiness development to be expressed (Peace et al. 2005a, 2006). CNMF (and CNSF) fruit are effectively resistant to mealiness, in that they remain too firm to be classified as mealy and don't exhibit the partial expression of endoPG in cold storage. Thus the *F-M* locus is epistatic in the genetic control of mealiness because some of its alleles mask the expression of other loci conditioning mealiness susceptibility. Avoiding mealiness in the fresh market peach industry can therefore be achieved by using non-melting or non-softening types, but if the buttery texture that consumers tend to prefer is to remain, the genetic basis of mealiness susceptibility must be identified in melting flesh types.

Within melting flesh cultivars, susceptibility to mealiness appears to be a quantitative trait with a significant genetic component (Peace et al. 2005a). Heritability was calculated as 0.25–0.30 within melting flesh progeny of two peach populations, with duplicated trees and observations conducted over three years (Peace et al. 2006), indicating that unless this genetic component is controlled by many small-effect loci, it should be feasible to discover and exploit loci conditioning mealiness susceptibility. Genetic models based on phenotypic segregation in controlled crosses suggest that in melting flesh types, mealiness is controlled by as few as two loci with dominant gene action (Peace et al. 2006). Genome-wide QTL analysis in one population identified at least three stable QTLs collectively accounting for almost 50% of the genotypic variation in melting flesh progeny. These QTLs did not always combine additively; some were compensatory, suggesting that if used in marker-assisted selection, one can be selected for in the absence of another to achieve the same level of resistance (Ogundiwin et al. 2007). The QTLs are being further targeted, via map saturation and verification in larger populations, while simultaneously taking a candidate gene approach (see below) to develop diagnostic genetic tests (Peace et al. 2005a, 2006; Ogundiwin et al. 2007). EndoPG may also have a further role to play. Given the major effect of the absence of the *Melting flesh* endoPG gene on mealiness, absence of the *Freestone* gene or other, less extreme, alleles of the *F-M* locus may be expected to quantitatively affect mealiness susceptibility. Indeed, in a melting flesh population segregating only for presence of the *Freestone* endoPG gene, a QTL for mealiness susceptibility co-located with the *F-M* locus (E. Ogundiwin et al., manuscript in prep.). A comprehensive microarray analysis is currently underway to identify additional candidate genes associated with mealiness and elucidate the functional relationships (Ogundiwin et al. 2008).

In conclusion, mealiness susceptibility is a heritable quantitative trait for which an understanding of its genetic basis would be valuable for crop improvement. Genomic analysis is dissecting its complexity into specific elements, and it appears likely that with available resources and technologies, the controlling genes will soon be identified.

As shown in the examples above, highly heritable single gene traits are the most amenable to revealing their genetic architecture. Dirlewanger et al. (2004) summarized the locations of 28 qualitative traits in the *Prunus* genome, for example. Closer examination of such loci may reveal further complexity, and thus additional efforts are useful to understand further effects and interactions. Nevertheless, identification of such major loci allows their use in crop improvement in Rosaceae through genotyping and other approaches (described later). Because quantitative traits require an understanding of potentially many interacting genetic and environmental factors, they are more difficult to elucidate. Ideally, quantitative traits are dissected into their individual components, as attempted for susceptibility to cold storage disorders in peaches (Ogundiwin et al. 2007) which led to the identification of a possible controlling gene for a large proportion of the genetic component of susceptibility to browning (Ogundiwin et al. 2008), in addition to a major heritable role for endoPG in mealiness and bleeding (Peace et al. 2006). For many other quantitative traits, their complexity has not yet been untangled. Highly heritable quantitative traits, while easier to dissect, are also readily improved by phenotypic selection, as the better performing individuals in a breeding program tend to carry the alleles for that superiority. However, there still remains much value in elucidating the genetic architecture of these traits, such as in parent and cross selection, gene pyramiding (combining a series of positive alleles from multiple genes, e.g. for durable disease resistance), and in saving time in progeny selection for crops with extended juvenility. Valuable traits with low heritability are the greatest challenge for genetic architecture dissection, and yet would benefit most from such genomic knowledge as phenotypic selection results in slow genetic gain over generations. Advances in genomics tools and technologies may address even these historically recalcitrant traits.

## 1.2 Genomic Approaches for Crop Improvement

Genomics approaches fall into various categories and go under various names. Structural, functional, and comparative genomics describe three basic categories of knowledge that researchers gather as they ultimately seek to discover the genetic basis of biological processes and important agronomic traits. Within, and often spanning each of these fields of study, are interconnected technologies and techniques that can be brought to bear in such scientific endeavors, and form an expanding toolkit that the modern Rosaceae genomicist (or geneticist) can employ for their fundamental or applied research (Fig. 2). Some of these approaches are described below, with examples of their application in the Rosaceae family.

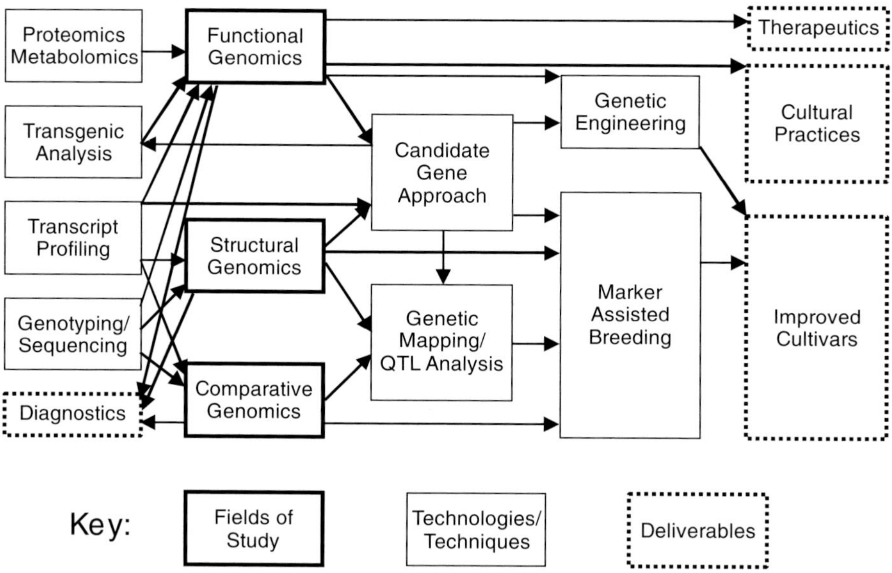


**Fig. 2** Genomic approaches for crop improvement use various technologies and techniques to address three general fields of genomic study: functional, structural, and comparative genomics. Genomic research in Rosaceae crops is commonly directed at understanding the genetic control of important agronomic traits with the aim of producing "deliverables" to improve product quality and reduce production costs

### 1.2.1 Fields of Study

Structural Genomics

Structural genomics is concerned with the physical structure and organization of individual genomes. Genome maps, both genetic linkage maps and physical maps, provide an informative description of the chromosomes of Rosaceae crops that can be used to localize important loci and determine interactions between them that ultimately produce phenotypes of interest. Genetic linkage maps, which describe the degree of co-inheritance between genetic loci across the genome of an organism, are abundant for *Prunus* crops (stone fruit and almond), pome fruit crops (apple and pear, and under development for others such as loquat), cane berries (raspberry and blackberry), rose, and diploid strawberry (Dirlewanger et al. 2004; Shulaev et al. 2008; Jung et al. 2008). Such maps are usually produced for the purposes of determining the genetic control of specific traits. However, reference maps of *Prunus* (Aranzana et al. 2003), apple (Silfverberg-Dilworth et al. 2006), and strawberry (Sargent et al. 2006) have been developed for use as a general resource within crop groups. The bin-mapping approach (described later) can simplify the general placement of any DNA sequence within an existing genetic linkage map, particularly when combined with a heterozygous reference linkage map (Howad et al. 2005).

More recently, physical maps have been created (peach: Zhebentyayeva et al. 2008; apple: Han et al. 2007) to allow more precise genomic placement of any DNA sequence. Physical maps describe the location of hundreds or thousands of overlapping identifiable sequence landmarks, such as BACs (large segments of the genome cloned into bacteria) or ESTs (expressed sequence tags), which have been anchored to a genetic linkage map. In contrast to linkage maps, physical maps measure distance in base pairs rather than relative genetic linkage measured in centiMorgans (cM) and the number of base pairs contained within a cM will vary across the genome. A fully-saturated physical map would contain BACs (or other DNA clones) tiled across the entire genome and allow physical base-pair distances between features of a chromosome to be readily calculated.

The ultimate genome map would fully integrate genetic linkage markers, physical location of cloned genomic DNA and complete genomic sequence for all the chromosomes of an organism. Complete genome sequencing is currently underway for one reference individual each for apple, peach, and strawberry, placing the Rosaceae plant family at the cutting edge of the genomics revolution. These complete maps are expected to function as publicly accessible resources for each crop group by 2010, greatly enhancing the efficiency of identifying gene networks controlling important traits for rosaceous crop production.

Structural genomics is also applicable to a finer scale, in determining the physical structure of individual loci. The structural organization of the self-incompatibility locus of *Prunus*, for example, as described by Ushijima et al. (2004), deepens our understanding of the functioning, diversity, and manipulation of the evolutionarily and commercially important traits of cross-compatibility and self-fertility.

## Functional Genomics

Functional genomics addresses biological questions by studying the function of individual genes and the interactions among groups of genes. It uses both "forward" genomic (or genetic) approaches that start from a phenotype or function and work toward the identification of DNA sequence, and "reverse" approaches that start from DNA sequence and work back to a function. Functional genomics typically relies upon a complement of forward approaches to discover genes associated with a trait and reverse approaches to confirm and study the role of specific genes in biological function. Global or genome-wide analysis is used to identify various genes or gene networks associated with a trait. These include methodologies such as phenotypic screening of mutant "libraries" and many types of transcriptional profiling which are described below. BLAST analysis is often relied upon to compare the nucleic acid sequence homology of the genes or transcripts identified in the analysis against large databases of previously annotated sequences from a wide array of organisms. The supposition of function from sequence homology is a valuable shortcut in functional genomics, and effective in many cases. However, BLAST annotation does not determine gene function, it merely suggests it. In addition, many sequences lack significant homology to functionally characterized proteins or domains, and functionality cannot be implied. For example, Horn et al. (2005) reported that 24.3% of almost 10,000 peach EST sequences had no known homology. Functionality is usually determined or verified using "reverse" genomic approaches, such as RNAi, over-expression of a transgene, and some types of transcript profiling. Given the large cost of one-gene-at-a-time functional testing, such studies are usually limited to genes with known or at least strongly suspected horticulturally significant function.

A major constraint in functional genomics is the availability of high-throughput phenotyping and functional assays to facilitate the analysis of hundreds or thousands of mutants and genes. The long juvenility of many Rosaceae species and the importance of flower and fruit traits in these crops make this a significant challenge for Rosaceae genomics. Altering the expression of the MADS box genes regulating juvenility through genetic engineering provides a possible mechanism to overcome this obstacle and has recently been demonstrated to reduce juvenility in apple and plum (Flachowsky et al. 2007; Kotoda et al. 2006).The development of high-throughput gene function testing platforms are also being developed for Rosaceae (http://strawberrygenomics.com/).

## Comparative Genomics

Comparative genomics connects studies in structural or functional genomics across crops by seeking commonalities within and between Rosaceae genera and subfamilies. Comparative genomics in Rosaceae is based on the assumption, originating from taxonomic classification by morphology, that rosaceous species are connected through a shared ancestry and thus have genomic similarities. Researchers investigating Rosaceae evolution seek to identify how, when, and where the family split from others, and member species and higher taxonomic clades (e.g. subgenera, genera, and subfamilies) differentiated from each other (e.g. Potter et al. 2007). Evolutionary genomics in Rosaceae is beginning to provide a solid foundation for comparative genomics, facilitating the transfer of genetic knowledge between species and shedding light on the genomic basis of the diversity of form and function in this family. Genetic maps constructed for specific experimental populations are readily aligned with others from the same crop where common markers are used, and such efforts between crops have identified an almost identical genome structure between apple and pear (Pierantoni et al. 2004) and between all crop members of *Prunus* (Dirlewanger et al. 2004). Common elements have also been identified across more distantly related Rosaceae crops, such as strawberry and *Prunus* (Vilanova et al. 2007) and apple and *Prunus* (Dirlewanger et al. 2004). Once whole genome sequences are available for apple, peach, and strawberry, as well as others as sequencing and assembly technologies advance and become cheaper, it should be possible to reconstruct the ancestral "Rosaceae genome" – a reference genome map for the Rosaceae family that will greatly facilitate the transfer of genetic knowledge across crops. Large-scale functional annotation of ESTs in Rosaceae relies on comparative genomics extending beyond this family, with gene function described from GenBank sequences that rarely originate from Rosaceae species. Comparative genomics on a gene-specific scale is illustrated in the characterization of a powdery mildew resistance locus across *Malus*, *Prunus*, and *Rosa* (Xu et al. 2007).

### 1.2.2 Technologies and Techniques

#### Genetic Mapping and QTL Analysis

Genetic loci are assembled into linkage groups and ordered relative to each other within groups using the technique known as genetic linkage mapping. Linkage groups represent chromosomes, initially partial segments and eventually whole chromosomes, once enough loci are included to span intervening gaps between the segments. Distances between loci are described in recombination units, usually cM. Genetic maps are created to serve either of two purposes: (1) to develop a general resource that locates new loci or "markers" on the genome of an organism, frequently referred to as a reference map or (2) to determine the number, location, and effects of loci controlling specific traits of interest. As described in *Structural genomics* above, genetic maps exist for all of the major rosaceous crops, including reference maps for each of the three crop subfamilies. The public Genome Database for Rosaceae displays many of these maps to allow users to focus in on regions of interest, or align maps for comparative purposes (Jung et al. 2008). Genetic maps in Rosaceae began with qualitative morphological traits under monogenic control, expanded to the use of isozyme, restriction fragment length polymorphism (RFLP), and random amplified polymorphic DNA (RAPD) markers, and now tend to rely on simple sequence repeat (SSR) markers due to their ability to facilitate alignment with other maps. Readily generated, but population-specific, dominant markers such as amplified fragment length polymorphisms (AFLPs) are used to saturate the regions between SSRs. Markers derived from ESTs and candidate genes are increasingly being included in genetic maps (see *Candidate gene approach* below).

Traditional QTL discovery approaches utilize software that require specific population types, usually $F_2$ or backcross which have relatively simple statistics for QTL analysis underlying them. However, experimental mapping germplasm of Rosaceae crops are typified by $F_1$ populations, because widespread self-incompatibility and high heterozygosity, coupled with long generation times for the tree crops, make it difficult to construct the simpler $F_2$ genetic populations. The development of the double pseudo-testcross strategy (Grattapaglia and Sederoff 1994) allowed researchers to use $F_1$ populations by mimicking the backcross model for each heterozygous parent of the cross. Once software became available for QTL analysis of $F_1$ populations that considered possible effects from all four alleles segregating at a locus in a diploid cross of QTLs (MapQTL, van Ooijen 2005), it quickly became popular in Rosaceae QTL studies. Available QTL mapping software applies to diploid species, and QTL analysis in polyploids remains problematic. Thus, genetic maps in polyploid crops of Rosaceae (strawberry, plum, rose, and tart cherry) are the least developed, and QTL discovery in these crops the least advanced.

To "map" or estimate the location of a marker on a given chromosome, a marker must be polymorphic within the population so that the frequency of recombination among the progeny can be measured. Reference maps therefore benefit from the use of wide crosses between unrelated parents to maximize the chance that any marker is polymorphic. The *Prunus* reference map relies on a cross between peach and almond and most markers screened on it are polymorphic (Aranzana et al. 2003).

Bin-mapping, described by Howad et al. (2005), is emerging as an efficient approach to locate any marker or DNA sequence in an organism's genome. By screening markers on a subset of the mapping population pre-selected to represent widely separated cross-over events on all chromosomes (referred to as a bin-set), a general map location can be rapidly estimated for polymorphic markers. The bin-set for the TxE (almond "Texas" × peach "Earlygold") reference population for *Prunus* consists of just eight plants (one of the parents, the $F_1$ hybrid, and six of $F_2$ progeny plants), and any polymorphic codominant marker will fall into one of 67 intervals on the map ("bins"), each equivalent to approximately an eighth of a chromosome (Howad et al. 2005). Bin-sets are also being developed for specific mapping populations, such as apple (Celton et al. 2009), aiding in the marker saturation of targeted regions and the rapid placement of candidate gene markers within a genetic map.

To determine the genetic control of specific traits, parents are chosen that contrast in expression of the trait, in order that the mapping population segregates for the trait, and in general genetic background, so that markers are likely to be polymorphic. For example, in genetic investigations of fruit size in sweet cherry, a large-fruited elite cultivar was crossed with a small-fruited wild variety (Olmstead et al. 2008). With genotypic and phenotypic data collected for the population, statistical procedures are then used to associate allelic variation in genetic markers with performance differences, and position influencing loci on the genetic map. Locations of many qualitative and quantitative traits, for quality, productivity, resistance to diseases, pests, environmental stresses, and physiological disorders are known for Rosaceae crops (described in the later crop-specific chapters). Loci controlling single gene traits can be readily located with simple linkage analysis, where the locus is treated as just another marker. Dirlewanger et al. (2004) summarized the locations of numerous qualitative traits in *Prunus*. The most common method used to identify associations between markers and quantitative traits in Rosaceae is standard QTL analysis, which uses genotypic and phenotypic data collected from large mapping populations. A popular software package used for QTL analysis is MapQTL (van Ooijen 2005). This QTL identification software is compatible with the map construction software JoinMap that enables the joint analysis of markers inherited from each parent. While inheritance from multiple parental sources complicates genetic analyses, this is a common feature for outcrossing Rosaceae species, where $F_1$ experimental populations are used. The use of MapQTL therefore allows the simultaneous detection and joint influence determination of QTLs with multiple alleles from each parent. Association mapping is a technique to establish gene-trait associations that is useful where mapping populations have not been established, but germplasm from a diverse collection of unrelated individuals is available (Oraguzie and Wilcox 2007). Use of association mapping in Rosaceae has only just begun.

### Transcript Profiling

Transcript profiling has become a cornerstone of functional genomics because it provides a high-throughput forward (function to sequence) genomic approach for gene discovery that does not require high-throughput functional assays. Transcripts are isolated from a group of plant samples and identified by a variety of methods that take advantage of advances in DNA sequencing and/or the accumulated DNA sequence information available for a given species. In general, the genomicist designs treatments to be applied to plants prior to sample harvest and that provide insight into the particular function to be studied. There are currently several good methodologies for transcript profiling and the development of new methodologies is a rapidly evolving field. The most recent advances in transcript profiling may be considered obsolete within a few years. However, all methodologies have their own advantages and disadvantages, and an understanding of these can help the genomicist select the methodology most appropriate for their crop and goals. Often, no single method is "best" for a specific experimental system, and a variety of methods can overcome the disadvantages of any one method, to provide a more complete or robust analysis. The most appropriate methodology is also dependent upon the amount of EST and genomic sequence information available for the crop of interest. Hence, the most appropriate methods will change as more sequence information becomes available and new technologies are developed.

Transcript profiling within a crop often begins with expressed sequence tag (EST) profiling studies. ESTs are transcribed, spliced nucleotide sequences that are derived from a specific tissue under a specific set of conditions that provide a crude inventory of the genes expressed under those conditions. Often, the tissue used for transcript isolation is the sole treatment or condition in the study. ESTs are usually produced by high-throughput single pass sequencing of cDNA resulting in low quality sequence information of relatively short sequence length, with a relatively high sequencing error rate. The major advantage of EST profiling is the low cost generation of coding (gene) sequence information in species with little genomic sequence information. EST profiling studies have been conducted in Rosaceous crops and there are currently 416,000 Rosaceae ESTs sequences available in GenBank, with over 255,500 ESTs for *Malus* × *domestica*, over 71,100 for *Prunus persica,* over 45,400 for *Fragaria vesca* and over 5,500 for hybrid *Rosa* (Lazzari et al. 2005; NCBI EST Database 2008; Newcomb et al. 2006). Park et al. (2006) used publicly available EST data to predict genes expressed in apple during fruit growth and development and to predict biochemical pathways involved in biosynthesis of precursors for volatile esters important to fruit flavor. The primary disadvantages of EST profiling are: (1) the genes identified include both genes associated with experimental treatments and housekeeping genes, (2) rare transcripts are difficult to detect, and (3) because EST sequence information is often derived from several different laboratories using different cultivars grown under different environmental conditions, associating trends in gene expression with specific biological functions can be difficult. To facilitate the identification of rare transcripts, cDNA libraries are often "normalized" to equalize the relative abundance of all transcripts (Soares et al. 1994). Because normalization eliminates information on EST abundance, these libraries should not be included in studies attempting to use EST profiling data to establish trends in gene expression with specific biological functions.

Suppression subtractive hybridization (SSH) and cDNA-amplified fragment length polymorphism (cDNA-AFLP) are transcript profiling techniques that can efficiently identify both abundant and rare transcripts differentially up- or down-regulated under specific experimental conditions. Both techniques are useful in species with little or no sequence information. A limitation of both techniques is that they usually do not yield a complete inventory of gene expression. Genes regulated in response to specific treatments are selected in SSH by sequential nucleic acid hybridizations in which the reference treatment cDNA, designated as the "driver", is present in a molar excess compared to "tester" cDNA, in which changes in gene expression are being investigated (Diatchenko et al. 1996). As the mechanics of the assay also include amplification of differentially expressed sequences by PCR that favor the normalization of up- and down-regulated sequences, regardless of the relative abundance of the original mRNA in the cell, a primary advantage of SSH is its ability to detect rare transcripts often missed by general EST profiling methods. Disadvantages of SSH are that it is not quantitative and requires careful control of non-treatment variation between samples. SSH analysis has been successfully used to characterize apple's response to several abiotic and biotic stresses, including short days, cold temperature, UV irradiation, fire blight, apple scab and phyllosphere colonization (Ban et al. 2007; Bassett et al. 2006; Degenhardt et al. 2005; Kuerkcueoglu et al. 2007; Norelli et al. 2009). cDNA-AFLP is another PCR based methodology that uses restriction enzymes to cut cDNA, followed by subsequent ligation of adaptors to facilitate PCR amplification and visualization of fragments on polyacrylamide gels to identify differentially expressed transcript. The primary advantage of cDNA-AFLP is that it facilitates direct, side by side comparison of transcript fragments from different cultivars under different experimental conditions for cross cultivar comparisons. Because cDNA-AFLP requires extraction of individual DNA fragments from gels to obtain sequence information, it is labor intensive and is not amenable to high-throughput data recovery. cDNA-AFLP has been widely used to characterize transcriptional responses in the Rosaceae (Balogh et al. 2005; Campalans et al. 2001; Geuna et al. 2007; Jensen et al. 2003)

The generation of EST and genome sequence information within the Rosaceae makes the use of more comprehensive transcript profiling techniques, such as microarray analysis, possible. Microarray analysis is a hybridization-based technique in which thousands of gene probes designed to match predicted open reading frames are arrayed on a solid surface and hybridized to transcripts labeled with a fluorescent dye. In general, RNA is isolated from biological samples, used as template for the synthesis of cDNA that is either labeled with a fluorescent dye or used as template for the synthesis of RNA that is labeled. After hybridization the array surface is laser scanned to determine the amount of transcript hybridized to each probe. It is a quantitative method that allows the researcher to obtain a "snap shot" of the expression of thousands of gene in specific tissues under a specific set of conditions. There are many different platforms used for microarray analysis due to the numerous options in manufacturer, method of fabrication, probe type and array design. Microarrays can be printed with fine-pointed pins on glass slides, often referred to as "printed arrays", or produced by various photolithography and electrochemical printing methods, sometimes referred to as "biochips". Early microarrays often used cDNA probes that do not require extensive genome sequence information for design but tend to have higher cross-hybridization between gene family members, greater difficulty detecting splice variants, require PCR synthesis of hundreds of genes that is time consuming and prone to error, and tend to have lower quality control than oligonucleotide probes. Oligonucleotide probes require extensive genome sequence information and bioinformatics analysis for proper design, but they have largely replaced cDNA probes because they can overcome the problems associated with cDNA probes and they facilitate high density arrays. Short oligonucleotide probes (20–30 mers) are cheaper to produce and facilitate ultra-high density arrays, whereas long oligonucleotide probes (60–70 mers) tend to have greater specificity to individual gene family members. Two-channel arrays allow direct comparison of two different treatment samples on a single array by labeling each cDNA template with a different fluorophore. Although an absolute level of gene expression can be obtained from two-channel designs, results from these arrays are often presented in relative difference or the ratio of gene expression among the various probes. One-channel arrays are designed to estimate the absolute level of gene expression from single-dye hybridization. This makes it easier to compare microarray results from different experiments, but requires twice as many arrays. The design and analysis of microarray experiments is complex due to the multiple variables in microarray technology and the large number of gene probes tested. Multiple levels of replication are necessary to account for the variation among biological samples, arrays, transcript labeling and dye detection, resulting in large, costly experiments utilizing many individual arrays. The vast amount of data generated and multiple comparisons between thousands of probes make statistical analysis and data interpretation challenging. Because the technique is quantitative, statistical analysis is necessary to draw valid conclusions regarding changes in gene expression. Although an array may contain tens of thousands to a couple of hundred thousand probes, only a small percent of them may show statistically significant results. The large number of platforms, the number of independent users, the varying data formats and the varying methods of analysis used in microarray experiments make standardization and comparison of results difficult. Despite these limitations, microarrays are a powerful tool for transcript profiling because they are capable of simultaneously detecting changes in the expression of many genes which facilitates the association of specific signaling and enzymatic pathways to complex biological functions.

One of the first *Malus* microarrays was a 15,720 oligonucleotide probe, printed array developed at Plant and Food Research in 2005 that was based on a subset of non-redundant EST contigs (unigenes) derived from Plant and Food Research's apple EST database (Newcomb et al. 2006). This array has been used to study the environmental effects on tree-to-tree variability in the orchard and the genesis of fruit aroma (Pichler et al. 2007; Schaffer et al. 2007). More recently, a 40,000 feature *Malus* array was developed in the laboratory of Dr. Schuyler Korban at University of Illinois that contains 548 control probes and 39,412 long-oligonucleotide (70 mer) probes designed to *Malus* unigenes derived from publicly available EST data and approximately184,000 *Malus* EST*s* (154,000 5′ reads and 30,000 3′ reads) identified in an NSF-funded project from different tissues, genotypes, developmental stages and stress conditions (Gasic et al., submitted). These *Malus* microarrays have also been successfully used for transcript profiling during stone development in peach fruit (Callahan et al. 2008). Microarrays have also been developed as diagnostic tools to detect pathogen development within Rosaceae host species (Schneider and Sherman 2007; Sholberg et al. 2005).

### Proteomics and Metabolomics

System-wide technologies in molecular biology extend to detection and analyses of the entire protein and metabolite array in organisms, although often focused on particular tissues at particular developmental stages as for transcriptomics. Proteomics (Pandey and Mann 2000) and metabolomics (Fiehn 2002) are the disciplines concerned with the application of such technologies. They can be employed to better understand the molecular physiological processes underlying traits (complementing forward genetics approaches), or the downstream effects of gene expression (complementing reverse genetics approaches). While a rosaceous plant may have in total tens of thousands of different genes and slightly more transcripts (although only a fraction in any given tissue), it may contain hundreds of thousands of different proteins (including enzymes and structural units) and metabolites (particularly secondary metabolites). The interactive networks of these gene-environment products are therefore likely to be extremely complex. Proteomics and metabolomics have great potential to elucidate biological processes, but are recent arrivals on the molecular biology scene and their associated toolboxes are still mostly under development. Challenges remain in large-scale identification of proteins and metabolites (Fridman and Pichersky 2005), in addition to associating networks and specific proteins and metabolites with horticultural traits. Furthermore, while individual proteins can be readily connected to their encoding gene, connecting specific metabolites with their underlying genetic sources is difficult (Schauer and Fernie 2006). As yet, there are only a few applications of these disciplines in Rosaceae. Grimplet et al. (2004) used proteomics to connect expressed genes with their translated products in apricot. Alm et al. (2007) examined hundreds of proteins in strawberry to study allergen content. In apple flesh, Guarino et al. (2007) detected 303 distinct proteins, of which 44 were identified and associated with 28 different genes. Metabolic profiling of apple peel detected more than 200 components, of which 78 were identified (Rudell et al. 2008).

### Candidate Gene Approach

Narrowing the vast array of information resulting from genomic research to specific genes is fundamental to the application of genomics for crop improvement. The candidate gene approach attempts to utilize knowledge generated by structural, functional, and/or comparative genomics, as well as classical molecular biology, physiology and genetics, to identify "candidate" genes with a high likelihood of playing an important role in the phenotype of a specific trait. Once candidates are identified, DNA markers, such as simple sequence repeat (SSR) and single nucleotide polymorphism (SNP), are developed for the genes. These gene-specific markers are then mapped, and their locations compared to known loci for the trait of interest. Co-localization of candidate gene markers with either known qualitative or quantitative trait loci identifies candidates that warrant functional verification, and provides a rational approach to maximize limited resources for greatest impact. If further functional analysis determines a causative role for the gene in the trait of interest, functional allele-specific markers for the trait are established that test for the causative DNA sequence differences underlying the functional differences. Allele-specific gene markers are also known as "perfect markers", as they avoid the possibility of recombination that can occur when a marker is only genetically linked to the gene. Gene-specific markers are usually very robust and can be useful across different genera of the Rosaceae. Etienne et al. (2002) described a candidate gene study of peach to identify genes underlying major loci and QTLs for acidity and sugar content. Eighteen candidate genes were chosen, twelve were mapped, and a gene involved in solute accumulation co-located with a QTL for soluble solids concentration. Other examples of significant progress with the candidate gene approach in Rosaceae include associating the gene for flavanone 3-hydroxylase with yellow fruit color in strawberry (Deng and Davis 2001), genes for ethylene biosynthesis and cell wall modification genes of Md-ACS1, Md-ACO1, and Md-Exp7 (an expansin) with firmness and/or storability in apple (Oraguzie et al. 2004; Costa et al. 2005, 2008), PpLDOX (leucoanthocyanidin dioxygenase) with cold storage-induced browning in peach (Ogundiwin et al. 2008), the transcription factor MdMYB10 with red flesh color in apple (Chagne et al. 2007), and endoPG with *Freestone*, *Melting flesh*, and mealiness in peach (Peace et al. 2005b).

In some cases, the identification of specific candidate genes can be based on prior biological research that established the association of specific enzymes or proteins with a biological process. The association of endoPG with fruit softening and the establishment of it as a marker for melting flesh is an example of such a case. In other cases, a specific class of protein may be associated with a trait that can serve as a means of identifying "candidates". For example, major resistance (R) genes often encode nuclear binding site (NBS) – leucine rich repeat (LRR) protein kinases, and NBS-LRR resistance gene analogs (RGAs) can be used for the identification of candidate disease resistance genes (Baldi et al. 2004; Samuelian et al. 2008). A "resistance gene map" was presented by Lalli et al. (2005) that described the genomic location of such gene sequences putatively involved in pathogen resistance in *Prunus*.

For many complex traits of importance, genomic analysis can lead to the identification of several hundred genes associated with a specific trait. In such cases, bioinformatics combined with inference drawn from the scientific literature can be used to narrow the focus to a smaller number of candidate genes. For example, transcript profiling of fire blight-challenged apple leaf tissue resulted in the identification of 650 *Malus* expressed sequence tags (ESTs) associated with fire blight disease (Norelli et al. 2009; Malnoy et al. 2008a). Bioinformatics was used to identify fire blight-associated ESTs that (1) appeared unique when compared with ESTs isolated from apple tissues that were not challenged with the fire blight pathogen (Baldo et al. 2007), (2) had significant BLAST similarities to 2,800 *Arabidopsis* genes known to be regulated in response to bacterial challenge (Thilmony et al. 2006) or systemic acquired resistance, and (3) had been identified by both suppression subtractive cDNA hybridization (SSH) and cDNA-AFLP transcript profiling. The ESTs identified by bioinformatics were then ranked for their potential importance in resistance based upon inferences from the scientific literature. SSR and SNP markers derived from highly ranked fire blight-associated ESTs were mapped in a "M.9" × "Robusta 5" population in which a major QTL for fire blight resistance has been located on linkage group 3 (Peil et al. 2007). Markers for heat shock protein 90 (Hsp81-2), a secretory class III peroxidase, and a serine/threonine-protein kinase mapped to the LG3 fire blight resistance QTL and reduced the QTL's size from 12 to 4 cM (S. Gardiner, manuscript in preparation). Markers for a "putative disease resistance protein" (NCBI AY347778) and Skp1 (SCF-type E3 ubiquitin ligase) mapped to positions corresponding to the location of two QTLs reported in other populations (Calenge et al. 2005; Khan et al. 2006). To date, of 28 candidate fire blight resistance gene markers that have been mapped, six have co-located to or near known fire blight resistance QTLs (Norelli et al. 2008). As whole genome sequence becomes available for Rosaceae species, similar approaches could be used to scan coding regions within established QTLs for potential candidate genes, further improving the efficiency of the candidate gene approach.

Co-localization of a candidate gene marker and a specific genetic locus does not prove a causative role for the gene in a specific trait phenotype; co-localization could be the result of coincidental linkage. Further functional analysis is necessary to establish a causative role. Furthermore, if the candidate gene does not co-localize in a specific segregating population, it cannot be concluded that the gene is not associated with the trait of interest. Complex traits can be affected by multiple biological mechanisms controlled by genes at several locations of the genome that may be of importance in other populations. To be able to detect all functional alleles of all causative genes for a trait of interest in a crop, or at least in the germplasm to be improved in a breeding program, it is therefore important to survey individuals that fully represent the germplasm and apply appropriate bioinformatics tools such as pedigree based analysis (see *A uniting statistical approach* below) or association mapping.

## Transgenic Analysis

Efficient *Agrobacterium*-mediated plant transformation technology has been developed within the three Rosaceae subfamilies containing the majority of rosaceous crops: Amygdaloideae (Srinivasan et al. 2005), Maloideae (Chevreau and Bell 2005; Dandekar 2002), and Rosoideae (Folta 2006; Martin 2002; Oosumi et al. 2006). However, some important crop species within the Rosaceae, such as peach, remain difficult to transform. In crops where the preservation of cultivar identification is desirable, such as apple and pear, genetic engineering provides a means to correct specific trait defects, such as disease susceptibility, in desirable cultivars of economic importance (Malnoy et al. 2004, 2008b). Genetic engineering can also produce novel phenotypes that may not occur in nature, such as blue colored roses (Katsumoto et al. 2007). In cases where desired phenotypic variation occurs in wild species with deleterious agronomic traits, such as poor fruit quality, genetic engineering can bypass the several generations of breeding crosses that may be required to incorporate the trait into a favorable genetic background (Malnoy et al. 2008b). Gene transfer technology also provides a powerful tool for the analysis of gene function, which has been difficult by classical genetic methods in much of the Rosaceae, due to the extended juvenility, large plant size, and self incompatibility that occurs within the family. Although improved transgenic cultivars can result directly from functional analysis, intellectual property rights associated with the technology often necessitate separate tracks for functional analysis and cultivar improvement.

Observing the effects of altered candidate gene expression on biological processes is a proven approach for using sequence information to study biological function (Dandekar et al. 2004; Malnoy et al. 2004). Candidate gene expression can be increased by transgenic expression (over-expression) or reduced by gene silencing. Over-expression requires cDNA or genomic sequence for the entire coding region, frequently referred to a "full-length" sequence. Because the cauliflower mosaic virus (CaMV) 35S promoter will usually result in high levels of gene transcription in most plant tissues, it is frequently the promoter of choice for over-expression studies. However, somaclonal variation and secondary effects caused by the CaMV 35S promoter can complicate analysis of gene function by this method. The 35S promoter can activate expression from other cis-located promoters

(Zheng et al. 2007), complicating analysis by the increased expression of more than one gene and thus making this promoter an especially poor choice for the functional analysis of transcription factors. The high levels of transcription resulting from use of the 35S promoter can also trigger gene silencing (Mishiba et al. 2005). Somaclonal variation, which can arise during transformation and tissue culture procedures, results from several causes including gene inactivation (or activation) mediated by transfer DNA (T-DNA) insertion, polyploidy, chromosomal translocations and physiological changes resulting from tissue culture (Brown et al. 1992; Filipecki and Malepszy 2006). Observed biological differences between a transgenic line and the parent cultivar are the combined result of transgene expression and line-specific somaclonal events, thus requiring the comparison of many transgenic lines to either (1) establish a statistically significant correlation between the level of gene expression and the level of biological function, or (2) separately estimate the effect of somaclonal variation in several transgenic lines transformed with an empty vector and the effect of transgene expression in several lines containing the transgene. The use of a chemically inducible promoter for over-expression studies (Malnoy et al. 2006; Norelli et al. 2007; Zuo et al. 2000) can overcome most of these problems by allowing comparison of the same transgenic line under conditions of non-induced and induced transgene expression, thus overcoming the problem of line-specific somaclonal effects by direct biological comparisons within a single transgenic background. Additionally, constitutive expression of candidate genes can frequently result in deleterious effects on plant growth and/or pleiotropic phenotypes. For example, expression of an endochitinase gene in apple under the control of the 35S promoter both increased resistance to apple scab and stunted growth, making it difficult to conclude what part of the change in resistance was due to endochitinase versus reduced growth (Bolar et al. 2000). An inducible promoter system can overcome these problems by limiting gene expression to a short period of time during biological analysis. A drawback to chemically inducible promoters is that they are usually not applicable for cultivar improvement, thus necessitating separate development tracks for functional analysis and cultivar improvement.

Gene expression can be down-regulated by both transcriptional gene silencing mechanisms, such as transposon insertion, and post-transcriptional gene silencing (PTGS) that is RNA-mediated and also known as RNA interference (RNAi). RNAi is mediated by double-stranded RNA (dsRNA) and is homology-dependent gene silencing (Eamens et al. 2008). RNAi does not require full length coding sequence for gene silencing, thus facilitating functional analysis from EST data or in situations where full-length cDNA clones are not available. RNAi also has some advantages over knock-out, or insertional, mutants when conducting reverse analysis (sequence to function). First, RNAi constructs directly target a specific gene, which overcomes the problem of having to generate a large population of lines with knock-out mutations in order to have a high degree of certainty of disrupting the function of any given gene (Helliwill et al. 2002). RNAi constructs will also give rise to plants with different degrees of gene silencing; this can result in viable, partially-silenced lines for genes that are lethal when completely knocked out. RNAi can be induced by transgenes containing an inverted repeat of DNA sequence that will result in the direct synthesis of dsRNA. It can also be induced by aberrant RNA molecules, either native or transgenic, that are converted to dsRNA by an RNA-dependent RNA polymerase (RdRP). Although the requirement for RdRP in initiating RNAi is undisputed, the substrate for this enzyme and what defines an "aberrant' mRNA is not fully understood (Eamens et al. 2008). RNAi can therefore be induced by various transgene designs in a sense (5' to 3'), antisense (complementary DNA strand) or inverted-repeat orientation. Because sense and antisense constructs are dependent upon RdRP activity for the production of dsRNA, they tend to be less efficient in inducing RNAi than transgenes containing inverted repeats (Wesley et al. 2001). Inverted repeats occurring in the 3' untranslated region of the transcript can also efficiently induce RNAi (Brummell et al. 2003), which may allow the use of RNAi in forward genomic approaches (function to sequence) (RNAi News, 2005). Currently, the most efficient transgene design for the induction of RNAi are "hairpin" constructs in which the 2 inverted repeat DNA sequences are separated by a transposable element. Hairpin designs are difficult to construct de novo, however several vectors have been developed to facilitate these designs (Wesley et al. 2001). Additionally, hairpin RNAi vectors that utilize lambda phage recombination, or GATEWAY™ technology (Helliwill et al. 2002; Mathews 2004), are amenable to high-throughput approaches.

The high cost of developing transgenic over-expressing and RNAi-silenced lines limits their use to the genomics analysis of a limited number of candidate genes of horticultural importance. Transient RNAi expression by agroinfiltration has been demonstrated in *Fragaria* and provides a rapid, low-cost alternative to the selection of stable transgenic lines for the analysis of gene function (Hoffmann et al. 2006). High-throughput reverse genetic analysis would also be greatly facilitated by the development of a virus-induced gene silencing (VIGS) system for the Rosaceae (Constantin et al. 2008; Godge et al. 2008). In comparison to agroinfiltration, in which gene silencing is restricted to the area of infiltration, VIGS can provide systemic gene silencing. A VIGS system is being developed for apple (Li et al. 2004; Yaegashi et al. 2007).

Random insertional mutagenesis has been a powerful tool for the analysis of complex biological traits in model systems because it allows a forward approach (function to sequences) that makes no a priori assumptions regarding the genetic control of a trait. Transformation technology facilitates random mutagenesis by transposon or T-DNA insertion and a large collection of T-DNA insertion mutants and *AcDs* activation tag lines are under development for *Fragaria vesca* (Oosumi et al. 2006; Shulaev et al. 2008). Mutagenesis by transposon or T-DNA insertion in the genome provide DNA "tags" to facilitate rapid identification of disrupted sequence, thus eliminating the need for extensive genetic analysis in gene identification. Because of the large number of mutants that must be screened in this approach, forward genomic analysis by mutagenesis relies upon efficient high-throughput phenotypic assays. Random insertional mutagenesis is less effective in organisms with large genomes, such as apple, due to a larger amount of non-coding DNA and therefore a lower frequency of gene disruption per insertion. Although the insertion of some transposons in non-coding regions can alter the regulation of down- and up-stream coding regions, these regulatory mutants can be difficult to analyze and complexity greatly increases with larger amounts on non-coding DNA. Similarly, the methodology has not been very effective in polyploid species.

### Genotyping and Marker-Assisted Breeding

The most commonly touted channel for using genomics in crop improvement is through genotyping (i.e. the application of genetic tests) of cultivars and breeding germplasm. Genotyping can be applied to existing cultivars in production and advanced selections in breeding programs to better understand and monitor their field performance (diagnostics), to potential breeding parents to better understand their breeding value (parent selection), and to seedling populations in breeding programs to improve efficiency of selection (seedling selection). Genotyping requires the development of marker "tool kits", which are sets of robust markers that can be readily screened on the germplasm of interest. Robustness refers to verification that the marker-trait associations are maintained over a wide range of germplasm and production conditions, or at least verification in the germplasm and conditions for which the markers are to be specifically applied. Ready screening refers to the availability of genotyping protocols and technologies suited to the number and condition of plants to be tested. Markers flanking a QTL region following QTL analysis, or functional markers representing the genes themselves following candidate gene analysis, are used as the predictive genetic tests of performance by screening DNA obtained from the plants of interest. Marker-assisted breeding (MAB) refers to the use of markers to assist in one or more operations of breeding programs, such as parent selection, family size planning, parentage verification, seedling selection, performance evaluation of advanced selections, and cultivar commercialization. Marker-assisted selection (MAS) refers just to the use of markers for selection in breeding – both of parents and seedlings, but usually referring to seedlings. The development of marker-trait associations, i.e. the experimental stage, is often erroneously included in MAB and MAS, often through the ambiguous term "marker development" which can mean the generation of new markers such as for map construction, the search for marker-trait associations, or the conversion of an experimental association into a robust marker for practical application.

Marker-trait associations must be verified to ensure they are applicable in the material to be tested. There are several reasons that associations may be lost during this verification step. First, the association may be a false positive arising from experimental conditions. Second, linkage disequilibrium (the association between a particular marker allele and trait allele) may be lost due to too much historical recombination. The more closely linked a marker is to the functional sequence difference itself (e.g. a specific mutation in a gene), the greater the likelihood that functional association is maintained. Researchers therefore seek these functional sequence differences even if linked markers are available, although the latter are often able to adequately serve breeding purposes. Use of flanking markers for a QTL increases the likelihood of successful performance prediction, as the specific QTL allele targeted will only lose its association with marker alleles if the very rare case of recombination between both markers and the QTL has occurred. Third, functional alleles identified in an experimental population may not be frequent in wider germplasm and therefore not detected in unrelated plants, limiting the extent of germplasm to which the genetic test is applicable. For example, markers for a newly-introgressed resistance allele are not applicable in the bulk of crosses where neither parent carries the resistance allele. However, verification of marker-trait associations may detect additional functional alleles that do not exist in the experimental population. The purposeful search for available functional alleles is known as allele mining. Allele mining includes describing the alleles present in the plants of interest such as the parents of a breeding program, and may extend to wider genepools such as germplasm collections.

For diagnostics and parent selection, methods of DNA extraction and genotyping can be low-throughput, i.e. at the scale of tens to hundreds of samples at time, as the numbers of plants under investigation are correspondingly limited. The actual marker types used for these low-throughput purposes can be isozymes and RFLPs through to the latest automated technologies. Genotyping for diagnostics and parent testing have therefore advanced the furthest in Rosaceae, and soon after marker-trait associations are discovered in experimental material, the genotypic profiles of cultivars are often reported to indicate the robustness of the associations and to describe the genetic character of each cultivar. Self-incompatibility (SI) groups to which cultivars belong greatly influences orchard design for most Rosaceae tree crops. In almond, cherry, plum, and apricot, uncovering the genes controlling this trait at the *S* locus enabled the development of simple PCR tests to place cultivars into SI groups, which are used to determine cross-compatible combinations and identify self-fertile cultivars (Tamura et al. 2000; Sonneveld et al. 2003; Sutherland et al. 2004; Halász et al. 2005). Discovery of an allele of an ACS (1-aminocyclopropane-1-carboxylic acid [ACC] synthase) gene in apple conferring low ethylene production and longer storage life led to genotyping of cultivars to characterize their ethylene genotype (Oraguzie et al. 2007). This ACS gene was also genotyped in combination with another gene in the ethylene biosynthetic pathway, ACO (ACC oxidase), to characterize cultivars and advanced selections of an apple breeding program (Zhu and Barritt 2008).

Application of genetic tests in breeding programs to reduce the squandering of resources on low-value seedlings requires the implementation of high-throughput DNA extraction and genotyping. Every year, Rosaceae breeders produce hundreds to many thousands of seeds, which are germinated, grown, field-planted, and eventually mostly eliminated, all the while undergoing phenotypic evaluation for traits of importance, to arrive at a tiny proportion of selected individuals ("selections") that are worthy of proceeding to more intensive performance evaluations. Marker-assisted seedling selection (MASS) involves integrating genotyping into these routine operations, augmenting the selection process by substituting genetic marker tests for sensory or instrumental phenotypic tests wherever it is determined to be more efficient in cost and/or time. Implementation of MASS for thousands of seedlings in a season requires the development of a streamlined process for sampling, extracting DNA, genotyping, and timely supply and application of results that is relevant to the idiosyncrasies of a breeding program. This infrastructure is an obstacle that most public breeding programs of Rosaceae crops have yet to overcome, as robust markers for numerous traits exist but very few are in operation. MASS for resistance to the diseases of scab and powdery mildew in apple, reported by Kellerhals et al. (2004), represents a rare case of real world implementation.

## *1.3 Outlook*

Unlike field crops such as wheat, corn, or soybean, most individual Rosaceae crops are supported by relatively small industries. The dozen major rosaceous crops represent a very diverse group of plants with assorted attributes and challenges for genetic improvement. Yet this diversity is also the strength of the family. Having a shared ancestral "Rosaceae genome" predicts that the controlling genes of common traits will often be the same, and underlying biological mechanisms may not be as different as appearances suggest. Comparisons between Rosaceae crops provide contrasts that can reveal the controlling gene networks and speed genetic improvement. For example, comparisons between cherry and plum or strawberry and raspberry may reveal the genetic basis of fruit size, apple and pear for fruit shape, and across Rosaceae for disease resistance mechanisms. Similarly, basic biological processes can be uncovered within Rosaceae, for example by comparing plant form between strawberry and apple, fruit ethylene response between climacteric peach and non-climacteric cherry, and fruit development between strawberry, raspberry, rose, apple, almond, and peach. Research funds offered by individual industries are both inappropriate and insufficient to address such fundamental yet far-reaching issues. Studies spanning two or more Rosaceae crops, particularly those across subfamily borders, will require an unprecedented level of coordination and collaboration. Fortunately, the international Rosaceae genomics, genetics, and breeding community has taken enormous strides in this direction, exemplified by several exciting initiatives.

### 1.3.1 A Centralized Web Portal and Database

The Genome Database for Rosaceae (GDR) was created in 2003 in response to rapidly expanding volumes of genomic data in the public domain. EST libraries and genetic maps were the first genomic resources to be hosted on the GDR, followed by transcript and physical maps. Frequent access to such resources has made the GDR an information hub for the Rosaceae network of scientists, breeders, and allied professionals, displaying community announcements, highlighting projects, providing bioinformatics tools for data analyses, and storing ever-increasing genomic data in a readily-accessible and public database. By collecting and processing these structural, functional, and comparative genomics data in one open location, the GDR has enabled the development of an active Rosaceae genomics community in which members operate beyond the limits of single crops. The GDR continues to take on a greater role as a community information hub. A series of twelve

USDA-funded Rosaceae genomics projects that started in 2005 uploaded their data and other project outcomes to the GDR, ensuring wide community dissemination. The diversity of genomic and genetic information represented by these twelve projects is large, and beyond that traditionally housed at this site. Indeed, the GDR is expanding to incorporate genotypic, phenotypic, QTL, pedigree, and gene expression data. Plans are underway to develop education and extension modules, to better inform stakeholders – from researchers and breeders to policy makers, industry, and the general public – about the activities, concerns, breakthroughs, and promise of Rosaceae genetics and genomics.

### 1.3.2 Shared Mapping Resources

The Rosaceae Consortium of Mapping Populations (RosPOP, www.bioinfo.wsu.edu/gdr/community/international/rospop.php) is an initiative designed to facilitate access to plant materials and information from segregating progeny populations of Rosaceae for researchers other than the population owners. Participation in RosPOP requires a formal, although essentially a goodwill, agreement between consortium members that specifies the resources to be shared. Material supplied includes access to the plants themselves and derived materials (e.g. budwood, pollen, fruit, leaves, DNA, and RNA) and data collected from those individuals (phenotypic and genotypic). Traditionally, individual researchers create and study their own experimental mapping populations, focusing on the traits they are most interested in and have the resources to collect data for. Constructing mapping populations is a time-consuming and expensive endeavor in itself, requiring for the tree crops four or more years from making the crosses between the desired parents until fruit production from the resulting seedlings. RosPOP intends to make maximum use of these valuable genetic resources by bringing to bear additional funds, labor, and scientific expertise for a broader scope and increased efficiency of genetic analyses. This approach fosters new and strengthened collaborations between researchers and institutions, and reduces redundancy in worldwide efforts toward Rosaceae genetic improvement. The advent of RosPOP represents a new era in international collaboration for Rosaceae genetic mapping and gene-trait association research.

### 1.3.3 Standardized Phenotyping

Another recent advance in community coordination is the concept of standardized phenotyping across Rosaceae genetic resources. Various studied sets of Rosaceae germplasm, such as individuals from genetic experiments, cultivars, breeding populations, or ex situ germplasm collections, tend to be phenotypically characterized according to immediate needs of individual investigations. Lack of consistency between studies limits the utility of collected phenotypic data. In contrast, DNA genotypes can be readily compared between studies. Standardized phenotyping offers an opportunity to align the characterization of germplasm collection accessions, such as those of the USDA's National Plant Germplasm System, more closely with the needs of breeding programs and the interests of genomics researchers.

In a wider context, the ability to directly compare both genotypic and phenotypic data across germplasm sets will greatly enhance Rosaceae-wide efforts to establish gene-trait associations by increasing the size of datasets available for analyses. For example, standardized phenotyping could facilitate direct comparisons between populations from two or more breeding programs to obtain more accurate estimates of heritability and genotype × environment interaction for priority traits. Another example application could be determining similarities in genetic factors underlying phytonutrient composition within and among apple, cherry, and raspberry cultivars.

Standardized phenotyping is a challenging approach, requiring considerable coordination and agreement between researchers to establish both a comprehensive set of descriptors and trust in the validity of the resulting data. Descriptors used must be fairly heritable, efficient to use, and relevant to both industry priorities and biological questions. Although challenging, standardized phenotyping will be necessary to address the extent to which the same gene networks control similar traits across Rosaceae crops or their functional divergence from the ancestral genome. As such, this approach will require and foster unity in international Rosaceae genomics, genetics, and breeding.

### 1.3.4 A Uniting Statistical Approach

Pedigree Based Analysis (PBA) is a powerful statistical approach able to simultaneously identify marker-trait associations, validate their robustness and applicability to individual breeding programs, and mine alleles for functional diversity. While traditional QTL discovery approaches rely on experimental populations which are usually created for the specific purpose of identifying or fine-mapping QTLs, the PBA approach avoids the need for such dedicated populations. Furthermore, although genomics-assisted breeding requires validation and allele mining in breeding germplasm, traditional QTL approaches stop at the discovery stage. The versatility of PBA is achieved by analyzing genotypic and phenotypic data of breeding germplasm itself. This approach is well suited to the multiple pedigree-linked populations of variable size that typify Rosaceae breeding germplasm (van de Weg 2004). PBA identifies networks of major genes and QTLs that determine genetic variation in horticulturally important traits, elucidating their interactions and mining their functional allelic diversity (van de Weg 2004). The strategy integrates marker and phenotypic data over past, current, and future generations within and across breeding programs, thus creating a flexible and continuously expanding platform for marker identification, validation, and use (van de Weg 2004). The PBA approach is based on two complementary statistical approaches. The first identifies QTL regions based on Markov chain Monte Carlo simulations and Bayesian statistics. The second is based on "Identity By Descent" values of each allele of a genotype, taking the different alleles of founding cultivars as factors in statistical analysis (Bink et al. 2008). PBA was the underlying and unifying scheme for the European HiDRAS project, concluding in late 2007, that aimed to identify genetic factors controlling apple fruit quality (including texture components) for increasing the acceptability of disease resistant apples (Gianfranceschi and Soglio 2004; Kellerhals and Eigenmann 2006).

### 1.3.5 Team Building

Because the journey from investment in genomic science to profitable fruit production spans a tremendous range of expertise, teams of specialists functioning as collaborative units are necessary to ensure that genomic research will impact crop improvement. Effective team building starts with direct and two-way communication between the scientific community and the fruit industry (Fig. 3A). Communication goals include: (1) making the project more responsive to industry needs, (2) improving the dissemination of genomic research information to the industry community, and (3) fostering the efficient integration of industry needs, research objectives, and the development of new cultivars. Industry-research communication should take place during both project planning and execution, receiving input on industry needs during project planning and identifying possible extension "deliverables" resulting from the research during project execution.

All three fields of genomics will not necessarily be involved in all projects, particularly in smaller projects with limited resources. However, communication and collaboration among researchers in the various fields of genomics will facilitate the project's ability to capitalize on new community resources developed in other fields of genomics as they become available (Fig. 3B). Because of the tremendous size

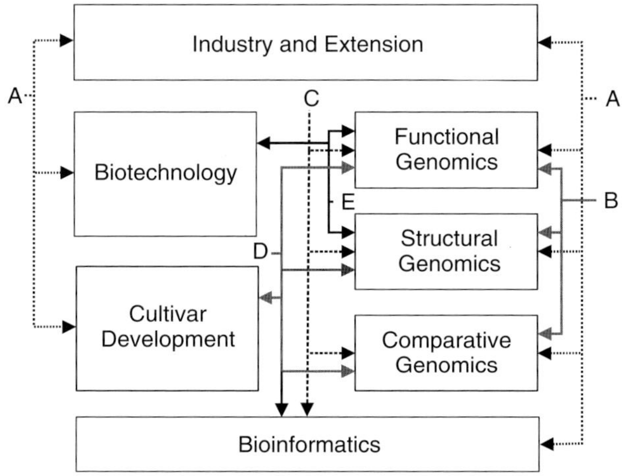


**Fig. 3** Team building is necessary for investment in genomic research to lead to increased profitability for Rosaceae industries and improved products for consumers. **A)** Effective team building starts with direct and two-way communication between the scientific community and Rosaceae industries. **B)** The development of community resources in each specific field of genomics fosters the development of knowledge in all fields. **C)** Collaboration with computational biologists strengthens projects and leads to development of useful bioinformatic tools. **D)** Collaboration and two-way communication between scientists working in cultivar development, genomics, and bioinformatics fosters the timely development of new cultivars that meet the needs of industry. **E)** Similarly, collaborations between biotechnologists and genomicists lead to the development of genetically engineered cultivars, therapeutics, and diagnostic tests that meet industry needs

of many genomic databases and the need to connect them into effective matrixes, the inclusion of computational biologists or bioinformaticists will strengthened the project team. Their involvement as a collaborating scientist, rather than a support consultant, increases the likelihood that the project will result in innovative computational approaches and useful bioinformatic tools (Fig. 3C).

Similarly, when scientists that will apply advances in genomics to specific horticultural practices are involved in project planning and execution, the likelihood that a project will have a significant impact on crop improvement is increased (Fig. 3D and E). Potential collaborators include geneticists (molecular mapping), plant breeders, horticulturalists, plant physiologists, plant pathologists, entomologists, genetic engineers, and chemists. Respectful two-way communication between scientists, rather than an arrogant assumption that genomics research is superior, will facilitate a synergistic collaboration between disciplines. In summary, the vertical integration of genomic research with industry needs and other scientific disciplines increases the likelihood that funds invested in genomic research will result in significant impact on crop improvement and increased profitability for the fruit industry.

## 2 Conclusions

The potential impact of genomics on Rosaceae crop improvement is enormous. Just as past breeding and research has delivered varied and valuable genetic products, the science of genomics will contribute to the ongoing advances in cultivar improvement necessary to keep up with new challenges to production and the demands of the marketplace. In-depth understanding of Rosaceae genomes and their functional components will not only impact cultivar improvement, but also foster the development of new diagnostics, therapeutics, and cultural practices. Genomic advances will need to address agricultural sustainability by reducing environmental impact, reducing land and water use requirements, and reducing chemical and energy input. It will need to address consumer desires for high quality products that are beautiful, tasty, healthy, consistent, and convenient, and enhance our quality of life. It will need to address the needs of the Rosaceae crop industries to reduce production costs in order to remain viable in the world market. The genomic tools, technologies, and basic knowledge developed in the short term will provide the foundation for addressing many of these challenges in the long term as they are directed to practical benefit. Advances in Rosaceae systems will also aid under-researched fruit, nut, and perennial flower crops for which Rosaceae crops often provide the unofficial model. Working collaboratively with industry and other scientific disciplines, the opportunity exists to anticipate future needs and, with current genomics capabilities, to pro-actively develop solutions for sustained supply of the many Rosaceae products that improve human health and well-being.

## References

Alm R, Ekefjärd A, Krogh M, Häkkinen J, Emanuelsson C (2007). Proteomic variation is as large within as between strawberry varieties. J. Proteome Res. 6:3011–3020

Aranzana MJ, Carbo J, Arus P (2003). Microsatellite variability in peach [*Prunus persica* (L.) Batsch]: cultivar identification, marker mutation, pedigree inferences and population structure. Theor. Appl. Genet. 106:1341–1352

Bailey JS, French AP (1949). The inheritance of certain fruit and foliage characteristics in the peach. Mass. Agr. Expt. Sta. Bul. 452

Baldi P, Patocchi A, Zini E, Toller C, Velaso R, Komjanc M (2004). Cloning and linkage mapping of resistance gene homologues in apple. Theor. Appl. Genet. 109:231–239

Baldo AM, Bassett CL, Malnoy M, Korban SS, Gasic K, Farrell RE, Aldwinckle HS, Norelli JL (2007). Computational identification of candidate genes involved in response to fire blight in apples. Plant Animal Genomes XV Conference, Abstract: http://intl-pag.org/15/abstracts/PAG15˙P08c˙886.html Cited 9 Jan 2009

Balogh A, Koncz T, Tisza V, Kiss E, Heszky L (2005). The effect of 1-MCP on the expression of several ripening-related genes in strawberries. HortScience 40:2088–2090

Ban Y, Honda C, Bessho H, Pang X-M, Moriguchi T (2007). Suppression subtractive hybridization identifies genes induced in response to UV-13 irradiation in apple skin: isolation of a putative UDP-glucose 4-epimerase. J. Exp. Bot. 58:1825–1834

Bassett CA, Wisniewski ME, Artlip TS, Norelli JL, Renaut J, Farrell RE Jr (2006). Global analysis of genes regulated by low temperature and photoperiod in peach bark. J. Am. Soc. Hort. Sci. 131:551–563

Beavis WD, Grant D, Albertsen M, Fincher R (1991). Quantitative trait loci for plant height in four maize populations and their associations with qualitative genetic loci. Theor. Appl. Genet. 83:141–145

Bink MCAM, Boer MP, ter Braak CJF, Jansen J, Voorrips RE, van de Weg WE (2008). Bayesian analysis of complex traits in pedigreed plant populations. Euphytica 161: 85–96

Bolar JP, Norelli JL, Wong K-W, Hayes CK, Harman GE, Aldwinckle HS (2000). Expression of endochitinase from *Trichoderma harzianum* in transgenic apple increases resistance to apple scab and stunts growth. Phytopathology 90: 72–77

Brown PTH, Lang FD, Lorz H (1992) Molecular changes in tissue culture-derived plants. Adv. Mol. Genet. 5:171–195

Brummell DA, Balint-Kurti, Harpster MH, Payly JM, Oeller PW, Gutterson N (2003) Inverted repeat of a heterologous 3'-untranslated region for high-efficiency, high-throughput gene silencing. Plant J. 33:793–880

Buescher RW, Furmanski RJ (1978). Role of pectinesterase and polygalacturonase in the formation of woolliness in peaches. J. Food. Sci. 43:264–266

Calenge F, Drouet D, Denance C, Van de Weg WE, Brisset M-N, Paulin JP, Durel CE (2005). Identification of a major QTL together with several minor additive or epistatic QTLs for resistance to fire blight in apple in two related progenies. Theor. Appl. Genet. 111:128–135

Callahan AM, Scorza R, Bassett C, Nickerson M, Abeles FB (2004). Deletions in an endopolygalacturonase gene cluster correlate with non-melting flesh texture in peach. Func. Plant Biol. 31:159–168

Callahan AM, Dardick CD, Chiozzotto R, Schaffer RJ, Piagnani MC, Scorza R (2008) Gene Expression Profiling Of Peach Fruit During Stone Development. Plant & Animal Genomes XVI Conference, 2008 http://www.intl-pag.org/16/abstracts/PAG16_P07a_716.html

Campalans A, Pages M, Nesseguer R (2001) Identification of differentially expressed genes by the cDNA-AFLP technique during dehydration of almond (*Prunus amygdalus*). Tree Physiol. 21:633–643

Celton J-M, Tustin DS, Chagn Gardiner SE (2009) Construction of a dense genetic linkage map for apple rootstocks using SSRs developed from *Malus* ESTs and *Pyrus* genomic sequences. Tree Genet. Genomes. 5:93–107

Chagne D, Carlisle CM, Blond C, Volz RK, Whitworth CJ, Oraguzie NC, Crowhurst RN, Allan AC, Espley RV, Hellens RP, Gardiner SE (2007). Mapping a candidate gene (MdMYB10) for red flesh and foliage colour in apple. BMC Genomics 8:212

Chevreau E, Bell R (2005) *Rosaceae: Pyrus* spp. Pear and *Cydonia* spp. Quince. In: Litz RE (ed) Biotechnology of fruit and nut crops, CABI Publishing, Wallingford and Cambridge, MA, pp 543–565

Constantin GD, Gronlund M, Johansen IE, Stougaard J, Lund OS (2008) Virus-induced gene silencing (VIGS) as a reverse genetic tool to study development of symbiotic root nodules. Mol. Plant Microbe Interact. 21:720–727

Costa F, Stella S, Van de Weg WE, Guerra W, Cecchinel M, Dallavia J, Koller B, Sansavini S (2005). Role of the genes *Md-ACO1* and *Md-ACS1* in ethylene production and shelf life of apple (Malus domestica Borkh). Euphytica 141:181–190

Costa F, van de Weg WE., Stella S, Dondini L, Pratesi D, Musacchi S, Sansavini S (2008). Map position and functional allelic diversity of Md-Exp7, a new putative expansin gene associated with fruit softening in apple (*Malus* × *domestica* Borkh.) and pear (*Pyrus communis*). Tree Genet. Genomes 4:575–586

Crisosto CH, Mitchell FG, Ju Z (1999) Susceptibility to chilling injury of peach, nectarine, and plum cultivars grown in California. HortScience 34: 1116–1118

Dandekar AM (2002) Introduction and expression of transgenes in apples. In: Khachatourians GC, McHughen A, Scorza R, Nip W-K, Hui YH (eds) Transgenic plants and crops, Marcel Dekker, Inc., New York and Basel, pp 327–344

Dandekar AM, Teo G, Defilippi BG, Uratsu SL, Passey AJ, Kader AA, Stow JR, Colgan RJ, James DJ (2004). Effect of down-regulation of ethylene biosynthesis on fruit flavor complex in apple fruit. Transgenic Res. 13:373–384

Degenhardt J, Al-Masri AN, Kurkcuoglu S, Szankowski I, Gau AE (2005). Characterization by suppression subtractive hybridization of transcripts that are differentially expressed in leaves of apple scab-resistant and susceptible cultivars of *Malus domestica*. Mol. Gen. Genet. 273: 326–335

Deng C, Davis TM (2001) Molecular identification of the yellow fruit color (c) locus in diploid strawberry: a candidate gene approach. Theor. Appl. Genet. 103:316–322

Diatchenko L, Lau YFC, Campbel AP, Chenchik A, Moqadam F, Huang B, Lukyanov S, Lukyanov K, Gurskaya N, Sverdlov E, Siebert PD (1996). Suppression subtractive hybridization: a method for generating differentially regulated or tissue-specific cDNA probes and libraries. Proc. Natl. Acad. Sci. 93: 6025–6030

Dirlewanger E, Graziano E, Joobeur T, Garriga-Caldere F, Cosson P, Howard W, Arus P (2004). Comparative mapping and marker assisted selection in Rosaceae fruit crops. Proc. Natl. Acad. Sci. USA 101:9891–9896

Eamens A, Wang M-B, Smith NA, Waterhouse PM (2008) RNA silencing in plants: yesterday, today and tomorrow. Plant Physiol. 147:456–468

Etienne C, Rothan C, Moing A, Plomion C, Bodenes C, Svanella-Dumas L, Cosson P, Pronier V, Monet R, Dirlewanger E (2002). Candidate genes and QTLs for sugar and organic acid content in peach [*Prunus persica* (L.) Batsch]. Theor. Appl. Genet. 105: 145–159

Fiehn O (2002). Metabolomics – the link between genotype and phenotype. Plant Mol. Biol. 48: 155–171

Filipecki M and Malepszy S (2006) Unintended consequences of plant transformation: A molecular insight. J. Appl. Genet. 47:277–286

Flachowsky H, Peil A, Sopanen T, Elo A, Hanke V (2007). Overexpression of BpMADS4 from silver birch (*Betula pendula* Roth.) induces early-flowering in apple (*Malus* × *domestica* Borkh.). Plant Breeding 126:137–145

Folta, KM, Dhingra A (2006). Transformation of strawberry: the basis from translation genomics in *Rosaceae*. In Vitro Cell. Devel. Biol. Plant 42:482–490

Fridman E, Pichersky E (2005). Metabolomics, genomics, proteomics, and the identification of enzymes and their substrates and products. Curr. Opin. Plant Biol. 8: 242–248

Gasic K, Gonzales DO, Thimmapuram J, Malnoy M, Gong G, Han Y, Vodkin LO, Liu L, Aldwinckle HS, Carroll N, Orvis K, Goldsbrough P, Clifton S, Pape D, Fulton L, Martin J, Theising B, Wisniewski ME, Fazio G, Korban SS (*submitted*) Comparative analysis and functional annotation of a large expressed sequence taf collection of apple (*Malus* × *domestica*)

Geuna F, Banfi R, Bassi D (2007). Identification and characterization of transcripts differentially expressed during development of apricot (*Prunus armeniaca* L.) fruit. Tree Genet. Genomes 1:69–78

Gianfranceschi L, Soglio V (2004). The European project HiDRAS: innovative multidisciplinary approaches to breeding high quality disease resistant apples. Acta Horticult. 663:327–330

Godge MR, Purkayastha A, Dasgupta I, Kumar PP (2008) Virus-induced gene silencing for functional analysis of selected genes. Plant Cell. Rep. 27:209–219

Grattapaglia D, Sederoff R (1994). Genetic linkage maps of *Eucalyptus grandis* and *Eucalyptus urophylla* using a pseudo-testcross mapping strategy and RAPD markers. Genetics 137: 1121–1137

Grimplet J, Romieu C, Sauvage FX, Lambert P, Audergon JM, Terrier N (2004). Transcriptomics and proteomics tools towards ripening markers for assisted selection in apricot. Acta Horticult. 663: 291–296

Guarino C, Arena S, De Simone L, D'ambrosio C, Santoro S, Rocco M, Scaloni A, Marra M (2007). Proteomic analysis of the major soluble components in Annurca apple flesh. Mol. Nut. Food Res. 51 : 255–262

Hadfield KA, Bennett AB (1998). Polygalacturonases: many genes in search of a function. Plant Physiol. 117:337–343

Halász J, Hegedus A, Hermán R, Stefanovits-Bányai É, Pedryc A (2005). New self-incompatibility alleles in apricot (*Prunus armeniaca* L.) revealed by stylar ribonuclease assay and *S*-PCR analysis. Euphytica 145:57–66

Han Y, Gasic K, Marron B, Beever JE, Korban S (2007). A BAC-based physical map of the apple genome. Genomics 89:630–637

Helliwill CA, Wesoey SV, Wielopolska AJ, Waterhouse PM (2002) High-throughput vectors for efficient gene silencing in plants. Funct. Plant Biol. 29:1217–1225

Hoffmann T, Kalinowski G, Schwab W (2006) RNAi-induced silencing of gene expression in strawberry fruit (*Fragaria* × *ananassa*) by agroinfiltration: a rapid assay for gene function analysis. Plant J. 48:818–826

Horn R, Lecouls AC, Callahan A, Dandekar A, Garay L, McCord P, Howad W, Chan H, Verde I, Main D, Jung S, Georgi L, Forrest S, Mook J, Zhebentyayeva T, Yu Y, Kim HR, Jesudurai C, Sosinski B, Arus P, Baird V, Parfitt D, Reighard G, Scorza R, Tomkins J, Wing R, Abbott AG (2005). Candidate gene database and transcript map for peach, a model species for fruit trees. Theor. Appl. Genet. 110:1419–1428

Howad W, Yamamoto T, Dirlewanger E, Testolin R, Cosson P, Cipriani G, Monforte AJ, Georgi L, Abbott AG (2005). Mapping with a few plants: using selective mapping for microsatellite saturation of the *Prunus* reference map. Genetics 171:1305–1309

Jensen P, Rytter J, Detwiler EA, Travis JW, NcNellis TW (2003). Rootstock effects on gene expression patterns in apple tree scions. Plant Mol. Biol. 53:493–511

Jung S, Staton M, Lee T, Blenda A, Svancara R, Abbott A, Main D (2008). GDR (Genome Database for Rosaceae): integrated web-database for Rosaceae genomics and genetics data. Nucleic Acids Res., 36(Database issue): D1034–D1040

Katsumoto Y, Fukuchi-Mizutani M, Fukui Y, Brugliera F, Holton TA, Karan M, Nakamura N, Yonekura-Sakakibara K, Togami J, Pigeaire A, Tao G-Q, Nehra NS Lu CY, Dyson BK, Tsuda S, Ashikar R, Kusumi T, Mason JG, Tanaka Y (2007) Engineering of the rose flavanoid biosynthetic pathway successfully generated blue-hued flowers accumulating delphinidin. Plant Cell Physiol. 48:1589–1600

Kellerhals M, Eigenmann C (2006). Evaluation of apple fruit quality within the EU project HiDRAS. In: M Boos (ed) Ecofruit – 12th International Conference on Cultivation Technique and Phytopathological Problems in Organic Fruit-Growing: Proceedings to the Conference, pp 165–171

Kellerhals M, Sauer C, Guggenbuehl B, Gantner S, Frey B, Frey JE, Patocchi A, Gessler C (2004). Apple breeding for high fruit quality and durable disease resistance. Acta Horticult. 663: 751–756

Khan MA, Duffy B, Gessler C, Patocchi A (2006). QTL mapping of fire blight resistance in apple. Mol. Breed 17:299–306

Koornneef M, Alonso-Blanco C, Peeters AJM, Soppe W (1998). Genetic control of flowering time in *Arabidopsis*. Annu. Rev. Plant Physiol. Plant Mol. Biol. 49:345–370

Kotoda N, Iwanami H, Takahashi S, Abe K (2006). Antisense expression of MdTFL1-like gene, reduces the juvenile phase in apple. J. Am. Soc. Horticult. Sci. 131:74–81

Kuerkcueoglu S, Degenhardt J, Lensing J, Al-Masri A-N, Gua AE (2007). Identification of differentially expressed genes in *Malus domestica* after application of the non-pathogenic bacterium *Pseudomonas fluorescens* Bk3 to the phyllosphere. J. Exp. Bot. 58:733–741

Lalli DA, Decroocq V, Blenda AV, Schurdi-Levraud V, Garay L, Le Gall O, Damsteegt V, Reighard GL, Abbott AG (2005). Identification and mapping of resistance gene analogs (RGAs) in *Prunus*: a resistance map for *Prunus*. Theor. Appl. Genet. 111:1504–1513

Lazzari B, Caprera A, Vecchietti A, Stella A, Milanesi L, Pozzi C (2005). ESTree db: a tool for peach functional genomics. BMC Bioinformatics 6(Supple 4):S16 doi:10.1186/1471-2105-6-S4–S16

Lester DR, Speirs J, Orr G, Brady CJ (1994). Peach (*Prunus persica*) endopolygalacturonase cDNA isolation and mRNA analysis in melting and nonmelting peach cultivars. Plant Physiol. 105: 225–231

Lester DR, Sherman WB, Atwell BJ (1996). Endopolygalacturonase and the melting flesh (M) locus in peach. J. Am. Soc. Hort. Sci. 121: 231–235

Li C, Sasaki N, Isogai M, Yoshikawa N (2004). Stable expression of foreign proteins in herbaceous and apple plants using Apple latent spherical virus RNA2 vectors. Arch Virol. 149:1541–1558

Mackay TFC (2004). The genetic architecture of quantitative traits: lessons from *Drosophila*. Curr. Opin. Genet. Dev. 14:253–257

Malnoy M, Borejsza-Wysocka E, Jin Q-L, He SY, Aldwinckle HS (2004). Overexpresssion of the apple MpNRP1 gene confers increased disease resistance in *Malus* × *domestica*. Acta Horticult. 663:463–468

Malnoy M, Borejsza-Wysocka EE, Aldwinckle HS, Baldo AM, Carlisle CM, Bowatte D, Gardiner SE, Norelli JL, Bassett CL, Farrell RE Jr (2008a). Identification and mapping of genes differentially expressed during interaction of resistant and susceptible apple cultivars *(Malus* × *domestica)* with *Erwinia amylovora*. Acta Horticult. 793:195–199

Malnoy M, Reynoird JP, Borejsza-Wysocka E, Aldwinckle HS (2006). Activation of the pathogeninducible Gst1 Promoter of potato after elicitation by and *Erwinia amylovora* in transgenic apple *(Malus* × *domestica)*. Transgenic Res. 15:83–93

Malnoy M, Xu M, Borejsza-Wysocka E, Korban S, Aldwinckle HS (2008) Two receptor-like genes, Vfa1 and Vfa2, confer resistance to the fungal pathogen *Venturia inaequlais* inciting apple scab disease. Mol. Plant Microbe Interact. 21: 448–458

Martin RR (2002). Genetic engineering of strawberries and raspberries. In: Khachatourians GC, McHughen A, Scorza R, Nip W-K, Hui YH (eds) Transgenic plants and crops, Marcel Dekker, Inc., New York and Basel, pp 449–464

Mathews L (2004). RNAi for plant functional genomics. Comp. Funct. Genom. 5: 240–244

Mishiba KI, Nishihara M, Nakatsuka T, Abe Y, Hirano H, Yokoi T, Kikuchi A, Yamamura S (2005) Consistent transcriptional silencing of 35S-driven transgenes in gentian. Plant J. 44: 541–556

NCBI EST Database (2008). National Center for Biotechnology Information, Bethesda. http://www.ncbi.nlm.nih.gov/sites/entrez?db=nucest Cited 6 Jul 2008

Newcomb RD, CrowHurst RN, Gleave AP, Rikkerink HA, Allan AC, Beuning LL, Bowen JH, Gera E, Jamieson KR, Janssen BJ, Laing WA, McArtney S, Nain B, Ross, GS, Snowden KC, Souleyre EJF, Walton EF Yauk Y-K (2006). Analyses of expressed sequence tags from apple. Plant Physiol. 141: 147–166

Norelli JL, Bassett, C, Artlip T, Aldwinckle HS, Malnoy M, Borejsza-Wysocka EE, Gidoni D, Glaishman M (2007) Inducible DNA promoters for use in apple. Acta Horticult. 738: 329–334

Norelli JL, Farrell RE Jr, Bassett CL, Baldo AM, Lalli DA, Aldwinckle HS, Wisniewski ME (2009) Rapid transcriptional response of apple to fire blight disease revealed by cDNA suppression subtractive hybridization analysis. Tree Genet. Genomes 5:27–40

Norelli JL, Gardiner SE, Malnoy M, Aldwinckle HS, Baldo AM, Borejsza-Wysocka E, Farrell RE Jr, Lalli DA, Celton J-M, Bassett CL, Korban SS, Wisniewki ME (2008). Using functional genomics to develop tools to breed fire blight resistant apple. Plant & Animal Genomes XVI Conference 2008 http://intl-pag.org/16/abstracts/PAG16_P05h_455.html

Ogundiwin EA, Peace CP, Gradziel TM, Dandekar AM, Bliss FA, Crisosto CH (2007). Molecular genetic dissection of chilling injury in peach fruit. Acta Horticult. 738:633–638

Ogundiwin EA, Peace CP, Nicolet CM, Rashbrook VK, Gradziel TM, Bliss FA, Parfitt D, Crisosto CH (2008). Leucoanthocyanidin dioxygenase gene (PpLDOX): a potential functional marker for cold storage flesh browning in peach. Tree Genet. Genomes 4:543–554

Olmstead JW, Sebolt AM, Cabrera A, Sooriyapathirana SS, Hammar S, Iriarte G, Wang D, Chen CY, van der Knaap E, Iezzoni AF (2008). Construction of an intra-specific sweet cherry (*Prunus avium* L.) genetic linkage map and synteny analysis with the Prunus reference map Tree Genet. Genomes 4:897–910

Oosumi T, Gruszewski HA, Blischak LA, Baxter AJ, Wadl PA, Shuman JL, Veilleux RE, Shulaev V (2006) High-efficiency transformation of the diploid strawberry (*Fragaria vesca*) for functional genomics. Planta 223:1219–1230

Oraguzie NC, Iwanami H, Soejima J, Harada T, Hall A (2004). Inheritance of the *Md-ACS1* gene and its relationship to fruit softening in apple *(Malus × domestica Borkh.)*. Theor. Appl. Genet. 108:1526–1533

Oraguzie NC, Wilcox PL (2007). An overview of association mapping. In: N Oraguzie, EHA Rikkerink, SE Gardiner, and HN De Silva (eds) Association Mapping in Plants. Springer Verlag, New York, pp. 1–9

Oraguzie NC, Volz RK, Whitworth CJ, Bassett HCM, Hall AJ, Gardiner SE (2007). Influence of *Md-ACS1* allelotype and harvest season within an apple germplasm collection on fruit softening during cold air storage. Postharvest Biol. Technol. 44:212–219

Pandey A, Mann M (2000). Proteomics to study genes and genomes. Nature 405: 837–846

Park S, Sugimoto N, Larson MD, Beaudry R, van Nocker S (2006). Identification of genes with potential roles in apple fruit development and biochemistry through large-scale statistical analysis of expressed sequence tags. Plant Physiol. 141: 811–824

Peace CP, Ahmad R, Gradziel TM, Dandekar AM, Crisosto CH (2005a). The use of molecular genetics to improve peach and nectarine post-storage quality. Acta Horticult. 682:403–410

Peace CP, Crisosto CH, Gradziel TM (2005b). Endopolygalacturonase: a candidate gene for *Freestone* and *Melting flesh* in peach. Mol. Breeding 16:21–31

Peace CP, Crisosto CH, Garner DT, Dandekar AM, Gradziel TM, Bliss FA (2006). Genetic control of internal breakdown in peach. Acta Horticult. 713:489–496

Peace CP, Callahan AM, Ogundiwin EA, Potter D, Gradziel TM, Bliss FA, Crisosto CH (2007). Endopolygalacturonase genotypic variation in *Prunus*. Acta Horticult. 738:639–646

Pichler FB, Walton EF, Day M, Triggs C, Janssen B, Wunsche JN, Putterill J, Schaffer RJ (2007). Relative developmental, environmental, and tree-to-tree variability in buds from field-grown apple trees. Tree Genet. Genomes 3:329–339

Peil A, Garcia-Libreros T, Richter K, Trognitz FC, Trognitz B, Hanke M-V, Flachowsky H (2007). Strong evidence for a fire blight resistance gene of *Malus robusta* located on linkage group 3. Plant Breeding 126:470–475

Pierantoni L, Cho KH, Shin IS, Chiodini R, Tartarini S, Dondini L, Kang SJ, Sansavini S (2004). Characterisation and transferability of apple SSRs to two European pear F1 populations. Theor. Appl. Genet. 109:1519–1524

Pflieger S, Lefebvre V, Causse M (2001). The candidate gene approach in plant genetics: a review. Mol. Breeding 7:275–291

Potter D, Eriksson T, Evans RC, Oh S, Smedmark JEE, Morgan DR, Kerr M, Robertson KR, Arsenault M, Dickinson TA, Campbell CS (2007). Phylogeny and classification of Rosaceae. Plant Syst. Evol. 266:5–43

RNAi News, Technology for Functional Genomics (2005). Rich Jorgensen discusses sense RNAi and forward genetics. http://www.rnainews.com/issues/3_23/rnainsider/112622-1.html. Cited June 17, 2008

Robertson DS (1989). Understanding the relationship between qualitative and quantitative genetics. p. 81–87. In: Helentjaris, T. and Burr, B. (eds), Development and Application of Molecular Markers to Problems in Plant Genetics. Cold Spring Harbor Laboratory, Cold Spring Harbor, New York

Rudell DR, Mattheis JP, Curry EA (2008). Prestorage ultraviolet-white light irradiation alters apple peel metabolome. J. Agr. Food Chem. 56:1138–1147

Samuelian SK, Baldo AM, Pattison JA, Weber CA (2008). Isolation and linkage mapping of NBS-LRR resistance gene analogs in red raspberry (*Rubus idaeus* L.) and classification among 270 Rosaceae NBS-LRR genes. Tree Genet. Genomes 4:881–896

Sargent DJ, Clarke J, Simpson DW, Tobutt KR, Arús P, Monfort A, Vilanova S, Denoyes-Rothan B, Rousseau M, Folta KM, Bassil NV, Battey NH (2006). An enhanced microsatellite map of diploid *Fragaria*. Theor. Appl. Genet. 112:1349–1359

Schaffer RJ, Friel EN, Souleyre EJF, Bolitho K, Thodey K, Ledger S, Bowen JH, Ma J-H, Nain B, Cohen D, Gleave AP, Crowhurst RN, Janssen BJ, Yao J-L, Newcomb RD (2007). A genomics approach reveals that aroma production in apple is controlled by ethylene predominantly at the final step in each biosynthetic pathway. Plant Physiol. 144: 1899–1912

Schauer N, Fernie AR (2006). Plant metabolomics: towards biological function and mechanism. Trends Plant Sci. 11:508–516

Schneider WL, Sherman DJ (2007). A comprehensive *Prunus* pathogen detection array. Phytopath. 97:S105

Sholberg P, O'Gorman D, Bedford K, Leveswue CA (2005). Development of a DNA macroarray for detection and monitoring of economically important apple diseases. Plant Disease 89: 1143–1150

Shulaev V, Korban SS, Sosinski B, Abbott AG, Aldwinckle HS, Folta KM, Iezzoni A, Main D, Arus P, Dandekar AM, Lewers K, Brown SK, Davis TM, Gardiner SE, Potter D, Veilleux RE (2008). Multiple models for Rosaceae genomics. Plant Physiol. 147:985–1003

Silfverberg-Dilworth E, Matasci CL, van de Weg WE, van Kaauwen MPW, Walser M, Kodde LP, Soglio V, Gianfranceschi L, Durel CE, Costa F, Yamamoto T, Koller B, Gessler C, Patocchi A (2006). Microsatellite markers spanning the apple (*Malus* × *domestica* Borkh.) genome. Tree Genet. Genomes 2:202–224

Soares M, Bonaldo M, Jelene P, Su L Lawton L (1994). Construction and characterization of a normalization cDNA library. Proc. Natl. Acad. Sci. USA 91:9228–9232

Sonneveld T, Tobutt KR, Robbins TP (2003). Allele-specific PCR detection of sweet cherry self-incompatibility (S) alleles $S_1$ to $S_{16}$ using consensus and allele-specific primers. Theor. Appl. Genet. 107:1059–1070

Srinivasan C, Padilla IMG, Scorza R (2005) *Rosaceae: Prunus* spp. almond, apricot, cherry, nectarine, peach and plum In: Litz RE (ed) Biotechnology of fruit and nut crops, CABI Publishing, Wallingford and Cambridge, MA, pp 512–542

Sutherland BG, Tobutt KR, Robbins TP (2004). Molecular genetics of self-incompatibility in plums. Acta Horticult. 663:557–562

Tamura M, Ushijima K, Sassa H, Hirano H, Tao R, Gradziel TM, Dandekar AM (2000). Identification of self-incompatibility genotypes of almond by allele-specific PCR analysis. Theor. Appl. Genet. 101:344–349

The United States Rosaceae Genomics, Genetics, and Breeding Initiative White Paper (2006). http://www.rosaceaewhitepaper.com Cited June 18, 2008

Thilmony R, Underwood W, He SY (2006) Genome-wide transcriptional analysis of the *Arabidopsis thaliana* interaction with the plant pathogen *Pseudomonas syringae* pv. *tomato* DC3000 and the human pathogen *Escherichia coli* O157:H7. Plant J. 46:34–53

Ushijima K, Yamane H, Watari A, Kakehi E, Ikeda K, Hauck NR, Iezzoni AI, Tao R (2004). The *S* haplotype-specific F-box protein gene, *SFB*, is defective in self-compatible haplotypes of *Prunus avium* and *P. mume*. Plant J. 39:573–586

van de Weg WE, Voorrips RE, Finkers HJ, Kodde LP, Jansen J, Bink MCAM (2004). Pedigree genotyping: a new pedigree-based approach of QTL identification and allele mining. Acta Horticult. 663:45–50

van Ooijen JW (2005). MapQTLÒ5, Software for the mapping of quantitative trait loci in experimental populations. Kyazma B.V., Wageningen, The Netherlands

Vilanova S, Arús P, Sargent DJ, Monfort A (2007). Map comparison between two distant genomes of the Rosaceae: *Prunus* and *Fragaria*. Poster 454 in Plant & Animal Genome XV, San Diego, CA

Wesley SA, Helliwell CA, Neil AS, Wang M-B, Liu Q, Gooding PS, Singh, SP, Abbott D, Stourtjesdijk PA, Robinson SP, Gleave AP, Green AG, Waterhouse PM (2001) Construct design for efficient, effective and high-throughput gene silencing in plants. Plant J. 27: 581–590

Xu Q, Wen X, Deng X (2007). Phylogenetic and evolutionary analysis of NBS-encoding genes in Rosaceae fruit crops. Mol. Phylogeneti Evol. 44:315–324

Yaegashi H, Yamatsuta T, Takahashi T, Li C, Isogai M, Kobori T, Ohki S, Yoshikawa N (2007). Characterization of virus-induced gene silencing in tobacco plants infected with apple latent spherical virus. Arch Virol. 152:1839–1849

Zhebentyayeva TN, Swire-Clark G, Georgi LL, Garay L, Jung S, Forrest S, Blenda AV, Blackmon B, Mook J, Horn R, Howad W, Arús P, Main D, Tomkins JP, Sosinski B, Baird WV, Reighard GL, Abbott AG (2008). A framework physical map for peach, a model Rosaceae species. Tree Genet. Genomes 4:745–756

Zheng, X, Deng W, Luo K, Duan H, Chen Y, McAvoy R, Song, S, Pei Y, Li Y (2007). The cauliflower mosaic virus (CaMV) 35 S promoter sequence alters the level and patterns of activity of adjacent tissue- and organ-specific gene promoters. Plant Cell. Rep. 26:1195–1203

Zhu Y, Barritt B (2008). Md-ACS1 and Md-ACO1 genotyping of apple (*Malus* × *domestica* Borkh.) breeding parents and suitability for marker-assisted selection. Tree Genet. Genomes 4:555–562

Zuo J, Qi-Wen N, Chua N-H (2000). An estrogen receptor-based transctivator XVE mediates highly inducible gene expression in transgenic plants. Plant J. 24:265–273

# 3. Genomics Opportunities, New Crops and New Products

**Nahla Bassil and Kimberly Lewers**

## 1 Introduction

The ultimate goal of genomic technology is to benefit consumers by generating new or better crops and products. Such technology transfer is possible through direct transfer of genes that encode desirable traits. The two techniques that effect gene transfer include transgenic biotechnology and marker assisted selection (MAS) using molecular markers associated with economically important traits. In the Rosaceae family, durable genetic resistance to Plum pox virus (PPV, genus *Potyvirus*) that causes the devastating sharka disease of *Prunus* has been demonstrated through genetic engineering (reviewed by Scorza and Ravelonandro, 2006). However, due to limited public acceptance of transgenic fruits, the most immediately deployable technologies will emphasize the development of new and improved Rosaceae germplasm and possible products obtained through MAS, if any. In this chapter we will also discuss the potential trends in genomic technology transfer to breeding programs and provide an overview of marker applications in crop species discussed later in this volume.

Because the release of a Rosaceae product as a result of MAS has never been reported, we sought to assess the current level of MAS use by Rosaceae breeders worldwide through a questionnaire sent by e-mail to the Rosaceae Breeder's List of the International Society of Horticultural Science (ISHS) and listserve for the fruit breeders working group and the ornamental plant breeding working group of the American Society of Horticultural Sciences (ASHS). In the first section of this chapter, we discuss the results obtained from this questionnaire while the second part gives a brief update of the available markers reported to be linked to important traits in the various Rosaceae crops. For further information, in-depth discussion

N. Bassil (✉)
USDA ARS, National Clonal Germplasm Repository, Corvallis, OR USA

K.M. Folta, S.E. Gardiner (eds.), *Genetics and Genomics of Rosaceae*, Plant Genetics and Genomics: Crops and Models 6, DOI 10.1007/978-0-387-77491-6_3,

of the important economic traits under study for each rosaceous crop is presented previously in this text.

## 2 Current Use of MAS in Rosaceae Breeding Programs

The following questions were asked to obtain as yet unpublished information about new crops and products that are being generated as a result of MAS in Rosaceae crops:

For each Rosaceae crop, please answer the following questions:

1. Do you use Marker Assisted Selection for selecting desirable progeny in your Rosaceae crop of interest?
2. If you use MAS, what type of marker? What trait? Please list citation(s) if available.
3. For what traits can you justify the use of MAS?
4. If there are markers linked to these traits, would you use MAS? Why or why not?
5. Would identifying genes underlying traits of interest result in concrete benefits to your breeding program? How?
6. What are the most important traits that you screen for in your crop of interest?
7. What type of new crops or products, if any, do you anticipate to produce as a consequence of the genomics era?
8. When would you expect these new crops or products to become available to the public?

The brief two-month duration of the requested response window limited the replies to twenty-three breeders representing thirty-eight fruit and two rose breeding programs. Such a restricted response is reasonable when compared to the 79 replies obtained in a period spanning one year by David Byrne when surveying fruit and ornamental breeders for molecular marker use in perennial plant breeding (Byrne, 2007). Most breeders led programs in North America and consisted of 12 breeders from the United States and four from Canada. A limited number of answers was obtained from other continents and included five from Europe (Germany, Norway, Serbia, Spain and the Netherlands) and two from Asia (China and the Republic of Korea). Out of these 23 breeding programs represented, 19 were public, three were private programs (2 in the US, one in the Netherlands) and one semi-private (in Norway). Breeding programs from the majority of important Rosaceae crops were represented and included 10 in strawberry, 10 in red raspberry, 5 in blackberry, 3 in apple, 2 each in cherries, peaches, pears, red raspberry and roses, one each in almond, apricot and cloudberry.

Despite the small number of programs surveyed, the trend in MAS use in breeding indicated that it is mostly limited to private breeding programs. Two of three private companies included in this study reported using MAS in six out of seven breeding programs (two each in red raspberry and strawberry, one each in apple and blackberry). A semi-private company reported using allele-specific primers for detecting self-compatible almond individuals. One public cherry breeding program also reported using MAS for selecting parents with the desirable self-compatibility/incompatibility alleles. Private breeding programs did not divulge the traits of interest used in early selection by MAS with the exception of the apple breeding program where Sequence Characterized Amplified Region (SCAR) and Sequence Tagged Site (STS) markers are used to choose parents with pyramidized apple scab (*Venturia inaequalis*) resistance. None of the public breeding programs reported using MAS for early selection in progeny.

The more widespread use of MAS in private Rosaceae breeding programs, in contrast to public breeding programs, mirrors other negative factors that have weakened plant breeding programs at public institutions (Baenziger, 2006; Guner and Wehner, 2003). Reduced financial support to public breeding programs in universities and the public sector has pushed current public plant breeders to shift their activities toward basic/fundamental studies that can be supported by federal grants and the private industry. More plant breeders now work in the private (65–75%) vs. public sector (25–35%) (Frey, 1996). Investment in public research and development (R&D) has decreased overall while funds dedicated to private R&D have increased (Morris et al., 2006). In addition to the pressure on public budgets, other factors that have shifted the balance of plant breeding from the public to the private sector include globalization, intellectual property protection and the ability of private firms to earn returns on their R&D investments. One public breeder surveyed in this study expressed his disillusion when asked to anticipate the type of new crops or products that will be produced as a consequence of the genomic era: 'I don't see the point of giving my best ideas to the world where they will be taken up by commercial enterprises with the funds to exploit them before I get the research done.'

In all of the Rosaceae crops surveyed except for cloudberry, the most important category that is selected for or used in screening parents and progeny consisted of disease resistance (20 breeding programs) followed by fruit quality (19), productivity in fruit crops or floriferousness in roses (10), plant architecture / development and other traits (9) (Fig. 1). Resistance to abiotic stress was mentioned in only seven breeding programs and included winter hardiness and heat tolerance. Important diseases were listed for each crop type. Size was the most important fruit quality trait. Other fruit quality traits included color, flavor, firmness, shelf life, soluble solids, anthocyanin levels and appearance or shape. In addition to crop-specific plant architecture, the plant architecture / development category also included fruiting season mostly in fruits harvested for fresh fruit consumption. Other traits consisted of compatibility in almond and cherries, machine harvestability or ease of harvest in such processed fruits as strawberries, blackberries and red and black raspberries, tree size control and induction of desirable scion tree traits in pear rootstocks. Suitability for the Immediately Quick Frozen fruit market (IQF) was listed as a selection criterion in red raspberry.

Rosaceae breeders were asked to list the traits for which they would use MAS. Again disease resistance was the most important category followed by fruit quality. Traits for which MAS was justified consisted of simply inherited traits that are

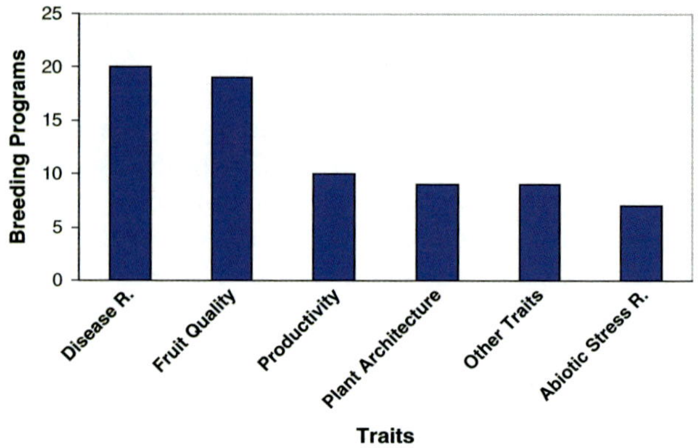


**Fig. 1** Six trait categories evaluated by Rosaceae breeding programs that include: Disease resistance, fruit quality, plant productivity in fruit crops and floriferousness in roses, plant architecture and development, other traits, and abiotic stress resistance

difficult to screen for. Traits that are difficult to screen for would include: Traits expressed in the mature phase of long-lived species with long juvenile periods; and traits that require high added costs for conventional screening in the target population of environments (as was concluded by Luby and Shaw, 2001).

Breeders agreed that identification of the number, relative importance, mode of action (recessive, dominant, additive) and interactions of genes would greatly facilitate our understanding of the inheritance /genetic control of economically important traits and allow us to manipulate these traits to benefit the growers and the consumers. Marker assisted selection was reported to allow more rational choice of parents, complementation of parents used for breeding, rapid release of disease resistant and profitable fruits and ornamental crops, production of fruits that contain higher levels of bioactive phytochemicals, and introduction of new traits to those crops. Use of molecular markers to select and match parents is an interesting contrast to early ideas of using molecular markers to select progeny from breeding crosses.

While private breeders are actively using MAS for early selection and for rapid release of cultivars possessing economically desirable traits, many public Rosaceae breeders are at the development stage for markers that are linked to their trait of interest. Examples of traits mentioned in this survey for which marker assisted selection is being sought include: Raspberry resistance to aphids, Raspberry Bushy Dwarf Virus (RBDV) and to raspberry root rot caused by *Phytophthora fragariae* var. *rubi*; repeat fruiting of strawberry, blackberry, and raspberry; sex control of strawberry; and resistance to bacterial angular leafspot disease of strawberry caused by *Xanthomonas fragariae* (Kennedy and King, 1962).

None of the public Rosaceae breeders surveyed in this study reported a new or improved product that has resulted from using MAS in their program. One cherry breeder anticipated releasing an improved fruit cultivar in 10 years. Private breeders did not respond to this question.

# 3 Traits for Which Markers are Currently Available for Use in Developing New Crops

While no new products or cultivars have yet to be released from MAS, progress has been made in identifying quantitative trait loci (QTL) that control economically important quantitative traits and in tagging major candidate genes responsible for important traits in the Rosaceae. QTLs associated with many traits have been identified, however, the utility of QTL markers must be validated over different genotypes, years, growth sites, generations and genetic backgrounds (epistasis) before co-segregating markers can be implemented in marker assisted selection schemes. Therefore, markers linked to QTLs have not been used in marker assisted selection of Rosaceae crops and will not be described in this chapter. We will, however, briefly describe markers that are linked to qualitative or single gene traits that have been used or that are ready for use in this family. This treatment provides the reader with a sense of the state of the art of marker application in these crops. A more comprehensive discussion follows in the relevant chapters within this volume.

Most of the current marker suite was developed in apple. Disease resistance traits for which efficient markers exist include: root knot nematode (RKN) resistance in peach and plum (Lecouls et al., 1999; Yamamoto and Hayashi, 2002), scab resistance and powdery mildew in apple (reviewed in Gardiner et al., 2007; Bus et al., 2008a); a major scab resistance gene (*Vnk*) in Japanese pear (Terakami et al., 2006); red stele and anthracnose in strawberry (Haymes et al., 2000; Hokanson and Maas, 2001; and Lerceteau-Köhler et al., 2005); resistance to sharka disease caused by the plum pox virus (PPV) in apricot (Soriano et al., 2008); resistance to woolly apple aphid (Bus et al., 2008b), rosy leaf-curling aphid (Cevik and King, 2002), rosy apple aphid and the leaf curling aphid in apple (Stoeckli et al., 2008). Markers for fruit quality traits were developed for: the *Rf* locus that is responsible for the red/yellow skin color polymorphism in apple (Cheng et al., 1996); the green skin phenotype in Japanese pear (Inoue et al., 2006); the low acid character in apple (Liebhard et al., 2003); genes controlling fruit firmness or softening in apple (Costa et al., 2005, 2008; Oraguzie et al., 2007; Zhu and Barritt, 2008), pear (Itai et al., 2003), and peach (Peace et al., 2005; reviewed by Peace et al., 2008). The columnar habit in apple is the only tree architecture trait that can be easily identified with molecular markers (Tian et al., 2005). Many marker systems were developed in the Rosaceae family for S-allele genotyping and are briefly summarized in this section.

## *3.1 Markers for Resistance to Nematode Pathogens*

Clones P.2175, P.1079 and P.2980 of the Myrobalan plum (*Prunus cerasifera*), section Euprunus of the Prunophora subgenus, an outbreeding diploid species, carry one dominant allele of a single resistance gene, designated *Ma1, Ma2* and *Ma3,* respectively (Esmenjaud et al., 1996; Rubio-Cabetas et al., 1998). Each of these *Ma* resistance alleles confers a high and wide-spectrum resistance to root knot nematode (RKN) caused by *M. arenaria, M. incognita, M. javanica* and *M.* sp. Florida which overcomes the resistance of the Amygdalus sources (Lecouls et al., 1997; Rubio-Cabetas et al., 1999) and to the minor species *Meloidogyne mayaguensis* (Fargette et al., 1996; Rubio-Cabetas et al., 1999). The *Ma* gene is being introgressed into the genome of new *Prunus* rootstocks by interspecific hybridization (e.g. Myrobalan × Amygdalus) (Lecouls et al., 2004). Molecular studies identified two reliable Sequence Characterized Amplified Region (SCAR) markers, $SCAL19_{690}$ and $SCAFLP2_{202}$, that are linked in coupling to the dominant resistance alleles *Ma1* and *Ma3* (Lecouls et al., 1999; Bergougnoux et al., 2002). SCAL19 is located less than 1 cM from *Ma* and SCAFLP2 co-segregates with *Ma* (Claverie et al., 2004). These markers appear to be powerful tools for selecting *Prunus* interspecific rootstocks (Lecouls et al., 2004). *Ma* and $R_{jap,}$ a wide-spectrum dominant resistance gene identified in Japanese plum (*Prunus salicina*), mapped very close to the SSR marker pchgms6 on LG7 of the reference *Prunus* map T×E (Claverie et al., 2004). Peach RKN resistance to *M. incognita* and *M. arenaria* was found in the Nemared rootstock ($R_{MiaNem}$) and Shalil sources ($R_{Mia557}$ through GF.557, an almond × Shalil peach hybrid), and in Japanese RKN resistant ornamental peach Juseitou (*Mia*) (Yamamoto and Hayashi, 2002). $R_{Mia557}$ and $R_{MiaNem}$ were co-localized in a subtelomeric position on linkage group 2 and were flanked by STS markers obtained by Yamamoto and Hayashi (2002) for the resistance gene *Mia*, thus suggesting that these three independent peach RKN resistance sources, 'Shalil', 'Nemared' and 'Juseitou', share at least one major gene resistance to *M. incognita* located in this subtelomeric position. Plum and peach RKN resistance genes are independent and, thus, can be pyramided into interspecific hybrid rootstocks based on the plum and peach species.

### *3.2 Markers for Resistance to Diseases*

In apple (see also Chapter "Apple"), molecular markers that are tightly linked to different major scab resistance genes have been identified and are used in pyramiding these genes (reviewed in Gardiner et al., 2007; Bus et al., 2008a). The *Vf* gene, has been analyzed in detail and a cluster of four *Vf* paralogs were cloned from 'GoldRush' and its original source in crabapple, *Malus floribunda* 821 (Xu and Korban, 2002). Recently, *Vfa1* and *Vfa2* were shown to be responsible for eliciting disease resistance to five races of the fungal pathogen, *Venturia inaequalis*, when introduced through *Agrobacterium*-mediated transformation into two susceptible apple cultivars, 'Galaxy' and 'McIntosh' (Malnoy et al., 2008). *Vf* resistance has been widely used in generating scab resistant cultivars and is found in more than 70 scab-resistant cultivars (Janick et al., 1996). However, new races of *Venturia inaequalis* have appeared including races 6, 7 and 8 that can overcome the resistance conferred by *Vf*. A number of additional major scab resistance genes have been identified and mapped in apple including *Va* (LG1), *Vbj*, *Vh2*, *Vh4*, *Vh8* and *Vr2* (LG2); *Vd* (LG10); *Vg* and *Vb* (LG12); and *Vm* (LG17) (reviewed in

Gardiner et al., 2007). Many groups are using molecular markers to introgress scab resistance from different sources into *Vf*-containing cultivars. One source in particular, a *Malus pumil*a Russian seedling (RS) apple R12740-7A has been used and appears to involve three major genes, one of which is race nonspecific (*Vr*) while the other two are specific to races 2 (*Vh2*) and 4 (*Vh4*) and map 40 cM apart on LG2 (an example is given in Chapter 18, Bus et al., 2008a). Identification of molecular markers linked to scab resistance in pear has lagged behind. However, sequence-tagged site (STS) markers that are linked to a major resistance gene (*Vnk*) to *Venturia naschicola* in a Japanese pear cultivar Kinchaku, were recently developed (Terakami et al., 2006). In European pear, two major QTLs on linkage groups 3 and 7 that are associated with resistance to *Venturia pyrina* were also recently identified (Pierantoni et al., 2007).

In contrast to scab resistance genes, genes conferring resistance to other diseases like powdery mildew and other pests do not show characteristic distinct phenotypes. Therefore, selection of individuals that carry more than one major gene to the fungal agent of powdery mildew, *Podosphaera leucotricha*, in apple was reported as almost impossible before the advent of linked molecular markers (James et al., 2004). Markers that are closely linked to five sources of resistance to the fungal agent of powdery mildew have been identified (reviewed by Gardiner et al., 2007). While *Pl1* from *Malus robusta* and *Pl2* from *Malus zumi* have been widely used in apple breeding programs, it is increasingly important to introgress other genes from different sources including *Plw* from the ornamental crabapple 'White Angel', *Pld* from the D12 clone and *Plm* from 'Mildew Immune Seedling'. The major resistance gene *Pl2* in apple was overcome by virulent strains of the powdery mildew fungus (Caffier and Parisi, 2007). Gardiner et al., 2007 reported pyramiding of *Pl2* and *Plm* in a resistance population at Plant and Food Research. Whole genome selection in a background selection 'fast breeding' approach is a promising new technique developed to quickly select progeny that contains high contribution from the high quality grandparent(s) ('Royal Gala') and lower proportion of the low quality *Pl2*-containing ancestor (A698-24) (Bus et al., 2008a).

In strawberry (see also Chapter "Strawberry"), molecular markers linked to resistance to two important diseases, red stele (caused by *Phytophthora fragariae*) and anthracnose (*Colletotrichum acutatum* pathogenicity group 2), were identified (Haymes et al., 2000; Hokanson and Maas, 2001; and Lerceteau-Köhler et al., 2005). A SCAR marker linked to the *Rpf1* gene (Haymes et al., 2000) and two RAPD markers linked to the *Rpf1* and *Rpf3* red stele resistance genes were reported (Hokanson and Maas, 2001). Two out of four AFLP markers linked to the *Rca2* anthracnose resistance gene were converted to SCAR markers (STS-Rca2_417 and STS_Rca2_240) and effectively predicted resistant and susceptible strawberry cultivars (Lerceteau-Köhler et al., 2005).

In apricot (Chapter "Genomics Opportunities in Apricot" herein), an SSR marker, ssrPaCITA5, that proved effective for MAS for resistance to sharka disease caused by plum pox virus was recently reported (Soriano et al., 2008). This marker allowed preservation of >90% (in $F_1$ populations) to >95% (in $F_2$ populations) resistant seedlings based on selection of the seedlings that contained resistance-linked alleles. Furthermore, PPV-resistance-linked alleles at ssrPaCITA5 and ssrPaCITA17 were present in all resistant cultivars studied containing four different sources of PPV resistance including North American cultivars 'Stark Early Orange', 'Sunglo,' and 'Reliable,' as well as *P. mandchurica* sp. A common origin for the PPV resistance from North Chinese cultivars, particularly those that contain *P. sibirica* or/and *P. mandshurica,* into North American germplasm was recently supported by targeted SSR analysis (Zhebentyayeva et al., 2008). Additional SSR markers linked to resistance-gene candidates were also recently reported (Sicard et al., 2008).

### 3.3 Markers for Resistance to Pests

Molecular markers linked to three major apple genes, *Er1* from 'Northern Spy', *Er2* from 'Robusta 5' and *Er3* from *Malus sieboldii* 'Aota 1' conferring woolly apple aphid (*Eriosoma lanigerum* Hausn.) were identified (Bus et al., 2008b). *Er1* and *Er3* mapped to LG8 while *Er2* was located on LG17. These genes are used in apple rootstock and scion cultivar breeding. A SNP marker, NZsn_O05, linked to both *Er1* and *Er3* was effective in identifying individuals carrying both genes from a 'Northern Spy' (*Er1*) × S26R01T053 (*Er3*) population (Bus et al., 2008a, b). However, a breeding strategy is needed to maintain gene combinations involving such linked resistance genes as the resulting selections will carry the genes in repulsion phase and will dissociate in further crossings. Markers to two additional sources of woolly apple aphid resistance (*Erm* and *Erl*) that do not map to markers for *Er1* or *Er3* were also reported (Gardiner et al., 2007). The *Sd1* gene for resistance to rosy leaf-curling aphid (*Dysaphis devecta* Wlk.) was fine mapped on LG7 of apple and co-localized with RFLP marker MC064 within a 1.3 cM interval between the SSR marker SdSSRa and RFLP marker 2B12a (Cevik and King, 2002). These molecular markers were also tightly linked to *Sd2*, indicating that *Sd1* and *Sd2* are tightly linked and possibly allelic (Cevik and King, 2002). Another SSR marker, CH-Sd1 was also developed from the BAC clone 49N23 containing the contig spanning the Sd1 locus (Khan et al., 2007). Alleles 255 (bp) of Hi07h02 and 216 (bp) of Hi03a10 SSR markers were closely associated with QTLs for resistance to the rosy apple aphid *Dysaphis plantaginea* and the leaf curling aphid *Dysaphis devecta* in apple, respectively (Stoeckli et al., 2008). These two SSR markers appear useful for marker assisted breeding as Hi07h02-255 was found in *D. plantaginea*-resistant 'Wagener' and Hi03a10-216 was present in *D. devecta*-resistant 'Cox's Orange Pippin' apples (Stoeckli et al., 2008).

### 3.4 Markers for Fruit Quality Traits

A single dominant gene at the *Rf* locus is responsible for the red/yellow skin color polymorphism in apple where a universal primer pair, BC226, can differentiate between the two phenotypes (Cheng et al., 1996). Amplification of alleles $A^1$ (1160 bp) and $A^2$ (1180 bp) is associated with red color while alleles $a^1$ (1230 bp) and $a^2$ (1320 bp) co-segregate with yellow fruit color in apple (Cheng et al., 1996; Melounová et al., 2005). A RAPD marker associated with the green skin phenotype in Japanese pear was recently identified (Inoue et al., 2006).

Acidity in apple fruit is due to malic acid and the low-acid character (pH 3.8 and above) is determined by the presence of recessive alleles *ma ma* for the *Ma* gene. In apple, the *Ma* gene was located on LG 16 in the cross 'Prima' × 'Fiesta' (Maliepaard et al., 1998). Markers that are linked to fruit acidity include RAPD OPT161000 (Maliepaard et al., 1998), AFLP marker E31M38-0193 and SSR marker CHO5e04z (Liebhard et al., 2003).

Fruit shelf-life and storability is a desirable factor that impacts the economic value of a cultivar. In apple, functional markers in three critical candidate genes associated with fruit softening were recently identified and include ethylene biosynthesis genes 1-aminocyclopropane-1-carboxylate (*Md-ACS1*) and 1-aminocyclopropane-1-oxidase (*Md ACO-1*) (Costa et al., 2005, 2008; Oraguzie et al., 2007; Zhu and Barritt, 2008), and an expansin gene, *MdExp7* (Costa et al., 2008). Earlier studies showed that *ACS* had stronger effect on ethylene production than *ACO* and that genotypes homozygous for the *ACS1*-2 allele produced less ethylene and had firmer fruit than *ACS1*-1/2 and *ACS1*-1/1 (Costa et al., 2005; Oraguzie et al., 2007; Zhu and Barritt, 2008). Recent data however suggests a higher impact of *Md-Exp7* and *Md-ACO1* in the control of firmness loss in apple (Costa et al., 2008). Difference between the unfavorable and favorable allele of *Md-ACS1* causes a difference of only 17% in fruit softening, while the difference rose up to 60 and 72.6% in *Md-ACO1* and *Md-Exp7*, respectively. The relative effects of each of these three genes will always depend on the germplasm under investigation. Although the *Md-Exp7* gene may show strong effects in some genotypes, the 214 allele (good firmness) and 198 allele (low firmness) are rare in modern cultivars, while the 202 (neutral firmness) is present in 71% of the cultivars tested by Costa et al. (2008). *Md-ACS1* has been investigated the most so far, and in independent studies, was shown to be associated with softening to an extent in modern cultivars that makes its use in MAS worthwhile (Cameron Peace, personal communication). In pear, cleaved-amplified polymorphic sequence (CAPS) markers A and B respectively of two 1-aminocyclopropane-1-carboxylate (ACC) synthase genes (*PPACS1* and *PPACS2*) were associated with the amount of ethylene produced (Itai et al., 2003). Marker A was associated with high ethylene producers and marker B with moderate ethylene producers while the absence of these two markers was characteristic of low ethylene producers. These CAPS markers were effective in identifying ethylene genotypes for 40 commercially important Japanese pear cultivars and two Chinese pear (*P. bretschneideri*). The apple Md-Exp$7_{SSR}$ marker appears to be also associated with softening in 41 individuals of a Passe Crassane × Harrow Sweet pear mapping population (Costa et al., 2008). In peach, endopolygalacturonase (endoPG) was identified as the gene controlling the major fruit firmness and texture traits of *Melting flesh* (*M*) and *Freestone* (*F*) (Peace et al., 2005; reviewed by Peace et al., 2008). Two endoPG genes underlie the Freestone-Melting flesh (F-M) locus Allelic variation at a microsatellite locus in the 5'UTR of a peach endoPG was associated with four functional phenotypic groups that include a null (n) allele. These groups correspond to haplotypes and include F (where F- is associated with the FMF phenotype), f (where ff, ffl, or fn gives CMF fruit), fl (where flfl or fln confers a CNMF phenotype), and n (nn leads to CNSF) (Peace et al., 2007). Association of endoPG genotypic variation with fruit firmness and flesh adhesion in other *Prunus* fruit crops like apricot and sour cherry is being investigated. Both Md-ACS1 in apple and endoPG in peach are ready for use in MAS. Still, due to our incomplete knowledge of all the genes/alleles involved in fruit softening and their interaction, caution should be exercised in applying these markers in marker assisted selection.

## 3.5 Markers for Tree Architecture

The columnar growth habit in apple is caused by a mutation at the *co* locus that is characterized by a reduced number of lateral shoots, an increased number of spurs and compact internodes. A single dominant gene controls the columnar habit and modifier genes might also be involved in the inheritance of this trait (Lapins, 1976). Columnar varieties are heterozygous for the *Co* gene (*Coco*) and include Telamon, Tuscan, Trajan, Maypole and Charlotte and the spontaneous sport of McIntosh where the columnar habit was initially identified. SSR and SCAR markers that are linked to the Co gene on LG 10 were identified and include $SSR^{CO}$ (Hemmat et al., 1997), $SCB82_{670}$ (Kim et al., 2003), $SCAR_{682}$ and $SCAR_{216}$ (Tian et al., 2005). When used in multiplex PCR, $SCAR_{682}$ and $SCAR_{216}$ were efficient in identifying individuals that exhibit the columnar habit (Tian et al., 2005).

## 3.6 Markers for Self-Incompatibility

Many Rosaceae species exhibit gametophytic self-incompatibility (GSI) that has been naturally selected to promote out-breeding. Consequently commercial fruit set requires the presence of mutually compatible cultivars, artificial pollination or self-compatible cultivars, if possible. Self-incompatibility has been extensively studied at the molecular level and is controlled by a single locus with multiple alleles. The gene encoding stylar specificity encodes a ribonuclease (S-RNase) (reviewed by McCubbin and Kao, 2000) and the gene controlling pollen specificity consists of an F-box protein (SLF) (reviewed by Kao and Tsukamoto 2004). These two genes are separate but tightly linked at the S-locus and their allelic form determines the S-haplotype. The progress made in uncovering the genetic and molecular basis of the self-incompatibility reaction in the Rosaceae has led to the development and application of PCR-based S-allele typing for two main aspects of Rosaceae fruit crop breeding: identification of self-compatible individuals; and assignment of S-alleles for determination of cross-compatible varieties. This paragraph is not meant to provide a review of the extensive literature describing S-allele typing in Rosaceae crops but to give a brief overview. S-allele typing was initially based on distinguishing S-RNase alleles by size following amplification with conserved or S-allele specific primers. Sometimes restriction endonuclease digestion of the resulting PCR fragments was necessary to differentiate between S-alleles that generated similar size fragments. S-RNase-based S typing was reported in apple (Broothaerts, 2003), European pear (Moriya et al., 2007), Japanese pear (Kim et al., 2006, 2007), Chinese pear (Tan et al., 2007), almond (López et al., 2006; Ortega et al., 2006), apricot (Halász et al., 2005; Jie et al., 2005), sweet cherry (Sonneveld et al., 2003, 2005), and sour cherry (Bošković et al., 2006; Hauck et al., 2006a, b; Tsukamoto et al., 2006). Molecular markers can now be used for early selection of self-compatible seedlings in many Rosaceae fruit crops including sweet cherry carrying the S4' genotype (Ikeda et al., 2004); almond expressing the $S_f$ allele (Lopez et al., 2006); and apricot with the $S_c$ allele (Halász et al., 2005). The pollen S-determinant for GSI, SFB/SLF was only recently identified in Rosaceae species including almond (Ushijima et al., 2003), sweet cherry (Yamane et al., 2003a), sour cherry (Yamane et al., 2003a, b), apricot (Romero et al., 2004), Japanese plum (Zhang et al., 2007), apple (Cheng et al., 2006; Sassa et al., 2007), and Japanese pear (Sassa et al., 2007). Consequently, new markers for S-typing and identification of self-compatible genotypes at the seedling stage are based on haplotype-specific polymorphism and were developed in Japanese pear (Kakui et al., 2007; Okada et al., 2008), and sour cherry (Tsukamoto et al., 2006), for example.

## 4 Conclusion

Genomic technology can provide valuable tools for faster release of new crops and products when used in combination with traditional genetic and breeding techniques. The most striking result of the survey we conducted was the disparity in MAS application in breeding between public and private breeding programs. Concern about the decline in the national breeding capacity over the last 20 years has led to the establishment in February 2007 of the Plant Breeding Coordinating Committee (PBCC). PBCC provides a forum for leadership regarding issues, problems and opportunities of long-term strategic importance to the contribution of plant breeding to national goals (Hancock and Stuber, 2008). Translating genomics research into a toolbox for breeder's use and public investment in implementing these tools are necessary for plant breeders to meet future consumer needs. It has been argued that marker assisted selection might be too expensive to use in breeding except under certain circumstances (Luby and Shaw, 2001), but breeders are easing into molecular marker use cost effectively by identifying and matching parents rather than identifying desirable progeny. This ability may help support breeding programs by attracting students who need to be proficient with both traditional breeding methods and molecular marker methods.

In summary, many factors have contributed to the limited use of genomic-based biotechnology in Rosaceae crops. Limited public acceptance of transgenic fruits, and strict regulations have hampered the release of plum pox resistance stone fruits. However, in roses, Australian researchers at Florigene Ltd., and the Japanese Suntory group of companies recently applied RNA interference (RNAi) technology for gene replacement in plants in developing the world's only blue rose (Katsumoto et al., 2007). It was used to remove the gene encoding the enzyme dihydroflavonol reductase (DFR) in transgenic roses that were expressing a pansy and an iris delphinidin enzyme which produces blue pigment. While many markers that are linked to single qualitative genes encoding various traits have been identified and are being used or ready for use in MAS in Rosaceae crops, markers linked to QTLs have lagged behind. Progress, however, is being made in identifying genes responsible for major QTLs as recently reported for qP-Brn5.$1^{m}$ affecting browning in peach (Ogundiwin et al., 2007, 2008). A gene encoding the leucoanthocanidin dioxygenase (PpLDOX) enzyme was identified as potentially responsible for this major QTL. SSR variation within the intron of this gene was associated with low browning incidence and might be useful for marker assisted breeding of peach and nectarine cultivars with low incidence of mealiness, browning and bleeding (Ogundiwin et al., 2008). Furthermore, as new technologies such as Pedigree Based Analysis (van de Weg et al., 2004) become more widely adopted for QTL validation and allele mining, new markers will be identified and used in generating new Rosaceae crops and products.

## References


Baenziger PS (2006) Plant breeding training in the U.S.A. HortScience 41: 28–29

Bergougnoux V, Claverie M., Bosselut N, Lecouls AC, Salesses G, Dirlewanger E, Esmenjaud D (2002) Marker-assisted selection of the *Ma* gene from Myrobalan plum for a complete-spectrum root-knot nematode (RKN) resistance in *Prunus* rootstocks. Acta Hort 592: 223–228

Bošković RI, Wolfram B, Tobutt KR, Cerović R, Sonneveld T (2006) Inheritance and interactions of incompatibility alleles in the tetraploid sour cherry. Theor Appl Gen 112: 315–326

Broothaerts W (2003) New findings in apple S-genotype analysis resolve previous confusion and request the re-numbering of some S-alleles. Theor Appl Gen 106: 703–714

Bus et al., 2008a Chapter 20 This book

Bus V, Chagné D, Bassett H, Bowatte D, Calenge F, Celton JM, Durel CE, Malone M, Patocchi A, Ranatunga A, Rikkerink E, Tustin D, Zhou J, Gardiner S (2008b) Genome mapping of three major resistance genes to woolly apple aphid (*Eriosoma lanigerum* Hausm.). Tree Genet Genomes 4: 223–236

Byrne D (2007) Molecular marker use in perennial plant breeding. Acta Hort 751: 163–167

Caffier V, and Parisi L (2007) Development of apple powdery mildew on sources of resistance to *Podosphaera leucotricha*, exposed to an inoculum virulent against the major resistance gene *Pl-2*. Plant Breeding 126: 319–322

Cevik V, and King G (2002) High-resolution genetic analysis of the *Sd-1* aphid resistance locus in *Malus* spp. Theor Appl Gen 105: 346–354

Cheng FS, Weeden NF, Brown SK (1996) Identification of co-dominant RAPD markers tightly linked to fruit skin color in apple. Theor Appl Gen 93: 222–227

Cheng J, Han Z, Xu X, Li T (2006) Isolation and identification of the pollen-expressed polymorphic F-box genes linked to the S-locus in apple (*Malus* × *domestica*). Sex Plant Reprod 19: 175–183

Claverie M, Dirlewanger E, Cosson P, Bosselut N, Lecouls AC, Voisin R, Kleinhentz M, Lafargue B, Caboche M, Chalhoub B, Esmenjaud D (2004) High-resolution mapping and chromosome landing at the root-knot nematode resistance locus *Ma* from Myrobalan plum using a large-insert BAC DNA library. Theor Appl Gen 109: 1318–1327

Costa F, Stella S, Van de Weg WE, Guerra W, Cecchinel M, Dallavia J, Koller B, Sansavini S (2005) Role of the genes *Md-ACO1* and *Md-ACS1* in ethylene production and shelf life of apple (*Malus domestica* Borkh). Euphytica 141: 181–190

Costa F, Van de Weg W, Stella S, Dondini L, Pratesi D, Musacchi S, Sansavini S (2008) Map position and functional allelic diversity of *Md-Exp7* , a new putative expansin gene associated with fruit softening in apple (*Malus* × *domestica* Borkh.) and pear (*Pyrus communis*). Tree Genet Genomes 4: 575–586

Esmenjaud D, Minot JC, Voisin R, Bonnet A, Salesses G (1996) Inheritance of resistance to the root-knot nematode *Meloidogyne arenaria* in Myrobalan plum. Theor Appl Gen 92: 873–879

Fargette M, Phillips MS, Block VC, Waugh R, Trudgill DL (1996) An RFLP study of relationships between species, populations, and resistance breaking lines of tropical *Meloidogyne*. Fundam Appl Nematol 19: 193–200

Frey K (1996) National plant breeding study: I. Human and financial resources devoted to plant breeding research and development in the United States in 1994. Spec. Rep. No. 98. Iowa Agric. Home Economics Exp. Stn., Ames, IA

Gardiner SE, Bus VGM, Rusholme RL, Chagné D, Rikkerink EHA (2007) Apple. In: Genome mapping and molecular breeding, vol. 4. Kole C (ed) Springer, Heidelberg, pp 1–62

Guner N, and Wehner TC (2003) Survey of U.S. land-grant universities for training of plant breeding students. Crop Sci 43: 1938–1944

Halász J, Hegedüs A, Hermán R, Stefanovits-Bányai É, Pedryc A (2005) New self-incompatibility alleles in apricot (*Prunus armeniaca* L.) revealed by stylar ribonuclease assay and S-PCR analysis. Euphytica 145: 57–66

Hancock JF and Stuber C (2008) Sustaining public plant breeding to meet future national needs. HortScience 43: 298–299

Hauck NR, Ikeda K, Tao R, Iezzoni AF (2006a) The mutated S1-haplotype in sour cherry has an altered S-haplotype-specific F-box protein gene. J Hered 97: 514–520

Hauck NR, Yamane H, Tao R, Iezzoni AF (2006b) Accumulation of nonfunctional S-haplotypes results in the breakdown of gametophytic self-incompatibility in tetraploid *Prunus*. Genetics 172: 1191–1198

Haymes KM, Van de Weg WE, Arens P, Maas JL, Vosman B, Nijs APMD (2000) Development of SCAR markers linked to a *Phytophthora fragariae* resistance gene and their assessment in European and North American strawberry genotypes. J Am Soc Hort Sci 125: 330–339

Hemmat M, Weeden NF, Conner PJ, and Brown SK (1997) A DNA marker for columnar growth habit in apple contains a simple sequence repeat. J Am Soc Hort Sci 122: 347–349

Hokanson SC, and Maas JL (2001) Strawberry biotechnology. Plant Breed Rev 21: 139–180

Ikeda K, Igic B, Ushijima K, Yamane H, Hauck N, Nakano R, Sassa H, Iezzoni A, Kohn J, Tao R (2004) Primary structural features of the S haplotype-specific F-box protein, SFB, in *Prunus*. Sex Plant Reprod 16: 235–243

Inoue E, Kasumi M, Sakuma F, Anzai H, Amano K, Hara H (2006) Identification of RAPD marker linked to fruit skin color in Japanese pear (*Pyrus pyrifolia* Nakai). Scientia Hort 107: 254–258

Itai A, Kotaki T, Tanabe K, Tamura F, Kawaguchi D, Fukuda M (2003) Rapid identification of 1-aminocyclopropane-1-carboxylate (ACC) synthase genotypes in cultivars of Japanese pear (*Pyrus pyrifolia* Nakai) using CAPS markers. Theor Appl Gen 106: 1266–1272

James C, Clarke J, Evans K (2004) Identification of molecular markers linked to the mildew resistance gene *Pl-d* in apple. Theor Appl Gen 110: 175–181

Janick J, Cummins JN, Brown SK, and Hemmat M (1996) Apples. In: Fruit Breeding, Tree and Tropical Fruits. J Janick and JN. Moore, eds. Wiley, New York, pp. 1–77

Jie Q, Shupeng G, Jixiang Z, Manru G, Huairui S (2005) Identification of self-incompatibility genotypes of apricot (*Prunus armeniaca* L.) by S-allele-specific PCR analysis. Biotechnol Lett 27: 1205–1209

Kakui H, Tsuzuki T, Koba T, Sassa H (2007) Polymorphism of SFBB $-\gamma$ and its use for S genotyping in Japanese pear (*Pyrus pyrifolia*). Plant Cell Report 26: 1619–1625

Katsumoto Y, Fukuchi-Mizutani M, Fukui Y, Brugliera F, Holton TA, Karan M, Nakamura N, Yonekura-Sakakibara K, Togami J, Pigeaire A, Tao G-Q, Nehra NS, Lu C-Y, Dyson BK, Tsuda S, Ashikari T, Kusumi T, Mason JG, and Tanaka Y (2007) Engineering of the rose flavonoid biosynthetic pathway successfully generated blue-hued flowers accumulating delphinidin. Plant Cell Physiol 48: 1589–1600

Kao T-h, Tsukamoto T (2004) The molecular and genetic bases of S-RNase-based self-incompatibility. Plant Cell 16: S72–83

Khan MA, Durel C-E, Duffy B, Drouet D, Kellerhals M, Gessler C, Patocchi A (2007) Development of molecular markers linked to 'Fiesta' linkage group 7 major QTL for fire blight resistance and their application for marker-assisted selection. Genome 50: 568–577

Kim MY, Song KJ, Hwang J-H, Shin Y-U, and Lee HJ (2003) Development of RAPD and SCAR markers linked to the *Co* gene conferring columnar growth habit in apple (*Malus pumila* Mill). J Hort Sci Biotechnol 78: 512–517

Kim H-T, Hirata Y, Kim H-J, Nou I-S (2006) The presence of a new S-RNase allele (S10) in Asian pear (*Pyrus pyrifolia* (Burm; Nakai)). Genet Resour Crop Evol 53: 1375–1383

Kim HT, Kakui H, Koba T, Hirata Y, Sassa H (2007) Cloning of a new *S-RNase* and development of a PCR-RFLP system for determination of the *S*-genotypes of Japanese pear. Breed Sci 57: 159–164

Lapins KO (1976) Inheritance of compact growth type in apple. J Am Soc Hort Sci 101: 133–135

Lecouls AC, Bergougnoux V, Rubio-Cabetas MJ, Bosselut N, Voisin R, Poessel JL, Faurobert M, Bonnet A, Salesses G, Dirlewanger E, Esmenjaud D (2004) Marker-assisted selection of *Prunus* rootstocks for the wide-spectrum root-knot nematode resistance conferred by the *Ma* gene from Myrobalan plum (*Prunus cerasifera*). Mol Breed 13: 113–124

Lecouls AC, Rubio-Cabetas MJ, Minot JC, Voisin R, Bonnet A, Salesses G, Dirlewanger E, Esmenjaud D (1999) RAPD and SCAR markers linked to the *Ma1* root-knot nematode resistance gene in Myrobalan plum (*Prunus cerasifera* Ehr.). Theor Appl Gen 99: 328–335

Lecouls AC, Salesses G, Minot JC, Voisin R, Bonnet A, Esmenjaud D (1997) Spectrum of the *Ma* genes for resistance to *Meloidogyne* spp. in Myrobalan plum. Theor Appl Gen 95: 1325–1334

Lerceteau-Köhler E, Guérin G, Denoyes-Rothan B (2005) Identification of SCAR markers linked to *Rca2* anthracnose resistance gene and their assessment in strawberry germplasm. Theor Appl Gen 111: 862–870

Liebhard R, Koller B, Gianfranceschi L, Gessler C (2003) Creating a saturated reference map for the apple (*Malus* × *domestica* Borkh.) genome. Theor Appl Gen 106: 1497–1508

López M, Vargas F, Batlle I (2006) self-(in)compatibility almond genotypes: A review. Euphytica 150: 1–16

Luby JJ and Shaw DV (2001) Does marker-assisted selection make dollars and sense in a fruit breeding program? HortScience 36: 872–879

Maliepaard C, Alston FH, van Arkel G, Brown LM, Chevreau E, Dunemann F, Evans KM, Gardiner S, Guilford P, van Heusden AW, Janse J, Laurens F, Lynn JR, Manganaris AG, den Nijs APM, Periam N, Rikkerink E, Roche P, Ryder C, Sansavini S, Schmidt H, Tartarini S, Verhaegh JJ, Vrielink-van Ginkel M, King GJ (1998) Aligning male and female linkage maps of apple (*Malus pumila* Mill.) using multi-allelic markers. Theor Appl Gen 97: 60–73

Malnoy M, Xu M, Borejsza-Wysocka E, Korban SS, Aldwinckle HS (2008) Two receptor-like genes, *Vfa1* and *Vfa2*, confer resistance to the fungal pathogen *Venturia inaequalis* inciting apple scab disease. Mol Plant Microbe Interact 21: 448–458

McCubbin AG, Kao T-h (2000) Molecular recognition and response in pollen and pistil interactions. Ann Rev Cell Develop Biol 16: 333–364

Melounová M, Vejl P, Sedlák P, Blažek J, Zoufalá J, Milec Z, and Blažková H (2005) Alleles controlling apple skin colour and incompatibility in new Czech apple varieties with different degrees of resistance against *Venturia inaequalis* CKE. Plant Soil Environ 51: 65–73

Moriya Y, Yamamoto K, Okada K, Iwanami H, Bessho H, Nakanishi T, Takasaki T (2007) Development of a CAPS marker system for genotyping European pear cultivars harboring 17 S alleles. Plant Cell Report 26: 345–354

Morris M, Edmeades G, and Pehu E (2006) Building capacity for international plant breeding: What roles for the public and private sectors? HortScience 41: 30–39

Ogundiwin EA, Peace CP, Nicolet M, Rashbrook VK, Gradziel TM, Bliss FA, Parfitt D, Crisosto CH (2008) Leucoanthocyanidin dioxygenase gene (PpLDOX): a potential functional marker for cold storage browning in peach. Tree Genet Genomes 4: 543–554

Ogundiwin EA, Peace CP, Gradziel TM, Dandekar AM, Bliss FA, Crisosto CH (2007) Molecular genetic dissection of chilling injury in peach fruit. Acta Hort 738: 633–638

Okada K, Tonaka N, Moriya Y, Norioka N, Sawamura Y, Matsumoto T, Nakanishi T, Takasaki-Yasuda T (2008) Deletion of a 236 kb region around S4- RNase in a stylar-part mutant S4 sm -haplotype of Japanese pear. Plant Mol Biol 66: 389–400

Oraguzie NC, Volz RK, Whitworth CJ, Bassett HCM, Hall AJ, Gardiner SE (2007) Influence of *Md-ACS1* allelotype and harvest season within an apple germplasm collection on fruit softening during cold air storage. Postharvest Biol Technol 44: 212–219

Ortega E, Bošković R, Sargent D, Tobutt K (2006) Analysis of S-RNase alleles of almond (*Prunus dulcis*): characterization of new sequences, resolution of synonyms and evidence of intragenic recombination. Mol Genet Genomics 276: 413–426

Peace et al. (2008) Chapter 3- This book

Peace CP, Callahan A, Ogundiwin EA, Potter D, Gradziel TM, Bliss FA, and Crisosto CH (2007) Endopolygalacturonase genotypic variation in *Prunus*. Acta Hort 738: 639–646

Peace CP, Crisosto CH, Gradziel TM (2005) Endopolygalacturonase: a candidate gene for Freestone and Melting Flesh in peach. Mol Breed 16: 21–31

Pierantoni L, Dondini L, Cho KH, Shin IS, Gennari F, Chiodini R, Tartarini S, Kang SJ, Sansavini S (2007) Pear scab resistance QTLs via a European pear (*Pyrus communis*) linkage map. Tree Genet Genomes 3: 311–317

Romero C, Vilanova S, Burgos L, Martínez-Calvo J, Vicente M, Llácer G, Badenes ML (2004) Analysis of the S-locus structure in *Prunus armeniaca* L. Identification ofS-haplotype specific S-RNase and F-box genes. Plant Mol Biol 56: 145–157

Rubio-Cabetas MJ, Lecouls AC, Salesses G, Bonnet A, Minot JC, Voisin R, Esmenjaud D (1998) Evidence of a new gene for high resistance to *Meloidogyne* spp. in Myrobalan plum (*Prunus cerasifera*). Plant Breed 117: 567–571

Rubio-Cabetas MJ, Minot JC, Voisin R, Esmenjaud D, Salesses G, Bonnet A (1999) Response of the *Ma* genes from Myrobalan plum to *Meloidogyne hapla* and *M. mayaguensis*. HortScience 34: 1266–1268

Sassa H, Kakui H, Miyamoto M, Suzuki Y, Hanada T, Ushijima K, Kusaba M, Hirano H, Koba T (2007) S locus F-box brothers: Multiple and pollen-specific F-box genes with S haplotype-specific polymorphisms in apple and Japanese pear. Genetics 175: 1869–1881

Scorza R, and Ravelonandro M (2006) Control of Plum pox virus through the use of genetically modified plants. EPPO Bullet 36: 337–340

Sicard O, Marandel G, Soriano J, Lalli D, Lambert P, Salava J, Badenes M, Abbott A, Decroocq V (2008) Flanking the major Plum pox virus resistance locus in apricot with co-dominant markers (SSRs) derived from candidate resistance genes. Tree Genet Genomes 4: 359–365

Sonneveld T, Tobutt KR, Robbins TP (2003) Allele-specific PCR detection of sweet cherry self-incompatibility (S) alleles S1 to S16 using consensus and allele-specific primers. Theor Appl Gen 107: 1059–1070

Sonneveld T, Tobutt KR, Vaughan SP, Robbins TP (2005) Loss of pollen-S function in two self-compatible selections of *Prunus avium* is associated with deletion/mutation of an S haplotype-specific F-box gene. Plant Cell 17: 37–51

Soriano J, Vera-Ruiz E, Vilanova S, Martínez-Calvo J, Llácer G, Badenes M, Romero C (2008) Identification and mapping of a locus conferring plum pox virus resistance in two apricot-improved linkage maps. Tree Genet Genomes 4: 391–402

Stoeckli S, Mody K, Gessler C, Patocchi A, Jermini M, Dorn S (2008) QTL analysis for aphid resistance and growth traits in apple. Tree Genet Genomes: Online first

Tan XF, Zhang L, Wuyun TN, Yuan DY, Cao YF, Jiang AF, Liang WJ, Zeng YL (2007) Molecular identification of two new self-incompatible alleles (S-alleles) in Chinese pear (*Pyrus bretschneideri*). Zhi Wu Sheng Li Yu Fen Zi Sheng Wu Xue Xue Ba 33(1): 61–70

Terakami S, Shoda M, Adachi Y, Gonai T, Kasumi M, Sawamura Y, Iketani H, Kotobuki K, Patocchi A, Gessler C, Hayashi T, Yamamoto T (2006) Genetic mapping of the pear scab resistance gene *Vnk* of Japanese pear cultivar Kinchaku. Theor Appl Gen 113: 743–752

Tian Y-K, Wang C-H, Zhang J-S, James C, Dai H-Y (2005) Mapping Co, a gene controlling the columnar phenotype of apple, with molecular markers. Euphytica 145: 181–188

Tsukamoto T, Hauck N, Tao R, Jiang N, Iezzoni A (2006) Molecular characterization of three non-functional S-haplotypes in sour cherry (*Prunus cerasus*). Plant Mol Biol 62: 371–383

Ushijima K, Sassa H, Dandekar AM, Gradziel TM, Tao R, Hirano H (2003) Structural and transcriptional analysis of the self-incompatibilityl of almond: Identification of a pollen-expressed F-box gene with haplotype-specific polymorphism. Plant Cell 15: 771–781

Van de Weg WE, Voorrips RE, Finkers R, Kodde LP, Jansen J and Bink MCAM (2004) Pedigree genotyping: A new pedigree-based approach of QTL identification and allele mining. Acta Hort 663: 45–50

Xu ML, and Korban SS (2002) A cluster of four receptor-like genes resides in the *Vf* locus that confers resistance to apple scab disease. Genetics 162: 1995–2006

Yamamoto T, and Hayashi T (2002) New root-knot nematode resistance genes and their STS markers in peach. Scientia Hort 96: 81–90

Yamane H, Ikeda K, Hauck NR, Iezzoni AF, Tao R (2003a) Self-incompatibility (S) locus region of the mutated S6-haplotype of sour cherry (*Prunus cerasus*) contains a functional pollen S allele and a non-functional pistil S allele. J Exp Bot 54: 2431–2437

Yamane H, Ikeda K, Ushijima K, Sassa H, Tao R (2003b) A pollen-expressed gene for a novel protein with an f-box motif that is very tightly linked to a gene for S-RNase in two species of cherry, *Prunus cerasus* and *P. avium*. Plant Cell Physiol 44: 764–769

Zhang SL, Huang SX, Kitashiba H, Nishio T (2007) Identification of S-haplotype-specific F-box gene in Japanese plum (*Prunus salicina* Lindl.). Sex Plant Reprod 20: 1–8

Zhebentyayeva T, Reighard G, Lalli D, Gorina V, Krška B, Abbott A (2008) Origin of resistance to plum pox virus in apricot: what new AFLP and targeted SSR data analyses tell. Tree Genet Genomes 4: 403–417

Zhu Y, Barritt B (2008) Md-ACS1 and Md-ACO1 genotyping of apple (*Malus* ×*domestica* Borkh.) breeding parents and suitability for marker-assisted selection. Tree Genet Genomes 4: 555–562

# Part I
# Apples
# (Chapters 4 – 7)

# 4. Introduction to Apple (*Malus* × *domestica*)

**Markus Kellerhals**

## 1 Origin and History

Apple is the most important temperate fruit crop and has been cultivated in Asia and Europe from antiquity (Janick et al., 1996). The genus *Malus* has, according to most authorities, 25–30 species and several subspecies of so-called crabapples. The cultivated apple is supposed to be the result of interspecific hybridization. The denomination *Malus* × *domestica* has been generally accepted as the appropriate scientific name (Korban and Skirvin, 1984). The main progenitor of the domestic apple is considered to be *Malus sieversii* which grows wild in the Heavenly Mountains (Tien Shan) at the boundary between western China and the former Soviet Union to the edge of the Caspian sea (Morgan and Richard, 1993; Forsline and Aldwinckle, 2004). Central Asia is the area of greatest diversity and the center of origin. The apple was carried by humans to the Middle East, Europe and eventually to North America. Several collection trips to Central Asia have shown that *M. sieversii* is very diverse and has all qualities present in *M.* × *domestica*. *Malus sieversii* is endemic to the Republic of Kazakhstan (Harris et al., 2002; Hokanson et al., 1998; Luby et al., 2001). In his book 'The Story of the Apple' from 2006, B.E. Juniper refers to both the domestic apple and the wild Central Asian apple as *Malus pumila*. The USDA has also adopted this convention. However, Coart et al. (2006) calls into question this hypothesis and declares *Malus sieversii* to be the correct denomination for the wild Central Asian apple, and *Malus* × *domestica* for the domestic apple.

Way et al. (1990) stated that while the genus *Malus* is very diverse genetically, cultivated varieties of apple have a rather narrow genetic base, the two cultivars 'Delicious' and 'Golden Delicious' being predominant. The word's most important commercially produced apple cultivars belong to the species *Malus* × *domestica*

M. Kellerhals (✉)
Research Station Agroscope Changins-Wädenswil, Wädenswil, Switzerland

K.M. Folta, S.E. Gardiner (eds.), *Genetics and Genomics of Rosaceae*, Plant Genetics and Genomics: Crops and Models 6, DOI 10.1007/978-0-387-77491-6_4,

Borkh. Some other species also have significance in commercial apple production and almost all scab resistant cultivars commercially available have *M.* × *floribunda* Siebold ex Van Houte in their ancestry. Genes for disease resistance have also been obtained from a wide spectrum of other *Malus* species such as *M. micromalus* Makino, *M.* × *atrosanguinea* (Spaeth) C. Schneider, *M. baccata jackii* Rehder and *M. sargentii* Rehder (Korban and Skirvin, 1984).

The history of apple is documented in many civilisations. Each country has fostered its own range of cultivars adapted to its growing conditions and to the requirements of their people. The discovery of 'McIntosh' (1796, USA), 'Jonathan' (1826, USA), 'Rome Beauty' (1848, USA), 'Cox Orange' (1850, UK), 'Granny Smith' (1868, Australia), 'Red Delicious' (1880, USA) and 'Golden Delicious' (1890, USA) were of historic importance. Moreover, 'Red Delicious' and 'Golden Delicious' are still among the most important apple cultivars worldwide.

## 2 Apple Production Worldwide

More than 50 million tons of apples are produced worldwide annually (www.faostat.fao.org). Apples are adaptable to various climates, but best adapted to the cool temperate zone from about 35–50° latitude. They have a more northern range than many other tree fruits, due to relatively late blooming and cold hardiness. World apple production has slightly increased in recent years and is supposed to further increase (Table 1). However, experts estimate a significant increase in production and consumption in the coming years, mainly in Asia. For North America and Europe the production is predicted to remain stable (World Apple Report, 2006). In China, the main varieties are 'Fuji' (60.4%, 'Red Delicious' (9.7%), 'Golden Delicious' (6.2%), 'Jonagold' 3.0%, 'Gala' 2.8% and others 1.9%. (Youngbin, 2006). The primary destination markets for China are Southeast Asia and Russia. As the emerging economies surrounding China and economic developments in China itself contribute to a considerable increase of apple consumption in Asia, it is not likely that large quantities of apples from China will be exported to Europe or North America in the near future.

According to the World Apple Report (Desmond O. Rourke) China is the major apple supplier to Asia and much of the growth in the apple trade in Asia is from imports from China. Fresh apples imported from China into Asian countries in 2005

**Table 1** Trends in world apple production (× 1.000 tons)

| Area | 2000 | 2005 | 2010* | 2015* |
|---|---|---|---|---|
| Europe | 14.104 | 12.394 | 13.739 | 14.245 |
| North America | 5.671 | 5.187 | 5.487 | 5.521 |
| Asia | 23.638 | 28.383 | 29.788 | 32.718 |
| Southern Hemisphere | 4.301 | 4.978 | 5.795 | 6.330 |
| World | 59.199 | 63.489 | 68.441 | 72.820 |

Source: World Apple Report, 2006.
* forecast.

showed that China supplied 68.2% of all apples in SE Asia and 54.1% in South Asia. Apple imports from China for selected countries included India (19.8%), Hong Kong (30.3%), Bangladesh (44.6%), Singapore (56.9%), Malaysia (59.6%), Thailand (63%), Indonesia (69.2%) and the Philippines (92.9%).

## 3 Breeding

The first controlled apple crosses are attributed to Thomas A. Knight in 1806. According to Alston and Spiegel-Roy (1985), for most tree fruit crops the selection of superior phenotypes and their subsequent random mating, followed by mass selection, is the most effective means of increasing the number of favorable alleles, because there is a relatively high additive variance governing the inheritance of most traits. Fruit breeders usually select superior genotypes in the F1 progenies. Modern breeding goals in apple are determined by market and consumer requirements. The markets are becoming more and more global, leading to the worldwide cultivation of apples such as 'Golden Delicious', 'Delicious', 'Gala', 'Fuji' and 'Braeburn' in addition to cultivars that are locally produced and desired by consumers. As part of sustainable cultivation systems, cultivars also need to satisfy economical and ecological demands. Towards addressing ecological concerns, there is an intensive effort to develop disease resistant, high quality cultivars. In many apple production areas of the world the major fungal diseases are apple scab (*Venturia inaequalis*) and powdery mildew (*Podosphaera leucotricha*). The bacterial disease fire blight (*Erwina amylovora*) is a considerable threat to apple production and is difficult to control. Disease resistant varieties allow a significant reduction of orchard pesticide inputs in temperate climates, where a greater number of treatments are typically needed in comparison to other crops. The production of apples in adapted and sustainable systems could improve the image of apples as a healthy food and contribute to increased food safety. However, none of the disease resistant cultivars has yet achieved a breakthrough in the marketplace comparable to varieties such as 'Golden Delicious' or 'Gala'. However the success of disease resistant varieties at the point of sale is not related to their disease resistance attributes, as the successful introduction of a new cultivar is primarily determined by wholesalers and retailers who perceive fruit quality as the principal consumer driver and design their marketing strategy accordingly. A recent consumer test performed by Agroscope Changins-Wädenswil, Switzerland, in a local supermarket revealed fruit quality criteria to be more important to consumers than the attribute 'from organic production' (Table 2). On the production side, disease resistant cultivars need to be durable with respect to resistance attributes.

### *3.1 Breeding Strategies for Durable Disease Resistance*

The challenge of breeding varieties durably resistant to disease can be approached in different ways, including marker assisted selection. Gessler et al. (2006) performed a comprehensive review on *Venturia inaequalis* resistance in apple, including the

**Table 2** Importance of different quality and production criteria to consumers in a survey in Switzerland, June 2004, 200 consumers, in percent

| | Firmness | Crispness | Flavor | Organic production |
|---|---|---|---|---|
| Not important | 10 | 6 | 5 | 57 |
| Important | 51 | 43 | 47 | 25 |
| Very important | 39 | 51 | 48 | 18 |

development of molecular markers for these resistances. The availability of molecular markers and genetic linkage maps enables the detection and the analysis of major resistance genes, as well as of quantitative trait loci (QTL) contributing to the resistance of a genotype (Liebhard et al., 2003b). A promising route to reduction of the risk of resistance breakdown is the combination of several functionally different resistances in a cultivar. Examples of molecular selection of genotypes with such pyramided genetic resistance against scab (*Venturia inaequalis*) and powdery mildew (*Podosphaera leucotricha*) are promising. Molecular markers are available nowadays that allow the detection of the major scab resistances *Vf*, *Vr* and *Vbj* (Tartarini et al., 1999; Hemmat et al., 2002; Gygax et al., 2004) and the $Pl_1$, $Pl_2$, *Pld* and *Plw* mildew resistances (Markussen et al., 1995; Seglias and Gessler, 1997; James and Evans, 2004). At Agroscope Changins-Wädenswil, progress has been achieved in establishing a system for marker assisted selection (Frey et al., 2004) whereby up to eight markers are analysed in a single multiplex reaction, substantially reducing costs. Systems have been established that allow a microsatellite based screening on an automated fragment analyzer.

In the framework of the European DARE project (Durable Apple Resistance in Europe), local European cultivars were examined as sources for durable scab resistance in apple (Laurens et al., 2004). It is known that the *Vf* scab resistance widely used in apple breeding programs can be overcome by specific races or strains of the fungus (Parisi et al. 2002). During this research very diverse and complex resistance behaviors were found: the cultivars which showed the widest range of resistances were mostly local cultivars as well as some newly selected hybrids that combine major genes for resistance with partial resistances.

To achieve a comparable level of resistance to that conferred by major genes, several quantitative resistance loci would have to be combined. Liebhard et al. (2003a) performed a QTL analysis based on a genetic linkage map that was constructed by using a segregating population of the cross between the apple cultivars ‘Fiesta’ and ‘Discovery’. The progeny was observed for three years at three different sites in Switzerland and field resistance against apple scab was assessed. The QTL analysis revealed 8 genomic regions whereby six conferred resistance against leaf scab and two resistance against fruit scab. However, the effectiveness of these QTLs has to be confirmed at higher disease levels, and in other genetic backgrounds.

There is also scope for breeding fire blight resistant apple cultivars by exploiting genetic variation in germplasm and by developing QTL markers. Fire blight, caused by the bacterium *Erwinia amylovora*, is the most serious bacterial disease of pipfruit. Forsline and Aldwinckle (2002) screened the USDA Apple Collection at

Geneva N.Y., including apple germplasm from Asia and Europe, for natural occurrence of fire blight and found no major resistance genes to this disease. However, QTLs for resistance to fire blight have been found in the cultivated apple *Malus* × *domestica* (cultivars 'Fiesta' and 'Nova Easygro'; Khan et al., 2006, Khan et al., 2007) as well as in wild *Malus* species (*Malus robusta* 5, Peil et al., 2007).

## 3.2 Pest Resistance

Evidence for a genetic basis of pest resistance in apple has been given for several herbivore species, for example for the woolly apple aphid (*Eriosoma lanigerum*) (Knight et al., 1962), and the brownheaded leafroller (*Ctenopseustis obliquana*) (Wearing et al., 2003). An indication of the genetic regions associated with pest resistance has been provided for *Dysaphis devecta* (Roche et al., 1997) and *Eriosma lanigerum* (Bus et al., 2007).

## 3.3 Cisgenics in Fruit Trees

Introgression of desired traits from wild germplasm and specific donors into top quality apple cultivars is a challenge that can be approached efficiently by the new approach of cisgenesis (Schouten et al., 2006), whereby only the allele of interest is inserted. Cisgenesis is defined as the genetic modification of a plant, inserting genes of the same plant species or from other crossable relatives. A cisgenic plant does not contain genes from outside the gene pool of the classical breeder.

## 3.4 Recurrent Breeding Strategies

The Horticulture and Food Research Institute of New Zealand Ltd (Plant and Food Research) has established an Apple Genetics Population to maintain biodiversity for cultivar development (Noiton and Alspach, 1996) and to provide genetic information on important apple characteristics (Oraguzie et al., 2000, 2001). The population is anticipated to provide novel fruit characteristics for application in a long-term breeding strategy based on recurrent selection. Families were derived from open-pollinated seed from a wide range of apple cultivars, as well as crab apples (*Malus spp.*), contributed from repositories from different countries (Noiton et al., 1999). Adapted recurrent selection strategies such as those applied in New Zealand can be considered amongst the most promising strategies to keep and increase genetic variability.

## 3.5 Genetics of Apple

The majority of apples are diploids (2n = 34). However, some triploids and tetraploids exist. Genomics is now being used to gain a better understanding of the genetic control of, as well as the interactions among traits. Whole genome sequences are being currently determined for *Malus* (see Chapter "Rosaceous Genome Sequencing: Perspectives and Progress").

Genetic linkage maps allow the identification of quantitative trait loci (QTL), which can identify chromosomal regions controlling phenotypic traits (Collard et al., 2005). Such a linkage map should be densely covered with molecular markers, in order to obtain the maximum probability of identifying a QTL (Liebhard et al., 2003a). The saturation of linkage maps with molecular markers (AFLP, RAPD, SSR, SCAR markers) has been considerably improved during recent years (Liebhard et al., 2003a; Maliepaard et al., 1998; Silfverberg-Dilworth et al., 2006; Celton et al., 2008; Fernández-Fernández et al., 2008).

## 3.6 Sources for Increased Genetic Diversity

Some years ago, efforts were made to enlarge the genetic basis in apple by collecting material in the centers of origin. An expedition to Kazakhstan and Kyrgyzstan by Forsline (1995) was successful and *Malus* collections included 65 accessions (18,000 seeds) representing 3 species endemic to that area. Collection of cuttings of elite materials was kept to a minimum due to restrictions in the USA quarantine facilities. Some unique germplasm from areas that had not been previously explored was collected. Seven ecosystems were explored in 12 expeditions. In the meantime, this genetic material is being screened for a wide range of fruit and tree characters. The Fruit Genebank at Dresden-Pillnitz (GER) collected about 7000 seeds from 55 accessions, as well as scions from 28 accession of *Malus hupehensis, Malus kansuensis, Malus prattii, Malus sieboldii, Malus transitoria* and *Malus toringoides* in 2001 during an expedition to 6 sites in the Chinese provinces Sichuan and Chongqing (Geibel and Hohlfeld, 2003).

A national inventory of top and small fruit genetic resources in Switzerland completed in spring 2005 (Kellerhals and Egger, 2004) is the basis for a complete and secure conservation of fruit genetic resources in Switzerland. In the course of the inventory, a fundamental project of the national plan of action, information was collected regarding the origin, abundance and frequency of accessions.

Inventorying, collecting, characterizing, evaluating and utilizing the fruit genetic resources e.g. in a breeding program are of great public relevance. It enables retention and utilization of a heritage for future generations, broadening the genetic basis in breeding programs and the meeting of consumers' evolving demands for healthy new and innovative products. The inventory of the national collection has demonstrated the rich genetic fruit diversity in Switzerland. We have used classical pomological knowledge to verify and determine varietal trueness-to-type. However, many samples remained undetermined, probably being unnamed chance seedlings or unknown varieties. It might be worth considering modern molecular techniques to determine the correct varieties and to highlight synonyms and homonyms (King et al., 1998).

## 4 Genetic Resources

The United Nations Conference on Environment and Development was held in Rio de Janeiro in 1992. A key agreement adopted was the Convention on Biological Diversity (CBD). The convention has fostered international activities for the conservation and sustainable use of plant genetic resources for food and agriculture. Three main goals are followed: the conservation of biological diversity, its sustainable use, and the sharing of benefits from their use. The conservation of plant genetic resources (PGR) has gained significantly in importance and is now accepted as an essential responsibility of national governments (Engels, 2002) This situation is demonstrated by the impressive number of nations which have ratified the CBD, endorsed the International Undertaking on Plant Genetic Resourcers, or both. Bioversity is the world's largest international research organization dedicated solely to the conservation and use of agricultural biodiversity and is non-profit and independently operated. In 2006, IPGRI (International Plant Genetic Resources Institute) and the International Network for the Improvement of Banana and Plantain (INIBAP) became a single organization. IPGRI and INIBAP changed their name to Bioversity International. The new name reflects an expanded vision of its role in the area of biodiversity research for development, seeking to advance the conservation and use of plant genetic diversity for the well-being of present and future generations. The European Cooperative Program for Plant Genetic Resources (ECPGR) is a collaborative program including most European countries and is aimed at facilitating the long-term conservation and increased utilization of plant genetic resources in Europe. It was founded in 1980 on the basis of the recommendations of the United Nations Development Program (UNDP), the Food and Agriculture Organization of the United Nations (FAO) and the Genebank Committee of the European Association for Research on Plant Breeding (EUCARPIA).The program is entirely financed by the member countries and coordinated by IPGRI. The ECPGR Documentation and Information Network, with the creation of crop specific and multi crop databases has contributed to making information available on ex situ conserved germplasm. As a result of an EU-funded project and ECPGR support, the EURISCO catalogue (http://eurisco.ecpgr.org/) provides on-line passport information on accessions conserved in European collections. The ECPGR network on fruit comprises working groups on *Malus/Pyrus*, *Prunus* and *Vitis*. Special central databases for each species are established and currently a European *Malus/Pyrus* collection is being established. As in many crops, the genetic diversity in fruit species has considerably decreased in the last decades and a few fruit species and cultivars grown worldwide have become predominant. ECPGR decided to establish an effective, efficient and rational European conservation system, with an initial focus on existing ex situ genebank collections in European countries. The goal is to create **A** **E**uropean **G**enebank **I**ntegrated **S**ystem (AEGIS) for plant genetic resources for food and agriculture, aimed at conserving the genetically unique and important accessions for Europe and making them available for breeding and research. Such material will be safely conserved under conditions that ensure genetic integrity and viability in the long term.

The United States Department of Agriculture (USDA) Agricultural Research Service (ARS) maintains a collection of apple germplasm in Geneva NY that includes over 8500 accessions representing at least 50 species. Of these, some 2600 accessions are clonally propagated cultivars, 3100 are seedlings mostly representing species collections, 1600 are in the form of seed, and 1250 are wild by elite hybrids that were generated specifically for genetic studies (Simon et al., 2008). The core mission of this collection includes the acquisition, maintenance, characterization and distribution of the diversity of *Malus*. The collection was primarily characterized with 154 descriptors including pomological, pathological, anatomical and physiological characteristics which are recorded in the Germplasm Resources Information Network (GRIN: www.ars-grin.gov). A set of microsatellite markers was recently included and the data are publically available.

## 4.1 Cryoconservation

The long-term storage of germplasm under cryogenic conditions is an efficient approach (Towill et al., 2004). The USDA – ARS, Plant Genetic Resources Unit at Geneva, NY (USA) has processed 1915 accessions of *Malus* representing 48 species using a winter vegetative bud method for cryopreservation. Overall the method has been successful with 91% of the lines tested having viability after cryo-exposure of 40% or more. Genotypic differences in survival were observed and the success varied from year to year. For *M.* × *domestica* 95% of the accessions tested have been cryopreserved. For species other than *M.* × *domestica*, 83% met the criterion. As an internal control, eight lines have been collected, cryopreserved and recovered through grafting each year. While genomics and bioinformatics are essential tools to understand and exploit variation of genes, genomes and genepools, cryopreservation with its minimal requirements of space and maintenance continues to gain importance as a means of long term physical storage of genetic materials (Forsline et al., 2003).

Breeding strategies most often consider the most advanced selections and cultivars and thus lead to a narrowing of the genetic base. The introgression of traditional varieties and accessions into the genepool is often feared due to undesirable characteristics that might be incorporated. However, there is scope for considering a wider genetic basis in apple breeding for sustainable fruit production. The directed use of genetic resources in genetic apple improvement programs could address many components of sustainable apple growing systems.

## 4.2 Nutritional Value and Health Issues

Regular consumption of fruits and vegetables in general are considered to decrease the risk of chronic diseases such as cancer and cardiovascular diseases. Apples are low in energy, but good sources of vitamins, minerals, pectin as well as secondary plant metabolites. Based on epidemiological studies, it seems that apples may play a role in maintaining a healthy lifestyle in general. Apple consumption may be associated with a reduced risk of cancer, heart disease and Type II diabetes as well as a increased weight loss when compared to other fruits and vegetables (Boyer and Rui 2004). Pectin, as a soluble fibre has a positive impact on satiety and may attenuate blood glucose and blood lipid levels when ingested in high amounts.

Apples contribute importantly to the flavanoid intake in humans. Thus they act as major antioxidants that may scavenge and neutralize free radicals, which in turn play a role in the onset of degenerative diseases (Biedrzycka and Amarowicz 2008). However, their contents vary considerably with the variety, storage conditions and other pre- and postharvest conditions. Stushnoff et al. (2003) examined juice and fruit tissue from 321 Malus species, selections and cultivars from the USDA Plant genetic Resources Unit at Geneva N.Y. A huge diversity in total phenolics and antioxidant capacity was detected with the aim of supplying breeders with data on antioxidant composition as a guide to parental selection. Information is being gathered on phenols and flavonols to benefit postharvest physiology studies. Data on the compositional diversity of apple germplasm is retained to aid medical epidemiological and dietary intervention studies, as well as identify bioextracts for supplemental application. Fruit were collected from the core collection at Geneva, NY, from wild apples collected in Central Asia, and from apple trees in the University of Minnesota germplasm collection.

Approximately 2% of the West-European population has an apple allergy. The use of molecular markers for allergy research is new. Apple allergy is dominated by IgE antibodies against Mal d 1 in areas where birch pollen is endemic. Apples with significantly decreased levels of Mal d 1 would allow most patients in these areas to eat apples without allergic reactions (Gilissen et al. 2005). Mal d 1 expression was successfully reduced by RNA interference. This translated into significantly reduced in vivo allergenicity. These observations support the feasibility of the production by gene silencing of apples hypoallergenic for Mal d 1 (Gilissen et al. 2005).

## 5 Looking Forward

Apple's unique cultivation history and genetic resources have led to the production of outstanding cultivars through many breeding programs. Today's present day challenges will require implementation of these same resources in new and clever ways. The increasing genomics tools in the species will accelerate genomics-enabled breeding strategies, ensuring a continued supply of apples that contribute to their sustained cultivation as well as human health and nutrition.

## References

Alston F.H., Spiegel-Roy P., 1985. Fruit Tree Breeding: Strategies, Achievements and Constraints. In: Attributes of Trees as Crop Plants, 49–67.

Biedrzycka, E. and Amarowicz, R. (2008). Dieta and health: apple polyphenols as antioxidants. Food Rev. Int. 24: 235–251.

Boyer, J. and Rui, H.L. (2004). Apple phytochemicals and their health benefits. Nutr. J. 3: 5.

Bus V., Chagne D., Bassett H., Bowatte D., Calenge F., Celton J., Durel C., Malone M., Patocchi A., Ranatunga A., Rikkerink E., Tustin D., Zhou J., Gardiner S. 2007. Genome mapping of three major resistance genes to woolly apple aphid (*Eriosoma lanigerum* Hausm.) Tree Genetics & Genomes 4: 233–236.

Celton J.-M., Tustin D.S., Chagné D., Gardiner S.E. 2008. Construction of a dense genetic linkage map for apple rootstocks using SSRs developed from *Malus* ESTs and *Pyrus* genomic sequences. Tree Genetics and Genomes Online First 10.1007/s11295-008-0171-z.

Coart E., Van Glabeke S., De Loose M., Larsen A.S., Roldán-Ruiz I. 2006. Chloroplast diversity in the genus *Malus*: new insights into the relationship between the European wild apple (*Malus sylvestris (L.) Mill.*) and the domesticated apple (*Malus domestica Borkh.*). Molecular Ecology 15: 2171–2182.

Collard B.C.Y., Jahufer M.Z.Z., Brouwer J.B., Pang E.C.K. 2005. An introduction to markers, quantitative trait loci (QTL) mapping and marker-assisted selection for crop improvement: the basic concepts. Euphytica 142:169–196.

Engels, J.M.M. (2002). Genebank management: an essential activity to link conservation and plant breeding. Plant Genet. Resour. Newsl. 129: 17–24.

Fernández-Fernández F., Evans K.M., Clarke J.B., Govan C.L., James C.M., Marić S., Tobutt K.R. 2008. Development of an STS map of an interspecific progeny of *Malus*. Tree Genetics & Genomes 4: 469–479.

Forsline P.L. 1995. Report of an Expedition to Kazakhstan to collect genetic resources of wild apple. August 23 – September 16, 1995. www.ars-grin.gov/gen/kaz95.html.

Forsline P.L., Aldwinckle H.S. 2002. Natural occurrence of fire blight in USDA apple germplasm collection after 10 years of observation. Acta Horticulture 590: 351–357.

Forsline P.L., Aldwinckle H.S., 2004. Evaluation of *Malus sieversii* populations for disease resistance and horticultural traits. Acta Horticulture 663: 529–534.

Forsline P.L., Fideghelli C., Knuepffer H., Meerow A.W., Nienhuis J., Stoner A.K., Thorn E., Tombolato A.F., Williams D. 2003. XXVI International Horticultural Congress: Plant Genetic Resources, The Fabric of Horticulture's Future. International Horticultural Congress. Acta Horticulture. 623–355.

Frey J.E., Frey B., Sauer C., Kellerhals M. 2004. Efficient low-cost DNA-extraction and multiplex fluorescent PCR method for high-throughput marker-assisted selection (MAS) in apple breeding. Plant Breeding 123:554–557.

Geibel M., Hohlfeld B. 2003. Malus Germplasm from Asia and its evaluation at the German Fruit Genebank. Acta Horticulture 623: 275–282.

Gessler C., Patocchi A., Sansavini S., Tartarini S., Gianfranceschi L. 2006. *Venturia inaequalis* Resistance in apple. Critical Reviews in Plant Science 25: 473–503.

Gilissen, L.J.W.J., Bolhaar, S.T.H.P., Matos, C.I., Rouwendal, G.J.A., Boone, M.J., Krens, F.A., Zuidmeer, L., van Leeuwen, A., Akkerdaas, J., Hoffmann-Sommergruber, K., Knulst, A.C., Bosch, D., van de Weg, W.E. and van Ree, R. (2005). Silencing the major apple allergen Mal d 1 by using the RNA interference approach. J. Allergy Clin. Immunol. 115: 364–369.

Gygax M., Gianfranceschi L., Liebhard R., Kellerhals M., Gessler C., Patocchi A. 2004. Molecular markers linked to the apple scab resistance gene Vbj derived from *Malus baccata jackii.* Theoretical and Applied Genetics 109: 1702–1709.

Harris S.A., Robinson J.P., Juniper, B.E. 2002. Genetic clues to the origin of the apple. Trends in Genetics 18:426–430.

Hemmat M., Brown S.K., Weeden N.F. 2002. Tagging and mapping scab resistance genes from R12740-7A apple. Journal of the American Society for Horticultural Science 127: 365–370.

Hokanson S.C., Szewc-McFadden A.K., Lamboy W.F., McFersen J.R. 1998. Microsatellite (SSR) markers reveal genetic identities, genetic diversity and relationships in a *Malus* × *domestica* Borkh. Core collection. Theoretical and Applied Genetics 97:671–683.

James C.M., Evans K. 2004. Identification of molecular markers linked to the mildew resistance genes Pl-d and Pl–w in apple. Acta Horticulture 663: 123–128.

Janick J., Cummins J.N., Brown, S.K., Hemmat M. 1996. Apples. In Fruit Breeding, Volume I: Tree and Tropical Fruits, ed. J. Janick and J.N. Moore, J. Wiley & Sons, Inc.

Kellerhals M., Egger S. 2004. Inventory on fruit genetic resources in Switzerland. Acta Horticulture, 663, 585–588.

Khan M.A., Duffy B., Gessler C., Patocchi A. 2006. QTL mapping of fire blight resistance in apple. Molecular Breeding 17:299–306.

Khan M.A., Duffy B., Durel C.E., Denancé C., Kellerhals M., Patocchi A., Gessler C. 2007. Development of markers linked to the 'Fiesta' 7 major QTL for fire blight resistance and their application for marker-assisted selection. Genome 50: 568–577.

King G., Brown L., Ryder C., Periam N. 1998. Microsatellite markers for accession identification, pedigree analysis and assessment of allelic diversity in Malus genetic resources. ECP/GR Report of a Working Group on Malus/Pyrus, IPGRI Rome, 82–89.

Knight R.L., Briggs J.B., Massee A., Tydeman H.M. 1962. The inheritance of resistance to woolly aphid, *Eriosoma lanigerum* in the apple. Journal of Horticulture Science 3:207–218.

Korban S.S., Skrivin R.M. 1984. Nomenclature of the cultivated apple. Hort Science 19:177–180.

Laurens F., Chevalier M., Dolega E., Genari F., Goerre M., Fischer C., Kellerhals M., Lateur M., Lefrancq B., Parisi L., Schouten H., Tartarini S. 2004. Local European cultivars as sources of durable scab resistance in apple. Acta Horticulture 663, 115–122.

Liebhard R., Koller B., Gianfranceschi L., Gessler C. 2003a. Creating a saturated reference map for the apple (*Malus* × *domestica* Borkh.) genome. Theoretical and Applied Genetics 106: 1497–1508.

Liebhard R., Koller B., Patocchi A., Kellerhals M., Pfammatter W., Jermini M., Gessler C. 2003b. Mapping quantitative field resistance against apple scab in a fiesta × discovery' progeny. Phytopathology 93:493–501.

Luby J., Forsline P., Aldwinckle H., Bus V., Geibel M. 2001. Silk road apples-Collection, evaluation, and utilization of *Malus sieversii* from Central Asia. HortScience 36:225–231.

Maliepaard C., Alston F.H., Van Arkel G., Brown L.M., Chevreau E., Dunemann F., Evans K.M., Gardiner S., Guilford P., Van Heusden A.W., Janse J., Laurens F., Lynn J.R., Manganaris A.G., Den Nijs A.P.M., Periam N., Rikkerink E., Roche P., Ryder C., Sansavini S., Schmidt H., Tartarini S., Verhaegh J.J., Vrielink-Van Ginkel M., King G.J. 1998. Aligning male and female linkage maps of apple (*Malus pumila* Mill.) using multi-allelic markers. Theoretical and Applied Genetics 97:60–73.

Markussen T., Krüger J., Schmidt H., Dunemann F. 1995. Identification of PCR-based markers linked to the powdery mildew resistance gene Pl1 from *Malus robusta* in cultivated apple. Plant Breeding 114: 530–534.

Morgan J., Richards A. 1993. The book of apples. Ebury Press, London.

Noiton D.A.M., Alspach P.A. 1996. Founding clones, inbreeding, coancestry, and status number of modern apple cultivars. Journal of the American Society for Horticultural. Science 121: 773–782.

Noiton D., Hofstee M.E., Alspach P.A., Brewer L.R., Howard C. 1999. Increasing genetic diversity for apple breeding: a preliminary report. Acta Horticulture 484, 105–107.

Oraguzie N.C., Hofstee M.E., Alspach P.A., Brewer L.R., Howard C. 2000. Quantitative genetic studies in a recurrent selection program in apple. Acta Horticulture 538:191–195.

Oraguzie N.C., Hofstee M.E., Brewer L.R., Howard C. 2001. Estimation of genetic parameters in a recurrent selection program in apple. Euphytica 118: 29–37.

Parisi, L., Fouillet, V., Schouten, H. J., Groenwold, R., Laurens, F., Didelot, F., Evans, K., Fischer, C., Gennari, F., Kemp, H., Lateur, M., Patocchi, A., Thissen, J. and Tsipouridis, C. (2004). Variability of the pathogenicity of *Venturia inaequalis* in Europe. ISHS. Acta Hort. 663: 107–114.

Peil A., Garcia-Libreros T., Richter K., Trognitz F.C., Trognitz B., Hanke M.V., Flachowsky H. 2007. Strong evidence for a fire blight resistance gene of *Malus robusta* located on linkage group 3. Plant Breeding 126: 470–475.

Roche P., Alston F.H., Maliepaard C., Evans K.M., Vrielink R., Dunemann F., Markussen T., Tartarini S., Brown L.M., Ryder C., King G.J. 1997. RFLP and RAPD markers linked to the rosy leaf curling aphid resistance gene (*$Sd_1$*) in apple. Theoretical and Applied Genetics 94:528–533.

Schouten H.J., Krens F.A., Jacobsen E. 2006. Do cisgenic plants warrant less stringent oversight? Nature Biotechnology 24(7):9.

Seglias N.P., Gessler C. 1997. Genetics of apple powdery mildew resistance derived from *Malus zumi* (Pl2). IOBC/WPRS Bull 20(9): 195–208.

Silfverberg-Dilworth, E., Matasci, C., Van de Weg, W., Van Kaauwen, M., Walser, M., Kodde, L., Soglio, V., Gianfranceschi, L., Durel, C., Costa, F., Yamamoto, T., Koller, B., Gessler, C. and Patocchi, A. (2006). Microsatellite markers spanning the apple (*Malus* × *domestica* Borkh.) genome. Tree Genet. Genomes 2:202–224.

Simon, C.J., Volk G.M. and Seewe-McFadden A.K. (2008). Genomic characterization of the us apple germplasm collection. Mitt. Julius Kühn-Institut 416: 96.

Stushnoff C., McSay A.E., Luby J.J., Forsline P.L. 2003. Diversity of phenolic antioxidant scavenging capacity in the apple germplasm collection. Acta Horticulture 623: 305–312.

Tartarini S., Gianfranceschi L., Sansavini S., Gessler C. 1999. Development of reliable PCR markers for the selection of the Vf gene conferring scab resistance in apple. Plant Breeding. 118: 183–186.

Towill L.E., Forsline P.L., Walters C., Waddell J.W., Laufmann J. 2004. Cryopreservation of *Malus* germplasm using a winter vegetative bud method: results from 1915 accessions. CryoLetters 25:323–334.

Way R.D., Aldwinckle H.S., Lamb R.C., Rejman A., Sansavini S., Shen T., Watkins R., Westwood M.N., Yoshida Y. 1990. Apples (*Malus*). In, Genetic Resources of Temperate Fruit and Nut Crops, eds. Moore J.N. and J.R. Ballington, ISHS, p 3–62.

Wearing C.H., Colhoun K., McLaren G.F., Attfield B., Bus V.G.M. 2003. Evidence for single gene resistance in apple to brownheaded leafroller, *Ctenopseustis obliquata*, and implications for resistance to other New Zealand leafrollers. Entomologia Experimentalis et Applicata 108: 1–10.

Youngbin, Y. (2006). China's apple production and exports: Current situation and trends. Congress Papers, Interpoma 2006. K. Werth ed., 74–79.

# 5. Apple Structural Genomics

**Schuyler S. Korban and Stefano Tartarini**

## 1 Introduction

A primary focus of apple genetics is the elucidation of genes influencing diverse phenotypes of economically important horticultural traits. Most of these phenotypes are genetically complex; i.e., controlled by multiple genes occupying chromosomal positions referred to as quantitative trait loci (QTL). Mapping of QTLs has become a common first step toward understanding the molecular basis of complex genetic traits, and it has provided the impetus for developing detailed genome maps. These genome maps are built with the aid of various biochemical and molecular markers such as isozymes, restriction fragment length polymorphisms (RFLPs), random amplified polymorphic DNAs (RAPDs), amplified fragment length polymorphisms (AFLPs), and simple sequence repeats (SSRs), among others. Dominant markers, such as RAPDs, can be used for map alignment if these markers are heterozygous in both parents, but their transferability to other maps is limited. While, co-dominant markers such as SSRs are also useful in map alignment, but they are also transferable between mapping populations. More recently, single nucleotide polymorphisms (SNPs) have taken hold as SNPs can occur in both coding (gene) and noncoding regions of the genome. Those SNPs found within a coding sequence are of particular interest as they are more likely to alter the biological function of a protein. SNPs are major contributors to genetic variation, comprising approximately 80% of all know polymorphisms, and their density in plants is variable depending on the species, while in the human genome it is estimated to be on average of 1 per 1000 base pairs. Although SNPs are mostly biallelic (less informative than short tandem repeats), they are more frequent and mutationally stable, making them suitable for association studies in which linkage disequilibrium (LD) between markers and an unknown variant is used to map mutations in complex traits. SNP maps will help

S.S. Korban (✉)
Department of Natural Resources and Environmental Sciences, University of Illinois, Urbana, Illinois USA 61801

K.M. Folta, S.E. Gardiner (eds.), *Genetics and Genomics of Rosaceae*, Plant Genetics and Genomics: Crops and Models 6, DOI 10.1007/978-0-387-77491-6_5,

in identifying multiple genes associated with such complex traits influencing tree architecture, fruit quality, and disease resistance. These associations are difficult to establish with conventional gene-hunting methods because a single altered gene may only be a small contributor to such a trait.

Expressed sequence tag (EST) sequence datasets are most suited to redundancy based SNP discovery. The highly redundant nature of EST data sets permits the selection of polymorphisms that occur multiple times within a set of aligned sequences.

Nevertheless, various structural genomics tools have become available in recent years that allow for sequence/structure analysis and data management. Genetic linkage maps have contributed to the dissection of complex inherited traits, QTLs, and for positional cloning of traits of economic importance. These linkage maps have made inroads into the use of molecular markers in breeding. With the development of new marker systems, this has allowed for the construction of genetic maps for tree species, including apple. Comparative mapping based on the co-alignment of common molecular markers among genetic linkage maps allows us to correlate linkage information from different genetic maps and to validate the accuracy of locus ordering from the different mapping strategies. With coalignment of genetic maps form different experiments, it is possible to compare QTLs from different genetic and environmental backgrounds. Comparative mapping also allows for comparisons of genomic structures within the genus, and thus, helps in studying chromosomal evolution by detecting chromosome rearrangements.

With successful efforts in whole genome sequencing of such higher plants as *Arabidopsis*, rice, and *Populus*, there has been an explosion in sequencing efforts of other groups of higher plants including a recent effort in sequencing the apple genome by Dr. Riccardo Velasco and his team in Italy.

Structural genomics has been extremely successful at increasing the scope of our structural knowledge of protein families. Although this requires a large number of processing steps to convert sequence information into a 3D protein structure, a high percentage of proteins coded in genomes sequenced so far have unknown function and either minimal or undetectable sequence homology to proteins of known structures (Service 2002). Thus, the majority of new protein structure determinations would remain very labor-intensive using conventional methods. However, high-throughput technological advances are now changing this facet of structural biology. Chandonia and Brenner (2006) reported that the structure of a previously solved protein in a different conformation or with a different binding partner could provide insight into its functional mechanisms. It was predicted that by using standard sequence comparison techniques such as BLAST and PSI-BLAST, to avoid targeting homologs of known structures, information on gene function can be elucidated.

Instead of picking targets known to be biologically interesting, structural genomics researchers can now scan genome databases for stretches of DNA encoding genes of completely unknown function, hunt down their proteins, study the results, and perhaps in the process discover entirely new realms of biology.

# 2 Use of DNA Markers for Evaluating Genetic Diversity in Apple

## 2.1 Genotyping of Apple Cultivars

The first efforts to genotype apple cultivars relied on the use of isozyme systems. Weeden and Lamb (1985) indicated that 6-phosphogluconate dehydrogenase (6-PGD; EC 1.1.1.44) and aspartate aminotransferase (AAT; EC 2.6.1.1), also known as glutamate oxaloacetate transaminase (GOT) were useful in distinguishing among various apple cultivars. Soon after, Bournival and Korban (1987) reported that three isozyme systems, including phosphoglucomutase (PGM; EC 5.4.2.2), GOT, and peroxidase (PER; EC 1.11.1.9), were useful in differentiating among a set of nine apple cultivars. Early on, isozymes, such as (6-PGD), were also used to identify the contributing 2n gamete parent of triploid apple cultivars (Chyi and Weeden 1984). A detailed analysis of the inheritance patterns of seven isozyme systems, including an endopeptidase (ENP; EC 3.4.9.9), acid phosphatase (AP; EC 3.1.3.2), isocitrate dehydrogenase (Idh; EC 1.1.1.42), alcohol dehydrogenase (Adh; EC 1.1.1.1), phosphoglucoisomerase (PGI; EC 5.3.1.9), PGM, and esterase (EST; EC 3.1.1.11), in pollen from apple genotypes and segregating progenies were used to investigate and confirm the allopolyploid origin of the apple genome (Chevreau et al. 1985). Some of these isozyme systems, including PER, EST, AP, and indoleacetic acid oxidase (IAA-O; EC 1.2.3.7) were also found to be useful in distinguishing among eleven apple rootstocks (Menendez et al. 1986). Although isozymes were deemed useful in cultivar and rootstock identification as well as in establishing genetic relationships, they provided a limited set of robust and informative markers as they were influenced by environmental conditions and variations in plant development.

Later, RFLPs were the first DNA-based marker system used to characterize and identify apple cultivars, rootstocks, and seedlings (Nybom and Schall 1990; Watillon et al. 1991). Ishikawa et al. (1992) used RFLPs to detect chloroplast and mitochondrial DNA variations among 18 apple cultivars and three rootstocks. This was followed by a PCR-based marker system whereby RAPDs allowed for differentiation of apple cultivars using commercially available arbitrary primers (Koller et al. 1993; Mulcahy et al. 1993). A group of 14 RAPD markers were found useful in distinguishing among 11 apple cultivars (based on presence or absence of bands) (Koller et al. 1993); while, in another study by Mulcahy et al. (1993), two sets of decamer random primers were capable of distinguishing among a group of eight cultivars. Interestingly, the latter set of RAPD markers could not distinguish among individual sports of such cultivars as 'Red Delicious', 'Gala', and 'Golden Delicious', among others (Mulcahy et al. 1993). Previously, Nybom and Schall (1990) reported that using a cultivar-specific RFLP probe, M13 minisatellite DNA probe, was useful in fingerprinting different apple cultivars, but it could not detect differences in fingerprints among 15 different sports of 'Red Delicious' (Nybom 1990a). Later, another RFLP probe, a rDNA spacer region probe from the crabapple 'White Angel' and designated pAR72, was found useful in confirming paternity in offspring using the non-radioactive chemiluminescence to detect hybridized DNA gel fragments (Nybom et al. 1992).

Gianfranceschi et al. (1998) identified 16 reliable SSR markers, containing (AG)/(CT) repeats, that can be used to distinguish among apple cultivars as they amplified all alleles, with an average of 8.2 alleles per locus and an average heterozygosity index of 0.78. These SSRs were deemed highly polymorphic, and two selected SSRs could distinguish all 19 cultivars as well as *M. floribunda*, except for two mutant sports of 'Starking' and 'Red Delicious'. Moreover, for two triploid cultivars included in this study, 'Boskoop' and 'Jonagold', it was possible to detect three alleles in nine and ten cases, respectively. Using eight SSRs, Hokanson et al. (1998) were able to clearly distinguish among 66 apple cultivars, but for seven pairs of genotypes. High levels of variation were detected with a mean of 12.1 alleles per locus and a mean heterozygosity of 0.693 across all eight loci.

Since then, over 250 SSRs have been identified (Liebhard et al. 2002; Silfverberg-Dilworth et al. 2006) along with over 140 expressed sequence tag (EST)-SSRs (Naik et al. 2006; Silfverberg-Dilworth et al. 2006). It was reported that markers derived from sequences containing dinucleotide repeats were generally more polymorphic than sequences containing trinucleotide repeats (Silfverberg-Dilworth et al. 2006).

From the above findings, it is clear that differentiation among sports or strains of cultivars such as those of 'Golden Delicious', 'Red Delicious', 'Gala', and 'Fuji', is very difficult with standard molecular techniques (RFLP, RAPD, SSR, and AFLP). However, a recent technique, S-SAP, seems to be promising for intra-cultivar differentiation (Venturi et al. 2006).

## 2.2 Evaluating Genetic Diversity in *Malus* Species

A collection of 23 ornamental apple trees of unknown origin, likely derived from various species hybridizations and possibly with *M. floribunda* and *M. sikkimensis*, among others, were subjected to fingerprinting using the minisatellite M13 repeat RFLP probe (Nybom 1990b). When DNA from these individuals and their seedling offspring were restricted with different restriction enzymes and run on agarose gels, their banding patterns suggested presence of 8–10 morphotypes in this collection, while each seedling exhibited a unique banding pattern. Genetic diversity in this material appeared to be associated with known levels of genetic relatedness, and paternity could be estimated in some of the derived open-pollinated seedlings (Nybom 1990b).

Harada et al. (1993) used an 18-mer arbitrary primer P-T, 5'-CACTTAGAA CAGCGGTAC-3', to distinguish and confirm the identity of several *Malus* species, including *M. hupehensis*, *M. sargentii*, *M. baccata*, *M. asiatica*, and *M. sheideckeri* as well as various crabapples, among others. It is interesting to note that based on the banding patterns of these various species, they were able to correct the identity of one of the plant introductions received from the U.S.

Using decamer random oligonucleotide primers, from Operon Technologies, Inc. (Alameda, Calif.), combined with bulk segregant analysis, a single RAPD marker, OPA15/900, was found useful in identifying chromosomal regions introgressed from a wild *Malus* species, *M. floribunda* clone 821, into the cultivated apple (Durham and Korban 1994). Other RAPD markers were identified that were also useful in detecting chromosomal regions introgressed from *M. floribunda* 821 and were linked to the *Vf* locus for resistance to the fungal disease apple scab that was derived from this species (Gardiner et al. 1996a, b) and these will be discussed further in a later section.

## 2.3 Proposal of a Common Set of SSR Markers for Genetic Diversity Studies

Large collections of apple germplasm are available at various locations around the world (USA, Canada, UK, France, Germany, Italy, Japan, and in many other countries). However, there is limited molecular knowledge available on the genetic diversity of these germplasm resources. It is anticipated that the richness of this diversity, in terms of useful genes, is invaluable and yet to be understood.

As mentioned above, most fingerprinting studies have been conducted using different molecular marker systems, primarily to demonstrate their feasibility in distinguishing among different apple cultivars. This has proven rather straightforward, and a few markers are commonly needed to distinguish among various apple genotypes.

Although genetic diversity studies in apple can be efficiently performed by using different types of markers, SSRs are probably still preferable even if AFLPs are known to be more efficient for these sorts of studies. The choice of using SSRs is mainly attributed to the observed high reproducibility among different laboratories even when different detection techniques (radioactivity, silver staining, or fluorescence) are employed. Moreover, SSRs are preferably selected for their wide transferability not only for within species, but also among related species (such as apple and pear) as well.

Screening of these large germplasm collections will have a great impact on coordination and optimization efforts of germplasm conservation and in overcoming problems of mislabelling of plant material among the different sites, and thereby creating a worldwide fingerprinting database. To this end, it is critical to ask how many SSRs are necessary for properly fingerprinting large apple collections? However, the answer to this question is not very obvious for the following reasons. To adequately characterize a large apple germplasm collection, both the genomic distribution and the degree of polymorphisms have to be considered along with the fingerprinting costs incurred. Regarding the issue of genome distribution, it is best to select SSRs that are well-distributed within the genome along with the provision of including 1–2 SSRs per linkage group, possibly those mapping along the two extreme ends of the linkage group. This, of course, will result in a large number of SSRs, 17–34, thus adding to the cost of the fingerprinting analysis. Those highly polymorphic SSRs, known to be multi-locus SSRs, can be very useful in reducing the amount of work needed and thereby lowering cost, as different alleles would usually map on homeologous chromosomes (e.g., chromosomes 4–12 and

**Table 1** A recommended list of 17 highly polymorphic SSRs, spanning the apple genome, for apple germplasm characterization

| SSR | Linkage group | SSR | Linkage group |
|---|---|---|---|
| Hi02c07 | 1 | CH02c11 | 10 |
| CH02c06 | 2 | CH02d08 | 11 |
| G12 | 3 | CH01f02 | 12 |
| NZ05g08 | 4 | GD147 | 13 |
| CH5f06 | 5 | CH04c07 | 14 |
| CH03d07 | 6 | CH02c09 | 15 |
| CH04e05 | 7 | CH04f10 | 16 |
| CH01h10 | 8 | CH01h01 | 17 |
| CHCH01f03b | 9 | | |

5–10) (Silfverberg-Dilworth et al. 2006). A listing of 17 SSRs spanning the apple genome have been identified and tested on a set of standard *Malus* accessions by the ECPGR group (Evans et al. 2007). This list includes a single SSR for each linkage group (Table 1).

Again, SSRs must be selected on the basis of the degree of polymorphisms and the number of different alleles that can be efficiently recognized. To this end, there are more than 70 SSRs that are capable of revealing at least 8 different alleles on a set of eight to nine cultivars, as estimated based on the findings of Liebhard et al. (2002) and Silfverberg-Dilworth et al. (2006). Of course, the high number of alleles is not always the only driving factor for consideration as only clean and reproducible amplification would make it possible for easy identification of the different alleles present in a wide germplasm collection.

To date, there are only a few examples of wide germplasm fingerprinting efforts underway. A collection of 142 accessions from 23 *Malus* species from the USDA-ARS Plant Genetic Resources Unit's core collection was screened with eight SSR markers. A high level of variation was detected with a mean of 26.4 alleles per locus (Hokanson et al. 2001). Recently, in the framework of an EU project (HiDRAS 2003–2007), about 350 different apple cultivars have been fingerprinted with about 80 SSRs well-distributed within the genome; however, the results are not yet published (Silfverberg-Dilworth et al. 2006). This set of SSRs could serve as a basis for establishing a common protocol for germplasm analysis, particularly as coordination of efforts and exchange of information among all researchers in the field become more commonplace.

## 2.4 Usefulness of Apple SSRs in Other Species

Dirlewanger et al. (2002) reported that SSRs have a good rate of transferability among closely related Rosaceae species, but this was mainly observed in the genus *Prunus*. Yamamoto et al. (2001) utilized nine apple SSRs, including 02b1, 05g8, 28f4 (Guilford et al. 1997), CH01B12, CH01E12, CH01F02, CH01H01, CH01H10,

CH02B12 (Gianfranceschi et al. 1998), to characterize 36 pear accessions, including 19 Japanese pears (*Pyrus pyrifolia*), seven Chinese pears (*P. bretschneideri*, *P. ussuriensis*), five European pears (*P. communis*), three wild relatives (*P. calleryana*), and two hybrids between *P. pyrifolia* and *P. communis*. All tested apple SSR primers produced discrete amplified fragments in all pear accessions. Nucleotide repeats were detected in amplified bands by both Southern blotting and sequencing analysis, and nucleotide sequences of pear were compared with those of apple. Differences in fragment sizes among pear or between pear and apple were primarily attributed to differences in repeat numbers. Moreover, the DNA sequence of flanking regions in apple was highly conserved in pear. Hybrids of *P. pyrifolia* × *P. communis* showed a single fragment was inherited from each parent in all scorable cases, which suggested that each primer pair amplified fragments originating from the same locus. More than 70% of apple SSRs (among 112) were found to be polymorphic in two different European pear progenies confirming the very high SSR transferability between apple and European pear (Pierantoni et al. 2004). Moreover, apple SSRs were also mapped on genetic linkage maps of pear, and it was reported that both their positions and linkages were conserved between apple and pear (Yamamoto et al. 2002; Pierantoni et al. 2004).

Liebhard et al. (2002) indicated that of 15 apple SSRs tested, they all could amplify fragments in various other genera in the subfamily Maloideae, including *Amelanchier*, *Cotoneaster*, *Cydonia*, *Pyrus*, among others, but only a single SSR, amplified bands in the subfamily Amygolaideae. Yamamoto et al. (2004) also used 78 apple SSRs to characterize genetic diversity in quince (*Cydonia oblonga*) and pear cultivars, and reported that 57 and 65 apple SSRs could amplify bands in quince and pear, respectively. Silfverberg-Dilworth et al. (2006) were able to determine the location of previously published SSRs of apple (GD 147), pear (HGA8b, KA4b, NB102a, NH009b, NH029a, and NH033b), and *Sorbus torminalis* (MSS6), whose position on the apple genome was unknown. They also indicated that the frequency of transferability of SSRs across *Maloideae* species was ~41%.

Due to the large public apple EST database, efforts are now underway to utilize EST-SSRs in various other Rosaceae species. This will help in identifying homologous loci in different species by comparing positions of homologous markers in linkage maps of different species, and should prove a highly valuable resource.

## 3 Linkage and Physical Maps

Apple is a highly heterozygous species that is characterized by a long juvenile period. Therefore, it is well know that breeding efforts can greatly benefit from the development of linkage maps based on molecular markers that will help in speeding up the selection process.

Early genetic maps were mainly based on either RAPDs or RFLPs (Hemmat et al. 1994; Conner et al. 1998; Maliepaard et al. 1998). The usefulness of these maps was rather limited due to difficulties in cross-transferability and/or technical complexities of these types of markers. In fact, one of the main problems in molecular breeding efforts is that the number of segregating traits that can be efficiently analyzed in a progeny is limited, thus making it necessary to develop new maps in different progenies segregating for various traits of interest.

SSRs are the markers of choice to build maps because of their reliability and transferability. Therefore, an SSR-based map can serve as an ideal reference map as it is possible to find information that can be easily transferred to different progenies, thus speeding-up the development of new maps. The value of a reference map is determined by the number of mapped SSRs, but more importantly is the distribution of these markers along the chromosomes to guarantee good genome coverage. In fact, by choosing only a few, e.g., 4–5, but well-distributed SSRs on a reference map, it is possible to build the backbone map of a new progeny. The gaps in the SSR backbone can be then readily filled by using other forms of markers, such as AFLPs, thus promptly allowing one to determine map position of new traits in a limited amount of time with minimal incurred costs.

As reported above, many SSRs have been developed in the last decade (Guilford et al. 1997; Gianfranceschi et al. 1998; Hokanson et al. 1998; Liebhard et al. 2002; Hemmat et al. 2003; Vinatzer et al. 2004; Silfverberg-Dilworth et al. 2006). More than 400 SSRs are now available for mapping, and for most their position is known within the apple genome.

The most extensive SSR work have been undertaken by Liebhard et al. (2002) and Silfverberg-Dilworth et al. (2006), publishing a list of about 300 SSRs that have been mostly used to build a reference map for the 'Fiesta' × 'Discovery' progeny. All this information has been made publicly available on a web site (http://www.hidras.unimi.it/) that has been created in the framework of the HiDRAS EU project (2002–2007). The web site includes all available SSR information (sequences, primers, annealing temperatures, fragment sizes in a small set of cultivars, map position, and amplification profiles) along with published maps. Whole genome maps have been reported by Liebhard et al. (2002), Kenis and Keulemans (2005), and Silfverberg-Dilworth et al. (2006). Partial genomic maps have also been produced for a limited number of linkage groups in order to map specific traits (Gygax et al. 2004; Gao et al. 2005a, b, c; Costa et al. 2005; Tian et al. 2005; Boudichevskaia et al. 2006; Chagné et al. 2007).

Recently, two maps for apple rootstocks 'M.9' and 'Robusta 5' have been developed using 224 SSRs, 18 SCARs, 14 SNPs, and 42 RAPDs (S. Gardiner, Pers. Comm.). In addition a new set of 47 polymorphic SSRs was developed from apple EST sequences containing di-, tri- and tetra-nucleotide repeats, and used for construction of this rootstock maps. All 17 linkage groups have been identified and aligned to existing apple genetic maps using markers in common, and each linkage group carries a minimum of three markers. These maps span 1175.7 cM for 'M.9' and 1086.7 cM for 'Robusta 5', which is comparable with the latest apple cultivar genetic maps of Silfverberg-Dilworth et al. (2006) as mentioned above (S. Gardiner, Pers. Comm.).

Molecular maps have also been used to map sequences putatively related to resistance gene fragments known as resistance gene analogs or RGA (Baldi et al. 2004; Van der Linden et al. 2004; Calenge et al. 2005a; Naik et al. 2006).

Although information on molecular maps has increased exponentially in the last decade thus making it easier to construct new maps, some problems remain and must be resolved. While in some linkage groups, SSRs are very abundant and are evenly distributed along a chromosome (at less than 20 cM), in others large gaps of 20–40 cM remain. A set of 86 SSRs covering ~85% of the apple genome has been presented by Silfverberg-Dilworth et al. (2006), but a total of 16 genomic regions still lack good polymorphic SSRs markers.

Identifying new SSRs in specific genomic regions in order to fill in the gaps is not easily achieved by employing standard random approaches that have been thus far used to develop current SSRs (i.e., SSRs from enriched libraries) now that the number of gaps has been reduced. Therefore, other and better suited strategies must be identified. To this extent, screening of a BAC library screening with specific markers (AFLPs, RFLPs, or RAPDs), known to map on SSR-deficient regions, will increase the odds of finding new useful SSRs on the identified BAC clones. A PCR-based protocol for extracting SSRs from BACs has been already developed and tested in apple (Vinatzer et al. 2004). More hints for the development of new SSRs in specific genomic regions will probably become available in a couple of years once the whole sequencing of the apple genome is completed.

### *3.1 BAC Libraries*

There are a handful of apple BAC libraries that have been constructed in the past few years. Early on, Vinatzer et al. (1998) constructed a BAC library for cultivar Florina in the BAC cloning vector pECBAC1 following partial digestion of high molecular weight (HMW) apple DNA with *Eco*RI and *Eco*RI methylase. This library consisted of 36,864 recombinant clones with an average insert size of ~120 kb and representing ~5× apple haploid-genome equivalents. Later, Xu et al. (2001) constructed a BAC library from the wild species *M. floribunda* 821 in the BAC cloning vector pBeloBAC11 following partial digestion of HMW apple DNA fragments with *Bam*HI. The library consisted of 31,584 BAC clones with an average insert size of 125 kb, representing approximately ~5× *M. floribunda* haploid-genome equivalents. Yet another BAC library was later constructed from the apple cultivar GoldRush, consisting of 35,328 clones with an average insert size of ~110 kb, using a similar strategy (Xu et al. 2002). All the above BAC libraries have been used to isolate and clone genes of interest, such as the *Vf* gene, controlling resistance to the fungal disease apple scab (Vinatzer et al. 2001; Xu and Korban 2002), *Sbe* genes coding for starch branching enzymes (Han et al. 2007a, b), COMT genes coding for caffeic acid *O*-methyltransferases (Han et al. 2007c), among others.

In an effort to develop a genome-wide physical map for the apple using BAC fingerprinting, a second BAC library for 'GoldRush' was also constructed; however, this time HMW apple DNA was restricted with *Hind*III instead of *Bam*HI (Han et al. 2007d). The resulting library consisted of 46,791 clones, and was equivalent to ~7× haploid genome equivalents with an average insert size of 125 kb.

Another BAC library was constructed from the apple rootstock Geneva 41 or G.41, a cross of 'Malling 27' × 'Robusta 5'. This library has a total of 41,472 clones with an average insert size of ~120 kb cloned into the pECBAC1 vector. The partial digest enzyme was *Mbo*1 (G. Fazio, Pers. Comm.). Yet another BAC library from 'Pinkie' has also been constructed by Plant and Food Research in New Zealand. This BAC library ~7x genome coverage and consisting of 56,400 clones, containing an average insert size of 110 kb. This library was constructed in pCLD04541 vector with HMW partially digested with *Hin*dIII (E. Rikkerink, Pers. Comm.).

A cosmid library from 'Pinkie' was also constructed by Plant and Food Research in SuperCos1 whereby apple DNA was partially digested with *Sau*3AI. This cosmid library has a coverage of ~7x of the apple genome with nearly 170,000 clones containing an average insert size of 35 kb (E. Rikkerink, Pers. Comm.).

## 4 Genomic Location of Some Monogenic and Polygenic Traits

### *4.1 Scab Resistance*

Major efforts have focused on identifying molecular markers linked to genes controlling apple scab, incited by the fungal pathogen *Venturia inaequalis*. Several major and polygenic genes controlling apple scab have been identified. Among the major genes for apple scab, the *Vf* gene from the small-fruited crab apple species *M. floribunda* 821 received the most attention as it was heavily used in apple breeding programs all over the world. Initially, RAPD markers were identified that are closely linked to the *Vf* gene. These included OPD20/ 600, OPM18/900, OPU01/400, OPAL07/580, OPC08/1100, OPC09/900, OPAM19/2200, OPK16/1300, OPAR4/1400, S5/2500, S29/1150, B505/1700, B398/480, P198/750 (Yang and Krüger 1994; Koller et al. 1994; Tartarini 1996; Gianfranceschi et al. 1996; Gardiner et al. 1996a; Yang et al. 1997a, b; Hemmat et al. 1998). Later, 15 AFLP markers were found linked to this *Vf* locus (Xu and Korban 2000). Some of these RAPD and AFLP markers were then converted into sequence characterized amplified regions (SCARs) or cleaved amplified polymorphic sequences (CAPs) (Yang and Korban 1996; Tartarini 1996; Tartarini et al. 1999; Gianfranceschi et al. 1996; Yang et al. 1997a, b; Xu et al. 2001). The *Vf* locus has been further characterized and a cluster of four gene paralogs have been identified within this locus using a map-based cloning strategy (Vinatzer et al. 2001; Xu and Korban 2002). At first, one of the four paralogs, *HcrVf2*, has been determined to be functional (Belfanti et al. 2004), and later this gene, also known as *Vfa2*, along with another paralog, *Vfa1*, have been confirmed to confer resistance to scab in transgenic apple lines (Malnoy et al. 2008).

Three RAPD markers, OPB08/710, OPK08/848, and OPZ13/869, linked to another scab resistance locus *Vbj* introgressed from the species *M. baccata jackii* were identified and converted into co-dominant SCARs (Gygax et al. 2004). By discarding plants showing genotype-phenotype incongruence (GPI), a linkage map for

*Vbj* was developed, and mapped between markers CH05e03 (SSR) and T6-SCAR, at 0.6 cM from and 3.9 cM distances from these markers, respectively (Gygax et al. 2004). Without removal of GPI plants, *Vbj* was placed 15 cM from the closest markers. Using segregation data of alleles in coupling with the *Vbj* locus, all seven markers were linked to a single linkage group with a total length of 10.3 cM (Gygax et al. 2004).

Hemmat et al. (2002) used a SCAR marker developed from the RAPD marker OB18/620 along with the SSR CH02B10 to map the *Vr* gene for scab resistance from the Russian seedling R12740-7A; while, another SCAR marker from the RAPD marker S22/1300 was used to map another gene, *Vx*, also from R12740-7A.

Recently, the *Vr1* (Boudichevskaia et al. 2004) and *Vr2* (Patocchi et al. 2004) genes from different Russian apple accessions were mapped to the proximal end of LG2 in apple. Bus et al. (2005) characterized hosts 2 and 4 derived from the Russian apple R12740-7A in the PRI breeding program, and demonstrated that *Vr-A* and *Vh2* were in fact the same gene. They also established genetic maps for regions around the *Vh2* and *Vh4* genes, and positioned them on the apple genome using SCARs of the RAPD markers OPL19 and OPB10, previously found to be closest to these genes (Bus et al. 2000), as well as S22SCAR (Hemmat et al. 2002), Z13SCAR, and four SSR markers including CH02b10, CH02c02a, CH05e03, and CH03d10 for LG2 (Liebhard et al. 2002).

Gardiner et al. (2003) mined candidate resistance genes from an apple EST database on the basis of homology to genes from five resistance (*R*) gene classes whereby these were screened as RFLP probes over Southern blots of DNA from seedling populations segregating for five selected resistances to apple scab, powdery mildew, and woolly apple aphid infection. These probes were converted to PCR-based markers, including SCARs or single nucleotide polymorphisms (SNPs).

Hemmat et al. (2003) identified SCAR markers linked to other scab resistance genes, including *Vb* from Hansen's baccata #2 and *Va* from 'Antonovka' PI 172633. They reported that B398/480 was the closest marker segregating with the resistance response from descendants of 'Antonovka' PI 172633 and UBC220/700 was the closest marker for resistance from 'Hansen's baccata #2. *Va* and *Vb* also shared flanking markers with each other and also with *Vf* on linkage group 8 of 'Rome Beauty' × 'White Angel'. A molecular marker linked to *Vm* has also been identified (Cheng et al. 1998) and its location on LG17 was determined by Patocchi et al. (2005). Two other apple scab resistance genes, *Vx* and *Vr2*, and molecular markers associated to them have also been identified (Hemmat et al. 2002; Patocchi et al. 2004).

Naik et al. (2006) utilized a total of 229 markers, including 46 apple EST-derived STSs (E-STSs), 8 resistance gene analogs (RGAs), 85 SSRs from apple and peach, and 88 RAPDs, and assigned them to 17 linkage groups covering 832 cM of the apple genome based on 52 individuals originating from the cross 'Antonovka debnicka' (Q12-4) × 'Summerred'. Clusters of E-STS and RGA loci were located in linkage groups previously identified to carry resistance genes, some of which confer resistance to apple scab disease.

Liebhard et al. (2003a) mapped quantitative trait loci (QTL) for scab resistance in progeny of 'Fiesta' × 'Discovery' and found eight genomics regions, six conferring resistance against leaf scab and two conferring fruit scab resistance. Although cv. Discovery showed a much stronger resistance against scab in the field, most identified QTLs were attributed to the more scab-susceptible parent 'Fiesta'. This indicated a high degree of homozygosity at scab resistance loci in 'Discovery'. Later, Durel et al. (2003) identified five QTLs for scab resistance in a 'Prima' × 'Fiesta' progeny inoculated with *V. inaequalis* isolates 302 and EUD42 (race 6). One of these QTLs was identified on LG1 with isolate 302 in 'Prima' around *Vf*, while another QTL for resistance to both isolates EU-D42 and 302 was detected on LG17 in 'Fiesta'. Calenge et al. (2004) working with an $F_1$ progeny derived from a cross between the partially scab-resistant apple cv. Discovery and the apple hybrid 'TN10-8', inoculated in the greenhouse with eight *V. inaequalis* isolates, found a single major resistance gene, *Vg*, along with seven QTLs for resistance to these isolates. Three QTLs on LG12, LG13, and LG15 were isolate-specific; while, a single QTL on LG5 was detected with two isolates. In addition, three QTLs on LG1, LG2, and LG17 were identified with most tested isolates, but not with every isolate. A single QTL on LG2 displayed alleles conferring different specificities, and it co-localized with the major scab resistance genes *Vr* and *Vh8*; whereas, the QTL on LG1 co-localized with *Vf*. These results further confirmed the involvement of these genomic regions on LG1 and LG17 in partial resistance to scab. Resistance to scab in apple was also recently reviewed by Gessler et al. (2006).

## 4.2 Other Diseases and Pest Resistance

### 4.2.1 Powdery Mildew

It is one of the most important apple diseases which attacks foliage and young shoots with marked effects on the quantity and quality of the fruit produced. The intensity of mildew attacks caused by the fungus *Podosphaera leucotricha* varies accordingly to climate conditions, but the disease is widespread wherever apples are grown. Some apple cultivars possess a high degree of field resistance, such as 'Delicious' and 'McIntosh', but most commercial cultivars are susceptible to this disease. Monogenic mildew immunity was reported only in some wild apple species or in interspecific hybrids of unknown origin (reviewed by Alston et al. 2000).

At least six major genes for resistance to mildew have been identified in apple germplasm. *Pl-1* from *M. robusta*, *Pl-2* from *M. zumi* (Knight and Alston 1968), *Pl-w* from White Angel (Gallot et al. 1985), *Pl-d* from the D12 accession (Visser and Verhaegh 1976), *Pl-8* from *M. sargenti* 843 (Korban and Dayton 1983) and *Pl-m* from the Mildew Immune Seedling or MIS (Dayton 1977). Breeders have concentrated mostly on using the resistance from *M. robusta* and *M. zumi* (Knight and Alston 1968) in spite of the fact that the levels of resistance conferred by *Pl-w* and *Pl-d* have been shown to be higher than those of *Pl-1* and *Pl-2* (Alston 1983).

Overall, the evaluation of powdery mildew resistance by using natural inocula is highly influenced by environmental conditions as well as plant age (seedlings vs. adult plants) making selection of resistant seedlings very difficult. This is also complicated by the existence of different physiological races of *P. leucotricha* that have been identified by pathogenicity tests with different fungal isolates (Lesemann et al. 2004; Urbanietz and Dunemann 2005; Dunemann et al. 2007). Finally, the breakdown of different resistance genes by some fungal isolates has been demonstrated as well (Korban and Dayton 1983; Lespinasse 1983, 1989; Caffier and Laurens 2005).

A lot of work has been done in recent years to characterize the different sources of mildew resistance at the molecular level, and several markers linked to the different mildew resistance genes have been identified (Table 2). Because of the complexity of the subject the information will be presented separately for each resistance source.

### 4.2.2 *Pl-1* Gene from *M. robusta*

The *Pl-1* resistance gene determines a necrotic reaction that is typical of this source of resistance (Alston 1983). Most of the molecular work on *Pl-1* has been conducted at the Federal Centre for Breeding Research on Cultivated Plants (Dresden, Germany) using the 78/18-4 accession ('A142/5' × 'Gloster') as source of mildew resistance. The pedigree of this selection includes the apple genotype 3762 (MAL59-1; *Malus robusta* 5 o.p.) as reported by Markussen et al. (1995). Two different progenies have been used for phenothypic and molecular analyses, progenies 93/9 (64 plants) and 99/2 (about 400 individuals). Different types of linked markers have been identified (Markussen et al. 1995; Dunemann et al. 2004, 2007), but surprisingly the marker-assisted selection (MAS) of *Pl-1* mildew-resistant seedlings was achieved by using one of these markers, SCAR AT20-450. Earlier, this marker has been reported to be incompletely successful in MAS in Switzerland and New Zealand (Kellerhals et al. 2000a; Dunemann et al. 2004) as the proportion of susceptible plants carrying the marker allele in coupling with *Pl-1* has been higher than expccted. A possible explanation for these discrepancies could reside in the origin of the source of resistance as different siblings of the progeny A142 have been used in different programs (A142-5 in Germany, A142-23 in Switzerland, and A142/8 in New Zealand). Also, a two-gene-model of *Pl-1* inheritance has been postulated to explain this discrepancy due to the fact that the two selections A142-8 and A142/23 are heterozygous for a second precursor gene, *Cc*, linked to *Pl-1*; while, A142-5 is homozygous (Dunemann et al. 2004).

Although the variability observed for different mildew resistance scores in greenhouse and field evaluations due to environmental conditions in different years, the position of the *Pl-1* gene at one end of LG12 is not in question. The *Pl-1* gene has been mapped on LG12 close to the Hi07f01 SSR and close to an NBS-LRR candidate resistance gene (Dunemann et al. 2007). Furthermore, this part on chromosome 12 seems to be very interesting as a cluster of mildew and scab resistance genes have been identified in this genomic region.

**Table 2** List of markers linked with different mildew resistance genes

| R gene | Source of resistance | Marker type and name | Reference(s) |
|---|---|---|---|
| *Pl-1* | A142-5, derived from MAL59-1 | OPD2-1000, OPAT20-450, AT20-450 SCAR | Markussen et al. (1995) |
| | A142-5 | AFLP2 and AFLP1 (the latter also transformed to AU-SCAR and AU-CAPS) | Dunemann et al. (2004) |
| | A142-5 | Hi07f01 SSR and 15G11 RGA | Dunemann et al. (2007) |
| *Pl-2* | A679-2 | Various RAPDs on LG6* including K14-1400, G4-1800, N18-1000 | Seglias and Gessler (1997) |
| | A679-2 | Various RAPDs on LG15* including AJ4-750, AD3-450, AR2-510 and AL07-400 | Seglias and Gessler (1997) |
| | SA572/2 | OPAT20-900 and UBC227-850 RAPDs | Dunemann et al. (1999) |
| | A689-24 | OPN18SCAR, OPAY17/OPAB16SCARa and b, OPU02SCARa and NBS 4 PCR | Gardiner et al. (2003) |
| | A689-24 | Various RFLPs with LRR-EST sequences (2d, 2e, 2c) and unknown sequences (2 g, 1f, 1d, 2b, 1 g, 2d, 1 h, 2e and 2 h) | Gardiner et al. (2003) |
| *Pl-w* | White Angel | LAP-2 isozyme | Manganaris (1989); Batlle and Alston (1996) |
| | White Angel | Acp-3 and Aat-1 isozymes | Hemmat et al. (1994) |
| | E295-4(= Gloster × White Angel) | CH01e12 and Ch05a02y SSRs | Evans and James (2003) |
| | E295-4 | EM M01 and EM M02 SCARs | Evans and James (2003) |
| *Pl-d* | A871-14 (= Worcester Permain × D12) | RAPD OPA01, AFLP ETA-CTC and SCAR EM DM01 | James et al. (2004) |
| | A871-14 | CH03c02 and CH01d03 SSRs | |
| *Pl-m* | MIS o.p. 93.051 | OPN18SCAR, OPAY17/OPAB16SCARa and b, OPU02SCARb and NBS 4 PCR | Gardiner et al. (2003) |

* The chromosome number is the one reported in the original paper, but after Maliepaard et al. (1998) and Liebhard et al. (2002), the LG15* corresponds to LG1 and LG6* corresponds to LG11, respectively.

Another mildew resistance source, probably related to *M. robusta,* is 'Novosibirski Sweet' OP. This accession also carries the SCAR AT20-450 marker, and its analysis in a segregating progeny (of about 200 seedlings) correlates well with the resistance phenotype. It has not yet been determined if this mildew resistance gene is either new, identical, or allelic to *Pl-1* (Dunemann et al. 2004).

### 4.2.3 *Pl-2* Gene from *M. zumi*

The presence of a monogenic resistance in *M. zumi* was first postulated by Knight and Alston (1968). This early hypothesis was later modified as the segregation for resistance could be better explained by the action of two complementary genes (Alston 1977). The resistance gene was derived from the apple genotype MAL58-1. While searching for markers linked to *Pl-2* gene derived from the A679-2 source, Seglias and Gessler (1997) observed a rather continuous distribution of resistance/susceptibility. Therefore, they postulated a polygenic inheritance, and found two main QTLs, one of which on the same chromosome carrying the *Vf* gene and probably derived from *M. floribunda* 821 (the A679-2 parent also carried the *Vf* gene). The second QTL, identified by the N18-1000 marker, was on LG6, LG11 after Maliepaard et al. (1998) as cited in Dunemann et al. (2007), of A679/2. This probably included the 'true' *Pl-2* gene derived from *M. zumi.*

The N18SCAR was also found to be linked to the *Pl-2* gene by Dunemann et al. (1999), Kellerhals et al. (2000b), and Gardiner et al. (2003). Using probes for candidate resistance genes, a series of RFLPs were found to be linked to the *Pl-2* gene (Gardiner et al. 2003).

### 4.2.4 *Pl-w* Gene from White Angel

The ornamental crabapple cv. White Angel is related to both *M. sargenti* and *M. sieboldii,* and appears to carry a monogenic source of mildew resistance (Gallot et al. 1985). 'White Angel' has been shown to provide a higher level of resistance than either *Pl-1* or *Pl-2* (Alston 1983), but the breakdown of mildew resistance from 'White Angel' has already reported (Korban and Dayton 1983).

The first marker linked to *Pl-w* was the isozyme leucine aminopeptidase (LAP-2). The observed segregation for mildew resistance in three different crosses was explained on the basis of complementary gene action, and both genes, *Pl-w* and *Rw,* were reported to be heterozygous in 'White Angel' (Manganaris 1989; Batlle and Alston 1996). The *Plw* gene was also found to be linked with the isozymes *Acp-3* and *Aat-p* (Hemmat et al. 1994), but because of the linkage with the *Lap-2* gene, *Plw* was believed to map onto LG8 (Maliepaard et al. 1998).

Recently, two SCARs developed from AFLP bands, EM M01 and EM M02, and two SSRs, CH01e12 and CH05a02y, were found to be linked to the *Plw* gene. The linkage with the two SSRs was not very tight (10–13 cM) but enough to confirm that the *Plw* gene was indeed on LG8 (Evans and James 2003). The maximum mildew score correlated well with the presence and absence of EM M01 and EM M02 markers. These markers were not completely specific for 'White Angel' as a band of the same size was amplified in some *M. zumi* derivatives carrying the *Pl-2* gene (A143-12 and A143-24). However, no amplification was observed on A679-2 (*Pl-2*), *Pl-1*, and *Pl-d* sources (Evans and James 2003).

### 4.2.5 *Pl-d* Gene from D12

The *Pl-d* is a mildew resistance gene from a seedling (D12) of the 'D series' derived from wild apples from South Tyrol (Visser and Verhaegh, 1976). Different types of markers (RAPDs, AFLPs, and SSRs) linked to the *Pl-d* gene have been identified (James et al. 2004). The RAPD marker OPA01 and the AFLP marker ETA-CTC are the closest to the gene, at distances of 4 and 5 cM, respectively. A SCAR marker, EM DM01, designed from a polymorphic AFLP band, has been mapped at a distance of about 9 cM from *Pl-d*. This 90 bp SCAR band is specific for the D12 source of resistance, while no amplification has been obtained from other resistance sources, including 'White Angel', 'MIS', some susceptible genotypes, some *M. robusta,* and *M. zumi* derivatives. Therefore, this marker is well suited for use in MAS. The linkage of *Pl-d* with the two SSRs CH03c02 and CH01d03 made it possible to assign this gene to LG12 at a position similar to that of the *Pl-1* gene (James et al. 2004), thus confirming the presence of a cluster of scab and mildew resistance genes in this chromosomal region.

### 4.2.6 *Pl-m* Gene from the Mildew Immune Seedling (MIS)

The Mildew Immune Seedling (MIS), carrying the *Pl-m* gene, is derived from open-pollinated 'Starking Delicious' (Dayton 1977). Using various markers previously identified for *Pl-2* made it also possible to develop a map of the genomic region controlling the *Pl-m* mildew resistance (Gardiner et al. 2003). The marker order and distances on LG11 are very similar except for a small inversion between NBS 4 PCR and OPU02SCARa, thus suggesting a possible extensive colinearity between these two sources of resistance. However, because of different sizes of the amplification products of OPU02SCAR (1700 for *Pl-2* and 2000 bp for *Pl-m*) a syntenic relationship seems to be more likely rather then identity (Gardiner et al. 2003). The N18SCAR and OPU02SCAR markers have also been amplified also in the mildew resistant rootstock Aotea 1 derived from *M. sieboldii,* but the amplified fragment is 2100 bp in size.

### 4.2.7 Other Mildew Resistance Sources

The clone U211 derived from an open-pollinated 'Primula' was selected in Poland because of its high level of mildew resistance that was also transmitted to its progeny (Stankiewicz et al. 2001). A preliminary analysis has made it possible to identify different QTLs, and one of these QTLs is on LG12 at a position where both *Pl-d* and *Pl-1* have been mapped (Stankiewicz et al. 2005). The RAPD marker OPAT20 and the two SCARs N18 and U02 do not amplify on U211. Therefore, the resistance of this clone is not attributed to the presence of either *Pl-1* from *M. robusta* nor the *Pl-2* from *M. zumi*.

A total of seven QTLs have been detected over five seasons in a cross between 'Discovery' × TN10-8, but two of these QTLs on LG2 and LG13 were consistently identified and were accounted together from 29 to 37% of the observed phenotypic variation, depending on the year of assessment. The other QTLs were identified during either one (LG1 and 14), two (LG10), or three (LG8 and 17) years (Calenge and Durel 2006). The strongest QTL effects were identified on LG groups 2, 13, and 8.

A QTL for polygenic field resistance to mildew in LG2 of 'Prima' × 'Fiesta' cross was also reported by Dunemann et al. (2000).

### 4.2.8 Fire Blight

This serious disease of both apples and pears is caused by the bacterium *Erwinia amylovora*. This bacterium, once confined to the United States, is spreading to most of the apple cultivation regions around the world. Immunity to fire blight disease has been reported for some *Malus* species (*M. robusta* 5 and *M. fusca* H-12), but there are also some cultivars ('Splendor', 'Delicious', and 'Winesap') as well as scab-resistant selections ('Prima', 'Priscilla', 'Enterprise', and 'Liberty') that are highly resistant to *E. amylovora* (Janick et al. 1996). Monogenic resistance to fire blight has not been identified in apple, and most of the molecular analysis has been conducted using a QTL approach on progenies derived from commercial cultivars.

Several QTLs were detected in two progenies sharing one common parent, 'Fiesta', that was crossed with each of 'Prima' and 'Discovery'. The three parental cultivars displayed different levels of fireblight resistance. 'Fiesta' and 'Prima' exhibited intermediate phenotypes, with 'Prima' being slightly more susceptible than 'Fiesta'; while, 'Discovery' was deemed as susceptible. In general, the average resistance scores of the two progenies were generally intermediate to the scores of their respective parents; however, transgressive segregation was also observed (Calenge et al. 2005b). In the 'Prima' × 'Fiesta' progeny, two additive QTLs were detected on LG7 and LG3. In the 'Fiesta' × 'Discovery', four additive QTLs were detected, two in 'Fiesta' (LG3 and LG7) and two in 'Discovery' (LG12 and LG13). The QTL associated with the strongest effects in both progenies, explaining about 40% of the phenotypic variation, was on LG7, and was derived from 'Fiesta'. In both progenies, the QTL peak on LG7 was close to the RAPD marker GE80-19-0550; while, the QTLs on LG3 of 'Prima' and 'Fiesta' were detected in different genetic positions (at about 40 cM distance).

The strong QTL on the same region of LG7 from 'Fiesta' was also detected in another progeny of 'Fiesta' × 'Discovery' (Khan et al. 2006). The phenotypic variation explained by the QTL on L7 was about 37–38% for the different scoring dates, 13–20 and 27 days following inoculation, and for the area under disease progress curve. The QTL peak was close to the AFLP marker E37M40-0400, and this marker was positioned between the two SSR markers CH04e05 and Hi05b09 (Silfverberg-Dilworth et al. 2006). The QTL on LG7 of 'Fiesta' could be considered a stable

QTL as it was detected in different genetic backgrounds, and following inoculations with two different strains of *E. amylovora* (Khan et al. 2006 and 2007); while, no other QTLs have been found in other genomic regions. Two RAPD markers bracketing the QTL have been transformed into SCAR markers, and an SSR marker specific for the region was developed. This stability and the availability of reproducible markers bracketing the QTL make this locus promising for use in MAS (Khan et al. 2007).

Recently, a fire blight resistance gene from *M. robusta* 5 was also identified on LG3. As the distribution of the symptoms in the progeny of 'Idared' × *M. robusta* 5 was continuous, a quantitative inheritance was postulated. However, this finding must be considered in light of a lack of identification of other QTLs and with previous evidence suggesting the presence of a dominant resistance gene in *M. robusta* 5 (Gardener et al. 1980). The QTL analysis made it possible to map the fire blight resistance gene close to the SSR CH03g07 and CH03e03 (Peil et al. 2007). Due to availability of common SSRs, the QTL on LG3 from *M. robusta* was located in the same genomic region of the QTL from 'Fiesta' as reported by Calenge et al. (2005b). The percentage of phenotypic variation explained by this QTL was very high (80%), and therefore a major gene was likely to be involved in the resistance response. The co-localization and the high proportion of phenotypic variation explained by the QTL suggested that different alleles conferring different levels of resistance must be present in this region of LG3 (Peil et al. 2007).

The modulation of defense responses in two *Malus* genotypes with contrasting susceptibility to fire blight was also demonstrated; i.e., delayed induction of several genes of the phenylpropanoid metabolism in tissues of the susceptible genotype (Venisse et al. 2002).

### 4.2.9 Aphids

Monogenic resistances to rosy apple aphid (*Dysaphis plantaginea*), rosy leaf-curling aphid (*Dysaphis devecta*) and woolly apple aphid (*Eriosoma lanigerum*) were found either in cultivated or wild apples.

### 4.2.10 *Dysaphis plantaginea*

A single dominant gene ($Sm_h$) for *D. plantaginea* resistance was identified in an open-pollinated selection of *M. robusta* (MAL59/9), but no molecular work has been conducted so far on this resistance source. Some genetic studies have been conducted on the scab-resistant cultivar 'Florina' that was also found to be resistant (or tolerant) to *D. plantaginea* (Lespinasse et al. 1985). A genetic control based on the action of at least two independent genes was postulated by the segregations observed in two crosses of 'Florina' with two different susceptible parents, 'Raxao' and 'Perico'. The tolerance to rosy apple aphid was also found to be independently inherited from the *Vf* scab resistance carried by 'Florina' (Minarro and Dapena 2004). A preliminary approach to identifying complex transcriptional changes occurring following *D. plantaginea* attack was conducted using a cDNA-AFLP approach. This allowed for the identification of 21 DNA fragments which were differentially expressed in the resistant cultivar 'Florina' following aphid infestation when compared to either the susceptible cultivar 'Topaz', mechanical wounding, or non-infested leaves (Qubbaj et al. 2005).

### 4.2.11 *Dysaphis devecta*

Resistance to three biotypes of the rosy leaf curling aphid have been identified in different apple cultivars (Alston and Briggs 1970). 'Cox's Orange Pippin' carries the *Sd1* gene, and it is resistant to biotypes 1 and 2. 'Northern Spy' has *Sd2,* and it is resistant to biotype 1; while, a selection from *M. robusta* carries the *Sd3* gene which confers resistance to all biotypes.

Molecular studies have been conducted mainly on the *Sd1* gene derived from 'Fiesta' ('Cox's Orange Pippin' × 'Idared'). The resistance gene was initially mapped on LG7 of 'Fiesta' and close to three RFLPs (Roche et al. 1997a). One of these RFLPs, 2B12a, was transformed to an easy-to use PCR marker (Roche et al. 1997b). A fine mapping of the *Sd1* locus with an AFLP bulked segregant analysis along with analysis of additional progenies (759 additional individuals) was later done. Three new AFLP tightly linked to *Sd1* were then identified, two of which were found to carry a GA repeat that was used to develop an easy-to-use SSR marker (SdSSR). The *Sd1* gene was then located within a 1.3 cM interval flanked by the SdSSR and the 2B12 RFLP. The RFLP MC064 was confirmed to co-localize with the *Sd1* gene (Cevik and King 2002a). A BAC contig of about 600 kb spanning the *Sd1* region was identified (Cevik and King 2002b).

### 4.2.12 *Eriosoma lanigerum*

The woolly apple aphid (WAA), *Eriosoma lanigerum,* is a major economic pest in most apple growing areas, particularly for susceptible rootstocks as it is difficult to prevent infestations in below-ground parts of trees. Roostock breeding efforts for WAA resistance began in 1920s and 1930s by using 'Northern Spy' as source of WAA resistance. The Malling-Merton (MM) series of rootstocks was specifically bred for resistance to WAA (Crane et al. 1936). Other sources of WAA resistance have been identified in the *Malus* germplasm (Knight et al. 1962; Cummings et al. 1981). At least three different major genes have been reported. The *Er1* gene derived from 'Northern Spy' and the *Er2* gene from *M. robusta,* although there was some confusion regarding this latter source of resistance. Therefore, because of its resistance, *M. robusta* 5 has been proposed as the source of WAA resistance gene from *M.* × *robusta* (Bus et al. 2007). A third gene for WAA resistance, *Er3*, was identified in *M. sieboldii* 'Aotea 1. The ability of WAA to overcome the resistance of these three resistance sources was also reported (Bus et al. 2007).

Recently, these three genes for WAA resistance have been mapped. The *Er1* and *Er3* genes are located along the top of LG8 (Bus et al. 2007). As a result, a previous report suggesting a linkage between *Er1* with the *S* self-incompatibility locus (Knight et al. 1962) must be incorrect, and the observed segregation distortions of

WAA resistance in some progenies must be attributed to other reasons. This absence of linkage between *Er1* and *S* loci was also postulated by Tobutt et al. (2000). Although *Er1* and *Er3* map to the same genomic region, the identification of a new biotype of WAA capable of overcoming *Er3* but not *Er1* (Sandanayaka et al. 2003) suggests the presence of two closely linked genes or different alleles having different functions (Bus et al. 2007). The *Er2* gene has been mapped along the top of LG17, but at the opposite end of the *S* locus, thus indicating that segregation distortions are not due to self-incompatibility (Bus et al. 2007). The *Er1* and *Er3* genes are mapped close to the SSR CH01c06 and to the SNP NZsc_O05; while, the *Er2* gene is close to the SSR CH04c06 and the SSR NZms_EB145764. The usefulness of these markers in MAS has also been validated in other segregating progenies, and the prediction rate is reported as good (Bus et al. 2007).

## *4.3 Horticultural Traits*

### 4.3.1 Apple Tree Architecture

Apple tree architecture is an important factor in fruit tree management and production; however, most traits associated with plant growth are under polygenic control. Costes et al. (2006) has reviewed our knowledge of fruit tree architecture and its implications for tree management and production.

Lawson et al. (1995) were the first to report on the inheritance of some morphological and developmental traits in apple, such as branching habit, root suckering, vegetative and reproductive budbreak. Mapping of these traits was achieved using a RAPD-based map and a small subset of plants (56 individuals) derived from a 'Rome Beauty' × 'White Angel' cross. Therefore, transferability of this information to other populations is rather limited, and only major genes effects could have been detected.

Various studies have reported on mapping of the columnar growth habit trait in 'Wijcik', a mutant of the apple cultivar McIntosh. Conner et al. (1997) reported the first RAPD marker linked to the columnar gene (*Co*) that was later sequenced and transformed to an easy-to-score marker containing an SSR repeat ($SSR^{Co}$) (Hemmat et al. 1997). The estimation of the recombination frequency was estimated only by using extremes of the distribution of growth habit phenotypes and eliminating intermediate phenotypes due to the presence of minor genes modifying expression of *Co*. Although the columnar type has been reported to be simply inherited, the complexity of this architectural trait necessitated its further dissection using the cross 'Wijcik' × 'NY 75441-58' (Conner et al. 1998). One to eight significant QTLs were found for different traits influencing juvenile growth and development (height increment, internode number, internode length, base diameter increment, base diameter after 9 years, branch number, and leaf break). For most of these traits, different QTL regions have been identified on the genetic map, but in some cases many traits clustered together along the same genomic region.

The most important QTL cluster included the branch number, internode length (Year 1), internode number (Y1 and Y2), and base diameter increment (Y1 and Y2), and it was identified on LG10 in the position where the *Co* gene was mapped. Another large QTL cluster was identified on LG7 for height increment (Y1 and Y3), internode number (Y1), base diameter increment (Y3), and base diameter (Y9). The stability of the identified QTLs was also estimated, but most were found to vary depending on the year of determination. Therefore, this variation in QTL detection in different years could be a result of either genotype × environment interactions or chance fluctuations in datasets in different years.

The columnar trait was also investigated in a 'Braeburn' × 'Telamon' (Kenis and Keulemans 2004, 2007) progeny. A QTL clustering for many growth-related traits was confirmed in the *Co* gene region on TEL LG15, equivalent to LG10 of Maliepaard et al. (1998). This cluster included growth increments in years 1 and 2, proportion of total and Y2-sylleptic branches, total branch length, total branch number (proleptic and sylleptic in Y2), internode length (Y1 and Y2), growth rate (Y1 and Y2), and height increment (Y1 and Y2). The influence of the Co gene was more pronounced as seedlings matured. Then the influence on total branching characteristics was higher for branch length than for branch number. Another clustered QTL was observed on TEL LG11, equivalent to LG8 of Maliepaard et al. (1998). The QTL included growth increment (Y1), branch length, number and proportion of sylleptic branches (Y1), internode length and number (Y1), height increment (Y1), and growth rate (Y2). The observed QTL instability over years indicated that the genetic control of growth characteristics changed as trees matured, and it was influenced by environmental factors.

Tian et al. (2005) reported on the fine mapping of the *Co* region by adding nine new markers using a 'Spur Fuji' × 'Telamon' progeny. These markers were identified by a bulked segregant analysis, and the *Co* gene was mapped between the two SSRs CH3d11 and COL.

The results obtained in different independent experiments on the genetics of the columnar habit are rather consistent, and it is possible to confirm that this trait is under polygenic control even if a major gene on LG10 is responsible for most of the observed genetic variation. The dissection of the columnar trait into different growth variables has also confirmed the presence of minor genes, and to a certain degree the influence of environmental variation.

Recently, the *Dwarfing 1* (*Dw1*) gene, the first such reported locus controlling dwarf growth habit, has been located in a region (<2.5 cM) at the top of LG5 of the rootstock 'Malling 9' (Rushlomé-Pilcher et al. 2008). *Dw1* is mapped between the RAPD marker NZra AM18_700 and the SSR marker CH03a09; however, the presence of other loci conditioning dwarfing growth habit is also reported due to the identification of vigorous phenotypes, in a segregating 'Malling 9' × 'Robusta 5' population, carrying this gene.

Five QTLs for growth-related traits have been identified on own-rooted seedlings of a 'Fiesta' × 'Discovery' progeny (Liebhard et al. 2003b). Three were associated with stem diameter and two with leaf size, but none of these QTL positions have been detected on the same progeny after grafting. A total of nine QTLs for stem diameter were identified on grafted progeny. Although three of these QTLs were found on the same chromosomes identified on seedlings, their locations were clearly different. Six genomic regions were found to be associated with height increment, and four of these co-localized with QTLs for stem diameter suggesting a possible clustering of growth and developmental traits (Liebhard et al. 2003b).

Due to the complexity of tree architecture, a combination of topological and geometric traits measurements on trunks and long sylleptic axillary shoots (LSAS) have been recorded on 1-year old seedlings (Segura et al. 2006). Using these measurements, several QTLs controlling tree geometry have been identified for some integrated traits (tree surface and volume), traits related to the form of long sylleptic axillary shoots (bending and basis angle), and for other traits such as internode length of the trunk, and LSAS. Four QTLs were mapped for total number of sylleptic branching (Segura et al. 2007). Recently, a very interesting study was conducted to better understand the genetic determinisms of some ecophysiological traits as this knowledge would have a direct impact on the selection of more efficient trees. Leaf gas-exchange parameters, including stomatal conductance, net CO2 assimilation, and transpiration rates were scored on each tree of a segregating progeny, 'Starkrimson' × 'Granny Smith', grown under dry air stress condition in a controlled environment (Regnard et al. 2007). Three co-localized QTLs were detected for transpiration and stomatal conductance on LG 8, LG11, and LG17 at the intermediate leaf-to-air vapor pressure deficit. The QTL on LG 8 was also detected at the higher and lower leaf-to-air vapor pressure deficits.

### 4.3.2 Flowering Related Traits

A QTL for bloom time has been reported on LG1 of 'White Angel' by Lawson et al. (1995). This map has been mainly constructed using RAPD markers, but due to the presence of the Aat isozyme, the 'White Angel' LG1 would correspond to LG17 of Maliepaard et al. (1998). This finding is rather interesting because in this same region is also the *S* locus controlling self-incompatibility (Maliepaard et al. 1998). The bloom time has also been investigated by Liebhard et al. (2003b) in the 'Fiesta' × 'Discovery' progeny, and three QTLs have been identified on LG 7, LG 10, and LG 17. However, the position of the QTL on LG17 is on the other extreme end of that of the *S* locus. Two QTLs have also been identified for the number of flower bunches on LG8 and LG15, and two other QTLs for length of the juvenile period on LG3 and LG 15 (Liebhard et al. 2003b). Unfortunately, these QTLs explain only a small part of the genetic variation, and therefore we are still very far from understanding fruit tree flowering and juvenility traits, whose knowledge would have a significant impact on both on fruit tree breeding and production.

## *4.4 Fruit Quality*

Besides the main component of water, apple fruits are rich in sugar and starch. Unripe apples contain as much as 15% starch, and ripe apples consist of 8–14%

sugar (Reed 1975). When young fruitlet size reaches more than 20 mm in diameter, starch rapidly accumulates (Magein and Leurquin 2000). As apples ripen, starch stored in their flesh is gradually converted to sugar. Starch content changes from variety to variety. Apple starch has a great influence on eating and processing quality. First, starch together with acid, is known to make up the flavor. Apples with low content in starch and low levels of acid taste sweet, while those rich in both starch and acid really make you pucker. Apple starch is composed of amylose and amylopectin. Amylose is a continuous chain of up to 4000 D-glucose units joined by α-1, 4-glucosidic linkages. It is not completely linear, but rather lightly branched at α-1, 6-glucan branch points. Amylopectin is a highly branched polyglucan with α-1,6-glucosidic linkages that connect linear chains. Han et al. (2007a) have isolated a gene encoding for starch branching enzymes I from apple. This gene, designated *SbeI*, consists of 14 exons and 13 introns, covering 6,075 bp. Fingerprinting and genomic Southern blot analysis has indicated that *SbeI* is a single copy gene in the apple genome. RT-PCR analysis suggests that apple *sbeI* transcripts show very low levels at the early stage of fruit development when its diameter is less than 1 cm, and reach a steady maximum level at around 44 days after pollination. Besides fruits, apple *sbeI* is also expressed in buds and flowers, and very weakly in leaves. Later, Han et al. (2007b) have identified two genes encoding starch branching enzyme II (SBEII) in apple. These genes share 94% and 92% identity in coding DNA sequences and amino acid sequences, respectively; moreover, they have similar expression patterns. Both genes are expressed in vegetative and reproductive tissues, including leaves, buds, flowers, and fruits.

Yao et al. (2007) used cDNA-AFLP along with bulked segregant analysis (BSA) to screen differentially expressed genes between low- and high-acid apple fruits from hybrids of a 'Toko '× 'Fuji' cross. They identified a *Mal-DDNA* gene, present as a single copy gene in the apple genome. This gene was transcribed in low-acid fruits at both early and ripe stages of fruit development; but only in the ripe stage of development in both high- and mid-acid fruits.

Sorbitol, the primary photosynthate and translocated carbohydrate in apple, is converted to fructose by sorbitol dehydrogenase (SDH; EC 1.1.1.14) which is active in apple fruit throughout development. Although nine SDH genes have been isolated and their sequences characterized, Nosarzewski and Archbold (2007) have determined that five of these nine SDH genes are expressed in apple fruit. Two SDH genes, *SDH1* and *SDH3*, are expressed in both seed and cortex tissues, while, SDH2 expression is limited to cortex whereas, SDH6 and SDH9 are expressed in seed tissues only. SDH isomeric proteins of different pI values are detected in apple fruit. SDH isomers with pI values of 4.2, 4.8, 5.5, and 6.3 are found in seeds, and SDH isomers with pI values of 5.5, 6.3, 7.3, and 8.3 are found in cortex.

Han et al. (2007c) found two clusters of genes coding for caffeic acid O-methyltranfeases (COMT) in the apple genome. Three genes from one cluster and two genes from another cluster were isolated and characterized. These five genes encoding COMT, designated *Mdomt*1 to *Mdomt*5 were distinguished by a (CT)n microsatellite in the 5' UTR and two transposon-like sequences present in the promoter region and intron 1, respectively. The transposon-like sequence in intron 1 unambiguously traced the five *Mdomt* genes in the apple to a common ancestor.

Recently, Han and Korban (2007) identified a novel family of transposable elements, designated *Spring*, in the apple genome. By aligning *Mdomt*1 and *Mdomt*2, differences in DNA sequences upstream of their transcript start sites were noted; however, two small fragments, ~148 bp in size, exhibit 87% sequence identity, and these were deemed as transposon-like element, and were designated as *Spring-1* and *Spring-2*. Both *Spring-1* and *Spring-2* contained a terminal inverted repeat and a flanking direct repeat, suggesting they may be indeed transposable elements. Further to identification of these miniature inverted repeat sequences (MITEs), two additional *Spring* elements were then identified. The first, designated *Spring-3*, was detected in intron 3 of a gene encoding α-farnesene synthase in apple (Genbank AY805412), while the second, designated *Spring-4*, was identified in the 3' UTR of two genes encoding 1-aminocyclopropane-1-carboxylic acid (ACC) synthase (Genbank AB010102 and U89156) (Han and Korban 2007).

Liebhard et al. (2003b) were able to detect major genes and QTL contributing to various fruit quality traits using a segregating population of 'Fiesta' × 'Discovery'. A strong QTL for 'harvest date' was identified on LG 3, originating from the early ripening parent 'Discovery' and accounting for 16% of phenotypic variability; while, three genomic regions on LG5, LG15, and LG16 were identified for 'number of fruit', and at least 8 QTLs for 'fruit weight'. They also identified five QTLs for fruit flesh firmness with one QTL on LG 3, a second QTL on LG 6, a third on LG 11, a fourth on LG 12, and a fifth on LG 14, accounting for 27, 16, 11, 8, and 6%, respectively, of phenotypic variability. Sugar content in fruit juice was attributed to five genomic regions on LGs 3, 6, 8, 9, and 14; while fruit acidity was attributed to QTLs on LGs 8 and 16. Previously, Maliepaard et al. (1998) identified the *Ma* locus for fruit acidity at the top of LG 16 of the cross 'Prima' × 'Fiesta'. An AFLP marker, E31M38-0193, with an *aa* allele and an SSR marker, CH05e04z, with a 167 bp fragment were found to be tightly linked with the two identified loci for acidity on LGs 8 and 16, respectively (Liebhard et al. 2003b). The presence of both detected alleles confer earlier ripening, higher fruit numbers, higher fruit weight, firmer fruit, and more sugar and acid fruit contents (Liebhard et al. 2003b).

The main apple allergens, *Mal d 1*, *Mal d2*, *Mal d 3,* and *Mal d 4*, have been mapped onto different apple chromosomes (Gao et al. 2005a, b, c). The two *Mal d* 3 genes have been mapped onto two homoeologous chromosomes, LG4 and LG12 (Gao et al. 2005a). A total of 18 *Mal d 1* genes, subdivided into four subfamilies, were identified, and 17 of those were mapped in two main clusters on two homoeologous chromosomes on LG 13 (7 genes), LG 16 (9 genes), and on LG6 (Gao et al. 2005b). Two copies of the *Mal d 2* gene, differing for intron length, were mapped at an identical position on LG 9. The four *Mal d 4* genes were mapped onto LG 9 (two copies), LG2, and LG8 (Gao et al. 2005c). *Mal d 1* expression and apple allergenicity were successfully reduced by RNA interference, thus supporting the feasibility of production of hypoallergenic apples by gene silencing (Gilissen et al. 2005).

Earlier, Cheng et al. (1996) identified co-dominant RAPD markers linked to red color pigmentation in fruit skin. Recently, Chagné et al. (2007) used candidate gene mapping and identified the *Rni* locus, a major genetic determinant of the red foliage and red color pigmentation in the core of apple fruit. In a population segregating for the red flesh and foliage phenotype we have determined the inheritance of the *Rni* locus and DNA polymorphisms of candidate anthocyanin biosynthetic and regulatory genes. SSRs and SNPs in the candidate genes were also located on an apple genetic map. They have shown that the transcription factor *MdMYB10* gene co-segregates with the *Rni* locus, and it is located on LG 9 of the apple genome.

## 5 Marker-Assisted Selection

The New Zealand apple breeding at Plant and Food Research has been heavily involved in marker-assisted selection (MAS) by screening apple breeding populations to identify seedlings with a desired resistance gene profile, including seedlings with pyramided resistances to apple scab, powdery mildew, and woolly apple aphid, using various molecular markers developed from various laboratories (Bus et al. 2000, 2002; Gardiner et al. 2002).

Gardiner et al. (2003) mined the apple EST database for candidate ESTs on the basis of homology to genes from five resistance (*R*) gene classes. These were screened as RFLP probes over Southern blots of DNA from seedling populations segregating for five selected resistances to pest or pathogen infection. They have used this targeted strategy to efficiently obtain a set of additional markers for resistance genes that are candidates for marker assisted selection in their apple breeding program, namely for screening for *Vf, Pl2, PlMIS*, and *Er3* genes.

Tartarini et al. (2000) and Kellerhals et al. (2000a, b) reported that SCAR markers derived from RAPD markers linked to the *Vf* gene were effective in MAS. However, they reported some inconsistencies between phenotypic scoring and presence of theses SCARs in some scab-resistant seedlings. Huaracha et al. (2004) assessed the use of SCAR markers, derived from AFLPs tightly linked to the *Vf* gene, for MAS for scab resistance under field conditions, and found that none of the SCARs could be detected in scab-susceptible apple seedlings. However, presence of these SCARs in scab-resistant seedlings varied among the different progeny analyzed. This was attributed to phenotypic classification, races of the fungal pathogen in the field, and presence of QTL, among others.

Khan et al. (2007) identified a fire blight resistance QTL explaining 34.3–46.6% of the phenotypic variation that was recently identified on LG 7 of apple cultivar 'Fiesta', and designated a F7. However, AFLP and RAPD markers flanking this QTL were deemed unsuitable for MAS. Two RAPD markers bracketing the QTL have been transformed into SCAR (sequence-characterized amplified region) markers, and an SSR marker specific for the region was developed. Pedigree analysis of 'Fiesta' with these markers enabled tracking of the F7 QTL allele back to 'Cox's Orange Pippin'.

# 6 Current and Future Developments

## 6.1 Bin Mapping

The demanding efforts necessary to find the genomic position of new markers can nowadays be reduced by the application of a 'BIN' mapping approach that have been successfully developed and applied in *Prunus* (Howad et al. 2005). A set of 67 BINs covering the whole *Prunus* genome have been determined through the identification of segregation patterns on a limited set of recombinant key seedlings. Therefore, the position of new markers in *Prunus* can be estimated through the molecular analysis of only six plants of the TxE reference progeny. The choice of the restricted number of plants for BIN mapping can be better performed after the development of a map frame based on well-distributed SSR. This work had been recently undertaken in apple by testing the optimal number of plants (six to twelve) that must be used in BIN mapping experiments (Van Dyk et al. 2007).

## 6.2 Pedigree-Based Approaches for Mining and Detection of QTLs

To date, most of the QTL work conducted in apple have been done using the standard method of combined analysis of phenotypic and molecular data using a single progeny. In the framework of the EU HiDRAS project, a new method was evaluated to identify QTLs on multiple progenies linked by pedigrees (Bink et al. 2002). The first results were recently presented at the Eucarpia Symposium held in Zaragoza Spain in 2007. In this project, about 2,000 plants (including progenitors, parents and seedling progenies) were phenotyped and genotyped using a total of 83 SSRs with good coverage of the apple genome.

Multiple progenies are essential for building sound genetic models explaining the inheritance of quantitative traits and to draw proper conclusions on the number of genomic positions critical for the control of each trait, their contributions, their mode of action (additive, dominant, or epistatic), and the influence of the genetic background (Bink et al. 2007). The software FlexQTL has been developed to analyze for phenotypic and molecular data (Van de Weg et al. 2004). This approach has already been applied within HiDRAS to firmness and acidity traits (Van de Weg et al. 2007; Stankiewicz-Kosyl et al. 2007).

## 6.3 Whole Genome Physical Map

Han et al. (2007d) constructed a genome-wide physical map for the apple using restriction-based fingerprinting of BAC clones. Two BAC libraries, *Bam*H and *Hind* BAC libraries constructed from apple cv. GoldRush, were used for fingerprinting. A total of 74,281 BAC clones representing $\sim$10.5 $\times$ haploid genome equivalents were fingerprinted using the agarose gel-based restriction fingerprinting method.

The physical map consists of 2,702 contigs, and it is estimated to span ~927 Mb in physical length. The reliability of contig assembly was evaluated using several approaches, including assembling contigs using variable stringencies, assembling contigs using fingerprints from individual libraries, checking the consensus maps of contigs, and DNA markers. Altogether, the results demonstrated that the contigs were properly assembled. The apple genome-wide BAC-based physical map represents the first draft genome sequence.

Efforts are now underway to anchor the physical map onto an apple genetic map by anchor the fingerprint contigs to linkage groups. The integration of physical map and genetic map is now underway. This physical map will not only serve as a platform for large-scale genome sequencing efforts, but it is also very helpful for various other purposes such as development of DNA markers for a genomic region of interest, QTL fine mapping, effective positional cloning of genes, high-throughput EST mapping (functional genomics), and comparative genomics (synteny studies).

## *6.4 Current Plans for Whole Genome Sequencing*

Apple genome sequencing efforts are currently underway. Sequencing of the heterozygous cultivar 'Golden Delicious' at 15.4 × genome coverage has been completed at the Istituto Agrario San Michele all'Adige (IASMA) in Italy. In addition, this team is collaborating with groups at INRA (France) and Plant and Food Research (New Zealand) to generate a high density genetic map to anchor this genomic sequence. Moreover, sequencing of a double haploid 'Golden Delicious', provided by INRA, is underway at Washington State University. These two sequencing efforts are jointly collaborating to release the whole apple genome sequence in the near future. This will certainly be a landmark event for the dawn of a new genomic era for apple and for all Rosaceae cultivated species. It is expected that the entire Rosaceae community will participate in post-genomic efforts as manual annotation of the genome will also support comparative genomics approaches within several Rosaceae species.

## References


Alston, F.H. (1977) Practical aspects of breeding for mildew (*Podosphaera leucotricha*) resistance in apples. Proc. VII Eucarpia Fruit Section Symp., Top Fruit Breeding, Wageningen, 1976: 4–13

Alston, F.H. (1983) Progress in transferring mildew (*Podosphaera leucotricha*) resistance from *Malus* species to cultivated apple. IOBC/WPRS Bull. 6: 87–95

Alston, F.H., and J.B. Briggs (1970) Inheritance of hypersensitivity to rosy apple aphid *Dysaphis plantaginea* in apple. Can. J. Genet. Cytol. 12:257–258

Alston, F.H., K.L. Phillips, K.M. Evans (2000) A *Malus* gene list. Acta Hortic. 538: 561–570

Baldi, P., A. Patocchi, E. Zini, C. Toller, R. Velasco, and M. Komjanc (2004) Cloning and linkage mapping of resistance gene homologues in apple. Theor. Appl. Genet. 109:231–239

Batlle, I., and F.H. Alston (1996) Genes determining leucine aminopeptidase and mildew resistance from the ornamental apple, 'White Angel'. Theor. Appl. Genet. 93:179–182

Belfanti, E., E. Silfverberg-Dilworth, S. Tartarini, A. Patocchi, M. Barbieri, J. Zhu, B.A. Vinatzer, L. Gianfranceschi, C. Gessler, and S. Sansavini (2004) The *HcrVf2* gene from a wild apple confers scab resistance to a transgenic cultivated variety. Proc. Natl. Acad. Sci. USA 101: 886–890

Bink, M.C.A.M., P. Uimari, M.J. Sillanpää, L.L.G. Janss, and R.C. Jansen (2002) Multiple QTL mapping in related plant populations via a pedigree-analysis based approach. Theor. Appl. Genet. 104: 751–762

Bink, M., R. Voorrips, E. Van de Weg, and H. Jansen (2007) Statistical tools for QTL mapping in multiple, pedigreed populations. Eucarpia XII Fruit Section Symposium, September 16–20, 2007, Zaragoza, Spain: 59

Boudichevskaia, A., H. Flachowsky, C. Fischer, V. Hanke, and F. Dunemann (2004) Development of molecular markers for *Vr1*, a scab resistance factor from R12740-7A apple. Acta Hort. 663:171–176

Boudichevskaia, A., H. Flachowsky, A. Peil, C. Fischer, and F. Dunemann (2006) Development of a multiallelic SCAR marker for the scab resistance gene *Vr1/Vh4/Vx* from R12740-7A apple and its utility for molecular breeding.Tree Genet. Genomes 2: 186–195

Bournival, B.L. and S.S. Korban (1987) Electrophoretic analysis of genetic variability in the apple. Scient. Hort. 31:233–243

Bus V., C. Ranatunga, S. Gardiner, H. Bassett, and E. Rikkerink (2000) Marker assisted selection for pest and disease resistance in the New Zealand apple breeding programme. Acta Hort. 538: 541–547

Bus, V., A. White, S. Gardiner, R. Weskett, C. Ranatunga, A. Samy, M. Cook, and E. Rikkerink (2002) An update on apple scab resistance breeding in New Zealand. Acta Hort. 595:43–47

Bus, V.G.M., E.H.A. Rikkerink, W.E. van de Weg, R.L. Rusholme, S.E. Gardiner, H.C.M. Bassett, L.P. Kodde, L. Parisi, F.N.D. Laurens, E.J. Meulenbroek, and K.M. Plummer (2005) The *Vh2* and *Vh4* scab resistance genes in two differential hosts derived from Russian apple R12740-7A map to the same linkage group of apple. Mol. Breed. 15:103–116

Bus, V.G.M., D. Chagné, H.C.M. Bassett, D. Bowatte, F. Calenge, J.M. Celton, C.E. Durel, M.T. Malone, A. Patocchi, A.C. Ranatunga, E.H.A. Rikkerink, D.S. Tustin, J. Zhou, and S.E. Gardiner (2007) Genome mapping of three major resistance genes to woolly apple aphid (*Eriosoma lanigerum* Hausm.). Tree Genet. Genomes DOI10.1007/s11295-007-0103-3

Caffier, V. and F. Laurens (2005) Breakdown of *Pl2*, a major gene for resistance to apple powdery mildew, in a French experimental orchard. Plant Pathol. 54: 116–124

Calenge, F., A. Faure, M. Goerre, C. Gebhardt, W. E. Van de Weg, L. Parisi, and C.-E. Durel (2004) Quantitative Trait Loci (QTL) Analysis reveals both broad-spectrum and isolate-specific QTL for scab resistance in an apple progeny challenged with eight isolates of *Venturia inaequalis*. Phytoapthology 94: 370–379

Calenge, F., C.G. van der Linden, W.E. van de Weg, H.J. Schouten, G. Van Arkel, C. Denancé, and C.E. Durel (2005a) Resistance gene analogues identified through the NBS-profiling method map close to major genes and QTL for disease resistance in apple. Theor. Appl. Genet. 110:660–668

Calenge, F., D. Drouet, C. Denancé, W.E. Van de Weg, M.N. Brisset, J.P. Paulin, and C.E. Durel (2005b) Identification of a major QTL together with several minor additive or epistatic QTLs for resistance to fire blight in apple in two related progenies. Theor. Appl. Genet. 111:128–135

Calenge, F. and C.E. Durel (2006) Both stable and unstable QTLs for resistance to powdery mildew are delected in apple afler four years of field assessments. Molecular Breed. 17: 329–339

Cevik, V. and G.J. King (2002a) High-resolution genetic analysis of the *Sd1* aphid resistance locus in *Malus* spp. Theor. Appl. Genet. 105: 346–354

Cevik, V. and G.J. King (2002b) Resolving the aphid resistance locus *Sd1* on a BAC contig within a sub-telomeric region of *Malus* linkage group 7. Genome 45: 939–945

Chagné, D., C.M. Carlisle, C. Blond, R.K. Volz, C.J. Whitworth, N.C. Oraguzie, R.N. Crowhurst, A.C. Allan, R.V. Espley, R.P. Hellens, and S.E. Gardiner (2007) Mapping a candidate gene (MdMYB10) for red flesh and foliage colour in apple. BMC Genomics 8: 212. doi: 10.1186/1471-2164-8-212 Chandonia, J.-M. and S.E. Brenner (2006) The impact of structural genomics: expectations and outcomes. Science 311: 347–351

Cheng F.S., N.F. Weeden, and S.K. Brown (1996) Identification of co-dominant RAPD markers tightly linked to fruit skin color in apple. Theor. Appl. Genet. 93:222–227

Cheng, F.S., N.F. Weeden, S.K. Brown, H.S. Aldwinckle, S.E. Gardiner, and V.G. Bus (1998) Development of a DNA marker for *Vm*, a gene conferring resistance to apple scab. Genome 41:208–214

Chevreau, E., Y. Lespinasse, and M. Gallet (1985) Inheritance of pollen enzymes and polyploidy origin of apple (*Malus* × *domestica*). Theor. Appl. Genet. 71:268–277

Chyi, Y.S. and N.F. Weeden (1984) Relative isozyme band intensities permit the identification of the 2n gamete parent of triploid apple cultivars. HortScience 19:818–819

Conner, P.J., S.K. Brown, and N.F. Weeden (1997) Randomly amplified polymorphic DNA-based genetic linkage maps of three apple cultivars. J. Am. Soc. Hort. Sci. 122: 350–359

Conner, P.J., S.K. Brown, and N.F. Weeden (1998) Molecular-marker analysis of quantitative traits for growth and development in juvenile apple trees. Theor. Appl. Genet. 96:1027–1035

Costa, F., S. Stella, W.E. Van de Weg, W. Guerra, M. Cecchinel, J. Dallavia, B. Koller, and S. Sansavini (2005) Role of the genes *Md-ACO1* and *Md-ACS1* in ethylene production and shelf life of apple (*Malus domestica* Borkh). Euphytica 141: 181–190

Costes, E., P.É. Lauri, and J. L. Regnard (2006) Implications for Tree Management and Fruit Production. Horticultural Reviews, J. Janick (ed), John Wiley & Sons, 32: 1–61

Crane, M.B., R.M. Greenslade, A.M. Massee, and H.M. Tydemans (1936) Studies on the resistance and immunity of apples to the wooly apple aphid, *Eriosoma lanigerum* (Hausm.). J. Pomol. Hortic. Sci. 14: 137–163

Cummings, J.N., P.L. Forsline, and J.D. Mackenzie (1981) Woolly apple aphid colonization on *Malus* cultivars. J. Am. Soc. Hortic. Sci. 106: 26–30

Dayton, D.F. (1977) Genetic immunity to apple mildew incited by *Podosphaera leucotricha.* HortScience 12:225–226

Dirlewanger, E., P. Cosson, M. Tavaud, M. Aranzana, C. Poizat, A. Zanetto, P. Arús, and F. Laigret (2002) Development of microsatellite markers in peach [*Prunus persica* (L.) Batsch] and their use in genetic diversity analysis in peach and sweet cherry. Theor. Appl. Genet. 105: 127–138

Dunemann, F., G. Bräcker, T. Markussen, and P. Roche (1999) Identification of molecular markers for the major mildew resistance gene *Pl2* in apple. Acta Hort 484:411–416

Dunemann, F., C. Fischer, B. Merkt, M. Thiermann, and A. Urbanietz (2000) Molecular approaches for breeding apple cultivars with durable powdery mildew resistance. Durable Resistance Symposium, Wageningen, SP26

Dunemann, F., A. Urbanietz, S. Gardiner, H. Bassett, W. Legg, R. Rusholme, V. Bus, and C. Ranatunga (2004) Marker assisted selection for *Pl-1* powdery mildew resistance in apple – old markers for a new resistance gene? Acta Hort. 663:757–762

Dunemann, F., A. Peil, A. Urbanietz, and T. Garcia-Libreros (2007) Mapping of the apple powdery mildew resistance gene *Pl1* and its genetic association with an NBS-LRR candidate resistance gene. Plant Breed. 126: 476–481

Durel, C.-E., L. Parisi, F. Laurens, E. van de Weg, R. Liebhard, B. Koller, and M.F. Jourjon (2003) Genetic dissection of partial resistance against two monoconidial strains of the new race 6 of *Venturia inaequalis* in apple. Genome 46:224–234

Durham, R.E. and S.S. Korban (1994) Evidence of gene introgression in apple using RAPD markers. Euphytica 79: 109–114

Evans, K.M. and C.M. James (2003) Identification of SCAR markers linked to *Pl-w* mildew resistance in apple. Theor. Appl. Genet. 106:1178–1183

Evans, K.M., F. Fernández, F. Laurens, L. Feugey, and W.E. Van de Weg (2007) Harmonising fingerprinting protocols to allow comparisons between germplasm collections. Eucarpia XII Fruit Section Symposium, September 16–20, 2007, Zaragoza, Spain: 57–58

Gallot, J.C., R.C. Lamb, and H.S. Aldwinkle (1985) Resistance to powdery mildew from some small fruited *Malus* cultivars. HortScience 20: 1085–1087

Gao, Z.S., W.E. van de Weg, J.G. Schaart, I.M. van der Meer, L. Kodde, M. Laimer, H. Breiteneder, K. Hoffmann-Sommergruber, L.J.W.J. Gilissen (2005a) Linkage map positions and allelic diversity of two *Mal d 3* (non-specific lipid transfer protein) genes in the cultivated apple (*Malus domestica*). Theor. Appl. Genet. 110: 479–491

Gao, Z.S., W.E. van de Weg, J.G. Schaart, H.J. Schouten, D.H. Tran, L.P. Kodde, I.M. van der Meer, A.H.M. van der Geest, J. Kodde, H. Breiteneder, K. Hoffmann-Sommergruber, D. Bosch, L.J.W.J. Gilissen (2005b) Genomic cloning and linkage mapping of the *Mal d 1* (PR-10) gene family in apple (*Malus domestica*). Theor. Appl. Genet. 111: 171–183

Gao, Z.S., W.E. van de Weg, J.G. Schaart, G. van Arkel, H. Breiteneder, K. Hoffmann-Sommergruber, L.J.W.J. Gilissen (2005c) Genomic characterization and linkage mapping of the apple allergen genes *Mal d 2* (thaumatin-like protein) and *Mal d 4* (profilin). Theor. Appl. Genet. 111: 1087–1097

Gardener, R.G., J. N. Cummins, and H.S. Aldwinckle, (1980) Inheritance of fire blight resistance in *Malus* in relation to rootstock breeding. J. Am. Soc. Hortic. Sci. 105: 912–916

Gardiner, S.E., H.C.M. Bassett, D.A.M. Noiton, V.G. Bus, M.E. Hofstee, A.G. White, R.D. Ball, R.L. Forster, and E.H.A. Rikkerink (1996a) A detailed linkage map around an apple scab resistance gene demonstrates that class 3A and 3B progeny both carry the *Vf* gene. Theor. Appl. Genet. 93:485–493

Gardiner, S.E., H.C.M. Bassett, C. Madie, and D.A.M. Noiton (1996b) Isoenzyme, RAPD and RFLP markers used to deduce a putative parent for the apple cv. Braeburn. J. Amer. Soc. Hort. Sci. 121:996–1001

Gardiner, S., J. Murdoch, S. Meech, R. Rusholme, H. Bassett, M. Cook, V. Bus, E. Rikkerink, A. Gleave, R. Crowhurst, G. Ross, and I. Warrington (2002) Candidate resistance genes from an EST database prove a rich source of markers for major genes conferring resistance to important apple pests and diseases. Acta Hort. 622: 141–151

Gessler, C., A. Patocchi, S. Sansavini, S. Tartarini, and L. Gianfranceschi (2006) V*enturia inaequalis* Resistance in Apple. Critical Rev. Plant Sci. 25: 473–503

Gianfranceschi, L., B. Koller, N. Seglias, M. Kellerhals, and C. Gessler (1996) Molecular selection in apple for resistance to scab caused by *Venturia inaequalis*. Theor. Appl. Genet. 93:199–204

Gianfranceschi, L., N. Seglias, R. Tarchini, M. Komjanc, and C. Gessler (1998) Simple sequence repeats for the genetic analysis of apple. Theor. Appl. Genet. 96: 1069–1076

Gilissen, L., S. Bolhaar, C. Matos, G. Rouwendal, M. Boone, F. Krens, L. Zuidmeer, A. van Leeuwen, J. Akkerdaas, and K. Hoffmann-Sommergruber (2005) Silencing the major apple allergen *Mal d 1* by using the RNA interference approach. J. Allergy Clin Immunol 115: 364–369

Guilford, P., S. Prakash, J.M. Zhu, E. Rikkerink, S. Gardiner, H. Bassett, and R. Forster (1997) Microsatellites in *Malus* × *domestica* (apple): abundance, polymorphism and cultivar identification. Theor. Appl. Genet. 94:249–254

Gygax, M., L. Gianfranceschi, R. Liebhard, M. Kellerhals, C. Gessler, and A. Patocchi (2004) Molecular markers linked to the apple scab resistance gene *Vbj* derived from *Malus baccata jackii*. Theor. Appl. Genet. 109: 1702–1709

Han, Y., K. Gasic, F. Sun, M.L. Xu, and S.S. Korban (2007a) A gene encoding starch branching enzyme I (SBEI) in apple (*Malus* × *domestica,* Rosaceae) and its phylogenetic relationship to *Sbe* genes from other angiosperms. Mol. Phylogenet. Evol. 43: 852–863

Han, Y., E. Bendik, F. Sun, K. Gasic, and S.S. Korban (2007b) Genomic isolation of genes encoding starch branching enzyme II (SBEII) in apple: Towards characterization of evolutionary disparity in *SbeII* genes between monocots and eudicots. Planta 226:1265–1276

Han, Y., K. Gasic, and S.S. Korban (2007c) Multiple-copy cluster-type organization and evolution of genes encoding *O*-methyltransferases in apple. Genetics 176: 2625–2635

Han, Y., K. Gasic, B. Marron, J.E. Beever, and S.S. Korban (2007d) A BAC-based physical map of the apple genome. Genomics 89: 630–637

Han, Y. and S.S. Korban (2007) *Spring*: A novel family of miniature inverted-repeat transposable elements in the apple genome. Genomics 90:195–200

Harada, T., K. Matsukawa, T. Sato, R. Ishikawa, M. Niizeki, and K. Saito (1993) DNA RAPDs detect genetic variation and paternity in *Malus*. Euphytica 65:87–91

Hemmat, M., N.F. Weeden, A.G. Manganaris, and D.M. Lawson (1994) Molecular marker linkage map for apple. J. Hered. 85:4–11

Hemmat, M., N.F. Weeden, P.J. Conner, and S.K. Brown (1997) A DNA marker for columnar growth habit in apple contains a simple sequence repeat. J. Amer. Soc. Hort. Sci. 122: 347–349

Hemmat, M., N.F. Weeden, H.S. Aldwinckle, and S.K. Brown (1998) Molecular markers for the scab resistance ($V_f$) region in apple. J. Amer. Soc. Hort. Sci. 123: 992–996

Hemmat, M., S.K. Brown, and N.F. Weeden (2002) Tagging and mapping scab resistance genes from R12740-7A apple. J. Amer. Soc. Hort. Sci. 127:365–370

Hemmat, M., S.K. Brown, H.S. Aldwinckle, and N.F. Weeden (2003) Identification and mapping of markers for resistance to apple scab from 'Antonovka' and 'Hansen's baccata #2'. Acta Hort. 622:153–161

Hokanson S.C., A.K. Szewc-McFadden, W.F. Lamboy, and J.R. McFerson (1998) Microsatellite (SSR) markers reveal genetic identities, genetic diversity and relationships in a *Malus* × *domestica* Borkh. core subset collection. Theor. Appl. Genet. 97: 671–683

Hokanson S.C., W.F. Lamboy, A.K. Szewc-McFadden, and J.R. McFerson (2001) Microsatellite (SSR) variation in a collection of Malus (apple) species and hybrids. Euphytica, 118: 281–294

Howad, W., T. Yamamoto, E. Dirlewanger, R. Testolin, P. Cosson, G. Cipriani, A.J. Monforte, L. Georgi, A.G. Abbott,and P. Arús (2005) Mapping With a Few Plants: Using Selective Mapping for Microsatellite Saturation of the Prunus Reference. Genetics 171: 1305–1309

Huaracha, E.M., M.L. Xu, K. Gasic, E. Pauwels, A. Van den Putte, J.W. Keulemans, and S.S. Korban (2004) Phenotypic reaction and genetic analysis using AFLP-derived SCARs for resistance to apple scab. J. Phytopathol. 152: 260–266

Ishikawa, S., S. Kato, S. Imakawa, T. Mikami, and Y. Shimamoto (1992) Organelle DNA polymorphism in apple cultivars and rootstocks. Theor. Appl. Genet. 83: 963–967

James, C.M., J.B. Clarke, and K.M. Evans (2004) Identification of molecular markers linked to the mildew resistance gene *Pl-d* in apple. Theor. Appl. Genet. 110:175–181

Janick, J., J.N. Cummins, S.K. Brown, and M. Hemmat (1996) Apples. In: Janick, J. and J.N. Moore (eds) Fruit Breeding: Tree and Tropical Fruits. Vol I. John Wiley, New York, USA, pp 1–77

Kellerhals, M., E. Dolega, B. Koller, and C. Gessler (2000a) Advances in marker-assisted apple breeding. Acta Hort. 538: 535–540

Kellerhals, M., L. Gianfranceschi, N. Seglias, and C. Gessler (2000b) Marker-assisted selection in apple breeding. Acta Hort. 521:255–265

Kenis K., and J. Keulemans (2004) QTL analysis of growth characteristics in apple. Acta Hortic. 663: 369–374

Kenis, K., and J. Keulemans (2005) Genetic linkage maps of two apple cultivars *(Malus × domestica* Borkh.) based on AFLP and microsatellite markers. Mol. Breed. 15:205–219

Kenis, K., and J. Keulemans (2007) Study of tree architecture of apple (*Malus × domestica* Borkh.) by QTL analysis of growth traits. Mol. Breed. 19: 193–208

Khan, M.A., B. Duffy, C. Gessler. and A. Patocchi (2006) QTL mapping of fire blight resistance in apple. Mol. Breed. 17: 299–306

Khan, M.A., C.E. Durel, B. Duffy, D. Drouet, M. Kellerhals, C. Gessler, and A. Patocchi, (2007) Development of molecular markers linked to the 'Fiesta' linkage group 7 major QTL for fire blight resistance and their application for marker-assisted selection. Genome 50: 568–577

Knight, R.L. and F.H. Alston (1968) Sources of field immunity to mildew *(Podosphaera leucotricha)* in apple. Can. J. Genet. Cytol. 10:294–298

Knight, R.L., J.B. Briggs, A.M. Massee. and H.M. Tydeman (1962) The inheritance of resistance to woolly aphid, *Eriosoma lanigerum* (Hsmnn.), in the apple. J. Hort. Sci. 37:207–218

Koller, B., A. Lehmann, J.M. McDermott, and C. Gessler (1993) Identification of apple cultivars using RAPD markers. Theor. Appl. Genet. 85: 901–904

Koller, B., L. Gianfranceschi, N. Seglias, J. McDermott, and C. Gessler (1994) DNA markers linked to *Malus floribunda* 821 scab resistance. Plant Mol. Biol. 26:597–602

Korban, S.S. and D.F. Dayton (1983) Evaluation of *Malus* germplasm for resistance to powdery mildew. HortScience 18:219–220

Lawson, D.M., M. Hemmat, and N. Weeden (1995) The use of molecular markers to analyze the inheritance of morphological and developmental traits in apple. J. Am. Soc. Hort. Sci. 120: 532–537

Lesemann, S., A. Urbanietz. and F. Dunemann (2004) Determining population variation of apple powdery mildew at the molecular level. Acta Hort. 663: 199–203

Lespinasse, Y. (1983) Amélioration du pommier pur la résistance à l'oïdium (*Podosphaera leucotricha*): premiers résultats concernant la virulence du champignon. IOBC/WPRS Bull. 6: 96–110

Lespinasse, Y. (1989) Breeding pome fruits with stable resistance to disease. Genes, resistance mechanisms, present work and prospects. IOBC Bullettin 2: 100–115

Lespinasse, Y., J.M. Olivier, J.M. Lespinasse, and M. Le Lezec (1985) 'Florina- Quérina'®: la résistance du pommier à la tavelure. Arboriculture fruitière 378: 43–47

Liebhard, R., L. Gianfranceschi, B. Koller, C.D. Ryder, R. Tarchini, E. Van De Weg, and C. Gessler (2002) Development and characterization of 140 new microsatellites in apple (*Malus* × *domestica* Borkh.). Mol. Breed. 10:217–241

Liebhard, R., B. Koller, A. Patocchi, M. Kellerhals, W. Pfammatter, M. Jermini, and C. Gessler (2003a) Mapping quantitative field resistance against apple scab in a Fiesta' × 'Discovery' progeny. Phytopathology 93: 493–501

Liebhard, R., M. Kellerhals, W. Pfammatter, M. Jermini, and C. Gessler (2003b) Mapping quantitative physiological traits in apple (*Malus* × *domestica* Borkh.). Plant Mol. Biol. 52: 511–526

Magein, H., and D. Leurquin (2000) Changes in amylose, amylopectin and total starch content in Jonagold apple fruit during growth and maturation. Acta Hortic. 517: 487–494

Maliepaard, C., F.H. Alston, G. Van Arkel, L.M. Brown, E. Chevreau, F. Dunemann, K.M. Evans, S. Gardiner, P. Guilford, A.W. van Heusden, J. Janse, F. Laurens, J.R. Lynn, A.G. Manganaris, A.P.M. den Nijs, N. Periam, E. Rikkerink, P. Roche, C. Ryder, S. Sansavini, H. Schmnidt, S. Tartarini, J.J. Verhaegh, M. Vrielink-van Ginkel, and G.J. King (1998) Aligning male and female linkage maps of apple (*Malus pumila* Mil.) using multi-allelic markers. Theor. Appl. Genet. 97: 60–73

Malnoy, M., M.L. Xu, E. Borejsza-Wysocka, S.S. Korban, and H.S. Aldwinckle. 2008. Two receptor-like genes, *Vfa1* and *Vfa2*, confer resistance to the fungal pathogen *Venturia inaequalis* inciting apple scab disease. Mol. Plant-Microbe Inter. 21: 448–458

Manganaris, A.G. (1989) Isoenzymes as genetic markers in apple breeding. PhD thesis, University of London, London, UK

Markussen, T., J. Kruger, H. Schmidt. and F. Dunemann (1995) Identification of PCR-based markers linked to the powdery-mildew-resistance gene *Pl1* from *Malus robusta* in cultivated apple. Plant Breed. 114:530–534

Menendez, R.A., F.E. Larsen, and R. Fritts Jr. (1986) Protein and isozyme electrophoresis and isoelectric focusing for the characterization of apple clones. Scient. Hort. 29: 211–220

Minarro, M., and E. Dapena (2004) Inheritance of tolerance to the rosy apple aphid of the cv. 'Florina'. Acta Hort. 663:261–264

Mulcahy, D.L., M. Cresti, S. Sansavini, G.C. Douglas, H.F. Liskens, G. Bergamini Mulcahy, R. Vignani and M. Pancaldi (1993) The use of random amplified polymorphic DNAs to fingerprint apple genotypes. Scient. Hort. 54: 89–96

Naik, S., C. Hampson, K. Gasic, G. Bakkeren, and S.S. Korban. 2006. Development and linkage mapping of E-STS and RGAs for functional gene homologues in apple. Genome 49: 959–968

Nosarzewski, M. and D.D. Archbold (2007) Tissue-specific expression of SORBITOL DEHYDROGENASE in apple fruit during early development . J. Exp. Bot. 174: 43–78

Nybom, H. (1990a) DNA fingerprints in sports of 'Red Delicious' apples. HortSci. 25: 1641–1642
Nybom, H. (1990b) Genetic variation in ornamental apple trees and their seedlings (*Malus*, Rosaceae) revealed by DNA 'fingerprinting' with the M13 repeat probe. Hereditas 113: 17–28
Nybom, H. and B.A. Schaal (1990) DNA 'fingerprints' applied to paternity analysis in apples (*Malus* × *domestica*). Theor. Appl. Genet. 79:763–768
Nybom, H., S. Gardiner, and C.J. Simon (1992) RFLPs obtained from an rDNA probe and detected with enhanced chemiluminescence in apples. HortScience 27: 355–356
Patocchi A., B. Bigler, B. Koller, M. Kellerhals, and C. Gessler (2004) *Vr2*: a new apple scab resistance gene. Theor. Appl. Genet. 109: 1087–1092
Patocchi A., M. Walser, S. Tartarini, G.A.L. Broggini, F. Gennari, S. Sansavini, and C. Gessler (2005) Identification by genome scanning approach (GSA) of a microsatellite tightly associated with the apple scab resistance gene *Vm*. Genome 48:630–636
Peil, A., T. Garcia-Libreros, K. Richter, F.C. Trognitz, B. Trognitz, M.V. Hanke, and H. Flachowsky (2007) Strong evidence for fire blight resistance gene of *Malus robusta* located on linkage group 3. Plant Breed. 126: 470–475
Pierantoni, L., K-H. Cho, I-S. Shin, R. Chiodini, S. Tartarini, L. Dondini, S-J. Kang, and S. Sansavini (2004) Characterisation and transferability of apple SSRs to two European pear F1 populations. Theor. Appl. Genet. 109:1519–1524
Qubbaj, T., A. Reineke, and C.P.W. Zebitz (2005) Molecular interactions between rosy apple aphids, *Dysaphis plantaginea*, and resistant and susceptible cultivars of its primary host *Malus domestica*. Entomol. Experiment. Applicata 115: 145–152
Reed, G. (1975) Enzymes inhabitation and activation. In G. Reed (Ed.), Enzymes in Food Processing (2nd ed., pp. 48–50). London: Academic Press
Regnard, J.L., V. Segura, N. Merveille, C.E. Durel, and E. Costes (2007) QTL analysis for leaf gas exchange in an apple progeny grown under atmospheric constraints. Eucarpia XII Fruit Section Symposium, September 16–20, 2007, Zaragoza, Spain: 36
Roche, P., F.H. Alston, C. Maliepaard, K.M. Evans, R. Vrielink, F. Dunemann, T. Markussen, S. Tartarini, L.M. Brown, C. Ryder, and G.J. King (1997a) RFLP and RAPD markers linked to the rosy leaf curling aphid resistance gene (*Sd1*) in apple. Theor. Appl. Genet. 94:528–533
Roche, P., G. van Arkel, and A.W. van Heusden (1997b) A specific PCR assay for resistance to biotypes 1 and 2 of the rosy leaf curling aphid in apple based on an RFLP marker closely linked to the *Sd1* gene. Plant Breed. 116: 567–572
Rushlomé-Pilcher, R.L., J-M. Celton, S.E. Gardiner, D.S. Tustin (2008) Genetic markers linked to the dwarfing trait of apple rootstock 'Malling 9'. J. Amer. Soc. Hort. Sci. 133: 100–106
Sandanayaka, W.R.M., V.G.M. Bus, P. Connolly, and R. Newcomb (2003) Characteristics associated with woolly apple aphid, *Eriosoma lanigerum*, resistance of three apple rootstocks. Entomol. Exp. Appl. 109: 63–72
Seglias, N.P., and C. Gessler (1997) Genetics of apple powdery mildew resistance from *Malus zumi* (*Pl2*). IOBC/WPRS Bull 20:195–208
Segura, V., C. Cilas, F. Laurens, and E. Costes (2006) Phenotyping progenies for complex architectural traits: a strategy for 1-year-old apple trees (*Malus* × *domestica* Borkh.) Tree Genet. Genomes 2: 140–151
Segura, V., C. Denancé, C.-E. Durel, and E. Costes (2007) Wide range QTL analysis for complex architectural traits in a 1-year-old apple progeny. Genome, 50: 159–171
Service, R.F. (2002) Tapping DNA for structures produces a trickle. Science 298: 948–950
Silfverberg-Dilworth, E., C.L. Matasci, W.E. Van de Weg, M.P.W. Van Kaauwen, M. Walser, L.P. Kodde, V. Soglio, L. Gianfranceschi, C.E. Durel, F. Costa, T. Yamamoto, B. Koller, C. Gessler, and A. Patocchi (2006) Microsatellite markers spanning the apple (Malus × *domestica* Borkh.) genome. Tree Genet. Genomes 2:202–224
Stankiewicz, M., E. Pitera, and S.W. Gawronski (2001) Genetic analysis of a polish apple selection U211 as a new source of high resistance to apple powdery mildew. Acta Hort. 546: 641–644
Stankiewicz-Kosyl, M., E. Pitera, and S.W. Gawronski (2005) Mapping QTL involved in powdery mildew resistance of the apple clone U 211. Plant Breed. 124:63–66

Stankiewicz-Kosyl, M., A. Novicka, P. Krajewski, K.Tomala, Soska A., F. Laurens, C. Govan, M. Lateur, F. Costa, S. Tartarini, W. Guerra, M. Lewandowski, K. Rutkowski, E. Zurawicz, L. Gianfranceschi, C.E. Durel, F. Mathis, E. Barbaro, D. Mott, A. Patocchi, D. Gobbin, F. Fernandez, K. Evans, F. Dunemann, A. Boudichevskaja, J. Jansen, and E. van de Weg (2007) QTL analysis of acidity in apple using pedigree-based approach. Eucarpia XII Fruit Section Symposium, September 16–20, 2007, Zaragoza, Spain: 60–61

Tartarini, S. (1996) RAPD markers linked to the *Vf* gene for scab resistance in apple. Theor. Appl. Genet. 92: 803–810

Tartarini, S., L. Gianfranceschi, S. Sansavini, C. Gessler (1999) Development of reliable PCR markers for the selection of the *Vf* gene conferring scab resistance in apple. Plant Breed. 118: 183–186

Tartarini, S., S. Sansavini, B. Vinatzer, F. Gennari, and C. Domizi (2000) Efficiency of marker assisted selection (MAS) for the *Vf* scab resistance gene. Acta Hort. 538: 549–552

Tian, Y.-K., C.-H. Wang, J.-S. Zhang, C. James, and H.-Y. Dai (2005) Mapping Co, a gene controlling the columnar phenotype of apple, with molecular markers. Euphytica 145: 181–188

Tobutt, K.R., R. Boskovic, and P. Roche (2000) Incompatibility and resistance to woolly apple aphid in apple. Plant Breed. 119: 65–69

Urbanietz, A., and F. Dunemann (2005) Isolation, identification and molecular characterization of physiological races of apple powdery mildew (*Podosphaera leucothrica*). Plant Pathol. 54: 125–133

Van der Linden, C.G., D.C.A.E. Wouters, V. Mihalka, E.Z. Kochieva, M.J.M. Smulders, and B. Vosman (2004) Efficient targeting of plant disease resistance loci using NBS profiling. Theor. Appl. Genet. 109:384–393

Van de Weg, W.E., R.E. Voorrips, R. Finkers, L.P. Kodde, J. Jansen, and M.C.A.M. Bink (2004) Pedigree genotyping: a new pedigree-based approach of QTL identification and allele mining. Acta Hortic. 663: 45–50

Van de Weg, E., H. Jansen, M. Bink, R.E. Voorrips, C.E. Durel, F. Laurens, F. Dunemann, K. Evans, A. Patocchi, W. Guerra, M. Komjanc, M. Lateur, M. Kellerhals, C. Ryder, S. Sansavini, K.Tomala, E. Zurawicz, and L. Gianfranceschi (2007) QTL mapping in multiple, pedigreed populations: the concept and the framework of the statistical procedures. Eucarpia XII Fruit Section Symposium, September 16–20, 2007, Zaragoza, Spain: 57–58

Van Dyk, M.M., J. Baison, P. Hove, C. Marondedze, A. Klein, K. Soeker, and D.J.G. Rees (2007) BIN mapping of EST-SSRs and EST-SNPs in apple (*Malus* × *domestica* Borkh.). Eucarpia XII Fruit Section Symposium, September 16–20, 2007, Zaragoza, Spain: 57–58

Venisse, J.S., M. Malnoy, M. Faize, J.P. Paulin, and M.N. Brisset (2002) Modulation of defense responses of *Malus* spp. during compatible and incompatible interactions with *Erwinia amylovora*. MPMI 15: 1204–1212

Venturi, S., L. Dondini, P. Donini, and S. Sansavini (2006) Retrotransposon characterisation and fingerprinting of apple clones by S-SAP markers. Theor. Appl. Genet. 112: 440–444

Visser, T., and J.J. Verhaegh (1976) Review of tree fruit breeding carried out at the Institute for Horticultural Plant Breeding at Wageningen from 1951–1976. Proc. Eucarpia Meeting of Tree Fruit Breeding, Wageningen, The Netherlands, pp 113–132

Vinatzer, B.A., H.B. Zhang, and S. Sansavini (1998) Construction and characterization of a bacterial artificial chromosome library of apple. Theor. Appl. Genet. 97: 1183–1190

Vinatzer B.A., A. Patocchi, L. Gianfranceschi, S. Tartarini, H. Zhang, C. Gessler, and S. Sansavini (2001) Apple contains receptor-like genes homologous to the *Cladosporium fulvum* resistance gene family of tomato with a cluster of genes cosegregating with *Vf* apple scab resistance. Mol. Plant-Microbe Inter. 14:508–515

Vinatzer, B.A., A. Patocchi, S. Tartarini, L. Gianfranceschi, S. Sansavini, and C. Gessler (2004) Isolation of two microsatellite markers from BAC clones of the *Vf* scab resistance region and molecular characterization of scab-resistant accessions in *Malus* germplasm. Plant Breed. 123: 321–326

Watillon, B., P. Druart, P. Du Jardin, R. Kettmann, P. Boxus, and A. Burny (1991) Use of random cDNA probes to detect restriction fragment length polymorphisms among apple clones. Scient. Hort. 46: 235–243

Weeden, N.F. and R.C. Lamb (1985) Identification of apple cultivars by isoenzyme phenotypes. J. Amer. Soc. Hort. Sci. 110:509–515

Xu, M.L. and S.S. Korban (2000) Saturation mapping of the apple scab resistance gene $V_f$ using AFLP markers. Theor. Appl. Genet. 101: 844–851

Xu M.L., and Korban S.S (2002) A cluster of four receptor-like genes resides in the *Vf* locus that confers resistance to apple scab disease. Genetics 162: 1995–2006

Xu, M., S.S. Korban, J. Song, and J. Jiang (2002) Constructing a bacterial artificial chromosome library of the apple cultivar GoldRush. Acta Hort. 595: 103–112

Xu, M., J. Song, Z. Cheng, J. Jiang, S.S. Korban (2001) A bacterial artificial chromosome (BAC) library of *Malus floribunda* 821 and contig construction for positional cloning of the apple scab resistance gene *Vf*. Genome 44:1104–1113

Yamamoto, T., T. Kimura, Y. Sawamura, K. Kotobuki, Y. Ban, T. Hayashi and N. Matsuta (2001) SSRs isolated from apple can identify polymorphism and genetic diversity in pear. Theor. Appl. Genet. 102:865–870

Yamamoto, T., T. Kimura, M. Shoda, T. Imai, T. Saito, Y. Sawamura, K. Kotobuki, T. Hayashi and N. Matsuta (2002) Genetic linkage maps constructed by using an interspecific cross between Japanese and European pears. Theor. Appl. Genet. 106: 9–18

Yamamoto, T., T. Kimura, J. Soejima, T. Sanada, Y. Ban, and T. Hayashi (2004) Identification of quince varieties using SSR markers developed from pear and apple. Breed. Sci. 54: 239–244

Yao, Y.-X., M. Li , Z. Liu , Y.-J. Hao , H. Zha (2007) A novel gene, screened by cDNA-AFLP approach, contributes to lowering the acidity of fruit in apple. Plant Physiol. Biochem. 173: 44–54

Yang, H. and S.S. Korban (1996) Screening apples for OPD20/600 using sequence-specific primers. Theor. Appl. Genet. 92:263–266

Yang, H.Y., S.S. Korban, J. Krüger, and H. Schmidt (1997a) The use of a modified bulk segregant analysis to identify a molecular marker linked to a scab resistance gene in apple. Euphytica 94: 175–182

Yang, H.Y., S.S. Korban, J. Krüger, and H. Schmidt (1997b) A randomly amplified polymorphic DNA (RAPD) marker tightly linked to the scab resistance gene $V_f$ in apple. J. Amer. Soc. Hort. Sci. 122: 47–52

Yang, H.Y. and J. Krüger (1994) Identification of an RAPD markers linked to the $V_f$ gene for scab resistance in apples. Plant Breed. 112:323–329

# 6. Apple Functional Genomics

**Andrew C. Allan, Ross Crowhurst, Andrew Gleave, Richard Newcomb, and Robert Schaffer**

## 1 *Malus* EST Datasets

Publicly available databases currently contain in excess of 250,000 *Malus* sequences (Park et al., 2006), the majority being derived from large-scale sequencing efforts of cDNA libraries from Illinois at Washington University, United States and Plant and Food Research, New Zealand (Korban et al., 2004, Newcomb et al., 2006). These *Malus* cDNA libraries have been sequenced to varying depths (Table 1), depending on library quality and novelty, to generate expressed sequence tags (ESTs). *Malus* cDNA libraries have been generated from material derived from numerous apple genotypes, many of which are cultivars of commercial significance including 'Braeburn', 'Elstar', 'Fuji', 'Golden Delicious', 'Goldrush', 'Granny Smith', 'Holsteiner Cox', 'Red Delicious', 'Royal Gala', 'Pinkie', 'Scirose/Pacific Rose™' and a number of dwarfing rootstocks. In addition to being derived from a range of genotypes, the *Malus* cDNA libraries originate from a wide variety of different tissues and developmental time points. For example, libraries have been generated from a staged series of developing and ripening 'Royal Gala' fruit, including flower, whole fruit, fruit cortex, skin, and seed samples (Newcomb et al., 2006). Such a series is a valuable resource of genes for experiments aimed at understanding important processes and transformations in fruit development, such as early cell proliferation, cell expansion, and ripening. This series is also of value in identification of genes encoding enzymes and transcription factors involved in the biosynthesis of health and flavor compounds from apple fruit. Other plant tissues sampled include buds, shoots, leaves, roots, phloem, and xylem. As many genes are only expressed in response to external effects, cDNA libraries have also been constructed from tissues, plants, and cell lines that were subjected to abiotic (e.g. fruit stored at high or low temperature

A.C. Allan (✉)
The New Zealand Institute for Plant and Food Research Limited, Mt Albert Research Centre, Private Bag 92169, 1142 Auckland, New Zealand

K.M. Folta, S.E. Gardiner (eds.), *Genetics and Genomics of Rosaceae*, Plant Genetics and Genomics: Crops and Models 6, DOI 10.1007/978-0-387-77491-6_6,

**Table 1** The degree of sequencing of publicly available *Malus* cDNA libraries

| Species | Code | Cultivar | Tissue | Developmental stage | Number of sequences | Number of singletons |
|---|---|---|---|---|---|---|
| *Malus* | ABIA | Unassigned sequences | | | 981 | 192 |
| *Malus angustifolia* | AGAA | | | | 1 | 0 |
| *Malus asiatica* | AGAB | | | | 3 | 0 |
| *Malus baccata* | AGAC | | Leaves | | 11 | 5 |
| *Malus coronaria* | AGAD | | | | 2 | 0 |
| | AGAE | | | | 2 | 0 |
| *Malus doumeri* | AGAG | | | | 2 | 0 |
| *Malus florentina* | AGAH | | | | 3 | 0 |
| *Malus floribunda* | AGAI | | | | 6 | 0 |
| | ABGA | | Leaf | | 141 | 20 |
| *Malus fusca* | AGAJ | | | | 9 | 0 |
| *Malus halliana* | AGAK | | | | 1 | 0 |
| *Malus honanensis* | AGAL | | | | 1 | 0 |
| *Malus hupehensis* | AGAM | | | | 3 | 0 |
| *Malus ioensis* | AGAN | | | | 3 | 0 |
| *Malus kansuensis* | AGAO | | | | 2 | 0 |
| *Malus mandshurica* | AGAP | | | | 2 | 0 |
| *Malus micromalus* | AGAQ | | | | 3 | 0 |
| *Malus niedzwetzkyana* | AGAR | | | | 4 | 0 |
| *Malus ombrophilia* | AGAS | | | | 1 | 0 |
| *Malus orientalis* | AGAT | | | | 4 | 0 |
| *Malus prattii* | AGAU | | | | 3 | 0 |
| *Malus prunifolia* | AGAV | | | | 18 | 13 |
| *Malus pumila* | AGAW | | | | 5 | 3 |
| *Malus sargentii* | AGAX | | | | 1 | 0 |
| *Malus sieboldii* | AGAY | | | | 4 | 0 |

**Table 1** (continued)

| Species | Code | Cultivar | Tissue | Developmental stage | Number of sequences | Number of singletons |
|---|---|---|---|---|---|---|
| *Malus sieboldii* | ABOA | Jack Humm | Fruit | Mature | 71 | 4 |
| | ABOC | Pomme Grise | Fruit | Mature | 1,138 | 116 |
| *Malus sieversii* | AGAZ | | | | 19 | 1 |
| *Malus sp.* | AGBA | | | | 8 | 0 |
| *Malus sylvestris* | AGBB | | | | 9 | 0 |
| *Malus toringoides* | AGBC | | | | 3 | 0 |
| *Malus transitoria* | AGBD | | Stylar | | 4 | 0 |
| *Malus trilobata* | AGBE | | | | 4 | 0 |
| *Malus tschonoskii* | AGBF | | | | 4 | 0 |
| *Malus × domestica* | AGBG | | | | 12 | 0 |
| | AGZZ | | | | 481 | 57 |
| | AGBX | | Fruit | | 133 | 14 |
| | ABFA | A172 | Leaf | | 94 | 13 |
| | AEAA | Aotea | Leaf | Expanding | 1,670 | 121 |
| | ABLA | Braeburn | Cell cultures | 3d post subculture | 12 | 12 |
| | ABLB | | | 3d post subculture | 105 | 1 |
| | ABLC | | | 3d post subculture | 4,705 | 395 |
| | ABNA | | Cultured fruit cells | | 45 | 10 |
| | ABNB | | Fruit cells | | 943 | 118 |
| | AGBP | Elstar | leaves | Young leaves | 30 | 8 |
| | AGCK | Evereste × MM106 | Leaves | | 91 | 36 |
| | AGBY | Fiesta | | | 78 | 53 |
| | AGBL | Florina | | One year old trees | 9 | 4 |
| | AGCJ | | Leaves | One year old trees | 17 | 7 |

**Table 1** (continued)

| Species | Code | Cultivar | Tissue | Developmental stage | Number of sequences | Number of singletons |
|---|---|---|---|---|---|---|
| | AGBZ | 'Freedom' | | | 27 | 15 |
| | AGBJ | 'Fuji' | Shoot | Shoots maintained in *in vitro* culture for 2 years | 448 | 65 |
| | AGBH | Gala | Hypanthium tissue | Immature | 200 | 60 |
| | AGCC | Geneva 3041 × *Malus sieversii* apple rootstock | Root | Phytophtora challenged Young root | 3,944 | 766 |
| | AGCD | 'Golden Delicious' | Root | | 162 | 31 |
| | AGCE | | Stem | | 17 | 3 |
| | AGBR | GoldRush | Composite bud | Bud, balloon, open and after pollination | 22,217 | 5,229 |
| | AGBM | | Composite bud | Dormant terminal bud, Dormant lateral bud, active lateral bud | 7,065 | 1,700 |
| | AGBN | | Composite fruit | Young fruitlet(<1 cm), Young fruitlet (1 cm dia.), Young fruitlet (12 cm dia.), Maturing fruit I, Maturing fruit II, Mature fruit | 26,912 | 3,347 |
| | AGBW | | Dormant shoot internodes | Dormant and active shoots | 7,155 | 1,517 |

**Table 1** (continued)

| Species | Code | Cultivar | Tissue | Developmental stage | Number of sequences | Number of singletons |
|---|---|---|---|---|---|---|
| | AGBV | | Leaves | Stage I | 2,847 | 549 |
| | AGBU | | | Stage II | 3,892 | 763 |
| | AGBS | | | Stage III | 3,620 | 653 |
| | AGBT | | | Stage III | 3,399 | 671 |
| | AGCF | | Leaves *V. inaequalis* challenged | | 3,894 | 337 |
| | AGCH | | Leaves *C. rosaceana* challenged | | 3,913 | 487 |
| | AGBO | | Young Fruitlet | Young Fruitlet < 1 cm | 7,730 | 1,358 |
| | AGBI | Granny Smith | Fruit cortical tissue | Mature | 1 | 0 |
| | AGBK | Holsteiner Cox | Leaves | Young leaves | 48 | 4 |
| | AAZA | M9 | Xylem | Mature wood | 4,982 | 757 |
| | AGCI | | Root | Young root | 2,107 | 436 |
| | ABPA | | Root tips (distal 1.5 cm) | | 38 | 0 |
| | ABPB | | | | 4,792 | 566 |
| | ABMA | | Shavings of phloem tissue | | 4,896 | 355 |
| | AENA | Northern Spy | Leaf | Expanding | 1,232 | 217 |
| | ABAA | Scirose/Pacific Rose™ | Spur buds ON trees | From ON trees | 791 | 105 |
| | ABAB | | | From ON trees | 485 | 51 |
| | ABBA | | Spur buds OFF trees | From OFF trees | 1,074 | 110 |
| | ABBB | | | From OFF trees | 496 | 55 |
| | AAGA | Pinkie | Leaf | Expanding | 532 | 78 |
| | AAHA | | | Expanding | 104 | 6 |
| | AAIA | | | Expanding | 115 | 19 |
| | AAJA | | | Expanding | 73 | 8 |
| | AAKA | | | Expanding | 521 | 88 |

**Table 1** (continued)

| Species | Code | Cultivar | Tissue | Developmental stage | Number of sequences | Number of singletons |
|---|---|---|---|---|---|---|
| | AEPA | | | Expanding | 4,832 | 607 |
| | ABHA | | | | 117 | 18 |
| | AGCG | 'Red Delicious' | Leaves fire blight challenged | | 3,900 | 476 |
| | AGBQ | 'Remo' | Leaf | Young leaves | 58 | 9 |
| | AAMA | 'Royal Gala' | Floral bud | Early, flower differentiation just starting | 989 | 218 |
| | AFBA | | | Pre-opening | 60 | 12 |
| | AFBB | | | Pre-opening | 44 | 11 |
| | AFBC | | | Pre-opening | 5,026 | 743 |
| | AAUA | | Fruit | Fruit cortex 87 days after full bloom | 2,461 | 81 |
| | AABA | | Fruit core | 126 days after full bloom | 1 | 1 |
| | AAXA | | | 126 days after full bloom | 5,075 | 346 |
| | AAYA | | Fruit cortex | 126 days after full bloom | 4,477 | 370 |
| | AALA | | | 150 days after full bloom | 5,317 | 852 |
| | AALB | | | 150 days after full bloom | 1,287 | 271 |
| | AAVB | | | Mature | 121 | 0 |
| | ABCA | | Fruit cortex and skin | Mature | 4,687 | 320 |
| | ABDA | | | Mature | 4,785 | 112 |
| | AAFA | | Fruit skin peel | 150 days after full bloom | 6,156 | 633 |
| | AAFB | | | 150 days after full bloom | 1,573 | 422 |
| | AARA | | Leaf | Partially senescing leaf | 8,837 | 1,778 |
| | AANA | | | Partially senescing leaves | 1,888 | 398 |
| | AELA | | | Young, expanding | 5,453 | 984 |
| | ABKA | | | Leaves less than six months | 1,073 | 87 |
| | ABEA | | | Seven-week-old seedlings | 4,813 | 560 |

**Table 1** (continued)

| Species | Code | Cultivar | Tissue | Developmental stage | Number of sequences | Number of singletons |
|---|---|---|---|---|---|---|
| | ABEB | | | Seven-week-old seedlings | 561 | 156 |
| | AAOA | | Phloem | Scrapings from inside of bark of mature wood | 4,512 | 1,159 |
| | AAWA | | Seed | 59 days after full bloom | 5,404 | 506 |
| | ASYA | | | 59 days after full bloom | 74 | 14 |
| | AACA | | Vegetative bud | Pre-opening | 52 | 11 |
| | AADA | | | Pre-opening | 33 | 12 |
| | AAEA | | | Pre-opening | 54 | 24 |
| | AVBB | | | Pre-opening | 5,573 | 822 |
| | AVBC | | | Vegetative bud, pre-opening | 17,940 | 1,352 |
| | ABQA | | Whole flowers | 0 DAFB | 924 | 57 |
| | AASA | | Young fruit | 10 days after full bloom | 4,861 | 504 |
| | AASB | | | 10 days after full bloom | 4,235 | 174 |
| | AASC | | | 10 days after full bloom | 405 | 20 |
| | AYFA | | | 10 days after full bloom | 8 | 5 |
| | AYFB | | | 10 days after full bloom | 627 | 174 |
| | AAPA | | | 24 days after full bloom | 2,647 | 596 |
| | AOFA | | | 24 days after full bloom | 1,076 | 272 |
| | AAAA | | | 59 days after full bloom, seeds removed | 7,026 | 1,421 |
| *Malus domestica* × *Pyrus communis* | AGAF | | Leaves | Mature | 1 | 0 |
| *Malus xiaojinensis* | AGCA | | | | 6 | 1 |
| *Malus yunnanensis* | AGCB | | | | 4 | 0 |

and/or under altered atmospheric conditions), and biotic stresses, e.g. infection with the causal agents of some of the most significant bacterial and fungal diseases to affect apple cultivars including *Erwinia amylovra* (fire blight), *Venturia inaequalis* (apple scab or black spot) and *Phytophthora sp.* (root rot).

For the 259,683 *Malus* sequences available, the average edited length is 478 bases. Clustering of the sequences using The Institute for Genomic Research (TIGR) gene indices clustering tools (http://www.tigr.org/tdb/tgi/software) set at a 95% threshold yields 30,393 tentative consensus (TC) sequences with 38,206 sequences remaining unclustered (singletons). TC sequences range in length from 61 to 9,332 bases with an average of 789 bases, whereas singletons range in size from 50 to 2,243 bases with an average of 403 bases. Together, the TC sequences and singletons yield an apple EST dataset of 68,599 sequences. Hereafter, the singletons and TC sequences are referred to collectively as the non-redundant (NR) gene set. The proportion of singletons compared with the total number of ESTs provides a measure of the overall contribution of the library to the dataset. No single library contains more than 13.7% (AGBR) of the total number of singletons. Two libraries (AGBR and AGBN) contributed 22.5% of singletons while nine libraries (AGBR, AGBN, AARA, AGBM, AGBW, AAAA, AGBO, AVBC, AAOA) contributed 49.4% of singletons. These figures hide the fact that both AGBR and AGBN represent composite libraries produced by pooling mRNA from several developmental stages. Consequently, much of the diversity is derived by sequencing different sources of tissue. The availability of EST sequences from a wide variety of different genotypes has also proved a useful tool for identifying new genes. An example of this is provided by a comparison of the NR clusters shared between two of the larger libraries, AELA from 'Royal Gala' leaf (represented within 2,120 NR sequences) and AEPA from 'Pinkie' leaf (represented within 2,778 NR sequences). These two libraries share only 20 NR sequences, which comprises 0.41% of the total NR sequences represented in the combined dataset of the two libraries. Similarly, a comparison between the smaller AEAA Aotea expanding leaf library (contributing to 437 NR sequences) and AELA shows only 7 common NR sequences. However, 86 common NR sequences are shared with AEPA. These differences are not solely due to genotype-specific expression profiles, but also will include differences introduced by the two separate cloning procedures involved with making the libraries.

Tissues where further sequencing would be useful can be predicted by examining the ratio of singletons relative to total sequences within a library, for libraries with a low to moderate existing sequencing depth (500 to several thousand sequences). Ranking by these criteria suggest that libraries such as ABEB (0.278 ratio), AYFB (0.277), AAFB (0.268), AAOA (0.257), and AGBM (0.241) could be targeted for sampling for further genes from apple. The 68,599 NR sequences can be expected to be an overestimate of the number of protein-coding transcripts represented in apple, since TC and singleton sequences representing the same gene may not overlap, and distinct TCs or singletons may represent mRNAs that are alternatively transcribed or processed, but originate from the same gene. In addition, given that the EST dataset is derived from sequences from numerous apple cultivars and cultivated apple is highly heterozygous (Chevreau et al., 1985), it is likely that at the 95% stringency level used in the cluster analysis of the dataset, there are numerous instances where the degree of polymorphisms between distinct alleles of the same gene results in them being assigned to different NRs. It is likely that additional sequencing, both of the cDNAs sampled and novel cDNAs from apple, would reduce this number of NR sequences.

Other EST projects undertaken in fruit crops that are of a significant size in terms of total number of ESTs collected have reported lower numbers of NR sequences. For example, a study of 152,635 tomato ESTs produced 31,012 NR sequences (Fei et al., 2004), whereas a collection of 146,075 grape ESTs gave rise to 25,746 NR sequences (Goes da Silva et al., 2005). This is likely to be the result of the lower clustering threshold (90%) used in tomato and grape studies compared with the 95% clustering threshold of the *Malus* sequences presented here. The *Arabidopsis* unigene set, estimated from all *Arabidopsis* ESTs, produces a 35% overestimate of the actual number of protein-coding genes estimated from the genome sequence. Therefore, one might expect that the actual number of apple genes may be approximately 45,000, which is similar to the number of poplar (*Populus trichocarpa*) genes predicted for the genome sequence (Tuskan et al., 2006).

## 2 GC Content and Codon Usage

Knowledge of GC content of a genome and codon usage is useful when devising PCR-based strategies for mapping and gene isolation, as well as for hybridization studies and microarrays. The GC ratio of singletons ranges from 6 to 78%, with an average of 43.8%, whereas that for TC sequences ranges from 14 to 74%, with an average of 44.1%. Codon usage assessment has been carried out on the complete sequence of a set of 545 apple cDNA sequences predicted to contain full-length coding regions and devoid of introns and frameshift errors as determined by manual inspection of BLASTx versus the non-redundant sequence database (NRDB90) reports. From these data, the 545 open reading frames were defined and a codon usage table developed from the 203,267 codons (Table 2). All codons are found in the full-length cDNA dataset, with the least frequent codon represented over 100 times. The GC content in the third base position of the full-length cDNA sampled (52% GC) is higher than the overall GC ratio of 43.9% from the sequences of the NR sequences. This indicates some pressure towards a more balanced GC ratio in coding regions compared with UTRs. Similar GC ratios in coding regions are found in grape (51%) and pear (52%). Overall, the codon usage of apple shares many similarities with that of other dicotyledonous plants represented in the codon usage database (http://www.kazusa.or.jp/codon/). Apple codon usage differs markedly from *Arabidopsis* for 12 amino acids. Further comparisons with grape (*Vitis vinifera*), pear (*Pyrus communis*), peach (*Prunus persica*), loblolly pine (*Pinus taeda*), poplar (*Populus trichocarpa*), tomato (*Solanum lycopersicum*), citrus (*Citrus sinensis*), potato (*Solanum tuberosum*), and tobacco (*Nicotiana tabacum*) showed that apple codon preference is most similar to that of grape and pear, differing from grape only in its preference for His (CAC), Leu (TTG), and Ser (TCT), and to pear in its preference for an additional three codons, Arg (AGG), Val (GTG), and

**Table 2** Codon usage calculated using 545 full-length apple cDNA sequences[a]

| Codon | Amino acid | Fraction[b] | per/1,000[c] | No. |
|---|---|---|---|---|
| GCA | A | 0.27 | 18.92 | 3,846 |
| GCC | A | 0.26 | 18.42 | 3,745 |
| GCG | A | 0.14 | 9.82 | 1,997 |
| GCT | A | 0.33 | 23.36 | 4,748 |
| TGC | C | 0.59 | 10.38 | 2,110 |
| TGT | C | 0.41 | 7.20 | 1,464 |
| GAC | D | 0.44 | 23.91 | 4,860 |
| GAT | D | 0.56 | 29.98 | 6,095 |
| GAA | E | 0.45 | 28.70 | 5,834 |
| GAG | E | 0.55 | 34.67 | 7,047 |
| TTC | F | 0.53 | 22.14 | 4,500 |
| TTT | F | 0.47 | 19.95 | 4,056 |
| GGA | G | 0.29 | 19.85 | 4,034 |
| GGC | G | 0.23 | 16.09 | 3,270 |
| GGG | G | 0.22 | 15.40 | 3,130 |
| GGT | G | 0.25 | 17.43 | 3,542 |
| CAC | H | 0.51 | 13.34 | 2,712 |
| CAT | H | 0.49 | 12.67 | 2,575 |
| ATA | I | 0.21 | 9.98 | 2,028 |
| ATC | I | 0.37 | 17.49 | 3,556 |
| ATT | I | 0.42 | 19.74 | 4,012 |
| AAA | K | 0.39 | 23.77 | 4,832 |
| AAG | K | 0.60 | 36.10 | 7,337 |
| CTA | L | 0.09 | 7.77 | 1,579 |
| CTC | L | 0.21 | 18.83 | 3,828 |
| CTG | L | 0.17 | 15.33 | 3,116 |
| CTT | L | 0.21 | 18.76 | 3,813 |
| TTA | L | 0.09 | 8.09 | 1,644 |
| TTG | L | 0.23 | 21.11 | 4,291 |
| ATG | M | 1.00 | 24.87 | 5,056 |
| AAC | N | 0.51 | 23.02 | 4,679 |
| AAT | N | 0.49 | 21.98 | 4,467 |
| CCA | P | 0.30 | 17.37 | 3,530 |
| CCC | P | 0.20 | 11.60 | 2,358 |
| CCG | P | 0.21 | 11.84 | 2,406 |
| CCT | P | 0.29 | 16.37 | 3,328 |
| CAA | Q | 0.48 | 19.08 | 3,879 |
| CAG | Q | 0.52 | 20.52 | 4,172 |
| AGA | R | 0.26 | 13.02 | 2,646 |
| AGG | R | 0.27 | 13.60 | 2,764 |
| CGA | R | 0.11 | 5.28 | 1,073 |
| CGC | R | 0.12 | 6.17 | 1,254 |
| CGG | R | 0.12 | 6.17 | 1,255 |
| CGT | R | 0.10 | 5.14 | 1,044 |
| AGC | S | 0.16 | 14.65 | 2,977 |
| AGT | S | 0.14 | 12.18 | 2,476 |
| TCA | S | 0.19 | 16.63 | 3,381 |
| TCC | S | 0.19 | 16.99 | 3,453 |
| TCG | S | 0.13 | 11.21 | 2,279 |
| TCT | S | 0.19 | 17.28 | 3,512 |

**Table 2** (continued)

| Codon | Amino acid | Fraction[b] | per/1,000[c] | No. |
|---|---|---|---|---|
| ACA | T | 0.26 | 12.97 | 2,636 |
| ACC | T | 0.30 | 15.15 | 3,079 |
| ACG | T | 0.14 | 7.23 | 1,469 |
| ACT | T | 0.29 | 14.68 | 2,985 |
| GTA | V | 0.11 | 7.27 | 1,478 |
| GTC | V | 0.23 | 14.47 | 2,942 |
| GTG | V | 0.33 | 20.81 | 4,231 |
| GTT | V | 0.33 | 20.57 | 4,181 |
| TGG | W | 1 | 14.01 | 2,847 |
| TAC | Y | 0.57 | 14.83 | 3,015 |
| TAT | Y | 0.43 | 11.16 | 2,269 |
| TAA | * | 0.34 | 0.91 | 185 |
| TAG | * | 0.25 | 0.66 | 135 |
| TGA | * | 0.41 | 1.11 | 225 |

[a] Codon usage calculated using the CUSP programme from EMBOSS.
[b] Proportion of usage of a given codon among its redundant set (i.e. the set of codons which code for this codon's amino acid).
[c] Codon frequency normalised per 1,000 bases.

the stop codon (TGA). CpG suppression is also evident in apple with a XCG:XCC ratio of 0.65, similar to that of tomato (0.58). This modest level of suppression of the CpG dinucleotides differs markedly from that of nearly no suppression in *Arabidopsis* (0.92) to the high level found in grape (0.35). This may well reflect different levels of methylation in the coding sequences used by different species of plants.

## 3 MicroRNAs and Their Targets Identified from EST Databases

MicroRNAs (miRNAs) are small, non-coding RNAs that play important regulatory roles by down-regulating target transcripts in a sequence specific manner. The miRBase Registry (Release 8.2) lists 732 miRNAs from flowering plant species, with the majority identified from *Arabidopsis*, rice and poplar where genome sequences are available (http://microrna.sanger.ac.uk/sequences/). It is apparent that some plant miRNAs are highly conserved even amongst distantly related plant species and EST databases have provided a useful resource for the discovery of plant miRNAs (Zhang et al., 2005, 2006).

Gleave et al. (2008) used computational analysis of approximately 120,000 ESTs from 'Royal Gala', to identify ten distinct apple sequences which can be classified as being representatives of seven conserved plant miRNA families (miR156, miR159, miR162, miR167, miR172, miR393 and miR398). Secondary structure predictions have shown these sequences have the characteristic fold-back structures of precursor miRNAs (Fig. 1). Northern analysis has validated the presence of these miRNA families within 'Royal Gala' tissues, with miR159, miR162 and miR172

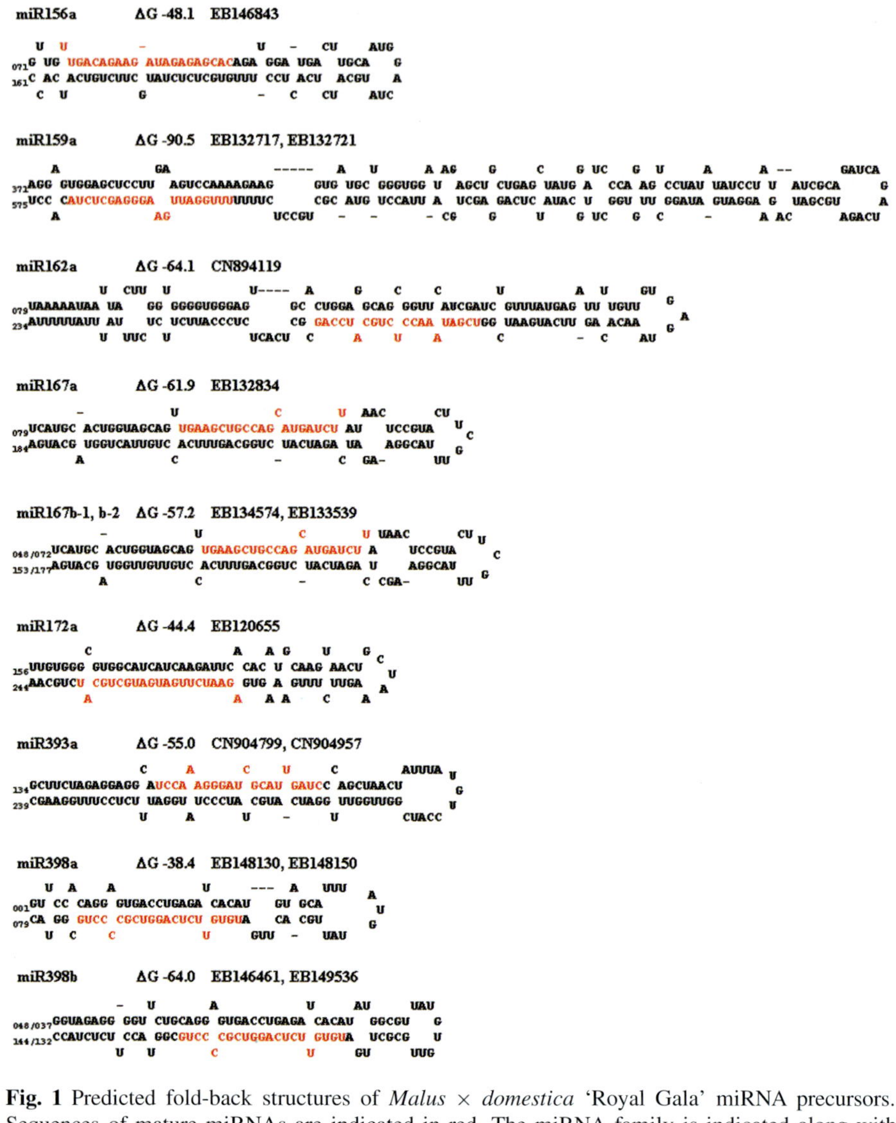

**Fig. 1** Predicted fold-back structures of *Malus* × *domestica* 'Royal Gala' miRNA precursors. Sequences of mature miRNAs are indicated in red. The miRNA family is indicated along with the free energy value (ΔG) kcal mol$^{-1}$ of the fold-back structure. Genbank accession numbers of the ESTs are shown and nucleotide coordinates relative to the start of the EST sequence shown at both the 5' and 3' ends of each fold-back structure

being expressed constitutively in leaves, roots, flower buds and developing fruit, whilst miR398 is primarily expressed in leaf tissue. The miR156 and miR167 are expressed in leaf and floral bud tissue and down-regulated during fruit development, whereas miR393 expression appears to be negligible in most tissues. However, it is significantly up-regulated in leaf tissue infected with the fungal pathogen *Venturia inaequalis*.

Plant miRNAs have few or no mismatches to their target mRNA, allowing the opportunity to use computational analysis to predict potential targets. Gleave et al. (2008), employing the miRNA/target rules of Schwab et al. (2005), identified potential targets for a number of the 'Royal Gala' miRNAs within a dataset of 120,000 'Royal Gala' ESTs and 1,312 complete 'Royal Gala' cDNA sequences. In general, the potential 'Royal Gala' mRNA targets encode proteins that have been shown to be the targets of corresponding miRNAs in other plant species, with the conserved miRNA/target relationship maintained for the 'Royal Gala' miR156 (*SPL*-like proteins), miR167 (ARFs), miR172 (AP2 domain proteins), miR393 (F-box auxin receptor *TIR1*) and miR398 (superoxide dismutase [Cu-Zn]). Furthermore, miR156- and miR398-mediated cleavage of their predicted conserved targets has been demonstrated experimentally.

## 4 Global Changes in Apple Gene Expression

Microarray technology has had a major impact in the study of non-model organisms. Using sequence knowledge from an EST sequencing program, information on gene expression can be achieved on many genes simultaneously. This has resulted in microarray technology being quickly embraced by researchers working on non-model organisms, such as apple. In apple, there are currently two published apple microarrays. Lee et al. (2007) used a 3,484 feature cDNA array to identify 192 apple cDNAs for which expression changes during early fruit development, and Schaffer et al. (2007) used a 15,726 feature oligonucleotide array to identify genes that change during apple fruit ripening.

Although microarray data provides an insight into global changes of gene expression, the greatest challenge is processing the often overwhelming amount of data obtained using this technology. As an example, a very simple dataset, comparing fruit 111 days after full bloom with mature leaves harvested, is presented (Fig. 2A). These hybridizations were performed at the workshop held before the 3rd International Rosaceae Genomics Conference (RGC3) at Plant and Food Research, New Zealand in 2006. Four microarrays, comparing labeled RNA from fruit to labeled RNA from leaf, were analyzed. The microarray and hybridization conditions used were the same as described in Schaffer et al. (2007). Differentially expressed genes with an adjusted p-value of less than 0.05 and 0.01 were selected following normalization using the limma package in Bioconductor (Smyth and Speed, 2003). Using these two cut-off values of the 15,726 genes on the array, 4,472 differentially expressed genes were identified with a p-value of less than 0.05, and 3,128 with a p-value of less than 0.01. Even at the higher stringency p-value, the number of genes selected is a challenge to assess in detail. To address this challenge, the lists of genes are often consolidated into groups.

A common method of consolidation is to examine gene ontology (GO; Ashburner et al., 2000). GO annotation works well with characterized model systems

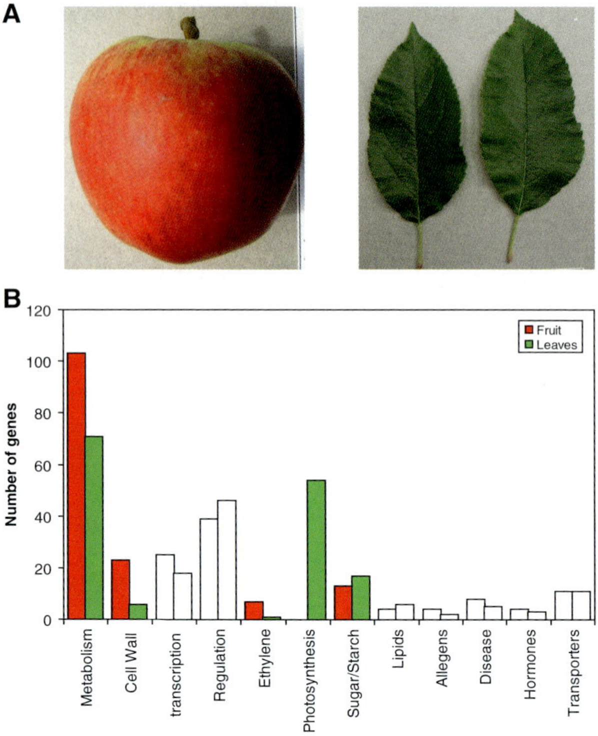


**Fig. 2 A**) Tissue used for microarray analysis. Fruit 111 days after full bloom was compared with mature leaves harvested at the same time (hybridizations were done at the RGC3 Rosaceae pre-conference workshop held at Plant and Food Research in 2006). **B**) Putative gene function of the top 500 fruit-selected genes and the top 500 leaf-selected genes, based on general categories generated by BLAST matches

and well annotated gene lists. However, with species such as *Malus*, it is less informative as an additional step is needed to align all the genes being assessed with a species where GO annotation is available. In the case of *Malus*, the closest species is *Arabidopsis*. This step results in some *Malus* genes not being aligned at all, and some being aligned with more than one function, causing a distortion in the output. A pseudo-GO ontology can be achieved by parsing the best hit blast matches for each gene and manually assigning a global function to that gene. From the data generated from the fruit-leaf comparison, the top 500 fruit genes and the top 500 leaf genes, based on p-value and fold change, were assigned function using this parsing technique (Fig. 2B). Of the top candidates that are more highly expressed in the leaves, there are a disproportionately high number of photosynthetic genes. In fruit, cell wall genes and ethylene-related genes are over-represented in expression data. However, examining the sugar and starch metabolism genes, the differences in numbers becomes less informative. Similar numbers of sugar and starch metabolism genes are seen in both fruit and leaves, but in the leaf sample there are many sugar and starch synthesis genes. In apple fruit, there appears to be more starch metabolism and sugar transport genes. This can be addressed by refining the lists to be more specific, but this can end up with an explosion in the number of categories presented.

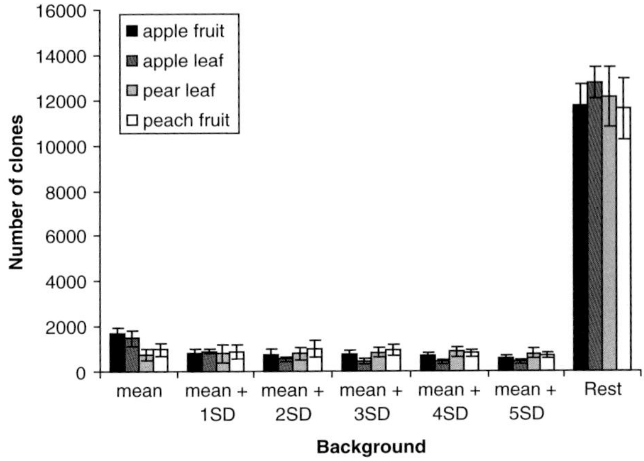


**Fig. 3** Heterologous hybridizations of other Rosaceous species on the apple oligo-microarray. The intensity of signal was measured from 20 negative control spots (background). Numbers of array features were measured for the mean plus 1 standard deviation of the mean, to mean plus 5 standard deviations of the mean, for apple (*Malus* × *domestica*), pear (*Pyrus communis*) and peach (*Prunus persica*)

An alternative method for examining microarray data is to link biology with candidate genes, using pathway maps such as KEGG (Kanehisa and Goto, 2000) to establish positioning of each gene. It quickly becomes obvious that there are many candidate genes for individual steps in each pathway. This method not only identifies potential candidate genes that are involved from the gene family, but also potential regulatory steps. This method was used by Schaffer et al. (2007), where ethylene-induced volatiles were analyzed in ripening apples. Candidate volatile biosynthetic genes were identified and their profiles matched with volatile production. Candidates for a number of steps in the biosynthesis of aroma compounds including esters, terpenes, and phenylpropanoid-derived volatiles were identified for the final step, and to a lesser extent the initial step, in each pathway that were ethylene up-regulated. The expression of these candidates can then be examined in the fruit versus leaf experiment to identify fruit predominant and leaf predominant expression patterns. It is of interest that comparing leaf to fruit, rather than different fruit time points, leads to the identification of further genes in these pathways (Table 3). While the apples in this study were mature, they were not yet undergoing ethylene-induced ripening, implying that some of the biosynthetic genes are expressed throughout fruit development, with the later regulatory time points controlling the final steps. In the ester biosynthetic pathway, the ethylene induction dataset identified four genes that are candidates for volatile production. These four genes (*LIPOXYGENASE 1* (*LOX1*), *LIPOXYGENASE 7* (*LOX7*), *PYRUVATE DECARBOXYLASE 1* (*PD1*), *ACYL TRANSFERASE 1* (*AT1*)) were also identified in the fruit versus leaf comparison in which 15 additional ester biosynthetic genes were identified. The two fruit *LOX* genes from apple are both members of one of the three lineages of *LOX* genes within plants (see Fig. 4). Interestingly, the *LOX* gene from tomato that is involved in flavor biosynthesis (Chen et al., 2004) is part of a separate lineage, suggesting that co-option into flavor pathways by members of the *LOX* multigene family has happened multiple times within the angiosperms.

**Table 3** The fruit or leaf predominant expression of potential apple aroma related genes (as in Schaffer et al. (2007). Genebank accessions are given (GB Acc). The fold change represents a fold (log base-2) change relative to leaves (negative for fruit, positive for leaves). Whether the gene was selected by Schaffer et al. (2007), as activated by ethylene in fruit is indicated

| Fruit up regulated | | | | | Leaf up regulated | | | |
|---|---|---|---|---|---|---|---|---|
| Step | Gene name | GB Acc | Fold change | Ethylene induced | Step | Gene name | GB Acc | Fold change |
| **Ester biosynthesis** | | | | | | | | |
| Es1 | LOX1 | CN851706.1 | −2.52 | Y | Es1 | LOX2 | CN876635.1 | 1.25 |
| Es1 | LOX7 | CN919697.1 | −0.82 | Y | Es1 | LOX4 | EB151028.1 | 2.54 |
| Es3a | BCDa2 | CN895673.1 | −1.52 | | Es1 | LOX5 | EB139610.1 | 1.61 |
| Es3c | BCDHC2 | EB144717.1 | −1.63 | | Es4 | ADH1 | CN915191.1 | 2.10 |
| Es3d | DLDH2 | CN918596.1 | −0.92 | | Es4 | ADH6 | CN939251.1 | 1.60 |
| Es3e | PD1 | EB142575.1 | −0.84 | Y | Es5 | ALDH1 | EB129749.1 | 1.27 |
| Es4 | ADH5 | EB115074.1 | −0.79 | | Es5 | ALDH8 | EB123767.1 | 2.53 |
| Es5 | ALDH2 | CN887130.1 | −0.89 | | Es6 | AT2 | EG631326.1 | 1.00 |
| Es5 | ALDH3 | EB140371.1 | −0.75 | | Es6 | AT8 | EB114837.1 | 0.73 |
| Es5 | ALDH6 | CN897114.1 | −0.97 | | Es7 | CXE1 | EB110805.1 | 0.99 |
| Es5 | ALDH8 | CN905672.1 | −1.85 | | Es7 | CXE10 | EB137884.1 | 1.73 |
| Es5 | AT1 | EG631323.1 | −3.47 | Y | Es7 | CXE12 | EE674220.1 | 0.50 |
| Es6 | AT3 | EE663747.1 | −2.15 | | Es7 | CXE4 | EB144756.1 | 1.03 |
| Es6 | AT6 | EG631195.1 | −0.68 | | Es7 | CXE5 | EB128672.1 | 1.20 |
| Es6 | AT7 | EG631324.1 | −2.72 | | | | | |
| Es6 | AT12 | EB121803.1 | −1.94 | | | | | |
| Es7 | CXE14 | CN916498.1 | −1.10 | | | | | |
| Es7 | CXE3 | EG631244.1 | −3.25 | | | | | |
| Es7 | CXE7 | EB129934.1 | −1.04 | | | | | |
| **Sesquiterpine biosynthesis** | | | | | | | | |
| T1 | ACOAAT1 | EB175518.1 | −0.48 | | T7 | IDDI1 | EE663731.1 | 1.48 |
| T3 | HMGR2 | EB111185.1 | −1.04 | | T8 | PPS4 | EB109638.1 | 2.24 |
| T8 | PPS7 | EB110788.1 | −0.70 | Y | | | | |
| T9 | aFS1 | EG631361.1 | NS | Y | | | | |

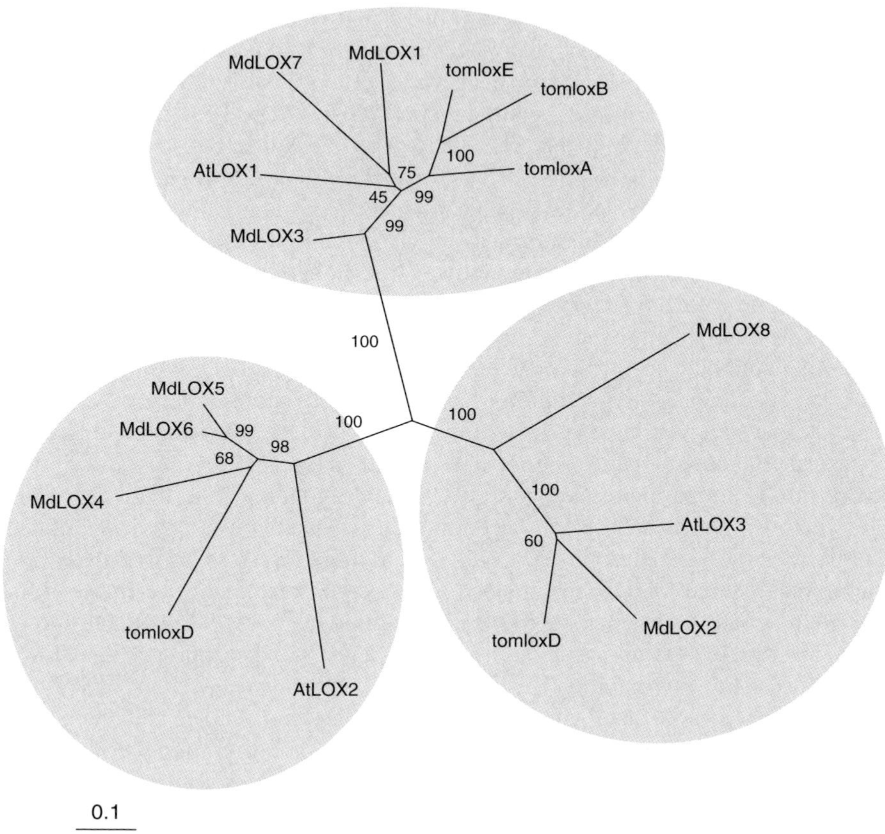


**Fig. 4** Un-rooted phylogram of plant lipoxygenase protein sequences from apple, tomato and *Arabidopsis*. The phylogram was constructed using Protdist and Fitch implemented within PHYLIP. Sequences include MdLOX1 (ES790104), MdLOX2 (ES790038), MdLOX3 (ES790007), MdLOX4 (ES790040), MdLOX5 (ES790006), MdLOX6 (ES790094), MdLOX7 (ES790041), MdLOX8 (ES790036) from apple, tomloxA (U63117), tomloxB (U63118), tomloxC (U37839), tomloxD (U37840), tomloxE (AY008278) from tomato, and AtLOX1 (AY093104), AtLOX2 (AY062611), AtLOX3 (AY075625) from *Arabidopsis*. Numbers on branches are percentage bootstrap values estimated from 1,000 bootstrap replicates

In the leaf versus fruit comparison, other members of these aroma pathways showed a fruit bias in their expression, even though they are not ethylene-induced. It should be noted that while apple volatiles are strongly induced with ethylene, there are detectable amounts of esters present before ethylene-induced ripening occurs, suggesting residual expression of this pathway or ester production is limited to availability of precursors. In this leaf-fruit dataset, a fruit-predominant alcohol dehydrogenase (*ADH5*) that does not appear to be ethylene induced was identified. *ADH1* was identified in the ethylene dataset as being repressed by the addition of ethylene, but in the leaf fruit dataset *ADH1* appears to be more highly expressed in the leaves.

In the ethylene dataset, *AT1* was found to be strongly induced by ethylene and has been implicated as a major enzyme involved in ester biosynthesis in apple (Souleyre et al., 2005). In this dataset, there appears to be a number of ATs with fruit-biased expression (ATs 1,3,6,7 and 12), while others such as *AT2* and *8* are preferentially expressed in leaves. Some of the ATs such as *MdAT6* fall into a second clade of ATs specializing in using alcohols as substrates (see Figure 8 in Newcomb et al., 2006) and therefore may also be involved in ester biosynthesis in apple. No aldehyde dehydrogenases were identified as ethylene-induced (Schaffer et al., 2007), while four (*ALDH 2,3,6*) are preferentially expressed in the fruit. Again, these data suggest the genes are not ethylene regulated but are more likely tissue regulated and highly expressed in fruit tissue.

Another method to consolidate gene lists is to assess the data in transcriptional units. The regulation of transcription is achieved by transcription factors, which in many instances are regulated at the transcriptional level themselves. Potential transcriptional regulators of gene groups can be identified by co-clustering transcription factors with a similar expression pattern. The selected genes from the fruit versus leaf experiment contained 25 transcription factors whose transcripts accumulated to higher steady-state levels in fruit, and 18 that more highly expressed in leaves. Among these, three MADS box transcription factors were identified in the fruit, and two in leaves. Additionally, a MYB transcription factor (*Md*MYB8) that shows high expression in the fruit was identified. Using a gene family approach, the *Malus* MYB transcription factor family has been targeted for more detailed analysis (Allan et al unpublished, more detail later in this chapter). This work showed that there was no apparent phenotype when this gene was over-expressed in *Arabidopsis*.

Finally, while the assessment of pathways and transcription factors are productive, these methods rely heavily on established data. When the lists of leaf-fruit genes are scrutinized, even when generously assigning function based on loose homology, the largest category of genes is those of unknown function. Of the top 500 genes with changes in expression level, there were 132 genes in fruit that only have homology with genes of unknown function (or more rarely, no homology at all with any other genes) and 158 such genes from leaves. The characterization of such genes is still a big challenge for molecular biologists, especially when working with non-model organisms.

## 5 Heterologous Microarrays

A further use of microarrays is to hybridize RNA from different species onto an array (heterologous hybridizations). We have successfully hybridized pear (*Pyrus communis*) and peach (*Prunus persica;* Chiozzotto et al. in preparation) RNA to the apple array. In both these experiments, the hybridization stringency used was identical to the apple as described in Schaffer et al. (2007). The efficiency of hybridization was assessed by taking six apple microarrays hybridized with apple leaf RNA, six with apple fruit RNA, six with pear leaf RNA and six with peach fruit RNA. For each array, the background level of hybridization was measured using negative control human clones located at different locations on the microarray and represented by 162 spots. The average signal and Standard Deviation were measured for all these control clones (excluding two that seem to cross hybridize). From this we then measured the number of spots hybridizing at background level, and then 1 to 5 standard deviations from the background (Fig. 3). The rest of the features that showed intensities of more than 5SD from the mean were also plotted, these were taken to be significant levels above background. The number of clones that showed a significant level of hybridization above background was similar in apple, pear and peach, suggesting meaningful data can be extracted from these heterologous arrays. Heterologous arrays do require the additional step, once differentially expressed genes are chosen, of identifying the native homologous gene, and this may take some time with more distant species.

## 6 Heterologous Expression of Apple Genes in Arabidopsis

The approach of transforming an apple gene into *Arabidopsis*, is both attractive (since transformation is relatively simple and plants easily grown), and fraught with danger in that the gene is certainly in a heterologous background. Nevertheless, several exciting and conclusive studies have used this approach.

The understanding of biosynthetic pathways, and their control, has benefited from transformation of genes into *Arabidopsis*. Genes involved in the phenylpropanoid pathway have been tested in *Arabidopsis*. A gene encoding an R2R3 MYB transcription factor, designated *MdMYB1*, was isolated from 'Cripps' Pink' and shown to control apple skin color. The expression of *MdMYB1* in *Arabidopsis* plants induced the synthesis of anthocyanin, particularly in the seeds (Takos et al., 2006). Two isolated spermine (Spm) synthases from apple (*MdACL5-1* and *MdACL5-2*, with high homology with *AtCL5*, and *MdSPMS* has high homology with *AtSPMS*) both complemented Spm biosynthesis in a yeast mutant deficient in Spm synthase, and ectopic expression of MdACL5-1 in the *Arabidopsis* dwarf mutant acl5 allowed recovery of the normal phenotype (Kitashiba et al., 2005).

Apple flowering-related genes can also be effective in *Arabidopsis*. In several studies, an *Arabidopsis* mutant has been complemented by an apple gene. For example, in *Arabidopsis*, *LIKE HETEROCHROMATIN PROTEIN1* (*LHP1*) is involved in silencing of flowering time genes (Mimida et al., 2007). Two apple *LHP1* homolog genes, called *MdLHP1a* and *MdLHP1b*, expressed constitutively in *Arabidopsis* could compensate for the pleiotropic phenotype of *lhp1/tfl2* mutant (Mimida et al., 2007). Transgenic *Arabidopsis* have been used to test the function of two apple orthologues of *FLORICAULA/LEAFY AFL1* and *AFL2* (apple *FLO/LFY;* Wada et al., 2002). Plants transformed with *AFL2* exhibited accelerated flowering and gave rise to several solitary flowers from rosette axils directly. *AFL1* had similar effects, but the phenotypes of the transgenic *Arabidopsis* with *AFL1* were weaker than those with *AFL2*.

As part of a gene discovery program at Plant and Food Research, over 120 *Malus domestica* genes have been transformed into *Arabidopsis* by the floral dip method (Clough and Bent, 1998). These genes have been selected by researchers in areas such as transcriptional regulation (62 genes), biosynthetic pathways (36), phytoremediation (14), and promoter testing (9). In the most part, a 35S promoter has been chosen to drive transgene expression, with over 20 extreme phenotypes produced, including changes in architecture (Foster et al., 2007) and color (Espley et al., 2007). It has been shown that transgenic *Arabidopsis* expressing *MdRGL2a*, one of six genes encoding the DELLA motif identified within the apple EST databases, have smaller leaves and shorter stems, take longer to flower in short days, and exhibit a reduced response to exogenous $GA_3$, indicating functional conservation of gene function between DELLA proteins from apple and *Arabidopsis*. *MDH1*, a homeobox gene similar to that of *BEL1*, which is involved in regulation of ovule development in *Arabidopsis*, was transformed into *Arabidopsis* driven by a 35S promoter. The resulting transgenics showed dwarfing, reduced fertility and changes in carpel and silique shape (Dong et al., 2000).

## 7 Transient Assays of Apple Genes in Tobacco

As a high-through-put assay of gene function, over 160 apple genes were overexpressed in tobacco (either *Nicotiana benthamiana or N. tabacum*) by the transient transformation technique (Hellens et al., 2005). These genes are involved in catalyzing key biosynthetic steps (e.g. activation of *Malus* × *domestica* chalcone synthase CHS1, an enzyme in the phenylpropanoid pathway: Hellens et al., 2005), or in studies of anthocyanin production (Espley et al., 2007).

The assays employing a set of versatile vectors have been built for the transient assay of promoter activity. Promoter sequences, for example apple chalcone synthase CHS1, were inserted into the multiple cloning site of pGreenII 0800-LUC, which then drives luciferase expression. This is ratio-ed against a 35S-Renilla. Tobacco leaves are then infiltrated with *Agrobacterium* carrying the promoter-reporter, and further *Agrobacterium* strains with other genes. Thus, any activation of the promoter (for example, an addition of a transcription factor, or treatment of the plant exogenously with ethylene) will result in a change in ratio between promoter-of-choice-luciferase to 35S-Renilla. Patches of tobacco leaf are then tested routinely 3–4 days later in a multi-plate luminometer. This enabled the identification, from a pool of 100s of transcription factors, those TFs likely to be involved in the regulation of the apple CHS gene (Hellens et al., 2005), and DFR gene (Espley et al., 2007).

The transient expression of MYB transcription factors in *N. tabacum* also drives colored, anthocyanic patches. *MdMYB10* in the presence of a bHLH such as *MdbHLH3* or *MdbHLH33* or *Arabidopsis TT8*, will develop such a color patch within 6 days (Espley et al., 2007). Souleyre et al. (2005) also used the system to show an acyl transferase is involved in aroma biosynthesis. Such fast screens of gene function are certainly good preliminary tests to employ before making stable transgenics, for further testing.

# 8 Apple Functional Genomics

Apple now represents a very well researched fruit in the genomics discipline. Several large EST databases, oligo-based microarrays. and new molecular tools which can confirm functional predictions from sequence homology, mean that apple is now a less intransigent system for plant biologists to study. In addition, future tools to aid functional genomics in apple are in development. A whole genome sequence project is in progress, and the sequence of other genomes will add to our understanding of apple.

## References


Ashburner M, Ball CA, Blake JA, Botstein D, Butler H, Cherry JM, Davis AP, Dolinski K, Dwight SS, Eppig JT, Harris MA, Hill DP, Issel-Tarver L, Kasarskis A, Lewis S, Matese JC, Richardson JE, Ringwald M, Rubin GM, Sherlock G (2000) Gene ontology: tool for the unification of biology. The Gene Ontology Consortium. Nat Genet 25(1): 25–29.

Chen G, Hackett R, Walker D, Taylor A, Lin Z, Grierson D (2004) Identification of a specific isoform of tomato lipoxygenase (TomloxC) involved in the generation of fatty acid-derived flavour compounds. Plant Physiol 136: 2641–2651.

Chevreau E, Lespinasse Y, Gallet M (1985) Inheritance of pollen enzymes and polyploidy origin of apple Malus × domestiva Borkh. Theor Appl Genet 71: 268–277.

Clough SJ, Bent AF (1998) Floral dip: a simplified method for Agrobacterium-mediated transformation of Arabidopsis thaliana. Plant J 16(6):735–743.

Dong YH, Yao JL, Atkinson RG, Putterill JJ, Morris BA, Gardner RC (2000) MDH1: an apple homeobox gene belonging to the BEL1 family. Plant Mol Biol 42(4): 623–633.

Espley RV, Hellens RP, Putterill J, Stevenson DE, Kutty-Amma S, Allan AC (2007) Red Colouration in Apple Fruit is Due to the Activity of the MYB Transcription Factor, MdMYB10. Plant J 49(3): 414–427.

Fei ZJ, Tang X, Alba RM, White JA, Ronning CM, Martin GB, Tanksley SD, Giovannoni JJ (2004) Comprehensive EST analysis of tomato and comparative genomics of fruit ripening. Plant J 40: 47–59.

Foster T, Kirk C, Jones W, Allan AC, Espley R, Karunairetnam S, Rakonjac J (2007) Characterisation of the DELLA subfamily in apple (*Malus* × *domestica* Borkh.). Tree Genet Genomes 3:187–197.

Gleave AP, Ampomah-Dwamena C, Berthold S, Dejnoprat S, Karunairetnam S, Nain N, Wang Y-Y, Crowhurst RN, MacDiarmid RM (2008) Identification and characterisation of primary microRNAs from apple (*Malus domestica* cv. 'Royal Gala') expressed sequence tags. Tree Genetics and Genomes (in press) doi 10.1007/s11295-007-0113-1.

Goes da Silva F, Iandolino A, Al-Kayal F, Bohlmann MC, Cushman MA, Lim H, Ergul A, Figueroa R, Kabuloglu EK, Osborne C, et al. (2005) Characterizing the grape transcriptome. Analysis of expressed sequence tags from multiple Vitis species and development of a compendium of gene expression during berry development. Plant Physiol 139: 574–597.

Hellens RP, Allan AC, Friel EN, Bolitho K, Grafton K, Templeton MD, Karunairetnam S, Laing WA (2005). Transient expression vectors for functional genomics, quantification of promoter activity and RNA silencing in plants. Plant Methods 1: 13 (doi:10.1186/1746-4811-1-13).

Kanehisa M, Goto S (2000) KEGG: Kyoto Encyclopedia of Genes and Genomes. Nucleic Acids Res 28: 27–30.

Kitashiba H, Hao YJ, Honda C, Moriguchi T (2005) Two types of spermine synthase gene: MdACL5 and MdSPMS are differentially involved in apple fruit development and cell growth. Gene 361:101–111.

Korban SS, Vodkin LO, Liu L, Aldwinkle HS, Carrol N (2004) Towards apple functional genomics: the EST project (abstract no.39. In International Plant and Animal Genomes XII Conference, Genome Sequencing and ESTs section, January 10–14, San Diego.

Lee Y-P, Yu G-H, Seo YS, Han SE, Choi Y-O, Kim D, Mok I-G, Kim WT, Sung S-K (2007) Microarray analysis of apple gene expression engaged in early fruit development. Plant Cell Rep 26(7): 917–926.

Mimida N, Kidou SI, Kotoda N (2007) Constitutive expression of two apple (Malus × domestica Borkh.) homolog genes of LIKE HETEROCHROMATIN PROTEIN1 affects flowering time and whole-plant growth in transgenic Arabidopsis. Mol Genet Genomics 2007 Jun 19; [Epub ahead of print].

Newcomb RD, Crowhurst RN, Gleave AP, Rikkerink EHA, Allan AC, Beuning LL, Bowen JH, Gera E, Jamieson KR, Janssen BJ, Laing WA, McCartney S, Nain B, Ross GS, Snowden KC, Soulyere EJF, Walton EF, Yauk Y-K (2006) Analyses of expressed sequence tags from apple. Plant Physiol 141:147–166.

Park S, Sugimoto N, Larson MD, Beaudry R, van Nocker S (2006) Identification of genes with potential roles in apple fruit development and biochemistry through large-scale statistical analysis of expressed sequence tags. Plant Physiol 141 811–824.

Schaffer RJ, Friel EN, Souleyre EJF, Bolitho K, Thodey K, Ledger, S, Bowen, JN, Ma J-H, Nain B, Cohen D, Gleave BJ, Crowhurst RN, Janssen BJ, Yao JL and Newcomb RD (2007) A genomics approach reveals that aroma production in apple is controlled by ethylene predominantly at the final step in each biosynthetic pathway. Plant Physiology 144: 1899–1912.

Schwab R, Palatnik JF, Riester M, Schommer C, Schmid M, Weigel D (2005) Specific effects of microRNAs on the plant transcriptome. Dev Cell 8: 517–527.

Souleyre EJF, Greenwood DR, Friel EN, Karunairetnam S, Newcomb RD (2005) An alcohol acyl transferase from apple (cv. Royal Gala), MpAAT1, produces esters involved in apple fruit flavour. FEBS J. 272: 3123–3144.

Smyth GK, Speed T (2003) Normalization of cDNA microarray data. Methods 31: 265–273.

Takos AM, Jaffe FW, Jacob SR, Bogs J, Robinson SP, Walker AR (2006) Light-induced expression of a MYB gene regulates anthocyanin biosynthesis in red apples. Plant Physiol 142(3): 1216–1232.

Tuskan G, DiFazio S, Jansson S, Bohlmann J, Grigoriev I, Hellsten U, Putnam N, Ralph S, Rombauts S, Salamov A et al. (2006) The genome of black cottonwood, *Populus trichocarpa* (Torr. & Gray ex Bradshaw) Science 313: 1596–1604.

Wada M, Cao QF, Kotoda N, Soejima J, Masuda T (2002) Apple has two orthologues of FLORICAULA/LEAFY involved in flowering. Plant Mol Biol 49(6): 567–577.

Zhang BH, Pan XP, Cannon CH, Cobb GP, Anderson TA (2006) Conservation and divergence of plant microRNA genes. Plant J 46: 243–259.

Zhang BH, Pan XP, Wang QL, Cobb GP, Anderson TA (2005) Identification and characterisation of new plant microRNAs using EST analysis. Cell Res 15: 336–360.

# 7. Apple Transformation and Translational Genomics

**Mickael Malnoy and Herb Aldwinckle**

## 1 Introduction

The apple genome sequencing, currently in progress at the *Istituto Agrario San Michele all'Adige* Italy and Washington State University USA, in combination of various computational and empirical approaches to sequence annotation, will make possible the identification of thousands of genes. At the time of completion of the apple genome sequence, few of these genes will have an experimental assigned function. Indeed, in *Arabidopsis thalinia* only 10% of the 25,500 unique genes that were initially predicted had an experimental assigned function (Arabidopsis Genome Initiative, 2000). Dertermination of the functions of the unknown genes will present a tremendous challenge, not only because of the large number of genes to be examined, but also because defining what constitutes a gene is itself a complex problem (Snyder and Gerstein 2003).

Recent improvements in genomic technology allow genome-wide capture of some basic information, such as the determination of gene or protein expression levels using microarrays, cDNA-AFLP or 2D gel electrophoresis (Fig. 1). Although informative, these types of data alone are typically not sufficient to define the function of a gene, as by its very nature this information is largely correlative. Mutant analysis or transformed lines provide an alternative and typically more reliable means to assign gene function. This phenotype-centric process is classically known as forward or reverse genetics. In apple forward genetics is a process that has not been yet used, instead reverse genetics was used for several years to improve some apple characteristics.

M. Malnoy (✉)
Istituto Agrario San Michele all'Adige, Italy

K.M. Folta, S.E. Gardiner (eds.), *Genetics and Genomics of Rosaceae*, Plant Genetics and Genomics: Crops and Models 6, DOI 10.1007/978-0-387-77491-6_7,

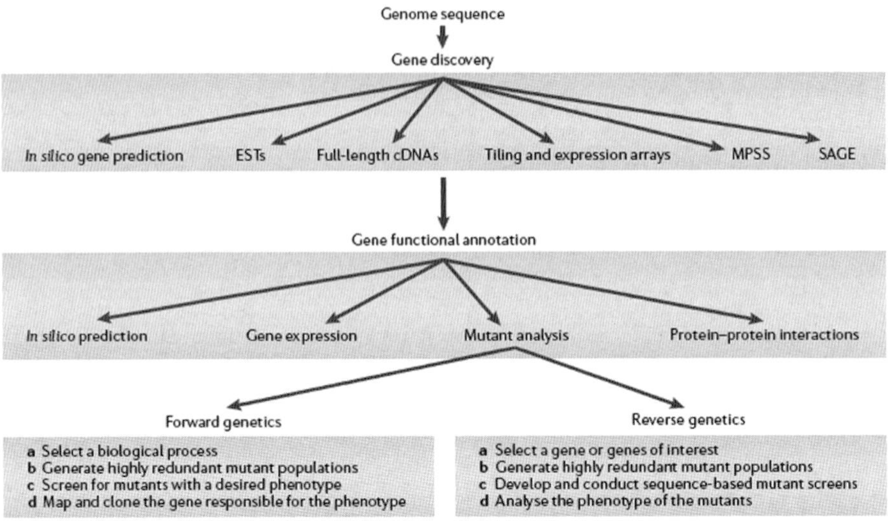


**Fig. 1** From genome sequence to gene function. Steps and experimental approaches that are used in the functional annotation of the genome. MPSS massively parallel signature sequencing; SAGE, serial analysis of gene expression

## 2 Reverse genetic, Gene Function Validation

The large amount of sequence information that has been generated by EST sequencing in USA and in New Zealand during the last couple of years and the sequencing of the apple genome, together with the implementation of approches to study gene expression, protein-protein interaction and other aspects of plant biology, have resulted in an increased interest in reverse genetic methodologies in apple. Strategies have been developed in apple to validate gene function; they are based on the over-expression or the silencing of a candidate gene.

The first report of in vitro regeneration of adventitious shoots from apple was made by Liu et al. (1983a, b) using apple seedlings. Several later reports (Barbieri and Morini 1987; Welander 1988; Fasolo et al. 1989; Sriskandarajah et al. 1990; James et al. 1994; Korban et al. 1992) revealed critical factors affecting the frequency of leaf regeneration in apple scion cultivars, rootstocks, and seed explants. They include nitrogen source and concentration, growth regulators, incubation conditions, leaf origin, leaf maturity and position on the stem, mode of excision and explant orientation. ‘Greensleeves’ was the first apple cultivar to be transformed (James et al. 1989) with an efficiency of transformation between 0.1 and 0.5% on a per explant basis. The method used was based on a leaf disk transformation method using disarmed strains of *Agrobacterium tumefaciens*. The *GUS* gene in the transformed ‘Greensleeves’ plant produced in this way displayed stable patterns of expression in fruit and Mendelian segregation in the progeny (James et al. 1995, 1996). To date, variations of this method have been used to transform different

*Malus* × *domestica* cultivars (30) and rootstocks (7), *M. prunifolia* (2) and one *M. robusta* variety (Malnoy et al. 2008). *A. tumefaciens* was used for transformation but some attempts were made with *A. rhizogenes* (Yamashita et al. 2004), mainly driven by the interest in the *rol* gene (Pawlicki-Jullian et al. 2002). Most early reports on transformed apple described 'proof of concept' experiments involving the development of regeneration and transformation protocols and the choice of appropriate promoters and selectable markers (summarized in Malnoy and Aldwinckle 2007). More recently, attention has focused on functional testing of traits of scientific and potential commercial interest. These traits can be grouped into two categories: production and consumer traits. Production traits of interest include bacterial, fungal and pest resistance, dwarfing, propagation, stress resistance, precocity, storage life and self fertility. Examples of consumer traits include novel health properties, flavor, reduced browning, color, and reduced allergenicity. However, some of these traits have been improved by an integrated foreign gene; this chapter will focus on the over-expression and silencing of apple gene.

## *2.1 Disease Resistance (Fire Blight and Apple Scab)*

### 2.1.1 Bacterial Disease Resistance

Engineering or increasing resistance to fire blight is the focus of research in bacterial disease resistance (also reviewed recently in Malnoy and Aldwinckle 2007). Fire blight, caused by the necrotic bacterium *Erwinia amylovora*, is arguably the most devastating disease of apple, with some outbreaks resulting in the destruction of whole orchards. The disease was first reported in the northeast of USA in 1784, and from there it has spread through infected material throughout most global apple growing regions, except China, India, South America and Australia. Fire blight bacteria are capable of infecting blossom, fruit, vegetative shoots and rootstock crowns through natural openings or wounds. Current disease management strategies rely on a mix of cultural practice and application of copper compounds and antibiotics. The most effective antibiotic is streptomycin. However, streptomycin is not registered for commercial use in many countries and streptomycin-resistant *E. amylovora* strains have arisen where streptomycin was used frequently (Jones and Schnabel 2000). Fire blight resistant rootstocks have been commercially released (e.g. the Geneva® series); but conventional breeding of scions is hampered by the very poor quality of resistance sources.

In most of the reported approches to create a fire blight resistant plants, the resistance induction was entrusted to a non-*Malus* gene (reviewed Malnoy and Aldwinckle 2007). However, two strategies based on the silencing of endogenous apple receptor protein and over-expression of an apple transcription factor have been proposed to improve the resistance to fire blight (Table 1).

*E. amylovora* also produces the pathogenicity effector protein *dspE*, which interacts directly with four leucine-rich repeat (LRR) receptor-like serine/threonine kinases of apple (DIPM). If this interaction does not take place *E. amylovora* will

**Table 1** Traits expressed in transformed apple

| Trait | Apple cultivars | Gene origin | Gene introduced | Promoter | Principal results | Reference |
|---|---|---|---|---|---|---|
| **Bacterial resistance** Fire Blight (*Erwinia amylovora*) | ***Galaxy*** M.26 | Over-expression | *MpNPR1* | Pin2CaMV35S | – Significant reduction in susceptibility to *E. amylovora* of 33–86% for Galaxy; M.26 showed a less substantial reduction in susceptibility compared to Galaxy (0 to 70%). | Malnoy et al. (2007b) |
| | ***Galaxy*** | RNAi | *DIPM* 4 different genes | CaMV35S | – Some transgenic lines showed silencing of the DIPM genes and an increase in resistance to *Erwinia amylovora* | Boresjza-Wysocka et al. (2007) |
| **Fungal resistance** Apple scab (*Venturia inaequalis)* | ***Gala*** | Over-expression | *HcrVf2* | CaMV35S | – Hcrvf2 confers scab resistance to the susceptible apple cultivar Gala.– Acquired resistance is race-specific. | Barbieri et al. (2003) Belfanti et al. (2004) Silfverberg-Dilworth et al. (2005) |
| | ***Galaxy*** McIntosh | Over-expression | *Vfa1, Vfa2 and Vfa4* | Their own native promoter | – Transgenic lines expressing either *Vfa1* or *Vfa2* showed a significant increase in resistance to *Venturia inaequalis*.<br>– Transgenic lines expressing *Vfa4* gene were as, or more susceptible than control | Malnoy et al. (2008) |

**Table 1** (continued)

| Trait | Apple cultivars | Gene origin | Gene introduced | Promoter | Principal results | Reference |
|---|---|---|---|---|---|---|
| **Tree characteristics** | | | | | | |
| Tree height | ***Greensleves*** | Over-expression Silencing | *GA 20-oxidase* (sense or antisense) | CaMV35S | – Reducing the expression of *MpGa20ox1* using either sense or antisense suppression resulted in a dwarf phenotype. | Bulley et al. (2005) |
| Precosity | ***Orin*** | Over-expression | *MdTFL1* | CaMV35S | – Transgenic apple expressing MdTFL1 antisense RNA first flowered 8–22 months after transfer to the greenhouse, whereas non-transgenic plants flowered 69 months after transfer to the greenhouse. | Kotoda et al. (2006) |
| Self-fertility | ***Elstar*** | Silencing | *S3* gene silencing | CaMV35S | – Production of transgenic apple tree with true self-fertility.<br>– This self-fertility was stable for several years without any obvious adverse effects on tree growth or fruit appearance. | Broothaerts et al. (2004) |
| **Modified metabolism** | | | | | | |
| Decreased flesh browning | ***Orin*** 'Fuji' | Silencing | *Polyphenol oxidase* | CaMV35S | – Transgenic lines, in which PPO expression is reduced, had less flesh browning. | Murata et al. (2000, 2001) |

**Table 1** (continued)

| Trait | Apple cultivars | Gene origin | Gene introduced | Promoter | Principal results | Reference |
|---|---|---|---|---|---|---|
| Sugar accumulation | ***Orin*** | Over-expression | *Sorbitol 6 phosphate dehydrogenase (S6PDH)* | CaMV35S | – Transgenic lines showed different expression level of S6PDH.<br>– Lines with less activity contained only a low level of sorbitol but showed 6 to 7 fold increase in sucrose. The growth of this plant stopped early during the summer due a deficiency of sugar.<br>– Lines with increased amount of S6PDH activity had increased sorbitol and sucrose content. | Kanamaru et al. (2004) |
| | ***Greensleeves*** | Over-expression | *Aldose 6 phosphate (A6PR) antisense* also called *S6PDH* | CaMV35S | – Antisense inhibition of A6PR expression significantly decreased A6PR activity and sorbitol synthesis, but increased concentration of sucrose and starch at both dusk and predawn. | Cheng et al. (2005) |
| Down regulation of ethylene production | ***Royal Gala*** | Silencing | *Aminocyclopropane 1 carboxylic (ACC) synthase 2 antisense* | CaMV35S | – Production of apple tree with down regulation of ethylene production.<br>– Some lines had fruits with delayed softening. | Hrazdina et al. (2003) |

**Table 1** (continued)

| Trait | Apple cultivars | Gene origin | Gene introduced | Promoter | Principal results | Reference |
|---|---|---|---|---|---|---|
| | ***Greensleeves*** | Silencing | *ACC synthase e ACC oxidase* | CaMV35S | – Some transgenics were significantly suppressed in ethylene production.<br>– The fruit of these apple trees was firmer and displayed increased shelf-life.<br>– No difference was observed in sugar or acid accumulation in these fruit compared to the control. However, a significant and dramatic suppression of the synthesis of volatile esters was observed. | Dandekar et al. (2004) Defilippi et al. (2004, 2005a, b) |
| | ***Royal Gala*** | Silencing | *ACC oxidase* | CaMV35S | – No production d'ethylene in some transgenic lines.<br>– Microarray data revealed that ethylene controlled only the last biosynthetic steps of aroma biosynthesis | Schaffer et al. (2007) |
| Decreased flesh browning | ***Orin*** 'Fuji' | Silencing | *Polyphenol oxidase* | CaMV35S | – Transgenic lines, in which PPO expression is reduced, had less flesh browning. | Murata et al. (2000, 2001) |

**Table 1** (continued)

| Trait | Apple cultivars | Gene origin | Gene introduced | Promoter | Principal results | Reference |
|---|---|---|---|---|---|---|
| Cell Adhesion | ***Royal Gala*** | Over-expression | *Polygalacturonase* | CaMV35S | – Phenotypic modification of apple tree (silvery colored leaves and leaf shedding).<br>– Mature leaves with malfunctioning and malformed stomata. | Atkinson et al. (2002) |
| Colors | ***Royal Gala*** | Over-expression | MdMYB10 | CaMV35S | – Regenerated callus and transformant were highly pigmented | Espley et al. (2007) |
| **Apple allergen** | ***Elstar*** | RNai | *Mal d1* Rnai | CaMV35S | – Reduction of *Mal d1* expression in the transgenic silenced apple. This translated into significantly reduced in vivo allergenicity. | Gilissen et al. (2005) |

not be able to infect that host, i.e., in the absence of the host DIPM receptor proteins the specific plant is a nonhost. Constructs with complete or partial hairpin *DIPM* gene sequences controlled by the CaMV 35S promoter were made. Several GM lines were recovered and in most genotypes the corresponding *DIPM* genes were silenced. However, no line with all four *DIPM* genes silenced was recovered. Preliminary fire blight resistance tests in growth chambers indicate that some lines are more resistant to artificial infection of the shoot tips (Borejsza-Wysocka et al. 2006). As the authors point out, 'resistance due to silencing of a native apple gene(s) is likely to be more acceptable to regulators, growers, and consumers than the addition of any foreign genes.'

During the infection process pathogens secrete compounds that elicit a range of plant host defense responses that occur at different rates. Researchers have tried to induce plant defense responses by introducing elicitors or by speeding up the defense response with various promoters. One approach proposed by Aldwinckle, has been to overexpress master regulators of the plant disease response. The *NPR1* gene is a key mediator of SAR (Cao et al. 1994; 1997). The *NPR1* protein in *Arabidopsis* is linked to the defense reaction and when overexpressed enhances resistance to *Pseudomonas syringae* and *Peronospora parasitica* (Cao et al. 1998). An additional copy of the apple ortholog, *MpNPR1* with the promoter Ppin2 (*E. amylovora* induced), was introduced into 'Galaxy' and the rootstock M.26 (Malnoy et al. 2007a). In test chamber challenges with *E. amylovora*, the transformed 'Galaxy' clones had 17.5–35.5% infected shoots - compared to 80% in controls. In addition, there was increased resistance to two other pathogens [*V. inequalis* and *Gymnosporangium juniperi-virginianae (cedar apple rust)*] (Malnoy et al. 2007a). The increased, broad spectrum resistance produced by the introduction of an additional copy of a gene sourced within apple, makes the use of *MpNPR1* (and similar strategies) very attractive, as all the other genes for resistance employed previously have been of viral, bacterial, fungal or animal origin.

### 2.1.2 Fungal Disease Resistance: *Venturia inaequalis*

Work on resistance to fungal disease has focused on apple scab (*Venturia inaequalis*). Scab is the most serious fungal disease of apple in growing areas with wet springs as it attacks both the foliage and fruit, resulting in reduced yield, and un-saleable blemished fruit. Scab is mainly controlled by the use of fungicides, although this method of control is threatened by increasing fungicide resistance. Conventional breeding has relied on one major resistance gene (*Vf*). Race 6 of the *V. inaequalis* that can overcome this resistance occurs in Europe, but apparently not in North America. Researchers have also introduced anti-fungal proteins from fungal and plant sources with some interesting results (Bolar et al. 2000, 2001; Faize et al 2003, 2004). A different approach to abtain scab resistance plant was attempted by the introduction of the apple scab resistance gene *Vf* from a wild into a cultivated apple. The introduction of the *Vf* by classical breeding began as early as the 1950s (Schmidt and van de Weg, 2005). Despite more than 50 years of traditional breeding programs, the new apple cultivars carrying this gene have yet to acquire the same fruit quality in terms of taste and texture as the susceptible top cultivars, because of linkage drag. The genes at the *Vf* locus have recently been cloned (Vinatzer et al., 2001; Xu and Korban, 2002). One of these *Vf* paralogs *HcrVf2* under the control of the CaMV35S promoter was introduced into the susceptible cultivar Gala using the *nptII* gene for selection. First in vitro tests evaluating the progression of the scab infection (penetration and stroma formation) (Barbieri et al. 2003), and later greenhouse scab inoculations of lines showed that the four transformed lines carrying *hcrVf2* gene were at least as resistant as resistant cultivars Florina when inculated with a field inoculum (Belfanti et al. 2004). Inoculation of these lines with race 7 from *M. floribunda* 821 resulted in sporulation on the transformed lines and controls; however, the inoculum was less abundant than the field inoculum. Moreover, the *HcrVf2* transformed lines still retained some resistance, being slightly more resistant than the lines transformed with only *nptII* gene and the untransformed 'Gala' (Silfverberg-Dilworth et al 2005). Recently, Malnoy et al. (2008) undertook a study of the function of each of the three *Vfa* full-length paralogs at the *Vf* locus in the resistance response of apple to *V. inaequalis* by complementary tests. Intact genomic candidate genes of each of the *Vfa* paralogs under the control of their own promoter were integrated into the genome of the scab-susceptible apple cultivars Galaxy and McIntosh. They found that transformed lines expressing *Vfa4* are found to be as susceptible as the control to *venturia inaequalis*; whereas, those expressing either *Vfa1* or *Vfa2* confer partial resistance to apple scab. It is the first study conducted in apple that validates the distinct functions of each of the three gene which compose the *Vf* locus in apple.

## 2.2 Tree Characteristics (Tree Height, Precosity, Self-Fertility, Colors)

Amelioration of tree characteristics by classical breeding can require several years before any change is observed, especially for precosity or self-fertility. The use of genetic engeneering can improve a trait in a shorted time and without modification of the interesting traits of specific cultivars. Some of these traits were improved in apple by silencing or overexpressing different genes from diverse organisms and apple itself. Here we will report only the work done with apple genes.

### 2.2.1 Tree Size

The ability to control plant vigor made modern high intensity apple production systems possible. In general, sufficient vigor control of apple cultivars is imparted by the use of rootstocks. The vigor can be also controlled with the application of growth retardants such as inhibitors of gibberellin (GA) biosynthesis. With this in mind, dwarf (reduced internode length) 'Greensleeves' scions were produced by silencing an endogenous GA 20-oxidase (Bulley et al. 2005). The size of the transformed trees ranged between 50 and 80% of non-transformed control and more pertinently, the dwarfing effect was retained in the T0 plants after grafting onto normal vigorous rootstocks (M.25 and MM.106). Earlier work also reported obtaining drawf apple tree with expression of the *phytochrome B* gene (*PhyB*) (Holefors et al. 2000): the *rol A* genes from *A. rhizogenes* (Zhu et al. 2001a; Holefors et al. 1998; Zhu and Welander 2000), the rol C (Igarashi et al. 2002) or the rol B gene (Radchuck and Korkhovoy 2005; Welander et al. 1998; Zhu and Welander 2000, Zhu et al. 2001b; Pawlicki-ullian et al. 2002).

### 2.2.2 Self Fertility

The majority of apples cultivars and many other tree crops are self-incompatible due to a system which specifically prevents self-pollen from fertilizing its own egg cells. In apple, a key gene in the self incompatibility system resides at the *S*-locus which encodes S-RNAses (Broothaerts et al. 1995). One allele isolated from 'Elstar' (S3; Broothaerts et al. 1995) was over-expressed under the control of the CaMV35S promoter in 'Elstar' to silence the endogenous allele. Self-compatible transformed lines were obtained that were silenced for the S3 allele but also the S5 allele presumably due to sequence similarity or methylation of the S5 allele (Broothaerts et al. 2004). Over a three year trial the S3/5 silenced lines were equally fertile with self (32% set) and non-self pollen (31% set), and to controls pollinated with non-self pollen (30%). Fertility of self-pollinated controls was very low (4%) (Broothaerts et al. 2004). Self-fertile apples would be attractive to growers, which will not depend of the presence of bees and separate cultivars for pollination. Self-fertile apple cultivars could also be useful to apple breeders and researchers for producing homozygous breeding lines.

### 2.2.3 Precosity

An attractive feature of many clonal rootstocks is that they reduce the length of juvenility and promote heavier flowering of grafted scions when compared with scion cultivars grown on their own roots or on seedling rootstocks. Kotoda et al. (2006); Kotoda and Wada (2005) reported that transformed 'Orin' apple plants expressing the *MdTFL1* gene in antisense orientation (*Md*TFL is an endogenous transcription factor that delays flowering), flowered 8–22 months after transfer to the greenhouse, whereas non-transgenic plants flowered 69 months after transfer to the greenhouse. However, this magnitude of early flowering effect is not strong enough, since flowering of transgenic lines was obtained around 22 month after transformation (Aldwinckle unpublished results). This plant was subjected to some special growth conditions, promoting rapid flowering.

Early flowering was reported also in apple by Flachowsky et al. (2007) by expressing the *BpMADS4* gene from silver birch. The 'Pinova' transgenic lines develop flowers in vitro- 13 weeks after transformation, and the flowers are morphologically normal. Flowers were also produced on own-rooted plants of three *BpMADS4*-transgenic lines after 3–4 months growth in a greenhouse. These flowers produced viable pollen, and when pollinated set fruit with 8–10 seeds per fruit.

By introducing the apple genes corresponding of *BpMADS4*, it will be possible to produce a early flowering model apple lines which can used for accelerating the classical breeding or to study genomics of flowering regulation and fruit quality characters.

## 2.3 Modifaction Aplle Metabolism (Sugar Accumulation, Ethylene Production)

Among the many quality parameters defining the desirability of fruits and vegetables, there are important nonvisual characteristics such as texture, nutritional value, and flavor, which influence the final acceptance by the consumer (Awad and de Jager 2003). Flavor composition has been defined as a complex attribute of quality, in which the mix of sugars, acids, and volatiles plays a primary role (Baldwin 2002). Another interesting group of metabolites are phenolic compounds, which are important secondary metabolites contributing to overall fruit quality, including flavor, nutritional value, and appearance (Golding et al. 2001). During fruit development there are many changes in flavor metabolites caused by their synthesis, transport, or degradation. In climacteric fruits, ethylene plays an important role as a modulator of ripening. All of these fruit quality related metabolites may be directly regulated by ethylene (ethylene-dependent processes) or by other signals (ethylene-independent process) (Flores et al. 2001).

In addition to the flavor metabolites mentioned above, sugars and organic acids measured through total soluble solids (TSS) and titratable acidity (TA), respectively, are most commonly associated with fruit taste. The sugars sucrose, glucose, and fructose are responsible for the sweetness, with some minor contribution of sorbitol in apple (Baldwin 2002; Knee 1993). Sugars are transported from source organs and accumulate in fruit during their development, where they form starch. The hydrolysis of starch in the fruit is an important source of sugars in the last stages of fruit development and starts before the climacteric peak (Knee 1993).

The effect of the ethylene production has been studied by silencing critical ethylene biosynthetic genes in apple. Transformed 'Gala', 'McIntosh' and 'Greensleeves' down-regulated for either ACS (ACC synthase; 1-aminocyclopropane-1-carboxylic acid synthase) or ACO (ACC oxidase) have been produced (Hrazdina et al. 2003; Dandekar et al. 2004; Schaffer et al. 2007). These genes encode key enzymes in biosynthesis of ethylene (Gray et al. 1992), which is a key hormone initiating and mediating ripening in climacteric fruit. Both Dandekar et al. (2004) and Defilippi et al. (2004) analyzed fruit from the same or similar transformed antisense ACO and ACS apple lines, that ranged in ethylene production rate from 4 to 40% that of the 'Greensleeves' control. They both observed greatly increased shelf life and increased firmness after 12 days at 20°C. Dandekar et al. (2004) reported no significant difference in soluble solids (°Brix) and acidity at harvest and after storage, whereas sugar and acid composition was different from controls in fruit stored without ethylene treatment, but was no different from controls for ethylene-treated fruit (Defilippi et al. 2004). Production of volatile esters such as and hexyl butanoate, as well as α-farnesene was suppressed (total ester production was down to 29% of control and α- farnesene was down to 40% of control). The volatile esters are important components of the fruit flavor complex, but a similar level of suppression of volatile ester production was observed when control fruit was treated with 1 $\mu g.L^{-1}$ 1-MCP for 20 h at 20°C (Defilippi et al. 2005a, b). Ethylene treatment allowed ester and alcohol synthesis to recover to 70% of the control fruit values, whereas ester and alcohol synthesis only slightly recovered in MCP-treated fruit (Defilippi et al. 2005a, b).

In a separate study, Schaffer et al. (2007) found that in ACC oxidase down-regulated 'Royal Gala', ester and alcohol synthesis completely recovered after ethylene treatment. Their microarray data revealed that ethylene controlled only the last biosynthetic steps of aroma biosynthesis.

### 2.3.1 Flavor

Alteration of sorbitol biosynthesis can impact on sugar partitioning in fruit. Sorbitol is the principal sugar translocated from source to sink organs in apple (Webb and Burley 1962). Transformation of 'Orin' with a sense sorbitol-6- phosphate dehydrogenase gene (*S6PDH*) under the control of the CaMV35S promoter resulted in lines either silenced or up-regulated in S6PDH activity (Kanamaru et al. 2004). Plants that were silenced for S6PDH had reduced levels of sorbitol (to almost zero in some lines) and increased steady state concentrations of sucrose compared with controls. In lines with increased S6PDH activity, both sorbitol *and* sucrose were increased compared with controls (Kanamaru et al. 2004). Conversion of triose-3-phosphate in the translocated photosynthate (in this case sorbitol) to 3-phosphoglyceric acid, results in NADPH production, which in turn reduces mannose-6-phosphate via mannose-6-phosphate reductase, was proposed to drive an apparent increase in photosynthetic rate (Kanamaru et al. 2004). It will be interesting to see what effects there are on fruit constituents in the lines with increased S6PDH activity. In a similar experiment, transformed 'Greensleeves' down-regulated for S6PDH showed marked changes in overall sugar metabolism (Cheng et al. 2005; Teo et al. 2006). The concentration of sorbitol was reduced in mature leaves while the concentration of sucrose and starch was increased. Although partitioning of newly fixed carbon to sucrose was unchanged, partitioning was increased for starch in S6PDH down-regulated lines. The increase in sucrose concentration was suggested to be due to breakdown of starch to glucose and maltose in chloroplasts for synthesis of sucrose during the night (Cheng et al. 2005). Fruit of S6PDH down-regulated lines had reduced sorbitol and increased sucrose concentrations. The fruit also had altered patterns of glucose, fructose, starch, and malic acid accumulation. Net carbon assimilation was also decreased in S6PDH down-regulated lines. Zhou et al. (2006) reported that these plants had reduced sorbitol dehydrogenase and increased sucrose synthase activities. These studies have shown that altering expression of a key enzyme in sugar metabolism can have wide-ranging effects on fruit flavor such as the sugar/acid balance. It would be of interest to determine how these changes affect sensory perception.

Atkinson et al. (2002) overexpressed polygalacturonase and obtained a range of new phenotypes, altering leaf morphology, plant water relations, stomata structure and function, as well as leaf attachment. Underexpression of polyphenol oxidase (PPO) (catechol oxidase), the enzyme responsible for enzymatic browning of apples, by use of an antisense *PPO* gene clearly led to reduced calli browning (Murata et al. 2000) and shoots had a similarly lower tendency for browning through the PPO activity (Murata et al. 2001). Such fruit would be attractive to both the consumer and the food processing industry.

## 2.4 Health Related: Alerginicity

The prevalence of allergies to foods is lower than that of airborne pollen. Roughly 8% of children, particularly in the first 3 years of life, exhibit allergic reaction to certain foods, but this drops to around 2% in adults. The prevalence of allergy to apple is uncertain, but it varies by geographic region. In Northern Europe most apple allergic patients (>90%) present oral allergy symptoms due to cross reactive IgE to the birch pollen allergen Bet v 1 (Fernandez-Rivas et al. 2006). The protein in apple that they react to is MALD1 which is an orthologue of Bet v 1 (Ebner et al. 1995; Vanek-Krebitz et al. 1995). The allergic symptoms are typically far less severe than peanut allergy because the MALD1 proteis extremely labile (Astwood et al. 1996). In Spain, apple allergic patients are sensitised to MALD3 (non-specific lipid transfer protein) (Fernandez-Rivas et al. 2006), and can develop much more severe allergic reactions than to MALD1.

The MALD1 allergen belongs to a group of pathogenesis related proteins, more specifically the PR10 proteins (Puhringer et al. 2000). Many plant foods, in particular fruits and tree nuts, contain homologous proteins that are recognized by the same Bet V 1- specific IgE antibodies. In apple, this allergen was designated *Mal d 1* (Vanek-Krebitz et al. 1995). Approximately 70% of patients allergic to birch pollen have been reported to have adverse reactions to apple as a consequence of the cross-reactive IgE antibodies (Ebner et al. 1995). Although birch pollen-related apple allergy is almost exclusively mild and restricted to the oral cavity, most patients allergic to apples avoid the fruit in their diet (Gilissen et al. 2005). Related fruits of the Rosaceae family, such as pear, cherry and peach can also induce adverse reactions on the basis of the same cross-reactive IgE antibodies (Ortolani et al. 1988). Therefore, avoidance often results in deprivation in the diet of a wide range of common plant foods that have important nutritional value. Production of an apple with a significant reduction of the overall expression of *Mal d 1* from existing economically successful cultivars seems to be an attractive approach. Gilissen et al. (2005) chose the approach of RNA interference (RNAi) for post-transcriptional silencing of the gene *Mal d1*. They have isolated one *Mal d 1* gene from 'Gala' and transformed the 'Elstar' apple cultivar with an *Mal d 1* RNAi vector. Normally it takes 3–5 years to grow an apple fruit-producing tree from seed or in vitro culture. Because *Mal d*

*1* genes are expressed in leaves as well as in the apple fruit, Gilissen et al. (2005) were able to evaluate *Mal d 1* gene silencing in the leaves of young apple shoots growing in vitro. Their results showed a reduction of expression of the *Mal d 1* by immunoblotting. This translated into significantly reduced in vivo allergenicity (Gilissen et al. 2005). These observations support the feasibility of production, by gene silencing, of apple cultivars hypoallergenic for *Mal d 1* (Gilissen et al. 2005). These data will need to be confirmed further by analyzing the expression of *Mal d 1* in transgenic fruit and by testing their allergenicity. *Mal d 1* silenced plants must also be evaluated for undersirable reduction in disease resistance.

The commercial feasibility of producing hypoallergenic apples needs to consider the potential market (between 2 and 10% of potential consumers, who typically avoid allergenic foods). Allergic patients can also react to multiple proteins and so multiple genes would need to be silenced. Furthermore, not all allergenic proteins present in apple have been characterized or identified. Nevertheless, there is the possibility that low allergenic cultivars may be given preference by food producers and processors because of food health and safety regulations and thus set the benchmark for future cultivars.

## 3 Conclusions

During the last 20 years researchers have made important progress in genetic engeneering of apple by trying to improve important horticultural problems using genes from apple. It was proven that it is possible to engineer resistanec apple to the major bacterial and fungal diseases. Promissing efforts have been made to try to reduce the amount of chemicals used for controlling self sterility, storage life, vigor, and increasing precosity of apple. Recently, researchers have been focusing their research on interesting and desirable trais of apple for the consumer, by reducing browning, reducing allerginicity of apple and by increasing health properties.

Aside of this research, there are other challenges to the development and commercialization of genetic engeneered apple. On of these challenges will be to limit the integration of non apple genetic material. That can be done by producing transformed apple without selectable marker (Malnoy et al. 2007b) and by using plant or apple border sequences (Rommens et al. 2004, Conner et al. 2007). The transformation without selectable marker is an interesting strategy but it will be necessair to identify chimeric and vector backbone free transformed apple. On other challenge for the commercialization of transformed apple will be the cost of the registration of GE apple lines by the differente organization. The intellectual properties are also a major barriere for the commercialization of GE apple.

In summary, genetic engineering of apple have a great potential for the consumer and improving major horticultural apple problemes. However, before these benefits can relise, the perception of the public must change. That can be address if we researchers are explaining what we are doing and what can be the benefit of GE apple for them.

## References

Arabidopsis Genome Initiative. (2000) Analysis of the genome sequence of the flowering plant Arabidopsis thaliana. Nature 408: 796–815.

Astwood, J.D., Leach, J.N. and Fuchs, R.L. (1996) Stability of food allergens to digestion in vitro. Nature Biotechnology 14:1269–1273.

Atkinson, R.G., Schroder, R., Hallett, I.C., Cohen, D. and MacRae, E.A. (2002) Overexpression of polygalacturonase in transgenic apple trees leads to a range of novel phenotypes involved in cell adhesion. Plant Physiology 129:122–133.

Awad, M.A. and de Jager, A. (2003) Influences of air and controlled atmosphere storage on the concentration of potentially healthful phenolics in apples and other fruits. PostharVest Biology and Technology 27: 53–58.

Barbieri, C. and Morini, S. (1987) In vitro regeneration from somatic tissues and seed explants of apple. Advances in Horticultural Science 1:8–10.

Barbieri, M., Belfanti, E., Tartarini, S., Vinatzer, B., Sansavini, S., Silfverberg-Dilworth, E., Gianfranceschi, L., Hermann, D., Patocchi, A. and Gessler, C. (2003). Progress of map-based cloning of the *Vf* resistance gene and functional verification: Preliminary results from expression studies in transformed apple. HortScience 38:329–331.

Baldwin, E., Fruit flavor, volatile metabolism and consumer perceptions. (2002) In Fruit Quality and its Biological Basis; M. Knee (Ed), CRC Press, Boca Raton, FL, 89–106.

Belfanti, E., Silfverberg-Dilworth, E., Tartarini, S., Patocchi, A., Barbieri, M., Zhu, J., Vinatzer, B.A., Gianfranceschi, L., Gessler, C. and Sansavini, S. (2004) The HcrVf2 gene from a wild apple confers scab resistance to a transgenic cultivated variety. Proceedings of the National Academy of Sciences USA 101:886–890.

Bolar, J.P., Norelli, J.L., Wong, K-W., Hayes, C.K., Harman, G.E. and Aldwinckle, H.S. (2000) Expression of endichitinase from *Trichoderma harzianum* in transgenic apple increases resistance to apple scab and reduces vigor. Phytopathology 90:72–77.

Bolar, J.P., Norelli, J.L., Harman, G.E., Brown, S.K. and Aldwinckle, H.S. (2001) Symergistic activity of endochitinase and exochitinase from *Trichoderma atroviride* (*T. harzianum*) against the pathogenic fungus (*Venturia inaequalis*) in transgenic apple plants. Transgenic Research 10:533–543.

Borejsza-Wysocka, E.E., Malnoy, M., Meng, X., Bonasera, J.M., Nissinen, R.M., Kim, J.F., Beer, S.V. and Aldwinckle, H.S. (2006) The fire blight resistance of apple clones in which DspE-interacting proteins are silenced. Acta Horticulturae 704:509–513.

Broothaerts, W., Janssens, G.A., Proost, P., Broekaert, W.F. (1995) cDNA cloning and molecular analysis of two self-incompatibility alleles from apple. Plant Molecular Biology 27:499–511.

Broothaerts, W., Keulemans, J. and Van Nerum, I. (2004) Self-fertile apple resulting from S-Rnase gene silencing. Plant Cell Reports 22:497–501.

Bulley, S.M., Wilson, F.M., Hedden, P., Phillips, A.L., Croker, S. and James, D. (2005) Modification of gibberellin biosynthesis in the grafted apple scion allows control of tree height independent of the rootstock. Plant Biotechnology Journal 3:215–223.

Cao, H., Bowling, S.A., Gordon, S.A. and Dong, X. (1994) Characterization of an *Arabidopsis* mutant that is nonresponsive to inducers of systemic acquired resistance. Plant Cell 6:1583–1592.

Cao, H., Gladzebrook, J., Clark,e J.D., Volko, S. and Dong, X. (1997) The *Arabidopsis NPR1* gene that controls systemic acquired resistance encodes a novel protein containing ankyrin repeats. Cell 88:57–63.

Cao, H., Li, X. and Dong, X. (1998). Generation of broad-spectrum disease resistance by overexpression of an essential regulatory gene in systemic acquired resistance. Proc. Natl. Acad. Sci. U.S.A. 11:6531–6536.

Cheng, L., Zhou, R., Reidel, E.J., Sharkey, T.D. and Dandekar, A.M. (2005) Antisenses inhibition of sorbitol synthesis leads to up regulation of starch synthesis without altering $CO_2$ assimilation in apple leaves. Planta 220:767–776.

Conner, A.J., Barrell, P.J.,. Baldwin, S.J., Lokerse, A.S., Cooper, P.A, Erasmuson, A.K, Nap, J.P. and Jacobs, J.M.E. (2007) Intragenic vectors for gene transfer without foreign DNA Euphytica 154:341–353.

Dandekar, A.M., Teo, G., Defilippi, B.G., Uratsu, S.L., Passey, A.J., Kader, A.A., Stow, J.R., Colgan, R.J. and James, D.J. (2004) Effect of down regulation of ethylene biosynthesis on fruit flavor complex in apple fruit. Transgenic Research 13:373–384.

Defilippi, B., Dandekar, A.M. and Kader, A.A. (2004). Impact of suppression of ethylene action or biosynthesis on flavor metabolites in apple (*Malus* × *domestica* Borkh.) fruit. Journal of Agricultural and Food Chemistry 52: 5694–5701.

Defilippi, B., Dandekar, A.M. and Kader, A.A. (2005a) Relationship of ethylene biosynthesis to volatile production, related enzymes, and precursor availability in apple peel and flesh tissues. Journal of Agricultural and Food Chemistry 53:3133–3141.

Defilippi, B., Dandekar, A.M. and Kader, A.A. (2005b) Apple aroma: alcohol acyltransferase, a rate limiting step for ester biosynthesis, is regulated by ethylene. Plant Science 168: 199–1210.

Ebner C., Hirschwehr R., Bauer L., Breiteneder H., Valenta R., Ebner H., et al. (1995) Identification of allergens in fruits and vegetables: IgE crossreactivities with the important birch pollen allergens Bet v 1 and Bet v 2 (birch profilin). Journal of Allergy and Clinical Immunology 95(5):962–969.

Espley, R.V., Hellens, R.P., Putterill, J., Stevenson, D.E., Kutty-Ammal, S. and Allan, A.C. (2007) Red colouration in apple fruit is due to the activity of the MYB transcription factor, MdMYB10. The Plant Journal 49: 414–427.

Faize, M., Malnoy, M., Dupuis, F., Chevalier, M., Parisi, L. and Chevreau E. (2003) Chitinases of *Trichoderma atroviridae* induce scab resistance and some metabolic changes in two cultivars of apple. Phytopathology 93:1496–1504.

Faize, M., Sourice, S., Dupuis, F., Parisi, L., Gautier, M.F. and Chevreau E. (2004) Expression of wheat puroindoline b reduces scab susceptibility in transgenic apple (*Malus* × *domestica* Borkh.). Plant Science 167:347–354.

Fasolo, F, Zimmermann, R.H. and Fordham, I. (1989) Adventitious shoot formation on excised leaves of in vitro grown shoots of apple cultivars. Plant Cell, Tissue and Organ Culture 16: 75–87.

Fernandez-Rivas, M., Bolhaar, S., Gonzalez-Mancebo, E., Asero, R., van Leeuwen, A., Bohle, B., Ma, Y., Ebner, C., Rigby, N., Sancho, A.I., Miles, S., Zuidmeer, L., Knulst, A., Breiteneder, H., Mills, C., Hoffmann-Sommergruber, K. and van Ree, R. (2006) Apple allergy across Europe: How allergen sensitization profiles determine the clinical expression of allergies to plant foods. Journal of Allergy and Clinical Immunology 118:481–488.

Flachowsky, H., Peil, A., Sopanen, T., Elo, A. and Hanke, V. (2007) Overexpression of BpMADS4 from silver birch (Betula pendula Roth.) induces early-flowering in apple (Malus domestica Borkh.). Plant Breeding 126:137–145.

Flores, F., Ben Amor, M., Jones, B., Pech, J.C., Bouzayen, M., Latché, A. and Romojaro, F. (2001) The use of ethylene-suppressed lines to assess differential sensitivity to ethylene of the various ripening pathways in cantaloupe melons. Physiologia Plantarum 113:128–133.

Gilissen, L.J.W.J., Bolhaar, S.T.H.P., Matos, C.I., Rouwendal, G.J.A., Boone, M.J., Krens, F.A., Zuidmeer, L., Van Leeuwen, A., Akkerdaas, J., Hoffmann-Sommergruber, K., Knulst, A.C., Bosch, D., Van de Weg, W.E. and Van Ree, R. (2005) Silencing the major apple allergen Mal d1 by using the RNA interference approach. Journal of Allergy and Clinical Immunology 115:364–369.

Golding, J.B., McGlasson, W.B., Wyllie, S.G., Leach, D.N. (2001) Fate of apple peel phenolics during cool storage. Journal of **Agricultural** and **Food Chemistry** 49:2283–2289.

Gray, J., Picton, S., Shabbeer, J., Schuch, W. and Grierson, D. (1992) Molecular biology of fruit ripening and its manipulation with antisense genes. Plant Molecular Biology 19: 69–87.

Holefors, A., Xue, Z.T. and Welander, M. (1998) Transformation of the apple rootstock M26 with rolA gene and its influence on growth. *Plant Science* 136:69–78.

Holefors, A., Xue, Z.T., Zhu, L.H. and Welander, M. (2000) The *Arabidopsis* phytochrome B gene influences growth of the apple rootstock M26. Plant Cell Reports 19, 1049–1056.

Hrazdina, G., Kiss, E., Galli, Z., Rosenfield, C., Norelli, J.L. and Aldwinckle, H.S. (2003) Down regulation of ethylene production in Royal Gala apples. Acta Horticulturae 628:239–251.

Igarashi, M., Ogasawara, H., Hatsuyama, Y., Saito, A. and Suzuki, M. (2002) Introduction of rolc into Marubakaidou (*Malus prunifolia* Borkh. Var. ringo Asami Mo 84-A) apple rootsock via *Agrobacterium tumefaciens*. Plant Science 163:463–473.

James, D.J., Passey, A.J., Barbara, D.J. and Bevan, M.V. (1989) Genetic transformation of apple (*Malus pumila* Mill.) using disarmed Ti-binary vector. Plant Cell Reports 7: 658–666.

James, D.J., Passey, A.J., Webster, A.D., Barbara, D.J., Viss, P., Dandekar, A.M. and Uratsu, S. (1993) Transgenic apples and strawberries: Advances in transformation, introduction of genes for insect resistance and field studies of tissue cultured plants. Acta Horticulturae 336:179–184.

James, D.J., Passey, A.J. and Baker, S.A. (1994) Stable gene expression in transgenic apple tree tissues and segregation of transgenes in the progeny – preliminary evidence. Euphytica 77: 119–121.

James, D.J., Passey, A.J. and Baker, S.A. (1995) Transgenic apples display stable gene expression in the fruit and medelian segregation of the transgenes in the R1 progeny. Euphytica 85: 109–112.

James, D.J., Passey, A.J., Baker, S.A. and Wilson, F.M., (1996) Transgenes display stable patterns of expression in apple fruit and Mendelian segregation in the progeny. Bio/Technology 14, 56–60.

Jones, A.L. and Schnabel, E.L. (2000) The development of streptomycin-resistant strains of *Erwinia amylovora*. p 235–251 In: J. Vanneste(Ed), Fire blight, CAB Int., New York.

Kanamaru, N., Ito, Y., Komori, S., Saito, M., Kato, H., Takahashi, S., Omura, M., Soejima, J., Shiratake, K., Yamada, K. and Yamaki, S. (2004) Transgenic apple transformed by sorbitol-6-phosphate dehydrogenase switch between sorbitol and sucrose supply due to its gene expression. Plant Science 167:55–61.

Knee, M. (1993) Pome fruits. In Biochemistry of Fruit Ripening; G.B. Seymour, J.E. Taylor, and G.A. Tucker (Eds), Chapman and Hall, New York, 325–346.

Kotoda, N. and Wada, M. (2005) *MdTFL1*, a *TFL1* like gene of apple, retards the transition from the vegetative to production phase in transgenic *Arabidopsis*. Plant Science 168: 95–104.

Kotoda, N., Iwanami, H., Takahashi, S. and Abe, K. (2006) Antisense expression of *MdTFL1*, a *TFL1* like gene, reduces the juvenile phase in apple. Journal of the American Society of Horticultural Science 131:74–81.

Korban, S.S., O'Connor, P.A. and Elobeidy, A. (1992) Effects of thidiazuron, naphthaleneacetic acid, dark incubation and genotype on shoot organogenesis from *Malus* leaves. Journal of the American Society of Horticultural Science 67: 341–349.

Liu, J.R., Sink, K.C. and Dennis, F.G. Jr. (1983a) Adventive embryogenesis from leaf explants of apple seedling. HortScience 18:871–873.

Liu, J.R., Sink, K.C. and Dennis, F.G. Jr. (1983b) Plant regeneration from apple seedling explants of apple and callus cultures. Plant Cell, Tissue and Organ Culture 2, 293–304.

Malnoy, M. and Aldwinckle, H.S. (2007), Development of fire blight resistance by recombinant DNA technology. Plant Breeding Reviews 29:315–344.

Malnoy M., Jin Q., Borejsza-Wysocka E.E., He S.Y. and Aldwinckle, H.S. (2007a) Over-Expression of the Apple Gene MpNPR1 confers Increased Disease Resistance in Malus × Domestica MPMI. 20:1568–1580.

Malnoy, M., Borejsza-Wysocka, E.E., Abbott, P., Lewis, S., Norelli, J.L., Flaishman, M.A., Gidoni, D. and Aldwinckle, H.S. (2007b) Genetic transformation of apple without use of a selectable marker. Acta Horticulturae 663:319–322.

Malnoy, M., Korban, S.S., Borejsza-Wysocka E.E. and Aldwinckle H.S. (2008) Apple in Kole, C and Hall, T.C. (eds.). "Compendium of tramsgenic cropsd plants: Transgenic Temperate Fruits and Nuts" Blackwell Publishing, oxford UK, 2008, pp 1–52.

Malnoy, M., Xu, M., Borejsza-Wysocka, E.E., Korban, S.S. and Aldwinckle, H.S. (2008) Two receptor like genes, Vfa1 and Vfa2 confer resistance to the fungal pathogen *venturia inaequalis* inciting apple scab disease. MPMI 21:448–458.

Murata, M., Haruta, M., Murai, N., Tanikawa, N., Nishimura, M., Homma, S. and Itoh, Y. (2000) Transgenic apple (*Malus* × *domestica*) shoot showing low browning potentail. Journal of Agricultural and Food Chemistry 48:5243–5248.

Murata, M., Nishimura, M., Murai, N., Haruta, M., Homma, S. and Itoh, Y. (2001) A transgenic apple callus showing reduced polyphenol oxidase activity and lower browning potential. Bioscience, Biotechnology and Biochemistry 65:383–388.

Ortolani, C., Ispano, M., Pastorello, E., Bigi, A. and Ansaloni, R. (1988) The oral allergy syndrome. Annals Allergy 61, 47–52.

Pawlicki-Jullian, N. Sedira, M. and Welander, M. (2002) The use of Agrobacterium rhizogenes transformed roots to obtain transgenic shoots of the apple rootstock Jork 9. Kluwer Academic Publ. pp. 163–171.

Puhringer, H., Moll, D., Hoffmann-Sommergruber, K., Watillon, B., Katinger, H. and Machado, M.L.D. (2000) The promoter of an apple *Ypr10* gene, encoding the major allergen Mal d 1, is stress- and pathogen-inducible. Plant Science 152, 35–50.

Radchuck, V.V. and Korkhovoy, V.I. (2005) The *rolB* gene promotes rooting in vitro and increases fresh root weight in vivo of transformed apple scion cultivar 'Florina'. Plant Cell, Tissue and Organ Culture 81, 203–212.

Rommens, C.M., Humara, J.M., Ye, J., Yan, H., Richael, C., Zhang, L., Perry, R. and Swords, K. (2004) Crop improvement through modification of the plant's own genome. Plant Physiology 135: 421–431.

Schaffer, R.J., Friel, E.N., Souleyre, E.J.F., Bolitho, K., Thodey, K., Ledger, S., Bowen, J.H., Ma J.-H., Nain, B., Cohen, D., Gleave, A.P., Crowhurst, R.N., Janssen, B.J., Yao, J.L. and Newcomb, R.D. (2007) A genomics approach reveals that aroma production in apple is controlled by ethylene predominantly at the final step in each biosynthetic pathway. Plant Physiology 144:1899–1912.

Schmidt, H. and van de Weg, W.E. (2005) Breeding. In J. Tromp, A.D. Webster, and S.J. Wertheim (Eds) Fundamentals of Temperate Zone Tree Fruit Production. Backhuys, Leiden, The Netherlands, 136–155.

Silfverberg-Dilworth, E., Patacchi, A., Belfanti, E., Tartarini, S., Sansavini, S. and Gessler, C. (2005) *Hcrvf2* introduced into Gala confers race-specific apple scab resistance. In: Plant & Animal Genome XIII Conference, January 15–19, 2005, San Diego, CA, USA.

Snyder, M. and Gerstein, M. (2003) vGenomics. Defining genes in the genomics era. Science 300:258–260.

Sriskandarajah, S., Skirvin, R.M., Abu Qaoud, H. and Korban, S.S. (1990) Factors involved in elongation and growth of adventitious shoots from tree apple scion cultivars in vitro. Journal Horticultural Science 65:113–121.

Teo, G., Suzuki, Y., Uratsu, S.L., Lampinen, B., Ormonde, N., Hu, W.K., Dejong, T.M. and Dandekar, A.M. (2006) Silencing leaf sorbitol synthesis alters long-distance partitioning and apple fruit quality. Proccedings of the National Academy of Sciences USA 103:18842–18847.

Vanek-Krebitz, M., Hoffmann-Sommergruber, K., Machado, M.L.D., Susani, M., Ebner, C., Kraft, D., Scheiner, O. and Breiteneder, H. (1995) Cloning and Sequencing of Mal d 1, the Major Allergen from Apple (*Malus domestica*), and Its Immunological Relationship to Bet v 1, the Major Birch Pollen Allergen. Biochemical and Biophysical Research Communications 214:538–551.

Vinatzer, B.A., Patocchi, A., Gianfranceschi, L., Tartarini, S., Zhang, H.B., Gessler, C., Sansavini, S. (2001) Apple contains receptor-like genes homologous to the *Cladosporium fulvum* resistance gene family of tomato with a cluster of genes cosegregating with *Vf* apple scab resistance. Molecular Plant Microbe Interactions 14:508–514.

Xu, M.L. and Korban, S.S. (2002) A cluster of four receptor-like genes resides in the *Vf* locus that confers resistance to apple scab disease. Genetics 162:1995–2006.

Webb, K.L. and Burley, J.W.A. (1962) Sorbitol translocation in apple. Science 137:766.

Welander, M. (1988) Plant regeneration from leaf and stem segments of shoots raised in vitro from mature apple trees. Journal of Plant Physiology 132:738–744.

Welander, M., Pawlicki, N., Holefors, A. and Wilson, F. (1998) Genetic transformation of apple rootstock M26 with *rolB* gene and its influence on rooting. Journal of Plant Physiology 53: 371–380.

Yamashita, H., Daimon, H., Akasaka-Kennedy, Y. and Masuda, T. (2004). Plant regeneration from hairy roots of apple rootstock, *Malus prunifolia* Borkh. Var. *ringo* Asami, strain Nagano No. 1, transformed by *Agrobacterium rhizogenes.* Jpn. Soc. Hortic. Sci. 73:505–510.

Zhou, R., Cheng, L. and Dandekar, A.M. (2006) Down regulation of sorbitol dehydrogenase and up-regulation of sucrose synthase in shoot tips of the transgenicapple trees with decreased sorbitol synthesis. J. Exp. Bot. 57: 3647–3657.

Zhu, L.H. and Welander, M. (2000) Growth characteristics of apple cultivar Gravenstein plants grafted onto the transformed rootstock M.26 with *rolA* and *rolB* under non-limiting nutrient conditions. Plant Science 147:75–80.

Zhu, L.H., Ahlman, A., Li, X.Y. and Welander, M. (2001a) Intergration of the *rolA* gene into the genome of the vigorous apple rootstock A2 reduced plant height and shortened internodes. Journal of Horticultural Science and Biotechnology 76:758–763.

Zhu, L.H., Holefors, A., Ahlman, A., Xue, Z.T. and Welander, M. (2001b) Transformation of the apple rootstock M.9/29 with the *rolB* gene and its influence on rooting and growth. Plant Science 160:433–439.

# 8. Pear Genomics

**Toshiya Yamamoto and Elisabeth Chevreau**

## 1 Introduction

### *1.1 Origin, Speciation and Botanic Characteristics*

Pear, like the other pip fruit species apple and quince, belongs to the sub-family *Maloideae* in the Rosaceae, sharing a basic chromosome number of x = 17 which indicates a polyploid origin. The genus *Pyrus* is believed to have arisen during the Tertiary period in the mountainous regions of western China. Dispersal and speciation is believed to have followed the mountain chains both east and west (Rubzov, 1944; Zeven and Zhukovsky, 1975). Wild pears can be found in the entire Eurasian zone. In Europe they are mostly *Pyrus communis* L. subsp. *pyraster* (L.) and in the Caucasus, *P. caucasica* (Fed.) Browicz. These pear trees produce small fruits of variable characteristics, which were probably picked and preserved dried by early humans. Domestication occurred from the better-fruited trees. As for apple, grafting played a key role in the diffusion of improved genotypes in Central Asia and in Eastern Mediterranean area. According to Hedrick et al. (1921), European pear culture was well established in Greece and cultivars with distinct names were propagated as early as 300 B.C. Oriental pears, which arose independently, were also grown in China for more than 2000 years (Kikuchi, 1946).

The taxonomy of the genus *Pyrus* is complex, due to synonymy between taxa, and to frequent interspecific crosses. A summary of the main recognized *Pyrus* species has been published by Bell et al. (1996) and is reproduced in Table 1. Economic usage of *Pyrus* species has been reviewed by Bell et al. (1996). *Pyrus communis* L. is the main edible pear species grown in Europe, North America, South America, Africa and Australia. The snow pear, *P. nivalis* Jacq., is also grown to a limited extent in Europe for making perry. In Asia, *P. pyrifolia* (Burm.) Nakai is the main cultivated species in southern and central China, Japan, Taiwan and countries

T. Yamamoto (✉)
National Institute of Fruit Tree Science, 2-1 Fujimoto, Tsukuba, Ibaraki 305-8605, Japan

K.M. Folta, S.E. Gardiner (eds.), *Genetics and Genomics of Rosaceae*, Plant Genetics and Genomics: Crops and Models 6, DOI 10.1007/978-0-387-77491-6_8,

**Table 1** Primary species in the genus *Pyrus* (from Bell et al., 1996)

| Species | Distribution |
|---|---|
| European : | |
| *P. communis* L. | West to SE Europe, Turkey, Eurasia |
| *P. caucasica* Fed. | SE Europe, Greece, Turkey |
| *P. nivalis* Jacq. | West, Central and Southern Europe |
| *P. cordata* Desv. | SW England, W France, Spain, Portugal |
| Circum-Mediterranean : | |
| *P. amygdalyformis* Vill. | Mediterranean Europe, Asia Minor |
| *P. eleagrifolia* Pall. | SE Europe, Russia, Turkey |
| *P. syriaca* Boiss. | Tunisia |
| *P. longipes* Coss & Dur. | Algeria |
| *P. gharbiana* Trab. | Morocco, W Algeria |
| *P. mamorensis* Trab. | Morocco |
| Mid-Asian | |
| *P. glabra* Boiss. | Iran |
| *P. salicifilia* Pall. | NW Iran, NE Turkey, South Russia |
| *P. reqelii* Redh. | South Central Asia (Afghanistan) |
| *P. pashia* D. Don. | Pakistan, India, Nepal |
| East Asian | |
| *P. pyrifolia* (Burm.) Nak. | China, Japan, Korea, Taiwan |
| *P. pseudopashia* Yu. | BNW China (Yunnan, Kweichow) |
| *P. ussuriensis* Maxim. | Siberia, Mandchuria, N China, Korea |
| *P. calleryana* Decne. | Central & S China, Vietnam |
| *P. betulaefolia* Bunge | Central & S China, S Mandchuria |
| *P. fauriei* Sceid. | Korea |
| *P. hondoensis* Kik. & Nak. | Japan |
| *P. dimorphophylla* Mak. | Japan |
| *P. kawakamii* Hayata | Taiwan, SE China |

of Southeast Asia. Some cultivars of this species are also grown on a more limited basis in Europe, under their Japanese name, nashi. In northern China and Japan, *P. ussuriensis* Max. and *P.* × *bretschneideri* Rehd. are grown for edible pear production, as well as some selections of *P. pashia* P. Don. Various *Pyrus* species are also used as rootstocks, or as ornamentals.

Wild native trees are deciduous, generally medium sized trees, with few species being shrubs. The inflorescence is a loose corymb of 6–9 flowers. The fruit of pear is a pyriform (European) or round (Asian) pome. As in apple, the fleshy edible portion is derived from the receptacle and the base of the perianth. There are 5 central seed cavities, usually bearing 2 seeds each as in apple. The flesh contains grit cells which are thick-walled, lignified cells that give the characteristic European pear flesh texture.

## *1.2 Production and Utilization*

Pears rank second to apples in the amount of worldwide production of deciduous tree fruit species The world production of pear has reached 19 MT in 2005

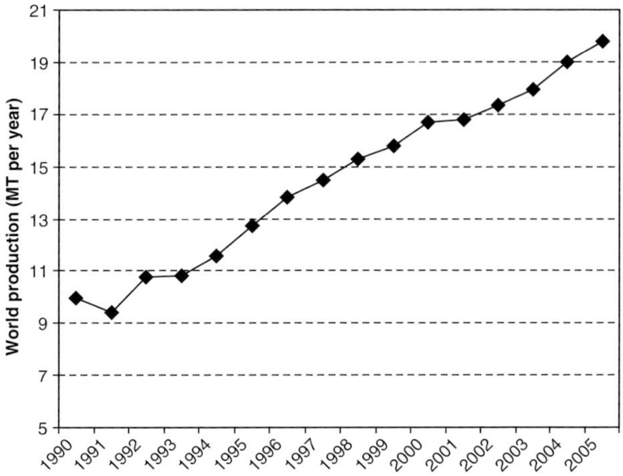


**Fig. 1** World pear production (FAO, 2007)

(FAOSTAT, 2007). This total production has doubled over the past 15 years (Fig. 1). It consists mostly of Asian pears (70%), China being the main producer, with more than 11 MT. European pears are produced mainly by Italy, United States, Argentina and Spain, each country producing between 0.5 and 1 MT. Most of the crop is utilized as fresh fruit, but significant amounts are processed and marketed as canned pears, puree, juice, or other products.

## *1.3 Traditional Breeding*

Pear breeding already has a very long history which deals with improvement of European as well as Asian pears. Breeding objectives are complex, but most pear breeders share a number of common purposes such as fruit quality, storage ability, consistent production, disease and pest resistance for cultivar improvement, and dwarfing ability, graft compatibility, iron chlorosis tolerance, disease resistance, cold hardiness for rootstock improvement (Bell et al., 1996). Information on objectives and recent introductions can be found in review by Bellini (1995).

Fruit tree breeding methods generally involve three major steps: creation of genetic variation, selection of elite material and extended experimentation of promising pre-selection before market release (Schmidt and van de Weg, 2005). Crossing is still by far the more largely used method of creation of genetic variability for pear. The genetic pool available for the pear breeder is wide. Most species of *Pyrus* are diploid ($2n = 34$; $x = 17$) and no significant interspecific cross-incompatibility is known to exist. Crossing programs rely on the availability of appropriate genetic resources, which can comprise recent breeding selections, old and new cultivars and various species of the genus *Pyrus*. Most pear varieties are diploid (2n=2x=34), with a few triploids (2n=3x=51) or tetraploids (2n=4x=68).

Most varieties are self-sterile, due to a strong gametophytic self-incompatibility controlled by a single S locus. This self-sterility leads to a high degree of heterozygosity.

Many new European pear scion cultivars have been developed from hybridization programs and released within the last 20 years with major improvements in the quality, storage and shelf life of early season pears, as well as in late season pears. Important progress has also been made in the field of disease resistance, particularly concerning fire blight. Countries with the highest number of pear scion releases since 1990 have been USA, Germany, France, Russia and Italy. However, very few of these novel varieties have the potential to replace the classical pear cultivars whose agronomical and commercial limits are already well known and which have gained a solid market (Bellini and Nin, 2002). Rootstock breeding programs are considerably longer than scion breeding programs, because no reliable laboratory technique is available so far to preselect for vigor and compatibility before testing rootstock/scion combinations in the orchard. Despite hybridization programs in several countries and release of several new rootstocks, there is still a lack of compatible dwarfing pear rootstocks of *Pyrus* type, combining a good propagation ability with disease resistance and an adaptation to difficult environmental conditions (cold, drought or lime-induced chlorosis) (Wertheim, 2002).

Mutation induction, although efficient for some characteristics such as skin color, compact growth or tetraploidy, never took a large place in pear breeding. The occurrence of chimeras, which are often unstable, is one of the major obstacles to the use of spontaneous or induced mutations for pear breeding (Chevreau et al., 1989). Recently, in vitro systems for pear mutation breeding have been developed to decrease the risk of obtaining chimeric plants, either by using adventitious regeneration or by applying rapid cycles of micropropagation to separate mutated from non-mutated sectors (Predieri and Zimmerman, 2001). So far, successes of induced mutagenesis for pear improvement are still limited. Three mutants of Asian pear ('Gold Nijisseiki', 'Kotobuki Shinsui' and 'Osa Gold') have been released. They were obtained after chronic $\gamma$-irradiation of plants and were selected for increased resistance to black spot disease, caused by *Alternaria alternata* (Yoshioka et al., 1998).

## 1.4 Biotechnical Approaches to Pear Improvement

The contribution of in vitro methods to create novel genetic variability in pear has advanced considerably during the two last decades. Haploidization via in situ parthenogenesis induced by irradiated pollen and in vitro rescue of the haploid plantlets has been successfully developed for pear (Bouvier et al., 1993). Techniques of adventitious bud regeneration from in vitro leaves have been developed for several genotypes of European and Asian pear and for quince. So far, applications of these techniques for the induction of somaclonal variation have been very limited. The occurrence of somaclonal variation has been demonstrated for fire blight resistance and iron-chlorosis tolerance, albeit at low frequency (Chevreau and Bell, 2005). Protoplast technology has also been applied to pear since 1986 by Ochatt and Caso. However, somatic hybridization between *Pyrus* and an incompatible genus, *Prunus*, has been reported only once (Ochatt et al., 1989).

Marketing of pears both within and among nations is characterized by limitations on the number of cultivars, because both producers and consumers prefer the old traditional cultivars, which are unique and clearly recognized. Therefore, acceptance of new hybrid cultivars is very slow. In this context, gene transfer offers pear breeders new tools to directly improve existing elite cultivars without changing their main recognizable characteristics. Transformation of pear is based on the co-culture of in vitro leaves with a disarmed *A. tumefaciens* strain carrying the gene(s) of interest in a binary vector. Since the first report of pear transformation on three European pear cultivars (Mourgues et al., 1996), several other genotypes have been transformed by various groups in Europe, United States and Asia.

Genetic engineering has already been applied to pear with the aim to modify important agronomical traits by several groups since 1999. These reports, mostly on European pears, are summarized in Table 2. Increased resistance to fire blight is the main objective of the pear genetic engineering program in France (Reynoird

**Table 2** Reports of introduction of transgenes in pear with the aim of modifying agronomical traits

| Country | Pear variety | Gene : trait of interest | Reference |
|---|---|---|---|
| USA | Beurre Bosc | Rol C / plant architecture | Bell et al. (1999) |
| France | Passe Crassane | Attacin / fire blight resistance | Reynoird et al. (1999) |
| France | Passe Crassane | Lysozyme / fire blight resistance | Malnoy et al. (2000) |
| USA | Bartlett | Lytic peptide DRC1 / fire blight resistance | Puterka et al. (2002) |
| Russia | Burakovka | Thaumatin II : taste improvement | Lebedev et al. (2002a) |
| Russia | Burakovka | Defensins Rs-AFP2 / resistance to fungal pathogens | Lebedev et al. (2002b) |
| Sweden | BP10030 | rolB / rooting efficiency | Zhu et al. (2003) |
| France | Passe Crassane | Lactoferrin / fire blight resistance | Malnoy et al. (2003a) |
| China | Fertility | Defensins Rs-AFP2 / resistance to fungal pathogens | Zhao et al. (2004) |
| Israel | Spadona | Stilbene synthase / health promoting | Flaishman et al. (2005) |
| France | Passe Crassane | Depolymerase / fire blight resistance | Malnoy et al. (2005a) |
| France | Passe Crassane | Harpin N / fire blight resistance | Malnoy et al. (2005b) |
| Japan | La France, Ballade | CiFT / flowering | Matsuda et al. (2006) |
| Japan | La France | ACC oxidase / fruit ripening | Gao et al. (2007) |
| China | Xueqing | Cry1Ac / insect resistance | Tang et al. (2007) |
| Japan | Ballade | Spermidine synthase / abiotic stress tolerance | Wen et al. (2008) |
| Israel | Spadona | TFL1 / flowering | Flaishman et al. (2007) |

et al., 1999, Malnoy et al., 2000, 2003a, 2005a, b). Promising strategies include the establishment of a competition for iron with the bacterial siderophores and the induction of plant defenses by expression of a bacterial effector. Transfer of exogenous transgenes for fungal and insect resistance has also been reported (Lebedev et al., 2002b, Zhao et al., 2004, Tang et al., 2007). Modification of polyamine levels has been attempted recently to confer abiotic stress tolerance (Wen et al., 2008). The *rol* B and C genes from *Agrobacterium rhizogenes* have been introduced into pear for modification of scion development (Bell et al., 1999) or for improvement of rootstock rooting ability (Zhu et al., 2003). A few reports indicate the use of transgenes related to fruit quality (Lebedev et al., 2002a, Flaishman et al., 2005, Gao et al., 2007). Finally, the development of juvenile-free pear has been accomplished by *CiFT* over-expression (Matsuda et al., 2006) and by a *TFL1*-RNAi strategy (Flaishman et al., 2007).

According to the APHIS Field Test Releases Database (updated October 9, 2007), two groups have already released transgenic pears for field trials in USA. The USDA-ARS in West Virginia has released transgenic pears carrying a cecropin gene for fire blight resistance, and the *rolC* gene for dwarfing. Exelixis in Oregon and Washington, has released transgenic pears carrying the *sam-k* gene for delayed fruit ripening. So far in Europe, only one field trial has been conducted by the Swedish University of Agriculture Sciences, with transgenic pear rootstocks BP10030 containing the *rolB* gene.

Practical applications of gene transfer for pear breeding are facing many obstacles, in particular the reluctance of the public in Europe to accept genetically modified fresh products such as fruits and the limited economical value of this fruit crop compared to the cost of intellectual property associated with the development of a GMO variety. Research projects to develop targeted expression of transgenes (Malnoy et al., 2003b) and marker-free transformation systems (Djennane et al., 2007) should contribute to minimize public concern. Even though the direct use of gene transfer for pear breeding seems a long distance target, the various possibilities offered by gene transfer to over-express or silence a precise gene constitutes a unique tool for the progress of genetic knowledge of this species. With the increased speed of gene discovery in fruit species, gene transfer will become a necessary tool to demonstrate the function of these genes.

# 2 Structural Genomics

## *2.1 Genetic Diversity*

### 2.1.1 Genetic Diversity in Asian Pears, European Pears, and Other Pyrus

The genus *Pyrus* contains at least 22 widely recognized primary species, all indigenous to Asia, Europe, and the mountainous area of North America. Presently, some pear species are cultivated commercially in the temperate regions of more than 50 countries around the world (Bell, 1990; Bell et al., 1996). It is considered that there are at least 10 naturally and artificially occurring interspecific hybrid species. The major edible species *P. communis* L. is used for cultivation in Europe, North America, South America, Africa and Australia. The other major edible species, *P. bretschneideri* Rehd., *P. ussuriensis* Maxim. and *P. pyrifolia* (Burm.) Nakai, are cultivated in East Asian countries. All the species of *Pyrus* are intercrossable and there are no major incompatibility barriers to interspecific hybridization in *Pyrus*, in spite of the wide geographic distribution of the genus (Westwood and Bjornstad, 1971). Although interspecific hybrids among *Pyrus* spp. have been tested to improve disease and pest resistances, fruit quality and adaptability were generally low (Bell et al., 1996). During the past decade, a lot of attempts have been tried to evaluate genetic diversity in Asian pears, European pears, and other *Pyrus*, using several types of DNA markers, i.e., RAPD, AFLP, SSRs (microsatellites) and ISSR.

Ten or more primary species are naturally distributed in East Asia, including *Pyrus pyrifolia*, *P. pashia*, *P. hondoensis*, *P. ussuriensis*, *P. kawakamii*, *P. calleryana*, *P. koehnei*, *P. fauriei*, *P. dimorphophylla* and *P. betulaefolia* (Bell, 1990, Bell et al., 1996). It is considered that some species such as *P. bretschneideri* and *P. phaeocarpa* are naturally occurring interspecific hybrids. Genetic diversity and genetic relatedness within species as well as between species in East Asian pears have been examined by DNA marker systems. Nineteen Japanese pear (*P. pyrifolia*) cultivars were successfully discriminated by 82 RAPD primers and 6 SCAR (sequence characterized amplified regions) markers converted from RAPD fragments (Kim et al., 2000a, b). Kim and Ko (2004) analyzed 33 Asian pears from 12 *Pyrus* species by 60 RAPD primers. Four groups were obtained based on a cladogram. A total of 118 *Pyrus* spp. and cultivars native mainly to east Asia were analyzed by 20 RAPD primers to evaluate genetic variation and relationships among the accessions (Teng et al., 2001, 2002). According to their reports, RAPD markers specific to species were identified, and the grouping of the species and cultivars by RAPD largely agrees with morphological taxonomy. Cultivars of *P. sinkiangensis* were suspected to be of hybrid origin involving *P. communis* and *P. bretschneideri* (Teng et al., 2001).

Kimura et al. (2002) identified 58 Asian pear accessions from 6 *Pyrus* species using 9 SSR markers with a total of 133 putative alleles. They obtained a phenogram based on the SSR genotypes, showing 3 major groups corresponding to the Japanese, Chinese and European groups. Bao et al. (2007) evaluated 98 pear cultivars native mainly to East Asia by 6 SSR markers. Chinese sand pear (*P. pyrifolia*) and Chinese white pear (*P. bretschneideri*) presented a large genetic diversity. Occidental pears generally had low affinities to Asian pears.

Edible European pears (*P. communis*) are derived from wild relatives native to the Caucasus Mountain region and eastern Europe. Thirteen SSR loci were used to determine the relationships among 145 wild and cultivated individuals of *P. communis* maintained in the National Plant Germplasm System (NPGS, USA) (Volk et al., 2006). Twelve clusters were obtained based on individual SSR genotypes by Bayesian clustering method. *Pyrus communis* ssp. *caucasica* which is native to the Caucasus Mountains can be genetically differentiated from *P. communis* ssp. *pyraster* native to eastern European countries. The domesticated pears cluster closely together and are most closely related to a group of genotypes that are intermediate to the *P. communis* ssp. *pyraster* and the *P. communis* ssp. *caucasica* groups. Dolatowski et al. (2004) studied the variability and genetic relationship of wild and semi-wild pears (*P. pyraster*) in Poland using AFLP markers.

Oliveira et al. (1999) investigated molecular characterization and phenetic similarities between several cultivars of *P. communis* and *P. pyrifolia* and several wild species by RAPD markers. Monte-Corvo et al. (2000) investigated the genetic relationships among 39 cultivars including 35 *P. communis* and 4 *P. pyrifolia* cultivars using AFLP and RAPD markers. They confirmed that AFLP markers were five times more efficient in detecting polymorphism per reaction. Although some differences can be noticed between the dendrograms resulting from AFLP and RAPD analyses, both techniques produced similar results. Lee et al. (2004) reported that RAPD, SCAR and the conserved 18S rDNA could be used to classify and identify cultivars of *P. pyrifolia* and *P. communis*.

### 2.1.2 Taxonomical Relationships in Pyrus Assessed by DNA Markers

Taxonomy in *Pyrus* is conducted mainly by morphology and geographic distribution. Classification of species of pears is very problematic and is often confused due to following reasons: (1) lack of wild populations, especially in cultivated species; (2) poor morphological diversity and lack of distinguishing characters among species; (3) widespread crossability and consequent interspecific hybridization and introgression among species. Recently, many efforts using DNA markers from the nuclear genome as well as chloroplast DNAs have been conducted, and reveal taxonomical relationships and the course of evolution in pears.

SSR markers are efficient tools to assess taxonomical relationships in pear, because of their advantages over other markers, i.e., co-dominant and typically neutral inheritance, large number of alleles per locus, abundance in genomes, and suitability for automation. Fifty-eight Asian pear accessions from 6 *Pyrus* species, 98 pear cultivars native mainly to East Asia, and 145 wild and cultivated individuals of *P. communis*, were analyzed by 9, 6 and 13 SSR loci, respectively (Kimura et al., 2002, Bao et al., 2007, Volk et al., 2006). SSR markers could reveal genetic and taxonomical relationships in pears and showed distinctive groups generally corresponding to species in taxonomy. It appears that exact identification of species may be difficult, for closely related varieties, and hybrid and introgressed germplasms.

Chloroplast DNA (cpDNA) usually shows maternal inheritance in angiosperms. Sequence conservation within cpDNA allows us to compare phylogenetic relationships at various taxonomic levels (Palmer et al., 1985). In spite of the conservation within the chloroplast genome, structural alterations such as insertions, deletions, inversions and translocations in cpDNA have been found in related plants by comparing the structure of cpDNAs. Tracing the mutational events in cpDNA provides useful tools to trace the course of evolution by reconstructing the plant phylogeny (Downie and Palmer, 1992). A physical map of cpDNA of pear was constructed using 5 restriction enzymes (Katayama and Uematsu, 2003). Pear cpDNA was found to be a circular molecule with a total size of about 156 kb in which 2 inverted repeats of 24.8 kb divide the molecule into small (17 kb) and large (90 kb) single-copy regions. RFLP analysis was carried out on cpDNAs from 5 *Pyrus* species (*P. pyrifolia*, *P. ussuriensis*, *P. calleryana*, *P. elaeagrifolia* and *P. communis*) and 2 mutations, a recognition-site mutation and a length mutation (deletion), were found only in the cpDNA of *P. pyrifolia* cultivars. This information will make it possible to investigate the phylogenetic relationships between *Pyrus* species.

Iketani et al. (1998) examined polymorphism of chloroplast DNAs of 106 accessions of mainly East Asian accessions. Four haplotypes were observed with the combination of 3 independent restriction site mutations and all 4 types appeared in the oriental pear accessions. This suggests that the oriental species of *Pyrus* and occidental ones may have evolved independently. The distribution of four haplotypes in the East Asian pear was quite incongruent with the species or infrageneric classification using mainly morphological characters. Considering the high crossability and frequent occurrence of suspected interspecific hybrids in wild populations, the dis-accordance is inferred to be the results of the hybridization and introgression between species.

Nucleotide sequences at 6 noncoding regions of cpDNAs were identified for 8 pear varieties from 5 species (Kimura et al., 2003). A total of 38 mutations such as nucleotide substitutions, deletions and insertions were found in more than 5.7 kbp of nucleotide sequences. Nucleotide sequences at the *trnL-trnF* were revealed for 33 pear varieties and 8 mutations were identified. A cladogram obtained from the data showed that Asian pear varieties were divided into 6 groups and that intraspecific as well as interspecific diversities existed in Asian pears, whereas European pear varieties were identical with respect to the *trnL-trnF* region.

## *2.2 DNA Markers*

### 2.2.1 DNA Marker Systems (Microsatellites, SNPs, AFLP, RAPD, etc.)

Microsatellites, or SSRs (simple sequence repeats), are polymorphic loci present in nuclear DNA that consist of repeating units of 1–4 base pairs in length. They are typically neutral and co-dominant, and show high degree of polymorphism and suitability for automation (Weber and May, 1989). SSR markers have several advantages over other molecular markers, which provide a more reliable method for DNA fingerprinting because of their co-dominant inheritance, large number of alleles per locus, and abundance in genomes. In addition, since SSR analysis is based on a PCR method, the technique is simple and only a small amount of DNA is required. More than 100 SSRs have been developed from European and Japanese pears (Yamamoto et al., 2002a, b, c; Sawamura et al., 2004; Fernandez-Fernandez et al., 2006; Inoue et al., 2007). These SSR markers have been used as molecular markers which have wide-ranging applications for the evaluation of genetic diversity (Kimura et al., 2002, Volk et al., 2006, Bao et al., 2007), cultivar identification, and the construction of genetic linkage maps (Yamamoto et al., 2002c, 2004, 2007).

RAPD stands for random amplification of polymorphic DNA and is the segments of DNA that are amplified are random. The RAPD reaction is performed with arbitrary, short primers (8–12 nucleotides). No knowledge of the DNA sequence for the targeted gene is required, as the primers will bind somewhere in the sequence, but it is not certain exactly where. Its resolving power is much lower than targeted, species specific DNA comparison methods, such as SSRs. In recent years, RAPD technology has been used to characterize, and trace, the phylogeny of diverse plant species. RAPDs have also been widely used on pear genetic studies because RAPDs have the advantage of being readily employed, requiring small amounts of genomic DNA. RAPD markers have been successfully used for identification and genetic relationships of pear (Oliveira et al., 1999, Teng et al., 2001, 2002).

AFLP (amplified fragment length polymorphism) markers are a highly sensitive method for detecting polymorphisms used in the study of genetics and in the practice of genetic engineering, which was developed in the early 1990's by Keygene (Vos et al., 1995). AFLP uses restriction enzymes to cut genomic DNA, followed by ligation of complementary double stranded adaptors to the ends of the restriction fragments. A subset of the restriction fragments are then amplified using 2 primers complementary to the adaptor and restriction site fragments. The fragments are visualized on denaturing polyacrylamide gels either through autoradiographic or fluorescence methodologies. AFLP has several advantages over the RAPD technique, like a higher number of loci analyzed and a higher reproducibility of banding patterns on genetic diversity study in pear (Monte-Corvo et al., 2000).

ISSR (inter-simple sequence repeat) is a general term for a genome region between microsatellite loci. The complementary sequences to 2 neighboring microsatellites are used as PCR primers. Sequences amplified by ISSR-PCR can be used for DNA fingerprinting. Since an ISSR may be a conserved or non-conserved region, this technique is not useful for distinguishing individuals, but rather for phylogeographical analyses or maybe delimiting species. Sequence diversity is lower than in SSRs, but still higher than in actual gene sequences. Monte-Corvo et al. (2001) reported that ISSR analysis was used for cultivar identification and the determination of phylogenetic relationship in pears (*P. communis*).

A single nucleotide polymorphism (SNP) is a DNA sequence variation occurring when a single nucleotide – A, T, C, or G – in the genome differs between members of a species. For example, 2 sequenced DNA fragments from different individuals, AAGCCTA to AAGCTTA, contain a difference in a single nucleotide. Single nucleotide polymorphisms may fall within coding sequences of genes, noncoding regions of genes, or in the intergenic regions between genes. SNPs within a coding sequence will not necessarily change the amino acid sequence of the protein that is produced, due to degeneracy of the genetic code. It is believed that SNP will be the most efficient tool for comprehensive genetic studies in near future. Since nucleotide sequences of pear genome remain very limited, SNP markers are not frequently utilized for pear at this moment.

### 2.2.2 DNA Markers Associated with Interest Genes

Several phenotypic traits, including tree and production characters, fruit quality, disease resistance, pest resistance and adaptability, have been studied in pears. Since it is believed that a lot of phenotypic traits are controlled by polygenes or QTLs, it is not easy to develop DNA markers associated with specific genes and various characteristics. Associated molecular markers have been isolated for some specific phenotypic traits, some of which were identified based on genetic linkage maps. Recently, resistance to major diseases for pear cultivation, fire blight, pear scab and black spot were analyzed and genes (or QTLs) controlling resistances were identified in genetic linkage maps (Table 3). Other genes of interest for self incompatibility and fruit ripening were also investigated (Table 3).

#### Disease Resistance (Fire Blight, Pear Scab, Black Spot)

Fire blight caused by *Erwinia amylovora* (Burr.) Winsl. is the most serious disease in pears in North America and Western Europe. In the 19th century, breeding for fire blight resistance started to introgress resistance of Asian pears into European pears. Some interspecific hybrid cultivars between Asian and European pears showed more resistant to fire blight than European pears, but they were inferior fruit quality. Large scale evaluation for fire blight disease showed that resistance in *P. communis* is relatively rare with 5–10% being rated as at least moderately resistant. Dondini et al. (2004) designed 2 genetic linkage maps of European pears 'Harrow Sweet' (resistant) and 'Passe Crassane' (susceptible) with F1 progeny crossed between them. Different types of DNA markers including SSRs, MFLPs (microsatellite-anchored fragment length polymorphisms), AFLPs, RGAs (resistant gene analogs) and AFLP-RGAs were applied for construction of their genetic maps. The 'Harrow Sweet' map consisted of 156 loci, for a total length of 912 cM divided into 19 linkage groups. Four putative QTLs related to resistance against fire blight were identified in LGs 2, 4 and 9 (2 QTLs in LG 2) of the map of 'Harrow Sweet'. No QTLs were found for susceptible cultivar 'Passe Crassane'. About 50% of the total variance was explained by 4 QTLs, and 2 QTLs found in LG 2 showed large LOD values and high variance explained. This study will lead to identify associated DNA markers with fire blight resistance in order to develop MAS approach.

Pear scab, caused by two species of *Venturia*, *V. nashicola* and *V. pirina*, is one of the most serious diseases of Asian and European pears, especially Japanese pear. *Venturia nashicola* infects Asian pears throughout their natural range, and *V. pirina* occurs in most regions where European pears are grown. The species are classified in the same genus as *V. inaequalis*, which causes apple scab. *Venturia nashicola* is pathogenic only on Asian pears and is not pathogenic on European pears (Bell et al., 1996, Ishii et al., 2002). In contrast, Japanese and Chinese pears are generally resistant to *V. pirina* (Bell et al., 1996, Ishii et al., 2002). None of the major commercial Japanese pear cultivars are resistant to scab disease caused by *V. nashicola* (Ishii et al., 1992, Bell et al., 1996), but no scab symptoms were observed on the indigenous Japanese pear cultivar 'Kinchaku', some Chinese pears, and many European

**Table 3** DNA markers associated with interest genes and their positions in genetic linkage map

| Characters | Gene symbol | Gene source | Associated DNA markers (gene name, QTL name) | Linkage group No. | References |
|---|---|---|---|---|---|
| scab resistance to *V. nashicola* | *Vnk* | Kinchaku | STS-OPW2, STS-CT/CTA | 1 | Terakami et al. (2006) |
| fire blight resistance | | Harrow Sweet | HS2A (QTL) | 2 | Dondini et al. (2004) |
| | | Harrow Sweet | HS2B (QTL) | 2 | Dondini et al. (2004) |
| | | Harrow Sweet | HS4 (QTL) | 4 | Dondini et al. (2004) |
| | | Harrow Sweet | HS9 (QTL) | 9 | Dondini et al. (2004) |
| black spot susceptibility | *A* | Osa Nijisseiki | CMNB41/2350 | unknown | Banno et al. (1999) |
| | *Ani* | Osa Nijisseiki | CH04h02, CH03d02 | 11 | Terakami et al. (2007) |
| | *Ana* | Nansui | CH04h02, CH03d02 | 11 | Terakami et al. (2007) |
| self-incompatibility | *S* | Japanese pear, European pear | S-RNase (gene name) | 17 | Yamamoto et al. (2002) |
| fruit skin color | | Niitaka | OPH-19-425 | unknown | Inoue et al. (2006) |
| fruit storage | | Japanese pear | ACC synthase (gene name) | unknown | Itai et al. (2003) |

pears. Inheritance analysis indicated that resistance of 'Kinchaku' is controlled by a single dominant gene (Abe and Kotobuki, 1998). Genetic linkage maps of the Japanese pear cultivars 'Kinchaku' and 'Kousui' were constructed using RAPD markers (Iketani et al., 2001). The 'Kinchaku' map consisted of 120 loci in 18 linkage groups covering a length of ca. 770 cM, in which 2 disease-related genes associated with resistance to pear scab and susceptibility to black spot (caused by *Alternaria alternata*) were mapped. The resistance gene *Vnk* of the Japanese pear 'Kinchaku' against pear scab disease was identified in the central region of LG 1 (Terakami et al., 2006). Six DNA markers (one SSR Hi02c07 and 5 STSs converted from AFLP and RAPDs) showed tight linkages to *Vnk*, being mapped with distances ranging from 2.4 to 12.4 cM. The SSR CH-Vf2, which was isolated from a BAC clone of the contig containing the apple scab gene *Vf*, was mapped at the bottom of linkage group 1 in 'Kinchaku', suggesting that the *Vnk* and *Vf* loci are located in different genomic regions of the same homologous linkage group.

Black spot disease, caused by *Alternaria alternata* (Fr.) Keissler Japanese pear pathotype, is one of the most serious diseases in Japanese pear cultivation. Large amount of costs and labors are required for bagging and spraying of fungicide in order to prevent infection of this disease (Kozaki, 1973). The AK-toxin, which is specifically produced by *A. alternata* Japanese pear pathotype, causes the necrosis, early leaf fall, and decrease of yield for Japanese pears (Nakashima et al., 1985). Many major Japanese pear cultivars 'Nijisseiki', 'Shinsui', and 'Nansui' show susceptibility to this disease, and the susceptibility to black spot is controlled by single dominant gene designated as *A* (Kozaki, 1973). Banno et al. (1999) tested 250 RAPD primers to screen a pair of bulked DNA samples derived from open-pollinated progeny of Japanese pear 'Osa Nijisseiki' to identify markers linked to the susceptible *A* gene. One RAPD marker CMNB41 was identified to show linkage to the susceptibility gene at a genetic distance of 3.1 cM. Iketani et al. (2001) reported that the susceptibility to black spot was identified in the genetic linkage map of 'Kinchaku'. More recently, the susceptibility genes to black spot disease of the Japanese pear cultivars 'Osa Nijisseiki' (designated as *Ani*) and 'Nansui' (*Ana*) were genetically identified and mapped in their genetic linkage maps, locating at the top region of a linkage group 11 (Terakami et al., 2007). Two SSR markers CH04h02 and CH03d02 showed tight linkages to *Ani* and *Ana*, using a genome scanning approach (GSA, Patocchi et al., 2005). This information about the position and molecular markers linked to the disease resistance genes will be useful for marker-assisted selection and for pyramiding resistances in pear breeding programs.

### Self Incompatibility

Most pear cultivars show self-incompatibility and the proposition of pollinizers inter-planted in the orchard is a requirement to get an economic crop from most of the cultivars (Sanzol and Herrero, 2002). In *Pyrus*, gametophytic self-incompatibility is controlled by a single locus, the S-locus. The S-locus harbors a multi-allelic gene, which encodes for S-RNase that blocks incompatible-tube growth through the style (Ushijima et al., 1998). In Japanese pear, cDNAs encoding S1- to S9-RNase have been isolated and sequenced (Sassa et al., 1997, Ishimizu et al., 1998, Takasaki et al., 2004). Ishimizu et al. (1999) established a PCR-RFLP system for S-genotype assignment in Japanese pear. Molecular techniques were used for the identification of S-genotypes in European pears (Sanzol and Herrero, 2002; Zuccherelli et al., 2002; Zisovich et al., 2004).

Recently, candidates of pollen S (S locus F-box brothers, SFBB) were identified in apple and pear (Sassa et al., 2007). Three SFBB genes were isolated in each of the Japanese pear S4 and S5 haplotypes. These SFBB genes in Japanese pear show S haplotype-specific sequence polymorphism, which can be used as CAPS markers for identifying self-incompatibility genotypes.

The self-incompatibility locus (S locus) was mapped in the Japanese pear 'Housui' and the European pear 'Bartlett' in their genetic linkage maps at the bottom of the linkage group 17 (Yamamoto et al., 2002c). The position of S locus in both pear and apple was identified in the same homologous LG 17 (Maliepaard et al., 1998).

### Other Phenotypic Traits

Ethylene production drastically varies during fruit ripening in cultivated Japanese pears. Climacteric-type fruits exhibit a rapid increase in ethylene production and show a low storage potential. Non-climacteric fruits show no detectable ethylene production and their fruit quality is maintained for over a month in storage. Itai et al. (1999, 2003a) cloned three ACC (1-aminocyclopropane-1-carboxylate) synthase genes (*PPACS1, 2, 3*) and showed that fruit storage potential was closely related to ethylene production and expression of ACC synthase genes during fruit ripening. It was identified that *PPACS1* was expressed in cultivars with high ethylene production, while *PPACS2* was specifically expressed in cultivars of moderate ethylene production. These two ACC synthase genes (*PPACS1, 2*) were identified as RFLP markers, differentiating to high, moderate and low ethylene production. Furthermore, CAPS markers were established, converted from RFLP markers (Itai et al., 2003b), which will be utilized for selection of Japanese pear cultivars with enhanced post-harvest storage ability.

Since pears are mainly served as fresh fruits and must have an attractive appearance, the fruit color is one of the most important factors contributing to appearance. In Japanese pears, yellow-green and brown russet fruits are preferred for consumers. Inoue et al. (2006) reported the RAPD marker linked to major genes controlling the fruit skin color in Japanese pear. Two F1 progenies from the cross of 'Kousui' × 'Kinchaku' and 'Niitaka' × 'Chikusui' segregated by fruit skin color were used for bulked segregant analysis. After 200 random primers were screened against bulks, the 425-bp band produced with OPH-19 primer (OPH-19-425) was selected in association with green bulks. The recombination rate between OPH-19-425 and the green skin phenotype was 7.3%.

## 2.3 Linkage and Physical Maps

### 2.3.1 Genetic Linkage Maps in European Pears and Japanese Pears

High-density genetic linkage maps are very useful for fundamental and applied genetic research. Linkage maps enable studies of the genome structure, the localization of interest genes, identification of quantitative trait loci (QTLs), and conduction of marker-assisted selection (MAS) and marker-assisted breeding (MAB). Several genetic linkage maps of pears have been reported in the European pear (*P. communis*) and the Japanese pear (*P. pyrifolia*) (Table 4). Recent pear genetic linkage maps contain linkage groups corresponding to its basic chromosome number (n = 17), and are sufficiently dense and saturated. In order to conduct MAS and to evaluate genome structures, it will be necessary to construct pear genetic maps covering entire genome regions with a large number of DNA markers.

Iketani et al. (2001) reported the construction of RAPD-based genetic maps of the Japanese pear varieties ‘Kinchaku’ and ‘Kousui’. The former map consisted of 120 loci in 18 linkage groups covering a length of ca. 770 cM. The map of ‘Kousui’ contains 78 RAPD loci in 22 linkage groups extending 508 cM. The resistance to pear scab disease (*Vn*) and the susceptibility to black spot disease (*A*) were identified in the genetic map of ‘Kinchaku’ and several RAPD markers were found to show significant linkages to pear scab resistance and black spot susceptibility.

Partial genetic linkage maps of the European pear cultivars ‘Passe Crassane’, ‘Harrow Sweet’, ‘Abbe Fetal’ and ‘Max Red Bartlett’ were constructed using apple SSRs, showing 3 linkage groups 10, 12 and 14 (Pierantoni et al., 2004). Dondini et al. (2004) reported on 2 genetic linkage maps that were made of the European pears ‘Passe Crassane’ and ‘Harrow Sweet’. The former map included 155 loci for a total length of 912 cM organized in 18 linkage groups. The ‘Harrow Sweet’ map consisted of 156 loci, for a total length of 930 cM divided into 19 linkage groups. Hemmat et al. (2003) suggested that many apple SSRs would be useful for genetic mapping in European pears in a preliminary experiment.

Integrated high-density genetic linkage maps were constructed for the European pear cultivar ‘Bartlett’ and ‘La France’, and the Japanese pear cultivar ‘Housui’ based on AFLPs, SSRs (from pear, apple and Prunus), isozymes, and phenotypic traits (Yamamoto et al., 2002c, 2004, 2007). The map of ‘Bartlett’ consisted of 447 loci including 58 pear SSRs, 60 apple SSRs and 322 AFLPs, which covered 17 linkage groups over a total length of 1,000 cM with an average distance of 2.3 cM between markers. Another genetic linkage map of ‘La France’ contained 414 loci including 66 pear SSRs, 68 apple SSRs and 279 AFLPs, on 17 linkage groups encompassing a genetic distance of 1,156 cM. Both maps consisted of more than 400 loci and covered 17 linkage groups, which corresponded to the basic chromosome number of pear (n=17). Both maps were well aligned using a total of 97 SSR markers in the 17 linkage groups. The map of ‘Housui’ contains 180 loci including 110 AFLPs, 64 SSRs (29 pear, 29 apple, 6 *Prunus* SSRs) on 20 linkage groups encompassing a genetic distance of 995 cM (Yamamoto et al., 2004). Three linkage groups or chromosomal regions could not be established because no

**Table 4** Characteristics of genetic linkage maps in pear

| Cultivar names | Species | Population | No. of linkage group | No. of loci | Map distance (cM) | Marker types | References |
|---|---|---|---|---|---|---|---|
| Kousui | *Pyrus pyrifolia* | Kinchaku × Kousui F1 (82 progeny) | 22 | 78 | 508 | RAPD | IKetani et al. (2001) |
| Kinchaku | *Pyrus pyrifolia* | Kinchaku × Kousui F1 (82 progeny) | 18 | 120 | 768 | RAPD | IKetani et al. (2001) |
| Housui | *Pyrus pyrifolia* | Bartlett × Housui F1 (63 progency) | 16 | 307 | 1,054 | SSR, AFLP | Yamamoto, unpublished |
| Passe Crassane | *Pyrus communis* | Pause Crassane × Harrow Sweet (99 progeny) | 18 | 155 | 912 | SSR, MFLP, AFLP, RGA | Dondini et al. (2004) |
| Harrow Sweet | *Pyrus communis* | Pasue Crassane × Harrow Sweet (99 progeny) | 19 | 156 | 930 | SSR, MFLP, AFLP, RGA | Dondini et al. (2004) |
| Bartlett | *Pyrus communis* | Bartlett × Housui F1 (63 progeny) | 17 | 447 | 1,004 | SSR, AFLP | Yamamoto et al. (2007) |
| La France | *Pyurs communis* | Shinsei × (Housui × La France) (55 progeny) | 17 | 414 | 1,156 | SSR, AFLP | Yamamoto et al. (2007) |

SSR and AFLP markers of these regions showed segregating alleles (fragments) for 'Housui'.

### 2.3.2 Genome Structure Between Species Based on Linkage Maps in PYRUS

Genetic linkage maps have been constructed for European pear cultivars 'Bartlett' and 'La France' (Yamamoto et al., 2004, 2007), 'Passe Crassane' and 'Harrow Sweet' (Dondini et al., 2004), and 'Passe Crassane', 'Harrow Sweet', 'Abbe Fetel' and 'Max Red Bartlett' (Pierantoni et al., 2004) (Table 4). Japanese pear maps were reported for 'Kinchaku' and 'Kousui' (Iketani et al., 2001), and 'Housui' (Yamamoto et al., 2002c, 2004). However, the saturated high-density genetic linkage maps of pear were constructed only for European pear cultivars 'Bartlett' and 'La France' (Yamamoto et al., 2007). At this moment, it is rather difficult to compare genome structure between species based on linkage maps in *Pyrus*. Yamamoto et al. (2007) described that all 17 linkage groups of 'Bartlett' and 'La France' could be connected together by using a total of 97 SSR loci with at least 1 SSR locus per linkage group. The linkage groups 10 and 14 of both maps were well consolidated by 10 and 11 anchor loci, respectively (Fig. 2). The positions and linkage groups of commonly mapped SSR loci were well conserved between 'Bartlett' and 'La France' except for a very few exceptions which may have been due to multi-locus SSRs, indicating that the genome structure is very well conserved within European pears.

Two saturated (or high-density) maps were published in apple (Maliepaard et al., 1998, Liebhard et al., 2003). The first one was based on the F1 progeny of a cross between the apple cultivars 'Prima' and 'Fiesta' (Maliepaard et al., 1998), and another one was based on 267 F1 progeny from a cross of 'Fiesta' × 'Discovery' (Liebhard et al., 2002, 2003). When pear genetic linkage maps of 'Bartlett' and 'La France' were compared with the maps of 'Discovery' and 'Fiesta', 66 apple SSR loci could be successfully identified into the same homologous linkage groups between pear and apple. Moreover, their positions within linkage groups were identified in almost the same regions between pear and apple. Since genetic relatedness within *Pyrus* is much closer than that between apple and pear, it is believed that genome structure should be very well conserved within *Pyrus*.

About 10–20% of SSR markers in pear are multi-loci as well as in apple (Yamamoto et al., 2004, Liebhard et al., 2002, 2003). Liebhard et al. (2002, 2003) pointed out duplication patterns of multi-locus SSRs in the linkage group pairs 1–7, 4–12, 5–10, 9–17, 12–13, 12–14 and 1–3. In pear, duplication of the linkage group pairs 2–5, 9–17, 5–10, 3–14, 2–15, 1–3, 10–17, 8–15, 13–16 and 12–14 were revealed by multi-locus SSRs between 'Bartlett' and 'La France'. Duplication of 1–3, 5–10, 12–14 and 9–17 were commonly observed in apple and pear. This information suggests that at least 2 homologous chromosomes or genomic regions exist in pear genome. Recent molecular genetic studies have provided supporting evidence that the subfamily Maloideae originated from autopolyploidy or hybridization between closely related members of a single lineage, with species of the Spiraeoideae subfamily being the most probable parental lineages (Morgan

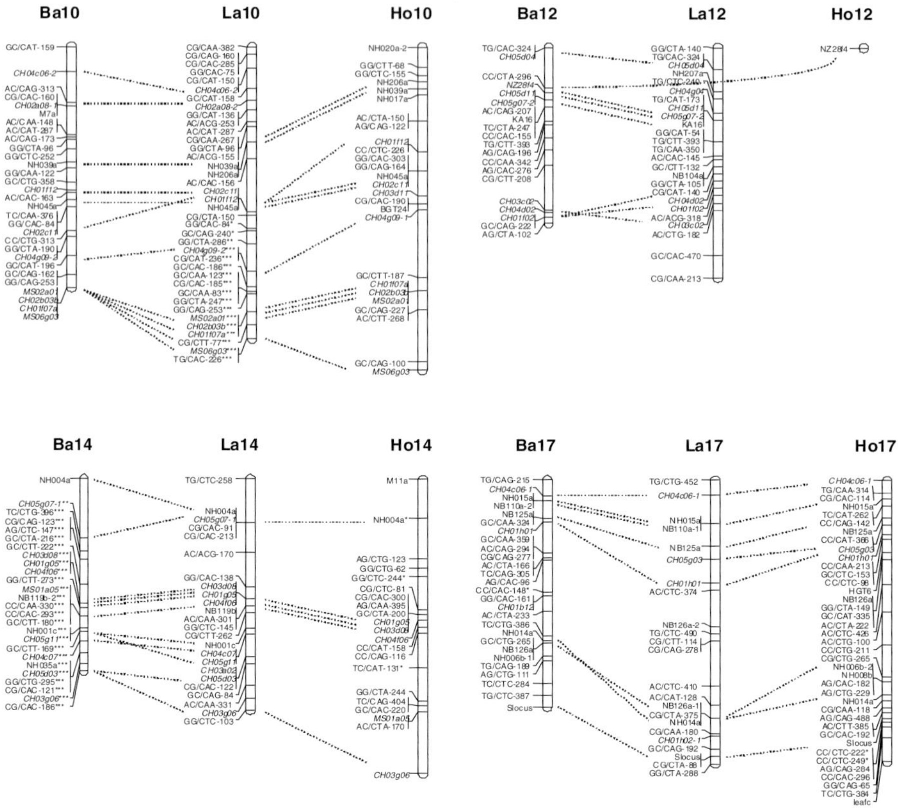


**Fig. 2** Comparative linkage groups 10, 12, 14 and 17 of pear cultivars 'Bartlett' (Ba), 'La France' (La) and 'Housui' (Ho) (Yamamoto et al., 2007, unpublished data). The SSR markers and S locus observed in both maps are indicated by the *dotted lines*

et al., 1994, Evans and Campbell, 2002). Linkage groups identified by multi-loci SSRs might reflect the polyploid nature of pear. It will be interesting to evaluate the genome structure of polyploid origin using detailed DNA markers.

### 2.3.3 Physical Map

Physical maps are particularly important when searching for interest genes by positional cloning strategies and for DNA sequencing. Physical mapping is the process of determining how DNA contained in a group of clones overlap without having to sequence all the DNA in the clones. Once the map is determined, we can use the clones as a resource to efficiently contain stretches of genome in large quantity. This type of mapping is more accurate than genetic maps. However, it appears that no or few information is available for physical map in pear.

## 2.4 Association Mapping

Variations in the DNA sequences of humans can affect how humans develop diseases, respond to pathogens, chemicals, drugs, etc. However, their greatest importance in biomedical research is for comparing regions of the genome between cohorts. The study of single nucleotide polymorphisms is also important in crop and livestock breeding programs. Genetic association studies including linkage disequilibrium (LD) are performed to determine whether a genetic variant is associated with a disease or trait: if association is present, a particular allele, genotype or haplotype of a polymorphism or polymorphism(s) will be seen more often than expected by chance in an individual carrying the trait. In pears, it appears that no or few information is available for association mapping. This approach will become a powerful tool to identify associated DNA markers for many important phenotypic traits controlled by polygenes or QTLs. Association mapping will lead to development of DNA markers associated with interest genes and characteristics such as tree and production characters, fruit quality, disease resistance, pest resistance and adaptability would be controlled by polygenes or QTLs in pears.

## References


Abe K, and Kotobuki K (1998) Inheritance of high resistance to *Venturia nashicola* Tanaka et Yamamoto in Japanese pear (*Pyrus pyrifolia* Nakai) and Chinese pear (*P. ussuriensis* Maxim.). J Japan Soc Hort Sci 67: 677–680

Banno K, Ishikawa H, Hamauzu Y, and Tabira H (1999) Identification of a RAPD marker linked to the susceptible gene of black spot disease in Japanese pear. J Japan Soc Hortic Sci 68: 476–481

Bao L, Chen K, Zhang D, Cao Y, Yamamoto T, and Teng T (2007) Genetic diversity and similarity of pear (*Pyrus* L.) cultivars native to East Asia revealed by SSR (simple sequence repeat) markers. Genet Resour Crop Evol 54: 959–971

Bell RL (1990) Pears (*Pyrus*). In: Moore JN, Ballington JR Jr. (eds), Genetic resources of temperate fruit and nut crops I. International Society for Horticultural Science, Wageningen, The Netherlands, pp655–697

Bell RL, Quamme HA, Layne REC, and Skirvin RM (1996) Pears. In: Janick J, Moore JN (eds), Fruit breeding, volume I: tree and tropical fruits. John Wiley and Sons Inc, UK, pp441–514

Bell RL, Scorza R, Srinivasan C, and Webb K (1999) Transformation of "Beurre Bosc" pear with the *rolC* gene., J Amer Soc Hortic Sci 124: 570–574

Bellini E (1995) Pear breeding and genetics (*Pyrus communis* L.). In Bellini E (ed), State of the art and perspectives of world genetic improvement of fruit tree species (peach, plum, apricot and pear), Societa Orticola Italiana, Firenze, pp99–157

Bellini E, and Nin S (2002) Breeding for new traits in pear. Acta Hort 596: 217–224

Bouvier L, Zhang YZ, and Lespinasse Y (1993) Two methods of haploidization in pear, *Pyrus communis* L., greenhouse seedling selection and in situ parthenogenesis induced by irradiated pollen. Theor Appl Genet 87: 229–232

Chevreau E, and Bell R (2005) *Pyrus* spp. pear and *Cydonia* spp. Quince. In Litz RE (ed), Biotechnology of Fruit and Nut Crops, C.A.B. International, Wallingford, UK, pp543–565

Chevreau E, Decourtye L, and Skirvin RM (1989) A review of pear chimeras: their identification and separation into pure types. HortSci 24: 32–34

Djennane S, Cesbron C, Sourice S, Loridon K, and Chevreau E (2007) Marker-free transgenic pear production by inducible expression of R-recombinase. ISHS International Symposium on Genetic transformations – challenges and possibilities for horticulture of the world. 16–20 September 2007, Ski, Norvège. (poster)

Dolatowski J, Nowosielski J, Podyma W, Szymanska M, and Zych M (2004) Molecular studies on the variability of Polish semi-wild pears (*Pyrus*) using AFLP. J Fruit Orna Plant Res 12: 331–337

Dondini L, Pierantoni L, Gaiotti F, Chiodini R, Tartarini S, Bazzi C, and Sansavini S (2004) Identifying QTLs for fire-blight resistance via a European pear (*Pyrus communis* L.) genetic linkage map. Mol Breed 14: 407–418

Downie SR, and Palmer JD (1992) Use of chloroplast rearrangements in reconstructing plant phylogeny. In: Soltis PS, Soltis DE, Doyle JJ (eds) Molecular systematics of plants. Chapman and Hall, New York, pp14–35

Evans RC, and Campbell CS (2002) The origin of the apple subfamily (*Maloideae; Rosaceae*) is clarified by DNA sequence data from duplicated GBSSI genes. Am J Bot 89: 1478–1484

FAOSTAT 2007 http://faostat.fao.org/

Fernandez-Fernandez F, Harvey NG, and James CM (2006) Isolation and characterization of polymorphic microsatellite markers from European pear (*Pyrus communis* L.). Mol Ecol Notes 6: 1039–1041

Flaishman MA, Shlizerman L, Cohen Y, Kerem Z, and Sivan L (2005) Expression of the health-beneficial stilbenes in transgenic 'Spadona' pear (*Pyrus communis* L.). Acta Hortic 671: 283–288

Flaishman M, Cohen Y, Freeman A, Golubowicz S, Korchinsky R, and Yablowicz Z (2007) Development of a fast and compact breeding system in pear by the use of juvenile free transgenic plants. EUCARPIA XII Fruit section Symposium, September 16–20, 2007, Zaragoza, Spain, Abstract book, p200

Gao M, Matsuta N, Murayama N, Toyomasu T, Mitsuhashi W, Dandekar A, Tao R, and Nishimura K (2007) Gene expression and ethylene production in transgenic pear (*Pyrus communis* cv. 'La France') with sense or antisense cDNA encoding ACC oxidase. Plant Sci 173: 32–42

Hedrick UP, Howe GH, Taylor OM, Francis EH, and Tukey HB (1921) The pears of new York. New York Dept. Agric. 29th Annu. Rep. Vol2, part2

Hemmat M, Weeden NF, and Brown SK (2003) Mapping and evaluation of *Malus* × *domestica* microsatellites in apple and pear. J Amer Soc Hort Sci 128: 515–520

Iketani H, Manabe T, Matsuta N, Akihama T, and Hayashi T (1998) Incongruence between RFLPs of chloroplast DNA and morphological classification in east Asian pear (*Pyrus* spp.). Genet Resour Crop Evol 45: 533–539

Iketani H, Abe K, Yamamoto T, Kotobuki K, Sato Y, Saito T, Terai O, Matsuta N, and Hayashi T (2001) Mapping of disease-related genes in Japanese pear using a molecular linkage map with RAPD markers. Breed Sci 51: 179–184

Inoue E, Kasumi M, Sakuma F, Anzai H, Amano K, and Hara H (2006) Identification of RAPD marker linked to fruit skin color in Japanese pear (*Pyrus pyrifolia* Nakai). Sci Hortic 107: 254–258

Inoue E, Matsuki Y, Anzai H, and Evans K (2007) Isolation and characterization of microsatellite markers in Japanese pear (*Pyrus pyrifolia* Nakai). Mol Ecol Notes 7: 445–447

Ishii H, Udagawa H, Nishimoto S, Tsuda T, and Nakashima H (1992) Scab resistance in pear species and cultivars. Acta Phytopathol Entomol Hung 27: 293–298

Ishii H, Watanabe H, and Tanabe K (2002) *Venturia nashicola*: Pathological specialization on pears and control trial with resistance inducers. Acta Hort 587: 613–621

Ishimizu T, Shinkawa T, Sakiyama F, and Norioka S (1998) Primary structural features of rosaceous S-RNase associated with gametophytic self-incompatibility. Plant Mol Biol 37: 931–941

Ishimizu T, Inoue K, Shimonaka M, Saito T, Terai O, and Norioka S (1999) PCR-based method for identifying the S-genotypes of Japanese pear cultivars. Theor Appl Genet 98: 961–967

Itai A, Kawata T, Tanabe K, Tamura F, Uchiyama M, Tomomitsu M, and Shiraiwa N (1999) Identification of 1-aminocyclopropane-1-carboxylic acid synthase genes controlling the ethylene level of ripening fruit in Japanese pear (*Pyrus pyrifolia* Nakai). Mol Gen Genet 261: 42–49

Itai A, Tanabe K, Tamura F, and Tomomitsu M (2003a) Cloning and characterization of a cDNA encoding 1-aminocyclopropane-1-carboxylate (ACC) synthase (*PPACS3*) from ripening fruit of Japanese pear (*Pyrus pyrifolia* Nakai). J Japan Soc Hort Sci 72: 99–106

Itai A, Kotaki T, Tanabe K, Tamura F, Kawaguchi D, and Fukuda M (2003b) Rapid identification of ACC synthase genotypes in cultivars of Japanese pear (*Pyrus pyrifolia* Nakai) using CAPS markers. Theor Appl Genet 106: 1266–1272

Katayama H, and Uematsu C (2003) Comparative analysis of chloroplast DNA in *Pyrus* species: physical map and gene localization. Theor Appl Genet 106: 303–310

Kikuchi A (1946) Speciation and taxonomy of Chinese pears. Collected Records of Hort. Res. 3, 1–8. Kyoto Univ. (Trans. by K. Park; ed. by M. N. Westwood)

Kim CS, Lee GP, Han DH, Ryu KH, and Lee CH (2000a) Classification and identification of *Pyrus pyrifolia* using RAPD. J Kor Soc Hortic Sci 41: 119–124

Kim CS, Lee GP, Han DH, Ryu KH, and Lee CH (2000b) SCARs markers derived from RAPD for cultivar identification in *Pyrus pyrifolia*. J Kor Soc Hortic Sci 41: 125–128

Kim D, and Ko KC (2004) Identification markers and phylogenetic analysis using RAPD in Asian pears (*Pyrus* spp.). J Kor Soc Hortic Sci 45: 194–200

Kimura T, Shi Y, Shoda M, Kotobuki K, Matsuta N, Hayashi T, Ban Y, and Yamamoto T (2002) Identification of Asian pear varieties by SSR analysis. Breed Sci 52: 115–121

Kimura T, Iketani H, Kotobuki K, Matsuta N, Ban Y, Hayashi T, and Yamamoto T (2003) Genetic characterization of pear varieties revealed by chloroplast DNA sequences. J Horti Sci Biotech 78: 241–247

Kozaki I. (1973) Black spot disease resistance in Japanese pear. I. Heredity of the disease resistance. Bull Hort Res Stn A12: 17–27. (in Japanese with English summary)

Lebedev VG, Taran SA, and Dolgov SV (2002a) Pear transformation by gene of supersweet protein Thaumatin II for fruit taste modification. Acta Hort 596: 199–202

Lebedev VG, Lavrova N, Lunin VG, and Dolgov SV (2002b) Plant-Defensin genes introduction for improvement of pear phytopathogen resistance. Acta Hort 596: 167–172

Lee GP, Lee CH, and Kim CS (2004). Molecular markers derived from RAPD, SCAR, and the conserved 18S rDNA sequences for classification and identification in *Pyrus pyrifolia* and *P. communis*. Theor Appl Genet 108: 1487–1491

Liebhard R, Gianfranceschi L, Koller B, Ryder CD, Tarchini R, Van De Weg E, and Gessler C (2002) Development and characterisation of 140 new microsatellites in apple (*Malus × domestica* Borkh.). Mol Breed 10: 217–241

Liebhard R, Koller B, Gianfranceschi L, and Gessler C (2003) Creating a saturated reference map for the apple (*Malus × domestica* Borkh.) genome. Theor Appl Genet 106: 1497–1508

Maliepaard C, Alston FH, van Arkel G, Brown LM, Chevreau E, Dunemann F, Evans KM, Gardiner S, Guilford P, van Heusden AW, Janse J, Laurens F, Lynn JR, Manganaris AG, den Nijs APM, Periam N, Rikkerink E, Roche P, Ryder C, Sansavini S, Schmidt H, Tartarini S, Verhaegh JJ, Vrielink-van Ginkel M, and J. King G (1998) Aligning male and female linkage maps of apple (*Malus pumila* Mill.) using multi-allelic markers. Theor Appl Genet 97: 60–73

Malnoy M, Reynoird JP, and Chevreau E (2000) Preliminary evaluation of new gene transfer strategies for resistance to fire blight in pear. Acta Hort 538: 635–638

Malnoy M, Venisse JS, Brisset MN, and Chevreau E (2003a) Expression of bovine lactoferrin cDNA confers resistance to *Erwinia amylovora* in transgenic pear. Mol Breed 12: 231–244

Malnoy M, Venisse JS, Reynoird JP, and Chevreau E (2003b) Activation of three pathogen-inducible promoters of tobacco in transgenic pears (*Pyrus communis* L.) after biotic and abiotic elicitation. Planta 216: 802–814

Malnoy M, Faize M, Venissse JS, Geider K, and Chevreau E (2005a) Expression of viral EPS-depolymerase reduces fire blight susceptibility in transgenic pear. Plant Cell Rep 23: 632–638

Malnoy M, Venisse JS, and Chevreau E (2005b) Expression of a bacterial effector, Harpin N, causes increased resistance to fire blight in *Pyrus communis*. Tree Genet Genomes 1: 41–49

Matsuda N, Ikeda K, Kurosaka M, Isuzugawa K, Endo T, Omura M, and Takashina T (2006) In vitro flowering of transgenic pears (*Pyrus communis* L.) expressing CiFT, a *Citrus* ortholog of the Arabidopsis FT gene. 3rd International Rosaceae Genomics Conference, Napier, New Zealand (poster)

Monte-Corvo L, Cabrita L, Oliveira C, and Leitao J (2000) Assessment of genetic relationships among *Pyrus* species and cultivars using AFLP and RAPD markers. Genet Resour Crop Evol 47: 257–265

Monte-Corvo L, Goulao L, and Oliveira C (2001) ISSR analysis of cultivars of pear and suitability of molecular markers for clone discrimination. J Am Soc Hortic Sci 126: 517–522

Morgan DR, Soltis DE, and Robertson KR (1994) Systematic and evolutionary implications of rbcL sequence variation in *Rosaceae*. Am J Bot 81: 890–903

Mourgues F, Chevreau E, Lambert C, and De Bondt A (1996) Efficient *Agrobacterium*-mediated transformation and recovery of transgenic plants from pear (*Pyrus communis* L.). Plant Cell Rep 16: 245–249

Nakashima T, Ueno T, Fukami H, Taga T, Masuda H, Osaki K, Otani H, Kohmoto K, and Nishimura S (1985) Isolation and structures of AK-toxin I and II, host-specific phytotoxic metabolites produced by *Alternaria alternata* Japanese pear pathotype. Agri Biol Chem 49: 807–815

Ochatt SJ, and Caso OH (1986) Shoot regeneration from leaf mesophyll protoplasts of wild pear (*Pyrus communis* var. *Pyraster* L.). J Plant Physiol 122: 243–249

Ochatt SJ, Patat-Ochatt EM, Rech EL, Davey MR, and Power JB (1989) Somatic hybridization of sexually incompatible top-fruit tree rootstocks, wild pear (*Pyrus communis* var. *pyraster* L.) and Colt cherry (*Prunus avium* × *pseudocerasus*). Theor Appl Genet 78: 35–41

Oliveira CM, Mota M, Monte-Corvo L, Goulao L, and Silva DL (1999) Molecular typing of *Pyrus* based on RAPD markers. Sci Hortic 79: 163–174

Palmer JD, Jorgensen RA, and Thompson WF (1985) Chloroplast DNA variation and evolution in *Pisum*; patterns of change and phylogenetic analysis. Genetics 109: 195–213

Patocchi A, Walser M, Tartarini S, Broggini GA, Gennari F, Sansavini S, and Gessler C (2005) Identification by genome scanning approach (GSA) of a microsatellite tightly associated with the apple scab resistance gene *Vm*. Genome 48: 630–636

Pierantoni L, Cho KH, Shin IS, Chiodini R, Tartarini S, Dondini L, Kang SJ, and Sansavini S (2004) Characterisation and transferability of apple SSRs to two European pear F1 populations. Theor Appl Genet 109: 1519–1524

Predieri S, and Zimmerman RH (2001) Pear mutagenesis, In vitro treatments with gamma-rays and field selection for productivity and fruit traits. Euphytica 117: 217–227

Reynoird JP, Mourgues F, Norelli J, Aldwinckle HS, Brisset MN, and Chevreau E (1999) First evidence for improved resistance to fire blight in transgenic pear expressing the attacin E gene from *Hyalophora cecropia*. Plant Sci 149: 13–22

Rubzov GA (1944) Geographical distribution of the genus *Pyrus* and trends and factors in its evolution. Am Naturalist 78: 358–366

Sanzol J, and Herrero M (2002) Identification of self-incompatibility alleles in pear cultivars (*Pyrus communis* L.). Euphytica 128: 325–331

Sassa H, Hirano H, Nishio T, and Koba T (1997) Style-specific self-compatible mutation caused by deletion of the S-RNase gene in Japanese pear (*Pyrus serotina*). Plant J 12: 223–227

Sassa H, Kakui H, Miyamoto M, Suzuki Y, Hanada T, Ushijima K, Kusaba M, Hirano H, and Koba T (2007) S locus F-box brothers: Multiple and pollen-specific F-box genes with S haplotype-specific polymorphisms in apple and Japanese pear. Genetics 175: 1869–1881

Sawamura Y, Saito T, Takada N, Yamamoto T, Kimura T, Hayashi T, and Kotobuki K (2004) Identification of parentage of Japanese pear 'Housui'. J Japan Soc Hort Sci 73: 511–518

Schmidt H, and van de Weg WE (2005) Breeding, In Tromp J, Webster AD, and Wertheim SJ (eds), Fundamentals of Temperate Zone Tree Fruit Production, Backhuys Publishers, Leiden, pp136–155

Takasaki T, Okada K, Castillo C, Moriya Y, Saito T, Sawamura Y, Norioka N, Norioka S, and Nakanishi T (2004) Sequence of the S9-RNase cDNA and PCR-RFLP for discriminating S1- to S9-alleles in Japanese pear. Euphytica 135: 157–167

Tang SH, Sun M, Lia ZH, Zhou QG, and Li DG (2007) Production of transgenic Xueqing pear plants with a synthetic Cry1Ac gene mediated by *Agrobacterium tumefaciens*. Acta Hortic Sinica 34: 56–62

Teng Y, Tanabe K, Tamura F, and Itai A (2001) Genetic relationships of pear cultivars in Xinjiang, China as measured by RAPD markers. J Hortic Sci Biotech 76: 771–779

Teng Y, Tanabe K, Tamura F, and Itai A (2002) Genetic relationships of *Pyrus* species and cultivars native to East Asia revealed by randomly amplified polymorphic DNA markers. J Am Soc Hortic Sci 127: 262–270

Terakami S, Shoda M, Adachi Y, Gonai T, Kasumi M, Sawamura Y, Iketani H, Kotobuki K, Patocchi A, Gessler C, Hayashi T, and Yamamoto T (2006) Genetic mapping of the pear scab resistance gene *Vnk* of Japanese pear cultivar Kinchaku. Theor Appl Genet 113: 743–752

Terakami S, Adachi Y, Iketani H, Sato Y, Sawamura Y, Takada N, Nishitani C, and Yamamoto T (2007) Genetic mapping of genes for susceptibility to black spot disease in Japanese pears. Genome 50: 735–741

Ushijima K, Sassa H, Tao R, Yamane H, Dandekar AM, Gradziel TM, and Hirano H (1998) Cloning and characterization of cDNAs encoding S-RNases from almond (*Prunus dulcis*): primary structural features and sequence diversity of the S-RNases in *Rosaceae*. Mol Gen Genet 260: 261–268

Volk GM, Richards CM, Henk AD, Reilley AA, Bassil NV, and Postman JD (2006) Diversity of wild *Pyrus communis* based on microsatellite analyses. J Am Soc Hortic Sci 131: 408–417

Vos P, Hogers R, Bleeker M, Reijans M, van de Lee T, Hornes M, Frijters A, Pot J, Peleman J, and Kuiper M (1995) AFLP: a new technique for DNA fingerprinting. Nucleic Acids Res 23: 4407–4414

Weber JL, and May PE (1989) Abundant class of human DNA polymorphism which can be typed using the polymerase chain reaction. Am J Hum Genet 44: 388–396

Wen XP, Pang XM, Matsuda N, Kita M, Inoue H, Hao YJ, Honda C, and Moriguchi T (2008) Over-expression of the apple *spermidine synthase* gene in pear confers multiple abiotic stress tolerance by altering polyamine titers. Transgenic Res 17: 251–263

Wertheim SJ (2002) Rootstocks for European pear : a review. Acta Hort 596: 299–307

Westwood MN, and Bjornstad HO (1971) Some fruit characteristics of interspecific hybrids and extent of self-sterility in *Pyrus*. Bull Torrey Bot Club 98: 22–24

Yamamoto T, Kimura T, Shoda M, Ban Y, Hayashi T, and Matsuta N (2002a) Development of microsatellite markers in Japanese pear (*Pyrus pyrifolia* Nakai). Mol Ecol Notes 2: 14–16

Yamamoto T, Kimura T, Sawamura Y, Manabe T, Kotobuki K, Hayashi T, Ban Y, and Matsuta N (2002b) Simple sequence repeats for genetic analysis in pear. Euphytica 124: 129–137

Yamamoto T, Kimura T, Shoda M, Imai T, Saito T, Sawamura Y, Kotobuki K, Hayashi T, and Matsuta N (2002c) Genetic linkage maps constructed by using an interspecific cross between Japanese and European pears. Theor Appl Genet 106: 9–18

Yamamoto T, Kimura T, Saito T, Kotobuki K, Matsuta N, Liebhard R, Gessler C, van de Weg WE, and Hayashi T (2004) Genetic linkage maps of Japanese and European pears aligned to the apple consensus map. Acta Hort 663: 51–56

Yamamoto T, Kimura T, Terakami S, Nishitani C, Sawamura Y, Saito T, Kotobuki K, and Hayashi T (2007) Integrated reference genetic linkage maps of pear based on SSRs and AFLPs. Breed Sci 57: 321–329

Yoshioka T, Masuda T, Kotobuki K, Sanada T, and Ito Y (1998) Gamma-ray-induced mutation breeding in fruit trees: breeding of mutant cultivars resistant to black spot disease in Japanese pear. Japan Agricultural Research Quarterly – Tsukuba Ibaraki 33: 227–234

Zeven AC, Zhukovsky M (1975) Dictionary of cultivated plants and their centres of diversity. Centre for Agricultural Publishing and Documentation, Wageningen

Zhao RH, Liu QZ, Sung QR, and Zhang XS (2004) Obtaining transgenic Fertility pear (*Pyrus communis* L.) plants with antifungal gamma-thionin Rs-afp1 gene. J Agric Biotech 12: 729–730

Zhu LH, Ahlman A, and Welander M (2003) The rooting ability of the dwarfing rootstock BP10030 (*Pyrus communis*) was significantly increased by introduction of the rolB gene. Plant Sci 135: 829–835

Zisovich AH, Stern RA, Shafir S, and Goldway M (2004) Identification of seven S-alleles form the European pear (*Pyrus communis*) and the determination of compatibility among cultivars. J Hort Sci Biotech 79: 101–106

Zuccherelli S, Tassinari P, Broothaerts W, Tartarini S, Dondini L, and Sansavini S (2002) S-allele characterization in self incompatible pear (*Pyrus communis* L.). Sex Plant Reprod 15: 153–158

# 9. Genomics of Almond

**Pere Arús, Thomas Gradziel, M. Margarida Oliveira, and Ryutaro Tao**

## 1 History and Evolution

The evolution of the cultivated almond closely parallels in both time and place, the emergence of agrarian societies in Asia and the adjoining Mediterranean regions. Archaeological studies in present day Israel have documented the collection and use of almond by early human communities as far back as 23,000 BC (Weiss et al., 2004). Legends and folklore containing references to almonds are common from the Levant to China (Rosengarten, 1984). One hypothesis has the cultivated almond resulting from selection from within a species, originally identified as *Amygdalus communis* L. (syn. *Prunus communis* Archang.), based on studies of two natural populations containing large numbers of sweet-kernel individuals rather than the bitter kernels typically found in the wild (Watkins, 1979). One population is located in the Kobet Dag mountain range in central Asia between present day Iran and Turkmenistan and the second population occurs on the lower slopes of the Tian Shan Mountains between Kyrgyzstan and western China. The natural range of *A. communis* was proposed to have extended across Iran, the Transcaucausus, and eastern Turkey, and into present-day Syria, and thus overlapped with known sites of early civilizations and almond cultivation (Denisov, 1988; Kester et al., 1991). According to this view, the distinction between cultivated and wild forms gradually disappeared with direct and indirect human selection. However, because the purportedly natural sweet-kernel populations closely resembles the phenotypes of present day cultivated almonds, it has been suggested that the Kobet Dagh and Tian Shan populations are, in fact, more recent remnants or escapes from later domesticated or semi-domesticated orchards (Ladizinsky, 1999). A more recent proposal is that cultivated almond represents a generalized kernel phenotype, probably derived from *Prunus fenzliana* but with contributions through natural interspecific hybridizations with a range of related species occurring naturally within this range, possibly including

P. Arús (✉)
IRTA, Centre de Recerca en Agrigenòmica CSIC-IRTA-UAB. Carretera de Cabrils Km 2. 08348 Cabrils, Spain

K.M. Folta, S.E. Gardiner (eds.), *Genetics and Genomics of Rosaceae*, Plant Genetics and Genomics: Crops and Models 6, DOI 10.1007/978-0-387-77491-6_9,

*P. bucharica, P. kuramica, P. triloba,* and *P. webbii* (Grasselly and Crossa-Raynaud, 1980; Kester et al., 1991; Godini, 2000; Socias i Company, 2002; Gradziel, 2009). While the cultivated almond and its wild relatives share basic botanical features and developmental patterns and are readily hybridized, the divergent selection pressures of the wide range of its ecological niches in the deserts, steppes and mountains of central Asia have resulted in an extensive variability in final tree and nut form (Grasselly, 1972; Niklasson, 1989; Felipe and Socias i Company, 1992). Within these regions the botanical structure defining horticultural quality was the edible kernel. High tree productivity, because it could over-extend plant resources, would be a liability in many of the marginal, dryland environments of both ancient and contemporary plantings. In these harsh environments, primary selection would be for tree survival with some level of consistent seed production regardless of final tree form or physiological pathways. Examples include the divergent photosynthetic strategies utilized by the cultivated almond, *P. dulcis* (syn. *P amygdalus*), and other wild relatives such as *P. fenzliana* and *P. scoparia* (Rouhi et al., 2007) and the pronounced differences in tree and fruit development between cultivated almond and the closely related cultivated peach (*P. persica*) (Ladizinsky, 1999; Gradziel, 2007).

While considerable genetic variability has been documented within cultivated almond, a much more extensive diversity among closely related and readily hybridized species, including the peaches (*P. persica*, *P. davidiana* and *P. mira*), is becoming known through genetic and morphological analysis. The genome of *P. dulcis,* as well as *P. fenzliana, P. nana* (syn. *P. tenella*), *P. bucharica, P. kotschyi,* and *P. scoparia* is diploid with $2n = 2x = 16$, which is the same as the peach species (Kester et al., 1991). Almond genome also has a small size (~300 Mbp), which is approximately twice that of Arabidopsis (Arumuganathan and Earle, 1991). The resulting genotypic and phenotypic variability represents a wealth of opportunities for cultivar improvement. The divergent development patterns based on unusually similar genomes also offer unique opportunities for the study of the fundamental regulation of plant ontogeny.

## 2 Breeding for Improved Cultivars

Almonds, either in cultivated orchards or as feral or wild seedlings, have been an important source of food for thousands of years. Almond's adaptation to harsh climates combined with an ability to develop a deep and extensive root system has allowed it to exploit a wide variety of ecological niches in its ancestral range in central Asia extending from the Takla Makan desert in western China to the Mediterranean (Kester et al., 1991; Ladizinsky, 1999). Almond is also well adapted to mild winter and dry, hot summer conditions due to its low chilling requirement for early bloom, rapid early shoot growth and high tolerance to summer heat and drought. It is the earliest temperate tree crop to bloom, which limits production to areas relatively free from spring frosts. Because almond is self-sterile, it requires cross-pollination which further acts to preserve genetic variability and so adaptability to new environments. Within different regions, promising wild seedlings have been selected for use by local communities (Grasselly, 1972). Continued human and natural selection led to greater adaptation to local environments, including resistance to many regionally important diseases and pests. The self-sterile nature of almond also insured a continuous exchange and mixing among cultivated as well as wild germplasm (Socias i Company, 2002). Although early almond improvement efforts mainly utilized natural genetic recombination within the species, the transfer of useful genes from wild species is also evident. For example, in the Apulia region of Italia, local accessions of *P. webbii* appear to be the source of self-fertility for the regionally important selection 'Tuono' (Godini, 2002).

Most modern cultivars in Asia, the Mediterranean area, and more recently in California, originated as such time-tested seedlings selections. The subsequent selection over hundreds of years and hundreds of thousands of clonal propagations has also identified improved clonal sources for many of these well established cultivars. Both genetic (deletions, point mutations, etc.), aneuploidy (Martínez-Gómez and Gradziel 2003), chromosomal (Jáuregui et al., 2001) and epigenetic (gene modulation/activation/silencing, etc.) changes would be selected, though because the subsequent selections are vegetatively propagated, the specific nature of control is rarely scrutinized. Most formal genetic improvement efforts involving controlled hybridizations and associated genetic analysis were established only in the last 50 years. Because of the long generation time and limitations on breeding population size resulting from the large tree size, these breeding programs have only recently begun to replace previously established seedling selections. Specific breeding objectives vary by region (see review by Kester and Gradziel, 1996), but typically target discrete improvements of regionally important seedling selections in the general areas of kernel quality, disease and pest resistance and yield consistency.

## 3 Development of Molecular Markers and Variability Analysis in Almond

All major types of markers such as isozymes (Hauagge et al., 1987), restriction fragment length polymorphisms (RFLPs; Viruel et al., 1995), random amplified polymorphic DNA (RAPDs; Bartolozzi et al., 1998), amplified fragment length polymorphisms (AFLPs; Bliss et al., 2002) and simple-sequence repeats (SSRs; Testolin et al., 2004), have been developed in almond for variability assessment, genome mapping, gene tagging and other applications. SSRs are currently the most utilized markers in almond due to their high quality (high polymorphism, codominance and reproducibility), the simplicity and robustness of the current analytical methods and the existence of large sets of publicly available markers. Almond SSRs have been obtained directly from genomic libraries (Testolin et al., 2004; Mnejja et al., 2005) and EST collections (Xu et al., 2004), but most SSRs used in almond come from the large SSR collection of other *Prunus* species particularly peach (Arús et al., 2005a). Most SSRs from other *Prunus* amplify a DNA fragment of the approximate expected size and are polymorphic and so are transferable to almond (Martínez-Gómez et al., 2003a, b). The first SSRs used for linkage analysis in *Prunus* were obtained in sour cherry and were mapped in the almond 'Ferragnès' × 'Tuono' progeny by Joobeur et al. (2000). Of 147 *Prunus* SSRs from six species 76% were transferable to almond (M. Mnejja and P. Arús, unpublished results).

Marker analyses has shown the level of variability and heterozygosity in almond genotypes to be very high. Arulsekar et al. (1986) found that 75% of the 12 almond isozyme genes surveyed were polymorphic compared to only 8% in peach. Byrne (1990) compared the levels of isozyme variability among different *Prunus* species and found that almond and Japanese plum had the highest levels, whereas apricot and peach were among the lowest. Self-incompatibility (SI) systems in almond and Japanese plum would promote outcrossing and so greater genetic variability, while the less variable species are either self-compatible (peach) or have a high proportion of self-compatible genotypes (apricot). Selfing and selection for inbred individuals, which is possible in the self-compatible genotypes, would have been a primary factor for the reduction of variability observed in peach and apricot. SSR variability is also consistent with earlier isozyme findings: Xie et al. (2006) analyzed 16 SSRs among 38 almond cultivars and found an average number of alleles per locus (A-value) of 7.7 and an average observed heterozygosity (Ho) of 0.65. A similar average A-value (7.3) but with approximately half the Ho (0.35) was reported by Aranzana et al. (2003) after examining the same number of SSRs in a much larger sample (212) of peach cultivars, which illustrates the enormous differences in variability levels and genetic structure between these two species. Similarly, Martínez-Gómez et al. (2003a) compared peach and almond variation using a common set of 18 SSRs in agarose gels, and reported average A-values of 6 in almond vs. A = 3.6 in peach and an Ho value of 0.59 in almond vs. 0.19 in peach. This high variability in almond contributes to a high marker efficacy for cultivar identification.

Isozyme genes have also been used to establish phylogenetic relationships among *Prunus* species (Mowrey and Werner, 1990) with findings in general agreement with previous morphology-based classifications, in which almond and peach were closely related and other *Prunus* species being more distant. These results have been confirmed with DNA sequence analysis (Potter, 2003). Almond and peach could be clearly distinguished from each other using SSRs (Martínez-Gómez et al., 2003a, b). SSR variability could also distinguish among almond related species including *P. kuramica*, *P. argentea*, *P webbii*, *P. bucharica*, *P. petunnikowii*, *P. pedunculata*, *P. mongolica*, *P. tangutica*, *P. ledebouriana* and *P. triloba* (Martínez-Gómez et al., 2003a, b, Xu et al., 2004). SSR differences between cultivars from Chinese and Mediterranean origins were reported by Xu et al. (2004) and Xie et al. (2006). The Chinese collection had a slightly higher level of variation, suggesting a possible founder role in the evolution of the cultivated almond. A more thorough evaluation of differences in genetic composition between the two origins may allow the understanding of the selective forces shaping cultivar evolution from environmentally distinct origins.

## 4 Marker Maps: Major Gene and QTL Mapping

Isozyme analysis provided the first opportunity to study linkage in almond allowing the early characterization of three separate linkage groups after analysis of ten different genes (Arús et al., 1994). Linkage studies were difficult before the availability of isozymes because the low number of Mendelian genes. Socias i Company (1998) had identified only four single-locus genes: kernel taste (sweet vs. bitter), shell hardness (soft vs. hard), delayed blooming time, and self-incompatibility (including the self-compatibility allele). With the development of DNA-based approaches capable of detecting large numbers of markers, whole genome linkage maps were possible, Viruel et al. (1995) published one of the first maps in *Prunus*, analyzing essentially transferable and codominant (RFLP and isozyme) markers in F1 offspring of a cross between almond 'Ferragnès' and 'Tuono'. Additional almond × almond maps have since been constructed from the cross between 'Felisia' (formerly named 'D-3-5') and 'Bertina' (Ballester et al., 2001) and from the cross between 'R1000' and 'Desmayo Largueta' (Sánchez-Pérez et al., 2007). The two latter maps were produced to analyze important agronomic characters and used a framework approach, i.e. selecting markers that covered the whole genome at distances of 15–25 cM from the *Prunus* reference map (Arús et al., 2005b).

A similar analysis of interspecific almond × peach populations has allowed development of some of the most complete maps within *Prunus*. Almond and peach are inter-compatible and hybrids are often fertile allowing the development of highly segregating of F2 populations. This high polymorphism and the F2 structure are optimal for linkage mapping, particularly when the objective is the construction of saturated or high density maps. Almond × peach maps allow an analysis of the peach genome by avoiding the low polymorphism associated with peach intraspecific progenies, and of the almond genome by avoiding the SI barriers to selfing. While of minor importance when compared to their advantages, interspecific F2 progenies have less attractive aspects, including low or no fruit set in some of the progenies, segregation for characters that are not the immediate objective of breeding programs and segregation distortions of loci in various genome regions due to sporophytic or gametophytic selection, (as when pollen containing almond SI allele is selected against when selfing an almond × peach F1 hybrid containing the corresponding pistil-SI allele). Four maps have been constructed with almond × peach progenies, which has allowed mapping several major genes and quantitative trait loci (QTL) of interest and a large number of markers. The main characteristics of these maps are summarized in Table 1.

One of the interspecific F2 maps, constructed from 'Texas' (syn. 'Texas Prolific', 'Mission') almond × 'Earlygold' peach population (Dirlewanger et al., 2004a; Howad et al., 2005), is generally accepted as the reference map for the genus and currently contains 826 loci, all of which are transferable including 361 RFLPs, 449 SSRs, 11 isozymes and 5 STS markers). This map distinguishes all expected eight linkage groups (G1–G8), has a total distance of 524 cM and has an average density of 0.63 cM/marker. Methods for fast mapping ('bin mapping') have been developed in this population that require only six plants of the progeny plus

**Table 1** Main characteristics of linkage maps constructed for almond

| Cross | Type (Size)[1] | No. markers | Anchors with TxE | Total size (cM) | % coverage[2] TxE map[2] | % common distance[3] | Paired t test[4] | References |
|---|---|---|---|---|---|---|---|---|
| Texas × Earlygold | F2 (82) | 826 | 826 | 519 | 100 | 100 | – | Dirlewanger et al. (2004a); Howad et al. (2005) |
| Garfi × Nemared | F2 (113) | 51 | 51 | 474 | 78 | 117 | 3.57** | Jáuregui et al. (2001) |
| Padre × 54P455 | F2 (64) | 161 | 24 | 1144 | 53 | 141 | 2.18 | Bliss et al. (2002) |
| Felinem | F1 (101) | 166 | 87 | 716 | 66 | 163 | 5.47** | Dirlewanger et al. (2004b) |
| Ferragnès | F1 (60) | 126 | 53 | 415 | 64 | 100 | 0.01 | Joobeur et al. (2000) |
| Tuono | F1 (60) | 99 | 41 | 416 | 48 | 111 | 0.76 | Joobeur et al. (2000) |
| Felisia | F1 (134) | 45 | 32 | 386 | 57 | 111 | 0.80 | Ballester et al. (2001) |
| Bertina | F1 (134) | 39 | 28 | 349 | 57 | 112 | 1.23 | Ballester et al. (2001) |
| R1000 | F1 (167) | 48 | 42 | 389 | 46 | 131 | 3.88** | Sánchez-Pérez et al. (2007) |
| Desmayo Largueta | F1 (167) | 45 | 39 | 395 | 51 | 132 | 2.97* | Sánchez-Pérez et al. (2007) |

[1]Progeny type. In parenthesis progeny size.
[2]Percentage of the TxE map covered by the other map: TxE distance covered * 100/total TxE distance.
[3]Relative size of the other map compared to that of TxE: other map distance covered* 100/common distance covered in TxE.
[4]Paired t test of the comparison between the distances of the two most separated common loci of each linkage group of TxE and the other map. *$P \leq 0.05$, **$P \leq 0.01$. The comparisons were done for all eight linkage groups in all progenies except for Padre × 54P455 where it was done with only six groups.

two parental individuals to detect the position of any segregating marker with an average precision of 7.8 cM (Howad et al., 2005). Markers from this map (abbreviated TxE) have been used to construct framework maps in many other *Prunus* species (Arús et al., 2005b), have proven useful for establishing the positions of a range of genes and QTLs of interest, and have been utilized as anchors between the peach genetic and physical maps (see Genome database for Rosaceae: http://www.bioinfo.wsu.edu/gdr/). A map with the position of the major genes and QTLs mapped with almond intraspecific or almond × peach populations is provided in Fig. 1.

*Prunus* map comparisons based on common TxE markers have established that the genome of all *Prunus* species analyzed is syntenic and collinear, i.e. that all *Prunus* species share the same genome (Arús et al., 2005b). However, differences exist among different *Prunus* genotypes used for mapping with respect to recombination rates, with TxE being the most compact map (Arús et al., 2005b). Some almond maps ('Ferragnès' × 'Tuono' and 'Felisia' × 'Bertina') have similar sizes as TxE (Table 1) while others, including the 'Garfi' × 'Nemared' F2 (Jáuregui et al., 2001), the 'Felinem' map (Dirlewanger et al., 2004b), and those obtained from the 'R1000' × 'Desmayo Largueta' F1 (Sánchez-Pérez et al., 2007) are significantly longer. 'Felinem' is a hybrid between 'Garfi' and 'Nemared' (Gómez-Aparisi et al.,

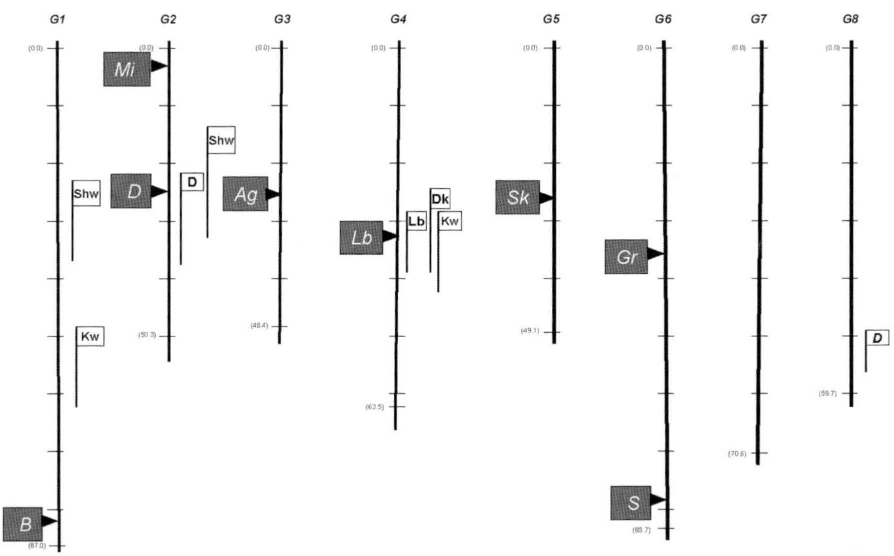


**Fig. 1** Positions of major genes (in dark background) and QTLs (in white background) on the TxE map (Dirlewanger et al., 2004a) based on the data of Joobeur (1998), Ballester (1998), Jáuregui et al. (2001), Ballester et al. (2001), Bliss et al. (2002), Dirlewanger et al. (2004b), Sánchez-Pérez et al. (2007). Major genes are: *B*, white vs. pink petal color; *Mi*, nematode resistance; *D*, shell hardness; *Ag*, anthocyanic vs. yellow anther color; *Lb*, late vs. early bloom time; *Sk*, sweet vs. bitter kernel taste; *Gr*, red vs. green leaf color and *S*, self-incompatibility. QTLs are: Shw, in-shell weight; Kw, kernel wheight; Dk, double kernels; D, shell hardness and Lb, bloom time

2001), demonstrating that while both maps were developed from different offspring of the same cross, the recombination data remained consistent. The observed differences in recombination rates were attributed by Arús et al. (2005b) to differences in individual genes affecting recombination rate. The map sizes observed in the almond suggest variability for these genes in different almond backgrounds.

## 5 Gene Cloning and EST Collections

The almond extensin gene that was extracted from an almond root cDNA library was the first gene to be reported in *Prunus* (Garcia-Mas et al., 1992). The cDNAs that were abundantly expressed during seed development were characterized by Garcia-Mas et al. (1995) who identified several sequences coding for proteins involved in storage processes. These proteins included two almond globulins (prunins), encoded by two genes (*Pru1* and *Pru2*) that are tightly linked in G8 (Viruel et al., 1995; Joobeur et al., 1998), and oleosin, a protein located on the surface of oil bodies in plant seeds, whose gene (*Ole1*) maps to G2 (Joobeur et al., 1998). Other genes that have been characterized and mapped include mandelonitrile lyase (Suelves and Puigdomènech, 1998a), a candidate gene for the bitter taste character of the almond kernel, O-methyltransferase (Suelves and Puigdomènech, 1998b), which appears to be related to lignin or flavonoid biosynthesis, and three lipid transfer proteins (Suelves and Puigdomènech, 1997), which may correspond to allergenicity factors in *Prunus*. Several genes strongly expressed in response to dehydration were also identified by Campalans et al. (2001) using the cDNA-AFLP technique. Almond is a highly drought tolerant species and so may provide a useful model for further molecular analysis.

From the total of 91,273 expressed sequence tags (ESTs) held in the Rosaceae Genome Database (Jung et al., 2004), a modest 3,864 are from almond, the majority (2,794, which correspond to 734 putative unigenes) from an almond developing seed cDNA library. Most of these sequences (3,863), were used by Xu et al. (2004) to identify 1,482 almond unigenes from which 178 different microsatellites (12.1% of the sequences examined) were obtained, with an estimated density of one EST-SSR every 5 kb.

## 6 Genetic Transformation

Prerequisites for successful plant transformation include the capacity to transform plant cells and the subsequent regeneration of plantlets from those cells. In almond, the ability to induce adventitious shoot regeneration has been demonstrated from both juvenile (seedling) tissues (Mehra and Mehra, 1974; Seirlis et al., 1979; Bimal and Jha, 1985; Rugini, 1986; Miguel, 1998; Ainsley et al., 2001b) and, to a lesser extent, adult tissues (Miguel et al., 1996; Ainsley et al., 2000). However, regeneration difficulties remain a major barrier to the use of genetic transformation for direct cultivar improvement.

Low regeneration success from transformed cells is a major problem, as in many fruit tree species, making optimized transformation protocols targeting the regeneration-competent cells highly desirable. There are a few reports of transgenic plant recovery in Prunoideae, and only three in almond (Miguel and Oliveira, 1999; Costa et al., 2006; Ramesh et al., 2006) using either a seed derived clone from the Portuguese 'Boa Casta' or the North-American 'Ne Plus Ultra'. Transformation protocols for the most important cultivars ('Nonpareil', 'Marcona', etc.) are not yet available.

All successful transformation protocols used *Agrobacterium*-mediated transformation. Pre-culture of leaf explants for three days on shoot induction medium before wounding and *Agrobacterium* infection resulted in higher regeneration rates and leading to a nearly seven-fold increase in the percentage of GUS positive explants (Miguel and Oliveira, 1999). The *Agrobacterium* functions of T-DNA targeting to the nuclear genome and the more targeted delivery of T-DNA in the regenerating cells (close to the wounds made by scalpel) may be important for the success of almond transformation.

To improve transformation efficiency, acetosyringone (AS) is often added during co-cultivation, aiming to stimulate transcription of *Agrobacterium* virulence genes. In almond, AS was used in the *Agrobacterium* growth medium prior to infection (Miguel, 1998; Miguel and Oliveira, 1999; Ramesh et al., 2006). However, in the experiments of Costa et al. (2006) AS was added to the plant culture medium at a concentration of 150 μM during the whole pre-culture, co-cultivation and induction period, with a clear positive effect. The combination of AS treatment with this modified selection scheme allowed the transformation efficiency increase by 100-fold (Costa et al., 2006).

Transgenic roots have been induced by infection of microcuttings with *A. rhizogenes* 1855 NCPPB and the transfer of *rol* genes was confirmed by the induction of the distinguishing hairy root phenotype and by Southern blot analysis (Damiano et al., 1995). The adventitious roots also displayed the typical plagiotropic emergence, but after transfer to soil they grew vigorously and the plants did not have any special requirements.

The *Agrobacterium* strains (EHA105, AGL1) most frequently used in the successful recovery of transgenic almond plants (see Miguel and Oliveira, 1999; Costa et al., 2006; Ramesh et al., 2006) are of the succinamopine super-virulent type. These strains are derived from the C58 strain and carry the virulence plasmid pTiBo542 which has been previously demonstrated as having high transformation ability in a number of plants, due to a higher expression of the *vir* genes, especially *vir*G (Liu et al., 1992; Ghorbel et al., 2001).

Reporter genes used to assess almond transformation efficacy include the $\beta$-glucuronidase gene (*gus*A, in p35SGUSint plasmid), which was chosen by a number of authors (Archilletti et al., 1995; Miguel and Oliveira, 1999; Ainsley et al., 2001a; Costa et al., 2006), the luciferase (*luc*, in plasmid pAL23) (Miguel, 1998) and the green fluorescent protein gene with modified codon usage designed to be targeted to the endoplasmic reticulum (*mgfp-5-ER*, in plasmid pBI121mgfp-5-ER) (Ramesh et al., 2006).

The selectable marker gene most often used to facilitate regeneration of transformed almond cells has been the neomycin phosphotransferase (*npt*II) gene conferring kanamycin resistance (Archilletti et al., 1995; Miguel and Oliveira, 1999; Ainsley et al., 2001a, Costa et al., 2006). The use of the herbicide resistance gene phosphinotricine acetyltransferase (*bar*) together with hygromycin phosphotransferase (*hpt*) (in plasmid pWRG2426) has been reported by Miguel (1998) and more recently, the use of the phosphomanose isomerase gene (*pmi*) (in plasmid pNOV2819 manA) has also shown success (Ramesh et al., 2006).

Although *Agrobacterium*-mediated transformation has been the method of choice in this species, particle bombardment was also tested (see Oliveira et al., 2008). Transient expression has been observed after particle bombardment, but no stable transformants were obtained even though the regeneration ability of leaf explants did not appear to be affected by the bombardment process (Miguel, 1998).

Kanamycin selection was utilized in the first successful report of almond transformation and regeneration (Miguel and Oliveira, 1999). Kanamycin was introduced immediately after co-cultivation with *Agrobacterium* but with a low concentration (17 or 26 μM), which was maintained during the induction period, then increased to 86 μM in the elongation phase. The surviving shoots were then propagated in 51 μM kanamycin for two subcultures, and then transferred to a kanamycin-free micropropagation medium. In the report of Costa et al. (2006), selection was also applied immediately after co-cultivation with 26 μM kanamycin, which was maintained during shoot elongation. Kanamycin concentration was increased in micropropagation, first to 51 μM for two cycles and to 86 μM thereafter. This latter strategy allowed a 40-fold increase in transformation efficiency (from 0.1 to up to 4%) when compared to the previously reported work of Miguel and Oliveira (1999). Indeed, all shoots surviving five subcultures on micropropagation medium (including the three subcultures at 86 μM kanamycin) were confirmed as transformed by Southern blotting (Costa et al., 2006). Three years after having initial transformation, transgenic shoots were transferred to medium containing 172, 344 or 516 μM kanamycin, for 4 subcultures, and were found not only to survive these high concentrations, but were even able to initiate axillary bud growth.

Ramesh et al. (2006) have reported that improved regeneration of *npt*II PCR-positive transgenic plants was achieved by delaying kanamycin selection (15 μM) until 70 days after co-cultivation with *Agrobacterium*. According to these authors, initiation of kanamycin selection (15 μM) 3 days after co-cultivation completely blocked shoot initiation, while kanamycin selection applied 21 days after co-cultivation (15 or 20 μM) resulted in the formation of a small number of buds, which developed into small shoots but remained stunted and did not grow further. The 70-day delay allowed the recovery of transgenic shoots with an efficiency of 5.6%, based on PCR screens. The authors further report that six PCR-positive transgenic shoots were recovered with Southern analyses confirming transgene integration.

Recently, Ramesh et al. (2006) have reported successful positive selection based on the inhibition of growth of non-transformed tissue through carbohydrate starvation when the phosphomannose isomerase structural gene (*pmi*) was used as the selection marker. This gene allows transgenic tissues to utilize mannose as a carbohydrate source. In nontransgenic tissue mannose accumulates in plant cells as mannose-6-phosphate which blocks glycolysis and so prevents further growth (Goldsworthy and Street, 1965). In tissue culture, almond can readily utilize sucrose, fructose or glucose as carbon source but grows poorly on mannose, making the mannose/PMI selection protocol an attractive selection option. At 20 g/l mannose, no callus formation was observed from almond leaf explants while mannose concentrations as low as 2.5 g/l severely reduced callus growth and shoot regeneration (Ramesh et al., 2006). The authors reported the regeneration of transgenic shoots from 'Ne Plus Ultra' on medium containing 2.5 g/l mannose together with 15 g/l sucrose when selection pressure was initiated 21 days after transformation.

Costa et al. (2006) have further observed that nine years after their initial recovery, transgenic almond shoots propagated *in vitro* in the absence of selection factors remained transgenic as assessed by Southern blot analyses. Ramesh et al. (2006), reported the successful rooting and transfer to soil of the transgenic plants which were able to flower less than 18 months after planting and eventually produced kernels (Almond Board of Australia, http://www.aussiealmonds.com).

# 7 Main Agronomic Characters, and Their Genetic Basis

## 7.1 Self-Incompatibility

Most of the fruit tree species in *Prunus*, including almond, sweet cherry (*P. avium*), Japanese apricot (*P. mume*), and Japanese plum (*P. salicina*), show gametophytic self-incompatibility (GSI). In GSI, the compatibility of a cross is determined by the haploid genome of the pollen and the diploid genome of the pistil (de Nettancourt, 2001). A pollen grain carrying an *S* allele that is also present in the style is recognized as 'self' pollen and its pollen tube growth is inhibited. The specificity of the GSI response is determined by at least two genes, one controlling pistil specificity and the other controlling pollen specificity. The former and the latter genes are referred to as the pollen *S* and the pistil *S*, respectively.

The presence of a minimum of two specificity genes has been confirmed through the identification and characterization of mutations exhibiting independent loss of pollen or pistil specificity. Because the specificity genes, pollen *S* and pistil *S*, segregate as if they were a single gene, the term '*S* haplotype' has been adopted to describe the set of these genes and the term '*S* allele' to describe variants of a given polymorphic gene at the *S* locus (McCubbin and Kao, 2000).

In the GSI system of Rosaceae, Solanaceae, and Plantaginaceae, the pistil *S* is known to be a glycoprotein with ribonuclease activity called *S-RNase* (Kao and Tsukamoto, 2004). The molecular basis of GSI has been studied most extensively in solanaceous plants. cDNAs associated with the *S* locus of GSI were first cloned from *Nicotiana alata* (Anderson et al., 1986). The deduced amino acid sequence strongly implicated stylar RNase involvement in the recognition and rejection reaction within the style (McClure et al., 1989). From these and other studies, it has been shown that the *S* gene products in pistils of Solanaceae are highly basic glycoproteins containing sequence motifs characteristic of the active site of the fungal RNase T2 (Kawata et al., 1988) and Rh (Horiuchi et al., 1988), called S-RNase.

Several years after the discovery of *S-RNase* in Solanaceae, Sassa et al. (1992) found the RNase activity bands that corresponded to the *S* haplotype of Japanese pear (*Pyrus pyriforia*). Later, cDNA encoding S-RNase of Japanese pear (Sassa et al., 1996) and apple (*Malus domestica*) (Broothaerts et al., 1995) were cloned. Since Rosaceae and Solanaceae are phylogenetically remote, it was an unexpected finding that they share the same type of molecules for the pistil *S* determinant. At around the same time that rosaceous *S-RNase* was cloned, the *S-RNase* of Plantaginaceae was also cloned (Xue et al., 1996).

The RNase activity bands corresponding to *S* haplotypes of almond (Tao et al., 1997) and sweet cherry (Bošković and Tobutt, 1996) were reported. The isozyme electrophoresis method developed by Bošković and Tobutt (1996) allowed the establishment of a correspondence between allozymes of stylar ribonucleases and *S* alleles in *Prunus*. This method was used to characterize a collection of almond cultivars (Bošković et al., 1997) and later to map the *S-RNase* gene in the almond 'Ferragnès' × 'Tuono' population to the distal part of G6 (Ballester et al., 1998).

*Prunus S-RNase* was cloned several years after the cloning of apple and pear *S-RNase*. As it was plausible to assume that the pistil *S* of *Prunus*, which belongs to the same family as apple and pear, shared a very high sequence homology with maloid *S-RNase*, researchers were trying to clone *S-RNase* based on the DNA sequence information from Japanese pear and apple *S-RNases*. Those attempts, however, were unsuccessful. Finally, *Prunus S-RNase* was cloned based on the N-terminal amino acid sequence information obtained from the almond (Tao et al., 1997; Ushijima et al., 1998) and sweet cherry (Tao et al., 1999) S-RNase spots on 2D-PAGE gel. It appeared that the DNA sequence identities between *Prunus S-RNase* and maloid *S-RNase* were as low as 30% while intra-subfamilial sequence identities of rosaceous *S-RNase* were over 70%. This is one of the reasons why the PCR-based cloning efforts were unsuccessful. Since the cloning of the first *S-RNase* sequence in almond and sweet cherry, more than 150 different alleles of *Prunus S-RNase* have been cloned. Interestingly, phylogenetic analyses showed that maloid and *Prunus S-RNase* form subfamily-specific clades in the phylogenetic tree while there is no species-specific clade in either the *Prunus* or maloid *S-RNase* clade (Igic and Kohn, 2001).

The pollen *S* of the three families eluded its identification for about 15 years after the discovery of *S-RNase* in Solanaceae. Since *S-RNase* and the unidentified pollen *S* were supposed to be physically linked, chromosome walking on the *S* locus region starting from *S-RNase* was first conducted in Solanaceae. The chromosome walking, however, was very difficult because the solanaceous *S* locus is located in the subcentromeric region of the chromosome and, thus, is very large in size and rich in repetitive sequences. Finally, it was shown that $S^2$ haplotype region of *Antirrhinum hispanicum* (Plantaginaceae) contained a male tissue-specific F-box protein gene, *AhSLF-*$S^2$ (Lai et al., 2002). At this point, however, it was unclear whether

*AhSLF-$S^2$* encodes pollen *S* because none of its alleles were cloned and furthermore, a gene with very high DNA sequence identity to *AhSLF-$S^2$* in other *S* haplotypes was cloned.

The first indication that F-box gene is the pollen *S* was shown in almond and Japanese apricot at about the same time by two independent research groups in Japan. Based on the assumption that *S* locus contains more DNA sequence polymorphism than outside of the *S* locus, Ushijima et al. (2001) delimited the *S* locus of the $S^c$ haplotype of almond to about 70 kbp through genomic DNA blot analyses. Surprisingly, the physical size of *S* locus of almond appeared to be very small as compared to the solanaceous one, whose size appears to be well over 1 Mbp (Wang et al., 2003). DNA sequencing of the 70 kbp region of almond $S^c$ locus revealed a good candidate gene for the pollen *S*, named *SFB* for *S* haplotype-specific F-box gene. *SFB* shows haplotype-specific sequence diversity, male tissue-specific expression, and physical linkage to *S-RNase*. At about the same time, DNA sequencing of the *S* locus of Japanese apricot revealed several pollen-expressed F-box genes including *S locus F-box*, *SLF*, which is a synonym of *SFB* in almond (Entani et al., 2003). *Prunus SFB/SLF* fulfilled necessary conditions for the pollen *S*, and thus it was presumed to be the pollen *S* (Ikeda et al., 2004 and 2005). Later, *SFBs* of pollen-part mutant self-compatible (SC) *S* haplotypes of cherries and Japanese apricot were shown to be mutated, providing additional evidence that *SFB* is the pollen part determinant (Ushijima et al., 2004).

The direct evidence that the pollen-expressed F-box gene is the pollen *S* was shown by the transformation experiments in *Petunia inflata* (Solanaceae) (Sijacic et al., 2004) and *Antirrhinum hispanicum* (Qiao et al., 2004). Recently, the sequencing analysis of the *S* locus of the subfamily Maloideae of the Rosaceae revealed multiple pollen-specific F-box genes with *S* haplotype-specific polymorphisms called *S locus F-box brothers* (*SFBB*) (Kakui et al., 2007; Sassa et al., 2007), a synonym of *SLF* by Cheng et al. (2006). It is still unclear whether all *SFBB* work together to give pollen specificity or only one of them is the pollen *S*.

Several working models for self-recognition have been proposed upon the identification of the pollen *S*. Most working models describe the *S* haplotype-dependent S-RNase degradation based on the fact that F-box protein is a component of the E3 complex (ubiquitin ligase) in the 26S proteasome pathway for protein degradation (Ushijima et al., 2003 and 2004; Takayama and Isogai, 2005), while another model based on the compartmentalization of S-RNase in the pollen tube has also been proposed (Goldraij et al., 2006; McClure and Franklin-Tong, 2006). So far, no conclusive results have been obtained from biochemical and histological studies.

From the practical point of view, identification of the pistil *S* and pollen *S* allows the development of molecular typing methods for *S* haplotypes. Fertilization is essential in almond production because the edible part is the kernel. In commercial orchards, cross-compatible (i.e. have different *S* haplotypes) cultivars that flower simultaneously are inter-planted, and honeybees are introduced to ensure cross-pollination. Thus, determination of correct *S* haplotypes of cultivars is essential., Classically, *S* haplotypes were determined through controlled pollination and pollen tube growth tests. These methods, however, are strongly influenced by environmental conditions. Furthermore, the methods can be applied only to the adult plants during the flowering period.

Almond cultivars are largely SI, although some SC cultivars exist (Tufts and Philp, 1922). Using controlled pollination and pollen tube growth tests, Crossa-Raynaud and Grasselly (1985) initially identified six different *S* haplotypes ($S^1$ to $S^4$, and $S^7$ and $S^8$), whereas Kester et al. (1994) identified four *S* haplotypes ($S^a$ to $S^d$) and assigned 31 cultivars to seven cross-incompatibility groups (CGI). Traditionally, European almond cultivars were designated by numerical code while Californian ones were designated by alphabetical code, leading to some confusion.

Since the *Prunus* pistil *S* (*S-RNase*) and pollen *S* (*SFB*) have been cloned, the DNA sequence information has been successfully used to develop a molecular marker for the *S* haplotype in SI fruit tree species in *Prunus* including almond (Tao et al., 1999; Tamura et al., 2000). The use of molecular markers has made it possible to type *S* haplotypes regardless of the time of the year and the age of plants, leading to the discovery of many new *S* haplotypes and unification of the numerical and alphabetical codes for *S* haplotypes (Table 2).

Most SC almond selections are considered to have originated from interspecific hybrids to SC species (Socias i Company, 1990). Herrero and Felipe (1975) conducted controlled pollination and pollen tube growth tests and reported two SC cultivars, 'Tuono' and 'A-S-1' with Italian and Spanish origins, respectively. The SC $S^f$ haplotype of 'Tuono' was presumed to be derived from *P. webbii*, which grows wild in southern Italy. Although partial DNA sequence of $S^f$-*RNase* has been reported, $SFB^f$ sequence remains to be reported (Ma and Oliveira, 2002). The molecular basis SC in $S^f$ haplotype is intriguing and may provide useful information for SC breeding programs in almond.

## 7.2 Bloom Time

Almond is the temperate fruit tree species with the earliest bloom time, typically blooming in late winter or early spring. Later bloom would avoid damage from winter/spring frosts, and improve conditions for honey-bee cross-pollination. In almond, bloom time is generally measured as a quantitative trait and has been usually found to be highly heritable (Kester et al., 1977a; Dicenta et al., 1993a). However, in the progeny of the late-blooming bud-sport 'Tardy Nonpareil', the blooming dates of the seedlings showed a clear bimodal distribution with a single dominant gene detected (Grasselly, 1978; Socias i Company et al., 1996). The late blooming allele, designed *Lb*, is dominant over early bloom, and confers an average bloom delay of 15 days. The gene has been mapped using the 'Felisia' × 'Bertina' cross (Ballester et al., 2001), and is located in the central part of linkage group 4 (G4).

Bloom time was also studied as a quantitative character in an intraspecific peach $F_2$ population (Dirlewanger et al., 1999), in an interspecific (*Prunus persica* × *P. ferganensis*) backcross (Verde et al., 2002) and in the TxE almond × peach $F_2$ (Joobeur, 1998). Two QTLs (in G2 and G7) were detected by Dirlewanger et al. (1999) and one in G4 by Verde et al. (2002). Joobeur (1998), using the TxE

**Table 2** DDBJ/GenBank/EMBL accession numbers for 22 S-RNase alleles of almond (Taken largely from Ortega et al. (2006) and López et al. (2006))

| S-RNase alleles | Synonimous S-RNase alleles from European cultivars | Synonimous S-RNase[‡] alleles from[‡] California cultivars[‡] | DDBJ/GenBank/EMBL accession no. |
|---|---|---|---|
| $S^{1}$ | $S^{16}$, $S^{17}$ | $S^{b}$ | AB011469 |
| $S^{2}$ | | | AF454000 |
| $S^{3}$ | | | AF490505 |
| $S^{4}$ | $S^{20}$ | | AM231656 |
| $S^{5}$ | $S^{15}$ | $S^{a}$ | AB026836 |
| $S^{6}$ | | $S^{e}$ | AM231657 |
| $S^{7}$ | | $S^{c}$ | AB011470 |
| $S^{8}$ | | $S^{d}$ | AB011471 |
| $S^{9}$ | | | AF454001 |
| $S^{10}$ | | | AM231658, AM231659 |
| $S^{11}$ | | | AM231660 |
| $S^{12}$ | $S^{k*}$ | | AM231661 |
| $S^{13}$ | $S^{19}$ | $S^{g}$ | AM231662 |
| $S^{14}$ | | $S^{j}$ | AM231663 |
| $S^{18}$ | | $S^{h}$ | AM231667 |
| $S^{21}$ | $S^{g*}$ | | AM231670 |
| $S^{22}$ | $S^{j*}$ | | AM231671 |
| $S^{23}$ | $S^{i*}$ | | AF454002 |
| $S^{24}$ | $S^{25\dagger}$ | | AM231672 |
| $S^{25}$ | | | AM231674 |
| $S^{26}$ | | | AM231674 |
| $S^{27}$ | | | AM231675 |
| $S^{28}$ | | | AM231676, AM231677 |
| $S^{29}$ | | | AM231678 |
| $S^{f}$ | | | AF510414 |

* Defined by Ma and Oliveira (2002).
† $S^{25}$-allele assigned to 'Gabaix' by López et al. (2004).
‡ Defined by Barckley et al. (2006).

progeny, found QTLs affecting bloom time in four linkage groups of the *Prunus* reference map (G1, G4, G6 and G7). Additional data are available from various peach intraspecific crosses, where three major QTLs were consistently located on G1, G4 and G7 (Howad and Arús, 2007). Some of these QTLs, although originating from different species (almond, peach and *P. ferganensis*), were found to locate at similar map positions (G1, G4 and G7), suggesting that the same genes may affect bloom time in different species. The co-location of the QTL on G4 with the major gene *Lb*, suggests that the phenotype of the major gene may be due to the action of an allele with major effects on the character, whereas other alleles have milder effects and can only be measured as quantitative variation. Aiming to identify the *Lb* gene and the genes within the QTLs, a candidate gene approach was used by Silva et al. (2005). Two almond partial cDNAs and eight *Prunus* expressed sequence tags (ESTs) showing sequence homologies with flowering regulatory genes previously characterized in other species, were used as candidate genes (CGs). The CGs were amplified from both the almond and peach parental lines of the mapping population using specific primers. DNA polymorphisms based on insertions/deletions, but mainly single nucleotide polymorphisms (SNPs), were identified and used to create differential restriction enzyme recognition sites and to develop co-dominant cleaved amplified polymorphic sequence (CAPS) markers to map the candidate genes on the *Prunus* reference map. The ten CGs were assigned to six linkage groups in the *Prunus* genome (Silva et al., 2005) and it was found that positions of two of these (*PrdFAR1* in G7 and *PrdAP2* in G6) were compatible with the regions where two QTLs for bloom time were detected although no associations with the *Lb* gene were observed.

In further studies, specific primers were used for rapid amplification of cDNA ends (RACE) and six full-length cDNA sequences were obtained representing different sequences (Silva, 2005; Silva et al., 2007). Three of the identified sequences code for MADS-box proteins (*PrdMADS1*, *2* and *3*), one for a putative homologue of the *Floricaula-leafy* gene (*FL*) gene of Arabidopsis (*PrdFL*) and another for a protein identical to a 'Constans-like' protein of Arabidopsis (*PrdCOL*). Additionally, another cDNA was identified coding for an enzyme putatively involved in the biosynthesis of active gibberellins, GA20-oxydase (*PrdGA20ox*). Expression studies revealed the involvement of these genes in molecular mechanisms underlying flowering in almond, and point for a conservation of these mechanisms between annual and perennial plants. However, to date, both the sequence and the specific function of *Lb* and of the genes responsible for the bloom time QTLs remain unknown. A MADS-box gene (*PrdMADS1*) co-located with the position of the *Evergrowing* (*Evg*) gene, which determines a non-deciduous behavior in peach (Silva et al., 2005), and with the QTL for bloom time in G1, suggesting a possible relationship between the two genes and the QTL.

In the Northern hemisphere, early vegetative growth in almond continues until roughly the end of June with flower initiation occurring during July and August. Around the beginning of October, almost all flower buds have entered arrested development and dormancy. Bud swelling typically occurs in January and February and flower development then proceeds until full bloom. Expression studies of the almond cDNAs putatively involved in flowering found similarities with transcription patterns of the corresponding putative apple homologues (Silva et al., 2007). *PrdMADS1* transcript was found to be absent from the leaves, seedlings and petals (but sepals and stamens were not tested for the presence of this transcript). Among the different floral buds tested *PrdMADS1* (homologue of apple *MdMADS10*, which is reported as the *Agamous-like 11* (*Agl-11*) gene of Arabidopsis, is expressed in carpels only at late developmental stages (late February) (Silva et al., 2007). Expression of *PrdMADS2* seemed to be ubiquitous in all examined tissues including leaves, seedlings and floral buds harvested in June, September and late February, as well as sepals, petals, stamens and carpels, and is similar to its close relative *MdMADS2* from apple (Sung et al., 1999). At the sequence level, *PrdMADS3* is closely related to the apple *MdMADS8* and *MdMADS9*. However, *PrdMADS3* transcripts were found in sepals, petals, stamens and carpels from mature flower buds harvested in late February. *MdMADS1* was found in all floral organs and young apple fruit (Sung and An, 1997). However, *PrdMADS3* was not found in earlier stages of floral development.

The almond putative homologue of *Leafy* (*PrdFL*) was found to be expressed in early flower bud development but was absent in mature buds, similar to *FLO* in *Antirrhinum* and *LFY* in Arabidopsis (Silva et al., 2007) and of homologues in other woody perennials such as grapevine (Carmona et al., 2002) and kiwifruit (Walton et al., 2001). The putative almond homologue of *LFY* thus may play a role in flower induction, although probably not directly in determining bloom time. The almond *constans-like (PrdCOL)* gene is expressed both in flowering and non-flowering tissues, though it appears to be stronger at the initial stages of floral bud development. It is therefore possible that in almond, as in Arabidopsis, the integration of photoperiodic and circadian signals is important in the regulation of floral induction (Silva, 2005; Silva et al., 2007). In Arabidopsis, *CO* (constans) is known to control flowering time and to be regulated by the circadian clock. *COL* (constans-like) genes do not have a significant effect on flowering time in Arabidopsis, but in plants with different photoperiodic responses (short-day plants) they seem to be mediators of flower induction (Yano et al., 2000; Liu et al., 2001).

An almond cDNA coding for a gibberellin 20-oxidase is expressed in floral buds about one month before their entrance in the winter dormancy period (Silva, 2005; Silva et al., 2007). In some perennial species, the regulation of winter dormancy seems to be mediated by photoperiod and it is possible that the presence of the almond GA 20-oxidase in pre-dormant buds is linked to a decrease in GA levels which, in turn, may promote dormancy.

## 7.3 Kernel Bitterness

Most cultivated almonds produce sweet kernels, but some have a slight amaretto to bitter flavor. Bitter kernels with a pronounced amaretto aftertaste are common in wild relatives of *P. dulcis* as well as kernels of most cultivated peaches. The bitter taste is due to the presence of the glucoside amygdalin, the end product of a metabolic pathway that includes prunasin and mandelonitrile as precursors. Variation in amygdalin content accounts for some cultivar flavor differences, particularly the distinct amaretto flavor associated with certain Mediterranean almonds (Dicenta and García, 1993; Vargas et al., 2001). Californian cultivars have amygdalin contents ranging from 0.33 to 0.84% with only 'Peerless' outside this range at 1.75% (dry weight). In contrast, Italian cultivars varied from 0.73 to 1.95% with only two cultivars below that range (Schirra, 1997). Even higher amygdalin levels occur in the wild-type bitter almond seeds which are often blended with sweet almonds to achieve desired levels of 'cherry' or 'amaretto' flavor.

Sweet vs. bitter kernel trait is inherited in a Mendelian manner, where sweet is dominant over bitter (Kester et al., 1977b; Dicenta and Garcia, 1993). Bitterness is determined by the presence of prunasin/amygdalin, which is synthesized in the bark and leaves of the tree and then transported to developing kernels. Consequently, all kernels of a given tree or cultivar are either sweet or bitter, depending on the genotype and irrespective of the pollen parent genotype. This gene (*Sk*) has been mapped in several populations. Joobeur (1998) located it on the central region of G5 of the TxE population. This position was later confirmed by Bliss et al. (2002), using also an almond × peach F2, and more recently by Sánchez-Pérez et al. (2007). In this map, constructed with a population of N=167, *Sk* was flanked by SSR markers BPPCT017 (at a distance of 8 cM) and PceGA025 (at 2 cM). Later, several additional markers located at this region in the TxE map have been found to be closer to this gene, such as BPPCT037 or CPDCT028 that are also flanking *Sk* (R. Sánchez-Pérez and W. Howad, pers. comm.) and that together provide a good resource for marker-assisted breeding for this character. Using EST sequences existing in the GDR, several candidate genes have been mapped in the TxE population and found to be in genome positions different from *Sk*, these are the mandelonitril lyase (Suelves and Puigdomènech, 1998a; Joobeur et al., 1998), amygdalin hydrolase, prunasin hydrolase and glucosyl transferase (R. Sánchez-Pérez and W. Howad, pers. comm.)

## 7.4 Shell Hardness

The structure of the almond shell is important for industrial processing. Hard shelled cultivars are generally grown in the Mediterranean region because they are more adapted to non-irrigated conditions, have better storage properties and are more resistant to damage by birds and insects. In contrast, soft-shelled cultivars are preferred in California and Australia. Shell hardness is usually measured with the kernel percentage, the proportion of the dry kernel weight on the total weight (kernel plus shell), and has been considered by some authors as a qualitative character determined by a single gene (*D/d*) (Grasselly, 1972) where the *D* allele would be dominant and code for a hard shell. Other researchers consider it as quantitative with high heritability (Kester et al., 1977b, Spiegel-Roy and Kochba, 1981; Dicenta et al., 1993b).

Results of genetic analyses in two populations segregating for shell hardness: 'Ferragnès' × 'Tuono' (Arús et al., 1998; Ballester, 1998) and 'Felisia' × 'Bertina' (Ballester, 1998) found that the major gene discovered by Grasselly (1972) was located in the middle of G3 between RFLP markers AG20 and AC27, separated 7-8 cM from each other. The character was scored as a single dominant gene since both populations segregated according to a 3:1 ratio, and the results on the position of the gene in G3 were consistent. Quantitative analysis of the data detected a QTL at the same position, which was responsible for most of the observed variability (approx. 80%) of this character in both populations. An additional QTL was observed in the 'Tuono' and the 'Bertina' maps Ballester (1998), which was consistently located in the distal part of G8 in both parental maps and explaining around 20% of the phenotypic variance, suggesting that at least another gene with minor effects is involved in the inheritance of this character. Kester and

Gradziel (1996) reported that both the hard (*D*-) and soft shell (*dd*) genotypes are affected by a range of modifier genes which are associated with different degrees of hardness.

## *7.5 Disease Resistance*

The most serious foliage diseases of almond include shothole caused by *Stigmina carpophila,* (syn. *Coryneum beijerinkii*), travelure *(Fusicladium amygdali),* polystigma (*Polystigma occhraceum*), fusiccocum (*Phomopsis amygdali* or *Fusicocum amygdali* as imperfect form) and anthracnose (*Gloeosporium amygdalinum* and *Colletotrichum acutatum*). Relative susceptibilities of important cultivars in different countries have been determined and potential sources of resistance have been identified (Kester et al., 1991). Blossom and twig blight, the major crop limiting fungal disease world wide, is caused by *Monilinia laxa* and *M. cinerea*. These fungi attack the flowers and are most serious in years when rain occurs with bloom. Other fungi, including *Botrytis cinerea* can also be a serious problem under these conditions. Aflatoxin producing *Aspergillus flavus* infections of the kernel can be a major problem, particularly where insect damage is common (Gradziel and Kester, 1994; Gradziel and Wang, 1994; Gradziel et al., 2000; Dicenta et al., 2003).

Almonds can be infected by the same range of viruses as other *Prunus* including the ALAR viruses (ringspot, prune dwarf, line pattern, calico, apple mosaic), and NEPO viruses (tomato black ring, tomato ring spot, yellow bud mosaic). However, many cultivars of almond appear to be immune to the plum pox virus which remains a serious problem for most other stone fruits (Martínez-Gómez et al., 2004).

Information on the inheritance and genomics of disease resistance is limited in almond. One research area that has yielded interesting results concerns the resistance to fusicoccum. This disease can be a major obstacle to nut production in the Mediterranean region. Fusicoccum affects mainly twigs of the lower part of the trees, causing cankers on shoots and necrosis on leaves (Barbé, 1993). The diffusion of the disease is very slow but its eradication from an orchard is difficult and expensive (Romero and Vargas, 1981). Additionally, in an infected plant, the symptoms may become visible only after several years, which makes the identification of molecular markers tightly linked to resistance traits a necessary tool to identify resistant plants at an early stage. Martins et al. (2001, 2005) identified three RAPD fragments related to tolerance/susceptibility to fusicoccum. These markers were first converted into sequence characterized amplified region (SCAR) markers, which were monomorphic, however the development of CAPS markers from these SCARs allowed recovery of the lost polymorphism (Martins et al., 2005) and the analysis of 15 almond cultivars using these CAPSs showed a high agreement with the respective phenotypes evaluated in field. These markers are being studied in progenies of controlled crosses between tolerant and susceptible plants to evaluate co-segregation rates (P. Barros and M.M. Oliveira, unpublished results). If these markers prove to be efficient they may be implemented in the certification system for routine analyses.

A set of resistance-gene candidate (RGC) sequences were cloned from one almond cultivar (described as resistant to several diseases, including fusicoccum) and two ecotypes of *P. webbii*, using a PCR-mediated approach with degenerated primers designed from conserved NBS (Nucleotide Binding Site) motifs of known plant disease resistance genes (Martins and Oliveira, 2005). Several of the cloned sequences could be translated into polypeptides, and the presence of conserved domains found in the NBS-LRR class of *R*-genes was observed. In phylogenetic analysis, the cultivated and wild almond RGCs clustered together in 5 different groups, independently of the species from which they were obtained (*P. dulcis* or *P. webbii*). This observation suggested a high similarity of genetic resistance backgrounds for both species. The candidate markers isolated are currently being mapped in the *Prunus* reference map.

## 7.6 Adventitious Regeneration

The main bottleneck in the regeneration of transgenic almond plants appears to be the fact that the host cells that become transgenic are often different from those possessing competence to re-enter the cell cycle and ultimately regenerate whole plants. To understand which cells become competent and by which mechanism, molecular studies have been conducted to identify genes/markers of early/late regeneration events (Santos et al., 2007). Two main strategies were used to screen and discover novel genes or markers: a candidate gene approach using the almond Knotted-1 (KN1) and CDK1:1 (putative markers for organogenesis events), and a transcriptomic approach with microarray technology, using two suppression subtractive hybribidization libraries constructed from two defined timeframes of organogenesis, early (E) and late (L), corresponding to 1–8 and 9–20 days of shoot induction in the dark, respectively (Santos et al., 2007). From the candidate gene approach, only the almond *KN1-like* showed a differential expression, occurring mostly in the second half of the shoot induction period (Santos et al., 2007). For the transcriptomics approach, the hybridization of three biological replicates represented by three induction assays produced a strong set of data that was statistically analyzed. One hundred and twenty eight DNA fragments showing an expression 2-fold higher in L vs E (70) or E vs L (58) were selected for sequencing and 64 were found to correspond to unique gene fragments.

After validation by Real-Time PCR, the expression of selected genes putatively involved in critical events of the adventitious organogenesis was assessed on a daily basis during the 21-day shoot induction period. The cDNA clones differentially expressed either in the early or in the late conditions revealed candidate genes with very interesting expression patterns during shoot induction. Genes putatively coding for lipid transfer protein and 1,3 $\beta$-glucanase and gibberelic acid stimulation giberellin-regulated proteins were identified and may be involved in distinct events of the almond regeneration process (A.M. Santos and M.M. Oliveira, unpublished results). The cDNA sequences of the genes currently known were deposited in the GenBank and will soon become public (EF640654-EF640717). In the future, these candidates may accurately mark de novo events occurring during organogenesis, which may be assisted by in situ techniques. The disclosure of a variety of known and unknown transcripts involved in adventitious regeneration may help to elucidate their putative regulatory roles in determining cell competence for shoot regeneration.

## *7.7 Characters Studied in Almond Species and Almond × Peach Crosses*

Almond is unique among the major cultivated crops as it appears to have originated from interspecific hybridizations with subsequent and continuing low levels of germplasm introgression from wild relatives occurring naturally along its wide-range of cultivation. The geographic range occupied by almond species is extensive with a wide diversity of traits (Kester and Gradziel, 1996; Gradziel et al., 2001a). Controlled crosses of *P. dulcis* with other almond species in Sections Euamygdalus and Spartiodes have been readily carried out (Gradziel et al., 2001a,b; Gradziel, 2003. Hybridization with Section Lycioides is somewhat more difficult and even more so with Chameamydalus. Despite their physical and developmental differences, crosses with peach (*P. persica, P. mira and P. davidiana*) can be readily achieved and have proven to be particularly valuable as rootstocks as well as sources of commercially useful traits (Gradziel et al., 2001a,b; Gradziel, 2003).

While several reports have documented the transfer of genes from related almond species through either natural or controlled crosses (Denisov, 1988; Socias i Company and Felipe, 1988; Socias i Company and Felipe, 1992; Gradziel and Kester, 1998; Felipe, 2000; Gradziel et al., 2002), only Rikhter (1969), Grasselly (1972), Socias i Company (1990) and Kester et al. (1991) have described the controlled introgression of genes from wild species to improve almond cultivars. The historical use of these species and their hybrids as almond rootstocks has facilitated such introgressions. More recently, crosses between almond and wild almond species have been evaluated as rootstocks (Gradziel and Kester, 1998; Gradziel et al., 2001a,b; Gradziel, 2003). While a wide variability in tree and branch architecture result, leaf and nut phenotypes of resultant hybrids are typically intermediate to the parents. In France, USA, Spain and Yugoslavia, controlled crosses between related species (mainly *P. persica* × almond but also others such as *P. webbii* × almond) initially directed towards almond rootstock development have more recently been utilized for cultivar improvement in the general areas of yield consistency, kernel quality, and disease/pest resistance (Rikhter, 1969; Grasselly, 1972; Denisov, 1988; Gradziel et al., 2001a,b).

Yield consistency is being pursued mainly through the transfer of self-compatibility originating from related species (Socias i Company, 1990). Genes introgressed from *P. mira*, *P. persica* and *P. webbii* have resulted in the highest levels of fruit set. Other introgressed traits showing promise for almond improvement include improved kernel quality, compact tree size, and resistance to aflatoxin producing *Aspergillus flavus* from peach; navelorange worm resistance and modified bearing-habit from *P. webbii*; and autogamy (self-pollinating flower structure) from *P. mira* (T.M. Gradziel, unpublished data). The very diverse almond species germplasm has also proven valuable for peach cultivar improvement, providing potential resistance genes for plum pox virus, mildew, leaf curl, green peach aphid and improved fruit texture and post-harvest storage life (Gradziel, 2007). The existence of several almond × peach genetic maps provides a useful starting point for inheritance studies of these characters and introgression of valuable genes into almond or peach.

# 8 Future Prospects

The last decade has witnessed unprecedented growth in our understanding of almond genetics. Eight major genes plus various QTLs for characters of agronomic importance are currently mapped. Hundreds of markers, most of them SSRs, are available and have been used for variability analysis, genotype fingerprinting and for the construction of high density maps. These maps have helped establish that all *Prunus* crops share a common genome (Dirlewanger et al., 2004a). Key genes for almond breeding have been studied in detail. One example is the SI locus, the sequence, map position and genetic basis of which have been established and are currently being used to determine the cross-compatibility group of new and old varieties and to breed for self-compatible cultivars (López et al., 2005, 2006). Major genes affecting kernel bitterness, shell hardness and bloom time have been mapped and markers in their neighborhood can be used for their selection. The first results are already available on inheritance and map position of genes affecting almond kernel size and shape, disease resistance and shoot regeneration capacity. Given the high similarity among *Prunus* genomes, and between *Prunus* and the Maloideae (apple, pear, etc.) (Dirlewanger et al., 2004a; Arús et al., 2005b), knowledge on marker or gene/QTL sequence and position from stone fruits and apple appear likely to also be applicable to almond improvement.

The genetic analysis of kernel quality requires special attention as it is a characteristic unique to almond with respect to other rosaceous crops. Socias i Company et al. (2008) describe in detail the different quality goals that breeders may target, from physical traits (kernel size, kernel shape, seed coat, double and twin kernels and fruit hull) to biochemical characteristics (cyanoglucosides, phenolics, vitamins and minerals, lipids, proteins, carbohydrates, fibers and allergens). Breeding strategies are now targeting specialized uses (i.e. processed, nutrient rich kernels, etc.) that will result in increased numbers if special use cultivars in the future. The development of markers associated with different aspects of kernel quality will not only accelerate breeding progress but will allow the efficient monitoring of kernel quality during marketing and processing.

In a long generation-time crop like almond, the continuing development of new populations segregating for targeted traits and their accurate phenotyping and genotyping provides the foundation for future genetic progress. The development of other population structures allowing finer QTL analyses should also be pursued, such as the construction of near-isogenic line (NIL) collections that have been so efficient for quantitative trait dissection in other species (Gur et al., 2004; Eduardo et al., 2005). The basis of genetic analysis with NILs is the comparison of each introgression line with the recurrent parent (usually an inbred line), where significant differences can be attributed to genes occurring in the introgressed fragment. In almond, SI and the resultant highly heterozygous nature of most individuals frustrates the development of inbred lines. Another approach is to use peach, a much more inbred species, or doubled-haploid peach lines, as the recurrent parent, with almond contributing with introgressed fragments to analyze the effect of almond specific alleles in a peach background. A peach-almond NIL collection based on the TxE cross is currently in its first stages of development (R. Picañol, W. Howad and P. Arús, unpublished results), and may prove very useful for the identification and introgression of valuable traits from almond to peach.

Association mapping, the association between a molecular marker and a phenotypic trait, depends on the conservation of linkage disequilibrium (non-random association between two loci). The degree to which this association will exist depends on the extent of linkage disequilibrium conservation (LD) in the population examined. Very high LD is anticipated in the typical mapping populations (F1s, F2s, backcrosses) and very low LD expected in collections of genotypes with a long history of random mating. Association mapping promises to be a powerful approach for genome analysis in almond or other *Prunus*. Preliminary results by Aranzana et al. (2007) indicate that collections of European and North American peach cultivars have a high level of LD, which can in part be attributable to the bottleneck produced by the use of a restricted set of founders by the first USA breeding programs during the last century (Hesse, 1975), and the relatively short number of generations from these founders to present. Since almond differs in both its ancient and recent history, it potentially has a much lower LD level, which needs to be assessed in comprehensive collections of materials from China, central and westerm Asia, and the Mediterranean. The existence of unstructured collections of *Prunus* with high LD (peach) and low LD (almond) may allow for a two-tiered approach where whole genome scans in high LD populations may help determining the positions of specific genes or QTLs and the detection of their alleles, while with low LD collections may be used to validate candidate genes and eventually clone specific genes.

In addition to greater cultivar development efficiency, marker-based selection also shows considerable promise for germplasm enhancement through interspecific gene introgression. Although related species have been shown to be valuable sources for improved crop quality, yield consistency and resistance, the great majority of introgressed genes appear to be suboptimal to deleterious in a cultivated almond background. While greater phenotypic plasticity is acceptable in tree crops such as almond because of the extensive cultural manipulations employed (for example, the common practice of budding almond trees onto the more disease resistant peach rootstock), this linkage drag has been a major impediment to the full exploitation of exotic gene introgression. The use of marker-assisted selection combined with continued progress in identifying genes and genetic mechanisms controlling key traits, will allow a much more efficient introgression of desired genes, facilitate the rapid selection of major genes or QTLs of interest, speeding up the introgression process by identifying the individuals with the lowest amount of the donor genome in early backcross generations, and limit the amount of donor DNA around the introgressed gene/QTL (Tanksley et al., 1989). These processes are necessarily long term, involving several generations and an as large a number of individuals as possible. Long-term introgression has to be viewed as an investment in fruit tree improvement from which new waves of cultivars with desired features can be obtained. The advantages of the use of molecular tools are that the time required should be lower than with traditional methods based only on phenotypic evaluation, and that the control of the introgressed and non-introgressed genome fragments can be monitored and the best genotype combinations selected.

Similar considerations can be made concerning the use of the exploitation of the enormous variability existing in almond and related species for the improvement of other *Prunus* crops, particularly those with less variability, such as peach. Valuable alleles for agronomical genes including resistances to plum pox virus, leaf curl, green leaf aphid and powdery mildew, or factors involved in fruit maturation and quality exist in almond which appear to be readily introgressed into peach or other *Prunus* species owing to their high genome similarity (Arús et al., 2005b) and ease of hybridization (Hesse, 1975).

Interestingly, while linkage drag would be expected to be more problematic as the source species become more divergent, one of the most useful species sources for almond improvement has been the more developmentally distinct peach group including *P. persica, P. mira* and *P. davidiana*. In spite of the ease of hybridization and subsequent gene introgression between almonds and peaches and the high level of synteny between these genomes, the divergent evolution of almonds in the harsh environments of central Asia vs. peach in the more temperate to subtropical climates of southeastern China have led to dramatically different ontogenetic and growth patterns. Almond and peach thus represent a unique situation in crop plant genetics, where very similar genomes are expressed as very different plant forms. This genetic-ontogenetic divergence between almond and the closely related peach may prove useful for the elucidation and eventually manipulation of the recently recognized epigenetic (i.e. brought about by an apparent change in gene activity rather than gene DNA sequence), mechanisms which are now recognized to have profound effects on fundamental plant developmental pathways and so final form and function.

Because epigenetic changes do not respond to traditional breeding methods designed to manipulate classic Mendelian genes, they are generally perceived as unpredictable and routinely rouged out, particularly in seed propagated crops. However, because both genetic and epigenetic composition can be captured through clonal propagation, these same methods can also be utilized to capture desirable epigenetic arrangements. An example would be the widespread practice among nurseries in selecting superior clonal sources of important vegetatively propagated cultivars (Hartmann et al., 2002). Epigenetic capture offers unique advantages to breeding programs utilizing wide crosses, since the interspecific hybridization process has been shown to increase the levels of epigenetic variability resulting in novel and transgressive phenotypes (Grant-Downton and Dickinson, 2006). Epigenetic-like changes also appear to be responsible for some of the differences within clonal sources of certain cultivars (Hartmann et al., 2002) and strategies for their management have been shown effective for the management of the epigenetic-like disorder Noninfectious Bud-failure (Kester et al., 2004). A similar breeding approach targeting the asexual 'capture' of novel phenotypes associated with specific interspecies genomic arrangements, has recently shown success for peach cultivar improvement where advanced processing peach selections derived from almond-peach hybrids, expressed useful fruit ripening patterns not evident in either species parent (Gradziel, 2003; Peace et al., 2005).

Epigenetic changes during transformation and plantlet regeneration also represent important challenges to maintaining cultivar integrity (Marcotrigiano and Gradziel, 1997). The capability to genetically engineer almond cultivars allows discrete improvements to otherwise elite genotypes, such as the elimination of self-sterility (using *S*-RNase gene silencing, antisense technology, RNA interference, etc.). This goal now appears attainable since the *S*-RNase gene sequences of various alleles are known and transformation protocols for 'Ne Plus Ultra' are available since 2006. Transformation experiments in 'Nonpareil', however, have not yielded transgenic plants yet (Ainsley et al., 2001a). For the moment, the use of transgenic breeding for almond improvement does not seem economically viable, except for specific cases in which elite cultivars of vast utilization are targeted (Oliveira et al., 2008). The ability to genetically engineer almond remains crucial, however, to validate gene function research.

Whole genome sequences of apple and peach are expected to be available within this decade. Information from these projects will inevitably accelerate progress of almond genetic improvement by facilitating the search for candidate genes, elucidating interactions between genes and providing markers and other tools to increase breeding efficiency. The new generation of high throughput sequencers that allow rapid and relatively inexpensive sequencing of large amounts of DNA, combined with the knowledge of the full peach genome sequence, will make the resequencing of other peach genotypes or related species such as almond feasible and affordable. The full sequence of one or more almond genotypes thus appears imminent and, given its close relatedness to peach, should facilitate our understanding of the *Prunus* genome in general and divergent speciation in particular. Results from high throughput sequencing will also identify a large number of DNA polymorphisms (Barbazuk et al., 2007) that can be used for the development of a new generation of markers allowing high resolution whole genome coverage. Resultamt maps would be particularly useful for the saturation of specific genome regions or for whole genome scans as part of future large-scale genetic studies.

## References

Ainsley P, Collins G, and Sedgley M (2000) Adventitious shoot regeneration from leaf explants of almond (*Prunus dulcis* Mill.). In Vitro Cellular and Developmental Biology 36: 470–474

Ainsley P, Collins G, and Sedgley M (2001a) Factors affecting *Agrobacterium*-mediated gene transfer and the selection of transgenic calli in paper shell almond (*Prunus dulcis* Mill.). Plant Cell Rep 76: 522–528

Ainsley PJ, Hammerschlag FA, Bertozzi T, Collins GG, and Sedgley M (2001b) Regeneration of almond from immature seed cotyledons. Plant Cell Tiss Org Cult 67: 221–226

Anderson MA, Cornish EC, Mau SL, Willams EG, Hoggart R, Atkinson A, Bönig I, Grego B, Simpson R, Roche PJ, Haley JD, Penschow JD, Niall HD, Tregear GW, Coghlan JP, Crawford RJ, and Clarke AE (1986) Cloning of DNA for a stylar glycoprotein associated with expression of self-incompatibility in *Nicotiana alata*. Nature 321: 38–44

Aranzana MJ, Carbó J, and Arús P (2003) Microsatellite variability in peach [*Prunus persica* (L) Batsch.]: cultivar identification, marker mutation, pedigree inferences and population structure. Theor Appl Genet 106: 1341–1352

Aranzana MJ, Howad W, and Arús P (2007) The extent of linkage disequilibrium in peach: a first approach. XII Eucarpia Symposium on Fruit Breeding and Genetics. Zaragoza (Abstract)

Archilletti T, Lauri P, and Damiano C (1995) *Agrobacterium*-mediated transformation of almond leaf pieces. Plant Cell Rep 14, 267–272

Arulsekar S, Parfitt DE, and Kester DE (1986) Comparison of isozyme variability in peach and almond cultivars. J Hered 77:272–274

Arumuganathan K, and Earle ED (1991) Nuclear DNA content of some important plant species. Plant Molec Biol Rep 9: 208–218

Arús P, Olarte C, Romero M, and Vargas F (1994) Linkage analysis of ten isozyme genes in F1 segregating almond progenies. J Amer Soc Hort Sci 119:339–344

Arús P, Ballester J, Jáuregui B, Joobeur T, Truco MJ, and de Vicente MC (1998) The European *Prunus* mapping project: update on marker development in almond. Acta Hortic 484: 325–330

Arús P, Howad W, and Mnejja M (2005a) Marker development and marker-assisted selection in temperate fruit trees. In: Proceedings of the International Congress "In the wake of the double helix: from the green revolution to the gene revolution". Avenue Media, Bologna, Italy, pp 309–325

Arús P, Yamamoto T, Dirlewanger E, and Abbott AG (2005b) Synteny in the Rosaceae. In: Plant Breed Reviews, Vol. 27. John Wiley & Sons Inc., New York, USA, pp 175–211

Ballester J (1998) Localització i anàlisi de caràcters d'interès agronòmic de l'ametller. Universitat Autònoma de Barcelona. Ph.D. Thesis

Ballester J, Bošković R, Batlle I, Arús P, Vargas F, and de Vicente MC (1998) Location of the self-incompatibility gene in the almond linkage map. Plant Breed 117: 69–72

Ballester J, Socias i Company R, Arús P, and de Vicente MC (2001) Genetic mapping of a major gene delaying blooming time in almond. Plant Breed 120: 268–270

Barbazuk WB, Emrich SJ, Chen HD, Li L, and Schnable PS (2007) SNP discovery via 454 transcriptome sequencing. Plant J 51: 910–918

Barbé L (1993) Favoriser le renouveau de la culture de l'amandier. Arboriculture Fruitière 358: 21–31

Barckley KK, Uratsu SL, Gradziel TM, and Dandekar AM (2006) Mutidimensional analysis of *S*-alleles from cross-incompatible groups of California almond cultivars. J Amer Soc Hort Sci 131:632–636

Bartolozzi F, Warburton ML, Arulsekar S, and Gradziel TM (1998) Genetic characterization and relatedness among California almond cultivars and breeding lines detected by ramdonly amplified polymorphic DNA (RAPD) analysis. J Amer Soc Hort Sci 123: 381–387

Bimal R, and Jha KK (1985) Role of cytokinin in shoot bud differentiation from endosperm of *Prunus amygdalus* Baill. J Plant Anat Morphol 2:63–66

Bliss FA, Arulsekar S, Foolad MR, Becerra V, Gillen AM, Warburton ML, Dandekar AM, Kocsisne GM, and Mydin KK (2002) An expanded genetic linkage map of Prunus based on an interspecific cross between almond and peach. Genome 45:520–529

Bošković R, and Tobutt KR (1996) Correlation of stylar ribonuclease zymograms with incompatibility alleles in sweet cherry. Euphytica 90: 245–250

Bošković R, Tobutt KR, Batlle I, and Duval H (1997) Correlation of stylar ribonuclease zymograms and incompatibility genotypes in almond. Euphytica 90:245–250

Broothaerts W, Janssens GA, Proost P, and Broekaert WF (1995) cDNA cloning and molecular analysis of two self-incompatibility alleles from apple. Plant Mol Biol 27:499–511

Byrne DH (1990) Isozyme variability in four diploid stone fruits compared with other woody perennial plants. J Hered 81(1):68–71

Campalans A, Pagès M, and Messeguer R (2001) Identification of differentially expressed genes by the cDNA-AFLP technique during dehydration of almond (Prunus amygdalus). Tree Phys 21:633–643

Carmona MJ, Cubas P, and Martínez-Zapater JM (2002) *VFL*, the grapevine *FLORICAULA/LEAFY* ortholog, is expressed in meristematic regions independently of their fate. Plant Phys 130:68–77

Cheng JH, Han ZH, Xu XF, and Li TZ (2006) Isolation and identification of the pollen-expressed polymorphic F-box genes linked to *S*-locus in apple (*Malus* × *domestica*). Sex Plant Reprod 19:175–183

Costa MS, Miguel CM, and Oliveira MM (2006) An improved selection strategy and the use of acetosyringone in shoot induction medium increase almond transformation efficiency by 100-fold. Plant Cell Tissue Organ Cult 85 :205–209

Crossa-Raynaud P, and Grasselly C (1985) Existence de groupes d'interstérilité chez l'amandier. Options Méditerranéennes 1985-I:43–45

Damiano C, Archilleti T, Caboni E, Lauri P, Falasca G, Mariotti D, and Ferraiolo, G (1995) *Agrobacterium*-mediated transformation of almond: in vitro rooting through localized infection of *A. rhizogenes.* Acta Hortic 392:161–169

de Nettancourt D (2001) Incompatibility and incongruity in wild and cultvated plants. Springer, Berlin, Heidelberg, New York

Denisov VP (1988) Almond Genetic Resources in the USSR and their use in production and breeding. Acta Hort 224: 299–306

Dicenta F, and García JE (1993) Inheritance of kernel flavour in almond. Heredity 70:308–312

Dicenta F, García JE, and Carbonell EA (1993a) Heritability of flowering, productivity and maturity in almond. J. Hortic Sci 68:113–120

Dicenta F, García JE, and Carbonell EA (1993b) Heritability of fruit characters in almond. J. Hortic Sci 68:121–126

Dicenta F, Martínez-Gómez P, Martinez-Pato E, and Gradziel T (2003) Screening for *Aspergillus flavus* resistance in almond. HortScience 38:266–268

Dirlewanger E, Moing A, Rothan C, Svanella L, Pronier V, Guye A, Plomion C, and Monet R (1999) Mapping QTLs controlling fruit quality in peach (*Prunus persica* (L.) Batsch). Theor Appl Genet 98:18–31

Dirlewanger E, Graziano E, Joobeur T, Garriga-Calderé F, Cosson P, Howad W, and Arús P (2004a) Comparative mapping and marker-assisted selection in Rosaceae fruit crops. Proc Nat Acad Sci USA 101:9891–9896

Dirlewanger E, Cosson P, Howad W, Capdeville G, Bosselut N, Claverie M, Voisin R, Poizat C, Lafargue B, Baron O, Laigret F, Kleinhentz M, Arús P, and Esmenjaud D (2004b) Microsatellite genetic linkage maps of myrobalan plum and an almond-peach hybrid – location of root-knot nematode resistance genes. Theor Appl Genet 109:827–838

Eduardo I, Arús P, and Monforte AJ (2005) Development of a genomic library of near isogenic lines (NILs) in melon (*Cucumis melo* L.) from the exotic accession PI161375. Theor Appl Genet 112:139–148

Entani T, Iwano M, Shiba H, Che FS, Isogai A, and Takayama S (2003) Comparative analysis of the self-incompatibility (*S*-)locus region of *Prunus mume*: Identification of a pollen expressed F-box gene with allelic diversity. Genes Cells 8:203–213

Felipe AJ (2000) El Almendro, I. El material vegetal., University of Zaragoza, Spain

Garcia-Mas J, Messeguer R, Arús P, and Puigdomènech P (1992) The extensin from *Prunus amygdalus*. Plant Physiol 100: 1603–1604

Felipe AJ, and Socias i Company R (1992) Almond Germplasm. HortScience 27: 718, 863

Garcia-Mas J, Messeguer R, Arús P, and Puigdomènech P (1995) Molecular characterization of cDNAs corresponding to genes expressed during almond (*Prunus amygdalus*) embryo development. Plant Molec Biol 27:205–210

Ghorbel R, La-Malfa S, López MM, Petit A, Navarro L, and Peña L (2001) Additional copies of *vir* G from pTiBo542 provide a super-transformation ability to *Agrobacterium tumefaciens* in citrus. Physiol Molec Plant Pathol 58:103–110

Godini A (2000) About the possible relationship between *Amygdalus webbii* Spach. and *Amygdalus communis* L. Nucis Newsl 9:17–19

Godini A (2002) Almond fruitfulness and role of self-fertility. Acta Hort 591:191–203

Goldraij A, Kondo K, Lee CB, Hancock CN, Sivaguru M, Vazquez-Santana S, Kim S, Phillips TE, Cruz-Garcia F, and McClure B (2006) Compartmentalization of S-RNase and HT-B degradation in self-incompatible *Nicotiana*. Nature 439:805–810

Goldsworthy A, and Street HE (1965) The carbohydrate nutrition of tomato roots: VIII. The mechanism of the inhibition by D-mannose of the respiration of excised roots. Ann Bot 29, 45–58

Gómez-Aparisi J, Carrera M, Felipe AJ, and Socias i Company, R (2001) 'Garnem', 'Monegro' y 'Felinem': Nuevos patrones híbridos almendro × melocotonero resistentes a nematodos y de hoja roja para frutales de hueso. Inf Técn Econ Agrar 97:282–288

Gradziel TM, and Wang D (1994) Susceptibility of California almond cultivars to aflatoxigenic *Aspergillus flavus*. HortScience 29:33–35

Gradziel TM, and Kester DE (1994) Breeding for resistance to *Aspergillus flavus* in almond. Acta Hort 373:111–117

Gradziel TM, and Kester DE (1998) Breeding for self-fertility in California almond cultivars. Acta Hort 470:109–117

Gradziel TM, Mahoney N, and Abdallah A (2000) Aflatoxin production among almond genotypes is not related to either kernel composition or *Aspergillus flavus* growth rate. HortScience 34:937–939

Gradziel TM, Martínez-Gómez P, and Dandekar AM (2001a) The use of S-allele specific PCR analysis to improve breeding efficiency for self-fertility in almond. HortScience 36: 440–440

Gradziel TM, Martínez-Gómez P, Dicenta F, and Kester DE (2001b) The utilization of related almond species for almond variety improvement. J Am Pomol Soc 55:100–109

Gradziel TM, Martínez-Gómez P, Dandekar A, Uratsu S, and Ortega E (2002) Multiple genetic factors control self-fertility in almond. Acta Hort 591:221–227

Gradziel TM (2003) Interspecific hybridizations and subsequent gene introgression within Prunus subgenus. Acta Hort 622:249–255

Gradziel TM (2009). Almond *(Prunus dulcis)* Breeding. In: S.M. Jain and M. Priyadarshan (eds). Breeding of Plantation Tree Crops. Springer Science, New York, pp 1–31.

Grant-Downton RT, and Dickinson HG (2006) Epigenetics and its Implications for Plant Biology 2. The Epigenetic Epiphany: Epigenetics, Evolution and Beyond. Ann Bot 97:11–27

Grasselly C (1972) L'Amandier; caractères morphologiques et physiologiques des variétés, modalité de leurs transmissions chez les hybrides de première génération. University of Bordeaux

Grasselly C (1978) Observations sur l'utilisation d'un mutant d'amandier á floraison tardive dans un programme d'hybridation Ann Amélior Plant 28:685–695

Grasselly C, and Crossa-Raynaud P (1980) L`amandier. G.P. Maisonneuve et Larose. Paris, XII 446 pp

Gur A, Semel Y, Cahaner A, and Zamir D (2004) Real Time QTL of complex phenotypes in tomato interspecific introgression lines. Trends Plant Sci 9:107–109

Hartmann HT, Kester DE, Geneve RL, and Davies FT (2002) Plant Propagation: Principles and Practices. Prentice Hall

Hauagge R, Kester DE, and Asay RA (1987) Isozyme variation among California almond cultivars: I. Inheritance. J Amer Soc Hort Sci 112(4):687–693

Herrero M, and Felipe AJ (1975) Pollinisation de ľamandier. Incompatibilité pollen-style. 2[e] Colloque GREMPA, Montepellier-Nimes, Septembre 1975

Hesse CO (1975) Peaches. In: Advances in Fruit Breeding. Temperate fruits. Purdue University Press, West Lafayette, USA, pp 285–335

Horiuchi H, Yanai K, Takagi M, Yano K, Wakabayashi E, Sanda A, Mine S, Ohgi K, and Irie M (1988) Primary strucuture of a base non-specific ribonuclease from *Rhizopus niveus*. J Biochem 103: 408–409

Howad W, Yamamoto T, Dirlewanger E, Testolin R, Cosson P, Cipriani G, Monforte AJ, Georgi L, Abbott AG, and Arús P (2005) Mapping with a few plants: using selective mapping for microsatellite saturation of the Prunus reference map. Genetics 171:1305–1309.

Howad W, and Arús P (2007) QTL analysis of blooming time in peach. XV Plant and Animal Genome Conference. San Diego. http://www.intl-pag.org/PAGarchives.html

Igic B, and Kohn JR (2001) Evolutionary relationships among self-incompatibility RNases. Proc Natl Acad Sci USA 98:13167–13171

Ikeda K, Igic B, Ushijima K, Yamane H, Hauck NR, Nakano R, Sassa H, Iezzoni AF, Kohn JR, and Tao R (2004) Primary strucutural features of the *S* haplotype-specific F-box protein, SFB, in *Prunus*. Sex Plant Reprod. 16:235–243

Ikeda K, Ushijima K, Yamane H, Tao R, Hauck NR, Sebolt AM, and Iezzoni AF (2005) Linkage and physical distances between the *S*-haplotype *S-RNase* and *SFB* genes in sweet cherry. Sex Plant Reprod 17: 289–296

Jáuregui B, de Vicente MC, Messeguer R, Felipe A, Bonnet A, Salesses G, and Arús P (2001) A reciprocal translocation between 'Garfi' almond and 'Nemared' peach. Theor Appl Genet 102:1169–1176

Joobeur T (1998) Construcción de un mapa de marcadores moleculares y análisis genético de caracteres agronómicos en *Prunus*. PhD thesis, Universitat de Lleida

Joobeur T, Viruel MA, de Vicente MC, Jáuregui B, Ballester J, Dettori MT, Verde I, Truco MJ, Messeguer R, Batlle I, Quarta R, Dirlewanger E, and Arús P (1998) Construction of a saturated linkage map for *Prunus* using an almond × peach F2 progeny. Theor Appl Genet 97:1034–1041

Joobeur T, Periam N, de Vicente MC, King G, and Arús P (2000) Development of a second generation linkage map for almond using RAPD and SSR markers. Genome 43:649–655

Jung S, Jesudurai C, Staton M, Du Z, Ficklin S, Cho I, Abbott A, Tomkins J, and Main D (2004) GDR (Genome Database for Rosaceae): integrated web resources for Rosaceae genomics and genetics research. BMC Bioinformatics. 5:130

Kao TH, and Tsukamoto T (2004) The molecular and genetic basis of S-RNase-based self-incompatibility. Plant Cell 16 [Suppl]: S72–S83

Kakui H, Tsuzuki T, Koba T, and Sassa H (2007) Polymorphism of *SFBB*$^{-\gamma}$ and its use for *S* genotyping in Japanese pear (*Pyrus pyrifolia*). Plant Cell Rep 26: 1619–1625

Kawata Y, Sakiyama F, and Tmaoki H (1988) Amino-acid sequence of ribonuclease T2 from *Aspergillus oryzae*. Eur J Biochem 176: 683–697

Kester DE, Gradziel TM, and Grasselly C (1991) Almonds (*Prunus*). In: Genetic resources of temperate fruit and nut crops. International Society for Horticultural Science, The Netherlands, pp 701–758

Kester DE, Gradziel TM, and Micke WC (1994) Identifying pollen incompatibility groups in California almond cultivars. J Amer Soc Hort Sci 119: 106–109

Kester DE, and Gradziel TM (1996) Almonds (*Prunus*). In: Fruit Breeding. John Wiley & Sons, Inc., New York, USA, pp 1–97

Kester DE, Raddi P, and Asay R (1977a) Correlation among chilling requirements for germination, blooming and leafing in almond (*Prunus amygdalus* Batsch.) within and among seedling populations in almond J Am Soc Hort Sci 102: 145–148

Kester DE, Hansche PE, Beres V, and Asay RN (1977b) Variance components and heritability of nut and kernel traits in almond. J Amer Soc Hort Soc 102: 264–266

Kester DE, Shackel KA, Micke WC, Viveros M, and Gradziel TM (2004) Noninfectious bud failure in 'Carmel' almond: I. Pattern of development in vegetative progeny trees. J Amer Soc Hort Sci 129: 244–249

Ladizinsky G (1999) On the origin of almond. Gen Resour Crop Evol 46: 143–147

Lai Z, Ma W, Han B, Liang L, Zhang Y, Hong G, and Xue Y (2002) An F-box gene linked to the self-incompatibility (*S*) locus of *Antirrhinum* is expressed specifically in pollen and tapetum. Plant Mol Biol 50: 29–72

Liu CN, Li XQ, and Gelvin SB (1992) Multiple copies of *vir*G enhance the transient transformation of celery, carrot and rice tissues by *Agrobacterium tumefaciens*. Plant Molec Biol 20: 1071–1087

Liu J, Yu J, McIntosh L, Kende H, and Zeevaart JAD (2001) Isolation of a *CONSTANS* ortholog from *Pharbitis nil* and its role in flowering. Plant Phys 125: 1821–1830

López M, Mnejja M, Rovira M, Collins G, Vargas FJ, Arús P, and Batlle I (2004) Self-incompatibility genotypes in almond re-evaluated by PCR, stylar ribonucleases, sequencing analysis and controlled pollinations. Theor Appl Genet 109: 954–964

López M, Mnejja M, Romero MA, Vargas FJ, Arús P, and Batlle I (2005) Use of Sf-specific PCR for early selection of self-compatible seedlings in almond breeding. Options Méditerraneeennes 63: 269–274

López M, Vargas FJ, and Batlle I (2006) Self-(in)compatibility almond genotypes: A review. Euphytica 150: 1–16

Ma RC, and Oliveira MM (2002) Evolution analysis of S-RNase genes from Rosaceae species. Mol Gent Genomics 267: 71–78

Marcotrigiano M, and Gradziel TM (1997) Genetic mosaics and plant improvement. In: Plant Breed Rev, Vol. 15 pp 43–84

Martínez-Gómez P, and Gradziel TM (2003) Sexual polyembryony in almond. Sex Plant Reprod 16: 135–139

Martínez-Gómez P, Arulsekar S, Potter D, and Gradziel TM (2003a) Relationships among peach, almond, and related species as detected by simple sequence repeat markers. J Amer Soc Hort Sci 128: 667–671

Martinez-Gomez P, Arulsekar S, Potter D, and Gradziel TM (2003b) An extended interspecific gene pool available to peach and almond breeding as characterized using simple sequence repeat(SSR) markers. Euphytica 131: 313–322

Martínez-Gómez P, Rubio M, Dicenta F, and Gradziel TM (2004) Resistance to Plum Pox Virus (Dideron isolate RB3.30) in a group of California almonds and transfer of resistance to peach. J Amer Soc Hort Sci 129: 544–548

Martins M, and Oliveira MM (2005) Cloning and characterization of NBS resistance gene candidate sequences in almond. Acta Hortic 663: 157–160

Martins M, Sarmento D, Batlle I, Vargas F, and Oliveira MM (2001) Search for molecular markers linked to *Fusicoccum* tolerance in almond. Acta Hortic 577: 87–90

Martins M, Sarmento D, Batlle I, Vargas F, and Oliveira MM (2005) Development of SCAR/CAPS markers linked to tolerance/sensitivity to *Fusicoccum* in almond. Options Méditerranéennes CIHEAM/IAMZ 63: 187–191

McClure BA, Haring V, Ebert PR, Anderson MA, Simpson RJ, Sakiyama F, and Clarke AE (1989) Style self-incompatibility gene products of *Nicotiana alata* are ribonucleases. Nature 342: 955–957

McClure BA, and Franklin-Tong V (2006) Gametophytic self-incompatibility: understanding the cellular mechanisms involved in "self" pollen tube inhibition. Planta 224: 233–245

McCubbin AG, and Kao TH (2000) Molecular recognition and response in pollen and pistil interactions. Annu Rev Cell Dev Biol 16: 333–364

Mehra A, and Mehra PN (1974) Organogenesis and plantlet formation in vitro in almond. Bot Gazette 135: 61–73

Miguel CM (1998) *Adventitious regeneration and genetic transformation of almond (Prunus dulcis Mill.).* Ph.D. Dissertation, Faculdade de Ciências, Universidade de Lisboa, Portugal

Miguel CM, and Oliveira MM (1999) Transgenic almond (*Prunus dulcis* Mill.) plants obtained by *Agrobacterium*-mediated transformation of leaf explants. Plant Cell Rep 18: 387–393

Miguel CM, Druart P, and Oliveira MM (1996) Shoot regeneration from adventitious buds induced on juvenile and adult almond (*Prunus dulcis* Mill.) explants. In Vitro Cellular Develop Biol-Plant 32: 148–153

Mnejja M, Garcia-Mas J, Howad W, and Arús P (2005) Development and transportability across *Prunus* species of 42 polymorphic almond microsatellites. Molec Ecol Notes 5: 531–535

Mowrey D, and Werner DJ (1990) Phylogenetic relationships among species of prunus as inferred by isozyme markers. Theor Appl Genet 80: 129–133

Niklasson M (ed.) (1989) The European almond catalogue. Alnarp, Nordic Gene Bank

Oliveira MM, Miguel CM, and Costa M (2008) Almond. In: Transgenics and Molecular Tailoring in Plants, Vol.4 – Temperate Fruits and Nuts. Blackwell Publishing. (in press)

Ortega E, Boskovic RI, Sargent DJ, and Tobutt KR (2006) Analysis of S-RNase alleles of almond (*Prunus dulcis*): characterization of new sequences, resolution of synonyms and evidence of intragenic recombination. Mol Gen Genomics 276: 413–426

Peace CP, Crisosto CH, and Gradziel TM (2005) Endopolygalacturonase: a candidate gene for freestone and melting flesh in peach. Molec Breed 161: 21–31

Potter D (2003) Molecular phylogenetic studies in Rosaceae. In: Plant Genome: Biodiversity and Evolution, Vol. 1, Part A: Phanerogams. Science Publishers, Inc. Enfield (SA), Plymouth (UK), pp 319–351

Qiao H, Wang F, Zhao L, Zhou J, Lai Z, Zhang Y, Robbins TP, and Xue Y (2004) The F-box protein *AhSLF-S*$_2$ controls the pollen function of S-RNase-based self-incompatibility. Plant Cell 16: 2307–2322

Ramesh S, Kaiser B, Franks T, Collins G, and Sedgley M (2006) Improved methods in *Agrobacterium*-mediated transformation of almond using positive (mannose/*pmi*) or negative (kanamycin resistance) selection-based protocols. Plant Cell Rep 25: 21–828

Rikhter AA (1969) Ways and methods of almond breeding (in Russian). Tr Gos Nikit Bot Sad 43:81–94

Romero MA, and Vargas FJ (1981) Contribution for the knowledge of almond sensitivity to *Fusicoccum amygdali* Del. Options Mediterranéennes, Série Études I: 143–147

Rosengarten F (1984) The book of edible nuts. Walker and Company, New York

Rouhi V, Samson R, Lemeur R, and van Damme P (2007) Photosynthetic gas exchange characteristics in three different almond species during drought stress. Environ Exper Bot 59: 117–129

Rugini E (1986) Almond. In: Handbook of Plant Cell Culture, Vol. 4. Macmillan, New York, USA, pp 574–611

Sánchez-Pérez R, Howad W, Dicenta F, Arús P, and Martínez-Gómez P (2007) Mapping major genes and quantitative trait loci controlling agronomic traits in almond. Plant Breed 126: 310–318

Santos AM, Oliver MJ, Sanchez AM, Payton P, Saibo N, and Oliveira MM (2007) Adventitious Shoot Regeneration in *Prunus dulcis* – a Molecular Approach to the Regeneration Process. Acta Hort 738: 663–668

Sassa H, Hirano H, and Ikehashi H (1992) Self-incompatibility-related RNases in styles of Japanese pear (*Pyrus serotina* Rehd.). Plant Cell Physiol 33: 811–814

Sassa H, Nishio T, Kowyama Y, Hirano H, Koba T, and Ikehashi H (1996). Self-incompatibility (*S*) alleles of the Rosaceae encodes members of a distinct class of the $T_2$/S rubonuclease superfamily. Mol Gen Genet 250: 547–557

Sassa H, Kakui H, Miyamoto H, Suzuki M, Hanada Y, Ushijima K, Kusaba M, Hirano H, and Koba K (2007) *S locus F-box brothers*: multiple and pollen-specific F-box genes with *S*-haplotype-specific polymorphisms in apple and Japanese pear. Genetics 175: 1869–1881

Schirra M (1997) Postharvest technology and utilization of almonds. Hort Rev 20: 267–292

Seirlis G, Mourras A, and Salesses G (1979) Tentatives de culture in vitro d'anthères et de fragments d'organes chez le *Prunus*. Ann Amélior Plantes 29: 145–161

Sijacic P, Wang X, Skirpan AL, Wang Y, Dowd PE, McCubbin AG, Huang S, and Kao TH (2004) Identification of the pollen determinant of S-RNase-mediated self-incompatibility. Nature 429: 302–305

Silva CM (2005) Mechanisms underlying flowering in almond (*Prunus dulcis* (Mill.) D.A. Webb): a molecular perspective. PhD thesis, ITQB-UNL, Lisboa, Portugal

Silva C, Garcia-Mas J, Sánchez AM, Arús P, and Oliveira MM (2005) Looking into flowering time in almond (*Prunus dulcis* (Mill) D. A. Webb): the candidate gene approach. Theor Appl Genet 110: 959–968

Silva C, Sanchez AM, and Oliveira MM (2007) Isolation of full-length cDNAs with putative functions in almond (*Prunus dulcis* Mill.) flowering. Acta Hort 738: 675–682

Socias i Company R, and Felipe AJ (1988) Self-compatibility in almond: Transmission and recent advances. Acta Hort 224: 307–317

Socias i Company R (1990) Breeding self-compatible almonds. Plant Breed Rev 8: 313–338

Socias i, Company R, Felipe AJ (1992) Almond: a diverse germplasm. HortScience 27:718, 863

Socias i Company R, Felipe AJ, and Gómez Aparisi J (1996) Genetics of late blooming in almond. Acta Hortic 484: 261–266

Socias i Company R (1998) Fruit tree genetics at a turning point: the almond example. Theor Appl Genet 96: 588–601

Socias i Company R (2002) The relationship of *Prunus webbii* and almond revisited. Nucis-Newsl 11: 17–19

Socias i Company R, Kodad O, Alonso JM, and Gradziel TM (2008) Almond Quality: A Breeding Perspective. In: Horticultural Reviews, Vol. 34. John Wiley & Sons Inc. New York, USA, pp 197–238

Spiegel-Roy P, and Kochba J (1981) Inheritance of nut and kernel traits in almond (*Prunus amygdalus* Batsch.) Euphytica 30: 167–174

Suelves M, and Puigdomènech P (1997) Different lipid transfer protein mRNA accumulate in distinct parts of *Prunus amygdalus* flower. Plant Sci 129: 49–56

Suelves M, and Puigdomènech P (1998a) Molecular cloning of the cDNA encoding for (R)-(+)-mandelinitrile lyase of *Prunus amygdalus*: temporal and spatial expression patterns in flowers and mature seeds. Planta 206: 388–393

Suelves M, and Puigdomènech P (1998b) Specific mRNA accumulation of a gene codig for an O-methyltransferase in almond (*Prunus amygdalus* Batsch.) flower tissues. Plant Sci 134: 79–88

Sung SK, and An G (1997) Molecular cloning and characterization of a MADS-box cDNA clone of the Fuji apple. Plant Cell Phys 38: 484–489

Sung SK, Yu GH, and An G (1999) Characterization of *MdMADS2* a member of the *SQUAMOSA* subfamily of genes, in apple. Plant Phys 120: 969–978

Takayama S, and Isogai A (2005) Self-incompatibility in plants. Annu Rev Plant Biol 56: 467–489

Tamura M, Ushijima K, Sassa H, Hirano H, Tao R, Gradziel TM, and Dandekar AM (2000) Identification of self-incompatibility genotypes of almond by allele-specific PCR analysis. Theor Appl Genet 101: 344–349

Tanksley SD, Young ND, Paterson AH, and Bonierbale MW (1989) RFLP mapping in plant breeding: new tools for an old science. BioTechnology 7: 257–263

Tao R, Yamane H, Sassa H, Mori H, Gradziel TM, Dandekar AM, and Sugiura A (1997) Identification of stylar RNases associated with gametophytic self-incompatibility in almond (*Prunus dulcis*). Plant Cell Physiol 38: 304–311

Tao R, Yamane H, Sugiura A, Murayama H, Sassa H, and Mori H (1999) Molecular typing of *S*-alleles trough identification, characterization and cDNA cloning for *S*-RNases in sweet cherry. J Amer Soc Hort Sci 124: 224–233

Testolin R, Messina R, Lain O, Marrazzo T, Huang G, and Cipriani G (2004) Microsatellites isolated in almond from an AC-repeat enriched library. Mol Ecol Notes 4: 459–461

Tufts WP, and Philp GL (1922) Almond pollination. California Agric Experimental Station Bull. 35

**Table 2** Sources of disease and pest resistance in species sexually compatible with peach

| Species | Traits | References |
|---|---|---|
| *P. persica* subsp. *ferganensis* | Powdery Mildew (*Sphaerotheca pannosa*) | Verde et al. (2004) |
| *P. dulcis* | Powdery Mildew (*S. pannosa*) | Gradziel (2003) |
| | Plum Pox | Pascal et al. (2002) |
| *P. kansuensis* | *Meloidogyne incognita* | Wang et al. (2002) |
| *P. davidiana* | Plum Pox Resistance | Pascal et al. (2002), Decroocq et al. (2005) |
| | Powdery Mildew (*S. pannosa*) | Foulongne et al. (2004) |
| | Green Peach Aphid Resistance *Myzus persicae* | Monet et al. (1998) |
| | Peach Leaf Curl (*Taprina deformans*) | Viruel et al. (1998) |

Stagnation or decrease in consumption of fresh peaches and nectarines in European Union and United States, in the last decade, has been attributed to low quality of fruit harvested at an immature stage for storage and shipment (Sansavini et al., 2006). This unfavorable trend has put an emphasis on importance of finding the right balance between fruit quality and immaturity at harvest. Most of the traits associated with fruit quality are quantitatively inherited and difficult to breed. There are 16 major traits associated with quantitative trait loci (QTLs), and 13 single gene traits described in peach that are associated with fruit characteristics (Hancock et al., 2008).

Significant work has been done on determining the genes and QTLs involved with various disorders that occur during prolonged cold storage, internal breakdown, mealiness, flesh browning (Peace et al., 2005; Ogundiwin et al., 2008); and in physical and chemical components of fruit quality, sugar and organic acid content (Dirlewanger et al., 1998; Etienne et al., 2002). Recent discovery of genes encoding endopolygalacturonase enzyme, that controls the freestone and melting characters (Peace et al., 2005), and leucoanthocanidin dioxygenase enzyme, co-located with the major QTL associated with flesh browning (Ogundiwin et al., 2008), show potential for marker assisted breeding of new varieties with lower susceptibility to internal breakdown. In addition, QTLs for fruit development period, fruit size, pH, soluble sugar and soluble solid contents have also been detected in peach (Abbott et al., 1998; Quarta et al., 2000).

## 2.3 Bloom and Ripening Times

Constant climatic changes and emerging abiotic stresses are affecting peach production worldwide. Hence, environmental adaptation has become a very important part in every breeding program. The most attention is given to winter cold hardiness, spring frost hardiness, and chilling requirement. Winter cold hardiness is important component of cultivars developed for colder climates. Most information on winter cold hardiness has been achieved through naturally occurring freezes and its complexity has been well documented (Stushnoff, 1972; Scorza and Sherman, 1996). Cold hardy genotypes have been reported in China (Scorza and Sherman, 1996) and North American germplasm (Layne, 1984). In general, genotypes with high chilling requirement tend to suffer less bud damage due to a winter cold. The degree of bud damage, observed in segregating populations, suggested quantitative inheritance (Mowry, 1964). Abundance of dehydrins has been associated with cold hardiness and its seasonal expression has been observed in woody plants (Wisniewski and Arora, 2000).

Late spring frosts are damaging factor in both cold and warm climates and their avoidance is achieved through alteration in blooming time via chill requirement reduction. Genetics of chill requirement has been largely unknown although segregation analyses suggested quantitative regulation involving few major genes with cumulative and similar effect on phenotype (Lesley, 1944; Sharp, 1961). A recessive gene responsible for absence of bud dormancy in peach was identified a few decades ago (Lammerts, 1945) and early observations suggested incomplete dominance with the heterozygotes exhibiting an intermediate phenotype (Rodriguez et al., 1994). Recently, this ‘evergrowing’ behavior of peach has been mapped (Wang et al., 2002) and explained as a deletion in a cluster of MIKC-type MADS-box transcription factors (Bielenberg et al., 2008).

Alterations in bloom time have always been associated with avoidance of late spring frosts and fulfillment of chilling requirement, and although it has been shown to exhibit moderate heritability (Hansche et al., 1972) its genetics is still mostly unknown. Early observations suggested that combination of cold and heat exposure is required for normal bloom time, and that chilling requirement influences bloom date. However, genotypes with low chilling requirement and late blooming have been observed (Yarnell, 1945; Scorza and Sherman, 1996).

Ripening date is probably the most observed character in active breeding programs. Desire to extend the growing season, with early and late ripening genotypes, lead to extensive evaluation of the time of fruit maturity. Majority of observations suggested additive polygenic control (Bailey and Hough, 1959; Hansche et al., 1972) and high heritability, 0.94 (de Souza et al., 1998) of ripening season. However, the evidence for few genes with relatively large effects was also available (French, 1951; Vileila-Morales et al., 1981).

### 2.3.1 Genetic Traits and Genomic Resources in Peach

Along with *Malus* × *domestica* and *Fragaria vesca* L., peach has been designated as one of three phylogenetically distinct genomic models within the Rosaceae and serves as the reference model species for stone-fruits (Shulaev et al., 2008). Among woody perennials, peach has a number of advantageous characteristics for genetic and genomic experiments. Peach is a diploid species with eight chromosomes ($2n = 2x = 16$) and does not appear to have a significant history of duplication within the genome. Peach trees are monoecious and do not display the self-incompatibility characteristic of most of species in the genus, allowing the creation of selfed $F_2$ populations for genetic mapping projects. Peach is reproductively mature within 2–3 years, short relative to other tree crops (Rieger, 2006). Hard and softwood cuttings can easily be used to clonally propagate populations for replicated plantings and maintenance of individuals (Okie, 1984). A large number of ‘simple’ segregating traits or mutations have been identified in peach in addition to the existence of thousands of cultivated varieties with well-described phenotypic variation (Scorza and Sherman, 1996).

Public molecular genetic resources for peach (and other closely related *Prunus* species) are abundant, including >80,000 EST sequences, multiple BAC genomic libraries, many genetic maps anchored to a common reference genetic map, and a physical map with anchored genetic markers and ESTs (Jung et al., 2004; Horn et al., 2005; Shulaev et al., 2008). The public resources for peach are centrally located in the Genome Database for Rosaceae, including many *Prunus* genetic maps with markers anchored to an interspecific almond × peach $F_2$ population (Jung et al., 2004).

The genome of peach is relatively small ($\approx$280 Mbp/C), less than twice the size of the *Arabidopsis* genome and smaller than the recently sequenced *Populus* genome (Baird et al., 1994; Tuskan et al., 2006). In 2007, The Joint Genome Institute announced the upcoming sequencing of the peach genome in with a targeted completion date of the end of 2008 (Shulaev et al., 2008), although it is likely that sequence may not be available until 2009. Integration of the genome sequence with existing genetic and physical maps will be a watershed for *Prunus* researchers as candidate genes are identified and made available for testing.

### 2.3.2 Translating the Potential of Genomics to Breeding Programs

The ultimate justification for the time and effort of genomics efforts is the translation of knowledge into crop improvement. Transgenic manipulation of significant agronomic traits is currently the paradigm for the improvement of crop characteristics and has been highly successful with many of our staple food crops, particularly with single gene manipulations for insect and herbicide resistance. Despite the successes seen in other *Prunus* species (Mante et al., 1991; Machado et al., 1992, 1995; Gutierrez-Pesce et al., 1998; Miguel and Oliveira, 1999; Gutierrez and Rugini, 2004; Petri et al., 2005; Padilla et al., 2006; Song and Sink, 2006; Dai et al., 2007; Maghuly et al., 2007), high efficiency transformation and regeneration of peach has proved elusive (Hammerschlag and Smigocki, 1991; Perez-Clemente et al., 2004; Padilla et al., 2006).

The recent report of transgenic success (Perez-Clemente et al., 2004) has not been replicated by other groups. Additionally, this technique made use of embryos as opposed to somatic tissues, which is less desirable for the improvement of superior commercial genotypes. Significant effort is being devoted to the development of a protocol with reasonable efficiency and applicability across genotypes at many institutions. In the United States, the USDA-ARS Appalachian Fruit Research Station group at Kearneysville, West Virginia is particularly active in the effort.

In peach, genomic data will most immediately be applied to crop improvement through the use of marker-assisted-selection (MAS). Markers associated closely with mature reproductive traits (flowering, fruiting, chilling requirement, etc.) could be used to cull undesirable offspring from crosses soon after germination, reducing the time, expense and effort of maintaining and evaluating large numbers of progeny. The use of MAS in peach breeding has not yet been reported, however, the available physical map should allow for marker saturation around important traits (Shulaev et al., 2008). It is foreseeable that a demonstration of successful MAS in peach will be performed in the next several years. Availability of the sequenced genome would speed the achievement of MAS for peach in traditional breeding programs.

## References

Abbott, A.G., Rajapakse, B., Sosinski, B., Lu, Z.X., Sossey-Alaoui, K., Gannavarapu, M., Reighard, G.L., Ballard, R.E., Baird, W.V., Scorza, R. and Callahan, A. (1998) Construction of saturated linkage maps of peach crosses segregating for characters controlling fruit quality, tree architecture and pest resistance. Acta Hort. 465:41–49.

Bailey, C.H. and Hough, L.F. (1959) An hypothesis for the inheritance of season of ripening in progenies from certain early ripening peach varieties and selections. Proc. Am. Soc. Hortic. Sci. 73: 125–133.

Baird, W.V., Estager, A.S. and Wells, J.K. (1994) Estimating nuclear-DNA content in peach and related diploid species using laser flow-cytometry and DNA hybridization. J. Am. Soc. Hortic. Sci. 119: 1312–1316.

Beckman, T.G. and Reilly, C.C. (2005) Relative susceptibility of peach cultivars to fungal gummosis (*Botryosphaeria dothidea*). J. Am. Pomol. Soc. 59:111–116.

Bielenberg, D., Wang, Y., Li, Z., Zhebentyayeva, T., Fan, S., Reighard, G.L., Scorza, R. and Abbott, A.G. (2008) Sequencing and annotation of the evergrowing locus in peach [*Prunus persica* (L.) Batsch] reveals a cluster of six MADS-box transcription factors as candidate genes for regulation of terminal bud formation. Tree Genet. Gen. 4:495–507.

Blenda, A.V., Verde, I., Georgi, L.L., Reighard, G., Forrest, S., Munoz-Torres, M., Baird, W.V. and Abbott, A.G. (2007) Construction of a genetic linkage map and identification of molecular markers in peach rootstocks for response to peach tree short life syndrome. Tree Genet. Gen. 81:281–288.

Byrne, D.H. (2005) Trends in stone fruit cultivar development. Horttechnology 15: 494–500.

Chang, L.S., Iezzoni, A., Adams, G. and Howell, G.S. (1989) *Lucostoma* tolerance and cold hardiness among diverse peach genotypes. J. Am. Soc. Hort. Sci. 114:482–485.

Chaplin, C.E. and Schneider, G.W. (1975) Resistance to the common peach tree borer (*Sanninoidea exitiosa* Say) in seedlings of 'Rutgers Redleaf' peach. HortScience 10:400.

Dai, W.H., Magnusson, V. and Johnson, C. (2007) Agrobacterium-mediated transformation of chokecherry (Prunus virginiana L.). Hortscience 42: 140–142.

Decroocq, V., Foulongne, M., Lambert, P., Gall, O., Mantin, C., Pascal, T., Schurdi-Levraud, V. and Kervella, J. (2005) Analogues of virus resistance genes map to QTLs for resistance to sharka disease in Prunus davidiana. Mol. Genet. Genom. 272(6):680–689.

de Souza, V.A.B., Byrne, D.H. and Taylor, J.F. (1998) Heritability, genetic and phenotypic correlations, and predicted selection response of quantitative traits in peach: II. An analysis of several fruit traits. J. Am. Hort. Sci. 123(4):604–611.

Dirlewanger, E., Pronier, V., Parvery, C., Rothan, C., Guye, A. and Monet, R. (1998) Genetic linkage map of peach [*Prunus persica* (L.) Batsch] using morphological and molecular markers. Theor. Appl. Genet. 97, 888–895.

Etienne, C., Rothan, C., Moing, A., Plomion, C., Bodénès, C., Svanella-Dumas, L., Cosson, P., Pronier, V., Monet, R. and Dirlewanger, E. (2002) Candidate genes and QTLs for sugar and organic acid content in peach [*Prunus persica* (L.) Batsch]. Theor. Appl. Genet. 105: 145–159.

Foulongne, M., Pascal, T., Pfeiffer, F. and Kervella, J. (2002) Introgression of a polygenic resistance to powdery mildew from a wild species *Prunus davidiana* into peach [*Prunus persica* (L.) Batsch], a case study of marker assisted selection in fruit tree. Acta Hort. 592: 259–265.

Foulongne, M., Pascal, T., Arus, P. and Kervella J (2003) The potential of Prunus davidiana for introgression into peach (Prunus persica L. Batsch) assessed by comparative mapping. Theor. Appl. Genet. 107:227–238.

Foulongne, M., Pascal, T., Pfeiffer, F. and Kervella, J. (2004) Re-examination of an introgression procedure of a polygenic resistance to powdery mildew from wild *prunus davidiana* into peach in the light of molecular data and Monte-Carlo simulations. Acta Hort. (ISHS) 663:849–854.

French, A.P. (1951) The peach: Inheritance of time of ripening and other economic characteristics. Massachusetts Agric. Exp. Sta. Bull. 462:2–31.

Gillen, A.M. and Bliss, F.A. (2005) Identification and mapping of markers linked to the Mi gene for root-knot nematode resistance in peach. J. Am. Soc. Hortic. Sci. 130:24–33.

Gradziel, T.M. (2003). Interspecific hybridizations and subsequent gene introgression within *prunus* subgenus *amygdalus*. Acta Hort. (ISHS) 622:249–255.

Gradziel, T.M., Thorpe, M.A., Bostock, R.M. and Wilcox, S.M. (1998) Breeding for brown rot (Monilinia fructicola) resistance in clingstone peache with emphasis on the role of fruit phenolics. Acta Hort. 465:161–170.

Gutierrez, P. and Rugini, E. (2004) Influence of plant growth regulators, carbon sources and iron on the cyclic secondary somatic embryogenesis and plant regeneration of transgenic cherry rootstock 'Colt' (*Prunus avium* × *P. pseudocerasus*). Plant Cell Tiss. Org. Cult. 79: 223–232.

Gutierrez-Pesce, P., Taylor, K., Muleo, R. and Rugini, E. (1998) Somatic embryogenesis and shoot regeneration from transgenic roots of the cherry rootstock Colt (Prunus avium × P-pseudocerasus) mediated by pRi 1855 T-DNA of Agrobacterium rhizogenes. Plant Cell Rep. 17: 574–580.

Hammerschlag, F.A. and Smigocki, A.C. (1991) Regeneration of plants from peach embryo cells infected with a shooty mutant strain of *Agrobacterium*. J. Am. Soc. Hortic. Sci. 116: 1092–1097.

Hancock, J.F., Scorza, R. and Lobos, G.A. (2008). Peaches. In: Hancock, J.F. (Ed.) Temperate fruit crop breeding, Springer, Netherlands, 265–298.

Hansche, P.E., Hesse, C.O., Beres, V. (1972) Estimates of genetic and environmental effects of several traits in peach. J. Am. Soc. Hortic. Sci. 97:76–79.

Horn, R., Lecouls, A.C., Callahan, A., Dandekar, A., Garay, L., McCord, P., Howad, W., Chan, H., Verde, I., Main, D., Jung, S., Georgi, L., Forrest, S., Mook, J., Zhebentyayeva, T., Yu, Y.S., Kim, H.R., Jesudurai, C., Sosinski, B., Arus, P., Baird, V., Parfitt, D., Reighard, G., Scorza, R., Tomkins, J., Wing, R. and Abbott, A.G. (2005) Candidate gene database and transcript map for peach, a model species for fruit trees. Theor. Appl. Genet. 110: 1419–1428.

Jung, S., Jesudurai, C., Staton, M., Du, Z.D., Ficklin, S., Cho, I.H., Abbott, A., Tomkins, J. and Main, D. (2004) GDR (Genome Database for Rosaceae): integrated web resources for Rosaceae genomics and genetics research. BMC Bioinformatics 5: Art. No. 130.

Lammerts, W.E. (1945) The breeding of ornamental edible peaches for mild climates, I. Inheritance of tree and flower characteristics. Am. J. Bot. 32:53–61.

Layne, R.E.C. (1984) Breeding peaches in North America for cold hardiness and perennial canker (*Leucostoma* ssp.) resistance: review and outlook. Fruit Varieties J. 38:130–136.

Lesley J.W. (1944) Peach breeding in relation to winter chilling requirements. Proc. Am. Soc. Hortic. Sci. 70:243–250.

Lu, Z.X., Sosinski, B., Reighard, G., Baird, W.V. and Abbott, A.G. (1998) Construction of a genetic linkage map and identification of AFLP markers for resistance to root-knot nematodes in peach rootstocks. Genome 41:199–207.

Machado, A.D., Puschmann, M., Puhringer, H., Kremen, R., Katinger, H. and Machado, M.L.D. (1995) Somatic embryogenesis of prunus-subhirtella autumno-rosa and regeneration of transgenic plants after agrobacterium-mediated transformation. Plant Cell Rep. 14: 335–340.

Machado, M.L.D., Machado, A.D., Hanzer, V., Weiss, H., Regner, F., Steinkellner, H., Mattanovich, D., Plail, R., Knapp, E., Kalthoff, B. and Katinger, H. (1992) Regeneration of transgenic plants of prunus-armeniaca containing the coat protein gene of plum pox virus. Plant Cell Rep. 11: 25–29.

Maghuly, F., Machado, A.D., Leopold, S., Khan, M.A., Katinger, H. and Laimer, M. (2007) Long-term stability of marker gene expression in Prunus subhirtella: A model fruit tree species. J. Biotechnol. 127: 310–321.

Mante, S., Morgens, P.H., Scorza, R., Cordts, J.M. and Callahan, A.M. (1991) agrobacterium-mediated transformation of plum (prunus-domestica l) hypocotyl slices and regeneration of transgenic plants. Bio-Technol. 9: 853–857.

Martinez-Gomez, P., Arulskar, S., Potter, D. and Gradziel, T.M. (2003) Relationships among peach, almond, and related species as detected by simple sequence repeat markers. J. Am. Soc. Hortic. Sci. 128:667–671.

Massonie, G., Maison, P., Monet, R. and Grasselly, C. (1982) Resistance to the green peach aphid *Myzus persicae* Sulzer (Homoptera: Aphididae) in *Prunus persica* (L.) Batsch and other Prunus species (in French). Agronomie 2:63–69.

Miguel, C.M. and Oliveira, M.M. (1999) Transgenic almond (*Prunus dulcis* Mill.) plants obtained by Agrobacterium-mediated transformation of leaf explants. Plant Cell Rep. 18: 387–393.

Monet, R. (1985) Heredity of the resistance to leaf curl (*Taphrina deformans*) and green aphid (*Myzus persicae*) in the peach. Acta Hortic. 173:21–23.

Monet, R., Guye, A. and Massonié, G. (1998). Breeding for resistance to green aphid *myzus persicae* sulzer in the peach. Acta Hort. (ISHS) 465:171–176.

Mowry, J.B. (1964) Inheritance of cold hardiness of dormant peach flower buds. Proc. Am. Soc. Hortic. Sci. 85:128–133.

Ogundiwin, E., Peace, C., Nicolet, C., Rashbrook, V., Gradziel, T., Bliss, F., Parfitt, D. and Crisosto, C. (2008) Leucoanthocyanidin dioxygenase gene (PpLDOX): a potential functional marker for cold storage browning in peach. Tree Genet. Gen. DOI 10.1007/s11295-007-0130-0

Okie W.R. (1984) Rapid multiplication of peach seedlings by herbaceous stem cuttings. Hortscience 19(2): 249–251.

Okie, W.R., Bacon, T. and Bassi, D. (2008) Fresh market cultivar development. In: Layne D.R. and D. Bassi (Eds.) The peach, botany, production and uses. CAB International Press, Wallingford, Oxon, UK. pp. 850.

Okie, W.R. and Rieger, M. (2003) Inheritance of venation pattern in *Prunus ferganensis* × *persica hybrids*. Acta Horticulturae 622: 261–264.

Padilla, I.M.G., Golis, A., Gentile, A., Damiano, C., and Scorza, R. (2006) Evaluation of transformation in peach Prunus persica explants using green fluorescent protein (GFP) and beta-glucuronidase (GUS) reporter genes. Plant Cell Tiss. Org. Cult. 84: 309–314.

Pascal, T., Kervella, J., Pfeiffer, F., Sauge, M.H. and Esmenjaud, D. (1998) Evaluation of the interspecific progeny *Prunus.persica* cv. Summergrand × *Prunus.davidiana* for disease resistance and some agronomic features. Acta Horticulturae 465:185–192.

Pascal, T., Pfeiffer, F. and Kervella, J. (2002) Preliminary observations on the resistance to sharka in peach and related species. Acta Hort. 592:699–704.

Peace, C.P., Cristoso, C.H. and Gradziel, T.M. (2005) Endopolygalacturonase: a Candidate Gene for Freestone and Melting Fleshin Peach. Mol. Breed. 16:21–31.

Perez-Clemente, R.M., Perez-Sanjuan, A., Garcia-Ferriz, L., Beltran, J.-P. and Canas, L.A. (2004) Transgenic peach plants (*Prunus persica* L.) produced by genetic transformation of embryo sections using the green fluorescent protein (GFP) as an in vivo marker. Mol. Breed. 14: 419–427.

Petri, C., Alburquerque, N. and Burgos, L. (2005) The effect of aminoglycoside antibiotics on the adventitious regeneration from apricot leaves and selection of nptII-transformed leaf tissues. Plant Cell Tiss. Org. Cult. 80: 271–276.

Petri, C., Alburquerque, N., Perez-Tornero, O. and Burgos, L. (2005) Auxin pulses and a synergistic interaction between polyamines and ethylene inhibitors improve adventitious regeneration from apricot leaves and Agrobacterium-mediated transformation of leaf tissues. Plant Cell Tiss. Org. Cult. 82: 105–111.

Quarta, R., Dettori, M.T., Sartori, A. and Verde, I. (2000) Genetic linkage map and QTL analysis in peach. Acta Hort. 521, 233–241.

Quilot, B., Wu, B.H., Kervella, J., Geńard, M., Foulongne, M. and Moreau, K. (2004) QTL analysis of quality traits in an advanced backcross between Prunus persica cultivars and the wild relative species P. davidiana. Theor. Appl. Genet. 109:884–897.

Rajapakse S, Belthoff LE, He G, Estager AE, Scorza R, Verde I, Ballard RE, Baird WV, Callahan A, Monet R, and Abbott AG (1995) Genetic mapping in peach using morphological, RFLP and RAPD markers. Theor Appl Genet 90: 503–510

Rieger M. (2006) Peach (*Prunus persica*). *In* Introduction to fruit crops. The Haworth Press, Inc., New York, pp. 311–323.

Rodriguez, A.J., Sherman, W.B., Scorza, R., Wisniewski, M. (1994) 'Evergreen' peach, its inheritance and dormant behavior. J. Am. Soc. Hortic. Sci. 119:789–792.

Rubio M, Dicenta F, Martinez-Gomez P (2003) Susceptibility to sharka (Plum pox virus) in Prunus mandshurica × P-armeniaca seedlings. Plant Breeding 122:465–466.

Sansavini, S., Gamberini, A. and Bassi, D. (2006) Peach breeding, genetics and new cultivar trends. Acta Hort. 713: 23–48.

Scorza, R. and Pusey, P.L. (1984) A wound-freezing inoculation technique for evaluating resistance to *Cytospora leucostoma* in young peach trees. Phytopathology 74:569–572.

Scorza, R. and Sherman, W.B. (1996) Peaches. *In*: Janick J., Moore J.N. (Eds.) Fruit breeding, vol 1.: Tree and tropical fruits. John Whiley & Sons, New York, pp.325–440.

Sharp, R.H. (1961) Developing new peach varieties for Florida. Proc. Fla. State Hortic. Soc. 74:348–363.

Shulaev, V., Korban, S.S., Sosinski, B., Abbott, A.G., Aldwinckle, H.S., Folta, K.M., Iezzoni, A., Main, D., Arus, P., Dandekar, A.M., Lewers, K., Brown, S.K., Davis, T.M., Gardiner, S.E., Potter, D. and Veilleux, R.E. (2008) Multiple models for Rosaceae genomics. Plant Physiol. DOI:10.1104/pp.107.115618.

Smykov, V.K., Ovcharenko, G.V., Perfilyeva, Z.N. and Shoferistov, E.P. (1982) Estimation of the peach hybrid resources by its mildew resistance against the infection background. Byull Gos Nikitsh Bot Sada 88:74–80.

Song, G.Q. and Sink, K.C. (2006) Transformation of Montmorency sour cherry (Prunus cerasus L.) and Gisela 6 (P-cerasus × P-canescens) cherry rootstock mediated by Agrobacterium tumefaciens. Plant Cell Rep. 25: 117–123.

Stushnoff C. (1972) Breeding and selection methods for cold hardiness in deciduous crops. HortScience 7:10–13.

Tsukanova Z.G., Sokolova S.A., Gatina E.S. and Smykov V.K. (1982) Inheritance of mildew resistance by peaches. Byull Gos Nikitsk Bot Sada 49: 72–75.

Tuskan, G.A., DiFazio, S., Jansson, S., Bohlmann, J., Grigoriev, I., Hellsten, U., Putnam, N., Ralph, S. Rombauts, S., Salamov, A., Schein, J., Sterck, L., Aerts, A., Bhalerao, R.R., Bhalerao, R.P., Blaudez, D., Boerjan, W., Brun, A., Brunner, A., Busov, V., Campbell, M., Carlson, J., Chalot, M., Chapman, J., Chen, G.L., Cooper, D., Coutinho, P.M., Couturier, J., Covert, S., Cronk, Q., Cunningham, R., Davis, J., Degroeve, S., Dejardin, A., Depamphilis, C., Detter, J., Dirks, B., Dubchak, I., Duplessis, S., Ehlting, J., Ellis, B., Gendler, K., Goodstein, D., Gribskov, M., Grimwood, J., Groover, A., Gunter, L., Hamberger, B., Heinze, B., Helariutta, Y., Henrissat, B., Holligan, D., Holt, R., Huang, W., Islam-Faridi, N, Jones, S., Jones-Rhoades, M., Jorgensen, R., Joshi, C., Kangasjarvi, J., Karlsson, J., Kelleher, C., Kirkpatrick, R., Kirst, M., Kohler, A., Kalluri, U., Larimer, F., Leebens-Mack, J., Leple, J.C., Locascio, P., Lou, Y., Lucas, S., Martin,

F., Montanini, B., Napoli, C., Nelson, D.R., Nelson, C., Nieminen, K., Nilsson, O., Pereda, V., Peter, G., Philippe, R., Pilate, G., Poliakov, A., Razumovskaya, J., Richardson, P., Rinaldi, C., Ritland, K., Rouze, P., Ryaboy, D., Schmutz, J., Schrader, J., Segerman, B., Shin, H., Siddiqui, A., Sterky, F., Terry, A., Tsai, C.J., Uberbacher, E., Unneberg, P., Vahala, J., Wall, K., Wessler, S., Yang, G., Yin, T., Douglas, C., Marra, M., Sandberg, G., de Peer, Y.V. and Rokhsar, D. (2006) The genome of black cottonwood, Populus trichocarpa (Torr. & Gray). Science 313: 1596–1604.

Verde, I., Vendramin, E., Dettori, M.T., Quarta, R., Wang, Y., Lecouls, A.C. and Abbott, A.G. (2004) Target SSR development in peach and ssr mapping in a peach $BC_1$ progeny. Acta Hort. (ISHS) 663:63–68.

Vileila-Morales, E.A., Sherman, W.B., Wilcox, C.J. and Andrews, C.P. (1981) Inheritance of short fruit development period in peach. J. Am. Soc. Hortic. Sci. 106:399–401.

Viruel, M.A., Madur, D., Dirlewanger, E., Pascal, T. and Kervella, J. (1998) Mapping quantitative trait loci controlling peach leaf curl resistance. Acta Hort. 465:79–87.

Wang, L., Zhu, G. and Fang, W. 2002. Peach germplasm and breeding programs at zhengzhou in china. Acta Hort. (ISHS) 592:177–182.

Weaver, D.J., Doud, S.L. and Wehunt, E.J. (1979) Evaluation of peach seedling rootstocks for susceptibility to bacterial canker, caused by *Pseudomonas syringae*. Plant Dis. Rep. 63: 364–367.

Werner, D.J., Ritchie, D.F., Cain, D.W. and Zehr, E.I. (1986) Susceptibility of peaches and nectarines, plant introductions, and other Prunus species to bacterial spot. HortScience 21:127–130.

Wisniewski, M.E. and Arora, R. (2000) Structural and biochemical aspects of cold hardiness in woody plants. *In*: Molecular biology of woody plants. In: Mohan Jain S. and Minocha S.C. (Eds.). Kluwer Academic Publishers, New York, pp.419–437.

Yarnell, S.H. (1945) Temperature as a factor in breeding peaches for a mild climate. Proc. Amer. Soc. Hortic. Sci. 44:239–242.

Yoon, J.H., Liu, D.C., Song, W.S. Liu, W.S., Zhang, A.M., Li, S.H. (2006) Genetic diversity and ecogeographical phylogenetic relationships among peach and nectarine cultivars based on simple sequence repeat (SSR) markers. J. Am. Soc. Hortic. Sci. 131:513–521.

# 11. Peach Structural Genomics

**Carlo Pozzi and Alberto Vecchietti**

Peach (*Prunus persica*) belongs to the *Prunus* genus and is a member of the Rosaceae family. It has been selected as a model species for genomics studies in virtue of several features (Abbott et al., 2002): it has a short juvenile phase (2–3 years) if compared to many other tree species; it has a small genome, just about twice the size of Arabidopsis ($5.9 \times 10^8$ bp; Baird et al., 1994); it is diploid with a base chromosome number of $x = 8$; it is the best characterized *Prunus* species, where a number of traits of agronomic interest are under the control of monogenic loci (recently reviewed in Dirlewanger et al., 2004). The recent development of genomics and functional genomics tools demonstrated throughout this volume is making possible to effectively implement comparative genomics strategies in Rosaceae, using peach as the "basic" genome (Abbott et al., 2002).

## 1 Molecular Markers and Breeding in Peach

Breeding in *Prunus* is challenged by the narrow genetic background of commercial cultivars (Scorza et al., 1985), the long juvenile period together with the large dimension of the plants. These facets make any breeding effort time and resource consuming. Peach is considered the least genetically variable crop in the genus (Byrne, 1990), and it is characterized by a level of inbreeding which has hampered efficient breeding prior to the development of molecular based tools.

Peach breeding thus benefits from the availability of highly efficient molecular marker systems for the identification of markers tightly linked to traits of interest. Most of the molecular markers developed so far are "anonymous", i.e. they are not based on the knowledge of the subtending DNA sequence or of its function. These markers recognize non-coding, or non-gene, sequences and traditionally have been used because the position and sequence of genes were unknown. Such markers have been used in the production of several maps in *Prunus*. The mapping populations

C. Pozzi (✉)
Parco Tecnologico Padano, via Einstein Loc. C.na Codazza, Lodi 26900, Italy

K.M. Folta, S.E. Gardiner (eds.), *Genetics and Genomics of Rosaceae*, Plant Genetics and Genomics: Crops and Models 6, DOI 10.1007/978-0-387-77491-6_11,

used derive from crosses within almond, peach, apricot, cherry, *P. davidiana*, *P. cerasifera*, and *P. ferganensis*. Our review focuses on maps constructed using peach populations, but it is worth mentioning the maps obtained from *P. persica* × *P. davidiana* (Dirlewanger et al., 1996; Foulongne et al., 2003a, b), (*P. persica* × *P. dulcis*) × *P. cerasifera* (Dirlewanger et al., 2003), and *P. persica* × *P. ferganensis* (Dettori et al., 2001).

The main molecular maps produced in peach are the result of coordinated efforts in the US and in Europe. The works of Chaparro et al. (1994), at NCSU Raleigh NC, Rajapakse et al. (1995), at Clemson University SC, and Foolad et al. (1995), at UC Davis CA, pioneered the field in the US. In Europe three highly interconnected initiatives took place at INRA (Bordeaux, France), ISF (Rome, Italy), and IRTA (Cabrils, Spain). A determining contribution to map expansion and saturation was the introduction of microsatellite markers at the end of the 90's (Cipriani et al., 1999; Sosinski et al., 2000; Dirlewanger et al., 2002; Aranzana et al., 2002).

## 1.1 Maps Derived from Peach × Almond Crosses

The interspecific cross peach × almond was used to produce dense maps (Foolad et al., 1995; Bliss et al, 2002; Jáuregui et al., 2001). Joobeur et al. (1998) and Aranzana et al. (2003) developed a saturated linkage map with 235 RFLPs and 11 isozyme markers using a $F_2$ cross almond ('Texas') × peach ('Early Gold') (TxE). This map is currently considered the reference map of the *Prunus* genus and is distributed over 8 linkage groups (LGs) which were aligned with the almond homologous groups originated from an F1 progeny from the cross 'Ferragnes' × 'Tuono' (Viruel et al., 1995). The total map length is 491 cM (average of 1 marker / 2.3 cM; Aranzana et al., 2003). The map is biased, to some extent, by the high level of heterozygosis and incompatibility of the two species: almost 46% of the mapped loci had a skewed segregation in the progenies and a major reciprocal translocation between LGs 6 and 8 was observed (Jáuregui et al., 2001). Despite these limitations, the TxE map provided a highly polymorphic population suitable for linkage studies, established a common terminology for linkage groups, and gave a set of transferable markers (anchor markers) of known map position to be used in comparative genomics (see below: Dirlewanger et al., 1998; Lu et al., 1998; Sosinski et al., 1998; Dettori et al., 2001; Jung et al., 2004; Gillen and Bliss, 2005; Yamamoto et al., 2001, 2005; Blenda et al., 2007). It was subsequently found that not all of the 88 individuals of the segregating population are necessary to map a new locus: in fact a sub-set of six plants can be used to define 65 different genotypes based on their SSR patterns (Howad et al., 2005). Sixty-three of these joint genotypes corresponded to a single chromosomal region (a bin) of the *Prunus* genome, and the two remaining corresponded to two bins each. Using this 'mapping sub-set' strategy, a total of 264 new SSRs were placed on the TxE map which has now a density of 0.63 cM/marker (826 markers distributed over 524 cM).

A total of 33 of the SSRs used in the TxE map have been placed on an intraspecific peach F2 (Yamamoto et al., 2005), and on a three-way interspecific progeny, involving almond, peach, and Myrobalan plum (Dirlewanger et al., 2004a).

## *1.2 Peach Maps Derived from Intraspecific Crosses*

Several maps of peach have been constructed using a large set of anonymous markers (RFLP/RAPD/SSR/AFLP; source: GDR, Genomic Database for Rosaceae; http://www.bioinfo.wsu.edu/gdr/; Dirlewanger et al., 1998; Lu et al., 1998; Sosinski et al., 1998; Dettori et al., 2001; Gillen and Bliss., 2005; Yamamoto et al., 2001, 2005; Blenda et al., 2007; Jung et al., 2004, 2008). All peach maps deriving from intraspecific crosses are anchored to the *Prunus* TxE reference map. We report here some of the maps which are currently enriched with anonymous (and/or functional) markers. A complete review of the molecular maps currently available is provided by Dirlewanger and Arùs (2004).

- *'Ferjalou Jalousia' × 'Fantasia' (JxF)*
  A single $F_1$ plant from a cross between a flat non-acid peach, 'Ferjalou Jalousia', and an acid round nectarine, 'Fantasia', (Dirlewanger et al., 1998) was selfed to obtain an $F_2$ population of 208 plants. The map locates 270 markers including four agronomic characters (peach/nectarine (*g*), flat/round fruit (*s*), acid/non-acid fruit (*d*), and pollen sterility (*PS*). Two hundred and forty-nine markers (1 isozyme, 50 RFLPs, 92 RAPDs, 8 inter-microsatellite amplification (IMA), and 115 AFLPs) were mapped to 11 LGs covering 712 cM, with an average density between pairs of markers of 4.5 cM. Twenty-six RFLPs are shared between JxF and the TxE maps is homologous positions. The density of the map allowed for markers to be found in close proximity to several Mendelian agronomic traits (see later), and were used as a basis for marker assisted breeding programs. This map is also used for the detection of quantitative trait loci (QTLs) controlling peach fruit quality (see later).
- *Prunus persica × (P. persica × P ferganensis) (PxF)*
  The PxF map is based on a BC1 progeny (Dettori et al., 2001) and was initially composed of 109 loci (74 RFLPs, 17 SSRs, 16 RAPDs, and two morphological traits, flesh adhesion (*F/f*) and leaf glands (*E/e*)) distributed in 10 LGs. The map was distributed on 521 cM of the peach genome and the average distance between adjacent loci was 4.8 cM. The PxF and TxE maps were aligned in virtue of 32 shared markers. Verde et al. (2005) subsequently enriched the PxF map which is now composed of 216 loci covering 665 cM with an average distance of 3.1 cM. Notably, the number of LGs was brought to eight and almost half of the 55 SSRs contained in this map are shared with the TxE map. Nine almond SSRs were mapped for the first time. A comparative analysis against the TxE map pointed out an almost complete synteny and co-linearity.

- *'Guardian' × 'Nemaguard' (GxN)*
  The GxN population was created to specifically address the mapping of resistance genes. The mapping population is an $F_2$ derived from the cross between 'Guardian BY520-9' selection 3-17-7 (tolerant to the peach tree short life syndrome, PTSL) and 'Nemaguard' (PTSL-susceptible) (Blenda et al., 2007). The map locates 151 AFLPs and 21 SSRs, including anchor loci from the TxE map. The markers covered 737 cM distributed on 11 LGs with an average interval of 4.7 cM between adjacent markers. The comparison of the distribution of resistance gene analogs (RGAs) on the *Prunus* resistance map (Lalli et al., 2005) with the distribution of the PTSL-associated markers mapped in the rootstock $F_2$ cross GxN revealed that some RGAs and several PTSL-associated markers were located in the same regions on LGs 1, 2, 4, 5, and 6.
- *'Akame' × 'Juseito' (AxJ)*
  The ornamental 'Juseito' was used in a cross with 'Akame' rootstock to produce an $F_2$ population of 126 plants (Yamamoto et al., 2001). Ninety two markers including 35 AFLPs, 31 RAPDs, 11 SSRs, 5 ISSRs, 1 STS traits encompassed 9 LGs. Thirteen morphological traits (nine major genes and 4 QTLs) were also mapped. A subsequent refinement of the map (Yamamoto et al., 2005) allowed the location of a further 178 markers on eight LGs. The map distance is of 571 cM with an average distance of 3.2 cM between each pair of loci. The AxJ peach genetic linkage map is completely aligned with the TxE map by using 42 common DNA markers. Interestingly, 55 SSRs that were unmapped in the TxE map could be positioned on the AxJ map. Two loci corresponding to flower color (*Fc*) and flesh color around the stone (*Cs*) were mapped on LG 3. Flesh adhesion (*F*) was mapped at the bottom of LG 4, similar to what found by Dettori et al. (2001).
- *'Harrow Blood' (HB) × 'Okinawa' (Oki) (PMP2)*
  Gillen et al. (2005) produced an $F_2$ population (PMP2) from a single $F_1$ plant derived from the cross between the rootstocks 'Harrow Blood (HB)' and 'Okinawa (Oki)', to locate the *Mi* locus for the root-nematode resistance. Two RFLPs (linked at 4.8 and 6.8 cM) and one RAPD marker (linked at 9.5 cM) were found in linkage to *Mi* on LG 2. The RAPD marker was cloned, sequenced, and converted to a cleaved amplified polymorphic sequence (CAPs) marker. RGAs developed from Oki were highly polymorphic when used as RFLP probes and mapped to four LGs. Even so, they may be useful as markers for disease resistance genes that occur in other populations. The PMP2 and TxE maps were colinear in specific regions, despite that a significant number of markers mapped to different LGs among the two populations. The locus for the blood-flesh trait (red-violet mesocarp) was mapped to the top of LG 4 of PMP2.
- *'New Jersey Pillar' × KV 77119 (NJxKV), 'Suncrest' × 'Bailey' (ScxB), 'Lovell' × 'Nemared' (LxN)*
  The NJxKV is the 'West Virginia' family developed by Sosinski et al. (1998) on which 65 markers were initially located. A highly saturated linkage map in peach was developed using this initial map and complemented with two other crosses (ScxB and LxN) (Lu et al., 1998; Sosinski et al., 1998): one segregating

**Table 2** Distribution of anchored markers on LGs of peach maps (source: GDR, Genomic Database for *Rosaceae*; http://www.bioinfo.wsu.edu/gdr/)

| Peach linkage group | Number of SSR on TxE map | Number of anchored SSR on peach maps | Maps with anchored SSR (number of anchored SSR on the map) |
|---|---|---|---|
| LG1 | 1 | 6 | JxF (4); PxF (2); AxJ (5); GxN (1) |
| LG2 | 3 | 4 | PxF (3); AxJ (4); PMP2 (1); LxN (1) |
| LG3 | 3 | 3 | PxF (1); AxJ (1); GxN (1) |
| LG4 | 3 | 5 | JxF (3); PxF (3); AxJ (2); GxN (1) |
| LG5 | 4 | 6 | JxF (4); PxF (4); AxJ (5); GxN (1) |
| LG6 | 1 | 4 | JxF (3); PxF (2); AxJ (4); GxN (1) |
| LG7 | 1 | 4 | JxF (4); PxF (1); AxJ (3) |
| LG8 | 4 | 5 | PxF (3); AxJ (2); GxN (1) |
| | Total: 20 | Total: 37 | |
| **Peach linkage group** | **Number of RFLP on TxE map** | **Number of anchored RFLP on peach maps** | **Maps with anchored RFLP (number of anchored RFLP on the map)** |
| LG1 | 10 | 10 | JxF (8); PxF (2) |
| LG2 | 5 | 5 | JxF (1); PxF (4) |
| LG3 | 6 | 6 | JxF (3); PxF (3) |
| LG4 | 7 | 7 | JxF (4); PxF (4) |
| LG5 | 6 | 7 | JxF (5); PxF (2); ScxB (1) |
| LG6 | 10 | 10 | JxF (5); PxF (6) |
| LG7 | 6 | 6 | JxF (3); PxF (3) |
| LG8 | 4 | 4 | PxF (4) |
| | Total: 54 | Total: 55 | |
| **Peach linkage group** | **Number of anchored EST (BAC contig) on TxE map** | **Number of anchored EST (BAC contig) on peach maps** | **Maps with anchored EST (BAC contig) (number of anchored EST (BAC contig) on the map)** |
| LG1 | 26 | 26 | JxF (26) |
| LG2 | 10 | 10 | JxF (10) |
| LG3 | 23 | 23 | JxF (23) |
| LG4 | 20 | 20 | JxF (20) |
| LG5 | 28 | 28 | JxF (28); ScxB (7) |
| LG6 | 9 | 9 | JxF (9) |
| LG7 | 8 | 8 | JxF (8) |
| | Total: 124 | Total: 124 | |

**Table 2** (Continued)

| Peach linkage group | Number of anchored EST (BAC) on TxE map | Number of anchored EST (BAC) on peach maps | Maps with anchored EST (BAC) (number of anchored EST (BAC) the map) |
|---|---|---|---|
| LG1 | 13 | 13 | JxF (13) |
| LG2 | 1 | 1 | JxF (1) |
| LG3 | 8 | 8 | JxF (8) |
| LG4 | 6 | 6 | JxF (6) |
| LG5 | 5 | 5 | JxF (5) |
| LG6 | 8 | 8 | JxF (8) |
| LG7 | 3 | 3 | JxF (3) |
| | Total: 44 | Total:44 | |
| Peach linkage group | Number of anchored RGA-I on TxE map | Number of anchored RGA-I on peach mapS | Maps with anchored RGA-I (number of anchored RGA-I on the map) |
| LG7 | 2 | 2 | JxF (2) |
| Peach linkage group | Number of anchored RGA-II on TxE map | Number of anchored RGA-II on peach mapS | Maps with anchored RGA-II (number of anchored RGA-II on the map) |
| LG2 | 1 | 1 | JxF (1) |
| LG5 | 5 | 5 | JxF (5) |
| LG7 | 1 | 1 | JxF (1) |
| | Total: 7 | Total:7 | |
| Peach linkage group | Number of anchored RGA-III on TxE map | Number of anchored RGA-III on peach maps | Maps with anchored RGA-III (number of anchored RGA-III on the map) |
| LG4 | 3 | 3 | JxF (3) |

## 3.1 QTLs for Peach Quality

The control of quality traits is usually genetically complex, for example the balance of acids and sugars in the fruit, although the non-acid character of the fruit juice is controlled by a single dominant gene *D* (Dirlewanger et al., 1998). The predominant organic acids in ripe peach fruit are malic and citric acids (Moing et al., 1998). The soluble sugars of peach are sucrose, fructose, glucose and, at lower levels, sorbitol. QTLs controlling hexose content were detected using a peach 'Suncrest' × 'Bailey' cross (Abbott et al., 1998). A population derived from peach × *Prunus ferganensis* allowed the mapping of one QTL for pH on LG 2 and QTLs for another compo-

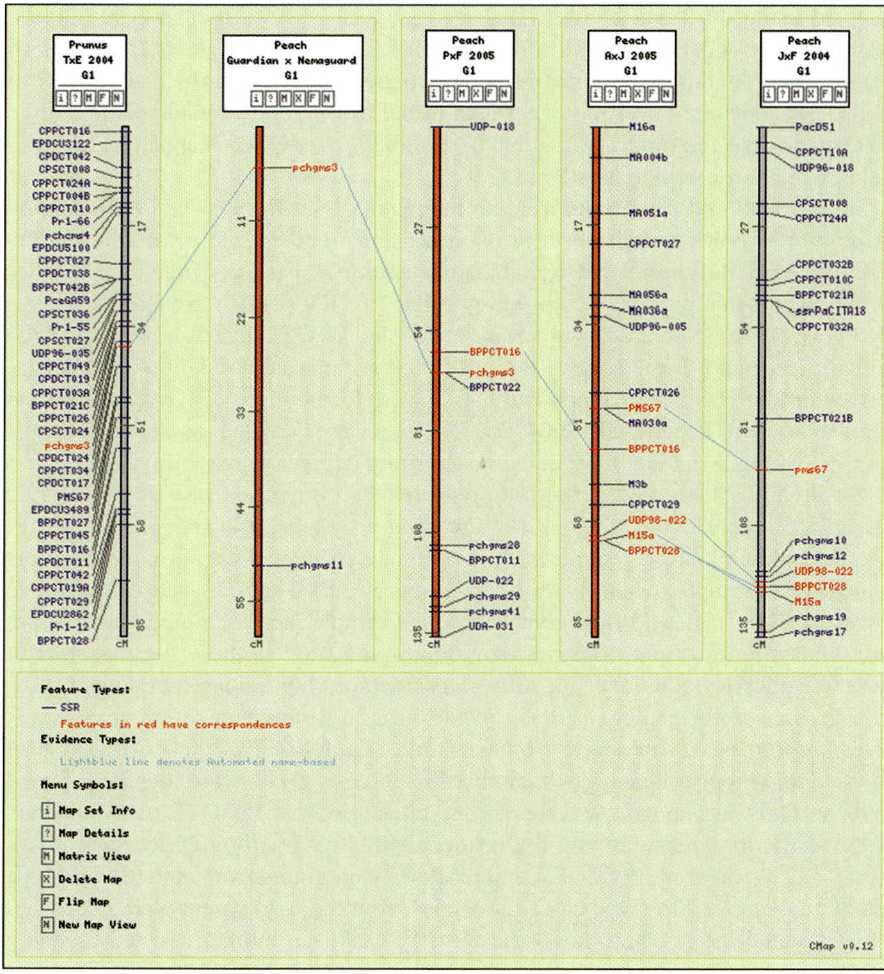


**Fig. 1** Distribution on anchored SSR on LG1 between TxE reference map and peach maps (source: GDR, Genomic Database for Rosaceae; http://www.bioinfo.wsu.edu/gdr/)

nent of the sugar/acid balance, the soluble solids content (SSC), were positioned on LGs 1 and 2 (Quarta et al., 2000). Further major QTLs involved with fruit quality were detected on the latter map: flowering time (mapping on LG 4); ripening time (on LGs 2 and 6); fruit skin color (on LGs 2 and 6); juice SSC (LG2 and 6). Further experiments conducted by Verde et al. (2002) led to the mapping of QTLs for ripening time, fruit skin color, and SSC of the fruit, confirming the positioning in narrow intervals of LGs 2 and 6. A partly overlapping set of QTLs was also mapped by Dirlewanger et al. (1999), and subsequently refined by Etienne et al. (2002) using the same mapping panel: QTLs for blooming, ripening, and fruit quality (fresh weight, color, pH, titratable acidity, SSC, acid and sugar contents) were detected using the JxF F2 cross (Dirlewanger et al., 1999; Etienne et al., 2002). Some were co-localized with the *D* locus on LG 5. Interestingly, QTLs for fresh weight and productivity were mapped on LG 6, near the *S* locus which controls fruit shape (flat fruits are lower in weight than round fruits). For titratable acidity, three QTLs were detected (on LGs 1, 5, and 6). For malic acid QTLs mapped on LG 1, 5, and 6, while citric acid mapped on LG 9.

Several fruit and plant traits were investigated by Sosinski et al. (1998): canopy shape, flower color and shape, fruit diameter and weight, flesh color and texture, SSC, juice pH, developmental period, and nematode resistance. Three F2 progenies were used for that purpose: 'New Jersey Pillar' × 'KV77119'; 'Suncrest' × 'Bailey'; 'Lovell' × 'Nemared' and seven out of the 12 QTL intervals were mapped in the joined map. Resistance to the post-harvest 'internal break down' pathological syndrome was also mapped and QTLs were located in the 'Georgia Belle' × 'Dr. Davis' map (Peace et al., 2004, 2005). QTLs were detected for all traits (mealiness, browning, bleeding, harvest date) at varying degrees or reliability.

As already stated, wild germplasm represents an important resource of genetic variability in peach. When using it in breeding programs, it is necessary to segregate out the wild parent genome which could be detrimental for agronomic performance. For example, when using *P. davidiana* in breeding for quality, a negative correlation between wild favorable traits (for example increased glucose and fructose concentrations) and negative agronomic traits such as those decreasing fruit mass was observed (Quilot et al., 2004). Despite those limitations, mapping of quality traits was performed in an advanced backcross breeding population involving *P. davidiana* (crossed with peach 'Summergrand'; Quilot et al., 2004). A total of 24 physical and biochemical traits were investigated and QTLs were identified for all of them. Only in one case was there good consistency of the QTL over the years of the study, in agreement with the results already obtained by Dirlewanger et al. (1999) and Verde et al. (2002). Favorable alleles were found from both the wild type and the cultivar and in some cases the alleles from the wild parent were not linked to detrimental factors, and thus were directly usable in marker assisted breeding programs.

## 3.2 QTLs for Resistance

The identification of QTLs and/or monogenic traits linked to resistance to the most important peach pathogens is of particular relevance to breeding programs. Powdery mildew resistance was extensively studied and 13 QTLs explaining up to a 65% of the related phenotypic variation have been identified (Foulongne et al., 2003b). Molecular markers are now available to be used in marker assisted selection (MAS), as they are able to discriminate between the favorable alleles for the QTLs on LGs 6 and 8. The mapping populations used (F1, F2, and BC2 populations; SD ('Summergrand' × *P. davidiana*); $SD40^2$ (a selected genotype from SD, selfed); and SD40 backcrossed with Summergrand) were derived from the cross between the resistant wild parent *P. davidiana* and the susceptible 'Summergrand' (Foulongne et al., 2003a; Quilot et al., 2004). For nine of the putative resistance loci identified, the allele from *P. davidiana* conferred higher level of resistance, as expected. Some QTLs could be confirmed only by virtue of usage of different mapping populations and only five were present in all populations. The "traditional" QTLs on LGs 6 and 8 were both confirmed in the same position in the three related crosses used. The co-localization of some QTLs for resistance and for quality traits (for example on LG 6) can be problematic if one wants to eliminate the negative wild allele deriving from *P. davidiana* while keeping the favorable resistance allele. Interesting physiological explanations for some of the co-localization observed are described in Quilot et al. (2004).

## 4 Functional Markers in Peach

Promising tools for peach breeding include the development of new types of molecular markers based on knowledge of the sequence and function of the subtended polymorphism (Andersen and Lubberstedt, 2003). The sequence is used to yield polymorphic, informative functional markers, indicative of genes of interest and highly transferable between species. Maps including this kind of marker are being used in Rosaceae for a candidate gene approach to gene cloning (Silva et al., 2005; Sargent et al., 2007). These markers can also be used in strategies based on synteny to identify agronomically relevant genes. The complement to these approaches are reverse genetics methodologies, including T-DNA tagging, RNA interference, which are also being developed for Rosaceae (see this book; Scorza and Ravelonandro, 2002.

The candidate gene approach is based on the *a priori* choice of genes which may be functionally related to the trait. A correlation between the trait under study and polymorphisms at the candidate gene is a strong argument in favor of the selected candidate gene. The bin mapping strategy recently designed by Howad et al. (2005) is particularly efficient in this approach, since it allows the fast screening of possible co-location of a large number of candidates with specific genes or QTLs. Technically, one of the most efficient strategies to generate functional markers is to exploit polymorphism in expressed sequences (ESTs). This allows direct tests of correlation between observed phenotypes of interest and genes of known or inferred function.

### *4.1 Mapping of Transcripts and Candidate Genes*

The peach maps are becoming more and more saturated with markers derived from known sequences (for example, in the TxE map 87% of loci correspond to a known DNA sequence and 37% of these sequences correspond to a putative protein). Maps based on ESTs are also being produced in peach. Barale et al. (2006) screened the ESTree database of 20,924 cDNA sequences (Lazzari et al., 2005; Lazzari et al.,

2007; Vecchietti A., unpublished results) in search of candidate genes (CGs) based on sequence similarity with genes relevant for fruit quality, already characterized in other related species like apple, apricot and strawberry. To rapidly identify polymorphisms, the database implements an algorithm for in silico single nucleotide polymorphism (isSNP) detection. Sequence-confirmed SNPs are then genotyped according to Howad et al. (2005) on the TxE population. Currently 50 ESTs have been mapped on this functional map (Barale et al., 2006). Further experiments are devoted to the identification of new fragrance genes in peach and targets were genes involved in the biosyntheses of lactones, shikimic acid derivates, branched-chain esters, alcohols, aldehydes, C6 compounds, and nor-isoprenoids (Vecchietti A., unpublished results).

A similar approach was undertaken by Etienne et al. (2002). Eighteen cDNAs encoding key proteins in soluble sugar and organic acid metabolism pathways as well as in cell expansion were isolated from peach fruit. Since no SNPs could be detected in the JxF population, gene mapping was performed at first on the TxE map. Twelve candidate genes were assigned to four linkage groups of the peach genome. Since the JxF population was used for QTL mapping, a further step concerned the "connection" of the TxE and JxE maps by anchor markers. Eventually, Etienne and co-workers were able to merge the mapping data and identify interesting co-location between candidate genes and mapped QTLs (addressed in greater detail later).

Bliss et al. (2002) have been working with some interesting genes for fruit quality. The catalase (*Cat1*) isozyme locus was mapped and shown to be linked with the *D* locus controlling fruit acidity, and preliminary studies point to an allele-specificity for this gene in cultivars with different juice pH.

The same approach was used to create maps of transcripts involved in resistance. Candidate genes representing analogs of major resistance genes, translation initiation factors known to be involved in recessive resistance to plant viruses, and defense response genes were hybridized to a peach BAC library. A "resistance map" for *Prunus* was generated which contains 42 map locations for putative resistance regions (Lalli et al., 2005). Bliss et al. (2002) positioned polygalacturonase inhibitor proteins on LG 7 (possibly a tandem duplication of the same gene), one of which is known to confer resistance to *Botrytis* fruit rot. One dehydrin (involved in abiotic stress tolerance) was mapped on LG 7. Several resistance genes analogs (RGAs) were co-localized with genes or QTLs that determine Sharka resistance or root-knot nematode resistance (Bliss et al., 2002), suggesting that they may belong to resistance gene clusters shared also by other Rosaceae (Dirlewanger et al., 2004a). Three RGAs mapped to the region of LG 7 containing the *Mai* gene, controlling resistance to root-knot nematode. RGAs were also mapped in the region hosting powdery mildew resistance on LG 6, but not on LG 8. RGAs were mapped in the region hosting Sharka resistance on LGs 1 and 7, but not on LG 6.

A similar approach was adopted to identify candidate genes for agronomically important traits related to fruit quality. Etienne et al. (2002) were able to demonstrate the co-localization of QTLs for fruit pH and sugars and the locus encoding for a vacuolar H+-pyrophosphatase, which controls solute accumulation from the cytosol into the vacuole, and thus possibly affects both pH and sugar concentration.

The study of mealiness was also approached in the same way: an initial literature survey led to the identification of 14 candidate genes and two were mapped, one endopolygalacturonase (endoPG), and a pectin methylesterase, which are the two most commonly implicated in the development of mealiness (Brummell et al., 2004).

EndoPG was also studied to understand flesh texture. EndoPG co-segregated completely with *F* (flesh adhesion) and *M* (melting flesh), suggesting that this enzyme controls one or both of these traits (Peace et al., 2005). Coincidence between the peak for mealiness QTL and the endoPG gene favor of this hypothesis (Peace et al., 2004). Markers were developed in absolute linkage with endoPG that can be used for distinguishing between freestone/melting flesh and clingstone/non-melting flesh and clingstone/melting flesh genotypes in marker assisted selection (Peace et al., 2005).

Based on similarity in predicted protein structures, candidate genes for plum pox virus (PPV) resistance were isolated and characterized from *Prunus* species (Decroocq et al., 2005). Fifteen candidate genes (belonging to the classes: R proteins, transcription factors, factors involved in RNA interference, translation initiation factors, and pathogenesis related (PR) proteins were isolated from plum, peach, and apricot. In the same study, a total of six *P. davidiana* genomic regions were identified as being involved in PPV resistance. Candidate genes were found in collinear regions of *P. davidiana* and of peach. Two regions (on LGs 2 and 7) were found to be linked to a trait involved in the restriction of PPV movement. The data do not support a linkage between QTLs and candidate genes, but open new perspective for MAS in PPV resistance. The PTSL syndrome was also studied using a candidate gene approach, and some RGAs were co-localized with AFLPs linked to the PTSL trait, on several LGs (Blenda et al., 2007).

## 5 Marker Assisted Selection (MAS) in Peach

MAS can shorten the number of generations required to select for the desired genes of the donor in backcrossing programs and is based on the availability of markers linked to the traits of interest. In *Prunus* this approach can very efficient and the linkage maps described can be used to locate genes of economic and biological value. The peach agronomic traits which are the objectives of current breeding programs are usually difficult to evaluate, as are resistances to biotic and abiotic stress (Quilot et al., 2004). Breeders generate large populations to recombine traits from different varieties, and tree breeding is still mostly based upon phenotypic selection. Examples of MAS in peach concern male sterility, peach/nectarine, non-acid/acid, and flat/round fruit (Dirlewanger et al., 1998), where tightly linked markers would allow the identification and selection of heterozygous individuals not expressing the trait.

MAS has been applied to other monogenic traits. The *F/f* (flesh adhesion) character, for example, is determined by a single gene pair with freestone dominant to clingstone. Jun et al. (2003) used the BSA strategy on F1 progenies of the "Yumyong" (clingstone) × "Baekhyang" (freestone) cross to identify four dominant coupling-phase RAPD markers which were converted into SCARs. Also the *G* ("nectarine" recessive character) and *S* ("saucer shaped" dominant character) traits were investigated by BSA. An AFLP was found that co-segregated with the *G* gene, and two markers (one RFLP and one AFLP) with the *S* gene (Dirlewanger et al., 2006). These results were obtained in the framework of a more complete mapping of monogenic characters: non-acid fruit (*D*); clingstone (*f*); pollen fertility (*Ps*), and non aborting fruit (*Af*) were mapped on the FxJ map. Markers were found associated with each of these characters and MAS was performed using markers linked to the *Af* character. A complete list of markers linked to principal monogenic traits in *Prunus* is given in Martinez-Gomez et al. (2005).

Another important set of agronomic characters where MAS can be used, are resistances to biotic stresses.

## 5.1 Root-Knot Nematode Resistance (RKN)

All the RKN resistance genes described so far were placed on LG 2 (even if not in the very same positions). This includes the two genes (*mi* and *Mj*, conferring resistance to *Meloidogyne incognita* and *M. javanica,* respectively), the *Mij* gene reported by Lu et al. (1998) which should confer resistance to both strains, and the $R_{MiaNem}$ gene (conferring resistance to *M. incognita* and *M. arenaria)* found by Dirlewanger et al. (2004a). The *Ma* gene (conferring resistance to *M. arenaria*, and derived from *P. cerasifera*; Lecouls et al., 2004) was mapped to LG 7. For resistance to *M. javanica*, a complex genetics with duplicate loci (*Mj1/mj1* and *Mj2/mj2*), and a single locus (*Mj2/mj2*) was reported by Yamamoto and Hayashi (2002).

Sources of resistance to RKN are difficult to find in peach, especially because of the narrow genetic base of commercial peach varieties; most rootstock material is susceptible and only the peach "Shalil" is resistant to *M. arenaria* and *incognita* but susceptible to *M. javanica* (to which "Nemared" is resistant), and to a population from Florida.

Current breeding programs are focusing on the exploitation of sources of resistance derived from the wild species *P. davidiana* and *P. ferganensis* (Foulongne et al., 2003a; Verde et al., 2002) *P. davidiana* was found to be resistant to powdery mildew, the green aphid, PPV, leaf curl virus, and can be used to introgress these resistances into peach. Several studies have initiated the development of markers to be used in advanced backcross breeding programs. Lu et al. (1998) developed STS and SSR markers linked to the *Mj* and *Mij* genes of "Nemared" using a "Juseitou" × "Akame" segregating population. A co-dominant AFLP was linked with the primary source of resistance to root-knot nematode (Lu et al., 1999). Markers tightly linked to the *Ma/ma* resistance gene from Myrobalan plum have also been identified. This gene and the *Mi/mi* from "Nemared" have been screened with markers to pyramid them in a three-way (peach, almond, and plum) progeny (Claverie et al., 2004a).

*Ma* represents the first precise mapping of a resistance gene in the *Prunus* genus: two SCARs were shown to be linked in coupling with the dominant resistance alleles *Ma1* and *Ma3* from Myrobalan (Bergougnoux et al., 2002). They have been identified by bulked segregant analysis (BSA) using intraspecific progenies involving parents with alternative genotypes at the trait. One marker is less than 1 cM from *Ma* and the other one is co-segregates with it (Claverie et al., 2004b). Current programs are using MAS to "collect" resistance genes from different species and "pyramid" them into a new generation of peach rootstocks. For example, Dirlewanger et al. (2003) undertook a MAS in a interspecific hybrid *P. cerasifera* × (*P. dulcis* × *P. persica*), carrying the resistance genes from each species. The breeding scheme was further enriched by adding resistance to chlorosis, drought, water logging, graft compatibility and rooting ability (Dirlewanger et al., 2004b).

RAPDs were used in combination with a BSA strategy to distinguish markers linked to resistance to ring nematode (*Mesocriconema xenoplax*) in peach (Blenda et al., 2002). The experiment was based on the cross between the "Guardian" (resistant rootstock) and the "Nemaguard" (susceptible parent) and allowed the identification of a linked AFLP. BSA in combination with AFLP technology was applied to identify diagnostic markers for the peach tree short life syndrome (PTSL) (Blenda et al., 2007).

## 6 Peach Physical Mapping

The use of the peach genome as a model for identification, cloning, characterization and manipulation of genes important to Rosaceae crop breeding requires the definition of structural and genomic resources. For this, physical and integrated genetic maps are necessary.

Few bacterial artificial chromosome (BAC) libraries are currently available. The BAC library constructed by Wang et al. (2001) using the "Jingyu" traditional Chinese variety, is composed of 20,736 clones with an average insert size is of 95 kb. The variety Okubo was used to produce a transformation-competent artificial chromosome based library composed of 41,000 clones, averaging in length 42 kb (Liang et al., 2004). The BAC library produced by Georgi et al. (2002) derived from "Nemared", consists of 44,160 BAC clones averaging 60 kb. A further library was developed using the haploid Lovell accession (Abbott A., unpublished results), and is composed of approximately 90,000 clones, derived from two separate libraries obtained using different restriction enzymes, averaging 100 kb in length. The latter libraries are being fingerprinted using a high information content fingerprinting (HICF) technique, based on capillary electrophoresis and the SNaPshot labeling kit. Altogether, 16,895 BAC fingerprints have been assembled using FPC software and the physical map (June 2008) is composed of 2,138 contigs containing 15,655 BAC. According to conservative estimate, the map covers 303 Mb of the peach genome. In total 2,636 markers (i.e. genetic markers, ESTs, cDNAs, AFLPs and overgo probes)

**Table 3** Statistics of the BAC physical map being developed at Clemson University, SC, as of June 2008

| | |
|---|---|
| Number of clones fingerprinted | 18,141 |
| Number of BACs used for map contig assembly | 16,895 |
| Number of clones in contigs | 15,655 |
| Number of singletons | 1,240 |
| Number of anchored contigs | 252 |
| Number of assigned contigs | 87 |
| Physical length of the contigs | 303 Mb |
| Physical length of the anchored contigs | 45 Mb |
| Physical length of assigned contigs | 15.9 Mb |

are integrated into the physical/genetic framework. ESTs mapped amount to 2,239. The current release of the peach physical map is displayed in GDR using WebAGCoL Package (Abbott and Zhebentyayeva, 2007; Jung et al., 2008). Table 3, derived from the GDR web site, represents the state of the art as of June 2008.

## 6.1 Use of the Peach Physical Map

The integration of the peach physical map with the genetic maps was initiated by Horn et al. (2005) who used 9,984 ESTs as a direct source of polymorphic markers or for the development of microsatellite markers. Their map location was in part determined by hybridizing them against previously marker-anchored contigs or BACs (i.e. placing them on an integrated physical/genetic map). Using this approach, 11.2% of the ESTs were assigned to locations on the reference *Prunus* genetic map (TxE). The BAC library developed by Georgi et al. (2002) was also used by Lecouls et al. (2002) who constructed contigs around the regions hosting major QTLs for pH and fructose content. From comparative analysis the putative conservation of QTLs affecting fruit quality in different crosses can be retrieved.

Wang et al. (2002b) developed an efficient method to search the peach genome for SSRs linked to traits of interest using AFLP-based genetic linkage maps and BAC library screening. Using this strategy, three separate contigs were developed at three AFLP marker positions that were close to the *evergreen* (*evg*) locus on the genetic linkage map. A chromosome walk in both directions was initiated from the BAC which contains the AFLP closest to the *evg* locus (Wang et al., 2002c). Bielenberg et al. (2004) approached the cloning of *evg* by comparative genomics. The region contains a tandem array of 6 MADS genes related to the short vegetative phase in Arabidopsis. Comparative sequencing of this gene region in apricot and plum demonstrates a preservation of gene composition and order. Comparisons are now being completed for cherry, strawberry, and raspberry (Abbott A., unpublished results).

Also the *Ma* locus was approached by a chromosome landing strategy. For this, a BAC library of approximately 30,000 clones was prepared from Myrobalan plum (Claverie et al., 2004b). AFLPs and derived SCARs allowed saturation of the region delimited by flanking markers around *Ma* leading to the isolation of a single 270 kb BAC clone supposedly containing the *Ma* locus. Sequencing of the BAC is in progress (Claverie et al., 2004b).

The availability of a complete physical map and the integration of it with a very dense genetic map is also a prerequisite for the sequencing of the genome. Substantial progress has been made in this effort to date, promising at least a draft genome sequence in the near future.

## References

Abbott A, Rajapaske S, Sosinski B, Lu ZX, Sossy-Alaoui K, Gannavarapu M, Reighard G, Ballard RE, Baird WV, Scorza R, and Callahan A (1998) Construction of saturated linkage maps of peach crosses segregating for characters controlling fruit quality, tree architecture and pest resistance. Acta Hort 465: 41–49

Abbott AG, Lecouls AC, Wang Y, Georgi L, Scorza R, and Reigherd G (2002) Peach: the model genome for rosaceae genomics. Acta Hort 592: 199–209

Abbott A, Zhebentyayeva T (2007) Physical Mapping of Peach, Department of Genetics& Biochemistry, Clemson University, Clemson, SC, 29634, USA http://www.bioinfo.wsu.edu/gdr/physical˙map.sht

Andersen JR, Lubberstedt T (2003) Functional markers in plants. Trends Plant Sci 11: 554–560

Aranzana MJ, Garcia-Mas, Carbo J, and Arus P (2002) Development and variability of microsatellite markers in peach. Plant Breed 121: 87–92

Aranzana MJ, Pineda A, Cosson P, Ascasibar J, Dirlewanger E, Ascasibar J, Cipriani G, Ryder CD, Testolin R, Abbott A, King GJ, Iezzoni AF, and Arùs P (2003) A set of simple-sequence repeat (SSR) markers covering the *Prunus* genome. Theor Appl Genet 106: 819–825

Baird WV, Estager As, and Wells J (1994) Estimating nuclear DNA content in peach and related diploid species using laser flow cytometry and DNA hybridization. J Am Soc Hort Sci 119: 1312–1316

Barale F, Lazzari B, Abbott A, Salamini F, and Pozzi C (2006) Steps towards the production of a function map in peach (*Prunus persica*). In PGEM Congress, Venice, Italy

Bergougnoux V, Claverie M, Bosselut N, Lecouls AC, Esmenjaud D, Dirlewanger E, and Salesses G (2002) Marker assisted selection of the Ma gene from Myrobalan plum for a complete-spectrum root-knot nematode (RKN) resistance in Prunus rootstocks. Acta Hort 592: 223–228

Bielenberg D, Wang Y, Fan S, Reighard G, Scorza R, and Abbott A (2004( A deletion affecting several gene candidates is present in the Evergrowing peach mutant. J Hered 95: 436–444

Blenda AV, Reighard GL, Baird WV, Georgi LL, and Abbott AG (2002) Molecular markers and candidate resistance genes: a genetic study of tolerance to ring nematode in peach. Plant, Animal & Microbe Genomes × Conference. San Diego,USA

Blenda AV, Verde I, Georgi LL, Reighard GL, Forrest SD, Muñoz-Torres M, Baird WV, and Abbott A (2007) Construction of a genetic linkage map and identification of molecular markers in peach rootstocks for response to peach tree short life syndrome. Tree Genet Genomes 3: 341–350

Bliss FA, Arulsekar S, Foolad MR, Becerra V, Gillen AM, Warburton ML, Dandekar AM, Kocsisne GM, and Mydin KK (2002) An expanded genetic linkage map of *Prunus* based on an interspecific cross between almond and peach. Genome 45: 520–529

Brummell DA, Dal Cin V, Lurie S, Crisosto CH, and Labavitch JM (2004) Cell wall metabolism during the development of mealiness in cold-stored peach fruit: association of mealiness with arrested disassembly of cell wall pectins. J Exp Bot 55: 2041–2052

Byrne DH (1990) Isozyme variability in four diploid stone fruits compared with other woody perennial plants. J Hered 81: 68–71

Calenge F, Van der Linden CG, Van de Weg E, Schouten HJ, Van Arkel G, Denance C, and Durel CE (2005) Resistance gene analogues identified through the NBS-profiling method map close to major genes and QTL for disease resistance in apple. Theor Appl Genet 110: 660–668

Chaparro JX, Werner DJ, O'Malley D, and Sederoff RR (1994) Targeted mapping and linkage analysis of morphological, isozyme, and RAPD markers in peach. Theor Appl Genet 87: 805–815

Cipriani G, Lot G, Huang WG, Marrazzo MT, Peterlunger E, and Testolin R (1999) AC/GT and AG/CT microsatellite repeats in peach: isolation, characterization and cross-species amplification in Prunus. Theor Appl Genet 99: 65–72

Claverie M, Bosselut N, Lecouls AC, Voisin R, Lafargue B, Poizat C, Kleinhentz M, Laigret F, Dirlewanger E, and Esmenjaud D (2004a) Location of independent root-knot nematode resistance genes in plum and peach. Theor Appl Genet 108: 765–773

Claverie M, Dirlewanger E, Cosson P, Bosselut N, Lecouls AC, Voisin R, Kleinhentz M, Lafargue B, Caboche M, Chalhoub B, and Esmenjaud D (2004b) High resolution mapping and chromosome landing at the root-knot nematode resistance locus Ma form Myrobalan plum using a large -insert BAC DNA library. Theor Appl Genet 109: 1318–1327

Decroocq V, Fave MG, Hagen L, Bordenave L, and Decroocq S (2003) Development and transferability of apricot and grape EST microsatellite markers across taxa. Theor Appl Genet 106: 912–922

Decroocq V, Foulongne M, Lambert P, Le Gall O, Mantin C, Pascal T, Shcurdi-Levraud V, and Kervella J (2005) Analogues of virus resistance genes map to QTLs for resistance to sharka disease in *Prunus davidiana*. Mol Genet Genom 272: 680–689

Dettori MT, Quarta R, and Verde I (2001) A peach linkage map integrating RFLPs, SSRs, RAPDs and morphological markers. Genome 44: 783–790

Dirlewanger E, Pascal T, Zuger C, and Kervella J (1996) Analysis of molecular markers associated with powdery mildew resistance genes in peach [*Prunus persica* (L.) Batsch] × *Prunus davidiana* hybrids. Theor Appl Genet 93: 909–919

Dirlewanger E, Pronier V, Parvery C, Rothan C, Guy A, and Monet R (1998) Genetic linkage map of peach [*Prunus persica* (L.) Batsch] using morphological and molecular markers. Theor Appl Genet 97: 888–895

Dirlewanger E, Moing A, Rothan C, Svanella L, Pronier V, Guye A, Plomion C, and Monet C (1999) Mapping QTLs controlling fruit quality in peach. Theor Appl Genet 98: 18–31

Dirlewanger E, Crosson A, Tavaud P, Aranzana MJ, Poizat C, Zanetto A, Arus P, and Laigret L (2002) Development of microsatellite markers in peach and their use in genetic diversity analysis in peach and sweet cherry. Theor Appl Genet 105: 127–138

Dirlewanger E, Poizat C, Cosson P, Lafargue B, Kleinhentz M, Claverie M, Bosselut N, Voisin R, Esmenjaud D, and Laigret F (2003) Genetic linkage maps of Myrobalan plum and of an almond-peach hybrid- Location of root-knot nematode resistance genes.7th International congress of plant molecular biology, ISPMB, Barcelona, Spain, 23–28 June 2003

Dirlewanger E, Arùs P (2004) Molecular markers in plant breeding and crop improvement: markers in fruit tree breeding: improvement of peach. In Biotechnology in agriculture and forestry, vol. 55 (eds. L. Hoerst, Gerhard W.): Springer.

Dirlewanger E, Cosson P, Howad W, Capdeville G, Bosselut N, Claverie M, Voisin R, Poizat C, Lafargue B, Baron O, Laigret F, Kleinhentz M, Arus P, and Esmenjaud D (2004a) Microsatellite genetic linkage maps of myrobalan plum and an almond-peach hybrid-location of root-knot nematode resistance genes. Theor Appl Genet 109: 827–838

Dirlewanger E, Kleinhentz R, Voisin R, Claverie M, Lecouls AC, Esmenjaud JL, Poessel M, Faurobert M, Arús P, Gómez-Aparisi J, Xiloyannis C, and Di Vito M (2004b) Breeding for a new generation of *Prunus* rootstocks: an example of MAS. Acta Hort 658: 581–590

Dirlewanger E, Graziano E, Joobeur T, Garriga-Caldere F, Cosson P, Howad W, and Arus P (2004c) Comparative mapping and marker-assisted selection in *Rosaceae* fruit crops. Proc Natl Acad Sci USA. 29: 9891–9896

Dirlewanger E, Cosson P, Boudehri K, Renaud C, Capdeville G, Tauzin Y, Laigret F, and Moing A (2006) Development of a second-generation genetic linkage map for peach (*Prunus persica*) and characterization of morphological traits affecting flower and fruit. Tree Genet Genomes 3: 1–3

Etienne C, Rothan C, Moing A, Plomion C, Bodnes C, Svanella-Dumas L, Cosson P, Pronier V, Monet R, and Dirlewanger E (2002) Candidate gene and QTLs for sugar and organic acid content in peach. Theor Appl Genet 105: 145–159

Foolad MR, Arulsekar S, Becerra V, and Bliss FA (1995) A genetic map of *Prunus* based on an interspecific cross between peach and almond. Theor Appl Genet 91: 262–269

Foulongne M, Pascal T, Arus P, and Kervella J (2003a) The potential of *Prunus davidiana* for introgression into peach [*Prunus persica* (L.) Batsch] assessed by comparative mapping. Theor Appl Genet 107: 227–38

Foulongne M, Pascal T, Pfeiffer F, and Kervella J (2003b) QTLs for powdery mildew resistance in peach × *Prunus davidiana* crosses: consistency across generations and environments. Mol Breed 12: 33–50

Georgi LL, Wang Y, Yvergniaux D, Ormsbee T, Inigo M, Reighard G, and Abbott AG (2002) Construction of a BAC library and its application to the identification of simple sequence repeats in peach. Theor Appl Genet 105: 1151–1158

Gillen AM, Bliss FA (2005) Identification and mapping of markers linked to the Mi gene for root-knot nematode resistance in peach. J Am Soc Hortic Sci 130: 24–33

Horn R, Lecouls AC, Callahan A, Dandekar A, Garay L, McCord P, Howad W, Chan H, Verde I, Main D, Jung S, Georgi L, Forrest S, Mook J, Zhebentyayeva T, Yu Y, Kim HR, Jesudurai C, Sosinski B, Arús P, Baird V, Parfitt D, Reighard G, Scorza R, Tomkins J, Wing R, and Abbott AG (2005) Candidate gene database and transcript map for peach, a model species for fruit trees. Theor Appl Genet 110: 1419–1428

Howad W, Yamamoto T, Dirlewanger E, Testolin R, Cosson P, Cipriani G, Monforte AJ, Georgi L, Abbott AG, and Arús P (2005) Mapping with a few plants: using selective mapping for microsatellite saturation of the *Prunus* reference map. Genetics 171: 1305–1309

Jáuregui B, de Vicente MC, Messeguer R, Felipe A, Bonnet A, Salesses G, and Arús P (2001) A reciprocal translocation between 'Garfi' almond and 'Nemared' peach. Theor Appl Genet 102: 1169–1176

Joobeur T, Viruel MA, de Vicente MC, Jáuregui B, Ballester J, Dettori MT, Verde I, Truco MJ, Messeguer R, Batlle I, Quarta R, Dirlewanger E, and Arús P (1998) Construction of a saturated linkage map for *Prunus* using an almond × peach $F_2$ progeny. Theor Appl Genet 97: 1034–1041

Jun JH, Chung KH, Jeong SB, and Lee HJ (2003) Development of RAPD and SCAR markers linked to the flesh adhesion gene in peach. Acta Hort 625: 89–96

Jung S, Jesudurai C, Staton M, Du Z, Ficklin S, Cho I, Abbott A, Tomkins J, and Main D (2004) GDR (Genome Database for *Rosaceae*): integrated web resources for *Rosaceae* genomics and genetics research. BMC Bioinformatics 9(5):130

Jung S, Main D, Staton M, Cho I, Zhebentyayeva T, Arus P, and Abbott A (2006) Synteny conservation between the *Prunus* genome and both the present and ancestral *Arabidopsis* genomes. BMC Genomics 4:7:81

Jung S, Staton M, Lee T, Blenda A, Svancara R, Abbott A, and Main D (2008) GDR (Genome Database for Rosaceae): integrated web-database for Rosaceae genomics and genetics data. Nucleic Acids Res: D1034-40

Ku HM, Liu J, Doganlar S, and Tanksley SD (2001) Exploitation of *Arabidopsis*-tomato synteny to construct a high-resolution map of the ovate containing region in tomato chromosome 2. Genome 44: 470–475

Lalli DA, Decroocq V, Blenda AV, Schurdi-Levraud V, Garay L, Le Gall O, Damsteegt V, Reighard GL, and Abbott AG (2005) Identification and mapping of resistance gene analogs (RGAs) in *Prunus*: a resistance map for *Prunus*. Theor Appl Genet 111: 1504–1513

Lazzari B, Caprera A, Vecchietti A, Stella A, Milanesi L, and Pozzi C (2005) ESTree db: a tool for peach functional genomics. BMC Bioinformatics 6: S16

Lazzari B, Caprera A, Vecchietti A, Merelli I, Barale F, Milanesi L, Stella A, and Pozzi C (2007) Version VI of the ESTree db: an improved tool for peach transcriptome analysis. BMC Bioinformatics 2:S9

Lecouls AC, Reighard AC, Abbott AG, and Dirlewanger E (2002) Physical mapping and integration of QTL intervals involved in fruit quality on peach fruit variety and rootstock molecular maps. Acta Hort 592: 273–278

Lecouls AC, Bergougnoux V, Rubio-Cabetas MJ, Bosselut N, Voisin R, Poessel JL, Faurobert M, Bonnet A, Salesses G, Dirlewanger E, and Esmenjaud D (2004) Marker-assisted selection for the wide spectrum resistance to root-knot nematodes conferred by the Ma gene from Myrobalan plum (*Prunus cerasifera*) in interspecific *Prunus* material. Mol Breed 13: 113–124

Lewers KS, Styan SMN, Hokanson SC, and Bassil NV (2005) Strawberry GenBank-derived and genomic simple sequence repeat (SSR) markers and their utility with strawberry, blackberry, and red and black raspberry. J Am Soc Hortic Sci 130: 102–115

Liang FS, Zhang KC, Yu ZW, Yang JL, Zhabg XM, Jin DM, and Wang B (2004) Construction, characterization and screening of a transformation-competent artificial chromosome library of peach. Plant Mol Biol Rep 22: 37–48

Lu ZX, Sosinski B, Reighard GL, Baird WV, and Abbott AG (1998) Construction of a genetic linkage map and identification of AFLP markers for resistance to root-knot nematodes in peach rootstocks. Genome 41: 199–207

Lu ZX, Sossey-Alaoui K, Reighard GL, Baird WV, and Abbott AG (1999) Development and characterization of a codominant marker linked to root-knot nematode resistance, and its application to peach rootstock breeding. Theor Appl Genet 99: 115–122

Martinez-Gomez P, Sanchez-Perez R, Rubio M, Dicenta F, Gradziel TM, and Sozzi GO (2005) Application of recent biotechnologies to Prunus tree crop genetic improvement Cien Inv Agr 32: 73–96

Moing A, Svanella L, Rolin D, Gaudillere M, Gaudillere JP, and Monet R (1998) Compositional changes during the fruit development of two peach cultivars differing in juice acidity. J Am Soc Hortic Sci 123: 770–775

Monet R, Guye A, Roy M, and Dachary N (1996) Peach mendelian genetics: a short review of results. Agronomie 16: 321–329

Peace CP, Ahmad R, Gradziel TM, Dandekar AM, and Crisosto CH (2004) The use of molecular genetics to improve peach and nectarine post-storage quality. Acta Hort 682: 403–410

Peace CP, Crisosto CH, and Gradziel TM (2005) Endopolygalacturonase: a candidate gene for freestone and melting flesh in peach. Mol Breed 16: 21–31

Quarta R, Dettori MT, Sartori A, and Verde I (2000) Genetic linkage map and QTL analysis in peach. Acta Hort 521: 233–241

Quilot B, Wu BH, Kervella J, Genard M, Foulogne M, and Moreau K (2004) QTL analysis of quality traits in an advanced backcross between *Prunus persica* cultivars and the wild relative species *P. davidiana*. Theor Appl Genet 109: 884–897

Rajapakse S, Belthoff LE, He G, Estager AE, Scorza R, Verde I, ballard RE, Baird WV, Callahan A, Monet R, and Abbott AG (1995) Genetic mapping in peach using morphological, RFLP and RAPD markers. Theor Appl Genet 90: 503–510

Sargent DJ, Rys A, Nier S, Simpson DW, and Tobutt KR (2007) The development and mapping of functional markers in *Fragaria* and their transferability and potential for mapping in other genera. Theor Appl Genet 114: 373–384

Scorza R, Mehlenbacher SA, and Lightner GW (1985) Inbreeding and coancestry of freestone peach cultivars of the eastern United States and implications for peach germplasm improvement. J Am Soc Hort Sci 110: 547–552

Scorza R., Ravelonandro M (2002) Gene silencing-based resistance to plum pox virus. Act Hort 622: 119–122

Silva C, Garcia-Mas J, Sanchez AM, Arus P, and Oliveira M (2005) Looking into flowering time in almond (*Prunus dulcis*): the candidate gene approach. Theor Appl Genet 110: 959–968

Sosinski B, Sossey-Alaoui K, Rajapakse S, Glassmoyer K, Ballard RE, Abbott AG, Lu ZX, Baird WV, Reighard G, Tabb A, and Scorza R (1998) Use of AFLP and RFLP markers to create a combined linkage map in peach [*Prunus persica* (L.) Batsch] for use in marker assisted selection. Acta Hort 465: 61–68

Sosinski B, Gannavarapu M, Hager LD, Beck GJ, Ryder CD, Rajapakse S, Baird WV, Ballard RE, and Abbott AG (2000) Characterization of microsatellite markers in peach [*Prunus persica* (L.) Batsch] Theor Appl Genet 101: 421–428

Tani E, Polidoros AN, and Tsaftaris AS (2007) Characterization and expression analysis of FRUITFULL- and SHATTERPROOF-like genes from peach (*Prunus persica*) and their role in split-pit formation. Tree Physiol 27: 649–659

Verde I, Quarta R, Cedrola C, and Dettori MT (2002) QTL analysis of agronomic traits in a BC1 peach population. Acta Hort 592: 291–295

Verde I, Lauria M, Dettori MT, Vendramin E, Balconi C, Micali S, Wang Y, Marrazzo MT, Cipriani G, Hartings H, Testolin R, Abbott AG, Motto M, and Quarta R (2005) Microsatellite and AFLP markers in the Prunus persica × P. ferganensis BC1 linkage map: saturation and coverage improvement. Theor Appl Genet 111: 1013–1021

Viruel MA, Messeguer R, de Vicente MC, Garcia-Mas J, Puigdomènech P, Vargas F, and Arús P (1995) A linkage map with RFLP and isozyme markers for almond. Theor Appl Genet 91: 964–971

Wang Q, Zhang K, Qu X, Jia J, Shi J, Jin D, and Wang B (2001) Construction and characterization of a bacterial artificial chromosome library of peach. Theor Appl Genet 103: 1174–1179

Wang Y, Georgi LL, Zhebentyayeva TN, Reighard GL, Scorza R, and Abbott AG (2002a) High-throughput targeted SSR marker development in peach (*Prunus persica*). Genome 45: 319–328

Wang Y, Georgi LL, Reighard GL, Scorza R, and Abbott AG (2002b) Genetic mapping of the evergrowing gene in peach [*Prunus persica* (L.) Batsch]. J Hered 93: 352–358

Wang Y, Garay L, Reighard GL, Georgi LL, Abbott AG, and Scorza R (2002c) Development of bacterial artificial chromosome contigs in the evergrowing gene region in peach. Acta Hort 592: 183–189

Xu Q, Wen XP, and Deng XX (2005) Isolation of TIR and NonTIR NBS-LRR resistance gene analogues and identification of molecular markers linked to a powdery mildew resistance locus in chestnut rose (*Rosa roxburghii* Tratt). Theor Appl Genet 111: 819–830

Xu Q, Wen X, and Deng X (2007) Phylogenetic and evolutionary analysis of NBS-encoding genes in *Rosaceae* fruit crops. Mol Phylogenet Evol 44: 315–324

Yamamoto T, Shimada T, Imai T, Yaegaki H, Haji T, Matsuta N, Yamaguchi M, and Hayashi T (2001) Characterization of morphological traits based on a genetic linkage map in peach. Breed Sci 51: 271–278

Yamamoto T, Hayashi T (2002) New root-knot nematode resistance genes and their STS markers in peach. Sci Hort 96: 81–90

Yamamoto T, Yamaguchi M, and Hayashi T (2005) An integrated genetic linkage map of peach by SSR, STS, AFLP and RAPD. J Japan Soc Hort Sci 74: 204–213

## 3 Functional Genomics Tools in Rosaceae

Within the Rosaceae research community, there is extensive collaboration and cooperation between laboratories, as is evidenced by the continuous data contributions to the GDR by those groups. Recently (March 2006), this community held a 3rd International Rosaceae Genome Mapping Conference in Napier New Zealand, and the community at large reached a number of conclusions as to how genomics in the family should proceed. A conference report for the meeting was generated, and this report served to mature both the National and International Rosaceae White Papers, both of which include a roadmap for future activities in genomics (conference reports and white papers available at www.bioinfo.wsu.edu/gdr. One important conclusion in the conference report and the white papers is that peach, due to the high level of genomic resources already in place, should serve as the first reference genome for the Rosaceae. Furthermore, at the "Technology Roadmap for Temperate Fruit Genomics Workshop" (Baltimore, October 18–19, 2004), one of the major focus areas identified was the development of microarray resources *for each Rosaceous species*. As a result, we have built upon the substantial genomic resources already in place for Rosaceae, to extend the functional genomics resources in this family, and worked to broaden access to data and research products. These resources include a large number of mapped ESTs, numerous genetic maps, and a substantial initial physical map projected to be completed by the end of 2007. All of these resources are readily available to the research community through our Genome Database for Rosaceae (GDR) website (www.bioinfo.wsu.edu/gdr).

### *3.1 Functional Genomics Resource Development*

As of August 2007, the public repository for EST sequences, NCBI dbEST (http://www.ncbi.nlm.nih.gov/dbEST/index.html) contained some 380,930 ESTs for Rosaceous plants. This marks an exponential growth in publicly available Rosaceae functional resources considering only 27,000 of these ESTs were present in May 2004. A preliminary unigene analysis was conducted using 200,068 Rosaceae ESTs. Employing relatively stringent criteria (–p90, –d60) to optimize the assembly, CAP3 (Huang and Madan, 1999) merged 80% of the ESTs into 23,548 contigs with an average length of 780 bp. Based on significant sequence similarity (EXP<1 e-6) using the fastx34 algorithm we have putatively identified 14,083 Arabidopsis gene homologues. Table 1 indicates the total number of homologous genes per species and the number of homologues genes that appear to be species specific (given the current EST data set). Rather interestingly, 92% of the contigs that did have an *Arabidopsis* homolog contained only single species ESTs. Further analysis is being conducted using the Swissprot and NCBI non redundant protein databases to identify other homologues of known or putative function for both the contigs and singlets (Sosinski, Folta and Main, unpublished results).

**Table 1** Summary of putative Arabidopsis genes identified in Rosaceae ESTs

| Species | Number of ESTs | % of total ESTs | Number of Arabidopsis genes | Number of unique Arabidopsis genes |
|---|---|---|---|---|
| Apple | 152,426 | 76.19 | 13,167 | 6,570 |
| Almond | 18,943 | 9.47 | 4,301 | 307 |
| Peach | 17,812 | 8.90 | 4,068 | 289 |
| Strawberry | 5,322 | 2.66 | 1,993 | 116 |
| Rose | 5,305 | 2.65 | 1,384 | 74 |
| Pear | 227 | 0.11 | 123 | 2 |
| Cherry | 12 | 0.01 | 11 | 1 |

**Table 2** Summary of Rosaceae and species specific unigenes available through GDR

| Organism | EST number | Unigene number | Contig number | Singlet Number |
|---|---|---|---|---|
| Rosaceae | 359,001 | 90,337 | 13,764 | 76,573 |
| *Fragaria* | 18,729 | 10,012 | 2,939 | 7,073 |
| *Malus* | 250,907 | 82,850 | 23,868 | 58,982 |
| *Prunus* | 83,751 | 23,721 | 8,818 | 14,903 |
| *Pyrus* | 330 | 271 | 35 | 236 |
| *Rosa* | 5,284 | 2,963 | 705 | 2,258 |

A more recent analysis of all Rosaceae ESTs has resulted in a combined unigene set for the family, as well as for each individual species with EST representation in the database (Table 2). This set was derived from a total of 369,106 ESTs representing 151 cDNA libraries and 17 species. Further curation of the data revealed that 20 tissue types were represented, and that over 40% of the tissues were from fruit. This unigene set was also compared to other available plant unigenes (*Medicago, Arabidopsis,* grape, soybean etc.) and a strong sequence similarity was shown. Additional screening of the contigs revealed in excess of 33,000 SSRs from the genera specific contigs and 27,260 were found in the Rosaceae contigs (Staton, 2007). To date, only a handful of published functional genomic studies employing microarrays have been conducted in Rosaceaous species. These include two studies in apple (Schaffer et al., 2007, Lee et al., 2007), three studies in strawberry (Aharoni et al., 2000, 2002, 2004), and one study from raspberry (Mazzitelli et al., 2007), but none of these employed the GDR unigenes. It is anticipated that future microarray projects within the Rosaceae will employ this useful resource thereby allowing interesting cross-species comparisons.

Given the scope and size of sequence conservation of ortholgous genes between Rosaceous species, as well as with other plant species, we believe that we will be able to generate a set of 60-mer oligonucleotides representing the Rosaceae unigene set, that will be useful for expression studies for any species within the Rosaceae. This microarray resource will be of great value for expression studies for the major crop plants within the family, but perhaps more importantly, this Rosaceae-wide microarray will be an invaluable tool for use in underrepresented and underfunded species including *Rubus* (brambles e.g. raspberry), rose, cherry and apricot. For many of these underrepresented species, where EST resources have not been developed, and may not be developed in the near future, a Rosaceae microarray will facilitate functional genomic analyses that would be otherwise unavailable.

The solidarity within the Rosaceae community and the availability of custom arrays coincides at a time when we can now standardize our functional-genomics efforts. This can be contrasted by work in the *Arabidopsis* or *Drosophila* communities that were performed on multiple, unrelated platforms that preclude interrelatabilty of data arising from well-conceived experimental design. This schism brings poor return on investment from thin federal resources, as the power of group bargaining and a common scheme of analysis are compromised. A strong suit of the community approach to functional genomics is that is unifies the collective interests of the community to generate a durable, cost-effective method to augment public resources with standardized functional data.

## 4 Functional Genomics Candidate Gene Databases

One of the critical components of family wide functional genomics effort is the development of a candidate gene database of at least one species to serve as a reference genome for the family. In this regard, peach [*Prunus persica* (L.) Batsch], is ideally suited as a Rosaceae genomics core species. Theoretically, peach could serve as the "rice of the Rosaceae", and provide a highly characterized "core" genome for subsequent comparative mapping and gene discovery. Additionally, this database would be invaluable to the Rosaceae research community for the purposes of linkage mapping, candidate gene/QTL associations, and functional genomics. The development of a candidate gene database in peach was the logical next step in development of genomic resources in this model genome species and capitalized on the major genomics advantages of the peach. The DNA content of peach, a diploid *Prunus* species with $n = 8$ (Jelenkovic and Harrington, 1972), is $5.9 \times 10^8$ bp or 0.61 pg/diploid nucleus (Baird et al., 1994) which is only about twice the value for *Arabidopsis thaliana* (Arumuganathan and Earle, 1991). In addition to its small genome size, peach is also a logical choice for the core species since currently peach is the most extensively mapped species in Rosaceae having extensive marker maps identifying genomic positions of a number of important simple as well as complex characters. Also important, peach is a fruit tree that shares many of the physiological features of other fruit trees, therefore the basic understanding in gene expression in different organs and throughout development, should provide important information that can be applied to other species.

Most of the efforts on functional genomics in peaches and nectarines have been focused on finding out what genes are expressed in the fruit. This has been performed by EST sequencing of cDNA libraries obtained from different stages of fruit development, ripening and postharvest. To date, more than 70,000 EST *Prunus persica* sequences are available at dbEST, Genebank. Most of these sequences come from fruit mesocarp; however, there are EST sequences also obtained from shoot, developing seeds, flower and fruit skin. Data from the Chilean Functional Genomics Innitiative, that sequenced 50,000 ESTs from mesocarp of fruit undergoing softening and postharvest, indicates that around 10,000 genes are expressed in peach mesocarp. We still do not know the actual number of genes present in *Prunus persica*; however, considering the information derived from *Arabidopsis thaliana*, *Oryza sativa* and *Populus trichocarpa* we would expect to find around 25,000–45,000 genes. Then, perhaps more than 25% of the genes are being expressed in the fruit. This is a significant proportion of the genome, suggesting that an important functional complexity is associated to softening and postharvest in peach fruit.

Most of the *Prunus persica* ESTs sequences have been annotated based on their sequence similarity to other species; however, there are a significant number of genes that have unknown function. An important challenge is to get evidence for the actual function of all these genes. The functional characterization of genes can be achieved by heterologous expression in other biological systems such as yeast, bacteria and/or Arabidopsis; however, in order to have a clear idea about the role that each of these genes is playing; it would ideal to express them in *Prunus persica*. Unfortunately, despite some attempts to get a reproducible transformation procedure (Pérez-Clemente et al., 2005), to date no good *Prunus persica* transformation procedure is available; therefore, more effort should be done to get a reliable transformation protocol that would allow the analysis of gene function in a high-throughput manner. An alternative to study the function of different genes in the fruit is to use agroinfiltration of this organ. This approach has been already successfully used in other members of the Rosaceae family (Hoffmann et al., 2006).

## 5 The Transcriptome

There are few studies on the *Prunus persica* trancriptome. Trainotti et al. (2003) used a matrix containing 800 genes to report that genes involved in cell wall metabolism exhibit changes during the ripening process but also before the climacteric rise. More recently, a chip containing 4,800 oligonucleotide probes was designed and used to study the transition from pre-climateric to climacteric phase in peach fruit (Trainotti et al., 2006). The results showed that 269 genes were up-regulated whereas 109 genes were down-regulated. This oligoarray chip was placed in the market and maybe a useful tool to study changes in gene expression in the fruit. More recently, Gonzalez-Aguero et al. (2008) developed an array containing 847 non-redundant expressed sequence tags and looked for the changes in gene expression between normal and fruit that has developed chilling injury due to the cold storage used to extend the postharvest life. 106 genes had a differential expression between juicy and chilling injured fruit, indicating that more than 10% of the transcriptome change during postharvest. These are initial studies and a more clear picture of the fruit transcriptome will be achieved once the whole *Prunus persica* genome is sequenced, allowing designed arrays to look at the expression of all genes present in the organism. This would have a great impact in the study not only in fruits but in every organ and during development.

The assignment and mapping of EST sequences on an integrated physical/genetic map for peach to creates a database resource for gene identification and characterization of use to all of the economically important Rosaceae species. If the genomic positions of EST sequences and qualitative or quantitative trait loci (QTLs) were available in a single core map, it would be possible to readily identify candidate gene sequences that could be immediately tested to determine whether that gene (EST) controls/ contributes to the phenotypic trait of interest. A genetically mapped EST database would provide framework, mapping probes for use in other Rosaceae species mapping efforts currently underway as well as provide the EST sequences for functional genomics studies (i.e., microarray technology).

## 6 The Peach Integrated Genetic/Physical/EST Functional Genomics Database

Physical maps of two rosaceous trees (peach, apple) are currently in progress (Han et al., 2007; Zhebentyayeva et al., 2008). Integrated physical/genetic maps are of critical importance for high-throughput EST mapping, QTL fine-mapping and effective positional cloning of genes (Zhang and Wing, 1997; Zhang and Wu, 2001; Green, 2001). To construct the physical map for peach we employed essentially the strategies utilized to develop the physical maps for the Arabidopsis thaliana and Drosophila melanogaster (Marra et al., 1999; Hoskins et al., 2000). The approach combines hybridization of the genetically mapped markers with BAC DNA fingerprinting and in our case, hybridization of EST sequences as well. Manual sequencing gel-based fingerprinting has been proven a reliable and cost-effective technique for BAC fingerprinting and under certain circumstances performs better than other traditional fingerprinting methods (Xu et al., 2004). An initial acrylamide gel-based physical/genetic map framework for peach was established and released recently (Zhebentyayeva et al., 2006). This framework was based on random fingerprinting of 3x peach genome equivalents, covered at least 50% of the genome and included hybridization data for 673 of 3,384 ESTs of the peach unigene set (PP_LE). On this framework, genetically anchored BAC contigs provided landmarks for a Prunus-Arabidopsis microsynteny study (Jung et al., 2006) and for further development of the *Prunus* transcript map by Horn et al. (2005). Since the first release, the peach physical framework underwent further enhancements. We have integrated global hybridization data that includes an additional 2,000 ESTs AFLPs (amplified fragment length polymorphism), SSRs (microsatellites), gene specific genomic probes and "overgo" probes derived from the BAC- end sequences. The total number of markers incorporated into the physical framework was increased up to 2,636 markers. Selectively, we have fingerprinted all hybridization positive BACs omitted during random fingerprinting. As a result, we incorporated an additional 1x peach genome equivalent composed of marker-positive BACs. Finally, we took an advantage of the HICF (a high-information content fingerprinting) technique along with improved FPC v8.5.2 software to increase an average number of bands per BAC clone and improve accuracy in contig assembly (Nelson et al., 2005). The initial HICF physical map for the peach consist of 2,138 contigs. Of these, 252 contigs are anchored to eight linkage groups of the *Prunus* reference map. The physical length of physical map contigs has been estimated at 303 Mb which is close to the estimated size of the peach genome (Baird et al., 1994). Due to the abundance of hybridization data the HICF physical map for peach is biased to the expressed genome regions and thus substantially covers the euchromatic portion of the peach genome.

## 7 The Prunus Resistance Gene Map as a Functional Genomics Tool

With the continued reduction in chemical means of controlling agricultural pests and the increased emphasis on integrated pest management strategies, fruit tree breeding programs are focusing major efforts on identification and characterization of natural resistance genes in the fruit tree germplasm. Once resistance genes have been identified in the fruit trees species, durable resistance may result from the introduction of naturally occurring disease resistance genes either by marker assisted breeding (MAS) or transgenic technologies. This approach is clearly a major goal of many breeding programs. Of course, introduction of natural resistance by MAS or transgenic technologies requires prior knowledge of either very tightly linked markers, in the case of MAS, or the gene itself, in the case of transgenic technologies. In the short term, using the physical map of peach for the development of genetically and physically anchored SSR sequences provides the tools necessary for MAS in *Prunus* resistance breeding programs. However, in the long term it is better to have identified the resistance gene itself for development of single nucleotide polymorphism (SNP) markers and for transgenic applications.

Currently the task of identification of specific resistance genes in *Prunus* species is daunting. As an example, plum pox (Sharka) is the most significant viral disease of fruit trees in Europe and has currently been identified in Northeastern United States. We have been working in cooperation with several European partners to genetically characterize resistance to the disease exhibited by several different apricot cultivars. We have currently mapped one resistance gene in apricot and are physically mapping the region conferring the resistance. However, fine structure genetic mapping is difficult since the cost and labor required to obtain and maintain high progeny numbers necessary to reduce the physical map intervals is prohibitive. However, in our favor is the fact that evidence from our physical mapping studies indicates that in some parts of the genome 1 cM could be as small as approximately 100 kb. In addition, BAC sequencing studies of the peach genome have revealed that within several 50–100 kb regions examined, the number of genes could be of the order of ~30 genes (Georgi et al., 2003). Therefore, it would be highly desirable to know if there are any potential resistance genes located in marked regions under study that could serve as candidates or at least landmarks for resistance gene clusters. One approach to expedite the process of identifying resistance genes, is to utilize conserved aspects of resistance gene structure previously documented in studies of plant resistance genes (for reviews, Hammond-Kosack and Jones, 1997; Hulbert et al., 2001), and to PCR amplify sets of resistance gene analogues using degenerate primers constructed to these conserved regions. This approach has been well documented in numerous plant species as a means to identify potential resistance genes (Resistance Gene Analogues, RGAs) and to enable the mapping of these RGAs onto genetic and physical maps. As documented by a number of laboratories (for review see, Hulbert et al., 2001), many resistance genes appear to cluster in plant genomes and thus identifying the position of one putative resistance gene can lead to identification of others in the clusters. If RGAs co-localize in areas of the genome where genetically mapped resistance has been positioned, these genes can serve as candidate genes or to help define more precisely a resistance gene-containing region on a physical map. This approach has been shown to assist in characterization of resistance gene containing regions in soybean (Kanazin et al., 1996, Yu et al., 1996); wheat and barley (Seah et al., 1998) potato, (Leister et al., 1996); Arabidopsis (Aarts et al., 1998, Speulman et al., 1998); citrus, (Deng et al., 2000); rice, (Mago et al., 1999); maize, (Collins et al., 1998); lettuce (Shen et al., 1998); tomato (Pan et al., 2000); and *Medicago* (Zhu et al., 2002).

A resistance gene map for *Prunus* (Lalli et al., 2005) was developed as a functional genomics database of candidate genes representing analogs of major resistance genes (NBS-LRR, kinase, and transmembrane domain classes), translation initiation factors (eIF4E) known to be involved in recessive resistance to plant viruses (Rodriguez et al., 1998; Duprat et al., 2002; Nicaise et al., 2003) and defense response genes. The resistance regions were mapped with the use of the peach physical map database and the Genome Database for Rosaceae (GDR) (Jung et al., 2004). Identifying and mapping RGAs with the use of the peach BAC libraries and physical map allowed location of resistance genes regions in *Prunus* with out the use of segregating populations. In this candidate resistance gene database, a total of 42 regions of resistance were mapped based on hybridization data obtained from 30 of 58 probes. Of the 30 probes that proved to be informative for mapping, only 17 returned BLAST results with sequence similarity to resistance genes of the NBS-LLR class; however, these 17 probes accounted for more than half of the total map locations identified. Several of these probes mapped to locations were QTLs for resistance mapped for traits such as, powdery mildew, (Foulongne et al., 2003; Dettori et al., 2001, I. Verde personal communication; Dirlewanger et al., 2004b), and Sharka resistance (Decroocq et al., 2005) . Moreover, 3 amplified RGAs mapped to the region of L7 that is known to contain the *Ma* gene or on L2 close to the *Mi* gene, both of which control resistance to root-knot nematodes (Claverie et al., 2004; Dirlewanger et al., 2004a). In the case of QTLs for sharka resistance in apricot, RGAs were positioned in regions associated with resistance QTLs in LG1 and 7 (Decroocq et al., 2005). Additionally, multiple RGA clones mapped in the same region of LG 1 as Vilanova et al. (2003) mapped Sharka resistance. RGAs were mapped in all of the linkage groups except linkage group three.

Thus, this candidate resistance map serves to integrate resistance gene regions of the genome with genetically mapped resistance and provides the gene substrate for functional genomics approaches to studying disease resistance in peaches and other *Prunus* species.

## 8 The Peach Proteome

The peach proteome has been poorly analyzed. One of the problems that play a role against this type of analysis is the difficulty on extracting proteins from plants tissues and specially fruits. In recent years this has been partly solved and a couple of papers dealing with protein analysis have been published. Recently, two studies on differential proteomics have been published. One of them (Chan et al., 2007) shows that infection with yeast or the treatment with salicylic acid show differences in the protein profile compared to the control. In addition, Uthairatanakij et al. (2005) performed a comparative analysis of the proteome from different nectarine varieties exposed to long term cold storage and the effect of controlled atmosphere. Eventhough 30 proteins were observed to change, the analysis was not quantitative. Finally, unpublished results (Nilo and Orellana) have shown that an important number of proteins change during postharvest and among different varieties.

These studies are based on the separation of proteins on two dimensional gels (2D-gels) followed by sequencing of the proteins using mass spectrometry. Fortunately, the database provided by the ESTs publicly available is allowing the match of the peptides sequence with genes that have been already annotated. Thus, most of the proteins sequenced so far have been related to a function. Although 2D-gels are quite informative, they are limited to the identification of the most abundant proteins and mostly towards hydrophilic proteins, therefore the low abundant and the hydrophobic proteins are still missing. Under good conditions, it is possible to identify around 1,000 proteins from fruit mesocarp, however; the number of genes that are expressed in the same tissue is larger than that. Therefore, other proteomic approaches needs to be implemented in order to get a broader picture of the peach proteome. Besides, protein extraction can also be a problem at the moment of a comparison of the proteome among different organs. However, there is no doubt that in order to get a global understanding of the role of the different genes it is necessary to look at the proteins since there are many reasons why the transcriptome data is not enough to get the actual picture of the molecular machinery involved in the physiological changes in a peach.

Finally, the peach metabolome is almost unknown. Beyond the analysis of solid solubles, ethylene and other parameters usually measured by fruit physiologist, our knowledge of the metabolome is almost zero. A great infrastructure (GC-MS, HPLC-MS, etc) is required to do a proper analysis of all metabolites that are produced in a peach. In addition, a robust bioinformatics platform is needed to store and analyze the data. Some of these capabilities are already available, thus hopefully in the near future we should begin to have information about the peach metabolome.

## 9 Perspectives

Functional genomics studies require significant gene information and technological advances within each plant family. In Rosaceae, due to the recent accumulation of EST sequences, physical maps and whole genome sequences, we are now in the period where functional genomics studies will increasingly dominate the research efforts of individual laboratories. Much of the work to this point has been directed at accumulation of these requisite tools. We expect in the near future to finally begin to understand the functional gene networks that give rise to the extensive plant diversity in this large and agriculturally important plant family and thus provide the tools to improve the performance in the marketplace of individually important species.

## References

Aarts MG M, te Lintel Hekkert B, Holub E.B, Beynon J.L, Stiekema W.J, Pereira, A. 1998. Identification of R-gene homologous DNA fragments genetically linked to disease resistance loci in Arabidopsis thaliana. MMPI. 11(4):251–258.

Aharoni A, Keizer LCP, Bouwmeester HJ, Sun Z, Alvarez-Huerta M, Verhoeven HA, Blaas J, van Houwelingen ML, De Vos RCH, van der Voet H, Jansem RC, Guis M, Mol J, Davis RW, Schena M, van Tunen AJ, O'Connell AP. 2000. Identification of the SAAT gene involved in strawberry flavor biogenesis by use of DNA microarray. Plant Cell. 12:647–661.

Aharoni A, Keizer LC, Van Den Broeck HC, Blanco-Portales R, Muñoz-Blanco J, Bois G, Smit P, De Vos RC, O′Connell AP. 2002. Novel insight into vascular, stress, and auxin-dependent and -independent gene expression programs in strawberry, a non-climacteric fruit. Plant Physiol. Jul;129(3):1019–1031.

Aharoni A, Giri AP, Verstappen FW, Bertea CM, Sevenier R, Sun Z, Jongsma MA, Schwab W, Bouwmeester HJ. 2004. Gain and loss of fruit flavor compounds produced by wild and cultivated strawberry species. Plant Cell. 2004 Nov;16(11):3110–3131.

Arumuganathan K, Earle E. 1991. Nuclear DNA content of some important plant species. Plant Mol Biol Rep. 9: 208–218.

Bachem CW, van der Hoeven RS, de Bruijn SM, Vreugdenhul D, Zabeau M, Visser RG. 1996. Visualization of differential gene expression using a novel method of RNA fingerprinting based on AFLP: analysis of gene expression during potato tuber development. Plant J 9: 745–753.

Baird WV, Estager AS, Wells J. 1994. Estimating nuclear DNA content in peach and related diploid species using laser flow cytometry and DNA hybridization. J Am Soc Hort Sci. 119: 1312–1316.

Casu RE, Dimmock CM, Chapman SC, Grof CP, McIntyre CL, Bonnett GD, Manners JM. 2004. Identification of differentially expressed transcripts from maturing stem of sugarcane by in silico analysis of stem expressed sequence tags and gene expression profiling. Plant Mol Biol. 54(4):503–517.

Chan Z, Qin G, Xu X, Li B, Tian S. 2007. Proteome approach to characterize proteins induced by antagonist yeast and Salicylic acid in peach fruit. J Proteome Res. 6: 1677–1688.

Claverie M, Bosselut N, Lecouls AC., Voisin R, Lafargue B, Poizat C, Kleinhentz M, Laigret F, Dirlewanger E, Esmenjaud D. 2004. Location of independent root-knot nematode resistance genes in plum and peach. Theor Appl Genet. 109(6):1318–1327.

Collins NC, Webb CA, Seah S, Ellis JG, Hulbert SH, Pryor A. 1998. The isolation and mapping of disease resistance gene analogs in Maize. MPMI 11(10):968–978.

Decroocq V, Foulongne M, Lambert P, Le Gall O, Mantin C, Pascal T, Schurdi-Levraud V, Kervella J. 2005. Analogues of virus resistance genes map to QTLs for resistance to sharka disease in Prunus davidiana. Mol Genet Genomics 272: 680–689.

Deng Z, Huang S, Ling P, Chen C, Yu C, Weber CA, Moore GA, Gmitter FG, Jr. 2000. Cloning and characterization of NBS-LRR class resistance-gene candidate sequences in citrus. Theor Appl Genet. 101: 814–822.

Dettori MT, Quarta R, Verde I. 2001. A peach linkage map integrating RFLPs, SSRs, RAPDs, and morphological markers. Genome 44, 783–790.

Dhingra A, Bies DH, Lehner KR, Folta KM. 2006. Green light adjusts plastid transcript accumulation during de-etiolation. Plant Physiol. 142(3):1256–166.

Diatchenko L, Lau Y-FC, Camplbell AP, Chenchik A, Moqadam F, Huang B, Lukyanof S, Lukyanov K, Gurskaya N, Sverdlov ED, Siebert PD. 1996. Suppression subtractive hybridization: a method for generating differentially regulated or tissue-spesific cDNA probes and libraries. Proc Natl Acad Sci USA. 93: 6025–6030.

Dirlewanger E, Cosson P, Howad W, Capdeville G, Bosselut N, Claverie M, Voisin R, Poizat C, Lafargue B, Baron O, Laigret F, Kleinhentz M, Arus P, Esmenjaud D. 2004a. Microsatellite genetic linkage maps of myrobalan plum and an almond-peach hybrid—location of root-knot nematode resistance genes. Theor Appl Genet 109: 827–838.

Dirlewanger E, Graziano E, Joobeur T, Garriga-Calderé F, Cosson P, Howad W, Arús P. 2004b. Comparative mapping and marker–assisted selection in Rosaceae fruit crops. Proc Natl Acad Sci USA. 101: 9891–9896.

Duprat A, Caranta C, Revers F, Menand B, Browning K, Robaglia C. 2002. The Arabidopsis eukaryotic initiation factor (iso)4E is dispensable for plant growth but required for susceptibility to potyviruses. Plant J. 32: 927–934.

Folta KM, Lieg EJ, Durham T, Spalding EP. 2003. Primary inhibition of hypocotyl growth and phototropism depend differently on phototropin-mediated increases in cytoplasmic calcium induced by blue light. Plant Phys. 133: 1464–1470.

Foulongne M, Pascal T, Pfeiffer F, Kervella J. 2003. QTLs for powdery mildew resistance in peach × Prunus davidiana crosses: consistency across generations and environments. Mol Breed 12: 33–50.

Georgi LL, Wang Y, Reighard GL, Mao L, Wing RA, Abbott AG. 2003. Comparison of peach and Arabidopsis genomic sequences: fragmentary conservation of gene neighborhoods. Genome. 46:268–276.

González-Aguero M, Pavez L, Ibanez F, Pacheco I, Campos-Vargas R, Meisel LA, Orellana A, Retamales J, Silva H, Gonzalez M, Cambiazo V. 2008. Identification of woolliness response genes in peach fruit after post-harvest treatments. J Exp Bot. 59(8): 1973–1986.

Girke T, Todd J, Ruuska S, White J, Benning C, Ohlrogge J. 2000. Microarray analysis of developing *Arabidopsis* seeds. Plant Physiol. 124: 1570–1581.

Green ED. 2001. Strategies for the systematic sequencing of complex genomes. Natl Rev Genet. 2: 573–583.

Hammond-Kosack KE, Jones JDG. 1997. Plant disease resistance genes. Annu Rev Plant Physiol Plant Mol Biol. 48:575–607.

Han J, Gasić K, Marron B, Beever JE, Korban SS. 2007 A BAC-based physical map of the apple genome. Genomics 89: 630–637.

Harmer SL, Hogenesch JB, Straume M, Chang H-S, Han B, Zhu T, Wang X, Kreps JA, Kay SA. 2000. Orchestrated transcription of key pathways in *Arabidopsis* by the Circadian clock. Science 290: 2110–2113.

Hoffmann T, Kalinowski G, Schwab W. 2006 RNAi-induced silencing of gene expression in strawberry fruit (Fragaria × ananassa) by agroinfiltration: a rapid assay for gene function analysis. Plant J. 48: 818–826.

Horn R, Lecouls A-C, Callahan A, Dandekar A, Garay L, McCord P, Howad W, Chan H, Verde I, Ramaswamy K, Main D, Jung S, Georgi L, Forrest S, Mook J, Zhebentyayeva TN, Yu Y, Kim HR, Jesudurai C, Sosinski BA, Arus P, Baird V, Parfitt D, Reighard G, Scorza R, Tomkins J, Wing R, Abbott AG. 2005. Candidate gene database and transcript map for peach, a model species for fruit trees. Theor Appl Genet. 110: 1419–1428.

Hoskins RA, Nelson CR, Berman BP, Laverty TR, George RA, Ciesiolka L, Naeemuddin M, Arenson AD, Durbin J, David RG, Tabor PE, Bailey MR, DeShazo DR, Catanese J, Mammoser A, Osoegawa K, de Jong PJ, Celniker SE, Gibbs RA, Rubin GM, Scherer SE. 2000. BAC-based physical map of the major autosomes of Drosophila melanogaster. Science 287 (5461): 2271–2274.

Huang X, Madan A. 1999. CAP3: A DNA sequence assembly program. Genome Res. 9:868–877.

Hulbert SH, Webb CA, Smith SM, Sun Q. 2001. Resistance gene complexes: evolution and utilization. Annu Rev Phytopathol. 39: 285–312.

Jelenkovic G, Harrington E. 1972. Morphology of the pachytene chromosomes in *Prunus persica*. Can J Genet Cytol. 14: 317–324.

Jung S, Jesudurai C, Staton M, Du Z, Ficklin S, Cho I, Abbott A, Tomkins J, Main D. Sept. 9, 2004. GDR genome database for rosaceae: integrated web resources for Rosaceae genomics and genetics research. BMC Bioinformatics 5:130.

Jung S, Main D, Staton M, Cho I, Zhebentyayeva T, Arus P, Abbott A. 2006. Synteny conservation between the Prunus genome and both the present and ancestral Arabidopsis genomes. BMC Genomics 7: 81.

Kanazin V, Marek LF, Shoemaker RC. 1996. Resistance gene analogs are conserved and clustered in soybean. Proc Natl Acad Sci USA 93:11746–11750.

Kieber JJ, Ecker JR. 1993. Ethylene gas: It's not just for ripening any more. Trends Genet. 9: 356–362.

Lalli DA, Decroocq V, Blenda AV, Schurdi-levraud V, Garay L, Le Gall O, Damsteegt V, Reighard GL, Abbott AG. 2005. Identification and mapping of resistance gene analogs (RGAs) in Prunus: A resistance map for Prunus. Theor Appl Genet Theor Appl Genet. 111:1504–1513.

Latorre M, Silva H, Saba J, Guziolowski C, Vizoso P, Martinez V, Maldonado J, Morales A, Caroca R, Cambiazo V, Campos-Vargas R, Gonzalez M, Orellana A, Retamales J, Meisel LA. 2006, JUICE: A data management system that facilitates the analysis of large volumes of information in an EST project workflow. BMC Bioinformatics, 7: 513.

Lee YP, Yu GH, Seo YS, Han SE, Choi YO, Kim D, Mok IG, Kim WT, Sung SK. 2007. Microarray analysis of apple gene expression engaged in early fruit development. Plant Cell Rep. 2007 Jul;26(7):917–926.

Leister D, Ballvora A, Salamini F, Gebhardt C. 1996. A PCR-based approach for isolating pathogen resistance genes from potato with potential for wide application in plants. Nat Genet. 14: 421–429.

Liang P, and Pardee AB. 1992. Differential display of eukaryotic messenger RNA by means of the polymerase chain reaction. Science 257:967–971.

Lin JF, Wu SH. 2004. Molecular events in senescing Arabidopsis leaves. Plant J. 39(4):612–628.

Lockhart DJ, Dong H, Byrne MC, Follettie MT, Gallo MV, Chee MS, Mittman M, Wang C, Kobayashi M, Horton H, Brown EL. 1996. Expression minitoring by hybridization to high-density oligonucleotide arrays. Nat Biotechnol. 14: 1675–1680.

Mago R, Nair S, Mohan M. 1999. Resistance gene analogues from rice: Cloning, sequencing and mapping. Theor Appl Genet. 99: 50–57.

Marra M, Kucaba T, Sekhon M, Hillier L, Martienssen R, Chinwalla A, Crokett J, Fedele J, Grover H, Gund C, McCombie WR, McDonald K, McPherson J, Mudd N, Parnell L, Schein J, Seim R, Shelby P, Waterston R, Wilson R. 1999. zA map for sequence analysis of the Arabidopsis thaliana genome. Nat Genet. 22: 269–270.

Mazzitelli L, Hancock RD, Haupt S, Walker PG, Pont SD, McNicol J, Cardle L, Morris J, Viola R, Brennan R, Hedley PE, Taylor MA. 2007. Co-ordinated gene expression during phases of dormancy release in raspberry (Rubus idaeus L.) buds. J Exp Bot. 58(5):1035–1045.

Narusaka Y, Narusaka M, Seki M, Ishida J, Nakashima M, Kamiya A, Enju A, Sakurai T, Satoh M, Kobayashi M, Tosa Y, Park P, Shinozaki K. 2003. The cDNA microarray analysis using an Arabidopsis pad3 mutant reveals the expression profiles and classification of genes induced by Alternaria brassicicola attack. Plant Cell Physiol. 44(4):377–387.

Nelson C. 1996. Qgene v2.26 Manual. Electronic address: greengenes.cit.cornell.edu port 70; directory "Software for genetics/qGene.

Nelson WM, Bharti AK, Butler E, Wei FS, Fuks G, Kim H, Wing RA, Messing J, Soderlund C. 2005. Whole-genome validation of high-information-content fingerprinting. Plant Physiol. 139: 27–38

NicaiseV, German-Retana S, Sanjuán R, Dubrana M, Mazier M, Maisonneuve B, Candresse T, Caranta C, LeGall O. 2003. The eukaryotic translation initiation factor 4E controls lettuce susceptibility to the potyvirus Lettuce mosaic virus. Plant Physiol. 132: 1272–1282.

Pan Q, Liu Y, Budai-Hadrian O, Sela M, Carmel-Goren L, Zamir D, Fluhr R. 2000. Comparative genetics of nucleotide binding site-leucine rich repeat resistance gene homologues in the genomes of two dicotyledons: tomato and Arabidopsis. Genetics 155: 309–322.

Pérez-Clemente RM, Pérez-Sanjuán A, García-Férriz L, Beltrán J, Cañas LA. 2005. Transgenic peach plants (Prunus persica L.) produced by genetic transformation of embryo sections using the green fluorescent protein (GFP) as an in vivo marker. Mol Breed. 14(4): 419–427.

Relógio A, Schwager C, Richter A, Ansorge W, Valcárcel J. 2002. Optimization of oligonucleotide-based DNA microarrays. Nucl Acids Res. 30(11): e51–e61.

Rodriguez CM, Freire MA, Camilleri C, Robaglia C. 1998 The Arabidopsis thaliana cDNAs coding for eIF4E and eIF(iso)4E are not functionally equivalent for yeast complementation and are differentially expressed during plant development. Plant J. 13: 465–473.

Schaffer RJ, Friel EN, Souleyre EJ, Bolitho K, Thodey K, Ledger S, Bowen JH, Ma JH, Nain B, Cohen D, Gleave AP, Crowhurst RN, Janssen BJ, Yao JL, Newcomb RD. 2007. A genomics approach reveals that aroma production in apple is controlled by ethylene predominantly at the final step in each biosynthetic pathway. Plant Physiol. 2007 Aug;144(4): 1899–1912.

Schena M, Shalon D, Davis RW & PO Brown (1995) Quantitative monitoring of gene expression patterns with a complementary DNA microarray. Science 270: 467–470.

Seah, S., Sivasithamparam K, Karakousis A, Lagudah ES. 1998. Cloning and characterization of a family of disease resistance gene analogs from wheat and barley. Theor Appl Genet. 97: 937–945.

Shen KA, Meyers BC, Islam-Faridi MN, Chin DB, Stelly DM, Michelmore RW. 1998. Resistance gene candidates identified by PCR with degererate oligonucleotide primers map to clusters of reisitance genes in lettuce. Mol Plant-Microbe Interact 11: 815–823.

Speulman E, Bouchez D, Holub EB, Beynon JL.1998. Disease resistance gene homologs correlate with disease resistance loci of Arabidopsis thaliana. Plant J. 14: 467–474.

Staton M. 2007. Bioinformatics Tool Development And Sequence Analysis Of Rosaceae Family Expressed Sequence Tags, Ph.D. Dissertation, Clemson University, April.

Tanksley SD, Ganal MW, Martin GB. 1995. Chromosome landing: a paradigm for map-based gene cloning in plants with large genomes. Trends Genet. 11:2, 63–68.

Trainotti L, Bonghi C, Ziliotto F, Zanin D, Rasori A, Casadoro G, Ramina A, Tonutti P. 2006. The use of microarray μPEACH1.0 to investigate transcriptome changes during transition from pre-climacteric to climacteric phase in peach fruit. Plant Sci. 170 (3): 606–613.

Trainotti L, Zanin D, Casadoro G. 2003. A cell wall-oriented genomic approach reveals a new and unexpected complexity of the softening in peaches. J Exp Bot. 54 (389) 1821–1832.

Uthairatanakij A, Holford P, McGlasson B. 2005. Changes in total proteins related to chilling injury of nectarine. Europ J Hort Sci. 70: 271–277.

Vilanova S, Romero C, Abbott AG, Llacer G, Badenes ML. 2003. An apricot (Prunus armeniaca L.) F2 progeny linkage map based on SSR and AFLP markers, mapping plum pox virus resistance and self-incompatibility. Theor Appl Genet. 107: 239–247.

Whitham SA, Quan S, Chang HS, Cooper B, Estes B, Zhu T, Wang X, Hou YM. 2003. Diverse RNA viruses elicit the expression of common sets of genes in susceptible Arabidopsis thaliana plants. Plant J. 33(2):271–283.

Xu Z, Sun S, Covaleda L, Ding K, Zhang A, Wu C, Scheuring C, Zhang HB. 2004. Genome physical mapping with large-insert bacterial clones by fingerprint analysis: methodologies, source clone genome coverage, and contig map quality. Genomics 84: 941–951.

Yu YG, Buss GR, Saghai-Maroof MA. 1996. Isolation of a super-family of candidate disease-resistance genes in soybean based on a conserved nucleotide-binding site. Proc Natl Acad Sci USA 93: 11751–11756.

Zhang HB, Wing RA. 1997. Physical mapping of the rice genome with BACs. Plant Mol Biol. 35: 115–127.

Zhang HB, Wu C. 2001. BAC as tools for genome sequencing. Plant Physiol Biochem. 39: 195–209.

Zhebentyayeva TN, Horn R, Mook J, Lecouls A, Georgi L, Abbott AG, Reighard GL, Swire-Clark G, Baird WV. 2006. A physical framework for the peach genome. Acta Hort. 713: 83–88.

Zhebentyayeva TN, Swire-Clark G, Georgi LL, Garay L, Jung S, Forrest S, Blenda, AV, Blackmon B, Mook J, Horn R, Howad W, Arús P, Main D, Tomkins JP, Sosinski B, Baird WV, Reighard GL, Abbott AG. 2008. A framework physical map for peach, a model Rosaceae species. Tree Genet Gen. 4(4):745–756.

Zhu T, Budworth P, Han B, Brown D, Chang H-S, Zou G, Wang X. 2001. Towards elucidation the global gene expression patterns of developing *Arabidopsis*: parallel analysis of 8300 genes by high-density oligonucleotide probe array. Plant Physiol Biochem 39:221–242.

Zhu H, Cannon SB, Young ND, Cook DR. 2002. Phylogeny and genomic organization of the TIR and Non-TIR NBS-LRR resistance gene family in Medicago truncatula. MPMI 15(6):529–539.

# 13. Genetic Engineering of Plum (*Prunus domestica* L.) for Plant Improvement and Genomics Research in Rosaceae

**Cesar Petri, Ralph Scorza, and Chris Dardick**

## 1 Introduction

Fruit trees are among the most recalcitrant of plants to regenerate adventitious shoots. In most woody fruit species, transformation and regeneration are difficult and often limited to a few genotypes or to seedlings (Petri and Burgos, 2005). This feature is the major limiting factor preventing the development of gene transfer technologies for fruit trees (Petri and Scorza, 2008). Such barriers slow the ability to relate the wealth of *Prunus* structural data, namely from peach, directly to testable questions of agricultural relevance. However, plum has been one of the more successful rosaceaous fruits to regenerate and transform, especially among *Prunus* species. A number of regeneration protocols have been reported from different tissues of European plum (*Prunus domestica* L.), such as leaf explants (Bassi and Cossio, 1991; Csányi et al., 1999; Escalettes and Dosba, 1993; Mikhailov and Dolgov, 2007; Nowak et al., 2004), or seed-derived tissues (Mante et al., 1989, 1991) (Table 1).

*Agrobacterium tumefaciens*-mediated transformation has been the principal technique applied to plum. One publication reported the use of *Agrobacterium rhizogenes*, but no transgenic shoots were recovered (Escalettes et al., 1994). There are few reports of regeneration of transformed plum shoots from clonal explants in European plum (Table 2). In these reports, transformation efficiencies are low and the varieties used as explant sources are generally of limited or local importance. Furthermore, in most cases, only marker genes were introduced into the plant genome, with few reports of modification of agronomically important traits (Escalettes et al., 1994; Mikhailov and Dolgov, 2007; Mikhailov et al., 2007; Yancheva et al., 2002).

The most successful transformation/regeneration protocol reported in plum has been developed in the authors' laboratory (USDA, Kearneysville), using embryonic

R. Scorza (✉)
Department of Genetic Improvement of Fruit Crops, USDA-ARS-Appalachian Fruit Research Station, 2217 Wiltshire Road, Kearneysville, WV 25430, USA

K.M. Folta, S.E. Gardiner (eds.), *Genetics and Genomics of Rosaceae*, Plant Genetics and Genomics: Crops and Models 6, DOI 10.1007/978-0-387-77491-6_13,

**Table 1** Adventitious shoot regeneration in plum (*Prunus domestica*)

| Variety/clone | Explant | Regeneration (%) | Shoots per explant[a] | Reference |
|---|---|---|---|---|
| Besztercei | Leaves | 32 | – | Csanyi et al. (1999) |
| Bluefree Susina di Dro | Leaves | 25 17 | 2.0 2.3 | Bassi and Cossio (1991) |
| P.1869 | Leaves | 12 | – | Escalettes and Dosba (1993) |
| Startovaya | Leaves | 80 | – | Mikhailov and Dolgov (2007) |
| Wegierka Zwykla | Leaves | 65 | 10.0 | Nowak et al. (2004) |
| Stanley | Hypocotyls | 63 | 6.8 | Mante et al. (1991) |
| B70173 | Cotyledons | 85 | 18.7 | Mante et al. (1989) |

[a] When not indicated it was non-specified by authors.

hypocotyl slices as the source of explants (Mante et al., 1991) (Fig. 1a). This protocol has been employed successfully for the introduction of agronomically useful genes into this species (Callahan and Scorza 2007; Gonzalez-Padilla et al., 2003; Hily et al., 2007; Scorza et al., 1994, 1995; Nagel et al., 2008; Petri et al., 2008). Although transformation of seed-derived material is not an ideal system for improving vegetatively propagated plum scion cultivars, it could have an impact on the development of new seed-propagated rootstock varieties and the introduction of novel genes into the plum germplasm. The high efficiency of the system will also enable functional genomics studies.

## 2 The Plum System

### *2.1 Description of Methodology*

Mante et al. (1991) developed an *Agrobacterium*-mediated transformation protocol in plum hypocotyls. This protocol was enhanced by improving the selection [80 mg $l^{-1}$ kanamycin (km) was added just after co-cultivation] as well as rooting and acclimatization steps and reached 4.2% transformation efficiency (Gonzalez-Padilla et al., 2003). Recently, the addition of 2,4-D during co-cultivation and the optimization of the timing of each step in the protocol has allowed transformation efficiencies up to 42% and enabled the production of self-rooted transgenic plants in the greenhouse in approximately 6 months (Petri et al., 2008).

Figure 1 represents a schematic of the Petri et al. (2008) procedure. Briefly, after the endocarp had been removed with a nutcracker, the seeds were surface-disinfected by immersion for 30 min in a 1% sodium hypochlorite solution containing approximately 20 μl Tween-20 per 100 ml solution and rinsed three times with sterile distilled water in a laminar flow bench. Disinfected seeds were soaked in sterile water overnight at room temperature and then the seed coats were removed with the aid of a scalpel. The radicle and the epicotyl were discarded, and the hypocotyl was sliced into three cross sections (0.5–1 mm), which were used for

**Table 2** Transformation of plum (*Prunus domestica*)

| Cultivar/clone | *Technique* | Genes | Explant | TE[a] (%) | Reference |
|---|---|---|---|---|---|
| Stanley | A. tumefaciens | *npt*II, *gus* | Hypocotyls | 3.3 | Mante et al. (1991) |
| Damas de Tolouse | A. rhizogenes | T-DNA (*ipt*) | Shoots[b] | 0 | Escalettes et al. (1994) |
| | | T-DNA (*ipt*), *PPV-CP* | | | |
| Marianna (GF8-1) | A. tumefaciens | *npt*II, *gus* | Leaves | – | |
| | | *npt*II, *gus, PPV-CP, hpt* | | | |
| Stanley | A. tumefaciens | *npt*II, *gus, PPV-CP* | Hypocotyls | 1.2 | Scorza et al. (1994) |
| B70146 | A. tumefaciens | *npt*II, *gus, PRV-CP* | Hypocotyls | 3.0 | Scorza et al. (1995) |
| Quetsche | A. tumefaciens | *npt*II, *gfp* | Leaves | 0.8 | Yancheva et al. (2002) |
| Kyustendilska sinya | | | | 2.7 | |
| Bluebyrd | A. tumefaciens | *nptII, gus* | Hypocotyls | 0.4 | Gonzalez-Padilla et al. (2003) |
| | | *nptII, PDV-CP* | | 1.4 | |
| | | *nptII,PNRSV-CP* | | 0.7 | |
| | | *nptII, gus, TomRSV-CP* | | 4.2 | |
| Stanley | | *nptII, gus, antisense ACO1* | | 2.0 | |
| Startovaya | A. tumefaciens | *nptII, gfp* | Leaves | 0.3 | Mikhailov and Dolgov (2007) |
| | | *hpt, gfp* | | 2.2 | |
| | | *pmi, gfp* | | 0.1 | |
| | | *nptII, PPV-CP* | Leaves | – | Mikhailov et al. (2007) |
| Stanley | A. tumefaciens | *nptII, ihpPPVCP* | Hypocotyls | – | Hily et al. (2007) |
| | | *nptII, gus, GAFP* | Hypocotyls | – | Nagel et al. (2008) |
| Bluebyrd | A. tumefaciens | *nptII, ihpPPV-CP* | Hypocotyls | 25.0 | Petri et al. (2008) |
| | | | | 20.0 | |
| | | | | 42.0 | |
| | | *nptII, ihpPDS* | | 15.0 | |

[a] Transformation efficiency. When not indicated, could not be deduced from the information provided by the authors.

[b] These shoots produced transgenic roots from which transformed shoots were regenerated.

**Fig. 1** Regeneration of transgenic plums. (**a**) Plum hypocotyl slices, where 1, 2 and 3 represents the explants used in this study. Epicotyl (E) and radicle (R) were not used. Vertical bar indicates 1 mm. (**b**) Explants co-cultivated with *Agrobacterium tumefaciens* in regeneration medium (SRM) supplemented 2 mg/l 2,4-D for 3 days (**c**) Adventitious regeneration from hypocotyl slices in selective medium. (**d**) Putative transgenic shoot cluster after cultivation in shoot growing medium (SGM) supplemented with timentin and kanamycin. (**e**) Shoots rooted in selective rooting medium. (**f**) Plant introduced in a plastic bag provided with zipper seal. (**g**) Rooted plantlet maintained in the culture room after the plastic bag is fully opened. (**h**) Transgenic plants grown in a greenhouse (Figure from Petri et al. (2008))

regeneration/transformation (Fig. 1a). Media used were described by Gonzalez-Padilla et al. (2003). They were MS medium-based with the following growth regulators: shoot regeneration medium (SRM) contained 7.5 μM thidiazuron (TDZ) and 0.25 μM indole butyric acid (IBA), shoot growing medium (SGM) contained 3.0 μM benzylaminopurine (BAP) and rooting medium (RM) contained 0.1 μM kinetin (K) and 5.0 μM α-naphthaleneacetic acid (NAA).

Immediately following the 3 day co-cultivation of hypocotyl segments with *A. tumefaciens* on SRM without antibiotics (Fig. 1b), the hypocotyl slices were transferred onto SRM with 300 mg $l^{-1}$ timentin (tim) and 80 mg $l^{-1}$ km. After 5–8 weeks, when shoots began to appear (Fig. 1c), explants were placed onto SGM supplemented with 300 mg $l^{-1}$ tim and 80 mg $l^{-1}$ km. Following three weeks on SGM, vigorous green shoot clusters were collected, labeled and placed onto fresh media (Fig. 1d). Shoot clusters were sub-cultured every 3 weeks on fresh SGM. When shoots reached 2–3 cm long they were separated from the cluster and transferred to RM supplemented with 40 mg $l^{-1}$ km and 300 mg $l^{-1}$ tim. In 3–4 weeks, roots started appearing and after an additional 1–3 weeks, shoots were ready for acclimatization (Fig. 1e).

For acclimatization, rooted shoots were washed in sterile water to eliminate agar residues and transferred to 3-inch-square peat pots containing Metro-Mix 510 potting medium (SUNGRO Horticulture, Bellevue, WA USA). The potted plantlets were introduced into plastic bags which were sealed and maintained in the tissue culture growth room with a 16/8 h light/dark cycle, light intensity of 45–50 μE $m^{-2}s^{-1}$, and a temperature of 24±1 °C (Fig. 1f). After 2 weeks, the plastic bags were fully opened (Fig. 1 g). Following 2 weeks in open plastic bags during which time plants were irrigated with deionized water as needed, the plants were sprayed with an aqueous solution (1:20, v/v) of Anti Stress 2000 (Polymer Ag, Fresno, Calif.) and cultured under ambient greenhouse conditions (Fig. 1 h).

The high transformation rates coupled with the rapid plant establishment methodology, make possible the application of this technique, not only for the introduction of agronomically useful genes into *P. domestica* but also for high-throughput functional genomic studies.

## 2.2 New Traits Introduced into Plum

Brief descriptions of successful attempts to genetically modify important traits in plum using the *Agrobacterium*-mediated plum hypocotyl transformation system are described below.

### 2.2.1 Honeysweet (C5)

In 1989 the USDA laboratory at Kearneysville (WV) began work on the development of resistance to plum pox virus (PPV) through genetic engineering. The first studies utilized the papaya ringspot virus (PRV) coat protein (CP) gene, which had been used to develop PRV resistant papayas (Gonsalves, 1998). This virus CP gene had significant homology to the *PPV-CP* gene and virus resistance was expected to be CP-mediated (Beachy et al., 1990). The heterologous protection against PPV in plum based on *PRV-CP* expression was effective for several years in greenhouse tests, but after 32 months resistance 'broke down', symptoms were evident and virus was detected throughout the plant (Scorza et al., 1995). The *PPV-CP* gene was then isolated, sequenced, and cloned (Ravelonandro et al., 1992) and used for *Agrobacterium*-mediated transformation of plum following the methods of Mante et al. (1991). Again, it was hypothesized that protection would be CP-based. Transferring the gene into plum, producing the genetically engineered (GE) plants, and propagating them for testing took two years. During the following two years, greenhouse tests for resistance were conducted at the USDA-ARS BSL3-P containment greenhouse at Ft. Detrick, Maryland. One transgenic plum plant that appeared highly resistant in greenhouse tests did not express *PPV-CP* (Ravelonandro et al., 1997; Scorza et al., 2001). This clone 'C5', since patented as 'HoneySweet', became the focus of research on the mechanism and stability of resistance to PPV. While the C5 clone appeared to be highly resistant in greenhouse tests, field testing was necessary in order to evaluate resistance under typical orchard conditions and in different growing environments. Testing of this resistant clone in areas where PPV was established was through collaboration with research partners in Europe (Poland, Romania and Spain). After appropriate field test permits were granted in each country, field trials were initiated in 1996–1997, which was six to seven years following the initial plum transformations. At USDA-ARS, Kearneysville a field trial was planted under a USDA-Animal and Plant Health Inspection Service (APHIS) permit. This was not to test for resistance since PPV was not present in the U.S. and plants could not be inoculated in the field, but rather to evaluate the trees for their horticultural traits including growth habit and fruit quality, as well as to initiate risk assessment studies. By 2002 the field tests in Europe clearly demonstrated the resistance of C5 to PPV infection through aphid vectors and by graft inoculation (Hily et al., 2004). Continuation of these tests through 2005 further confirmed these findings (Malinowski et al., 2006). Investigations of the mechanism of resistance in C5 showed low levels of transgene mRNA, methylation and production of siRNA specific to the silenced *PPV-CP*, indicating that resistance was through post-transcriptional gene silencing (PTGS) (Scorza et al., 2001; Hily et al., 2004, 2005).

'HoneySweet' was deregulated by APHIS in 2006 (Scorza et al., 2007), and approved it is by the U.S. Food and drug administration (FDA). It is currently being evaluated by the Environmental Protection Agency (EPA). Additional information on 'HoneySweet' can be found at www.agbios.com (http://www.agbios.com/dbase.php).

### 2.2.2 Intron Hairpin PPV-CP

Constructs with self-complementary sequences separated by an intron produce 'hairpin' RNA [intronhairpin-RNA (ihpRNA)] structures that efficiently elicit PTGS. In the case of 'Honeysweet' resistance was not produced by an ihpRNA vector, but rather PTGS developed as a result of peculiarities of the insertion event (Scorza et al., 2001)

Hily et al. (2007) designed different intron-hairpin RNA *PPV-Cp* constructs (Fig. 2c). Transgenic plum lines transformed with the ihpRNA-B14 and ihpRNA-E2 constitutively produced a short (21 nt) and a long (25–26 nt) class of siRNA (Fig. 2a, b). The siRNA accumulation levels in the transgenic clones were found to be similar to *P. domestica* C5 (Hily et al., 2005), suggesting that the clones would be highly resistant to PPV. Currently, these clones and additional *PPV-CP* hairpin transformed lines (Petri et al., 2008) are under evaluation in the greenhouse (Scorza, personal communication).

### 2.2.3 1-Aminocyclopropane Carboxylic Acid (ACC) Oxidase Antisense Plums

Fruit are harvested at a time determined by the handling properties that will yield the highest quality fruit that can withstand storage and transport. Genetic engineering efforts have been conducted to improve fruit quality in the market by manipulating genes that can affect the rate of softening. To this end, a project to reduce or delay the amount of ethylene produced by the fruit was undertaken to determine whether lower ethylene levels resulted in a firmer fruit that could remain in the tree longer to develop more tree-ripened flavors, yet resist damage incurred during harvesting, processing and shipping (Callahan and Scorza, 2007).

Plum hypocotyls were transformed with an antisense construct of a peach ACC oxidase (*ACCO*) gene (the enzyme responsible for the last step in ethylene synthesis) under the control of the CaMV35S promoter (Gonzalez-Padilla et al., 2003).

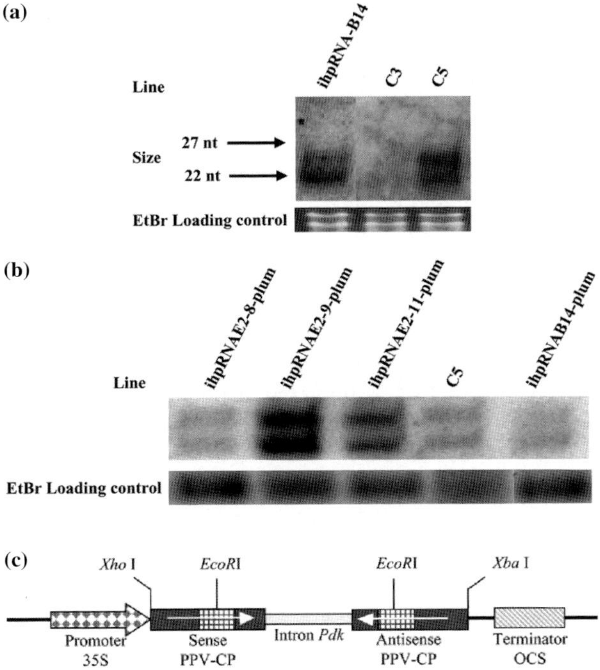


**Fig. 2** (**a, b**) Small interfering RNA (siRNA) production in leaves from a transgenic uninoculated *Prunus domestica* containing the intron-hairpin-RNA (ihpRNA)-B14 construct (b) and the ihpRNA-E2 construct (b). C5 or C3, a resistant and a sensitive transgenic plum, were used as positive and negative controls respectively for siRNA production. Nucleic acid preparations were separated on a 20% denaturing polyacrylamide gel and hybridized with radioactive DNA probes corresponding to both strands of the full-length plum pox virus–coat protein (PPV-CP) sequence (Scorza et al., 1994). Positions of the 22 and 27 nucleotides (nt) markers are indicated at the left. Relative quantification was performed on a 1% nondenaturing agarose gel using ethidium bromide (EtBr) staining. (**c**) ihp-RNA-B14 construct designed by Hily et al. (2007) based on the pHELLSGATE8 vector. (OCS, octopine synthase; Pdk, pyruvate orthophosphate dikinase intron). This construct carries the full length (1 kb) of the *PPV-CP* sequence (black rectangle) driven by the cauliflower mosaic virus (CaMV) (35S) promoter (Figure from Hily et al. (2007))

Eighteen lines were derived from hypocotyls from 'Bluebyrd'. DNA blots indicated that the majority of these lines had single insertions of the peach antisense *ACCO* gene (Callahan and Scorza, 2007). Analyses of the data suggested that in some transgenic lines ethylene production and softening was delayed. Sugar levels in most of these lines were, on the average, lower than in the non-transformed 'Bluebyrd' (Callahan and Scorza, 2007).

### 2.2.4 Gastrodianin Antifungal Protein

The *Gastrodia* anti-fungal protein (GAFP) is a monocot mannose-binding lectin isolated from the Asiatic orchid *Gastrodia elata* (Hu et al., 1988). This lectin has provided increased resistance in transgenic tobacco against root diseases from different phylogenetic lineages (Cox et al., 2006). *Agrobacterium tumefaciens* mediated transformation yielded three *GAFP-1* expressing plum lines (*Prunus domestica* 'Stanley') designated 4 J, 4I, and 5D (Nagel et al., 2008). These lines possessed one, two, and four copies of the *GAFP-1* gene, respectively, as demonstrated by DNA blotting. Lines 4 J and 4I were not phenotypically different from the non-transformed control line, but line 5D showed significant divergence in leaf morphology and growth habit. Compared to the inoculated control line, lines 4 J and 4I exhibited increased tolerance to *Phytophthora* root rot (PRR), caused by *P. cinnamomi.* When inoculated with the root-knot nematode, *Meloidogyne incognita,* the 4 J and 4I lines showed a significantly lesser degree of root galling than the inoculated control line, although nematode reproduction in these lines was not significantly different from the control line. The results of this study suggest that the expression of *GAFP-1* in the roots of a woody plant may confer some level of resistance to PRR and root-knot nematode. Long-term field trials will be necessary to confirm these findings in an applied context.

A new construct, where *GAFP* is under a potato ubiquitin promoter strongly expressed in roots, has been engineered. This construct has been introduced, via *A. tumefaciens*, into tobacco and plum, and currently, plants are under evaluation in the greenhouse (Schnabel, personal communication).

# 3 Future Applications

## *3.1 Functional Genomics*

Many aspects of tree biology are common to all plants, and therefore gene function studies for certain genes may be carried out in herbaceous model species such as *Arabidopsis* (http://www.nsf.gov/pubs/2002/bio0202/model.htm). Yet, there are many genes that are unique in their function or expression pattern that are of necessity, or are most appropriately, studied in the species of interest or a closely allied species. This is especially true of woody perennial fruit tree species, including *Prunus* spp., in which gene function and regulation in tree growth and fruit development are poorly understood. Research in Rosaceae genomics is generating a large body of gene sequence data and the analysis of gene function is critical for the application of this information to the genetic improvement of rosaceous species.

We developed two ihp constructs using a peach phytoene desaturase (*PDS)* gene (Fig. 3a, b), to show that the plum system provides a rapid, high throughput system for analyzing the function of endogenous *Prunus* genes (Petri et al., 2008). Silencing of the function of the *PDS* gene produces a readily visible photo-bleaching phenotype (albinism) due to the inhibition of chloroplast development. (Qin et al., 2007).

After *Agrobacterium*-mediated plum transformation, fifteen clusters displayed the albino phenotype (Fig. 4a). Gene-silencing efficiencies were 40 and 64% for PpPDS1.1 and PpPDS2.1 constructs, respectively. *PDS* mRNA level was highly reduced (Fig. 4b) in the albino lines and siRNA specific to the *PDS* sequence was detected (Fig. 4c).

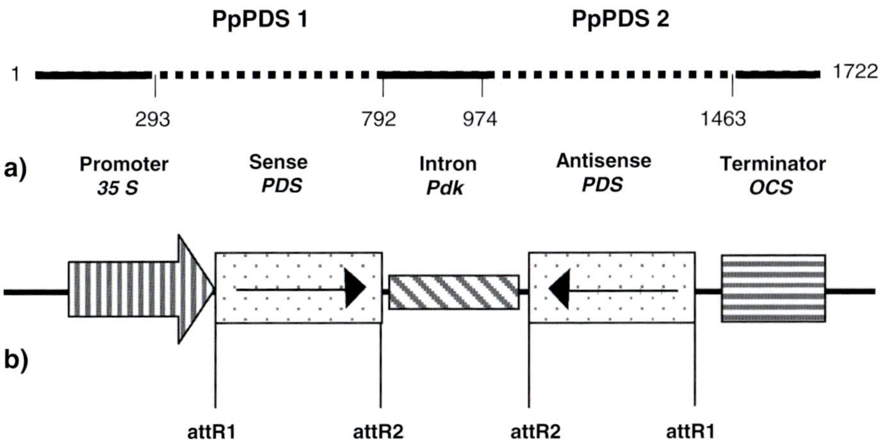


**Fig. 3** (**a**) Schematic of the peach (*Prunus persica*) (Pp) phytoene desaturase (PDS) sequence illustrating the positions of the two fragments – PpPDS1 and PpPDS2 that were used to produce intron hairpin (ihp) constructs. (**b**) Schematic of a PDS ihp contruct that was used to transform tobacco (*N. tabacum*) and plum (*P. domestica*) (Figure from Petri et al. (2008))

By demonstrating the functionality of peach *PDS* gene ihp constructs in plum, we have established that the plum system is amenable for the study of gene function in *Prunus* and could be used for other rosaceous species since a high level of synteny has been reported (Arús et al., 2006; Jung et al., 2006). The transformation

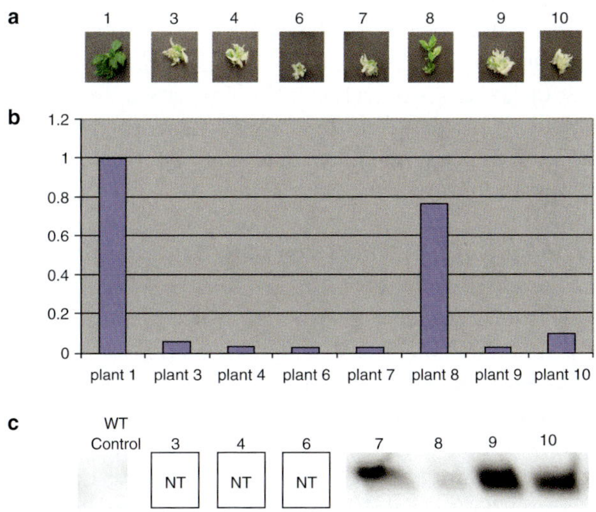


**Fig. 4** Silencing of plum phytoene desaturase (PDS) gene expression by peach-based PDS ihp constructs PpPDS1.1 and PpPDS2.1. (**a**) Plant 1 control; plants 3, 4, 6 7, 8, 9, 10 independents PDS silenced transgenic plum plants showing variable albino phenotype. (**b**) PDS mRNA accumulation. All transgenic lines showed lower PDS mRNA level than the control. (**c**) Detection of PDS siRNA in lines 7, 8, 9 and 10. Lines 3, 4 and 6 were not analyzed for PDS siRNA. NT = non tested (Figure from Petri et al. (2008))

efficiency and time are superior to those for many other tree species (Petri and Burgos, 2005) and comparable to those published for other model woody plants, such as poplar (*Populus*) (Cseke et al., 2007; Song et al., 2006). Moreover, transformation with genes that promote early flowering in other tree species, such as *LEAFY, APETALA 1* (Peña et al., 2001), *FT1* (Böhlenius et al., 2006) or *FT2* (Hsu et al., 2006), could reduce the time to flowering in the plum system, enabling rapid evaluation of flower and fruit specific gene functions.

Recently, a transformation protocol in Japanese plum (*Prunus salicina*) has been published (Urtubia et al., 2008). Transformation of *P. salicina* would allow for improvement of this species through the development of GE varieties and additionally, transformation of this diploid species would provide a platform for functional genomics studies that would be useful for all *Prunus* and Rosaceous species (Urtubia et al., 2008).

## 3.2 Regeneration of Marker-Free Transgenic Plums

In a transformation procedure only a few cells stably incorporate the transgene(s). Marker genes allow for the recognition and selection of transformed cells (Miki and McHugh, 2004). These marker genes are introduced with the gene of interest and are used for identifying the rare individuals that have taken up foreign DNA. Most plant transformation systems have relied on dominant selectable markers, such as antibiotic and herbicide resistance genes, to enable the recovery of transgenic plants. Unfortunately, their presence is also often problematic for commercial biotechnology products because of consumer concerns and regulatory requirements over the presence of ‘excess’ exogenous DNA. In Europe, release of plants carrying certain antibiotic resistance genes has been particularly problematic (Directive 2001/18/EEC of the European Parliament and the Council of the European Union). Therefore, development of procedures to avoid the use of antibiotic selection or to allow elimination of marker genes from the transformed plant should be a research priority in the coming years (Petri and Burgos, 2005).

Over the past decade, researchers have developed several approaches for excising marker DNA from transgenic plants and crops (See review Darbani et al., 2007). Only multi-autonomous-transformation (MAT)-vector system (Ebinuma et al., 1997) have been applied to fruit tree transformation procedures and some results have been published in apricot (López-Noguera et al., 2006, 2007) and citrus (Ballester et al., 2007, 2008).

The high-throughput plum transformation system that has been developed in our laboratory (Petri et al., 2008), represents an alternative system to obtain marker-free transgenic plum plants under non-selective conditions. Highly efficient transformation systems have been used to develop markerless transformation platforms in citrus species (Ballester et al., 2008; Dominguez et al., 2002; Ghorbel et al., 1999), with transformation efficiencies between 3 and 7% for the different species.

### 3.3 Cultivar Transformation

There are just few reports of regeneration of transformed plum shoots from clonal explants of *P. domestica*. Moreover, transformation efficiencies are low and the varieties used are generally of limited or local importance (Table 2).

In our laboratory a transformation/regeneration protocol is currently being developed from leaves of *P. domestica* 'Improved French'. 'Improved French', also known as 'French Prune', is perhaps the most economically important *Prunus domestica* variety in the world. It is particularly suited to drying and accounts for most of the world's commercial production of dried plums (or 'prunes'). 'Improved French' is particularly important in California where it accounts for 99% of the dried plum production, which reached 180,000 tons in 2006 with a production value of US $255,684,000 (USDA, 2006), consequently the potential impact of improving this cultivar by genetic engineering is significant.

Several factors affecting adventitious regeneration (including shoot proliferation medium, light, gelling agent, growth regulators, and ethylene inhibitors) were studied. Regeneration rate from 'Improved French' leaves reached up to ~50% (Petri and Scorza, submitted).

Some preliminary transformation experiments have been carried out with three different disarmed *Agrobacterium tumefaciens* strains (GV3101, EHA101 and LBA4404) harboring the pGA482GGi plasmid. This plasmid is a modification of the plasmid pGA482GG (An et al., 1985) that includes changes in the multiple cloning site and the addition of an intron in the *gus* gene for plant-specific expression. *Gus* assays were performed 4 weeks after *A. tumefaciens* infection in order to determinate stable transformation. *Gus* positive spots/zones were found in 90.5% of leaf explants, with an average of 6.8 *gus* spots/zones per transformed explant, opening the possibility for genetic engineering of the most important dried plum cultivar in the USA.

## 4 Conclusions

Substantial progress has been made in *Prunus* functional genomics particularly in peach where physical maps, sequenced EST libraries, and microarray platforms have been developed (reviewed by Shulaev et al., 2008). The peach genome sequence is also now underway and will lead to the development of even more sophisticated functional genomics tools. However, in the absence of an efficient transformation system, progress in determining gene function will remain slow. As long as peach transformation remains difficult and inefficient, the transgenic plum system offers a solution to bottlenecks in *Prunus* research. Whether the focus of study is peach, almond, apricot, or cherry, the plum system should serve as a useful model for assessing the functions of individual genes. By combining transformation with short generation cycles through the development of early flowering plum lines, it will be possible to further improve plum as a functional genomics model system for trees. As the plum system becomes further refined, the pace of functional discovery for key agronomic genes in *Prunus* will rapidly accelerate.

## References


An G, Watson BD, Stachel S, Gordon MP, Nester EW (1985) New cloning vehicles for transformation of higher plants. EMBO J 4: 277–284

Arús P, Yamamoto T, Dirlewanger E, Abbott AG (2006) Synteny in the Rosaceae. In: J Janick (ed.) Plant Breeding Reviews 27 (pp.175–211). John Wiley & Sons, NJ (USA)

Ballester A, Cervera M, Peña L (2007) Efficient production of transgenic citrus plants using isopentenyl transferase positive selection and removal of the marker gene by site-specific recombination. Plant Cell Rep 26:39–45

Ballester A, Cervera M, Peña L (2008) Evaluation of selection strategies alternative to *nptII* in genetic transformation of citrus. Plant Cell Rep (on line) Doi 10.1007/s00299-008-0523-z

Bassi G, Cossio F (1991) In vitro shoot regeneration on "Bluefre" and "Susina di Dro" prune cultivars (*Prunus domestica* L.). Acta Hort 289:81–82

Beachy RN, Loesh-Fries S, Tumer NE (1990) Coat protein-mediated resistance against virus infection. Ann Rev Phytopath 28: 451–474

Böhlenius H, Huang T, Charbonnel-Campaa L, Brunner AM, Jansson S, Strauss SH, Nilsson O (2006) *CO/FT* regulatory module controls timing of flowering and seasonal growth cessation in trees. Science 312:1040–1043

Callahan A, Scorza R (2007) Effects of a peach antisense ACC oxidase gene on plum fruit quality. In: RE Litz and R Scorza (eds.) Proceedings of the international symposium on biotechnology of temperate fruit crops and tropical species. Acta Hortic 738:567–573

Cox K, Layne D, Scorza R, Schnabel G (2006) Gastrodia anti-fungal protein from the orchid *Gastrodia elata* confers disease resistance to root pathogens in transgenic tobacco. Planta 224:1373–1383

Csányi M, Wittner A, Nagy A, Balla I, Vértessy J, Palkovics L, Balázs E (1999) Tissue culture of stone fruit plants basis for their genetic engineering. J Plant Biotechnol 1:91–95

Cseke LJ, Cseke SB, Podila GK (2007) High efficiency poplar transformation. Plant Cell Rep 26:1529–1538

Darbani B, Eimanifar A, Stewart CN Jr., Camargo WN (2007) Methods to produce marker-free transgenic plants. Biotechnol J 2:83–90

Dominguez A, Fagoaga C, Navarro L, Moreno P, Peña L (2002) Regeneration of transgenic citrus plants under non selective conditions results in high-frequency recovery of plants with silenced transgenes. Mol Genet Genomics 267:544–556

Ebinuma H, Sugita K, Matsubaga E, Yamakado M, Komamine A (1997) Principle of MAT vector. Plant Biotech 14:133–139

Escalettes V, Dahuron F, Ravelonandro M, Dosba F (1994) Utilisation de la transgénose pour l'obtention de pruniers et d'abricotiers exprimant le gène de la protéine capside du plum pox potyvirus. Bull OEPP/EPPO 24:705–711

Escalettes V, Dosba F (1993) In vitro adventitious shoot regeneration from leaves of *Prunus* spp. Plant Sci 90:201–209

Ghorbel R, Juarez J, Navarro L, Peña L (1999) Green fluorescent protein as a screenable marker to increase the efficiency of generating transgenic woody fruit plants. Theor Appl Genet 99: 350–358

Gonsalves D (1998) Control of papaya ringspot virus in papaya – A case study. Ann Rev Phytopath 36:415–437

Gonzalez-Padilla IM, Webb K, Scorza R (2003) Early antibiotic selection and efficient rooting and acclimatization improve the production of transgenic plum plants (*Prunus domestica* L.). Plant Cell Rep 22:38–45

Hily J-M, Scorza R, Malinowski T, Zawadzka B, Ravelonandro M (2004) Stability of gene silencing-based resistance to *Plum pox virus* in transgenic plum (*Prunus domestica* L.) under field conditions. Transgenic Res 13: 427–436

Hily J-M, Scorza R, Webb K, Ravelonandro M (2005) Accumulation of the long class of siRNA is associated eith resistance to *Plum pox virus* in a transgenic woody perennial plum tree. MPMI 18: 794–799

Hily J-M, Ravelonandro M, Damsteegt V, Bassett C, Petri C, Liu Z, Scorza R (2007) *Plum pox virus* coat protein gene intron-hairpin-RNA (ihpRNA) constructs provide resistance to *plum pox virus* in *Nicotiana benthamiana* Domin. and plum (*Prunus domestica* L.). J Am Soc Hortic Sci 132(6):850–858

Hsu CY, Liu Y, Luthe DS, Yuceer C (2006) Poplar *FT2* shortens the juvenile phase and promotes seasonal flowering. Plant Cell 18(8):1846–1861

Hu Z, Yang Z, Wang J (1988) Isolation and partial characterization of an antifungal protein from *Gastrodia elata* corm. Acta Botanica Yunnanica 10:373–380

Jung S, Main D, Staton M, Cho I, Zhebentyayeva T, Arus P, Abbott A (2006) Synteny conservation between the *Prunus* genome and both the present and ancestral *Arabidopsis* genomes. BMC Genomics 7:81

López-Noguera S, Petri C, Burgos L (2006) Production of marker-free transgenic plants after transformation of apricot cultivars. Acta Horticult 717:225–227

López-Noguera S, Petri C, Burgos L (2007) Using MAT vector system to produce marker-free transformed apricot plants. In: Litz R.E. and Scorza R. (eds.) Proceedings of the international symposium on biotechnology of temperate fruit crops and tropical species. Acta Horticult 738:607–612

Malinowski T, Cambra M, Capote N, Zawadzka B, Gorris MT, Scorza R, Ravelonandro M (2006) Field trials of plum clones transformed with the *Plum pox virus* coat protein (PPV-CP) gene. Plant Dis 90:1012–1018

Mante S, Morgens PH, Scorza R, Cordts JM and Callahan AM (1991) *Agrobacterium*-mediated transformation of plum (*Prunus domestica* L.) hypocotyl slices and regeneration of transgenic plants. Bio/Technol 9:853–857

Mante S, Scorza R, Cordts JM (1989) Plant regeneration from cotyledons of *Prunus persica*, *Prunus domestica*, and *Prunus cerasus*. Plant Cell Tiss Org Cult 19:1–11

Mikhailov RV, Dolgov SV (2007) Transgenic plum (*Prunus domestica* L.) plants obtained by *Agrobacterium*-mediated transformation of leaf explants with use of various selective agents. In: Litz R.E. and Scorza R. (eds.) Proceedings of the international symposium on biotechnology of temperate fruit crops and tropical species. Acta Hort 738:613–623

Mikhailov RV, Muratova SA, Dolgov SV (2007) Production of transgenic plum plants from vegetative tissues by means of positive selection. Acta Hort (ISHS) 734:129–138

Miki B, and McHugh S (2004) Selectable marker genes in transgenic plants: applications, alternatives and biosafety. J Biotechnol 107: 193–232

Nagel AK, Scorza R, Petri C, Schnabel G (2008) Generation and characterization of transgenic plum lines expressing the *Gastrodia*-anti fungal protein. HortScience 43(5):1514–1521

Nowak B, Miczynski K, Hudy L (2004) Sugar uptake and utilisation during adventitious bud differentiation on *in vitro* leaf explants of 'Wegierka Zwykla' plum (*Prunus domestica*). Plant Cell Tiss Org 76:255–260

Peña L, Martín-Trillo M, Juárez JA, Pina JA, Navarro L, Martínez-Zapater JM (2001) Constitutive expression of *Arabidopsis LEAFY* or *APETALA1* genes in citrus reduces their generation time. Nat Biotechnol 19:263–267

Petri C, Burgos L (2005) Transformation of fruit trees. Useful breeding tool or continued future prospect? Transgenic Res 14:15–26

Petri C, Scorza R (2008) Peach. In C Kole and T C Hall (eds.). "Compendium of Transgenic Crop Plants: Transgenic Temperate Fruits and Nuts", Blackwell Publishing, Oxford, UK. volume 4. pp 79–92.

Petri C, Webb K, Hily J-M, Dardick C, Scorza R (2008) High transformation efficiency in plum (*Prunus domestica* L.): a new tool for functional genomics studies in *Prunus* spp. Mol Breed 22(4):581–591

Qin G, Gu H, Ma L, Peng Y, Deng XW, Chen Z, Qu L-J (2007) Disruption of phytoene desaturase gene results in albino and dwarf phenotypes in *Arabidopsis* by impairing chlorophyll, carotenoid, and gibberellin biosynthesis. Cell Res 17:471–482

Ravelonandro M, Monsion M, Teycheney PY, Delbos R, Dunez J (1992) Construction of a chimeric viral gene expressing *plum pox virus* coat protein. Gene 120: 167–173

Ravelonandro M, Scorza R, Bachelier JC, Labonne G, Levy L, Damsteegt V, Callahan AM, Dunez J (1997) Resistance of transgenic *Prunus domestica* to plum pox virus infection. Plant Dis 81:1231–1235

Scorza R, Callahan A, Levy L, Damsteegt V, Webb K, Ravelonandro M (2001) Post-transcriptional gene silencing in plum pox virus resistant transgenic European plum containing the plum pox potyvirus coat protein gene. Transgenic Res 10:201–209

Scorza R, Hily J-M, Callahan AM, Malinowski T, Cambra M, Capote N, Zagrai I, Damsteegt V, Briard P and Ravelonandro M (2007) Deregulation of Plum Pox resistant transgenic plum 'HoneySweet'. In: RE Litz and R Scorza (eds.) Proceedings of the international symposium on biotechnology of temperate fruit crops and tropical species. Acta Hortic 738:669–673

Scorza R, Levy L, Damsteegt V, Yepes LM, Cordts J, Hadidi A, Slightom J, Gonsalves D (1995) Transformation of plum with the papaya ringspot virus coat protein gene and reaction of transgenic plants to plum pox virus. J Am Soc Hortic Sci 120: 943–952

Scorza R, Ravelonandro M, Callahan AM, Cordts JM, Fuchs M, Dunez J, Gonsalves D (1994) Transgenic plums (*Prunus domestica* L) express the plum pox virus coat protein gene. Plant Cell Rep 14:18–22

Shulaev V, Korban SS, Sosinski B, Abbott AG, Aldwinckle HS, Folta KM, Iezzoni A, Main D, Arús P, Dandekar AM, Lewers K, Brown SK, Davis TM, Gardiner SE, Potter D, Veilleux RE. (2008) Multiple models for Rosaceae genomics. Plant Phys. May 16, 2008; 10.1104/pp.107.115618

Song J, Lu S, Chen Z-Z, Lourenco R, Chiang VL (2006) Genetic transformation of *Populus trichocarpa* genotype Nisqually-1: A functional genomic tool for woody plants. Plant Cell Physiol 47(11):1582–1589

United States Department of Agriculture (2006) California prune (dried plum) acreage report. USDA, NASS, California Field Office, P.O. Box 1259, Sacramento, CA 95812

Urtubia C, Devia J, Castro A, Zamora P, Aguirre C, Tapia E, Barba P, Dell'Orto P, Moynihan MR, Petri C, Scorza R and Prieto H (2008) Agrobacterium-mediated genetic transformation of Prunus salicina. Plant Cell Reports. Doi 10.1007/s00299-008-0559-0

Yancheva SD, Druart P, Watillon B (2002) *Agrobacterium*-mediated transformation of plum (*Prunus domestica* L.). Acta Horticulturae 577: 215–217

# 14. Sweet and Sour Cherries: Linkage Maps, QTL Detection and Marker Assisted Selection

**Elisabeth Dirlewanger, Jacques Claverie, Amy F. Iezzoni, and Ana Wünsch**

## 1 Introduction

The cherry is one of the most popular temperate fruit crops despite its relatively high price. The fruits are attractive in appearance because of their bright shiny skin color, their subtle flavor and sweetness are appreciated by most consumers. Compared to other temperate fruits, such as apple and peach, breeding improvements for cherries have been slow. The long generation time and the large plant size of cherry trees severely limit classical breeding. Thus, the integration of molecular markers in breeding programs should be a powerful tool to hasten cultivar development. Only a few genetic linkage maps are available for sweet or sour cherry and quantitative trait loci (QTLs) have been reported only for sour cherry. Until now, most of the efforts have concentrated on the use of molecular markers in order to (i) identify the *S*-alleles controlling gametophytic self-incompatibility, (ii) characterize cultivars, and (iii) assess genetic diversity.

### *1.1 Brief history of the Crop*

*Prunus avium* L. includes sweet cherry trees cultivated for human consumption and wild cherry trees used for their wood, also called mazzards (Webster, 1996). The sweet cherry is indigenous to parts of Asia, especially northern Iran, Ukraine, and countries south of the Caucasus mountains. In Europe, the Romanian and Georgian wild cherry trees appeared to have significantly differentiated from those of central and western Europe (Tavaud, 2002). The Georgian wild cherry trees are the most genetically diverse, suggesting that this area could have been a main glacial refuge. The ancestors of the modern cultivated sweet cherries are believed to have originated around the Caspian and Black Seas, from where they have slowly spread. This

E. Dirlewanger (✉)
Institut National de la Recherche Agronomique, Centre de Bordeaux, Unité de Recherches sur les Espèces Fruitières et la Vigne, BP 81, 33883, Villenave d'Ornon Cedex. France

K.M. Folta, S.E. Gardiner (eds.), *Genetics and Genomics of Rosaceae*, Plant Genetics and Genomics: Crops and Models 6, DOI 10.1007/978-0-387-77491-6_14,

radiation was driven initially by birds. Sweet cherries are now cultivated commercially in more than 40 countries around the world, in temperate, Mediterranean, and even subtropical regions. Its natural range covers the temperate regions of Europe, from the North part of Spain to the Southeastern part of Russia (Hedrick, 1915). They prefer regions with warm and dry summers, but require adequate rainfall or irrigation during the growing season for production of fruit with appropriate size for marketing. Rainfall at harvest time may reduce the commercial potential of the production by inducing fruit cracking.

Fruit of *Prunus cerasus* L., the sour cherry tree, are mainly used for processed products such as pie filling, jam or liquor. Sour cherry originated from an area very similar to that of sweet cherry, around the Caspian Sea and close to Istanbul. While sour cherry is less widely cultivated than sweet cherry, large quantities of sour cherries are produced in many European countries and in the USA. Most of these are used in processing and processed cherry products are sold worldwide.

*Prunus fruticosa* Pall., the ground cherry tree, is sometimes used as rootstocks for other *Prunus* species. This species is widespread over the major part of central Europe, Siberia and Northern Asia (Hedrick, 1915).

The duke cherries, which result from crosses between *P. avium* and *P. cerasus*, are cultivated on a much smaller scale. Different names have been given to these interspecific hybrids are such as *Prunus acida* Dum, *Cerasus regalis*, *Prunus avium* ssp *regalis*, but the name used today is *P.* × *gondouinii* Rehd. (Faust and Suranyi, 1997; Saunier and Claverie, 2001). Duke cherry trees are intermediate for their tree and fruit characteristics compared to their progenitors.

### 1.1.1 Botanical Descriptions

All cherry species belong to the *Cerasus* subgenus of the *Prunus* genus, part of the Rosaceae family. The majority of cultivated cherry trees belong to *Prunus avium* L. and *Prunus cerasus* L. species. Together with *Prunus fruticosa* Pall., these species and their interspecific hybrids constitute the *Eucerasus* section of the *Cerasus* subgenus, based on morphological criteria (Krussmann, 1978; Rehder, 1947). This classification and the monophyletic origin of the *Eucerasus* clade have been confirmed by chloroplast DNA variation analysis (Badenes and Parfitt, 1995).

A large amount of morphological variation is observed among *P. avium*, *P. fruticosa* and *P. cerasus* species. Multivariate analysis on sour cherry revealed continuous variation between the *P. avium* and *P. fruticosa* traits throughout the geographic distribution of the species. In Western Europe, *P. cerasus* trees more closely resemble *P. avium* whereas in Eastern Europe, *P. cerasus* is closer to *P. fruticosa* (Hillig and Iezzoni, 1988; Krahl et al., 1991). This continuum of morphological characteristics makes species assignment difficult when considering only phenotypic traits. The sweet cherry is a deciduous tree of large stature, occasionally reaching almost 20 meters in height, with attractive peeling bark. The sour cherry is a small tree, or more often a deciduous bush, which suckers profusely from the base. It has smaller leaves and flowers than the sweet cherry. Sweet cherries are usually split into three groups on the basis of fruit characters: 1. Mazzards, often wild types with small inferior fruits of various shapes and colors, 2. Guignes, Hearts or Geans, with soft-fleshed fruit, and 3. the Bigarreaux with hard-fleshed, heart-shaped, light-colored fruit. Sour cherry cultivars are generally classified as Amarelles (or Kentishand) or as Griottes (or Morellos). Amarelles have pale red fruits flattened at the ends and uncolored juice. Griottes have, in contrast, dark spherical fruits and dark-colored juice. A third group of sour cherry cultivars, called Marasca, are characterized by small, very black-red colored and bitter fruit whose juice is of the best quality for making maraschino liquor. Marasca cultivars are sufficiently distinct to have been classified by early botanists as a subspecies of *P. cerasus* (*Prunus cerasus* Marásca (Reichb.) Schneid, Rehder, 1947).

## *1.2 Genome Content*

*Prunus avium* has a diploid genome (AA, 2n=2x=16) and small haploid genome size (338 Mb) (Arumuganathan and Earle, 1991), bigger than the genome of peach (290 Mb) which is the smallest *Prunus* genome evaluated to date. *Prunus fruticosa*, the ground cherry tree, is a tetraploid wild species (2n=4x=32) believed to be (FFFF). The genome size is still unknown.

*Prunus cerasus* is an allotetraploid species (AAFF, 2n=4x=32), with a genome size of 599 Mb, allegedly due to natural hybridization between *P. avium* (producing unreduced gametes) and *P. fruticosa* (Fig. 1). This origin was first suggested by Olden and Nybom (1968) who observed that artificial hybrids between tetraploid *P. avium* and *P. fruticosa* were very similar to *P. cerasus*. Isozyme analysis, genomic *in situ* hybridization and karyotype analysis further confirmed the hybrid origin of *P. cerasus* (Hancock and Iezzoni, 1988; Santi and Lemoine, 1990; Schuster and Schreiber, 2000). The patterns of inheritance of 7 isozymes in different crosses of sour cherry indicated that *P. cerasus* may be a segmental allopolyploid (Beaver and Iezzoni, 1993; Beaver et al., 1995). Studies based on cpDNA markers detected two distinct chlorotypes in *P. cerasus* which strongly suggest that crosses between *P. avium* and *P. fruticosa* have occurred at least twice to produce sour cherry (Badenes and Parfitt, 1995; Brettin et al., 2000; Iezzoni and Hancock, 1996). Moreover, these works showed that most of the time, *P. fruticosa* was the female progenitor of *P. cerasus,* but in few cases, *P. avium* was the female parent due to the formation of unreduced ovules. Tavaud et al. (2004) demonstrated that specific alleles in *P. cerasus* were not present in the A genome of *P. avium* and probably came from the F genome of *P. cerasus*. Recent analysis with cpDNA and microsatellite markers show that some *P. cerasus* share the same chloroplastic haplotype as some *P. fructicosa*, and that some microsatellite markers are share by both species (A. Horvath, personal communication). Triploid hybrids through the fusion of normal gametes of *P. avium* and *P. fruticosa* occur naturally but remain sterile. Due to this sterility and many unfavorable *P. fruticosa* traits, these triploids are not clonally propagated by humans (Olden and Nybom, 1968).

*P.* × *gondouinii* Rehd is an allotetraploid (AAAF, 2n=4x=32) species stemming from the pollinization of sour cherry by unreduced gametes of sweet cherry (Iezzoni et al., 1990). These hybrids are often sterile, due to disturbances during meiosis, but they are clonally propagated by human.

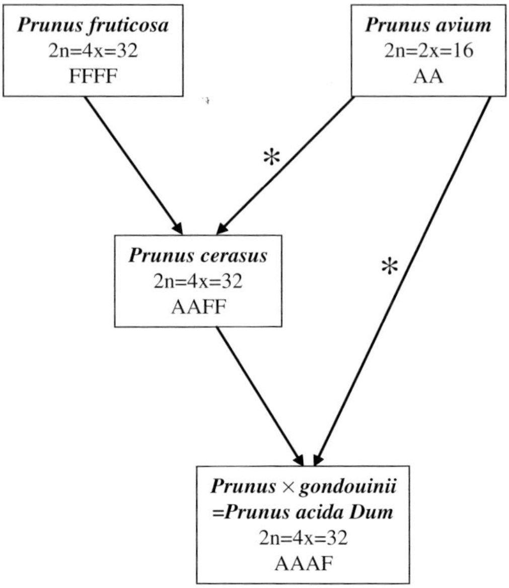


**Fig. 1** Relationships and genome constitution among the species of the Eucerasus section

## *1.3 Economic Importance*

Worldwide, 375,000 Ha of sweet cherry and 248,000 Ha of sour cherry are cultivated giving a total production of 1,896,000 Mt and 1,035,000 Mt respectively (FAO, 2005). The main production areas in the world for sweet and sour cherries are located in Europe (953,000 Mt and 711,000 Mt), Asia (653,000 Mt and 208,000 Mt) and North America (228,000 Mt for sweet cherry and 115,000 Mt for sour cherry) (FAO, 2005). However, a huge increase in sweet cherry production occurred 10 years ago in the Southern hemisphere, especially in Chile and Argentina. In Chile, the cultivated area increased by four times in two years and nearly all the production is exported to the USA and Europe. In the Northern hemisphere, sweet cherry production is mainly located in Europe but major shifts are occurring in European production. France was one of the main producers in Europe (100–120,000 Tonnes) but halved its production in 2003 and 2004 (57,000 Tonnes), and at the same time Spain doubled its production, especially with early maturing varieties. In the next following years, Turkey may become the leading world producer of sweet cherries.

## *1.4 Breeding Objectives*

The main breeding objectives for sweet cherry are:

- large, attractive and good-flavored fruits,
- reduced juvenile phase,

- large and constant yields,
- reduced susceptibility to fruit cracking,
- self-compatibility,
- improved resistance or tolerance to diseases, especially bacterial canker induced by *Pseudomonas* mors pv. *prunorum* and *P. syringae*.

Regular yields and superior fruit quality are the two main objectives of sour cherry breeding programs. Breeding for disease resistance in sour cherry is concentrated on resistance to cherry leaf spot caused by *Blumeriella japii.* When not properly controlled, CLS can cause leaf chlorosis and premature defoliation resulting in fruit that is poorly colored, low in soluble solids and softer than fruit on healthy trees (Keitt et al., 1937). Early defoliation can also result in reduced winter hardiness, potentially leading to flower bud loss and tree death (Howell and Stackhouse, 1973).

Yields per hectare vary by the country of production, the commercial use (for fresh market or for industry) and the training system. The average yield ranges from 8 to 10 T/Ha in classical orchards but can reach 30–40 T/ha for an intensive industrial orchard. The highest limitation to the development of cherry culture is the high cost required to manually pick the fruit, as manual picking may account for 70% of the production price. This has led to the selection in some breeding programs of new varieties that can be harvested partially with machines, such as 'Sweetheart' and 'Van' that can be harvested without the stem. At the same time, a better knowledge of the architecture of the tree has led to new approaches to orchard training.

Because of the efforts of classical breeding programs, a large number of cultivars are now available. Within the last 10 years, 20 new varieties have gained wide interest internationally such as 'Earlise' (early season), 'Summit' (middle season) and 'Sweetheart' (late season). Each of these should be widely cultivated in the next 15–20 years.

Classical breeding programs are time consuming because cherry trees take a minimum of 3–5 years of growth before they are capable of flowering and fruit production. Prior knowledge of linkage relationships between marker loci and important flower and fruit characteristics will facilitate and shorten the selection of promising individuals. Consequently, introduction of marker-assisted selection will be especially beneficial for sweet and sour cherry breeding.

## 2 Construction of Genetic Maps

The construction of genetic maps is useful for localisation of important genes controlling both qualitative and quantitative traits in numerous plant species and, then, for improving and shortening breeding selection (Tanksley et al., 1989). In the subgenus *Cerasus*, several maps have been published using five segregating populations. Until recently, only partial maps for sweet or sour cherry were available. The earliest of them was constructed in a sweet cherry using random amplified polymorphic DNA (RAPD) and allozyme analysis of 56 microspore-derived callus culture individuals of the cv. 'Emperor Francis' (Stockinger et al., 1996). Two allozymes and 89 RAPD markers were mapped to 10 linkage groups totalling 503 cM. Interestingly, another map integrating isozyme genes exclusively, was obtained using data from two inter-specific $F_1$ cherry progenies: *P. avium* 'Emperor Francis' × *P. incisa* E621 and *P. avium* 'Emperor Francis' × *P. nipponica* F1292 (Bošković and Tobutt, 1998). This map, one of the most exhaustive ever made with isozyme markers in plants, included a total of 47 segregating isozymes, of which 34 were aligned into seven linkage groups. The East Malling group has continued this research with the construction of an inter-specific cherry map from the cross *P. avium* ' Napoleon' × *P. nipponica* using microsatellite and gene-specific markers (Clark et al., 2009).

Another genetic linkage map is in progress for sweet cherry using an intra-specific $F_1$ progeny including 133 individuals from a cross between cultivars 'Regina' and 'Lapins' in INRA at Bordeaux (France). These cultivars were chosen as parents for their distinct agronomic characters and especially because they differ for resistance to fruit cracking which is a limiting factor in sweet cherry production ('Regina' is resistant and 'Lapins' is susceptible.) 'Lapins' is a self-compatible cultivar as opposed to 'Regina'. Moreover, they differ for several other characters: blooming and maturity dates, peduncle length, and fruit color, weight, firmness, titratable acidity and refractive index. Preliminary maps of each parent and their comparison with the reference *Prunus* map 'Texas' × 'Earlygold' (T×E) is described in Dirlewanger et al. (2004b). These maps include microsatellite markers, 30 of which are located in the 'Régina' map are anchors marker with T×E map, 28 located in the 'Lapins' map as anchor markers with the T×E map. Only one non-collinear marker was detected, but for all other markers the location was in the homologous linkage group. These results are in agreement with the high level of synteny within the *Prunus* genus (Arús et al. 2006).

An intra-specific sweet cherry genetic linkage map was also constructed at Michigan State University (US) from a $F_1$ progeny from a cross between a wild forest cherry with small (~2 g) highly acid dark-red colored fruit (NY54) and a domesticated variety with large (~6 g), yellow/ pink, sub-acid fruit 'Emperor Francis' (EF) (Olmstead et al., 2007, 2008 ). The 'EF' and 'NY' maps were 711.1 cM and 565.8 cM, respectively, with the average distance between markers of 4.94 and 6.22 cM (Fig. 2). A total of 82 shared markers between the 'EF' and 'NY' maps and the *Prunus* reference map supported previous findings that the cherry genome is collinear with other *Prunus* genomes. The $F_1$ population is composed of approximately 600 individuals, including 190 that were used for map construction and initial QTL analysis. The remaining progeny will be used for fine mapping of major QTLs. The objective of the study is to identify QTLs that control the fruit quality traits improved during domestication. In addition, this cross is fully compatible and progeny segregation for the *S*-locus fits the expected 1:1:1:1 ratio (Ikeda et al., 2005).

In sour cherry, linkage maps were constructed at Michigan State University (US) from 86 individuals from the cross of two cultivars, 'Rheinische Schattenmorelle' (RS) and 'Erdi Botermo' (EB). Since sour cherry is a tetraploid, informative restriction fragment length polymorphisms (RFLPs) were scored as single-dose restriction

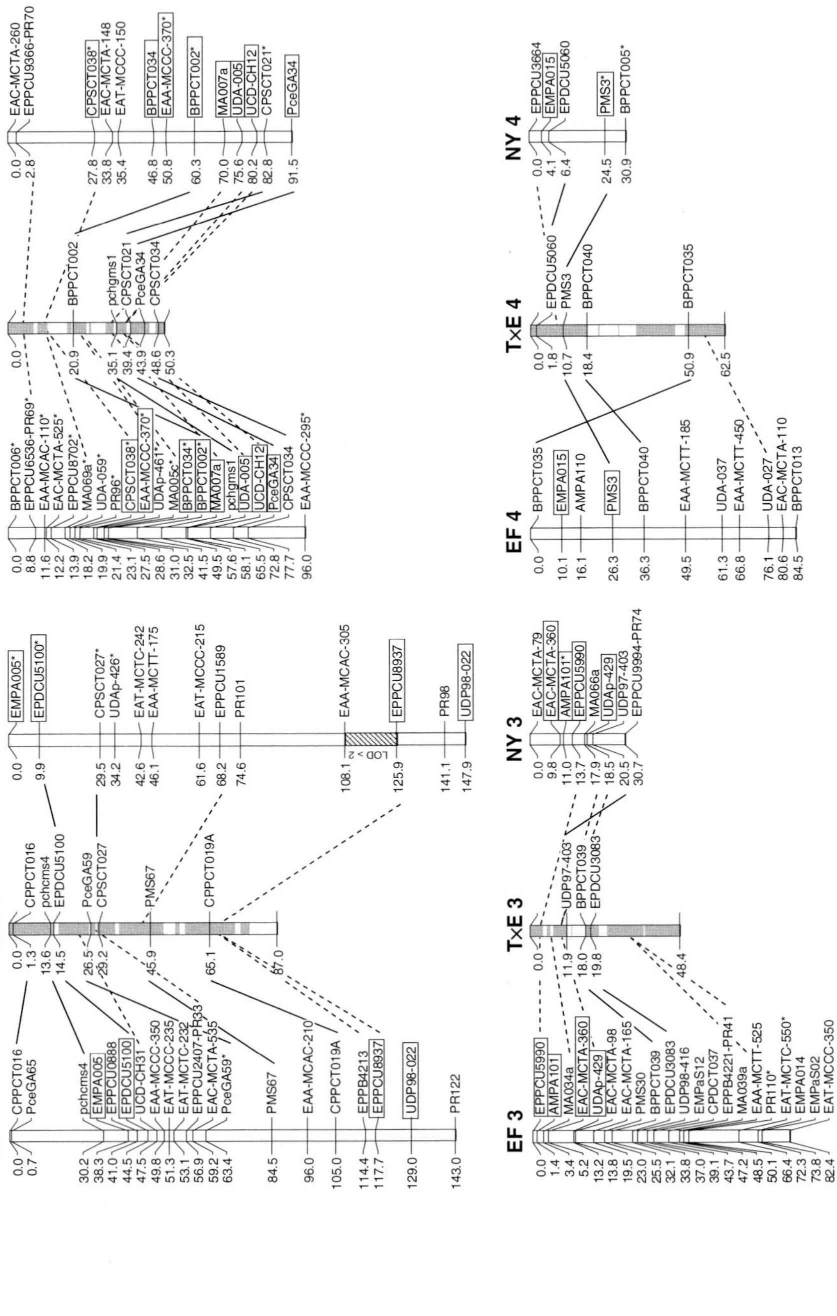


**Fig. 2** (continued)

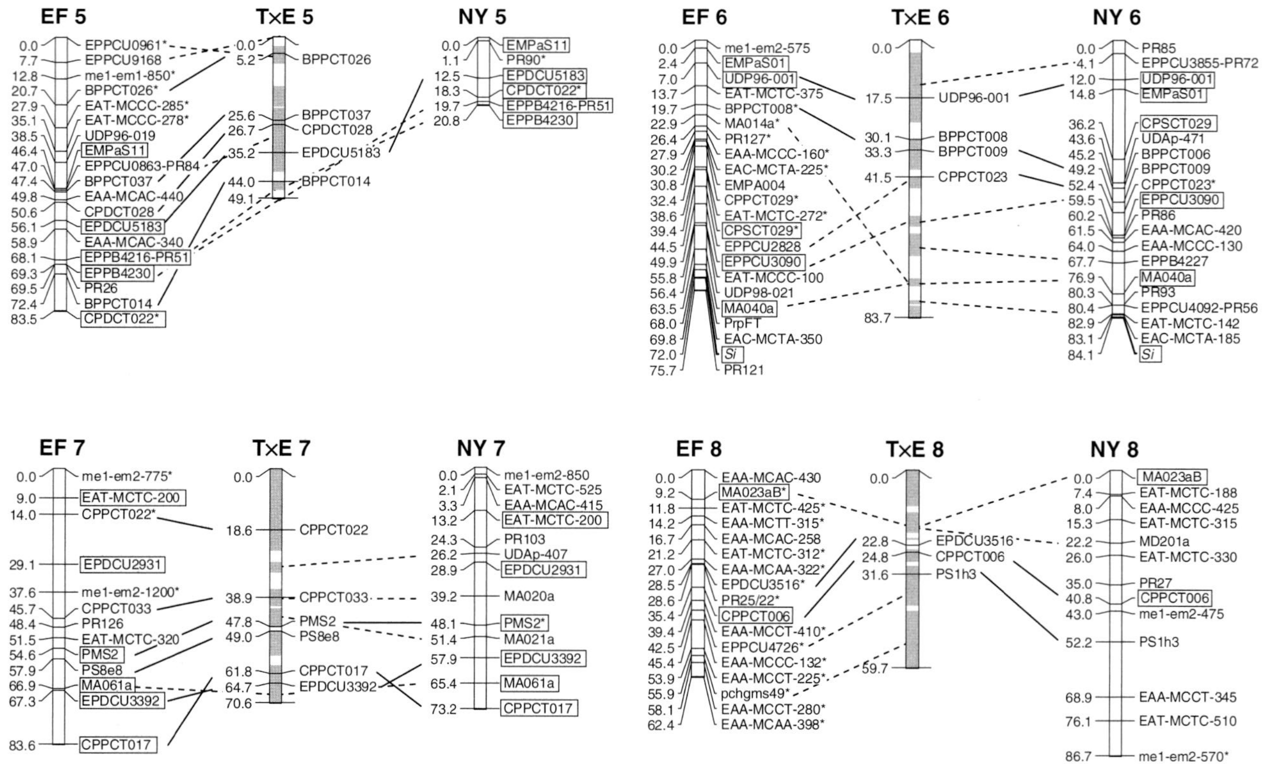


**Fig. 2** (continued) Alignment of 'Emperor Francis' (EF) and 'New York 54' (NY) sweet cherry parental maps with the *Prunus* reference and bin map ['Texas' almond × 'Earlygold' peach (T×E)] (Olmstead et al., 2008). Only common markers present on EF and NY maps are presented on the *Prunus* reference map. Shaded areas on the reference map linkage groups indicate bin locations. Solid lines indicate homology between mapped markers; dashed lines indicate markers present in *Prunus* bins. Boxed markers indicate anchor points between the EF and NY parental maps. Markers followed by an asterisk (*) indicate significant deviation ($P<0.05$) from the expected chi-square segregation ratio

fragments (SDRF) according to Wu et al. (1992). A genetic linkage map was constructed for RS that consists of 126 SDRF markers assigned to 19 linkage groups covering 461 cM (Wang et al., 1998). The EB linkage map had 95 SDRF markers assigned to 16 linkage groups covering 279 cM (Wang et al., 1998). Due to the limited number of shared markers between the RS × EB map compared to other *Prunus* maps, putative homologous linkage groups could only be identified in for the *Prunus* LGs 2, 4, 6, and 7. The other linkage groups were arbitrarily numbered from the longest to shortest and therefore the sour cherry linkage groups numbers have not been rigorously aligned with that of the *Prunus* consensus map. The RS × EB population was subsequently screened using 10 *Prunus* microsatellite primer pairs (Canli, 2004a) and a consensus map of 442 cM, less than the previously reported RS map of 461 cM, was constructed. A total of 16 microsatellite markers were added to 10 of the 19 linkage groups; however, the linkage groups were not re-numbered to reflect these markers. In addition, four of the microsatellite primer pairs identified duplicate linked markers. This 'double mapping' of a marker is due to the inclusion of progeny individuals exhibiting tetrasomic inheritance for that linkage group. If this correction had been done by Canli (2004a), it is likely that the number of microsatellite markers added to the map would be reduced to twelve.

The difficulty of identifying SDRFs and eliminating progeny that resulted from non-homologous pairing for the linkage group under study, illustrate the complexity of linkage mapping in a segmental allopolyploid. Hence, future work at Michigan State University will concentrate on linkage map construction in the diploid sweet cherry.

## 3 Gene Mapping and QTLs Detected

In sour or sweet cherries most of the agronomically important traits have complex inheritance. Only self-incompatibility (SI) in diploid sweet cherry is controlled by a single locus (*S*) with multiple alleles, and fertilization only takes place when the *S* allele in the haploid genome of the pollen is different from the two *S* alleles in the diploid tissue of the style. In contrast, blooming and ripening time, flower bud and pistil death and characters controlling fruit quality are quantitative traits. The self-incompatibility locus is located in the distal part of linkage group 6 in almond (Ballester et al., 1998; Bliss et al., 2002), apricot (Vilanova et al., 2003), and cherry (Olmstead et al., 2008).

Although in peach many major genes (Fig. 3) and QTLs involved in fruit quality (Dirlewanger et al., 1999; Etienne et al., 2002; Quilot et al., 2004) and diseases resistance (Quarta et al., 1998; Viruel et al., 1998; Foulongne et al., 2003) have been reported, the only QTL study published to date in cherry is a QTL analysis of flower and fruit traits using the sour cherry RS × EB linkage mapping population (Wang et al., 1998). Eleven QTLs (LOD>2.4) were identified for six traits (bloom time, ripening time, percent pistil death, percent pollen germination, fruit weight, and soluble solids concentration) (Wang et al., 2000, Fig 4). The percentage of phenotypic

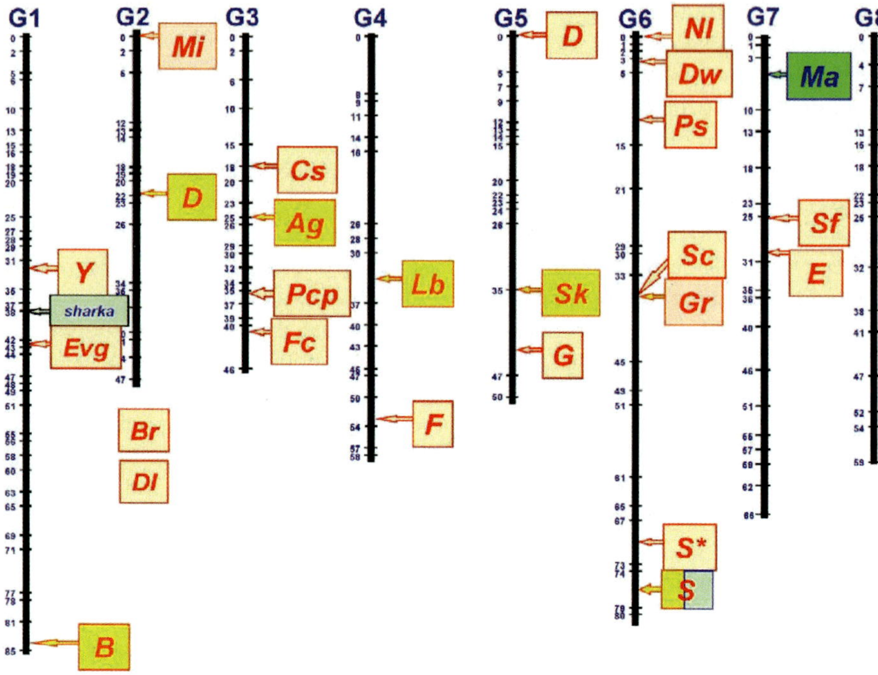


**Fig. 3** Approximate position of 28 major genes mapped in different populations of apricot (*blue background*), peach (*orange background*), almond or almond × peach (*yellow background*), and Myrobalan plum (green background) on the framework of the *Prunus* reference map (Dirlewanger et al., 2004b). Gene abbreviations correspond to: *Y*, peach flesh color; *B*, almond/peach petal color; *sharka*, plum pox virus resistance; *B*, flower color in almond × peach; *Mi*, nematode resistance from peach; *D*, almond shell hardness; *Br*, broomy plant habit; *Dl*, double flower; *Cs*, flesh color around the stone; *Ag*, anther color; *Pcp*, polycarpel; *Fc*, flower color; *Lb*, blooming date; *F*, flesh adherence to stone; *D*, non-acid fruit in peach, *Sk*, bitter kernel; *G*, fruit skin pubescence; *Nl*, leaf shape; *Dw*, dwarf plant; *Ps*, male sterility; *Sc*, fruit skin color; *Gr*, leaf color; *S**, fruit shape; *S*, self-incompatibility (almond and apricot); *Ma*, nematode resistance from Myrobalan plum; *E*, leaf gland shape; *Sf*, resistance to powdery mildew. Genes *Dl* and *Br* are located on an unknown position of G2

variation explained by a single QTL ranged from 12.9 to 25.9% (Wang et al., 2000). Subsequently, three microsatellite markers were identified that mapped within the putative location of the previously described QTLs (Wang et al., 2000) for bloom time (*blm2*), pistil death (*pd1*) and fruit weight (*fw2*), respectively (Canli, 2004a). Unfortunately these three microsatellite markers were not used in QTL analyses to determine their location relative to the previously published QTLs.

The identification of bloom time QTL is of particular interest for cherry breeding as the development of new cultivars with late bloom would significantly reduce the probability of spring freeze damage to the pistils (Iezzoni, 1996). Sour cherry exhibits extreme diversity for bloom time with many cultivars blooming exceedingly late in the spring (Iezzoni and Mulinix, 1992). This late bloom character in sour

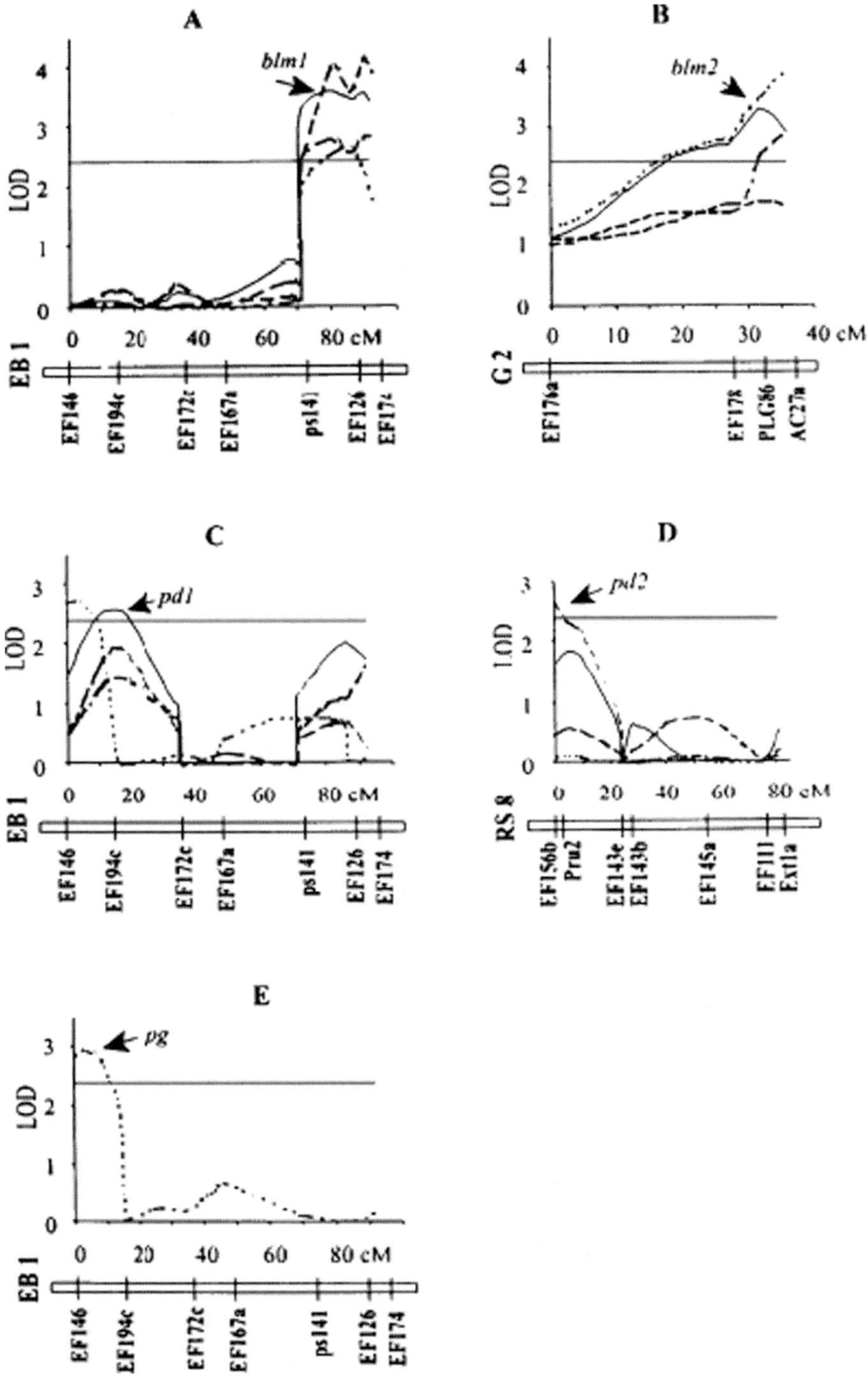


**Fig. 4** (continued)

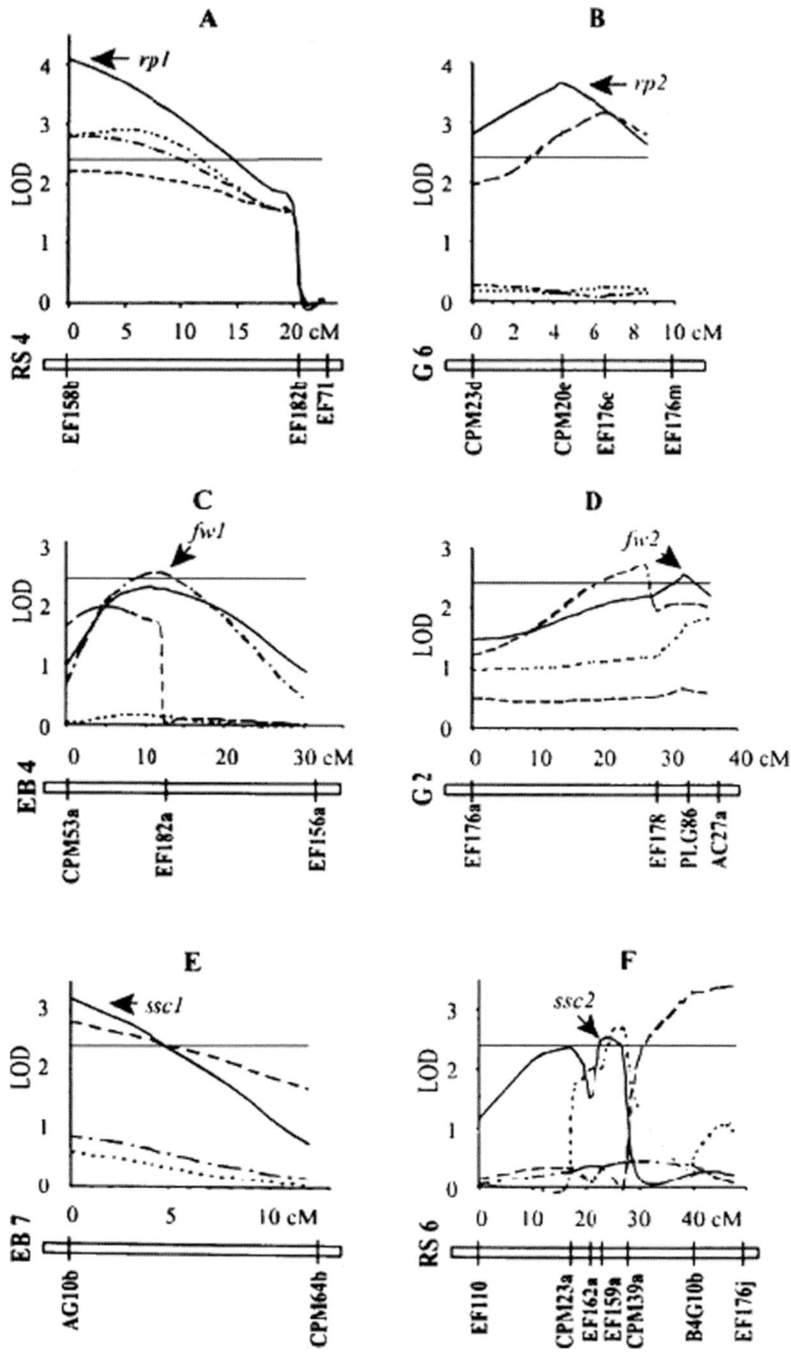


**Fig. 4** (continued)

cherry is likely due to the hybridization and continued introgression with the very late blooming ground cherry, *P. fruticosa*.

Bloom time in cherry is a quantitative trait; however its high broad sense heritability (0.91) led to the identification of two bloom time QTL, *blm*1 and *blm*2, in the 'RS' × 'EB' population (Wang et al., 2000). Unfortunately the genetic effects of these two QTL alleles from 'EB' were to induce early bloom. To identify QTL with alleles conferring late bloom time, a second mapping population between the mid-season blooming 'Ujfehertoi Furtos' and late blooming 'Surefire' has been developed at Michigan State University (US). The population exhibited transgressive segregation for bloom time permitting a bulk segregant approach to identify markers linked to bloom time QTL (Bond, 2004). To date, a third QTL for late bloom, named *blm*3, was identified using AFLP markers that is significantly associated with late bloom using an ANOVA. This QTL allele is present in 'Surefire' and confers late bloom time. Ongoing work attempts to determine the linkage map location of this QTL. Using this same mapping population, two AFLP markers were identified that differed between the early and late bulks (Canli, 2004b). However these markers were never screened over the 'Ujfehertoi Furtos' × 'Surefire' progeny population and the marker results described could not be repeated.

## 4 Analysis of Self Incompatibility

Sweet cherry, like other Rosaceae species, exhibits a strict self-incompatibility system that has been naturally selected to promote outbreeding (De Nettancourt, 2001). This mechanism disallows the fertilization of flowers of one genotype by its own pollen. As a consequence, commercial fruit set in this species depends upon the presence of other compatible pollinating genotypes or on the utilization of self-compatible cultivars. In sour cherry, self-incompatible as well as self-compatible genotypes have been identified (Lansari and Iezzoni, 1990; Yamane et al., 2001; Hauck et al., 2002). Sour cherry is a tetraploid hybrid of diploid sweet cherry and tetraploid ground cherry and the self-incompatibility mechanism seems to be conserved only in some genotypes.

**Fig. 4** (continued) (*Below*) QTLs detected for flower and fruit traits in sour cherry (Wang et al., 2000). LOD scores for bloom date on linkage groups EB 1 (*blm*1) (**A**) and Group 2 (*blm*2) (**B**); pistil death (*pd*) on linkage groups EB 1 (**C**) and RS 8 (**D**); pollen germination percentage (*pg*) on linkage group EB 1 (**E**). Peak LOD scores for each trait are indicated by *arrows*. Linkage groups are shown below the *x-axes*. The *horizontal line* indicates the level of significance at LOD=2.4. *Curves* represent individual years of 1995 (–––), 1996 (- - - -), 1997 (— - —) and over years (———). LOD scores for ripening date on linkage groups RS 4 (*rp1*) (**A**) and Group 6 (*rp2*) (**B**); fruit weight on linkage groups EB 4 (*fw1*) (**C**) and Group 2 (*fw2*) (**D**); soluble solids concentration on linkage groups EB 7 (*ssc1*) (**E**) and RS 6 (*ssc2*) (**F**)

The type of self-incompatibility operating in the Rosaceae is called gametophytic self-incompatibility (GSI) (De Nettancourt, 2001), and it is shared by other plant families like the Solanaceae and Plantaginaceae. The gametophytic self-incompatibility is controlled by different genes of one polymorphic locus (*S*) that determine the incompatibility response of the pollen and the style (McCubbin and Kao, 2000). In cherries the incompatibility phenotype of the style is determined by a ribonuclease called *S*-RNase (Bošković and Tobutt, 1996; Tao et al., 1999c; Yamane et al., 2001) and the specificity of the pollen is determined by the product of the F-box gene *SFB* (Yamane et al., 2003; Ushijima et al., 2004; Ikeda et al., 2004a). Together the RNAse and SFB protein would interact in an allele specific manner to confer the self-incompatibility reaction. The mechanism of this reaction is such that the growth of the pollen tube is inhibited in the style when the *S*-allele of the pollen factor matches either of the two *S*-alleles of the *S*-RNases expressed in the diploid style tissue. Several models have been proposed to explain how these factors mediate the incompatibility reaction of the *S*-RNase-based self-incompatibility (Luu et al., 2001, Kao and Tsukamoto, 2004; Ushijima et al., 2004; Goldraij et al. 2006; McClure 2006; Hua et al. 2008).

Like sweet cherry, sour cherry exhibits an *S*-RNase based GSI system (Yamane et al., 2001; Hauck et al., 2002; Tobutt et al., 2004; Bošković et al., 2006); however, natural sour cherry selections include both self-incompatible (SI) and self-compatible (SC) types (Redalen, 1984; Lansari and Iezzoni 1990). This genotype-dependent loss of self-incompatibility in sour cherry indicates that genetic changes, not polyploidy per se, cause the breakdown of SI. Instead the genetic control of SI and SC in sour cherry has been shown to be regulated by the accumulation of non-functional *S*-haplotypes according to the 'one-allele-match model' (Hauck et al., 2006b). In this model, the match between a functional pollen-*S* gene produced by the 2*x* pollen and its cognate functional *S*-RNase in the style results in an incompatible reaction. A similar reaction occurs regardless of whether the pollen contained a single functional pollen-*S* gene or two different pollen-*S* genes. The absense of a functional match results in a compatible reaction. Thus for successful fertilization, 2*x* sour cherry pollen must contain two non-functional *S*-haplotypes.

The progress made in the knowledge of the genetic and molecular basis of the self-incompatibility reaction has allowed the application of molecular techniques for two main aspects of sweet cherry breeding, the identification of cross-compatible combinations of different varieties by the identification their S-alleles and the selection of self-compatibility.

## *4.1 S-Allele Typing*

Self-incompatibility in sweet cherry prevents inbreeding but the same mechanism also prevents cross-pollination among varieties with the same *S* alleles. This means that it is necessary to know the *S* haplotypes of each variety to be able to establish which cultivar combinations are compatible and, thus, to select which varieties can be inter-planted. Varieties that have the same incompatibility alleles and are therefore cross-incompatible, form incompatibility groups. Until the molecular basis of self-incompatibility was characterized, *S* allele typing and incompatibility group assignment was carried out by controlled pollinations followed by recording fruit set (Crane and Brown, 1937; Matthews and Dow, 1969) or by the observance of pollen tube growth in the style by fluorescent microscopy. Since the style *S* factor in GSI was known to be a ribonuclease in Solanaceae (McClure et al., 1989), it was possible to identify *S* alleles in sweet cherry by correlating known *S* alleles with bands obtained from stylar proteins separated by isoelectric focusing and stained for ribonuclease activity (Bošković and Tobutt, 1996). Subsequently this biochemical assay would provide evidence that correlated well with the new incompatibility alleles ( Bošković et al., 1997).

The cloning and sequence characterization of the *S-RNases* of sweet cherry (Tao et al., 1999a, b) allowed the development of PCR and RFLP based methods of typing cherry *S*-alleles. Tao et al. (1999c) developed an *S*-allele typing method based in the utilization of two pairs of PCR primers, designed in the conserved regions of the sweet cherry *S-RNase* sequences. These *S-RNase* sequences have two introns varying in length for each different allele and, consequently, PCR amplification with those primers enables differentiation of the different *S*-alleles according to the size of the amplified fragments. Subsequently, other sweet cherry *S*-RNases were cloned and other PCR methods based in conserved sequence primers (Wiersma et al., 2001), allele specific primers (Sonneveld et al., 2001, 2003, 2006) or PCR followed by restriction fragment analysis (Yamane et al., 2000b) have been developed. RFLP profiles have also been used to assign self-incompatibility alleles to different sweet cherry genotypes (Hauck et al., 2001). The identification of the pollen-*S* (SFB) in sweet cherry (Yamane et al., 2003), has also been followed by the cloning and characterization of different cherry *SFB* alleles (Ikeda et al., 2004a; Vaughan et al., 2006; Yamane et al., 2003). The knowledge of the sequence and structure of these alleles has allowed the development of new *S*-allele PCR typing methods based in allele specific primer sets (Ikeda et al., 2005), in sequence conserved primers that distinguish *SFB* alleles by size polymorphisms (Vaughan et al. 2006), and dot-blot analysis using *SFB* sequence polymorphism (Kitashiba et al. 2008). The introduction of molecular methods in sweet cherry *S*-allele typing has allowed a rapid confirmation of the *S*-alleles and incompatibility groups of different cultivars reported previously, the identification of the *S*-genotype of new varieties and the identification of putative new *S* alleles by their correlation with new PCR products (Table 2, Tao et al., 1999c; Yamane et al., 2000a, b; Hauck et al., 2001; Sonneveld et al. 2001; Wiersma et al., 2001; Choi et al., 2002; Zhou et al., 2002; Sonneveld et al., 2003; Wunsch and Hormaza, 2004d; De Cuyper et al., 2005). S-allele typing has also become a useful tool for genetic studies of germplasm collections (Wünsch and Hormaza, 2004c; Marchese et al., 2007a; Schuster et al., 2007; Gisbert et al., 2008) and wild cherry populations (De Cuyper et al., 2005; Schueler et al. 2006).

To date, 31 functional *S*-haplotypes have been characterized in cherry, and due to overlapping studies and the use of different techniques, synonymous alleles have been subsequently detected and in some cases the number labeling does not follow a chronological order. These *S*-alleles are numbered $S_1 - S_7$, $S_9 - S_{10}$, $S_{12} - S_{14}$, $S_{16}$, as $S_8$, $S_{11}$, and $S_{15}$, appear to be synonyms of $S_3$, $S_7$ and $S_5$, respectively (Sonneveld et al., 2001, 2003). Three additional alleles, $S_{23}$, $S_{24}$ and $S_{25}$, were later characterized from Italian and Spanish cultivars (Wünsch and Hormaza, 2004a). Of these; $S_{23}$ seems to be synonymous to $S_{14}$ (Sonneveld et al. 2003; Vaughan et al., 2008). In wild sweet cherry populations six additional alleles, $S_{17}$ to $S_{22}$ were characterized (De Cuyper et al., 2005), and according to Vaughan et al. (2008) $S_{21}$ seems to be synonymous of $S_{25}$ (Wünsch and Hormaza, 2004a). Allele $S_{26}$ was reported in sour cherry (Hauck et al., 2006b), $S_{27} - S_{32}$ were described in wild sweet cherry (Vaughan et al., 2008), and finally, $S_{33}$ to $S_{36}$ were described in sour cherry (Tsukamoto et al., 2008).

## 4.2 Breeding for Self-Compatibility

The use of self-compatible varieties in sweet cherry orchards can limit some of the problems incurred from self-incompatibility, such as the cost derived from the need to use pollinator varieties and losses from erratic production (Tehrani and Brown, 1992). As a consequence, obtaining and introducing self-compatible varieties has been one of the main objectives of sweet cherry breeding (Brown et al., 1996). Self-compatibility was induced in sweet cherry by irradiation, giving rise to several self-compatible seedlings (Lewis, 1949). 'Stella' (Lapins, 1970), a descendent of one of these seedlings (JI2420), is self-compatible and has been widely used as a progenitor in self-compatible sweet cherry breeding. Most of the self-compatible varieties currently used derive from 'Stella'. Self-compatibility in these genotypes is caused by a pollen function mutation in the $S_4'$ allele ($S_4'$ standing for mutated $S_4$ allele), (Bošković et al., 2000). To carry on selection of self-compatible seedlings derived from these genotypes it is necessary to differentiate the genotypes that inherited the $S_4'$ allele. However, since the $S_4$-RNase in these genotypes is intact, it was not possible to differentiate genotypes with the $S_4'$ mutant allele from genotypes with a functional $S_4$ allele, using *S*-allele typing methods based on *S*-RNase sequence allele diversity. It was not until the finding of the pollen determinant (SFB) of GSI in *Prunus* (Yamane et al., 2003: Ushijima et al. 2004) that has been possible to establish a method that allows the identification of genotypes carrying the mutated $S_4'$ allele (Ikeda et al., 2004b). This method is based in the identification of a 4 bp deletion in the *SFB* sequence of the $S_4'$ allele when compared with the normal $S_4$ allele. This deletion has been used to design molecular markers that identify the $S_4'$ allele by PCR followed by polyacrylamide gel electrophoresis or restriction digestion (Ikeda et al. 2004b).

Additional sources of self compatibility, that can broaden the genetic base of cultivated germplasm and that can also be highly useful to understand the mechanism of GSI, are also being studied. Sonneveld et al. (2005) carried out molecular and genetic analysis of the two self-compatible accessions obtained by the radiation of pollen at the John Innes Institute, JI 2420 and JI 2434 (Lewis and Crowe,

1954). As determined by Ushijima et al. (2004), a 4 bp deletion was identified in *S4'-SFB* of JI 2420. On the other side, *S3'-SFB* (*S3'* standing for mutated *S3* haplotype) of accession JI 2434 appeared to be deleted (Sonneveld et al. 2005). Self-compatible progeny derived from JI 2434 can now be selected by detecting this *SFB* deletion through RFLP or PCR analysis of $S_3$-*SFB* (Sonneveld et al. 2005). Self-compatibility in the Spanish landrace 'Cristobalina' is also being investigated to identify markers that facilitate the introgression of this trait, as analysis in this genotype have shown that self-compatibility is not associated with the *S*-locus (Wünsch and Hormaza, 2004b). On the other hand, self-compatibility in the Sicilian sweet cherry 'Kronio' has been attributed to a pollen part mutation in $S_5$-*SFB* (thus called $S_{5'}$) caused by a premature stop codon that results in a truncated protein (Marchese et al., 2007b). The presence of a polymorphic microsatellite in the *S-RNase* intron of $S_5$ and $S_5'$ has allowed developing a marker to identify self-compatible genotypes carrying $S_{5'}$ (Marchese et al., 2007b).

Sour cherry selections that have two non-functional *S*-haplotypes are SC (Hauck et al., 2006b). These non-functional *S*-haplotypes can results from the loss of pollen function (termed pollen-part mutants) or loss of stylar function (termed stylar-part mutants), or both (Tsukamoto et al., 2006). Three of the *S*-haplotypes prevalent in sweet cherry ($S_1$, $S_6$ and $S_{13}$) have been shown to also have non-functional variants in sour cherry that have lost pollen or stylar function (Hauck et al., 2006a: Tsukamoto et al., 2006). Loss of function was due to structural alternations of the *S-RNase*, *SFB* or *S-RNase* upstream sequences.

## 5 Conclusion and Future Scope of Work

### *5.1 Genome Mapping and QTL Detection*

Genetic mapping and QTL detection will continue, especially in sweet cherry. Since sweet cherry is diploid, it is much easier to develop linkage maps when compared with sour cherry which is tetraploid with an in-complete disomic inheritance, and occasional intergenomic pairing and pre-or post zygotic selection. Because of the high level of synteny demonstrated within *Prunus*, results obtained in sweet cherry will be useful for sour cherry. For the same reason we can expect that cherry will take benefit of knowledge developed in other members of the Rosaceae family. The enormous progress made during the last decade on genetic characterization of the cultivated species of the Rosaceae, and particularly of peach as its more logical model, can be exploited for cherry.

### *5.2 Self-(in)Compatibility: Molecular Cloning and MAS*

Progress in the understanding of the RNase-based self-incompatibility, has allowed the development of molecular methods that accelerate two relevant aspects of cherry breeding: Incompatibility Group assignment through *S*-allele genotyping, and introgression of self-compatibility through marker assisted selection. At the same time, research in sour and sweet cherry self-incompatibility and self-compatible mutants is greatly contributing to the knowledge of the mechanism operating in the self-incompatibility reaction in the genus *Prunus*. A better understanding of the self-incompatibility reaction from future progress in Rosaceae GSI research, together with the increasing availability of genetic tools in cherry species will provide an appropriate ground for a more efficient cherry improvement.

## References


Arumuganathan K, and Earle ED (1991) Nuclear DNA Content of some important plant species. Plant Mol Biol Rep 9: 208–219.

Arús P, Yamamoto T, Dirlewanger E, and Abbott AG (2006) Synteny in the Rosaceae. *In:* Plant Breeding Reviews Vol 27, Wiley.

Badenes ML, and Parfitt DE (1995) Phylogenetic relationships of cultivated *Prunus* species from an analysis of chloroplast DNA variation. Theor Appl Genet 90: 1035–1041.

Ballester J, Boskovic R, Batlle I, Arús P, Vargas F, and de Vicente MC (1998) Localisation of the self-incompatibily gene on the almond linkage map. Plant Breed 116: 69–72.

Beaver JA, and Iezzoni AF (1993) Allozyme inheritance in tetraploid sour cherry (*Prunus cerasus* L.). J Am Soc Hortic Sci 118: 873–877.

Beaver JA, Iezzoni AF, and Ramn C (1995) Isozyme diversity in sour, sweet and ground cherry. Theor Appl Genet 90: 847–852.

Bliss FA, Arulsekar S, Foolad MR, Becerra V, Gillen AM, Warburton ML, Dandekar AM, Kocsisne GM, and Mydin KK (2002) An expanded genetic map of *Prunus* based on an interspecific cross between almond and peach. Genome 45:520–529.

Bond AM (2004) Bulk segregant analysis for bloom time QTL in sour cherry (Prunus cerasus L.) MS Thesis, Mich. State Univ. 54 pp.

Bošković R, and Tobutt KR (1996) Correlation of stylar ribonuclease zymograms with incompatibility alleles in sweet cherry. Euphytica 90: 245–250.

Bošković R, and Tobutt KR (1998) Inheritance and linkage relationships of isoenzymes in two interspecific cherry progenies. Euphytica 103:273–286.

Bošković R, Russell K, and Tobutt KR (1997) Inheritance of stylar ribonucleases in cherry progenies, and reassignment of incompatibility alleles to two incompatibility groups. Euphytica 95: 221–228.

Bošković R, Tobutt KR, Schmidt H, and Sonneveld T (2000) Re-examination of (in)compatibility genotypes of two John Innes self-compatible sweet cherry selections. Theor Appl Genet 101: 234–240.

Bošković R, Wolfram B, Tobutt KR, Cerovic R, and Sonneveld T (2006) Inheritance and interactions of incompatibility alleles in the tetraploid sour cherry. Theor Appl Genet 112:315–326.

Brettin TS, Karle R, Crowe EL, and Iezzoni AF (2000) Chloroplast inheritance and DNA variation in sweet, sour, and ground cherry. J Hered 91:75–79.

Brown SK, Iezzoni A, and Fogle HW (1996) Cherries. In: Moore JN (ed) Fruit breeding, Vol 1: Tree and tropical fruits. John Wiley & Sons, Inc., pp 213–255.

Canli FA (2004a) Development of a second generation genetic linkage map for sour cherry using SSR markers. Pak J Biol Sci 7:1676–1683.

Canli FA (2004b) A modified-bulk segregant analysis for late blooming in sour cherry. Pak J Biol Sci 7:1684–1688.

Choi C, Tao R, and Andersen RL (2002) Identification of self-incompatibility alleles and pollen incompatibility groups in sweet cherry by PCR based *S*-allele typing and controlled pollination. Euphytica 123: 9–20.

Clark JB, Sargent DJ, Boskovic RI, Belaj A, and Tobutt KR (2009) A cherry map from the interspecific cross *Prunus avium* 'Napoleon' × *P. nipponica* based on microsatellite, gene-specific and isoenzyme markers. Tree Genet Genomes 5: 41–51.

Crane MB, and Brown AG (1937) Incompatibility and sterility in the sweet cherry, *Prunus avium* L. J Pomol Hortic Sci 15: 86–116.

De Cuyper B, T Sonneveld, and KR Tobutt (2005) Determining self-incompatibility genotypes in Belgian wild cherries. Mol Ecol 14: 945–955.

De Nettancourt D (2001) Incompatibility and incongruity in wild and cultivated plants, 2nd edn. Springer, Berlin Heidelberg New York.

Dirlewanger E, Graziano E, Joobeur T, Garriga-Calderé F, Cosson P, Howad W, and Arús P (2004b). Comparative mapping and marker assisted selection in Rosaceae fruit crops. Proc Natl Acad Sci 101: 9891–9896.

Dirlewanger E, Moing A, Rothan C, Svanella L, Pronier V, Guye A, Plomion C, and Monet R (1999) Mapping QTL controlling fruit quality in peach (*Prunus persica* (L) Batsch). Theor Appl Genet 98: 18–31.

Etienne C, Rothan C, Moing A, Plomion C, Bodénès C, Svanella-Dumas L, Cosson P, Pronier V, Monet R, and Dirlewanger E (2002) Candidate genes and QTLs for sugar and organic acid content in peach [*Prunus persica* (L.) Batsch]. Theor Appl Genet 105: 145–159.

FAO (2005) FAOSTAT database 2004. Web site at http://faostat.fao.org

Faust M, and Suranyi D (1997) Origin and dissemination of cherry. Hort Rev 19: 263–317.

Foulongne M, Pascal T, Pfeiffer F, and Kervella J (2003) QTLs for powdery mildew resistance in peach × *Prunus davidiana* crosses: consistency across generations and environments. Mol Breed 12: 33–50.

Gisbert AD, Badenes ML, Tobutt KR, Llacer G and Romero C (2008) Determination of the *S*-allele composition of sweet cherry (*Prunus avium* L.) cultivars groen in the southeast of Spain by PCR analysis. J Hort Sci Biotech 83:246–252.

Goldraij A, Kondo K, Lee CB, Hancock CN, Sivaguru M, Vazquez-Santana S, Kim S, Phillips TE, Cruz-Garcia F and McClure B (2006) Compartmentalization of *S*-RNase and HT-B degradation in self-incompatible *Nicotiana*. Nature 435:805–810.

Hancock AM, and Iezzoni AF (1988) Malate dehydrogenase isozyme patterns in seven *Prunus* species. Hort Sci 23: 381–383.

Hauck NR, Iezzoni AF, Yamane H, and Tao R (2001) Revisiting the S-allele nomenclature in sweet cherry (*Prunus avium*) using RFLP profiles. J Am Soc Hortic Sci 126: 654–660.

Hauck NR, Yamane H, Tao R, and Iezzoni AF (2002) Self-compatibility and incompatibility in tetraploid sour cherry (*Prunus cerasus* L.). Sex Plant Reprod 15: 39–46.

Hauck NR, Ikeda K, Tao R, and Iezzoni AF (2006a) The mutated $S_1$-haplotype in sour cherry has an altered *S*-haplotype specific F-box protein gene. J Hered 97: 514–520.

Hauck NR, Yamane H, Tao R, and Iezzoni AF (2006b) Accumulation of non-functional *S*-haplotypes results in the breakdown of gametophytic self-incompatibility in tetraploid *Prunus*. Genetics 172: 1191–1198.

Hedrick UP (1915) The history of cultivated cherries. In: The cherries of New York., ed Albany, JB Lyon, NY,. pp. 39–64.

Hillig KW, and Iezzoni AF (1988) Multivariate analysis of a sour cherry germplasm collection. J Am Soc Hortic Sci 113: 928–934.

Howell GS, and Stackhouse SS (1973) The effect of defoliation time on acclimation and dehardening in tart cherry (*Prunus cerasus* L.). J Amer Soc Hortic Sci 98: 312–316.

Hua ZH, Fields A and Kao TH (2008) Biochemical Models for S-RNase-Based Self-Incompatibility. Molecular Plant Advance Access doi:10.1093/mp/ssn032.

Iezzoni AF (1996) Sour cherry cultivars: Objectives and methods of fruit breeding and characteristics of principal commercial cultivars: In Cherrries: Crop Physiology, Production and Uses (Eds. Webster AD and Looney NE), University press, Cambridge, UK, 223–241.

Iezzoni AF, and Hancock AM (1996) Chloroplast DNA variation in sour cherry. Acta Hortic 410: 115–120.

Iezzoni AF, and Mulinix CA (1992) Variation in bloom time in a sour cherry germplasm collection. Hort Sci 27: 1113–1114.

Iezzoni AF, Schmidt H, and Albertini A (1990) Cherries (*Prunus*). In: Moore JN, Ballington JR, Jr. (eds) Genetic Resources of Temperate Fruit and Nut Crops, Vol 1. I.S.H.S., Wageningen, The Netherlands, pp 111–173.

Ikeda K, Igic B, Ushijima K, Yamane H, Hauck NR, Nakano R, Sassa H, Iezzoni AF, Kohn JR and Tao R (2004a). Primary structural features of the *S* haplotype-specific F-box protein, SFB, in *Prunus*. Sex Plant Reprod 16: 235–243.

Ikeda K, Watari A, Ushijima K, Yamane H, Hauck NR, Iezzoni AF, and Tao R (2004b) Molecular markers for the self-compatible $S_4'$-haplotype, a pollen-part mutant in sweet cherry (*Prunus avium* L.). J Am Soc Hortic Sci 129: 724–728.

Ikeda K, Ushijima K, Yamane H, Tao R, Hauck N, Sebolt A, and Iezzoni AF (2005) Linkage and physical distances between the *S*-haplotype *S-RNase* and *SFB* genes in sweet cherry. Sex Plant Reprod 17: 289–296.

Kao TH, and Tsukamoto T (2004) The molecular and genetic bases of S-RNase-based self-incompatibility. Plant Cell 16: 572–583.

Keitt GS, Blodgett EC, Wilson EE, and Magie RO (1937) The epidemiology and control of cherry leaf spot. Univ Wisc Agric Exp Stn Res Bull 132.

Kitashiba H, Zhang SL, Wu J, Shirasawa K and Nishio T (2008) *S* genotyping and *S* screening utilizing *SFB* gene polymorphism in Japanese plum and sweet cherry by dot-blot analysis. Mol Breed 21:339–349.

Krahl KH, Lansari A, and Iezzoni AF (1991) Morphological variation within a sour cherry collection. Euphytica 52: 47–55.

Krussmann G (1978) Manual of cultivated broadleaved trees and shrubs. Vol. 3. PRU-Z. B.T. Batsford Ltd London, pp. 18–58.

Lansari A, and Iezzoni A (1990) A Preliminary-Analysis of Self-Incompatibility in Sour Cherry. Hort Sci 25: 1636–1638.

Lapins KO (1970) The Stella cherry. Fruit Varieties Horticultural Digest 24: 19–20.

Lewis D (1949) Structure of the incompatibility gene. II. Induced mutation rate. Heredity 3: 339–355.

Lewis D and Crowe LK (1954) Structure of the incompatibility gene. IV Types of mutation in *Prunus avium* L Heredity 8:357–363.

Luu DT, Qin XK, Laublin G, Yang Q, Morse D, and Cappadocia M (2001) Rejection of S-heteroallelic pollen by a dual-specific S-RNase in *Solanum chacoense* predicts a multimeric SI pollen component. Genetics 159: 329–335.

Marchese AKR, Tobutt KR, Raimondo A, Motisi A, Boskovic RI, Clarke J and Caruso T (2007a) Morphological characteristics, microsatellite fingerprinting and determination of incompatibility genotypes of Sicilian sweet cherry cultivars. J Hort Sci Biotech 82: 41–48.

Marchese A, Boskovic RI, Caruso T, Raimondo A, Cutuli M, and Tobutt KR (2007b) A new self-compatibility haplotype in sweet cherry 'Kronio', $S_{5'}$, attributed to a pollen-part mutation in the *SFB* gene. J Expt Bot 58: 4347–4356.

Matthews P, and Dow KP (1969) Incompatibility groups: sweet cherry (*Prunus avium*) En: Knight RL (eds) Abstract Bibliography of Fruit Breeding and Genetics to 1965: *Prunus*, pp 540–544. Commonwealth Agricultural Bureaux, Farnham Royal.

McClure BA, Haring V, Ebert PR, Anderson MA, Simpson RJ, Sakiyama F, and Clarke AE (1989) Style self-incompatibility gene products of *Nicotiana alata* are ribonucleases. Nature 342: 955–957.

McClure (2006) New views of S-RNase-based self-incompatibility. Curr Opin Plant Biol 9: 639–646.

McCubbin AG, and Kao TH (2000) Molecular recognition and response in pollen and pistil interactions. Ann Rev Cell Dev Biol 16: 333–364.

Olden EJ, and Nybom N (1968) On the origin of *Prunus cerasus* L. Hereditas 70: 3321–3323.

Olmstead JW, Iezzoni AF, and MD Whiting (2007) Genotypic differences in sweet cherry fruit size are primarily a function of cell number. J Amer Soc Hort Sci 132(5): 697–703.

Olmstead JW, Sebolt AM, Cabrera A, Sooriyapathirana SS, Hammar S, Iriarte G, Wang D, Chen CY, van der Knaap E, and Iezzoni AF (2008) Construction of an intra-specific sweet cherry (*Prunus avium* L.) genetic linkage map and synteny analysis with the *Prunus* reference map. Tree Genet Genomes (DOI 10.1007/s11295-008-0161-10).

Quarta R, Dettori MT, Verde I, Gentile A, and Broda Z (1998) Genetic analysis of agronomic traits and genetic linkage mapping in a BC1 peach population using RFLPs and RAPDs. Acta Hortic 465: 51–59.

Quilot B, Wu BH, Kervella J, Génard M, Foulongne M, and Moreau K (2004) QTL analysis of quality traits in an advanced backcross between *Prunus persica* cultivars and the wild related species *P. davidiana*. Theor Appl Genet 109: 884–897.

Redalen G (1984) Fertility in sour cherries. Gartenbauwissenschaft 49:212–217.

Rehder A (1947) Manual of cultivated trees and shrubs, 2nd edn. Macmillan Compagny, New-York pp. 452–481.

Santi F, and Lemoine M (1990) Genetic markers for *Prunus avium* L. 2. Clonal identifications and discrimination from *P. cerasus* and *P. cerasus* × *P. avium*. Annales des Sciences Forestières 47: 219–227.

Saunier R, and Claverie J (2001) Le cerisier : évolution de la culture en France et dans le monde. Point sur les variétés, les porte-greffe. Le fruit belge 490: 50–62.

Schuster M, Flachowsky H, and Kohler D (2007) Determination of self-incompatibility genotypes in sweet cherry (*Prunus avium* L.) accessions and cultivars of the German Fruit Genet Bank and from private collections. Plant Breed 126: 533–540.

Schuster M, and Schreiber H (2000) Genome investigation in sour cherry, *P. cerasus* L. Acta Hort 538: 375–379.

Schueler S, Tusch A, and Scholz F (2006) comparative analysis of within-population genetic structure in wild cherry (*Prunus avium L.*) at the self-incompatibility locus and nuclear microsatellites. Mol Ecol 15:3231–3243.

Sonneveld T, Robbins TP, Boskovic R, and Tobutt KR (2001) Cloning of six cherry self-incompatibility alleles and development of allele-specific PCR detection. Theor Appl Genet 102:1046–1055.

Sonneveld T, Tobutt KR, and Robbins TP (2003) Allele-specific PCR detection of sweet cherry self-incompatibility (*S*) alleles $S_1$ to $S_{16}$ using consensus and allele-specific primers. Theor Appl Genet 107:1059–1070.

Sonneveld T, Tobutt KR, Vaughan SP, and Robbins TP (2005) Loss of pollen-S function in two self-compatible selections of *Prunus avium* is associated with deletion/mutation of an S haplotype-specific F-box gene. Plant Cell 17: 37–51.

Sonneveld T, Robbins TP and Tobutt KR (2006) Improved discrimination of self-incompatibility S-RNase alleles in cherry and high throughput genotyping by automated sizing of first intron polymerase chain reaction products. Plant Breed 125:305–307.

Stockinger EJ, Mulinix CA, Long CM, Brettin TS, and Iezzoni AF (1996) A linkage map of sweet cherry based on RAPD analysis of a microspore-derived callus culture populations. J Hered 87: 214–218.

Tanksley S, Young N, Paterson A, and Bonierbale M (1989) RFLP mapping in plant breeding: new tools for an old science. Biotechnology 7: 257–264.

Tao R, Yamane H, and Akira H (1999a) Cloning of genomic DNA sequences encoding encoding S1-, S3-, S4- and S6-RNases (accession nos. AB031815, AB031816, AB031817, and AB0311818) from sweet cherry (*Prunus avium* L.). Plant Physiol 121: 1057.

Tao R, Yamane H, and Sugiura A (1999b) Cloning and sequences of cDNAs encoding S1- and S4-RNases (accession nos. AB028153 and AB028154) from sweet cherry (*Prunus avium* L.) (PGR99-121). Plant Physiol 120:1207.

Tao R, Yamane H, Sugiura A, Murayama H, Sassa H, and Mori H (1999c) Molecular typing of S-alleles through identification, characterization and cDNA cloning for S-RNases in sweet cherry. J Am Soc Hortic Sci 124: 224–233.

Tavaud M (2002) Diversité génétique du cerisier doux (*Prunus avium* L.) sur son aire de répartition : Comparaison avec ses espèces apparentées (*P. cerasus* et *P.* × *gondouinii*) et son compartiment sauvage. Thèse de l'ENSAM, 98p.

Tavaud M, Zanetto A, David JL, Laigret F, and Dirlewanger E (2004) Genetic relationships between diploid and allotetraploid cherry species (*Prunus avium*, *Prunus* × *gondouinii* and *Prunus cerasus*. Heredity 93: 631–638.

Tehrani G, and Brown SK (1992) Pollen–incompatibility and self-fertility in sweet cherry. Plant Breed Rev 9: 367–388.

Tobutt KR, Bošković R, Cerovic R, Sonneveld T, and Ruzic D (2004) Identification of incompatibility alleles in the tetraploid species sour cherry. Theor Appl Genet 108: 775–785.

Tsukamoto T, Hauck NR, Tao R, Jiang N, and Iezzoni AF (2006) Molecular characterization of three non-functional *S*-hapltypes in sour cherry (*Prunus cerasus*). Plant Mol Biol 62: 371–383.

Tsukamoto T, Potter D, Tao R, Vieira CP, Vieira J, and AF Iezzoni (2008) Genetic and molecular characterization of three novel *S*-haplotypes in sour cherry (*Prunus cerasus* L.). J Expt Bot 59: 3169–3185.

Ushijima K, Yamane H, Watari A, Kakehi E, Ikeda K, Hauck NR, Iezzoni AF, and Tao R (2004) The S haplotype-specific F-box protein gene, SFB, is defective in self-compatible haplotypes of *Prunus avium* and *P-mume*. Plant Journal 39: 573–586.

Vaughan SP, Russell K, Sargent DJ and Tobutt KR (2006) Isolation of *S*-locus F-box alleles in *Prunus avium* and their application in a novel method to determine self-incompatibility geneotype. Theor Appl Genet 112:856–866.

Vaughan SP, Boskovic RI, Gisbert-Climent, Russell K, and Tobutt KR (2008) Characterization of novel S-alleles from cherry (*Prunus avium* L.). Tree Genet Genomes 4:531–541.

Vilanova S, Romero C, Abbott A G, Llacer G, and Badenes ML (2003) An apricot (*Prunus armeniaca* L.) F2 progeny linkage map based on SSR and AFLP markers, mapping plum pox virus resistance and self-incompatibility traits. Theor Appl Genet 107: 239–247.

Viruel MA, Madur D, Dirlewanger E, Pascal T, and Kervella J (1998) Mapping quantitative trait loci controlling peach leaf curl resistance. Acta Hort 465: 79–88.

Wang D, Karle R, Brettin TS, and Iezzoni AF (1998) Genetic linkage map in sour cherry using RFLP markers. Theor Appl Genet 97: 1217–1224.

Wang D, Karle R, and Iezzoni AF (2000) QTL analysis of flower and fruit traits in sour cherry. Theor Appl Genet 100: 535–544.

Webster AD (1996) The taxonomic classification of sweet and sour cherries and a brief history of their cultivation. In: *Cherries: crop physiology, production and uses.* Cab international, Wallingford, pp. 3–23.

Wiersma PA, Wu Z, Zhou L, Hampson C, and Kappel F (2001) Identification of new self-incompatibility alleles in sweet cherry (*Prunus avium* L.) and clarification of incompatibility groups by PCR and sequencing analysis. Theor Appl Genet 102: 700–708.

Wu KK, Burnquist W, Sorrells ME, Tew TL, Moore PH, and Tanksley SD (1992) The detection and estimation of linkage in polyploids using single-dose restriction fragments. Theor Appl Genet 83: 294–300.

Wunsch A, and Hormaza JI (2004a) Cloning and characterization of genomic DNA sequences of four self-incompatibility alleles in sweet cherry (*Prunus avium* L.). Theor Appl Genet 108: 299–305.

Wunsch A, and Hormaza JI (2004b) Genetic and molecular analysis in Cristobalina sweet cherry, a spontaneous self-compatible mutant. Sex Plant Reprod 17: 203–210.

Wunsch A, and Hormaza JI (2004c) Molecular evaluation of genetic diversity and S-allele composition of local Spanish sweet cherry (*Prunus avium* L.) cultivars. Genet Res Crop Evol 51: 635–641.

Wunsch A, and Hormaza JI (2004d) S-allele identification by PCR analysis in sweet cherry cultivars. Plant Breed 123: 327–331.

Yamane H, Tao R, Murayama H, Ishiguro M, Abe Y, Soejima J, and Sugiura A (2000a) Determining S- genotypes of two sweet cherry (*Prunus avium* L.) cultivars, 'Takasago (Rockport Bigarreau)' and 'Hinode (Early Purple)'. J Jap Soc Hortic Sci 69: 29–34.

Yamane H, Tao R, Murayama H, and Sugiura A (2000b) Determining the S-genotypes of several sweet cherry cultivars based on PCR-RFLP analysis. J Hortic Sci Biotechnol 75: 562–567.

Yamane H, Tao R, Sugiura A, Hauck NR, and Iezzoni AF (2001) Identification and characterization of S-RNases in tetraploid sour cherry (*Prunus cerasus*). J Am Soc Hortic Sci 126: 661–667.
Yamane H, Ikeda K, Ushijima K, Sassa H, and Tao R (2003) A pollen-expressed gene for a novel protein with an F-box motif that is very tightly linked to a gene for S-RNase in two species of cherry, *Prunus cerasus* and *P. avium*. Plant Cell Physiol 44: 764–769.
Zhou L, Kappel F, MacDonald R, Hampson C, Bakkeren G, and Wiersma PA (2002) Determination of *S*-genotypes and self-fertility of sweet cherry in Summerland advanced selections. J Am Pomol Soc 56:173–179.

# 15. Genomics-Based Opportunities in Apricot

**Kevin M. Folta and Susan E. Gardiner**

## 1 Introduction

Apricot (*Prunus armeniaca* L.) is in many ways similar to peach. The two crops share common physical traits, such as highly familiar fruit forms, that point to their common center of origin and remarkably collinear genomes. At first glance, it may seem somewhat redundant to include a chapter on apricot adjacent to a substantial discussion of peach. However, while similar to peach, apricot features many important distinctions. Apricot has a discrete cultural history, its dissemination throughout Asia and Europe is unique, and the challenges faced by apricot breeders and growers discriminate it from other stonefruits. Unlike peach most apricot cultivars are not self compatible. Apricot varieties are much more diverse and they are more abundant in arid regions such as the Middle East. These facets justify an independent treatment of apricot in any discussion of rosaceous crop genomics.

### *1.1 Classification*

Apricot belongs to the section *Armeniaca* within the genus *Prunus* L. in the subfamily *Prunoideae*. Section *Armeniaca* comprises five separate species that are all recognized under the broad heading of apricots (Rehder 1940). The common apricot is *Prunus armeniaca*, and this group includes the Japanese apricot (*P. mume* Sieb. and Zucc.), the black apricot (*P. dasycarpa* Ehrh.), the Alpine plum (*P. brigantiaca* Vill.), and the Tibetan apricot (*P. holosericea*). Other variants once thought to be separate species were later classified as *P. armeniaca*, such as the Siberian apricot (*P. armeniaca* var. *sibirica* L.), Manchurian apricot (*P. armeniaca* var. *mandsurica* (Maxim) Koehne) and Ansu apricot (*P. armeniaca* var. *ansu* Komar) (Bailey 1927).

K.M. Folta (✉)
Horticultural Sciences Department and the Graduate Program in Plant Molecular and Cellular Biology, University of Florida, Gainesville, FL 32611, USA

K. M. Folta, S.E. Gardiner (eds.), *Genetics and Genomics of Rosaceae*, Plant Genetics and Genomics: Crops and Models 6, DOI 10.1007/978-0-387-77491-6_15,

## *1.2 Botanical Attributes*

Historically apricots have defied convenient botanical classification, as their extensive range and adaptations leave exceptions to the rules. While Bailey (1916) utilized leaf shape to discern various cultivars, Rehder (1940) scored plums separate from apricots based on ovary pubescence. A recent comprehensive comparison is provided by Ledbetter (2008) and illustrates the inapplicability of these hard botanical descriptors for the categorization of apricots. Kostina (1936) studied the extensive collections of apricot germplasm in Ukranian botanical gardens and grouped plants based on pubescent or glabrous fruit skin, cling or freestone, flesh color, fruit size and seed flavor. New genomic tools may further illuminate the genetic diversity underpinning the variability and perhaps allow a tighter description of relatedness.

Wild apricot trees are generally smaller in stature than other rosaceous tree crops, averaging between 3 and 12 m in height with a 40 cm diameter trunk. The leaves are ovate with a pointed tip and have serrations in the leaf margins. The flowers often emerge in spring before the leaves and feature five white petals, often with a pink cast. Fruit descriptors are presented in Layne et al. (1996). The fruit is a drupe with a pubescent or glaborous skin. The flesh may be sour or sweet, and apricots may be found with freestone or clingstone qualities. Wild fruits are small, between 3 and 20 g. Like peach, the fruits are pubescent and contain a stone (Faust et al. 1998) and they are easily mistaken for a miniature peach. Recent breeding advances have brought a considerable increase in fruit size compared with the wild fruit, the cultivation of which commenced over 3,000 years ago (Hormaza et al. 2007).

## *1.3 Genomic Architecture*

Apricots are diploid ($2n = 16$) with a genome that is intermediate in size ($5.9 \times 10^8$ bp/2*n*) between that of peach and cherry (Arumaganathan and Earle 1991). It shares a high degree of collinearity with other diploid *Prunus* species, as determined by the overlap of important makers on various genetic linkage maps (Dirlewanger et al. 2004; Hurtado et al. 2002). Apricot maintains a moderate degree of heterozygosity (Hurtado et al. 2002). Compared with other *Prunus* species, apricot is more heterozygous than peach, but not as much as cherry, plum or almond (Byrne and Littleton 1989). The degree of heterozygosity varies greatly, depending on breeding locale, which is low for some Spanish cultivars (namely 'Gitanos' and 'Pepito del Rubio'), but moderate for North American cultivars (Sanchez-Perez et al. 2006). While sharing substantial similarity with other *Prunus* crops, apricot remains the least specifically characterized.

## *1.4 Center of Origin and History of Cultivation*

Although thought to have originated in Armenia, at the junction of Eastern Europe and Western Asia (Hormaza et al. 2007), the history of apricot cultivation dates back thousands of years in Asia. By 600 BC apricots were a major fruit staple in China, eaten fresh, dried, salt-cured or smoked (Faust et al. 1998), and central Asia is its most likely origin. European cultivation is only a relatively recent phenomenon, dating back approximately two thousand years. Its native range is difficult to decipher, because apricots were extensively cultivated prior to any historical record keeping, but best estimates suggest that its range includes central Asia and western and northern China. Between three and six centers of origin have been recognized in the literature (Bailey and Hough 1975; Layne et al. 1996; Vavilov 1992). However, it is unlikely that the species truly originated in separate locales. These distinctions likely represent nth-order radiation of the species from its true center of origin, presumed to be in the Tien Shan Mountains. This is commonly referred to as the "Central Asian Center", and is based in the mountains of west-central China. The region known as the "Inner-Asiatic Center" is represented by a region recognized as modern day Afghanistan, Tajikistan, Uzbekistan and northwestern India. The "Asia Minor Center" is described as the third region, represented by Iran and Turkmenistan, as well as the South Caucasus region between the Caspian and Black Seas. This center includes Armenia, the namesake of the species. These classifications agree well with the delineations based on fruit traits presented by Kostina (1969).

Apricots were extensively cultivated in Armenia and eventually engrained in the culture, as evidenced by foods and folklore. Dried apricots were historically a valuable resource for trade. The expansion of apricots into Europe was fostered by Alexander the Great and a Roman general named Lucullus, who brought a series of fruit tree species to the European continent from adjacent Asian territories (reviewed in Faust et al. 1998). Apricots have long been a prized commodity in Persia, and modern day Iran remains one of the world's foremost apricot producers. Today many South Americans refer to the apricot as the "damasco", signifying its cultivation in the Middle East. Apricots were introduced to the New World by English Settlers who brought the crop to the colonies in eastern North America, but the crop never prospered. The species became a productive part of New World cultivation when Spanish missionaries ferried the apricot westward to settlements on the Pacific Ocean where it remains a valuable species to this day (Boriss et al. 2006).

Today, the vast majority of apricot production derives from Asia, Northern Africa and Europe, with Turkey (17% of world production) and Iran (11%) being the largest producers according to a UN FAO survey in 2005. That year, the worldwide area under cultivation was nearly 400,000 ha and the production was 2.6 million MT. Apricots are cultivated throughout temperate climates worldwide, but only thrive in Mediterranean climates where hot summers and cool, wet winters predominate (Boriss et al. 2006). In fact, apricot productivity is principally delineated by climate conditions. Apricots grow well in the Mediterranean as well as regions that experience cold winters. The apricot surpasses peach in coldhardiness, as the tree can survive temperatures as cold as –30°C. The Siberian apricot, *Prunus sibirica*, survives extremes to –50°C and may serve as a genetic source for cold tolerance. Cold-induced dormancy is a prerequisite for flowering and most cultivars benefit from the sustained chilling incurred in Asian and Mediterranean climates. The temperature fluctuations observed over parts of North America make it not as suitable for apricot production and limit its range of cultivation, although some cultivars are succesfully grown. Production is frequently hampered by untimely spring frosts that can affect developing flower buds and limit fruit set. This is exacerbated by the tendency of apricots to flower early relative to other tree fruit crops.

In South Australia, apricots are heavily cultivated near the town of Mypolonga in the Riverland region, along with stands in Tasmania, western Victoria and southwest New South Wales. In the United States, apricot cultivation is predominant in California, with minor contributions from Washington, Utah and Texas, although total production is declining from these regions.

Apricots are generally eaten fresh, canned or dried (Hormaza et al. 2007), although in some areas of Asia kernel production for edible seed and oil is more important than for fruit. Japanese apricots are eaten processed rather than fresh, mostly as pickles, and have been found to contain bioactive compounds (Chuda et al. 1999; Utsunomiya et al. 2002; Utsunomiya et al. 2005).

## *1.5 Traits of Economic Importance*

To understand how genomic tools may be implemented in apricot improvement, it is necessary to define the current suite of challenges to be resolved. Certainly the two major issues, self incompatibility and susceptibility to PPV, also known as "Sharka", are being addressed already. Adaptation to climatic conditions in specific production areas is a high priority for current breeding programs (Hormaza et al. 2007) and should be a target for future QTL studies. Apricots tend to flower early in the spring, and this attribute has significantly limited its range of cultivation. Many breeding programs have therefore considered later blooming as a desirable trait that reduces the risk of frost damage. The "Hunza" apricot was introduced to the United States in the late 1980s from Pakistan and was shown to flower 30 days later than the usual cultivars (Faust et al. 1998). However, in frost-free areas, early blooming is targeted, to produce early maturing cultivars (Hormaza et al. 2007).

In a study of tree growth habits in Mexico designed to inform breeding decisions, twenty different variables were scored in a population of apricot genotypes that exhibited a wide variation in growth habits and flowering time (Perezgonzales 1992). In a multivariate analysis of the data, yield potential was associated with bud number, flower number, and the number of fruits per unit length of fruiting spurs. In Europe, a study of morphological, phenological and fruit quality traits of important commercial clones from Spain, France and Italy revealed a narrower than expected range of variation (Badenes et al. 1998). However, a study within the larger germplasm resources of the Irano-Caucasian eco-geographical group found a wide variation in harvest season, fruit yield, total soluble solids, total acidity, fruit, and pit and kernel mass (Asma and Ozturk 2005). This germplasm, although mostly small-fruited, is a source of variation for breeders, especially the 'Levent' apricot that exhibits very late ripening.

## 2 Molecular Markers

Apricots, like all perennial tree crops, require a substantial investment in time, labor and space to propel genetic studies. Here, the application of molecular markers is very useful, especially considering the far-ranging effects of geographical and environmental conditions that may affect phenotype. Like other crops in the Rosaceae, molecular markers hold great promise for hastening the development of new cultivars and as a means for germplasm protection.

The development of molecular markers in apricot benefits from the high degree of synteny observed among *Prunus* species (Dirlewanger et al. 2004). The peach 'Earliglow' × almond 'Texas' (TxE) linkage map discussed in "Genomics of Almond" is an outstanding reference map for apricot, and loci for self-incompatibilty and Sharka resistance have been mapped in association with other markers that may prove useful in further studies.

A series of traditional molecular markers have been used to answer questions in apricot, typically in regard to relatedness of genotypes and to map genes associated with traits of interest. Early molecular evidence of genetic variability was observed using isozyme analyses. Patterns of inheritance were traced among five isozyme loci, primarily for the detection of hybrids and general characterization of germplasm (Byrne 1989; Byrne and Littleton 1989). Variation in aspartate aminotransferase banding patterns has also been used to analyze interspecific plum × apricot crosses (Pashkoulov et al. 1995). A report on the isozyme-based genetic diversity of apricot examined 94 accessions from North America, Europe and northern Africa for 10 isozyme systems (Badenes et al. 1996). The results revealed extensive polymorphism among North American genotypes that could be categorized into discrete subgroups, one of which was composed of PPV resistant germplasm (discussed later in this chapter) that appeared similar to Asian species.

De Vicente et al. (1998) employed 31 genomic and cDNA probes from almond to study a set of 52 European and North American apricot cultivars. The group of 25 Spanish cultivars had a lower level of polymorphism than the others, suggesting that bottlenecks may have occurred in the recent history of the apricot and have eroded its genetic variability. Restriction Fragment Length Polymorphism (RFLP) analyses generated a series of tools that could be used to delineate wild apricot germplasm from highly cultivated material (Uematsu et al. 1991).

Random Amplified Polymorphic DNA (RAPD) markers have utility in apricot because of the higher degree of heterozygosity relative to peach. They have been implemented for germplasm studies in *P. mume* and *P. armeniaca*. Shimada et al. (1994) separated 54 *P. mume* genotypes into seven groups using 95 primer pairs. A study by Mariniello et al. (2002) that sought a means to differentiate apricot cultivars (*P. armeniaca*) from the Campania region of Italy from those of other Italian regions, Europe and North America. They found that the OL9 primer (5' ACGCCAGTTC 3') produced a product specific to the Campania cultivars (Mariniello et al. 2002). Similarly, the two morphologically indistinguishable cultivars, 'Hajihalioulu' and 'Hasanbay', were genotyped with a set of RAPD markers (Takeda et al. 1998). The results confirmed suspicions that these visually identical plant lines were in fact the same, as no RAPD variability could be detected.

The ability of Amplified Fragment Length Polymorphisms (AFLP) markers to detect variability has been valuable to studies in apricot. Their high sensitivity to polymorphism and large number of alleles makes them useful for describing relatedness or genetic diversity as well as for linkage mapping. Diversity studies with AFLPs have been used to describe the relatedness of various cultivars and apricot species. A study of 47 cultivars representing a large geographical range was assayed with five AFLP primer pairs, yielding over 350 polymorphic products. Analysis allowed clustering of various cultivars that agreed well with known geographical origin, and confirmed that *P. mume*, *P. briganiaca* and *P. dasycarpa* were distant relatives, while *P. holosericia* was actually a variant of *P. armeniaca* (Hagen et al. 2002). The same applied to 14 *P. mume* cultivars from China and Japan (Fang et al. 2005). Similarly, while expanding on an earlier analysis based on simple sequence repeat (SSR) markers (Krichen et al. 2006), analysis of 31 cultivars of Tunisian apricots with an extensive set of AFLPs allowed clustering of Tunisian apricots into two distinct sub-types that correlated well with their geographical distribution (Krichen et al., (2008).

Similarly, analysis of Simple Sequence Repeats (SSRs) has been useful in study of apricot genetic variation. SSR analyses in apricots have benefited from well-defined primer sets designed for use in peach, especially those shown to be portable across *Prunus* species (Cipriani et al. 1999). Zhebentyayeva et al. (2003) showed that levels of polymorphism exhibited in a germplasm collection of 74 *P. armeniaca* accessions were greater with SSRs (from peach) than with other co-dominant marker systems (isoenzymes, RFLPs). Peach SSR primers were used to characterize a suite of cultivars from North America, Greece, Spain and France (Sanchez-Perez et al. 2005). The study revealed that peach-based primers could generate substantial evidence of variability, with an average number of alleles per locus of 4.1 and as many as 9. The results of this study validated the use of peach SSRs in apricot as the clustering of genotypes agreed well with plant pedigree and geographical distribution. Peach SSR markers were used to examine how populations of higher heterozygosity could be used to introduce genetic variability into less polymorphic populations in Spanish breeding programs. A study by Sanchez-Perez et al. (2006) surveyed several discrete populations of varying heterozygosity and explored how complementary crosses could enrich the genetic diversity of less heterozygous Spanish cultivars.

Heterologous SSR markers have also been used to assess diversity of native *P. armeniaca* apricot populations. Comprehensive SSR analysis of landrace cultivars of Tunisia (Krichen et al. 2006) preceded the AFLP study discussed above (Krichen et al., (2008), demonstrating the use of molecular fingerprinting as a tool for germplasm conservation. In another study, eight selected SSR primers from peach were used to determine the population genetic structure of *P. armeniaca* apricots from three populations growing in the wild in West China (He et al. 2007). Genetic variation was mainly found among individuals within the populations, however, the populations were also different at most SSR loci, and this information was used to devise a conservation strategy.

Gao et al. (2004) used SSRs from sweet cherry (5), sour cherry (10) as well as peach (9) to show that SSR primers developed in several other *Prunus* species could be implemented to study diversity in *P. mume*. Ahmad et al. (2004) confirmed the transferability of SSR markers among *Prunus* species in a study that involved 25 SSRs from cherry and three from peach to distinguish seven apricot, one plumcot and six pluot cultivars.

SSR markers have been developed from apricot genomic (Lopes et al. 2002; Messina et al. 2004) and cDNA libraries (Decroocq et al. 2003). Hagen et al. (2004) developed SSRs from fruit expressed sequence tags (EST) libraries as well as a genomic leaf cDNA library. Maghuly et al. (2006) used ten of the homologous primer combinations developed by Lopes et al. (2002) and Messina et al. (2004) to undertake the first demonstration of genetic variability in a large collection of *P. armeniaca* apricot cultivars (133) and related species (3). The homologous SSRs were selected for their high level of polymorphism and this was higher than reported with SSRs from peach (Romero et al. 2003; Zhebentyayeva et al. 2003). Most of the East European cultivars clustered together. American cultivars have European germplasm in their pedigree, and have been enriched with germplasm of Asian origin. The cluster that included accessions from Pakistan only exhibited a high allelic variability, consistent with its seedling rather than clonal origin.

A very elegant recent study further investigating the enrichment of American cultivars with germplasm of Asian origin was specifically related to PPV resistance and used a combination of AFLP and SSR technology (Zhebentyayeva et al. 2008). Twenty-seven AFLP markers found to be associated with resistant genotypes and not shared with European cultivars were used to trace putative introgressions into American breeding germplasm. The results indicated that native central Asian germplasm had contributed to the resistance of 'Harlayne' and 'Goldrich', and that Chinese material had been introgressed into American cultivars 'Stark Early Orange', LE 2904, LE 3276 and 'Vestar' used in PPV resistance breeding. Maximum likelihood analysis for genetic similarity of the PPV resistance region in resistant genotypes with five SSR markers linked to the PPV resistance on Linkage Group 1 (LG1) demonstrated the relatedness of resistant American cultivars 'Stark Early Orange', 'Goldrich' and 'Harlayne' with the Chinese cultivars 'Yuan-sin', 'Mai-che-sin' and 'In-ben-sin'. It was concluded that Chinese apricot cultivars and/or the wild species *P. mandshurica* and *P. sibirica* var. *davidiana* were important sources of PPV resistance in North American breeding material.

## *2.1 Genetic Linkage Maps*

A number of genetic maps have been constructed in *P. armeniaca*, principally with the goal of elucidating mechanisms of PPV resistance (see below for discussion of PPV resistance mapping). The first map was a partial 'Goldrich' (resistant) × 'Valenciano' (susceptible) (GxC) map constructed in a progeny of 81 F1 individuals using mainly AFLP (82) and RAPD (33) markers, with a few almond RFLP (4) and peach SSR (9) markers (Hurtado et al. 2002). Eight linkage groups were constructed for 'Goldrich' and seven for 'Valencia', with a total length of 511 cM, respectively 467 cM for two cultivars. Vilanova et al. (2003a) chose a $F_2$ population of 'Lito' ('Stark Early Orange' (resistant) × 'Tirynthos'), to enable studies of both PPV resistance and self-incompatibility. One hundred AFLPs and 29 peach and cherry SSRs mapped to 11 linkage groups covering 602 cM. Lambert et al. (2004) specifically chose markers previously used to construct the TxE map (Aranzana et al. 2003; Joobeur et al. 1998) to enable the alignment of the apricot map in 'Polonais' × 'Stark Early Orange' with those of almond and peach. The map construction utilized 142 individuals, and employed 88 RFLP probes used for the TxE map, with 20 SSR primer pairs from peach, 14 from apricot, and one from *P. davidiana*. Eighty-eight AFLP markers served to improve map density. The map length encompassed 538 cM and 699 cM for 'Polonais' and 'Stark Early Orange', respectively. Most markers in each of the eight apricot linkage groups aligned with those in the TxE map, indicating a high degree of collinearity among the genomes of apricot, peach and almond.

The maps developed by Dondini et al. (2007) in a progeny of 125 individuals from a 'Lito' × BO 81604311 cross were the first to include a substantial number of SSR markers derived from apricot. Of the total of 185 SSR markers mapped, 74 were from an apricot genomic library and the remainder were from peach, almond and cherry. Eight linkage groups were obtained for each parent and total map length was 504 cM and 620 cM for 'Lito' and BO 81604311, respectively. The maps aligned well with other published *Prunus* maps from almond, peach and 'Myrobalan', indicating that information from one genetic map will help breeders locate the same loci in related species. The progeny are expected to segregate for fruit size, firmness, flesh soluble solids and acid content, making it a valuable resource for trait mapping. Recently, a new set of peach and apricot SSR markers has increased the resolution of the existing GxC and $F_2$ 'Lito' maps, providing a strong association between discrete markers and PPV resistance loci that translates well to other breeding populations (Soriano et al. 2008). The addition of 43 SSR markers to mapping in GxC did not substantially change the length of the maps in 'Goldrich' or 'Currot' (designated 'Valencia' in Hurtado et al. 2002) and the authors believe that the maps may cover 38–58% and 37–56% of the parental genomes, respectively. The new 'Lito' map (designated LxL-98) remained unchanged in length after the addition of 37 new SSR markers, but may cover 55–77% of the nuclear genome. The latest map available was constructed in a $BC_1$ population of only 67 individuals from a cross between apricot selection LE-3246 ('Stark Early Orange' × 'Vestar') and 'Stark Early Orange', and represents a combined map for LE-3246 and 'Vestar'. It consists of 330 AFLPs and 26 SSRs assigned to eight linkage groups and extends over 523 cM. Bulked segregant analysis in a related population identified AFLPs segregating with PPV resistance and resulted in the development of SSR primers that detected loci mapping to the same linkage group as PPV resistance. No linkage maps have been reported to date in *P. mume*.

# 3 Significant Traits

## 3.1 Self-Incompatibility

Self-incompatibility is a mechanism in flowering plants that prevents self-fertilization, ensuring efficient outcrossing. Self-incompatible genotypes cannot be grown in mono-cultures, because of their need for other genotypes to function as pollinizers, which in turn requires bees for pollination. Cultivar mixtures limit production because the various genotypes must be treated individually with respect to flowering, harvest, pest management, tree maintenance and other cultural aspects. There is considerable interest from growers and breeders alike to understand the molecular-genetic basis of self-incompatibility. In apricot, European cultivars are generally described as self-compatible, while those of Asian or Iranian-Caucasian origin are typically self-incompatible (Kostina 1969). A mixture of self-compatible and self-incompatible cultivars have been identified in European and North American selections (Burgos et al. 1997).

The gametophytic mechanisms have been well elucidated in recent years and hinge on the diversity of a small gene family at the highly variable self-incompatibility (S)-locus. The genomic region of the S-locus contains principally two types of genes, alleles of the S-RNase and an F-Box protein (or several F-Box proteins). Polymorphism in this locus and interplay between components allows a plant to discriminate between its own pollen and that of a separate plant. Hence, incompatible pollen tubes cannot negotiate the style, precluding fertilization.

The precise mechanism lies in the S-RNase encoded within the S-locus. The S-RNases degrade the ribonucleic acid (RNA) within the extending pollen tube, inhibiting viable descent through the style. However, in order for outcrossing to occur, some pollen must be permitted to succeed on its quest to fertilization. To facilitate this process, specific pollen-resident F-box proteins (SFBs, for *S-haplotype-specifc F-Box*) target the non-self RNase for proteolytic degradation. F-box proteins are a diverse family of adapters that participate in the specific recognition and ubiquitination of target proteins. Targeted proteins are covalently tagged with ubiquitin chains that designate it a substrate for proteolysis via the 26S proteosome. In a compatible scenario, the F-box protein degrades the non-self RNase allowing the tube to grow, whereas in an incompatible situation the RNase remains active and cytotoxic. Thus, both stylar and pollen specific factors dictate the degree of compatibility.

The early description arose from study of the Japanese apricot, *P. mume*. Here the first pollen-part mutant S haplotypes were characterized (Ushijima et al. 2004). PCR and Southern blotting techniques can delineate self-compatible from self-incompatible genotypes in this species (Tao et al. 2000). The S-locus has been mapped in detail, revealing the presence of RNases and three F-box proteins, one that is solely expressed in pollen and therefore is like the gametophytic determinant (Entani et al. 2003). The sequences of the pistil and pollen components of self incompatibility have been identified and provide a basis for use of molecular markers (Tao et al. 2002; Yaegaki et al. 2001; Yamane et al. 2003).

Studies in *P. armeniaca* showed that the S-locus resides on LG6 (Vilanova et al. 2003a). Original descriptions indicated that the locus contained one allele for self-compatibility ($S_c$) and seven alleles for self-incompatibility ($S_1$-$S_7$) (Alburquerque et al. 2002; Burgos et al. 1998). A careful assessment of Hungarian germplasm revealed the presence of at least nine additional alleles (Halasz et al. 2005), where the variability arose from Asian genotypes that have contributed to the frost tolerance traits of regional cultivars. Other efforts have also expanded the resolution of the composition of the S-locus, noting the precise sequences of proteins encoded therein. These tools provide a basis for analysis of any apricot population and allow the construction of marker systems to select for self-compatible germplasm.

## *3.2 General Pathogen Resistance*

Nucleotide binding site-leucine-rich-repeat (NBS-LRR) proteins have been directly implicated in disease resistance (Belkhadir et al. 2004) and two studies have approached the issue of resistance in apricot via cloning of these genes. The first study identified more than 80 different sequences from 'Stark Early Orange' by cloning the PCR products of a single primer pair based on conserved NBS sequences (Dondini et al. 2007). All of the apricot RGA sequences were members of the toll-interleukin receptor (TIR) family. When 10 sets of primers based on hypervariable regions were used to amplify a panel of DNA from PPV resistant and susceptible genotypes, one primer pair was amplified in PPV resistant genotypes only.

In order to facilitate identification of genomic regions containing candidate resistance genes against a range of pathogens, AFLP-resistance gene analogue (RGA) markers were developed and mapped on the LxL map (Soriano et al. 2005). Development of the AFLP-RGA markers involved first identifying apricot resistance genes of the NBS-LRR class using degenerate PCR primers that targeted conserved NBS regions. These primers identified 43 unique proteins in 'Goldrich' and 'Lito', which could be aligned into 6 distinct groups of resistance gene analogs, all of which belonged to the TIR group of NBS-LRRs. Family-specific primers were then developed from non-consensus NBS regions and used to develop AFLP-RGA markers. These markers utilized standard AFLP primers coupled with the RGA family-specific primers and resulted in 27 polymorphic markers, of which 16 were mapped to five linkage groups on the $F_2$ 'Lito' map of Vilanova et al. (2003a). A degree of clustering was indicated on LG6 and LG8, and four markers were distributed over LG1. This map will complement a more extensive map constructed in peach (Lalli et al. 2005).

## *3.3 Sharka Disease*

Infection with PPV is the most limiting factor in apricot cultivation and the most economically devastating fruit disease in Europe (Hurtado et al. 2002). Description of the disease spectrum known as *Sharka* dates back to Bulgaria in the early 1900s, where patterns of susceptibility and manifestation were observed on plums. Also known as Plum Pox, the disease was then identified on specific *Prunus* genotypes as it crept throughout Eastern Europe during the first half of the 19$^{th}$ century. Sharka symptoms have not only been observed on apricots, but also on plums, peaches and nectarines. Infection may result in severe deformation of fruit and significant reduction in yield. There are no known controls aside from complete destruction of infected materials, as well as the adjacent trees. PPV disease has spread to five continents, showing up for the first time in the USA in a Pennsylvanian orchard in 1999.

The causative agent was determined to be a potyvirus, discernable as six specific sub-types. The first two were designated Dideron (PPV-D) and Marcus (PPV-M) (Candresse et al. 1998), and there are currently several additional sub-types that can be discriminated. PPV spreads via aphid vectors. In acute transmission, aphids feed on infected trees, the virus resides in their stylet and then can be transmitted to other adjacent uninfected trees. There also is evidence that some aphids may harbor the virus in their salivary glands, making them chronic, potent vectors of viral spread. Approximately 20 aphid species are capable of disease transmission, some of which may pass the virus directly to their offspring.

### 3.3.1 Symptoms

The presentation of PPV infection varies greatly between genotypes, and even clonal plants may exhibit contrasting symptoms, depending on ambient environmental conditions. Foliar symptoms may appear as chlorotic lesions, contrasting ring patterns and yellowing of the veins. Flowers of some *Prunus* species take on a speckled color pattern. The most economically relevant damage occurs on fruits, where necrotic rings cause severe deformation of apricot fruits. In some cases, infected trees show no obvious symptoms, until a dramatic decrease in yield is observed. Ultimately, the disease hinders productivity, leads to poorer grade fruit, and shortens the life of the tree.

### 3.3.2 PPV Detection

Evaluation of germplasm for PPV sensitivity is relatively simple. Buds containing the relevant serotype are grafted onto growing branches and then evidence of the virus can be detected in other parts of the tree by enzyme-linked immuno-sorbent assays (ELISA). Evidence of PPV is detectable within a year, is most prevalent in newly infected young trees, and varies in degree greatly within *Prunus* species. Polak (1998) observed that many apricot cultivars harbored some of the highest concentrations of Sharka, but 'Stark Early Orange' permits only low levels to accumulate.

The ELISA test has become a staple for germplasm evaluation in various breeding programs (Toma et al. 1998). Greenhouse-grown apricots grafted to infected peach scions were evaluated for PPV and compared against the same apricot genotypes as 5-year-old field grown plants (Rubio et al. 2008). The materials were evaluated by ELISA as well as by visual inspection, and generally the greenhouse and field detection methods agreed, suggesting that greenhouse evaluation may replace field screening and be a useful measure to curb the spread of the disease.

### 3.3.3 Sources of PPV Resistance

One of the major goals of PPV eradication programs has been the development of resistant germplasm. The PPV-M resistant apricot cultivars 'Stark Early Orange' and 'Stella' were identified among North American selections, however, their poor fruit characteristics and substantial chilling requirements meant that they were unsuitable for European cultivation. The basis for their resistance was therefore of interest, as if it could be identified it may be introduced into breeding populations. The consensus is that of the origin of PPV resistance is likely Asian, however, evidence has been presented that it is not conferred by *Prunus mandshurica*, a Central Asian apricot relative that was used as a source of frost-resistance genes in North American breeding programs (Rubio et al. 2003), though this view was not shared by Zhebentyayeva et al. (2008). In the latter study dendrograms generated from AFLP and SSR analysis of aboriginal European, Asian and North American cultivars provide evidence that *P. mandshurica* and other Chinese accessions were the source of PPV resistance in North American germplasm.

### 3.3.4 Genetic Analysis of Resistance Mechanisms

The genetic mechanisms of Sharka resistance have been the subject of much controversy, which can be explained by the complicating fact that even resistant materials may temporarily show symptoms.

Resistance to PPV has been shown to be conferred by various modes of inheritance, under the control of one (Dicenta and Audergon 1998; Dicenta et al. 2000), two (Rubio et al. 2007; Vilanova et al. 2003a), or three genetic loci (Guillet Bellanger and Audergon 2001; Salava et al. 2005), depending on the many environmental and experimental variables that could influence disease presentation. Genetic linkage maps generated with the resistant genotype 'Stark Early Orange' or 'Goldrich' did not clearly indicate the number of loci underlying resistance. A QTL-based approach using the susceptible parent 'Polonais' crossed with 'Stark Early Orange' defined four discrete genomic regions associated with resistance. One of these regions mapped to LG1 of 'Stark Early Orange', and accounts for 56% of the phenotypic variation (Lambert et al. 2007). Two other QTL provided a minor contribution to resistance, and a third was observable only in the first year of infection. A population developed from a cross between 'Stark Early Orange' and the susceptible cultivar 'Vestar' (Lalli et al. 2008) and segregating for PPV sensitivity was used to generate the genetic linkage map described in Section 2.1. PPV resistance segregated with markers on LG1. In a separate, contemporaneous study, Soriano et al. (2008) added additional SSR markers to the established GxC and the LxL-98 maps and obtained results that indicated control of the trait by at least one major dominant locus on LG1 in agreement with aforementioned QTL studies from Lambert et al. (2007) and Lalli et al. (2008). Several SSR markers resident to this interval were tested in two independent populations segregating for PPV resistance, which confirmed that they may have profound utility in marker assisted selection (Soriano et al. 2008), yet only one putative marker has been shown to faithfully segregate with Sharka resistance (Dondini et al. 2007).

To further describe the inheritance of PPV resistance, Karayiannis et al. (2008) performed a genetic analysis of PPV-M resistance in a large hybrid population (1178 trees, 12 crosses) over an extensive period, revealing that accessions NJA2, 'Harlayne', 'Veecot', 'Sunglo', 'Orangered' and 'Stark Early Orange' are heterozygous for resistance and pass the trait on as a monogenic dominant trait. The resistance in NJA2 and 'Orangered' originated from 'Scout', a Canadian selection raised from Chinese seeds (Karayiannis et al. 2008), again implicating the Asian origin. 'Harlayne' and 'Veecot' both share 'Reliable' as a common resistant ancestor.

Both 'Orangered' and 'Harlayne' have been described as immune to PPV, as neither exhibits post-inoculation evidence of infection, even with sensitive PCR-based assays (Fuchs et al. 2001). The precise mechanisms are open to speculation, but include the possibility of virus translocation, development of a hypersensitive response, or RNA-mediated gene silencing.

### 3.3.5 Transgenic Mitigation of Sharka

Transgenic technologies have been utilized in addressing PPV disease. These trials have been confined to plum, where a robust transformation system facilitates its application. This work is also described in chapter 13 within this volume. Briefly, the advantage is that direct incorporation of a transgene cassette may induce durable resistance to the virus, circumventing the years of selection, co-introgression of undesirable traits, and multigene effects experiences in traditional breeding. The first attempt involved introducing the PPV coat protein into plum using *Agrobacterium*-mediated transformation techniques (Scorza et al. 1994). The regenerated plants were challenged by grafting of infected buds onto them (Ravelonandro et al. 1997). The most resistant plant line expressed low levels of coat protein mRNA and had no detectable coat protein. Analysis of genomic DNA indicated that the transgene was methylated, consistent with the hypothesis that the expression was controlled by post-transcriptional gene silencing. Trees from one 'Honeysweet' transgenic line, the C5 genotype, remained resistant as they exhibited only mild symptoms to infection via naturally occurring aphid populations in the field, whereas nearby cultivars exhibited marked symptoms (Hily et al. 2004).

In safety trials, the transgenic plum trees showed no effects on the composition of, or numbers in, aphid or arthoropod populations. There was no evidence of recombination between the transgene and viruses from the environment. Overall, the risk assessment indicates that there was no evidence of engineered plums posing a higher risk than non-engineered trees (Capote et al. 2007).

### 3.4 Bitterness

Kernel taste is a trait that is significant to some breeders, as 'bitter' apricot seeds high in cyanogenic glucosides can be utilized for extraction of commercial amygdaline and 'sweet' seeds could potentially be used for flavoring pastes. It was believed that seed bitterness was a recessive monogenic trait, however, segregations determined in a multi-family study on inheritance of this trait have indicated that it is more likely to be based on five independent genes. Interactions among recessive alleles of three genes are postulated to control different steps in a pathway determining biosynthesis and/or transport of cyanogenic glucosides, while interactions between two dominant alleles would control catabolism of cyanogenic glucosides (Negri et al. 2008). This hypothesis could be tested using QTL mapping in suitably designed populations.

### 3.5 Color

An analysis of the genetic control of carotenoid levels in apricots was performed on the grounds that an improvement in carotenoid levels in pickled Japanese apricot could provide a health benefit, as well as enhance its attractiveness. Two cultivars of *P. mume* with different patterns of ripening were used for a RNA gel blot analysis of the genes involved in fruit carotenoid accumulation (Kita et al. 2007). In the fruit of a cultivar that exhibits carotenoid accumulation on the tree, both the phytoene synthase gene *PmPSY-1* and the downstream genes lycopene B-cyclase (*PmLCYb*), β-carotene hydroxylase (*PmHYb*), and zeaxanthin hypoxidase (*PmZEP*) were induced, while in fruit of a cultivar exhibiting ripening after harvest, only *PmPSY-1*was substantially induced. Ethylene was found to play a role in the induction of *PmPSY-1,* being a pre-requisite for its induction.

### 3.6 Flowering Time

A QTL analysis of the control of flowering time ('early' vs 'late') is underway at University of the Western Cape (South Africa) in collaboration with the CEBAS-CSIC (Spain). A back-cross design is being used, where a selection ('506/07') from the cross 'Orange Red' (late flowering) and 'Currot' (early flowering Spanish cultivar) has been crossed back to 'Currot' (Jose A. Campoy pers.comm.).

### 3.7 Genes Associated with Ripening

Several specific ripening-related genes have been isolated from apricots. Mbeguie-A-Mbeguie et al. isolated an ACC-oxidase clone (Mbeguie-A-Mbeguie et al. 1999) and two expansins (Mbeguie-A-Mbeguie et al. 2002) from a ripe apricot fruit cDNA library. Mita et al., (1999) documented the expression of ACC oxidase as well as ACC synthase in *P. mume*, and showed that expression patterns mirrored enzyme activity. Later the same group used fluorescent differential display to identify a set of 15 transcripts that were inducible by ripening or ethylene treatment (Mita et al. 2006). The transcripts encode an expansin, a water channel, pectate lyase, a series of metabolic enzymes, a transcription factor and a series of novel or hypothetical proteins. A subset was also induced by wounding, and all candidates are excellent targets for further study.

## 4 Genomics Resources

A 101,376 BAC library with an average insert size of 64 kb constructed of 'Goldrich', and representing 22x coverage of the apricot genome (Vilanova et al. 2003b) has been used to construct a contig around the S-RNase region.

The same library was used recently to develop SSR markers targeted to LG1. BAC libraries were probed with RFLP sequences mapping to this linkage group of the TxE *Prunus* map (Vilanova et al. 2003b). The selected BACs were then hybridized to repeat sequences that indicated the potential presence of SSR markers. Primers flanking these polymorphic regions were designed and resulted in development of eight polymorphic SSR markers (Vilanova et al. 2006), The markers aprigms 2, 18 and 24 have been demonstrated to map to LG1 of 'Goldrich' (Soriano et al. 2008).

## 5 Regeneration and Transformation

Definitive functional tests of gene activity require a transgenic system, and as illustrated throughout several chapters in this volume, *Prunus* transformation and regeneration is nothing short of challenging. Apricot regeneration from embryonic tissues was reported by several groups using Murashige and Skoog media supplemented with BA, TDZ and/or 2-4-D (Goffreda et al. 1995; Pieterse 1989). Stage 2 embryos were chosen as explants in both studies. A later study on regeneration of apricot from vegetative explant sources illustrated that regeneration frequency was highly genotype dependent and that each genotype had specific optima for light conditions, gelling agents, explant source and growth regulators (Perez-Tornero and Burgos 2000; Perez-Tornero et al. 2001; Perez-Tornero et al. 2000). Burgos and Alburquerque (2003) showed that the ethylene inhibitors silver thiosulfate and aminoethoxyvinylglycine had a positive effect on regeneration frequency. Surprisingly kanamycin had strong positive effects on regeneration frequency in specific cultivars. Strong effects of various gelling agents were observed. This study established 'Helena' as a useful in vitro system for efficient regeneration.

Machado et al. (1992) reported successful transformation and regeneration of embryonic apricot 'Kecskemeter' cotyledonary tissues with the PPV coat protein and β-glucuronidase enzyme. Over the next two years several studies by Petri and colleagues would carefully test a series of parameters likely to affect apricot transformation. The first report by Petri et al. (2004) demonstrated transient and stable transformation of a GFP construct in vegetative tissue. The study utilized the 'Helena' cultivar and *Agrobacterium*-mediated transformation. Careful consideration was paid to culture conditions, *Agrobacterium* type used, the use of acetosyringone during co-cultivation and overall co-cultivation time. Using this system the same research group later tested the effect of aminoethoxyvinylglycine, putrecine, spermidine and/or silver thiosulfate on transformation frequency, in combination of pulse treatments of 2-4-D (Petri et al. 2005b). Alternative aminoglycoside antibiotics were tested as plantlet selection agents. These careful tests revealed the high phytotoxicity of gentamycin and strong selection efficacy of paromomycin and kanamycin in the apricot system (Petri et al. 2005a). Although no reports have tested specific processes in transgenic apricot, the aforementioned optimization studies are invaluable in establishing a firm foundation for application of such technologies to relevant production questions.

## 6 Conclusions

Apricots are highly similar to other well-studied *Prunus* species on many levels. From the general fruit characteristics, to disease sensitivity, to the colinearity between genomes, to overarching production challenges, apricot is much like its *Prunus* cousins. However, its unique cultural attributes, novel traits that make it suitable for specific climates, and specific fruit qualities render apricots unique in the genus. Although sparse genomics resources are present for apricot relative to peach, cherry and plum, new technologies will allow rapid application of genomics tools to apricot breeding and productivity. The transformation capacities may speed resolution of gene function and ultimately assist in mitigating some of the barriers that threaten apricot production.

**Acknowledgment** The authors thank Vincent Bus, Emily Buck and Doron Holland for critical evaluation of this work.

## References

Ahmad R, Potter D, Southwick SM (2004) Identification and characterization of plum and pluot cultivars by microsatellite markers. Journal of Horticultural Science & Biotechnology 79: 164–169

Alburquerque N, Egea J, Perez-Tornero O, Burgos L (2002) Genotyping apricot cultivars for self-(in)compatibility by means of RNases associated with S alleles. Plant Breeding 121:343–347

Aranzana MJ, Pineda A, Cosson P, Dirlewanger E, Ascasibar J, Cipriani G, Ryder CD, Testolin R, Abbott A, King GJ, Iezzoni AF, Arus P (2003) A set of simple-sequence repeat (SSR) markers covering the Prunus genome. Theoretical and Applied Genetics 106:819–825

Arumaganathan K, Earle ED (1991) Nulcear DNA content of some important plant species. Plant Molecular Biology Reporter 9:208–218

Takeda T, Shimada T, Nomura K, Ozaki T, Haji T, Yamaguchi M, Yoshida M (1998) Classification of apricot varieties by RAPD analysis. Journal of Japanese Society Horticultural Science 67:21–27

Tao R, Habu T, Yamane H, Sugiura A (2002) Characterization and cDNA cloning for S-f-RNase, a molecular marker for self-compatibility, in Japanese apricot (Prunus mume). Journal of Japanese Society Horticultural Science 71:595–600

Tao R, Habu T, Yamane H, Sugiura A, Iwamoto K (2000) Molecular markers for self-compatibility in Japanese apricot (Prunus mume). Hortscience 35:1121–1123

Toma S, Isac M, Balan V, Ivascu A (1998) Detection of plum pox virus by enzyme-linked immunosorbent assay in some apricot and peach varieties and hybrids in Romania. Acta Virologica 42:276–277

Uematsu C, Sasakuma T, Ogihara Y (1991) Phylogenetic-relationships in the stone fruit group of Prunus as revealed by restriction fragment analysis of chloroplast DNA. Japanese Journal of Genetics 66:59–69

Ushijima K, Yamane H, Watari A, Kakehi E, Ikeda K, Hauck NR, Iezzoni AF, Tao RT (2004) The S haplotype-specific F-box protein gene, SFB, is defective in self-compatible haplotypes of Prunus avium and P-mume. Plant Journal 39:573–586

Utsunomiya H, Takekoshi S, Gato N, Utatsu H, Motley ED, Eguchi K, Fitzgerald TG, Mifune M, Frank GD, Eguchi S (2002) Fruit-juice concentrate of Asian plum inhibits growth signals of vascular smooth muscle cells induced by angiotensin II. Life Sciences 72:659–667

Utsunomiya H, Yamakawa T, Kamei J, Kadonosono K, Tanaka SI (2005) Anti-hyperglycemic effects of plum in a rat model of obesity and type 2 diabetes, Wistar fatty rat. Biomedical Research-Tokyo 26:193–200

Vavilov NI (1992) The phyto-geographical basis for plant breeding. In: Dorofeyev VF (ed) Origin and Geography of Cultivated Plants. Cambridge University Press, Cambridge, pp 316–366

Vilanova S, Romero C, Abbott AG, Llacer G, Badenes ML (2003a) An apricot (Prunus armeniaca L.) F2 progeny linkage map based on SSR and AFLP markers, mapping plum pox virus resistance and self-incompatibility traits. Theoretical and Applied Genetics 107:239–247

Vilanova S, Romero C, Abernathy D, Abbott AG, Burgos L, Llacer G, Badenes ML (2003b) Construction and application of a bacterial artificial chromosome (BAC) library of Prunus armeniaca L. for the identification of clones linked to the self-incompatibility locus. Molecular Genetics and Genomics 269:685–691

Vilanova S, Soriano JM, Lalli DA, Romero C, Abbott AG, Llacer G, Badenes ML (2006) Development of SSR markers located in the G1 linkage group of apricot (Prunus armeniaca L.) using a bacterial artificial chromosome library. Molecular Ecology Notes 6:789–791

Yaegaki H, Shimada T, Moriguchi T, Hayama H, Haji T, Yamaguchi M (2001) Molecular characterization of S-RNase genes and S-genotypes in the Japanese apricot (Prunus mume Sieb. et Zucc.). Sex Plant Reproduction 13:251–257

Yamane H, Ushijima K, Sassa H, Tao R (2003) The use of the S haplotype-specific F-box protein gene, SFB as a molecular marker for S-haplotypes and self-compatibility in Japanese apricot (Prunus mume). Theoretical and Applied Genetics 107:1357–1361

Zhebentyayeva TN, Reighard GL, Gorina VM, Abbott AG (2003) Simple sequence repeat (SSR) analysis for assessment of genetic variability in apricot germplasm. Theoretical and Applied Genetics 106:435–444

Zhebentyayeva TN, Reighard GL, Lalli D, Gorina VM, Krska B, Abbott AG (2008) Origin of resistance to plum pox virus in Apricot: what new AFLP and targeted SSR data analyses tell. Tree Genetics & Genomes 4:403–417

# Part III
# Rose
# (Chapters 16 – 19)

# 16. Introduction to *Rosa*

**Hilde Nybom**

## 1 Major Economic Use/Value of Roses and Limitations to Production

The ornamental value of rose flowers has been recognized and enjoyed since the dawn of civilization. A huge number of cultivars has been developed either as garden plants or for the cut rose market, and more recently as indoor pot plants. In addition, the fruits, i.e. the rose hips, have been identified as a source of ornamental value and are now being commercialised in several countries. Rose flowers are also used for rose oil production, and to a lesser extent, for direct consumption or making various types of food products like tea, jam and candy. Rose hips are similarly used for food products, and attention is now drawn also to their medicinal properties.

Commercial rose breeding and production used to take place only in the indigenous rose areas, i.e. mainly in the temperate regions of the Northern hemisphere. Most of the large rose breeding companies still have their centres in Europe or North America, but now they often have independent production facilities around the world, including the subtropics in both the Northern and Southern hemispheres.

## 2 Garden Roses (Including Landscape Roses)

Roses have been planted around human habitations for at least two millennia. In China, rose cultivation can be traced back to the Han dynasty (141–87 BC) when rose bushes adorned the gardens of the imperial palace (Guoliang, 2003). In West Asia and Europe, the culture of roses similarly goes far back in history, especially in the Orient where roses were grown for their beauty and the production of rose

H. Nybom (✉)
Swedish University of Agricultural Sciences, Kristianstad, Sweden

K.M. Folta, S.E. Gardiner (eds.), *Genetics and Genomics of Rosaceae*, Plant Genetics and Genomics: Crops and Models 6, DOI 10.1007/978-0-387-77491-6_16,

oil. From there, cultivation spread to Egypt, Greece and the Roman empire. Plants of *R. moschata* and *R. foetida* were probably grown and appreciated for their ornamental values, whereas *R. damascena* was grown both as an ornamental (recurrent blooming varieties) and for rose oil production (Joyeaux, 2003). At about 1200 AD, the rose was known in all of Europe, where it was regarded as an ornamental as well as a medicinal plant.

Presently, garden rose plants are produced commercially in almost all temperate and subtropical countries around the world, usually by grafting on rootstocks. Production of saleable plants requires 1–3 years. The most popular cultivars are currently derived from the Hybrid Tea and Floribunda groups, and to a smaller extent also from some Miniature and Shrub types. Improved ornamental value and increased fragrance are common breeding goals.

Lately, disease resistance has received increasing interest from both scientists and breeders. The causal organism of black spot, *Diplocarpon rosae* Wolf in its perfect stage and *Marssonina rosae* (Lib.) Died. in the imperfect, forms circular or irregularly coalescent spots on the upper leaf surfaces. Resistance is rare but has been found in wild species (Debener et al., 1998; Vries and Dubois, 2001) and some OGR cultivars (Carlson-Nilsson, 2002). Both a single, dominantly inherited resistance gene (Malek and Debener, 1998; Zhang, 2003) and polygenic mechanism (Xue and Davidson, 1998; Shupert et al., 2007) have been reported. Several other fungi also cause dark spots on the leaves, e.g. *Sphaceloma rosarum* (Pass.) Jenkins, often known as anthracnose.

*Podosphaera pannosa* (Wallr. ex Fr.) de Bary causes powdery mildew and grows on the surface of all green tissues, mainly on leaves but also on young stems, flower buds and rose hips. This is one of the most severe diseases in both field and glasshouse rose production (Linde and Debener, 2003). As with black spot, resistance to powdery mildew has been reported to be controlled by major genes (Linde and Debener, 2003; Linde et al., 2004; Lammerts, 1945; Xu et al., 2005) and QTLs (Schlosser, 1990; Zhang, 2003; Dugo et al., 2005; Linde et al., 2006; Xu et al., 2007). Another common mildew disease is downy mildew, caused by *Perenospora sparsa* Berk.

Nine different species of *Phragmidium* have been reported to cause rust on roses (Horst, 1983), a disease that is especially common in temperate regions with cool temperatures and moist conditions. Orange or brown discolorations of the tissue appear on the upper leaf surface, eventually leading to wilting and defoliation. A certain level of tolerance has been noted for some ornamental rose cultivars.

Roses are becoming increasingly popular not just in carefully tended gardens but also as landscape plants, thus adorning road sides and more rustic types of public parks and habitation areas. Obviously hardiness in all different respects is needed for these plants, i.e. the ability to grow well under adverse weather conditions and to withstand attacks from fungi and insects. In addition, these landscape plants should preferably offer ornamental qualities during a large part of the year, thus leaves, flowers and hips are all important, as well as a natural plant shape that does not require pruning to look good.

## 3 Cut Roses

The commercially important cut-rose industry started just before the year 1900, when the first specially constructed glasshouses were taken into use. Nowadays, the grafted rose plants for cut-flower production are usually grown in carefully climate-controlled glasshouses, where the plants can be harvested throughout the year after a growth cycle of 2–3.5 months depending on the quality of light, heating, irrigation, fertilization and $CO_2$ supplementation. Worldwide cut-rose production is estimated to cover about 8500 ha, with an annual production of 15–18 billion stems a year (Blom and Tsujita, 2003).

The most obvious characteristics of a successful cut-rose cultivar relate to its ornamental values, like color, size and shape of flower, leaves, neck and prickles (Chaanin, 2003). Red is still the most popular flower color, with more than 30% of the market. Other classic colors are yellow, pink and white. The so-called sunshine colors, apricot and orange, have been fashionable for some years, and efforts are also made to develop cultivars with even more unusual colors like brown, purple and green. These colors, as well as bicolored and striped patterns, have considerable novelty value but are not likely to ever capture a large share of the market.

Sufficient strength and length of the stem are important characteristics as is sufficient stability of the flower neck. Very large-sized flowers and long stems have been especially popular in Europe but cultivars with smaller flowers and shorter stems are now increasing their market share, while spray roses with five or more small flowers on each stem are very popular especially in Japan. Large and strong prickles are, of course, not desirable. Fragrance has received increasing attention lately but is usually associated with softer petals and thus a shorter vase life, often only 8–10 days. By contrast, cultivars with non-fragrant, double flowers often have vase lives reaching 18 days or more.

Resistance to plant diseases, and a high productivity even at low light and temperatures, are other desirable characteristics. Adaptation to climatic and cultural conditions has lately become increasingly important for the cut-rose cultivars since their production areas have expanded, especially in South America and Africa. To ensure optimal production results, cultivars are often bred and tested for different geographic areas and subsequently grown only in those areas where they are best adapted.

## 4 Pot Roses

Roses have been grown in pots for hundreds of years, but large-scale commercial production of especially developed pot roses did not start until the 1970s. Nowadays, pot roses are usually produced from softwood cuttings in heated glasshouses, around the year, in cycles that take 14–18 weeks.

The pot roses were originally intended mainly for outdoors use on patios but are now sold mostly as indoor pot plants. They are overall smaller in stature than the regular garden or cut flower roses. Specifically they have smaller flowers and leaves, and thinner shoots. While dwarf roses generally stay below 50 cm, a plant height of only 25 cm is most desirable especially when marketed as indoor pot plants. Sometimes application by length-reducing chemicals is used to ensure that the plants stay sufficiently small.

Pot rose cultivars usually originate from small genotypes (dwarfs) occurring naturally in one of the groups Miniatures (originating from *R. chinensis* var. *minima*), Koster roses, Dwarf Polyanthas and Compactas (Vries, 2003). Crosses have also been made with Hybrid Tea and Floribunda cultivars. The ornamental value is of prime importance, and efforts are now being made to increase flower size without raising overall plant size. Hardiness, disease resistance against black spot and powdery mildew, and prolonged shelf life are other common breeding goals. Especially the latter is important since potted roses usually succumb quickly when kept in the overly heated and poorly lit and aerated living conditions that human beings tend to favor.

## 5 Rose Oil Production

The use of flowers and leaves to produce desirable odors, to improve the smell of ourselves and our environments, has a long history. Among the most valuable and useful of the natural perfume oils are the essential oils extracted from roses. More than 400 volatile substances have been identified in rose oil but only about 20% of the wild species can be characterized as ‘fragrant’ (Schulz, 2003). Amount and composition of the essential rose oils vary strongly between species and cultivars. Presently, *R. damascena* (Bulgarian rose oil) and *R. centifolia* (Moroccan rose oil) are the most important species for production of rose oils (attar of roses) and a semisolid extract (rose concrete). Some local production in e.g. Turkey, Bulgaria and Russia is based on the less productive but more hardy and resistant *R. alba* and, to a smaller extent, on *R. moschata*, *R. rugosa* and *R. bourboniana*.

Rose plants grown for oil production are usually propagated with cuttings. Interestingly, several molecular marker-based studies (Baydar et al., 2004; Rusanov et al., 2005) suggest that the *R. damascena* plants used for essential oil production in Bulgaria and Turkey represent a single genotype. This genotype also dominates in the major production zone of Iran although much diversity was found throughout the production areas of Iran (Babaei et al., 2007; Kiani et al., 2007). It appears that the original oil bearing clones used in Turkey and Bulgaria were obtained from Iran and that this region is a center of diversity for *R. damascena*.

## 6 Ornamental Fruits

In the constant search for novelty, various kinds of decorative fruits now increase their share of the ornamental plants market. The attractive value and good shelf-life of rose hips make these very useful for wreath making and cut-branch arrangements.

A commercial production of rose hip branches has therefore recently emerged in Turkey and Italy. Ornamental rose hip cultivation is conducted outdoors and requires diligent control of pests and diseases to ensure the production of spot-free hips. Breeding and selection is carried out to produce rose hip branches with brightly colored, uniform and firm fruits in large clusters on long branches (Baktir et al., 2005). Early coloring of the rose hips enable growers to harvest the branches before they are fully ripe, thus extending storage ability and shelf-life. The plants should also be free from physiological disorders and show good resistance against fungi and insects.

## 7 Culinary and Medicinal Fruits

Rose hips have long been used in food products like tea, jam, marmalade and dessert soup, and for the production of some more specialized items like rose seed oil. Most of the raw material for commercial rose hip based products is harvested from wild or naturalized dogroses, especially in Chile, which is the world's largest exporter of rose-hip pulp.

Rose hips have more recently attracted attention because of their potential health benefits. The hips from taxa in sect. *Caninae* have a remarkably high content of carotenoids, vitamin C and phenolic compounds (Olsson et al., 2005) that can act as powerful antioxidants and scavengers of free radicals and thus prevent the lipid peroxidation associated with arteriosclerosis, cancer and chronic inflammation.

To date, the commercially most important medicinal effect of rose hips concern treatment of osteoarthritis with a standardized rose-hip powder, prepared from the seeds and husks of *R. canina* fruits (Warholm et al., 2003; Winther et al., 2005). Several commercial rose hip plantations have thus been implemented in Northern Europe for the production of rose-hip powder marketed as a health-promoting food additive. Several other health-beneficial effects of rose hips have been reported (Cinar and Colakoglu, 2005) and may lead to a future increase in commercial rose-hip production.

In south west China, a traditional drink called Cili is made from hips of the chestnut rose, *R. roxburghii*. This is thought to have senescence-retarding and cancer prevention properties related to its high vitamin C content and superoxide dismutase activity (Ma et al., 1997). It is seen as a promising fruit crop in China and is being bred to improve the commercial varieties available (Wen and Deng, 2005).

## 8 Rootstocks

Both garden roses and indoor-grown roses for cut-rose production are usually grafted on rootstocks. A rootstock should be easy to propagate and have good compatibility with scion cultivars. Since seedlings tend to be comparatively homogeneous in the dogrose species, many seed-propagated rootstocks, 'Edelcaninas', have been selected from this section. These rootstocks are still used for propagation of garden roses whereas indoor-grown roses have been grafted on clonally propagated rootstocks which show better vigour control. Some breeding of clonal rootstocks still takes place, but an increasing number of cut-rose and landscape rose production units have recently started to make use of own-rooted cultivars.

# 9 Taxonomy and Classification

## *9.1 Taxonomy of Wild Species*

The relatively well-defined genus *Rosa* contains perennial, often shrubby species with open flowers that have numerous stamens and free carpels, and develop into urn like fruits containing achenes.

Basic chromosome number is 7, with most taxa being diploid, tetraploid or pentaploid. Taxonomy within this genus has been quite controversial. According to Shepherd (1954), even Linnaeus realized some of these difficulties since he very succinctly stated that: 'The species of *Rosa* are very difficult to determine and those who have seen few species can distinguish them more easily than those who have examined many'. For many years, Rehder's classification system (Rehder, 1940) was widely accepted and it is still used as a basis for modern treatises (e.g. Wissemann, 2003), with the genus *Rosa* divided into four subgenera; *Hulthemia, Platyrhodon, Hesperhodos* and *Rosa*. The first three subgenera contain only one or two species each whereas subgenus *Rosa* contains approx. about 120 to 200 species. These species are usually divided into 10 sections; *Pimpinellifoliae, Rosa, Caninae, Carolinae, Cinnamomeae, Synstylae, Indicae, Banksiae, Laevigatae* and *Bracteatae*, with the four latter sections having only one to three species each. Actually, sect. *Rosa* also contains only one wild species, *R. gallica*, but several hybridogenous 'species' representing cultivated entities are commonly classified into this section.

In the last two decades, various DNA-based approaches have been used to improve the taxonomical treatment of *Rosa*. Jan et al. (1999) conducted a RAPD-based phenetic analysis of 119 accessions, representing 36 species (one species each from the subgenera *Platyrhodon* and *Hesperhodos*, and 34 species from 8 sections of subgenus *Rosa*). Their results suggested that the two subgenera *Platyrhodon* and *Hesperhodos* should be placed within subgenus *Rosa*. In a more recent study comprising phenetic as well as phylogenetic approaches, Koopman et al., manuscript submitted) used AFLPs to analyse 93 accessions from 47 species, comprising three subgenera and six sections. It was suggested that subgenera *Hulthemia* and *Platyrhodon* should be included in subgenus *Rosa*. The broadest sequencing study so far published was carried out with analysis of cpDNA IGS and nrDNA ITS in a total of 119 *Rosa* species representing all four subgenera, and all 10 sections within subgenus *Rosa* (Wissemann and Ritz, 2005). This study, as well as a study on the cpDNA *mat*K gene (Matsumoto et al., 1998) likewise suggest the lack of a sound genetic basis for a subgeneric division of the genus *Rosa*.

As for the division into sections, *Synstylae* appears to be monophyletic (Matsumoto et al., 2000; Wu et al., 2000; Koopman et al., 2008), except for its only European representative, *R. arvensis*, which probably does not belong in the section. The closest affinity of sect. *Synstylae* appears to be with sections *Indicae* (Wu et al., 2000; Wissemann and Ritz, 2005) and *Rosa* (Koopman et al., 2008).

Species of sect. *Carolinae* are embedded within sect. *Cinnamomeae* according to RAPD (Jan et al., 1999), AFLP (Koopman et al., 2008), SSR (Scariot et al., 2006) and sequencing data (Wissemann and Ritz, 2005), suggesting that these two sections should be merged. In addition, *Pimpinellifoliae* occurred in the same cluster as sections *Carolinae* and *Cinnamomeae* in an SSR study (Scariot et al., 2006). Various studies indicate that sect. *Pimpinellifoliae* is polyphyletic, and that *R. spinosissima* should be separated from the other species (Matsumoto et al., 1998; Wissemann and Ritz, 2005; Koopman et al., 2008).

In spite of its size (currently, about 50 species are acknowledged) and hybridogenous origination, the mainly European section *Caninae* (also known as dogroses) apparently constitutes a well-circumscribed monophyletic group, according to RAPD (Debener et al., 1996), AFLP (Koopman et al., 2008), SSR (Scariot et al., 2006) and ITS sequences (Matsumoto et al., 2000; Wissemann and Ritz, 2005). Its closest affinities are probably with sections *Rosa* and *Synstylae* (Wissemann and Ritz, 2005). The section *Caninae* is characterised by a unique, asymmetrical meiosis which results in heterogamy and matroclinal offspring (Nybom et al., 2006). Although sharing some ITS sequence types with species in other sections, thereby confirming their hybridogenous origin, the *Caninae* species also have one unique ITS sequence type which is further evidence of their monophyly (Ritz et al., 2005).

So far, few DNA-based studies have been conducted with the aim to unravel intra-sectional differentiation of taxa except in sect. *Caninae*. For this section, a number of subsections are recognized on morphological character combinations, with *Caninae, Rubigineae* and *Vestitae* being the largest. DNA marker analysis however tends to produce fewer and larger groups of taxa than the morphological characters do (Olsson et al., 2000). Possibly subsect. *Rubigineae* can still be treated as a monophyletic entity but the other subsections overlap considerably according to AFLP data (De Cock et al., 2008; Koopman et al., 2008). Moreover, species delimitations are weak, and genetic distances between sampled genotypes are sometimes more closely associated with geographic distances than with taxonomic distances (De Cock et al., 2008).

## 9.2 Classification and Identification of Cultivars

The wild ancestors of cultivated roses are mainly found in the sect. *Synstylae* species *R. moschata, R. wichurana* and *R. multiflora*; the sect. *Rosa* species *R. damascena* and *R. gallica*; the sect. *Indicae* species *R. chinensis* and *R. gigantea*; and the sect.

*Pimpinellifoliae* species *R. foetida* (Wylie, 1954). By selection and hybridization of these 8 species, a truly amazing variability in shape, color and fragrance has been achieved.

Rose cultivars are generally divided according to their age into two main categories, 'Old Garden Roses' (OGR) and 'Modern Roses' with 1867 as the cut-off point, i.e. the year that 'La France' was introduced on the market. Many of the OGR cultivars have large and fragrant flowers, but bloom only during a short period each year. In 1751, the first China roses reached Europe, and thus recurrent-blooming was introduced to European rose breeding, eventually resulting in the development of the Modern Roses. More than 10,000 rose cultivars have now been named and registered.

To classify rose cultivars further, numerous authors have constructed their own schemes with variable numbers of horticultural groups based on original parentage and morphological characteristics. In 2000, the World Federation of Rose Societies agreed on a system with a total of 34 groups of cultivars, 21 belonging to the category OGR and the remainder to Modern Roses (Cairns, 2003). Well-known groups among OGR are e.g. Alba, Bourbon, Centifolia, Damask, Hybrid China, Hybrid Gallica, Moss, Noisette, Portland and Tea. Especially popular and abundant among the Modern Roses are Hybrid Tea (of which 'La France' was the first cultivar), Grandiflora, Floribunda and Polyantha.

Recently, AFLP analysis showed that rose cultivars could be grouped into a European '*Rosa*' cluster related to *R. damascena* and *R. gallica*, and an Oriental '*Synstylae*' cluster related to *R. moschata, R. wichurana* and *R. multiflora* (Koopman et al., 2008). The '*Rosa*' cluster had a mainly European genetic background and included cultivars belonging to the Damask, Centifolia, Gallica, Alba, Moss, and Portland cultivar groups which all have their main ancestry in the old European garden roses in sect. *Rosa*, although many of them also include influences of the Hybrid China roses. The '*Rosa*' cluster also showed a (more distant) affinity with sect. *Caninae* from which the female parent of the Alba roses is thought to originate. The '*Synstylae*' cluster contained cultivars mainly belonging to the Moschata, Multiflora, Noisette, Bourbon, Tea, and Polyantha groups together with *R. moschata*, while a cluster with *R. wichurana* and *R. multiflora* was sister to this group, suggesting that the '*Synstylae*' cluster has a mainly Oriental genetic background. The Hybrid perpetuals, derived from crosses between Hybrid China roses and *R. gallica/R. damascena* hybrids, appeared to be closer to the *R. gallica/R. damascena* background.

An SSR-based investigation of both wild species and cultivars similarly resulted in a large and mainly European '*Rosa*' cluster which this time also contained the Hybrid China roses and one of the Noisette cultivars (Scariot et al., 2006). This '*Rosa*' cluster again showed affinity to the wild species in sect. *Caninae* as well as to the Alba roses. Species and cultivars affiliated with sections *Indicae*, *Carolinae* and *Cinnamomeae*, and *Synstylae* were further apart in the dendrogram.

Successful attempts to identify rose cultivars, and to distinguish between cultivars originating as sports or from genetic recombination, respectively, have been made with band profiles obtained with AFLPs (Vosman et al., 2004) and SSRs

(Esselink et al., 2003). Through quantification of ratios between band peak areas, exact allelic configuration can be achieved with many of the SSR markers also for tetraploids (Esselink et al., 2004).

## 10 Available Genetic Resources

New, improved rose cultivars are generally obtained from crosses between elite cultivars and subsequent selection among very large seedling progenies. However, this method has now been used so long that it becomes increasingly difficult to find sufficiently improved new genotypes, especially for characteristics like disease resistance and adaptation to adverse production and cultivation conditions. Therefore, it may become necessary to introduce 'new blood' from wild or naturalized species or from more or less forgotten old cultivars.

The genus *Rosa* is distributed in the temperate regions of the Northern hemisphere, in Eastern Asia, Europe and Western Asia, and North America. Roses also occur in warmer areas in New Mexico (USA), Iraq, Ethiopia, Bengal and southern China. Especially China is an important centre of genetic diversity, with many rose species growing in C and SW parts. Bushy, large-flowering, double and recurrent-blooming cultivars were developed in China about 1000 years ago. Many cultivated and highly disease-resistant old roses are still available in some areas of China and could be useful for modern breeding programs (Guoliang, 2003). Another important centre of diversity, especially for sect. *Caninae*, is found in Turkey (Ercisli, 2005).

Wild species and old cultivars are grown in rosaries around the world, but the origination and determination of this material is often insufficiently documented. New inventories may be needed for obtaining material with high variability and desirable genes. Knowledge about the amount and partitioning of genetic variability is unfortunately lacking for most rose species except for some taxa in sect. *Caninae* as well as a few European non-*Canina* species that have been analyzed with DNA markers (Olsson et al., 2000; De Cock et al., 2008).

## 11 Major Features of Rose Genetics

Wild rose species are most often diploid and have a regular meiosis with 7 bivalents. Usually these are ring bivalents with one cross-over in each chromosome arm. However, sections *Cinnamomeae* and *Carolinae* contain both 2x and 4x species together with a few 6x and 8x species in the former, and there are some polyploids also in other sections like *R. chinensis* (sect. *Indicae)* which has been denoted as 2x, 3x and 4x, and *R. gallica* (sect. *Rosa)* which is a tetraploid.

Species in sect. *Caninae* are characterized by the peculiar *canina* meiosis (Lim et al., 2005). Regardless of ploidy level (usually 5x, but some 4x and 6x taxa also occur), only 7 bivalents are formed in the first meiotic division. The remaining chromosomes occur exclusively as univalents and are not included in viable pollen grains, which contain only the 7 divided bivalent chromosomes. All the univalents are, however, transmitted to one of the daughter cells in the female meiosis, and are eventually included in the viable egg cells, which therefore contain 21, 28 or 35 chromosomes depending on ploidy levels. SSR-based analyses of different species and offspring from controlled crosses suggest that bivalent formation involves one biparentally inherited, highly homozygous diploid genome, whereas the remaining 2, 3 or 4 haploid and often highly differentiated genomes are maternally transmitted (Nybom et al., 2004, 2006).

## 12 Cross Breeding

Rose breeding is usually conducted by crossing different cultivars or genotypes and selecting among the resulting seedlings. Almost all presently grown rose cultivars are tetraploid and usually interfertile. Number of multivalents in these 4x hybrids differs among genotypes, but based on the most commonly encountered meiotic chromosome associations, segregation often appears to be intermediate between allotetraploid and autotetraploid (Byrne and Crane, 2003).

Crosses among diploids or between different ploidy levels are more problematic than on the tetraploid level, and embryo rescue may be needed. Some triploids produce a relatively high proportion of viable haploid and/or diploid gametes, with a potential to act as bridges in a breeding program. In crosses between triploids with tetraploids, using the triploids as pollen parents results in reasonably high seed set and mostly 4x offspring whereas the reciprocal crosses are less successful and produce more triploids (Huylenbroeck et al., 2005).

Since pollen produced by dogroses is haploid, these species behave as diploids when used as paternal parents. Unfortunately male fertility is considerably lower (pollen stainability c. 20–30%) than in diploid species (>75%). Using the dogroses as maternal parents produces more offspring but these become highly polyploid (6x–8x). For crosses involving dogrose species, the matroclinal inheritance makes direction of the cross very important even when the species are at the same ploidy level (Werlemark and Nybom, 2001).

Manipulation of ploidy levels has been attempted by in vitro culture of anthers and unfertilized egg cells. Successful haploidization has however been achieved only through in situ parthenogenesis using irradiated pollen and subsequent embryo rescue. Some of the resulting dihaploid plants have been used as seed parents in crosses with diploid wild species as male parents. Unfortunately, many dihaploids produce unreduced gametes, resulting in 3x or 4x offspring (El-Mokadem et al., 2002).

Another way to combine wild species with cultivars is to raise diploids to the tetraploid level. The most well-known rose amphidiploid, *R. kordesii*, arose spontaneously from the sterile diploid ‘Max Graf’ (*R. rugosa* × *R. wichurana*) and has had a large influence on rose breeding. Ploidy level can be increased also through the use of colchicine and oryzalin (Kermani et al., 2003). A second avenue is provided by the occasional production of 2x pollen in diploid rose hybrids through a first division restitution thus capturing much of the heterozygosity present in the diploid parent (Crespel et al., 2006).

## References

Babaei, A., Tabaei-Aghdaei, S.R., Khosh-Khui, M., Omidbaigi, R., Naghavi, M.R., Esselink, G.D., and Smulders, M.J.M. (2007). Microsatellite analysis of Damask rose (*Rosa damascena* Mill.) accessions from various regions in Iran reveals multiple genotypes. BMC Plant Biology 7, (8 March 2007).

Baktir, I., Hazar, D., Usyal, S., and Özel, S. (2005). Possible uses of dogrose branches and rose hips for ornamental purposes. Acta Horticulturae 690, 97–99.

Baydar, N.G., Baydar, H., and Debener, T. (2004). Analysis of genetic relationships among *Rosa damascena* plants grown in Turkey by using AFLP and microsatellite markers. Journal of Biotechnology 111, 263–267.

Blom, T.J., and Tsujita, M.J. (2003). Cut rose production. In Encyclopedia of rose science. Volumes 1–3, A.V. Roberts, T. Debener, and S. Gudin, eds (Oxford: Elsevier), pp. 594–600.

Byrne, D.H., and Crane, Y.M. (2003). Meiosis. In Encyclopedia of rose science, A.V. Roberts, T. Debener, and S. Gudin, eds (Oxford: Elsevier), pp. 273–279.

Cairns, T. (2003). Classification/Horticultural classification schemes. In Encyclopedia of rose science, A.V. Roberts, T. Debener, and S. Gudin, eds (Oxford: Elsevier), pp. 117–124.

Carlson-Nilsson, B.U. (2002). Variation in *Rosa* with emphasis on the improvement of winter hardiness and resistance to *Marssonina rosae* (blackspot) (Agraria: Acta Universitatis Agriculturae Sueciae), pp. 51 + papers I–X.

Chaanin, A. (2003). Breeding/Selection strategies for cut roses. In Encyclopedia of rose science, A.V. Roberts, T. Debener, and S. Gudin, eds (Oxford: Elsevier), pp. 33–41.

Cinar, I., and Colakoglu, A.S. (2005). Potential health benefits of rose hip products. Acta Horticulturae 690, 253–257.

Crespel, L., Ricci, S.C., and Gudin, S. (2006). The production of 2n pollen in rose. Euphytica 151, 155–164.

Debener, T., Bartels, C., and Mattiesch, L. (1996). RAPD analysis of genetic variation between a group of rose cultivars and selected wild rose species. Molecular Breeding 2, 321–327.

Debener, T., Drewes-Alvarez, R., and Rockstroh, K. (1998). Identification of five physiological races of blackspot, *Diplocarpon rosae* Wolf, on roses. Plant Breeding 117, 267–270.

De Cock, K., Van Der Mijnsbrugge, K., Breyne, P., Van Bockstaele, E., and Van Slycken, J. (2008). Morphological and AFLP-based differentiation within the taxonomical complex section *Caninae* (subgenus *Rosa*). Annals of Botany 102, 685–697.

de Vries, D.P. (2003). Breeding/selection strategies for pot roses. In Encyclopedia of rose science, A.V. Roberts, T. Debener, and S. Gudin, eds (Oxford: Elsevier), pp. 41–48.

Dugo, M.L., Satovic, Z., Millan, T., Cubero, J.I., Rubiales, D., Cabrera, A., and Torres, A.M. (2005). Genetic mapping of QTLs controlling horticultural traits in diploid roses. Theoretical and Applied Genetics 111, 511–520.

El-Mokadem, H., Crespel, L., Meynet, J., and Gudin, S. (2002). The occurrence of 2n-pollen and the origin of sexual polyploids in dihaploid roses (*Rosa hybrida* L.). Euphytica 125, 169–177.

Ercisli, S. (2005). Rose (Rosa spp.) germplasm resources of Turkey. Genetic Resources and Crop Evolution 52, 787–795.

Esselink, G.D., Smulders, M.J.M., and Vosman, B. (2003). Identification of cut rose (*Rosa hybrida*) and rootstock varieties using robust sequence tagged microsatellite site markers. Theoretical and Applied Genetics 106, 277–286.

Esselink, G.D., Nybom, H., and Vosman, B. (2004). Assignment of allelic configuration in polyploids using the MAC-PR (microsatellite DNA allele counting-peak ratios) method. Theoretical and Applied Genetics 109, 402–408.

Guoliang, W. (2003). History of roses in cultivation/Ancient Chinese roses. In Encyclopedia of rose science, A.V. Roberts, T. Debener, and S. Gudin, eds (Oxford: Elsevier), pp. 387–395.

Horst, R.K. (1983). Black spot. In Compendium of rose diseases (St. Paul, MN: The American Phytopathological Society), pp. vii + 50 pp.

Jan, C.H., Byrne, D.H., Manhart, J., and Wilson, H. (1999). Rose germplasm analysis with RAPD markers. HortScience 34, 341–345.

Joyeaux, F. (2003). History of roses in cultivation/European (Pre-1800). In Encyclopedia of rose science, A.V. Roberts, T. Debener, and S. Gudin, eds (Oxford: Elsevier), pp. 395–402.

Kermani, M.J., Sarasan, V., Roberts, A.V., Yokoya, K., Wentworth, J., and Sieber, V.K. (2003). Oryzalin-induced chromosome doubling in *Rosa* and its effect on plant morphology and pollen viability. Theoretical and Applied Genetics 107, 1195–1200.

Kiani, M., Zamani, Z., Khalighi, A., Fatahi, R., and Byrne, D.H. (2007). Wide genetic diversity of *Rosa damascena* Mill. germplasm in Iran revealed by RAPD analysis. Scientia Horticulturae 115, 386–392.

Koopman, W.J.M., Wissemann, V., Cock, K. de, et al. (2008). AFLP markers as a tool to reconstruct complex relationships: a case study in Rosa (Rosaceae). American Journal of Botany 95(3), 353–366.

Lammerts, W.E. (1945). The scientific basis of rose breeding. American Rose Annual 30, 70–79.

Lim, K.Y., Werlemark, G., Matyasek, R., Bringloe, J.B., Sieber, V., El-Mokadem, H., Meynet, J., Hemming, J., Leitch, A.R., and Roberts, A.V. (2005). Evolutionary implications of permanent odd polyploidy in the stable sexual, pentaploid of *Rosa canina* L. Heredity 94, 501–506.

Linde, M., and Debener, T. (2003). Isolation and identification of eight races of powdery mildew of roses (*Podosphaera pannosa*) (Wallr.: Fr.) de Bary and the genetic analysis of the resistance gene Rpp1. Theoretical and Applied Genetics 107, 256–262.

Linde, M., Mattiesch, L., and Debener, T. (2004). Rpp1, a dominant gene providing race-specific resistance to rose powdery mildew (*Podosphaera pannosa*): molecular mapping, SCAR development and confirmation of disease resistance data. Theoretical and Applied Genetics 109, 1261–1266.

Linde, M., Hattendorf, A., Kaufmann, H., and Debener, T. (2006). Powdery mildew resistance in roses: QTL mapping in different environments using selective genotyping. Theoretical and Applied Genetics 113, 1081–1092.

Ma, Y.X., Zhu, Y., and Wang, C.F. (1997). The ageing retarding effect of 'Long-Life-CiLi'. Mechanism of Ageing Development 96, 171–189.

Matsumoto, S., Kouchi, M., Fukui, H., and Ueda, Y. (2000). Phylogenetic analyses of the subgenus *Eurosa* using the ITS nrDNA sequence. Acta Horticulturae 521, 193–202.

Matsumoto, S., Kouchi, M., Yabuki, J., Kusunoki, M., Ueda, Y., and Fukui, H. (1998). Phylogenetic analyses of the genus *Rosa* using the matK sequence: molecular evidence for the narrow genetic background of modern roses. Scientia Horticulturae 77, 73–82.

Nybom, H., Esselink, G.D., Werlemark, G., and Vosman, B. (2004). Microsatellite DNA marker inheritance indicates preferential pairing between two highly homologous genomes in polyploid and hemisexual dog-roses, *Rosa* L. sect. *Caninae* DC. Heredity 92, 139–150.

Nybom, H., Esselink, G.D., Werlemark, G., Leus, L., and Vosman, B. (2006). Unique genomic configuration revealed by microsatellite DNA in polyploid dogroses, *Rosa* sect. *Caninae.* Journal of Evolutionary Biology 19, 635–648.

Olsson, A., Nybom, H., and Prentice, H.C. (2000). Relationships between Nordic dogroses (*Rosa* L. sect. *Caninae*, Rosaceae) assessed by RAPDs and elliptic Fourier analysis of leaflet shape. Systematic Botany 25, 511–521.

Olsson, M.E., Andersson, S., Werlemark, G., Uggla, M., and Gustavsson, K.E. (2005). Carotenoids and phenolics in rose hips. Acta Horticulturae 690, 249–252.

Rehder, A. (1940). Manual of cultivated trees and shrubs. (New York: MacMillan).

Ritz, C.M., Schmuths, H., and Wissemann, V. (2005). Evolution by reticulation: European dogroses originated by multiple hybridization across the genus *Rosa*. Journal of Heredity 96, 4–14.

Rusanov, K., Kovacheva, N., Vosman, B., Zhang, L., Rajapakse, S., Atanassov, A., and Atanassov, I. (2005). Microsatellite analysis of *Rosa damascena* Mill. accessions reveals genetic similarity between genotypes used for rose oil production and old Damask rose varieties. Theoretical and Applied Genetics 111, 804–809.

Scariot, V., Akkak, A., and Botta, R. (2006). Characterization and genetic relationships of wild species and old garden roses based on microsatellite analysis. Journal of the American Society for Horticultural Science 131, 66–73.

Schlosser, E. (1990). Horizontal resistance of some plant species to powdery mildew. Mededelingen van de Faculteit Landbouwwetenschappen, Rijksuniversiteit Gent 55, 203–206.

Schulz, H. (2003). Fragrance and pigments/odoriferous substances and pigments. In Encyclopedia of rose science, A.V. Roberts, T. Debener, and S. Gudin, eds (Oxford: Elsevier), pp. 231–240.

Shepherd, R.E. (1954). The history of the rose. (New York: Macmillan).

Shupert, D.A., Byrne, D.H., Pemberton, H.B. (2007). *Inheritance of flower traits, leaflet number and prickles in roses.* Proceedings of the Fourth International Symposium on Rose Research and Cultivation, California, USA, 18–22 September 2005. Acta Horticulturae 751, 331–335.

Van Huylenbroeck, J., Leus, L., and Bockstaele, E.v. (2005). Interploidy crosses in roses: use of triploids. Acta Horticulturae 690, 109–112.

Von Malek, B., and Debener, T. (1998). Genetic analysis of resistance to blackspot (*Diplocarpon rosae*) in tetraploid roses. Theoretical and Applied Genetics 96, 228–231.

Vosman, B., Visser, D., Van de Voort, J.R., Smulders, M.J.M., and Van Eeuwijk, F. (2004). The establishment of 'essential derivation' among rose varieties, using AFLP. Theoretical and Applied Genetics 109, 1718–1725.

Vries, D.P.d., and Dubois, L.A.M. (2001). Developments in breeding for horizontal and vertical fungus resistance in roses. Acta Horticulturae 552, 103–112.

Warholm, O., Skaar, S., Hedman, E., Molmen, H.M., and Eik, L. (2003). The effects of a standardized herbal remedy made from a subtype of *Rosa canina* in patients with osteoarthritis: A double-blind, randomized, placebo-controlled clinical trial. Current Therapeutic Research-Clinical and Experimental 64, 21–31.

Wen, X.P., and Deng, X.X. (2005). Micropropagation of chestnut rose (*Rosa roxburghii* Tratt) and assessment of genetic stability in in vitro plants using RAPD and AFLP markers. Journal of Horticultural Science and Biotechnology 80, 54–60.

Werlemark, G., and Nybom, H. (2001). Skewed distribution of morphological character scores and molecular markers in three interspecific crosses in *Rosa* section *Caninae*. Hereditas (Lund) 134, 1–13.

Winther, K., Apel, K., and Thamsborg, G. (2005). A powder made from seeds and shells of a rose-hip subspecies (*Rosa canina*) reduces symptoms of knee and hip osteoarthritis: A randomized, double-blind, placebo-controlled clinical trial. Scandinavian Journal of Rheumatology 34, 302–308.

Wissemann, V. (2003). Classification/Conventional taxonomy (wild roses). In Encyclopedia of rose science, A.V. Roberts, T. Debener, and S. Gudin, eds (Oxford: Elsevier), pp. 111–117.

Wissemann, V., and Ritz, C.M. (2005). The genus *Rosa* (Rosoideae, Rosaceae) revisited: molecular analysis of nrITS-1 and atpB-rbcL intergenic spacer (IGS) versus conventional taxonomy. Botanical Journal of the Linnean Society 147, 275–290.

Wu, S., Ueda, Y., He, H., Nishihara, S., and Matsumoto, S. (2000). Phylogenetic analysis of Japanese *Rosa* species using matK sequences. Breeding Science 50, 275–281.

Wylie, A.P. (1954). The history of garden roses, part I. J. Royal Horticultural Society 79, 555–574.

Xu, Q., Wen, X., and Deng, X. (2005). Isolation of TIR and nonTIR NBS-LRR resistance gene analogues and identification of molecular markers linked to a powdery mildew resistance locus in chestnut rose (*Rosa roxburghii* Tratt). Theoretical and Applied Genetics 111, 819–830.

Xu, Q., Wen, X., and Deng, X. (2007). Cloning of two classes of PR genes and the development of SNAP markers for powdery mildew resistance loci in chestnut rose (*Rosa roxburghii* Tratt). Molecular Breeding 19, 179–191.

Xue, A.G., and Davidson, C.G. (1998). Components of partial resistance to black spot disease (*Diplocarpon rosae* Wolf) in garden roses. HortScience 33, 96–99.

Zhang, L.H. (2003). Genetic linkage map in tetraploid and diploid rose (Clemson, SC: Clemson University).

# 17. Rose Structural Genomics

**David H. Byrne**

## 1 DNA Markers Available and Future Developments

The development of SSRs designed for rose has accelerated in the last several years and is now a focal point of the map development especially as we move towards developing a consensus map to combine the mapping data from the various maps that have been developed. In the past 5 years, there have been reports from Europe (Esselink et al., 2003; Debener, personal communication; Yan et al., 2005a; Hibrand-Saint Oyant et al., 2007), the USA (Zhang et al., 2006), and Japan (Kimura et al., 2006) describing the development of both genomic (256 primer pairs) and EST (44 primer pairs) based microsatellites. Of these, 168 of the genomic primer pair sequences are proprietary and are available via a MTA (material transfer agreement) with restrictions on their commercial use in breeding or in variety protection (Esselink et al., 2003; Yan et al., 2005a; Debener, personal communication, ConCipio) (Table 1).

Mapping work has shown that each of these major groups of microsatellites (Rh, Rw, and EST-SSRs) map to all 7 of thc linkage groups of rose (Yan et al., 2005a; Zhang et al., 2006; Hibrand-Saint Oyant et al., 2007). The only set of SSRs mapped to several maps are those developed in the USA (Zhang et al., 2006). These 30 SSRs have been mapped to the 90–69 map (Zhang et al., 2006), the 97/7 map (Zhang, 2003), and the Hw map (Hibrand-Saint Oyant et al., 2007). Using this data, a preliminary partial consensus map was developed with 5 of the 7 linkage groups (LG) by Sriyani Rajapakse (Byrne et al., PAG talk, Jan 2007) (Fig. 1). Current work in the USA, Europe and Israel is working towards rose maps with sufficient overlapping microsatellites to enable the development of a consensus map for the rose genome.

With a primary interest in the use of molecular markers to aid in the development of highly disease resistant plant varieties (Byrne, 2007) several research groups have developed, characterized, and mapped *RGAs (resistance gene analogues)* and PKs *(protein kinase)* as an approach to find markers for the loci controlling resistance

---

D.H. Byrne (✉)
Department of Horticultural Sciences, Texas A&M University, College Station, TX USA

K.M. Folta, S.E. Gardiner (eds.), *Genetics and Genomics of Rosaceae*, Plant Genetics and Genomics: Crops and Models 6, DOI 10.1007/978-0-387-77491-6_17,

**Table 1** Sources of SSRs developed for rose

| DNA source | Nomenclature | Number | Availability | Reference |
|---|---|---|---|---|
| Genomic DNA from hybrid tea cv Sonia, robust sequence tagged microsatellite site | Rh/MicD | 27 | MTA-PRI/BAFZ | Debener et al. (2001) |
| Genomic DNA from hybrid tea cv Sonia, robust sequence tagged microsatellite sites | Rh | 24 | MTA-PRI/BAFZ | Esselink et al. (2003) |
| Genomic DNA | Rh Rh MiCD | 37 21 4 | MTA-PRI/BAFZ Public-PRI/BAFZ | Yan et al. (2005a) |
| Genomic DNA | RMS | 107 | MTA-CC | Debener, personal communication |
| Genomic DNA | Rw | 30 | Public | Zhang et al. (2006) |
| *R. wichurana* cv Basye's Thornless | | | | |
| Genomic DNA | na | 13 | Public | Kimura et al. (2006) |
| Hybrid tea cv Asamii Red | | | | |
| 97/7 mapping | na | 4 | Mapped but no sequence reported | Linde et al. (2006) |
| Genomic DNA | Rw | 20 | Public | Hibrand-Saint Oyant et al. (2007) |
| *R. wichurana* cv Basye's Thornless | | | | |
| EST – cDNA library from vegetative and floral buds of *R. wichurana* and 'Black Baccara' | C, CL, H | 44 | Public | Hibrand-Saint Oyant et al. (2007) |

MTA = material transfer agreement, PRI = Plant Research International, BAFZ = Institute for Ornamental Plant Breeding, Ahrensburg, Germany, CC = Con/Cipio, Sangerhausen, Germany.

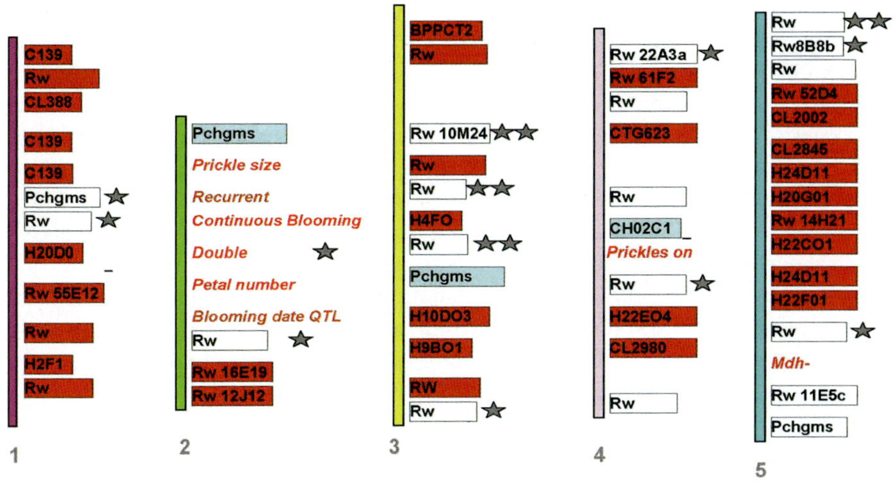


**Fig. 1** Partial Consolidated rose map combining the tetraploid 90–69 F2 map and the diploid 97–7 and Hw maps. Consolidation done by Dr. Sriyani Rajapakse

to powdery mildew and black spot (Xu et al., 2005; Yan et al., 2005a; Linde et al., 2006; Hattendorf and Debener, 2007b). Thus far, RGAs have been mapped onto 3 mapping populations. This includes a few selected RGA derived RFLPs near the CRPM1 locus on the *Rosa roxburghii* map (Xu et al., 2005) and about 50 RGA loci mapped on both the 94/1 and 97/7 populations (Yan et al., 2005a; Linde et al., 2006). In both of the latter populations the RGAs clustered heavily on LGs 1 and 7 with few or none on the LGs 3 and 6. Interestingly there were no RGAs located close to the major powdery mildew QTL clusters or the Rpp1 loci on LGs 3 and 4 although several co-localized with some minor QTLs on other LGs (Linde et al., 2006; Hattendorf and Debener, 2007b). The closest RGA locus to the Rdr1 locus was about 20 cM away (Hattendorf and Debener, 2007a). The 24 PK loci mapped on the 94/1 population were not associated with disease resistance either (Yan et al., 2005a).

*SNPs or single nucleotide polymorphisms* represent the third generation of molecular markers. SNPs are the most abundant co-dominant marker class. In comparison to SSRs, due to their abundance, they can easily be detected by direct sequencing of genomic regions or ESTs. However as they are mainly bi-allelic, they are less polymorphic than SSRs. In plants, they can be used for diverse applications such as marker assisted selection, germplasm fingerprinting, gene mapping, population structure analysis (as LD analysis), and genome wide analysis. Furthermore different techniques have been developed in human to genotype with high throughput SNPs protocol (SNPlex, Illumina®, chip based technologies, Amplifluor®, TaqMan®). As rose is a highly heterozygous plant, SNPs can be easily detected, and represent a large new source of markers. In rose, *SNPs* have been reported to exist every 64–68 bp in the PR2 and PR5 genes in *Rosa roxburghii* which is comparable to the frequency seen in rice (170 bp), maize (48 bp) (Xu et al., 2007) and *Prunus mume* (55 bp) (Fang et al., 2005). In a study to discover new markers for resistance to powdery mildew, SNPs were searched for within the PR2 (pathogenesis-related gene family class 2: *B*-1,3-glucanase) and PR5 (pathogenesis-related gene family class 5: osmotin) genes as created in rose via degenerate primers. The 23 SNPs discovered were converted to SNAPs (single nucleotide amplified polymorphism) to create a simple efficient marker that is scored as absent or present. The SNAP primer pairs were tested and 19 gave good products and 17 were polymorphic and mapped in the *R. roxburghii* mapping population. Sixteen mapped to the same linkage group and one (Glu7) was linked to a minor QTL for powdery mildew (Xu et al., 2007). This approach has promise as an additional procedure for finding markers for targeted regions of the genome.

## 2 Use of Markers for Identification and Studies of Rose Population Structure, Taxonomy, Domestication, and Diversity Studies

Since the early 1990s, the papers published using various molecular markers in the study of rose varieties, germplasm, and species has increased dramatically from 0 to 1 per year to 5–6 per year. This increase corresponded to the development of the PCR based markers which were less expensive per unit of information than either isozymes or genomic probe technology such as RFLPs, mini satellites or microsatellites.

Varietal identification has been an important field of inquiry especially as it pertains to patent rights of roses and other plants. UPOV (International Union for the Protection of New Varieties) introduced the concept of essentially derived varieties (EDV) in 1991 to protect the rights of breeders (UPOV, 1991). In the case of roses, this would be the right of the original breeder to any commercially useful sports (somatic mutations) of the varieties that they develop. It has been shown repeatedly that either RAPDs (Torres et al., 1993; Gallego and Martinez, 1996; Matsumoto and Fukui, 1996; Debener et al., 1997, 2000; Benedetti et al., 2001; Mohapatra and Rout, 2005) , AFLPs (Debener et al., 2000; Zhang et al., 2001; Vosman et al., 2004; Wen and Deng, 2005) or SSRs (Esselink et al., 2004; Rusanov et al., 2005; Scariot et al., 2006; Babaei et al., 2007) are very useful to distinguish among varieties and their seedlings but not necessarily between their sports . Consequently it is a straightforward process to determine whether a plant is a sport of a variety. Of these three markers, RAPDS is the least consistent among different laboratories (Jones et al., 1997) and consequently is the least desirable choice. The most consistent among laboratories and easiest to genetically interpret are SSRs and ISSRs which were recently improved (Laurent Crespel, personal communication). Nevertheless, AFLPs have been shown to be sufficiently consistent among laboratories to easily distinguish among sports and seedlings of rose varieties (Vosman et al., 2004) and thus useful for EDV evaluations.

These three markers, beyond being used for varietal identification are versatile and have been used effectively in assessing the segregation patterns in dog roses (Werlemark et al., 1999; Nybom et al., 2004, 2006), quantifying the heterozygosity of roses (Crespel et al., 2001), in bulked segregant analysis in search of a marker for the Rdr1 gene (Malek et al., 2000), the identification of outcrosses (Debener et al., 2003), and in a range of diversity and phenetic studies. The early studies used RAPDs and mostly examined a few varieties and/or species (Debener et al., 1996; Millan et al., 1996; Walker and Werner, 1997; Olsson et al., 2000). A few topics such as the relationships among rose species (Matsumoto et al., 1998; Jan et al., 1999; Wissemann and Ritz, 2005; Koopman et al., manuscript submitted) and the diversity of *Rosa damascena* (Iwata et al., 2000; Baydar et al., 2004; Rusanov et al., 2005; Tabaei-Aghdaei et al., 2006; Babaei et al., 2007; Kiani et al., 2008) have been studied with a variety of molecular markers all of which support the same conclusions confirming the usefulness of molecular markers.

## 3 Current Situation for Linkage Maps

Genetic mapping for rose has been done or is being developed in both diploid and tetraploid rose populations (Tables 2 and 3). The diploids consist of three groups of crosses. The German populations (94/1 and 97/7) that are based on an introgression population combining traits of *R. multiflora* with those of Garden roses (Debener and Mattiesch, 1999; Linde and Debener, 2003; Yan et al., 2005b; Linde et al., 2006). The second group involves various crosses with *R. wichurana*. This includes the population developed in France (Hw) (Crespel et al., 2002), in Spain (Dugo et al., 2005) and in the USA (WOB) (Shupert et al., 2007). It is also interesting to note that the *R. wichurana* genotype used in both the Spain and the USA rose mapping populations is the same genotype, 'Basye's Thornless'. The third mapping population is a population of *Rosa roxburghii*, developed to study the inheritance of powdery mildew resistance (Xu et al., 2005).

On the tetraploid level, a map has been developed with the 90-69 $F_2$ population which is a complex hybrid that involves parentage of cultivated roses and both tetraploid (*R. carolina* and *R. virginiana*) and diploid (*R. rugosa rubra* and *R. wichurana* Basye's Thornless') species in its recent breeding history (Rajapakse et al., 2001; Zhang et al., 2006). In the mapping efforts for the black spot resistance gene, Rdr1, three tetraploid populations, all crosses with the rose 91/100–5 which is heterozygous for the Rdr1 gene, were used in a bulked segregant analysis study to locate markers that could be mapped onto the diploid map. The third tetraploid population, a cross between the cut flower varieties 'Golden Gate' and 'Fragrant Cloud' was made to study the complex traits of fragrance and post harvest quality (Zamir, personal communication).

**Table 2** Maps developed for rose

| Name of population | Population | Markers | Linkage groups | Map length (cM/marker) | Map length (cM/marker) | % loci with distorted segregation | Reference |
|---|---|---|---|---|---|---|---|
| 94/1R | 93/1–117 × 93/1–119 (60 $F_1$) half sibs derived from R. multiflora × Garden rose progeny diploid | 157 RAPD<br>119 AFLP<br>Blfo – single/double corolla<br>Blfa – white/pink flower color | 7 | 326 cM (2.4 cM/marker) | 370 cm (2.6 cM/marker) | 16% | Debener and Mattiesch (1999) |
| 94/1R | Same as above, 60 additional markers | 321 markers<br>20 AFLP<br>27 microsatellites<br>8 RFLP<br>5 SCAR | 7 | na | na | 17% | Debener et al. (2001) |
| 94/1 | 93/1–119 × 93/1–117 (88 $F_1$) Reciprocal cross of 94/1R Parental and integrated maps | 360 AFLP<br>26 PK<br>88 SSR<br>29 RGA<br>41 others<br>Rdr1, Blfa-3, Blfo-2, | 7 | 487 cM (1.9 cM) Integrated map 545 cM | 490 cM (2.0 cM) | 22% | Yan et al. (2005a) and Hibrand-Saint Oyant et al. (2007) |

**Table 2** (continued)

| Name of population | Population | Markers | Linkage groups | Map length (cM/marker) | Map length (cM/marker) | % loci with distorted segregation | Reference |
|---|---|---|---|---|---|---|---|
| 97/7 | 95/13–39 × Sp3<br>270 $F_1$<br>Integrated map | 172 AFLP<br>50 RGAs<br>4 SSRs<br>1 CAP (Rd1)<br>1 Bac end derived marker<br>4 morphological (black spot, stem prickles, double flowers, white stripes)<br>QTL trait, powdery mildew resistance | 7 | 370 cM<br>(2.6)<br>Integrated map<br>418 cM<br>(1.8 cM/marker) | 354 cM<br>(2.4) | Na | Linde et al. (2006) |
| 97/7 | Same as above<br>Partially integrated | 95 (79 AFLP, 15 SSR, CAP), Dc, Rpm, Rbs<br>97 (91 AFLP, 6 SSR), Rb | 7 | 360 cM<br>3.7 cM | 376 cM<br>3.8 cM | 33% | Zhang (2003) |
| 90–69 $F_2$ | (82–1134 × 86–7)$F_2$<br>52 plants<br>tetraploid | 171, 167 AFLP<br>prickles/petiole<br>MDH | 15, 14 | 682 cM | 902 cM | ?? | Rajapakse et al. (2001) |

**Table 2** (continued)

| Name of population | Population | Markers | Linkage groups | Map length (cM/marker) | Map length (cM/marker) | % loci with distorted segregation | Reference |
|---|---|---|---|---|---|---|---|
| | Same as above<br>Partial map integration | 256, 286<br>19 rose SSRs<br>2 heterologous SSRs | 20, 14 | 920 cM | 770 cM | 15% | Zhang et al. (2006) |
| Hw | H190 × R. wichurana<br>91 plants diploid | 68, 108 AFLP<br>Recurrent/seasonal bloom<br>Double/single corolla<br>QTL prickle density | 8, 6 | 238 cM (3.7 cM) | 287 cM (3.1 cM) | 39% (29–50%) | Crespel et al. (2002) |
| | Same as above | 105, 136 SSRs<br>QTL traits (petal number, bloom date) | 7 | 432 cM (4.0) | 438 cM (3.0) | 26%, 10% | Hibrand-Saint Oyant et al. (2007) |
| Spain | Blush Noisette × *R. wichurana* 'Basye's Thornless' | 130 (85, 49) RAPD<br>2 heterologous SSRs<br>1 morphological traits (Blfo)<br>4 QTL traits (flower size, leaf size, resistance to powdery mildew, days to flowering) | 7 | 388 cM (5.6 cM/marker) | 260 cM (5.83 cM/marker) | 15% | Dugo et al. (2005) |

**Table 3** Details on rose mapping families

| Name ploidy | Female | Comments | Male | Comments | Reference |
|---|---|---|---|---|---|
| 94/1R 2x | 93/1–117 | Parents are selected from open pollinated population of 81/42–15 which derived from introgression program with *R. multiflora* and various garden roses | 93/1–119 | Same as female parent | Debener and Mattiesch (1999) |
| 94/1 2x | 93/1–119 | Same as above | 93/1–117 | Same as female parent | Yan et al. (2005a) |
| 97/7 2x | 95/13–90 | Parents are selected from open pollinated population derived from introgression program with *R. multiflora* and various garden roses | Sp3 (= 82/78–1) | Same as female parent | Linde et al. (2006) |
| 97/7 2x | 95/13–90 (R-pm) | (88/124–46 × 82/78–1) | 82/78–1 (S-pm) | Backcross segreagating for PM race 9 S/R | Linde and Debener (2003) |
| 95/1 95/2 95/3 4x | 91/100–5 | Has Rdr1 gene conferring resistance to broad range of black spot races | Caramba Charme Heckenzauber | Susceptible to black spot | Malek et al. (2000) – BSA to find marker for Rdr1 Yan et al. (2006) – marker association analysis for PM resistance |

**Table 3** (continued)

| Name ploidy | Female | Comments | Male | Comments | Reference |
|---|---|---|---|---|---|
| 90–69 $F_2$ 4x | 90–69 (= 86–7 × Basye's Blueberry) | 86–7 = (R. wichurana 'Bayse's Thornless' × *R. rugosa rubra*) chromosomes doubled with colchicine<br>Basye's Blueberry (82–1134) = [(*Rosa carolina* L. × 'Hugh Dickson') open pollinated] × 74–193. 74–193 = [(*Rosa carolina* L. × 'Hugh Dickson') open pollinated] × (*Rosa virginiana alba* Mill. × 'Betty Morse'). | Progeny produced by selfing the female | Same as female | Rajapakse et al. (2001) and Zhang et al. (2006) |
| Hw 2x | H190 | Dihaploid of 4x *R. hybrida* cv Zambra | *R. wichurana* hybrid | Original paper states it is R. wichurana but since heterozygous for recurrent it is probably a hybrid with China rose | Crespel et al. (2002) and Hibrand-Saint Oyant et al. (2007) |
| Spain 2x | Blush Noisette | Champney's Pink Cluster seedling = Pink China rose (= Parsons's Pink?) × *Rosa moschata* | *R. wichurana* 'Bayse's Thornless' | Same *R. wichurana* clone as used in the mapping work in Texas | Dugo et al. (2005) |
| Roxburghii 2x | Guinong No. 6 | Resistant to powdery mildew | Guinong No. 5 | Susceptible to powdery mildew | Xu et al. (2005, 2007) |
| WOB 2x | Old Blush | *Rosa chinensis* | WOB-26 | *R. wichurana* Basye's Thornless × Old Blush | Shupert et al. (2007) |
| GGFC 4x | Golden Gate | Excellent vase life but non fragrant cut flower variety. | Fragrant Cloud | Very fragrant but poor vase life cut flower rose variety | Zamir (personal communication) |

Given the highly heterozygous nature of the parents in most of these crosses, the general approach to their analysis has been using the pseudo test cross approach first described by Grattapaglia and Sederoff (1994) for use in forest trees. To further complicate mapping, from 10 to 39% of the markers have exhibited distorted segregation (Table 2). This segregation distortion could be caused by the interspecific nature of the crosses used, self incompatibility segregating in some of the populations, gametophytic selection by sub-lethal genes conferring low viability on the zygote, embryo, or seedling level or maybe competitive differences in the performance of genetically variable pollen. Whatever the reason, high levels of segregation distortion limit the quality of the map by limiting the number of markers that are used in making the initial frame work map as the markers with distorted segregation ratios are only added in the last mapping step.

The tetraploid mapping requires larger populations as compared to the diploids for the same quality map and are more difficult to map because the mapping programs are designed for the diploid situation. Thus the mapping is generally restricted to loci that are behaving as diploids.

The lengths of the diploid maps reported vary from 238 to 545 cM on 6 to 8 linkage groups and the tetraploid maps vary from 674 to 920 cM on 14 to 23 linkage groups (Table 2). Thus the total length of the diploid is about 500 cM with an average chromosome length between 70 and 80 cM (Yan et al., 2005a). The density of the maps ranges from low of 5.83 cM/marker (Dugo et al., 2005) to a high of 1.9 cM/marker (Yan et al., 2005a) with the majority being at medium density of 2.5–4.0 cM/marker (Table 2).

The earliest maps used RAPDs and AFLPs (Debener and Mattiesch, 1999; Rajapakse et al., 2001) and consisted of male and female parent maps as these markers do not lend themselves for map integration. Later as other markers such as RFLPs, SSRs, RGAs, PKs, SCARs, CAPs (Table 2) and most recently SNP derived SNAPs (Xu et al., 2007) were developed they have been and are being incorporated into the maps. At the present time there are several hundred microsatellites available for rose (Esselink et al., 2003; Yan et al., 2005a; Kimura et al., 2006; Zhang et al., 2006) and mapping of these have allowed the integration of the male and female maps (Yan et al., 2005a; Linde et al., 2006; Zhang et al., 2006; Hibrand-Saint Oyant et al., 2007). With microsatellite mapping data a preliminary consensus map for 5 of the 7 chromosomes has been constructed for the maps of 97/7, 90-69 $F_2$, and the Hw (Rajapaske, personal communication; Byrne et al., 2007 – PAG meeting abstract) (Fig. 1). As the various research groups place other SSRs and codominant markers on the maps, a rose consensus map will be developed.

The objective of mapping is to better understand the organization of the genome in question and to find markers for specific traits that might facilitate the breeding process. Consequently, a range of qualitative and quantitative trait loci have been placed on the maps. These include qualitative traits such as loci that control pink/white flower color, single/double flower corolla, white petal stripes, black spot resistance, powdery mildew resistance, recurrent flowering, and stem and petiole prickles. Among the QTLs mapped include those controlling powdery mildew resistance, plant vigor, stem prickle density and size, leaf size, flower size, days to flowering or to 50% bloom, and petal number (Table 4). Unfortunately, although these QTL studies are interesting, most used a small number of progeny which limits the accuracy of the mapping and thus the usefulness of the markers identified for the specific traits. In these mapping studies, a number of close linkages between a trait and a marker such as a RAPD or AFLP have been reported as well as a SCAR (SCM10) and two CAP (Rd1, RdY) markers for Rdr1 (Malek et al., 2000; Kaufmann et al., 2003; Zhang, 2003), two SCAR markers for Rpp1 (Linde et al., 2004) and two RGAs (RGA22C, Glu 7 RGA) for a major and minor QTL for powdery mildew resistance (Xu et al., 2005, 2007).

## 4 Inheritance of Known Characters

Roses have been cultivated for several thousand years but only since the late part of the 19th century have breeders been doing controlled crosses in the development of new varieties as opposed to planting out open pollinated seed and selecting among the seedlings (Krüssmann, 1981). At present, there are a few public breeding efforts, hundreds of small private breeders and scores of large breeding operations that name and release several hundred new varieties of roses every year throughout the world. In spite of all this breeding effort, the inheritance of the major traits is poorly understood due to the lack of public research into the breeding and genetics of the species. Nevertheless, in recent years in conjunction with the development of linkage maps of the rose this knowledge has increased several fold (Table 4).

The earliest reports on the inheritance of traits in the USA focussed on ornamental characteristics and were not necessarily well documented but revealed the breeder's conclusions from a few documented cases and his breeding experience (Hurst, 1941; Lammerts, 1945; Morey, 1954). This was followed with additional work in the1960s on the inheritance of red color (Lammerts, 1960, 1964), 1970s on the inheritance of the recurrent bloom trait (Semeniuck, 1971a, b), in the 1980s in Holland, Canada and India, and finally in the last 20 years a flurry of activity connected with the development of genetic linkage maps for a variety of rose mapping populations in the USA, Asia (China), and Europe (Spain, Germany, Holland, and France).

## 5 Flower Traits

The most important trait in roses is the *recurrent blooming* trait, a probable vernalization mutation which reduces the juvenility period and allows the plant to continuously bloom without any need for a vernalization period. This trait was introduced from China during the period of exploration and was initially integrated into European roses by random crosses from gardens where both types grew. This transformed the rose into the most popular ornamental plant in the world. As early as 1941, Hurst indicated that the recurrent trait was recessive to the seasonal bloom

**Table 4** Genetic control of phenotypic characters

| Character | Genetic control | Studied population | Remarks | References |
|---|---|---|---|---|
| **Adaptation traits** | | | | |
| Black spot resistance | Major genes<br>*Rdr1*<br>*Rbs*<br>Additive | 91/100–5, $F_2$, backcrossed to susceptible Caramba, $F_1$ with susceptible cv Hechenzauber, Parisier Charme, and Elina 97/7<br>Old Blush × WOB13/21/26 | Resistance dominant and race specific | Malek and Debener (1998), Malek et al. (2000), Debener et al. (2003), Yan et al. (2005a)<br>Zhang (2003)<br>Shupert et al. (2007) |
| Powdery mildew resistance | Major genes<br>Dominant<br>Rpp1<br>CRPM1<br>Rpm<br>QTLs<br>Pm1, Pm2<br>28 QTLS, 2 main regions<br>Rpm1, Rpm2<br>Minor QTL | Tetraploid (Crimson Glory × Captain Thomas and others)<br>97/9<br>Roxburghii<br>97/7<br>Spain<br>97/7<br>97/7<br>Roxburghii | Resistance dominant and in recent studies appear race specific | Lammerts (1945)<br>Zhang (2003), Linde et al. (2004), Xu et al. (2005)<br>Zhang (2003), Dugo et al. (2005), Linde et al. (2006), Xu et al. (2007) |
| Nematode resistance | Polygenic | Diallel using 3 *R. multiflora* and 2 *R. indica* rootstocks | Wang et al. (2004b) | |
| Cold hardiness | Additive | Diploid and tetraploid populations | Field evaluations, $h^2$ ranged from 51% to 92% | Svejda (1979) |

**Table 4** (continued)

| Character | Genetic control | Studied population | Remarks | References |
|---|---|---|---|---|
| **Flower traits** | | | | |
| Recurrent blooming | Major gene<br>R/r<br>r4 | *R. wichurana*,<br>Goldilocks × *R. wichurana*<br>Hybrid tea × *R. centifolia*, $F_1$ and $F_2$<br>Diploid *R .multiflora* populations<br>HW<br>Old Blush × WOB13/21/26 | Non recurrent dominant over recurrent | Semeniuck (1971a, b)<br>Vries and Dubois (1984)<br>Debener (1999)<br>Crespel et al. (2002)<br>Hibrand-Saint Oyant et al. (2007)<br>Shupert et al. (2007) |
| Repeat flowering ability | Major gene, Rb | 97/7 | Dominant, ability to flower in summer | Zhang (2003) |
| Flowering time | QTLs | Blush Noisette × *R. wichurana*<br>HW | ■ 50% of flowering<br>■ First flower to stage 1 | Dugo et al. (2005)<br>Hibrand-Saint Oyant et al. (2007) |
| Number of petals | Major gene<br>D/d<br>Blfo<br>d6<br>NP<br>Dc<br>QTLs | Tetraploid populations<br>Diploid *R. multiflora* populations<br>94/1R<br>Spain<br>Hw<br>97/7<br>Old Blush × WOB13/21/26<br>97/7<br>Hw | Dominant (single/double)<br>Lammerts (1945) indicates expression is also dose dependent | Lammerts (1945)<br>Debener (1999)<br>Debener and Mattiesch (1999)<br>Dugo et al. (2005)<br>Crespel et al. (2002)<br>Hibrand-Saint Oyant et al. (2007)<br>Zhang (2003)<br>Shupert et al. (2007)<br>Zhang (2003)<br>Hibrand-Saint Oyant et al. (2007) |

**Table 4** (continued)

| Character | Genetic control | Studied population | Remarks | References |
|---|---|---|---|---|
| Flower size | QTl | Spain | Flower diameter | Dugo et al. (2005) |
| | Additive | Old Blush × WOB13/21/26 | Flower diameter, petal length | Shupert et al. (2007) |
| Pink vs. white | Major gene, Blfa | Diploid *R. multiflora* populations | Shupert et al. (2007) report codominance | Debener (1999) |
| | | Old Blush × WOB13/21/26 | | Shupert et al. (2007) |
| Presence of anthocyanins | Major gene, A/a | Tetraploid | Presence of anthocyanins is dominant | Lammerts (1964) |
| Magenta red color | Major gene, M/m | 2 tetraploid populations | Expression of M is dose dependent | Lammerts (1960) |
| Individual pigments | Additive | Tetraploid populations | | Vries et al. (1980) and Marshall et al. (1983) |
| Pale, pink, dark pink and crimson | QTLs | 97/7 | | Zhang (2003) |
| Yellow | Major gene | *Rosa foetida* × Hybrid teas, F1 and back crosses | Dominant | Vries and Dubois (1978) |
| Light vs. dark yellow | | Tetraploid populations | Light dominant over dark yellow | Lammerts (1960) |
| Yellow and silver reverse | Major gene | Tetraploid populations | Recessive to normal | Lammerts (1945) |
| Male sterility | Major gene | $F_1$ : *R. setigera* × *R. brunoni* | Dominant | Lewis and Basye (1961) |
| Moss character | Major gene | Hybrid tea × *R. centifolia mucosa*, $F_1$ and $F_2$ | Dominant | Vries and Dubois (1984) |
| Fragrance | Major gene | Bo × Anna, tetraploids | Dominant gene control presence/absence of monoterpene alcohol and aldehyde pathway | Cherri-Martin et al. (2007) and Lammerts (1945) |
| | Quantitative | Tetraploid roses | | |

**Table 4** (continued)

| Character | Genetic control | Studied population | Remarks | References |
|---|---|---|---|---|
| **Plant traits** | | | | |
| Spreading vs. sprawling growth | Major gene | 90–69 $F_2$ | Codominant | Rajapakse et al. (2001) |
| Dwarf vs. non dwarf | Major gene, D | Polyantha × *R. chinensis minima* | Dominant | Dubois and Vries (1987) |
| Climbing vs dwarf or bush growth type | Major gene | Tetraploid populations | Dominant | Lammerts (1945) and Morey (1954) |
| Plant vigor (10 parameters) | QTLs | 94/1 | | Yan et al. (2007) |
| Leaf size | QTLs | Old Blush × WOB13/21/26 | Leaf width, length | Shupert et al. (2007) |
| | | 94/1 | Leaf area, weight | Yan et al. (2007) |
| | | Tetraploid populations | Leaf width | Lammerts (1945) |
| | | 97/7 | Leaf area | Zhang (2003) |
| Glossy vs dull foliage | Major gene | Tetraploid populations | | Lammerts (1945) |
| Stem prickles | Major gene | Diploid *R. multiflora* populations | Prickles dominant | Debener (1999) |
| | QTLs | Old Blush × WOB13/21/26 | | Shupert et al. (2007) |
| | Additive | HW | | Crespel et al. (2002) |
| | | Tetraploid populations | | Lammerts (1945) |
| | | 97/7 | | Linde et al. (2006) |
| Prickle size | QTLs | 97/7 | | Zhang (2003) |
| Petiole prickles | Major gene | 90–69 $F_2$, tetraploid | Independent of stem prickle locus | Rajapakse et al. (2001) |
| Malate dehydrogenase | Major gene | 90–69 $F_2$ | Codominant | Rajapakse et al. (2001) |

Old Blush × WOB13/21/26 = Old Blush × WOB13, × WOB21, and WOB26, all WOB are siblings from the cross *R. wichurana* 'Basye's Thornless' × Old Blush but it was not established until 1971 by Semeniuck (1971a, b) and then further confirmed by several others (Vries and Dubois, 1984; Debener, 1999; Crespel et al., 2002; Dugo et al., 2005). Recently it has been mapped on the Hw map (Crespel et al., 2002) (Tables 2 and 3).

The inheritance of the *flower form* (single vs. double flowers) was first reported in 1999 (Debener, 1999; Debener and Mattiesch, 1999) as a dominant allele in the locus conferring the double corolla/flower state and has since been confirmed in other rose progenies (Crespel et al., 2002; Yan et al., 2005b; Linde et al., 2006; Shupert et al., 2007). This locus has been mapped on several rose populations with different names: Blfo, d6, Dc, NP (Debener and Mattiesch, 1999; Crespel et al., 2002; Zhang, 2003; Linde et al., 2006; Hibrand-Saint Oyant et al., 2007). Preliminary data using SSR markers showed that *Blfo* and *NP/d6* are on the same homologous linkage group (Hibrand-Saint Oyant, personal communication). Within the double corolla/flower category there is a wide range of petals from 5 to several hundred. Genetic analysis in several populations indicates that the number of petals within the double flower is additive and controlled by several QTLs (Debener et al., 2001; Zhang, 2003; Hibrand-Saint Oyant et al., 2007).

*Flower size* (flower diameter, petal length) appears to be additively inherited with a high ($h^2 > 0.70$) heritability (Lal et al., 1982; Dugo et al., 2005; Shupert et al., 2007). Mapping of this trait indicated that there were 4 major QTL loci, 2 each on the first two linkage groups of each of the parents (Dugo et al., 2005).

*Flower production* has not been well studied. A small study in India indicated that floriferousness (number of flowers a plant produced) was highly heritable with a broad sense heritability of 0.96. More recent work focussed on flower production during the warm summer months and mapped a single gene (*Rb* = repeat bloom) on the D2 linkage group of the 97/7 progeny (Zhang, 2003).

*Flower color* in rose is extremely important and has been studied repeatedly beginning with early studies which indicated which colors were dominant over the others and suggestions on the breeding course to obtain specific colors (Lammerts, 1945; Morey, 1954). Subsequent studies by Lammerts (1960; 1964) with tetraploid roses concluded that there was a gene *A/a* which controlled the presence of cyanidin (red) and a gene *M/m* which controlled the magenta color. With the magenta color gene, the color was dependent on the gene dose with dark purple being homozygous dominant (*MMMM*), a solerino purple corresponding to a duplex condition (*MMmm*) and a homozygous recessive state (*mmmm*) giving a non fading currant red (assuming a red background with dominant A allele present). Since the pink to red color is controlled by a range of anthocyanins, the inheritance of specific anthocyanins (quercetin, kaempferol, cyanidin, peonin and pelargonidin) was examined in tetraploid roses. Both these studies came to the conclusion that these pigments were quantitatively inherited (Vries et al., 1980; Marshall et al., 1983). Recent work with diploid populations has identified a locus (*Bfla*) on the A2/B2 linkage group of the 94/1 map which controls pink and white color. In this case, pink is dominant over white although in one population (WOB), the heterozygous plants were a lighter pink as compared to the homozygous pink plants (Debener, 1999; Debener and Mattiesch, 1999; Yan et al., 2005b; Shupert et al., 2007). Finally in the 97/7 population, when the flower color was rated into 4 categories (Crimson, dark pink, pink, and pale) several QTLs were located on the 4th linkage group of this map (Zhang, 2003).

Beyond anthocyanins, the color in roses is determined by a range in carotenoids which give colors in the *yellow*/orange range. Lammerts (1945; 1960) hypothesized from his breeding experience that deep yellow was recessive to light yellow. The only other study describes a general breeding procedure to develop yellow everblooming roses utilizing *R. foetida* as the source of yellow flower color. Although the authors never hypothesize its mode of inheritance, the breeding procedure indicates that yellow is dominant as it is expressed in the $F_1$ and segregates in the back cross progenies (Vries and Dubois, 1978).

The presence or absence of *white stripes* on the petals, although it did not segregate as expected for a single dominant trait, was nevertheless mapped onto the 97/7 rose map on the 3rd linkage group along with the loci for double flowers and prickles (Linde et al., 2006).

QTLs for *flowering time* (50% bloom) were located on two chromosomes by Dugo et al. (2005) and another QTL (first bloom) was located on another chromosome by Hibrand-Saint Oyant et al., 2007). It has been suggested that the location of the QTLs were different because the flowering times were measured differently, and thus the two groups measured different aspects of this complex trait (Hibrand-Saint Oyant et al., 2007).

*Flower fragrance* is a very complex trait which is controlled by a multitude of chemical pathways as well as environmentally sensitive (Verhoeven et al., 2003) but has been bred for successfully over the years as evidenced by the presence of wonderfully fragrant varieties. The inheritance appears to be quantitative (Lammerts, 1945) but recent work looking at the monoterpene alcohol and aldehyde pathway that is responsible for the typical European rose fragrance indicates that there is also a major gene is responsible for turning the pathway on and off (Cherri-Martin et al., 2007) in addition to an array of genes that modify the quantity of the various scent components. Current work in the functional genomics of fragrance in rose is beginning to identify the key enzymes that create and regulate the various fragrance components and promises to lead to the development of better tools and understanding of this essential trait (Vainstein et al., 2003)

## 6 Plant Traits

In studies of *growth type*, the climbing growth types was found to be dominant over non-climbing growth types (Lammerts, 1945; Morey, 1954), and the dwarf type was found to be dominant to non dwarf types (Dubois and Vries, 1987). It is not known if these two traits are controlled at the same locus. In a recent mapping study, when a spreading bush growth type was crossed with a sprawling ground cover growth type the resultant $F_1$ plant was of an intermediate growth type. Upon selfing the $F_1$

plant, the $F_2$ segregated for all three types in a ratio to suggest that these forms were inherited co dominantly (Rajapakse et al., 2001).

Other growth type related traits such as the *length of the flowering stem* (Lammerts, 1945), *the branch number*, *plant height* and *plant spread* (Lal et al., 1982) all were inherited as polygenic characters with the last three with a high broad sense heritability ($h^2 = 0.88–0.95$) (Lal et al., 1982).

Plant vigor, an important trait in cut rose breeding, was measured using 10 parameters (number of nodes, stem thickness, shoot length, chlorophyll content, leaf area, specific leaf weight, leaf dry weight, stem dry weight, total dry weight and growth rate) in a replicated progeny at two sites. The additive $h^2$ for each individual trait ranged from moderate to moderately high (0.48–0.71). QTL analysis identified 42 QTLs which clustered mainly on three chromosomes (I-2, I-6, I-7 of 94/1 population). Very importantly it was shown that although a GxE interaction exists it was much smaller than the main genetic effect of the trait (Yan et al., 2007).

*Leaf size* whether measured as leaf width, leaf area, or leaflet number was found to be quantitative in nature (Lammerts, 1945; Zhang, 2003; Shupert et al., 2007)

There is a major locus controlling the presence or absence of *stem prickles* with the presence being the dominant allele (Debener, 1999; Linde et al., 2006; Shupert et al., 2007). Another independent locus controls the presence or absence of petiole prickles (Rajapakse et al., 2001). Among germplasm with the dominant gene for stem prickles, the density and size of the prickles is quantitatively inherited (Lammerts, 1945; Lal et al., 1982; Crespel et al., 2002; Zhang, 2003).

Additional traits that have been analyzed include the *male sterility*, *dwarfness*, the *moss* trait (dominant with modifiers), *MDH-2* (co-dominant), *foliage glossiness* (glossy dominant over dull), *flower neck strength* (quantitative), and the *ability of rose callus to form somatic embryos* (additive) (Lammerts, 1945; Lewis and Basye, 1961; Vries and Dubois, 1984; Rajapakse et al., 2001; Burrell et al., 2006).

## 7 Adaptation Traits

There have been numerous reports on the varietal differences in resistance to black spot (Wenefrida and Spencer, 1993; Carlson-Nilsson, 2000), powdery mildew (Mence and Hildebrandt, 1966; Atkiss, 1978; Ferrero et al., 2001), nematodes (Wang et al., 2004a), and cold hardiness (Lehmushovi, 1987; Karam and Sullivan, 1991). As might have been expected for black spot and powdery mildew both major gene and quantitative inheritance has been reported.

Two major genes (Rdr1, Rbs) have been mapped for black spot resistance within 10 cM on the same linkage group (Malek and Debener, 1998; Debener et al., 2003; Zhang, 2003; Yan et al., 2005b). The Rdr1 locus was found tightly linked to an AFLP marker (M10) which was converted into a SCAR (SCM10), several RFLPs (BMA1-4) (Malek et al., 2000) and several BAC end markers (Kaufmann et al., 2003). Subsequently one of the RFLP fragments (BMA3) was converted into

CAP marker (RdrY) (Zhang, 2003). These markers may be useful in MAS (marker assisted selection).

Powdery mildew resistance was first reported as a dominant gene in 1945 by Lammerts and more recently the dominant resistance genes *Rpp1* (Linde and Debener, 2003; Linde et al., 2004) and *Rpm* (Zhang, 2003) as well as a major resistance QTL, *CRPM1* (Xu et al., 2005) have been mapped. A SCAR marker has been developed for the *Rpp1* loci and a RGA (Resistance Gene Analogue) derived RFLP marker (RGA22C) has been located close to the *CRPM1* locus (Xu et al., 2005) which should facilitate MAS (Linde et al., 2004). Other QTLs for powdery mildew have also been located on maps of 3 rose mapping populations (Zhang, 2003; Dugo et al., 2005; Xu et al., 2005, 2007; Linde et al., 2006). Unfortunately, at this point it is not known how these powdery mildew resistance major genes or QTLs are related.

Resistance to both nematodes (*Meloidogyne hapla*) and winter cold have been reported to be polygenic (Svejda, 1977b, a, 1979; Wang et al., 2004b).

## 8 Physical Maps, BAC Libraries and Current Situation for Positional Cloning

The construction of BAC libraries is the first step towards the construction of a physical map and the positional cloning of genes of interest in rose. Thus far two BAC libraries have been constructed and one is under construction (Kaufmann et al., 2003; Hess et al., 2007). Thus far there is a BAC library constructed from a colchicine-induced tetraploid of *Rosa rugosa* var alba cv C5-23 which has been used for chromosome walking and contig assembly for the Rdr1 black spot resistance gene (Kaufmann et al., 2003). The other BAC library was constructed from the *Rosa chinensis* cv 'Old Blush' with the objective of using it for positional cloning of the recurrent blooming gene (Hess et al., 2007). The two libraries are complementary as they are constructed from genetically diverse genotypes and used different restriction enzymes (*Hind*II versus *BamH*I) for the partial digestion step. A third library from the multiflora derived black spot resistant 88/124–46 rose comprises 60,000 clones with an average insert size of 46 kb. This is being done with pCLC04541 cosmid that allows the transmission of small inserts via Agrobacterium. The objective of this library is to achieve complementation of the black spot resistance gene Rdr1(Kaufmann et al., 2003; Debener, personal communication).

In the construction of both BAC libraries, the basic procedure describes by Zhang (2003) was modified to minimize the interference of the high levels of polysaccharides and phenols in rose tissue which tend to lower the resultant DNA quality. Although both the German (Kaufmann et al., 2003) and the American (Hess et al., 2007) laboratories grew the plants in the greenhouse and put the leaves through a dark treatment for 3–8 days before extraction, they differed in the subsequent steps. With the *R. rugosa* library, high levels of PVP 40 (4%), 0.13% DIECA, 0.1% ascorbic acid, and 0.5% 2-mercaptoethanol was used in the extraction buffer to counteract the high phenolic concentration of the initial extraction (Kaufmann et al., 2003). The approach with the 'Old Blush' library was to heavily fertilize and prune the rose bushes to induce rapid growth in the dark followed by an increased number of wash/filter steps (from 1–2 to 6) of the nuclei before they were embedded into the LMP agarose plugs (Hess et al., 2007).

Although a contig of the Rdr1 locus was constructed and the several markers (AFLP, CAP and BAC end markers) were mapped within a cM of the Rdr1 locus, the high heterozygosity of the rose genome created the situation of having allelic forms of the same locus. As compared to creating a BAC library of the homozygous plant, this situation in a heterozygous plant essentially doubles the work involved in the development of a contig and positional cloning of a gene (Kaufmann et al., 2003).

## 9 Synteny to Other Rosaceous Crops

The comparative genome studies will allow the transfer of genetic and genomic information between species (e.g. orthologous gene controlling disease resistance or flowering) and may help to elucidate the genome relationships and evolution within the Rosoideae tribe (*Rosa*, *Fragaria* and *Rubus*, mainly) and more distant genera within the Rosaceae family (such *Malus*, *Pyrus* or *Prunus*).

The alignment of the genetic maps reveals a high synteny between species of a same genus, for instance, it was proposed that the *Prunus* genome can be treated as a single genetic entity (Arus et al., 2006), with a single exception of a reciprocal translocation. In rose, the first step was to develop consensus markers with other Rosaceae species. The first attempt was done using heterologous microsatellite primer pairs from *Malus*, *Prunus*, and *Fragaria*. Of the 31 apple, 22 peach and 8 sour cherry SSRs screened on rose DNA only a two apple, 6 peach, and none of the sour cherry SSRs were mapped to a rose map (Rajapakse et al., 2001; Dugo et al., 2005; Zhang et al., 2006; Hibrand-Saint Oyant et al., 2007). Recently, the screening of strawberry primers on rose DNA revealed a much higher rate of good amplification and polymorphism and seven new consensus polymorphic SSR markers were developed (Hibrand-Saint Oyant et al., 2007). Anonymous SSRs are not convenient markers for genome comparison in Rosaceae. Their transferability is too low (problem of amplification and present of the SSR) between the tribes. However, within the tribe, in the case of SSR, in the coding sequence (as EST-SSR) the transferability is higher, and those markers can be used in synteny studies (Rousseau et al., 2006). SSRs from *Fragaria* were transferred to *Rosa* (and vice versa) not by looking for SSR polymorphism in the homologue sequences but by detecting SNPs using the SSCP technique. By such a method, the transferability was increased to 10%. Furthermore, strawberry homologues of candidate genes for floral initiation in rose were isolated in strawberry (Rousseau et al., 2006). Twenty four candidate genes were tested and 25% can be amplified and were polymorphic in strawberry.

From these new consensus markers, Rousseau et al. (2006) started a macrosynteny study between *Rosa* and *Fragaria*. The 24 markers provided anchor points to link the 7 homologous linkage groups. Preliminary results indicate numerous rearrangements between Rosa and Fragaria (Hibrand Saint Oyant L. and Denoyes, unpublished). New consensus markers need to be developed, and those markers have also to be tested in other species from the Rosoideae tribe (*Rubus*) and Rosaceae family. Possible good consensus markers are marker based on coding sequences that are single or low copy. This is the case for COS (Conserved Orthologue Set) previously developed in tomato (Fulton et al., 2002) or the Gene Pair Haplotype developed in strawberry (Davis et al., 2008).

Interestingly, macrosynteny between the strawberry and rose diploids reveal that the locus controlling recurrent blooming and perpetual flowering in rose and strawberry respectively are homologous (Hibrand Saint Oyant L. and Denoyes B., unpublished). This raised the question whether the *RB* locus has the same function as genes for seasonal/recurrent flowering in strawberry (Albani et al., 2004) and primocane bearing in blackberry and raspberry (Lopez-Medina and Moore, 1999; Lim and Knight, 2000).

To date, QTL mapping has been the method of choice for localizing the genetic basis of ornamental traits in rose. However QTL mapping presents limitations as the need for large progenies (especially in perennial species as rose), the time and space necessary for their phenotyping, and the fact that traits depend on the environment. Now genetics are moving towards LD (Linkage Disequilibrium) mapping to identify genomic regions contributing to specific phenotypes. LD mapping identifies correlations between a trait and a marker (reviewed by Mackay and Powell, 2007). According to LD decay, LD mapping may help to validate candidate genes or detect new QTLs in genome wide associations. The major advantage of LD mapping is that you do not need crosses as it relies on population samples. Development of LD mapping will need an international effort for the development of adequate populations (determination of the number of individual, population structure) and high density marker map to identify genome wide associations. The development of high density, based on SSR or SNP map will be feasible with the complete genome sequencing of rose.

## References

Albani, M.C., Battey, N.H., and Wilkinson, M.J. (2004). The development of ISSR-derived SCAR markers around the SEASONAL FLOWERING LOCUS (SFL) in Fragaria vesca. Theoretical and Applied Genetics 109, 571–579.

Arus, P., Yamamoto, T., Dirlewanger, E., and Abbott, A.G. (2006). Synteny in the Rosaceae. Plant Breeding Reviews 27, 175–211.

Atkiss, L. (1978). Disease resistant rose varieties. American Rose Annual 63, 99–104.

Babaei, A., Tabaei-Aghdaei, S.R., Khosh-Khui, M., Omidbaigi, R., Naghavi, M.R., Esselink, G.D., and Smulders, M.J.M. (2007). Microsatellite analysis of Damask rose (Rosa damascena Mill.) accessions from various regions in Iran reveals multiple genotypes. BMC Plant Biology 7, (8 March 2007).

Baydar, N.G., Baydar, H., and Debener, T. (2004). Analysis of genetic relationships among Rosa damascena plants grown in Turkey by using AFLP and microsatellite markers. Journal of Biotechnology 111, 263–267.

Benedetti, L. de, Mercuri, A., Bruna, S., et al. (2001). Genotype identification of ornamental species by RAPD analysis. Proceedings of the International Symposium on Molecular Markers for Characterizing Genotypes and Identifying Cultivars in Horticulture, Montpellier, France, 6–8 March 2000. Acta Horticulturae 546, 391–394.

Burrell, A.M., Lineberger, R.D., Rathore, K.S., and Byrne, D.H. (2006). Genetic variation in somatic embryogenesis of rose. HortScience 41, 1165–1168.

Byrne, D.H. (2007). Molecular marker use in perennial plant breeding. Proceedings of the Fourth International Symposium on Rose Research and Cultivation, California, USA, 18–22 September 2005. Acta Horticulturae 751, 163–167.

Carlson-Nilsson, B.U. (2000). Resistance to Marssonina rosae in Rosa L. seedlings obtained from controlled crosses including germplasm L83. Acta Agriculturae Scandinavica. Section B, Soil and Plant Science 50, 176–182.

Cherri-Martin, M., Jullien, F., Heizmann, P., and Baudino, S. (2007). Fragrance heritability in Hybrid Tea roses. Scientia Horticulturae 113, 177–181.

Crespel, L., Zhang, D., Meynet, J., and Gudin, S. (2001). Estimation of heterozygosity in two botanic rose species using AFLP markers. Acta Horticulturae 546, 187–191.

Crespel, L., Chirollet, M., Durel, C.E., Zhang, D., Meynet, J., and Gudin, S. (2002). Mapping of qualitative and quantitative phenotypic traits in Rosa using AFLP markers. Theoretical and Applied Genetics 105, 1207–1214.

Davis, T.M., Folta, K.M., Shields, Z., and Zhang, Q. (2008). Gene Pair Markers: An Innovative Tool for Comparative Linkage Mapping. In North American Strawberry Symposium, F. Takeda, D.T. Handley, and E.B. Poling, eds. Proc. 2007. pp. 105–107. North American Strawberry Growers Association, Kemptville, ON Canada.

Debener, T. (1999). Genetic analysis of horticulturally important morphological and physiological characters in diploid roses. Gartenbauwissenschaft 64, 14–20.

Debener, T., and Mattiesch, L. (1999). Construction of a genetic linkage map for roses using RAPD and AFLP markers. Theoretical and Applied Genetics 99, 891–899.

Debener, T., Bartels, C., and Mattiesch, L. (1996). RAPD analysis of genetic variation between a group of rose cultivars and selected wild rose species. Molecular Breeding 2, 321–327.

Debener, T., Bartels, C., and Spethmann, W. (1997). Parentage analysis in interspecific crosses between rose species with RAPD markers. Gartenbauwissenschaft 62, 180–184.

Debener, T., Janakiram, T., and Mattiesch, L. (2000). Sports and seedlings of rose varieties analysed with molecular markers. Plant Breeding 119, 71–74.

Debener, T., Mattiesch, L., and Vosman, B. (2001). A molecular marker map for roses. Acta Horticulturae 547, 283–287.

Debener, T., Dohm, A., and Mattiesch, L. (2003). Use of diploid self incompatible rose genotypes as a tool for gene flow analyses in roses. Plant Breeding 122, 285–287.

Dubois, L.A.M., and Vries, D.P.d. (1987). On the inheritance of the dwarf character in polyantha × Rosa chinensis minima (Sims) Voss F1-populations. Euphytica 36, 535–539.

Dugo, M.L., Satovic, Z., Millan, T., Cubero, J.I., Rubiales, D., Cabrera, A., and Torres, A.M. (2005). Genetic mapping of QTLs controlling horticultural traits in diploid roses. TAG Theoretical and Applied Genetics 111, 511–520.

Esselink, G.D., Smulders, M.J.M., and Vosman, B. (2003). Identification of cut rose (Rosa hybrida) and rootstock varieties using robust sequence tagged microsatellite site markers. Theoretical and Applied Genetics 106, 277–286.

Esselink, G.D., Nybom, H., and Vosman, B. (2004). Assignment of allelic configuration in polyploids using the MAC-PR (microsatellite DNA allele counting-peak ratios) method. Theoretical and Applied Genetics 109, 402–408.

Fang, J., Devanand, P.S., and Chao, C.T. (2005). Practical strategy for identification of single nucleotde polymorphisms in fruiting mei (*Prunus mume* Sied. et Zucc.) from amplified fragment lenght polymorphism fragments. Plant Molecular Biology 23, 227–239.

Ferrero, F., Cadour-Marvaldi, P., Guilotea, E., Jacob, Y., Coudret, A., and Sallanon, H. (2001). Evaluation of the resistance to powdery mildew, *Sphaerotheca pannosa* var. rosae, of rose-tree species and hybrids. B. First exploration of the resistance biodiversity in a crossing program resistance. Acta Horticulturae 547, 379–381.

Fulton, T.M., Hoeven, R.v.d., Eannetta, N.T., and Tanksley, S.D. (2002). Identification, analysis, and utilization of conserved ortholog set markers for comparative genomics in higher plants. Plant Cell 14, 1457–1467.

Gallego, F.J., and Martinez, I. (1996). Molecular typing of rose cultivars using RAPDs. Journal of Horticultural Science 71, 901–908.

Grattapaglia, D., and Sederoff, R. (1994). Genetic linkage maps of Eucalyptus grandis and Eucalyptus urophylla using a pseudo-testcross: mapping strategy and RAPD markers. Genetics 137, 1121–1137.

Hattendorf, A., Debener, T. (2007a). NBS-LRR-RGAS in roses: diversity, genomic organization, expression and chromosomal location. Proceedings of the Fourth International Symposium on Rose Research and Cultivation, California, USA, 18–22 September 2005. Acta Horticulturae 751, 151–162.

Hattendorf, A., and Debener, T. (2007b). Molecular characterization of NBS-LRR-RGAs in the rose genome. Physiologia Plantarum 129, 775–786.

Hess, G., Byrne, D.H., Zhang, H.B. (2007). Toward positional cloning of everblooming gene (*evb*) in plants: a BAC library of *Rosa chinensis* cv. 'Old Blush'. Proceedings of the Fourth International Symposium on Rose Research and Cultivation, California, USA, 18–22 September 2005. Acta Horticulturae 751, 169–174.

Hibrand-Saint Oyant, L., Crespel, L., Rajapakse, S., Zhang, L., and Foucher, F. (2008). Genetic linkage maps of rose constructed with new microsatellite markers and locating QTL controlling flowering traits. Tree Genetics and Genomes 4, 11–23.

Hurst, C.C. (1941). Notes on the origin and evolution of our garden roses. Journal of Royal Horticultural Society 66, 73–82.

Iwata, H., Kato, T., and Ohno, S. (2000). Triparental origin of Damask roses. Gene 259, 53–59.

Jones, C.J., Edwards, K.J., Castaglione, S., Winfield, M.O., Sala, F., Wiel, C.v.d., Bredemeijer, G., Vosman, B., Matthes, M., Daly, A., Brettschneider, R., Bettini, P., Buiatti, M., Maestri, E., Malcevschi, A., Marmiroli, N., Aert, R., Volckaert, G., Rueda, J., Linacero, R., Vazquez, A., and Karp, A. (1997). Reproducibility testing of RAPD, AFLP and SSR markers in plants by a network of European laboratories. Molecular Breeding 3, 381–390.

Karam, F.H., and Sullivan, J.A. (1991). A rapid method for detection of cold hardiness in roses. HortScience 26, 59–60.

Kaufmann, H., Mattiesch, L., Lorz, H., and Debener, T. (2003). Construction of a BAC library of Rosa rugosa Thunb. and assembly of a contig spanning Rdr1, a gene that confers resistance to blackspot. Molecular Genetics and Genomics 268, 666–674.

Kiani, M., Zamani, Z., Khalighi, A., Fatahi, R., and Byrne, D.H. (2008). Wild genetic diversity of *Rosa damascena* Mill. germplasm in Iran revealed by RAPD analysis. Scientia Horticulturae, 115:386–392.

Kimura, T., Nishitani, C., Iketani, H., Ban, Y., and Yamamoto, T. (2006). Development of microsatellite markers in rose. Molecular Ecology Notes 6, 810–812.

Koopman, W.J.M., Wissemann, V., Cock, K. de, et al. (2008). AFLP markers as a tool to reconstruct complex relationships: a case study in Rosa (Rosaceae). American Journal of Botany 95(3), 353–366.

Krüssmann, G. (1981). The complete book of roses. (Portland, Ore.: Timber Press).

Lal, S.D., Seth, J.N., Yadav, J.P., and Danu, N.S. (1982). Genetic variability and correlation studies in rose. I. Phenotypic variability, heritability and genetic advance. Progressive Horticulture 14, 234–236.

Lammerts, W.E. (1945). The scientific basis of rose breeding. American Rose Annual 30, 70–79.

Lammerts, W.E. (1960). Inheritence of magenta color in roses. American Rose Annual 45, 119–125.

Lammerts, W.E. (1964). Inheritance of the scarlet-vermillon signal red colors. American Rose Annual 49, 167–172.

Lehmushovi, A. (1987). Rose varieties in the experimental filed 1981–1984. Annales Agriculturae Fenniae 26, 4.

Lewis, W.H., and Basye, R.E. (1961). Analysis of nine crosses between diploid Rosa species. Proceedings of the American Society for Horticultural Science 78, 572–579.

Lim, Y.K., and Knight, V.H. (2000). The successful transfer of primocane fruiting expression from raspberry to Rubus hybrid berry. Euphytica 116, 257–263.

Linde, M., and Debener, T. (2003). Isolation and identification of eight races of powdery mildew of roses (Podosphaera pannosa) (Wallr.: Fr.) de Bary and the genetic analysis of the resistance gene Rpp1. Theoretical and Applied Genetics 107, 256–262.

Linde, M., Mattiesch, L., and Debener, T. (2004). Rpp1, a dominant gene providing race-specific resistance to rose powdery mildew (Podosphaera pannosa): molecular mapping, SCAR development and confirmation of disease resistance data. Theoretical and Applied Genetics 109, 1261–1266.

Linde, M., Hattendorf, A., Kaufmann, H., and Debener, T. (2006). Powdery mildew resistance in roses: QTL mapping in different environments using selective genotyping. TAG Theoretical and Applied Genetics 113, 1081–1092.

Lopez-Medina, J., and Moore, J.N. (1999). Chilling enhances cane elongation and flowering in primocane-fruiting blackberries. HortScience 34, 638–640.

Mackay, I., and Powell, W. (2007). Methods for linkage disequilibrium mapping in crops. Trends in Plant Science 12, 57–63.

Malek, B.v., and Debener, T. (1998). Genetic analysis of resistance to blackspot (Diplocarpon rosae) in tetraploid roses. Theoretical and Applied Genetics 96, 228–231.

Malek, B.v., Weber, W.E., and Debener, T. (2000). Identification of molecular markers linked to Rdr1, a gene conferring resistance to blackspot in roses. Theoretical and Applied Genetics 101, 977–983.

Marshall, H.H., Campbell, C.G., and Collicutt, L.M. (1983). Breeding for anthocyanin colors in Rosa. Euphytica 32, 205–216.

Matsumoto, S., and Fukui, H. (1996). Identification of rose cultivars and clonal plants by random amplified polymorphic DNA. Scientia Horticulturae 67, 49–54.

Mence, M.J., and Hildebrandt, A.C. (1966). Resistance to powdery mildew in rose. Annals of Applied Biology 58, 309–320.

Millan, T., Osuna, F., Cobos, S., Torres, A.M., and Cubero, J.I. (1996). Using RAPDs to study phylogenetic relationships in Rosa. Theoretical and Applied Genetics 92, 273–277.

Mohapatra, A., and Rout, G.R. (2005). Identification and analysis of genetic variation among rose cultivars using random amplified polymorphic DNA. Zeitschrift fur Naturforschung 60, 611–617.

Morey, D. (1954). Observation on the genetics of the mutant climbing factor in hybrid tea roses. American Rose Annual 39, 89–97.

Nybom, H., Esselink, G.D., Werlemark, G., and Vosman, B. (2004). Microsatellite DNA marker inheritance indicates preferential pairing between two highly homologous genomes in polyploid and hemisexual dog-roses, Rosa L. sect. Caninae DC. Heredity 92, 139–150.

Nybom, H., Esselink, G.D., Werlemark, G., Leus, L., and Vosman, B. (2006). Unique genomic configuration revealed by microsatellite DNA in polyploid dogroses, Rosa sect. Caninae. Journal of Evolutionary Biology 19, 635–648.

Olsson, A., Nybom, H., and Prentice, H.C. (2000). Relationships between Nordic dogroses (Rosa L. sect. Caninae, Rosaceae) assessed by RAPDs and elliptic Fourier analysis of leaflet shape. Systematic Botany 25, 511–521.

Rajapakse, S., Byrne, D.H., Zhang, L., and erson, N., Arumuganathan, K., and Ballard, R.E. (2001). Two genetic linkage maps of tetraploid roses. Theoretical and Applied Genetics 103, 575–583.

Rousseau, M., Hibrand-Saint Oyant, L., Foucher, F., Barrot, L., Lalanne, L., Sargent, D., Simpson, D., Laigret, F., and Desnoyes-Rothan, B. (2006). Comparative mapping in the Rosoideae tribe: *Rosa* and *Fragaria*. Napier, NZ.

Rusanov, K., Kovacheva, N., Vosman, B., Zhang, L., Rajapakse, S., Atanassov, A., and Atanassov, I. (2005). Microsatellite analysis of Rosa damascena Mill. accessions reveals genetic similarity between genotypes used for rose oil production and old Damask rose varieties. TAG Theoretical and Applied Genetics 111, 804–809.

Scariot, V., Akkak, A., and Botta, R. (2006). Characterization and genetic relationships of wild species and old garden roses based on microsatellite analysis. Journal of the American Society for Horticultural Science 131, 66–73.

Semeniuck, P. (1971a). Inheritance of recurrent blooming in *Rosa wichuraiana*. The Journal of Heridity 62.

Semeniuck, P. (1971b). Inheritance of recurrent and non recurrent in 'Goldilocks' × *Rosa wichuraiana* progeny. The Journal of Heridity 62.

Shupert, D.A., Byrne, D.H., Pemberton, H.B. (2007). Inheritance of flower traits, leaflet number and prickles in roses. Proceedings of the Fourth International Symposium on Rose Research and Cultivation, California, USA, 18–22 September 2005. Acta Horticulturae 751, 331–335.

Svejda, F. (1977a). Breeding for improvement of flowering attributes of winterhardy Rosa rugosa hybrids. Euphytica 26, 697–701.

Svejda, F. (1977b). Breeding for improvement of flowering attributes of winterhardy Rosa kordesii Wulff hybrids. Euphytica 26, 703–708.

Svejda, F. (1979). Inheritance of winterhardiness in roses. Euphytica 28, 309–314.

Tabaei-Aghdaei, S.R., Monfared, H.H., Fahimi, H., Ebrahimzade, H., Jebelly, M., Naghavi, M.R., and Babaei, A. (2006). Genetic variation analysis of different populations of Rosa damascena in NW. Iran using RAPD markers. Iranian Journal of Botany 12, 121–127.

Torres, A.M., Millan, T., and Cubero, J.I. (1993). Identifying rose cultivars using random amplified polymorphic DNA markers. HortScience 28, 333–334.

Union for the Protection of New Varieties of Plants. 1991. 1991 Act Chapter 6, Article 20 Variety Denomination. 26 Feb. 2008. http://www.upov.int/en/publications/conventions/1991/w_up912_.htm#_20>.

Vainstein, A., Zamir, D., and Weiss, D. (2003). Fragrance and pigments/Functional genomics. In Encyclopedia of rose science, A.V. Roberts, T. Debener, and S. Gudin, eds (Oxford: Elsevier), pp. 263–265.

Verhoeven, H.A., Blass, J., and Brandenbourg, W.A. (2003). Fragrance profiles of wild and cultivated roses. In Encyclopedia of rose science, A.V. Roberts, T. Debener, and S. Gudin, eds (Oxford: Elsevier), pp. 240–248.

Vosman, B., Visser, D., Voort, J.R.v.d., Smulders, M.J.M., and Eeuwijk, F.v. (2004). The establishment of 'essential derivation' among rose varieties, using AFLP. Theoretical and Applied Genetics 109, 1718–1725.

Vries, D.P.d., and Dubois, L.A.M. (1978). On the transmission of the yellow flower colour from Rosa foetida to recurrent flowering hybrid tea-roses. Euphytica 27, 205–210.

Vries, D.P.d., and Dubois, L.A.M. (1984). Inheritance of the recurrent flowering and moss characters in F1 and F2 hybrid tea × R. centifolia muscosa (Aiton) Seringe populations. Gartenbauwissenschaft 49, 97–100.

Vries, D.P.d., Garretsen, F., Dubois, L.A.M., and Keulen, H.A.v. (1980). Breeding research on rose pigments. II. Combining ability analyses of variance of four flavonoids in F1 populations. Euphytica 29, 115–120.

Walker, C.A., Jr., and Werner, D.J. (1997). Isozyme and randomly amplified polymorphic DNA (RAPD) analyses of Cherokee rose and its putative hybrids 'Silver Moon' and 'Anemone'. Journal of the American Society for Horticultural Science 122, 659–664.

Wang, D., Fan, J., and Ranu, R.S. (2004a). Cloning and expression of 1-aminocyclopropane-1-carboxylate synthase cDNA from rosa (Rosa × hybrida). Plant Cell Reports 22, 422–429.

Wang, X., Jacob, Y., Mastrantuono, S., Bazzano, J., Voisin, R., and Esmenjaud, D. (2004b). Spectrum and inheritance of resistance to the root-knot nematode Meloidogyne hapla in Rosa multiflora and R. indica. Plant Breeding 123, 79–83.

Wen, X.P., and Deng, X.X. (2005). Micropropagation of chestnut rose (Rosa roxburghii Tratt) and assessment of genetic stability in in vitro plants using RAPD and AFLP markers. Journal of Horticultural Science and Biotechnology 80, 54–60.

Wenefrida, I., and Spencer, J.A. (1993). Marssonia rosae variants in Mississippi and their virulence on selected rose cultivars. Plant Disease 77, 246–248.

Werlemark, G., Uggla, M., and Nybom, H. (1999). Morphological and RAPD markers show a highly skewed distribution in a pair of reciprocal crosses between hemisexual dog rose species, Rosa sect. Caninae. Theoretical and Applied Genetics 98, 557–563.

Xu, Q., Wen, X., and Deng, X. (2005). Isolation of TIR and nonTIR NBS-LRR resistance gene analogues and identification of molecular markers linked to a powdery mildew resistance locus in chestnut rose (Rosa roxburghii Tratt). TAG Theoretical and Applied Genetics 111, 819–830.

Xu, Q., Wen, X., and Deng, X. (2007). Cloning of two classes of PR genes and the development of SNAP markers for powdery mildew resistance loci in chestnut rose (Rosa roxburghii Tratt). Molecular Breeding 19, 179–191.

Yan, Z., Bai, Y., and Silva, J.A.T.d. (2006). Molecular markers and their use in genetic studies in rose. In Floriculture, ornamental and plant biotechnology, pp. 498–503.

Yan, Z., Visser, P.B., Hendriks, T., Prins, T.W., Stam, P., and Dolstra, O. (2007). QTL analysis of variation for vigour in rose. Euphytica 154, 53–62.

Yan, Z., Denneboom, C., Hattendorf, A., Dolstra, O., Debener, T., Stam, P., and Visser, P.B. (2005a). Construction of an integrated map of rose with AFLP, SSR, PK, RGA, RFLP, SCAR and morphological markers. TAG Theoretical and Applied Genetics 110, 766–777.

Yan, Z.F., Dolstra, O., Hendriks, T., Prins, T.W., Stam, P., and Visser, P.B. (2005b). Vigour evaluation for genetics and breeding in rose. Euphytica 145, 339–347.

Zhang, D., Besse, C., Cao, M.Q., and Gandelin, M.H. (2001). Evaluation of AFLPS for variety identification in modern rose (Rosa hybrida L.). Acta Horticulturae 546, 351–357.

Zhang, L.H. (2003). Genetic linkage map in tetraploid and diploid rose (Clemson, SC: Clemson University).

Zhang, L.H., Byrne, D.H., Ballard, R.E., and Rajapakse, S. (2006). Microsatellite marker development in rose and its application in tetraploid mapping. Journal of the American Society for Horticultural Science 131, 380–387.

# 18. Functional Genomics in Rose

**Fabrice Foucher**

## 1 Introduction

Rose is in the infancy of genomics as the genome sequence is not available and only few tools have been developed. Here, we will present all the tools (cDNA libraries, EST, databases, microarrays) that have been developed and their uses in molecular approaches to study different ornamental traits as scent, color, flower development and senescence.

## 2 The Tools Available for the Genomic Approaches

### *2.1 cDNA Libraries*

Several cDNA libraries have been constructed from petals at different stages, and recently from vegetative and floral apices (Table 1).

### *2.2 EST Production*

From these cDNA libraries, ESTs have been sequenced. Today, in the public databases (NCBI sources, www.ncbi.nlm.nih.gov), 9,289 rose ESTs are present. This number represents a small numbers as compared to other Rosaceae species such as apple (260,000) or *Prunus* (90,000). However in the rosoid section, this EST production is significant compared to the 20,000 ESTs available in strawberry or 300 EST in *Rubus*.

Channeliere et al. (2002) presented the first overview of genes expressed in rose petals. Petals represent important organs as they influence flower form, determine

---

F. Foucher (✉)
Inra Angers, UMR Génétique et Horticulture (GenHort), Angers, France

K.M. Folta, S.E. Gardiner (eds.), *Genetics and Genomics of Rosaceae*, Plant Genetics and Genomics: Crops and Models 6, DOI 10.1007/978-0-387-77491-6_18, 

**Table 1** List of the cDNA libraries developed in rose

| Rose varieties | Tissues and stages | Number of ESTs | References |
|---|---|---|---|
| *R. hybrida* 'Kardinal' | Petal Stage 4 of senescing rose | – | Fukuchi-Mizutani et al. (1995) and Fukuchi-Mizutani et al. (2000) |
| *R. hybrida* 'Kardinal' | Petal Stage 8 of senescing rose | – | Wang et al. (2004a) |
| *R. hybrida* 'Ilseta' | Young flowers | – | Chmelnitsky et al. (2003) |
| *R. chinensis* 'Old Blush' (OB) | Petals at different stages | 1794 | Channeliere et al. (2002) |
| *R. hybrida* 'Fragrant Cloud' (FC) | Petal at stage 4 | 2466 | Guterman et al. (2002) |
| *R. hybrida* 'Golden Gate' (GG) | Petal at stage 4 | 1045 | Guterman et al. (2002) |
| *R. wichurana* | Vegetative apices | 1932 | Foucher et al. (manuscript submitted) |
| *R. hybrida* 'Black Baccara' (BB) | Terminal apices during the floral transition | 2041 | Foucher et al. (manuscript submitted) |

vase life and are the principal site of scent production. From a *Rosa chinensis* 'Old Blush' petal cDNA library, 1,794 5'ESTs were produced and clustered in 877 unique sequences. After gene annotation, a large proportion of the sequences had no known function. For the ESTs with an assigned function, the main categories are defense or stress function. Furthermore, a significant proportion of EST showed similarities with genes involved in primary and secondary metabolism, putatively involved in scent production (Channeliere et al., 2002).

In another study, Guterman et al (2002) used two different rose varieties: *R. hybrida* 'Fragrant Cloud' (FC) and 'Golden Gate' (GG). FC is an old garden variety with large red fragrant flowers that have a short vase life, whereas GG is a modern cut-flower rose with small yellow flowers and only faint scent, but it has a long vase life. From both cDNA libraries, ~3,500 ESTs were produced (Table 1). This set of ESTs represents after clustering 2,139 unique sequences (contigs). The distribution of unique genes among functional classes (based on the Munich Information Center Sequence) is identical between the two rose cultivars (Guterman et al., 2002). A large number (around 15%) are putatively unique sequences related to metabolism, and might be involved in scent production. Interestingly, a contig, with a strong similarity with a sesquiterpene synthase, contains ESTs only from FC. This gene may be involved in the synthesis of terpene, one important component of rose scent (Flament et al., 1993).

Two homologues of *AGAMOUS*, C type genes, were isolated (*MASAKO C1* and *D1*) in *R. rugosa* (Kitahara and Matsumoto, 2000). These two genes are expressed in the stamen and carpel, as is found for the *AGAMOUS* gene in Arabidopsis. Interestingly, the authors suggested the presence of alternative splicing for the *AG* homologue in rose. RAG, the homologue of *MASAKO C1* in *Rosa hybrida cv. Motrea*, might be involved in phyllody, the replacement of floral organs with leaf-like organs in flowers (Meyer, 1966). *RAG* expression in plants presenting phyllody such as *R. chinensis viridiflora* or *R. hybrida* cv. Motrea (Chmelnitsky et al., 2003) is delayed during the floral development, which may be responsible of the phyllody phenotype.

Similarly, using degenerated primers, the B-type floral identity homologues, *MASAKO BP* and *B3* of *PISTILLATA* and *APETALA3* respectively, were isolated in rose (Kitahara et al., 2001). Both genes are expressed in petals and stamens as expected for B-type genes, which are required, in interaction with A and C types, for the development of these floral organs (Weigel and Meyerowitz, 1994). Another rose B-type gene, *MASAKO euB3*, homologue to *APETALA3*, is expressed in all floral organs (Hibino et al., 2006), an expression pattern which is not expected for a B-type gene. According to phylogenetic analysis, *MASAKO euB3* belongs to the *euAP3* lineage, whereas *MASAKO B3* belongs to the *TM6* lineage (Kitahara et al., 2001; Hibino et al., 2006). Functional analysis has been conducted to precisely elucidate the involvement of these genes in rose floral development (see in translational genomics part, this chapter, T. Debener).

## 3 Ethylene Signaling and Metabolism and Its Involvement in Rose Opening and Senescence

The plant hormone ethylene is involved in the regulation of a number of physiological and developmental processes (Johnson and Ecker, 1998), and particularly in flower development, flower opening and senescence. Its role as a regulator of flower development is of particular economic importance in rose flower production. To study the involvement of ethylene in flower development, candidate genes were found by cloning in rose the homologues of genes known in Arabidopsis to play a role in ethylene synthesis and signaling.

## 4 Cloning of Genes Involved in Ethylene Signaling and Synthesis

Ethylene biosynthesis is well characterized and genes coding for the enzymes involved are well known in different plants. Ethylene is synthesized from methionine through the cyclic amino acid 1-aminocyclopropane-1-carbosylic acid (ACC) intermediates (Wang et al., 2002). The ACC synthase converts SAM (S-adenosyl methionine) to ACC, whereas the ACC oxidase converts ACC into ethylene. One ACC oxidases (*RhACO1*) and three ACC synthase (*RhACS1 to 3*) homologues were isolated in modern roses (Muller et al., 2000b; Wang et al., 2004a; Ma et al., 2006).

Ethylene synthesis, perception and signaling has been extensively studied in Arabidopsis (Bishopp et al., 2006). At low concentration of ethylene, ethylene receptors are active and can stimulate the negative regulator CTR1 (Constitutive Triple Response 1), which in turn shuts down ethylene signaling by allowing EIN3 (Ethylene Insensitive 3) degradation. Ethylene binding inactivates the receptors and therefore they fail to activate CTR1. As a result EIN3 is not degraded, which leads to EIN3 accumulation and activation of ethylene responsive genes. Using degenerated primers, four partial cDNAs with high similarity with ethylene receptors *RhETR1* to *4* (Muller et al., 2000a), two partial *CTR1* homologues, *RhCTR1* and *2* (Muller et al., 2002) and one *EIN3* homologue, *RhEIN3* (Muller and Stummann, 2003) were isolated in miniature roses. In addition a fifth ETR5 homologue (*RhETR5)* and a second *EIN3* homologue (*RhEIN3-2*) were isolated in *R. hybrida* cv Kardinal (Tan et al., 2006). By expression analysis, the involvement of theses genes was studied during flower opening and senescence.

## 5 Ethylene and Flower Opening

Roses respond differently to ethylene during flower development. For instance, in *R. hybrida* cv. 'Samantha', flower development is stimulated by ethylene, whereas in *R. hybrida* cv. 'Kardinal', the flower development is inhibited by ethylene (Ma et al., 2005). In 'Samantha', ethylene induced ethylene synthesis via upregulation of *RhACS3* and *RhACO1* (Ma et al., 2006). *RhETR1, RhETR3, RhCTR1* and *RhCTR2* are also induced by ethylene during floral opening and inhibited by ethylene inhibitors in 'Samantha'. Ethylene biosynthesis genes (*RhACS1, 2, 3* and *RhACO1*) are expressed similarly during flower development in 'Samantha' and 'Kardinal'. Differences between 'Kardinal' and 'Samantha' might be in ethylene response, as 'Kardinal' seems to be more responsive to ethylene (Ma et al., 2005) and produces only low levels of *RhETR* transcripts (Tan et al., 2006). However *RhCTRs* are expressed more in 'Kardinal' than in 'Samantha', and *RhEIN3* is constitutively expressed (Tan et al., 2006). These results are contradictory, and further experiments are needed to understand the differences in ethylene signaling between 'Samantha' and 'Kardinal'.

## 6 Ethylene and Flower Senescence

Flower senescence was studied by comparing two opposite rose varieties with respect to senescence behavior: *R. hybrida* cv. 'Vanilla' has a long vase life and is insensitive to ethylene, whereas *R. hybrida* cv. 'Bronze' is a short vase life and is sensitive to ethylene. The *RhACS* transcript increased during flower senescence in 'Vanilla', but remained at low level in 'Bronze' (Muller et al., 2000b), whereas *RhACO* transcript level increased more in 'Bronze' as compared to 'Vanilla'. In *R. hybrida* cv. 'Kardinal', the expression of the *RhACS1* increased dramatically as did the ethylene levels as the flower matured towards senescence (Wang et al., 2004a). Up regulation of this ethylene synthesis genes may be responsible for ethylene synthesis late in flower development and acceleration of petal senescence.

Interestingly, levels of ethylene receptors *RhETR1/2* transcripts in ‘Bronze’ (short vase life) increase more than in ‘Vanilla’ (long vase life) and *RhETR3* is induced during senescence in only ‘Bronze’ (Muller et al., 2000a). Therefore flower longevity may be regulated by the abundance of the ethylene receptors. However, this hypothesis is not in agreement with the negative regulation model of ethylene signaling (Bishopp et al., 2006). Furthermore, *RhCTR1-1* is slightly induced during senescence, whereas *RhCTR1-2* is constitutively expressed (Muller et al., 2002). Surprisingly, *RhEIN3* is constitutively expressed during flower development, and not up regulated during senescence by ethylene, as it would be expected (Muller et al., 2003). This might suggest that *EIN3* regulation in rose is not transcriptional. In conclusion, the role of ethylene in rose senescence needs more investigation to understand the perception and signaling of ethylene during this process. Functional approaches may help to better understand the role of these different genes.

Various additional genes have been isolated and proposed to be involved in rose senescence. A δ9 acyl lipid desaturase was cloned by a differential expression experiment designed to identify senescence-inducible genes (Fukuchi-Mizutani et al., 1995). The gene is expressed late during flower senescence and may play a role in the degradation of saturated fatty acids of membrane lipids during petal senescence. Another fatty acid modifying enzyme lipoxygenase, *Rlox1*, appears to play a role in rose senescence. It is expressed in petal tissue, induced by ethylene and strongly up regulated during senescence (Fukuchi-Mizutani et al., 2000). The authors suggest that the lipoxygenase may degrade phospholipids that will destabilize the membranes. Finally, an expansin gene, *RbEXPA1*, is expressed in the petal abscission zone during senescence and rapidly induced by ethylene (Sane et al., 2007). This expansin gene might be involved in cell wall modification during petal senescence.

## 7 Anthocyanin Biosynthesis in Flower Color

Anthocyanin is the principal pigment in flower conferring intense red-to-blue cyanic colors on petals. A rose homologue of dihydroflavonol 4-reductase (DHR), an enzyme involved in pink color production in Petunia, is developmentally regulated and parallels anthocyanin production in rose petals (Tanaka et al., 1995). Sepals, prickles and styles also contain anthocyanins and express *DHR*. Anthocyanin biosynthesis requires glycosylations of precursors that stabilize the molecule and are normally performed in two steps by different enzymes in plants. In rose, a new enzyme, RhGT1 that catalyzes glycosylation at two positions successively on the precursor molecule, was characterized (Ogata et al., 2005). *RhGT1* is temporally and spatially regulated during flower development, according to anthocyanin synthesis. This new glycosyltransferase represents a new pathway in plants for anthocyanin synthesis (Ogata et al., 2005).

Other developmental genes, such as a vacuolar $Na^+/H^+$ antiporter (Kagami and Suzuki, 2005) and a gene related to pathogen resistance as NBS-LRR (Lim et al., 2005; Hattendorf and Debener, 2007b) and PR (Pathogenesis Related) genes (Xu et al., 2007) were isolated but no expression data taken. Furthermore, only few of the previous genes have been validated by functional analysis. These will be described in the following chapter.

## References

Bishopp, A., Mahonen, A.P., and Helariutta, Y. (2006). Signs of change: hormone receptors that regulate plant development. Development 133, 1857–1869.

Channeliere, S., Riviere, S., Scalliet, G., Szecsi, J., Jullien, F., Dolle, C., Vergne, P., Dumas, C., Mohammed, B., Hugueney, P., and Cock, J.M. (2002). Analysis of gene expression in rose petals using expressed sequence tags. FEBS Letters 515, 35–38.

Chmelnitsky, I., Khayat, E., and Zieslin, N. (2003). Involvement of RAG, a rose homologue of AGAMOUS, in phyllody development of Rosa hybrida cv. Motrea. Plant Growth Regulation 39, 63–66.

Church, S.A., Livingstone, K., Lai, Z., Kozik, A., Knapp, S.J., Michelmore, R.W., and Rieseberg, L.H. (2007). Using variable rate models to identify genes under selection in sequence pairs: their validity and limitations for EST sequences. Journal of Molecular Evolution 64, 171–180.

Clark, J.I., Brooksbank, C., and Lomax, J. (2005). It's all GO for plant scientists. Plant Physiology 138, 1268–1279.

Flament, I., Debonneville, C., and Furrer, A. (1993). Volatiles constituents of roses: characterization of cultivars based on the headspace analysis of living flower emissions. In Bioactive volatile compounds from plants, R. Tenashi, R.G. Buttery, and H. Sugisawa, eds (Washington, DC: American Chemical Society), pp. 269–281.

Foucher, F., Chevalier, M., Corre, C., Soufflet-Freslon, V., Legeai, F., and Oyant, L.H.S. (2008). New resources for studying the rose flowering process. Genome 51, 827–837.

Fukuchi-Mizutani, M., Savin, K., Cornish, E., Tanaka, Y., Ashikari, T., Kusumi, T., and Murata, N. (1995). Senescence-induced expression of a homologue of DELTA 9 desaturase in rose petals. Plant Molecular Biology 29, 627–635.

Fukuchi-Mizutani, M., Ishiguro, K., Nakayama, T., Utsunomiya, Y., Tanaka, Y., Kusumi, T., and Ueda, T. (2000). Molecular and functional characterization of a rose lipoxygenase cDNA related to flower senescence. Plant Science 160, 129–137.

Guterman, I., Shalit, M., Menda, N., Piestun, D., Dafny-Yelin, M., Shalev, G., Bar, E., Davydov, O., Ovadis, M., Emanuel, M., Wang, J.H., Adam, Z., Pichersky, E., Lewinsohn, E., Zamir, D., Vainstein, A., and Weiss, D. (2002). Rose scent: genomics approach to discovering novel floral fragrance-related genes. Plant Cell 14, 2325–2338.

Hattendorf, A., and Debener, T. (2007b). Molecular characterization of NBS-LRR-RGAs in the rose genome. Physiologia Plantarum 129, 775–786.

Hibino, Y., Kitahara, K., Hirai, S., and Matsumoto, S. (2006). Structural and functional analysis of rose class B MADS-box genes MASAKO BP, euB3 and B3: paleo-type AP3 homologue MASAKO B3 association with petal development. Plant Science 170, 778–785.

Hibrand-Saint Oyant, L., Crespel, L., Rajapakse, S., Zhang, L., and Foucher, F. (2008). Genetic linkage maps of rose constructed with new microsatellite markers and locating QTL controlling flowering traits. Tree Genetics and Genomes 4, 11–23.

Johnson, P.R., and Ecker, J.R. (1998). The ethylene gas signal tranduction pathway: a molecular perspective. Annual Review of Genetics 32, 227–254.

Jung, S., Abbott, A., Jesudurai, C., Tomkins, J., and Main, D. (2005). Frequency, type, distribution and annotation of simple sequence repeats in Rosaceae ESTs. Functional & Integrative Genomics 5, 136–143.

Kagami, T., and Suzuki, M. (2005). Molecular and functional analysis of a vacuolar $Na^+/H^+$ antiporter gene of Rosa hybrida. Genes and Genetic Systems 80, 121–128.

Kitahara, K., and Matsumoto, S. (2000). Rose MADS-box genes 'MASAKO C1 and D1' homologous to class C floral identity genes. Plant Science (Limerick) 151, 121–134.

Kitahara, K., Hirai, S., Fukui, H., and Matsumoto, S. (2001). Rose MADS-box genes 'MASAKO BP and B3' homologous to class B floral identity genes. Plant Science 161, 549–557.

Lan, T.H., DelMonte, T.A., Reischmann, K.P., Hyman, J., Kowalski, S.P., McFerson, J., Kresovich, S., and Paterson, A.H. (2000). An EST-enriched comparative map of Brassica oleracea and Arabidopsis thaliana. Genome Research 10, 776–788.

Lavid, N., Wang, J.H., Shalit, M., Guterman, I., Bar, E., Beuerle, T., Menda, N., Shafir, S., Zamir, D., Adam, Z., Vainstein, A., Weiss, D., Pichersky, E., and Lewinsohn, E. (2002). O-methyltransferases involved in the biosynthesis of volatile phenolic derivatives in rose petals. Plant Physiology 129, 1899–1907.

Lim, K.Y., Werlemark, G., Matyasek, R., Bringloe, J.B., Sieber, V., El-Mokadem, H., Meynet, J., Hemming, J., Leitch, A.R., and Roberts, A.V. (2005). Evolutionary implications of permanent odd polyploidy in the stable sexual, pentaploid of Rosa canina L. Heredity 94, 501–506.

Ma, N., Cai, W., Lu, W., Tan, H., and Gao, J. (2005). Exogenous ethylene influences flower opening of cut roses (*Rosa hybrida*) by regulating the gene encoding ethylene biosynthesis enzymes. Science in China 48, 434–444.

Ma, N., Tan, H., Liu, X., Xue, J., Li, Y., and Gao, J. (2006). Transcriptional regulation of ethylene receptor and CTR genes involved in ethylene-induced flower opening in cut rose (Rosa hybrida) cv. Samantha. Journal of Experimental Botany 57, 2763–2773.

Meyer, V.G. (1966). Flower abnormality. The Botanical ***Review*** 32, 165–195.

Muller, R., and Stummann, B.M. (2003). Genetic regulation of ethylene perception and signal transduction related to flower senescence. Journal of Food, Agriculture & Environment 1, 87–94.

Muller, R., Stummann, B.M., and Serek, M. (2000a). Characterization of an ethylene receptor family with differential expression in rose (Rosa hybrida L.) flowers. Plant Cell Reports 19, 1232–1239.

Muller, R., Lind-Iversen, S., Stummann, B.M., and Serek, M. (2000b). Expression of genes for ethylene biosynthetic enzymes and an ethylene receptor in senescing flowers of miniature potted roses. Journal of Horticultural Science and Biotechnology 75, 12–18.

Muller, R., Owen, C.A., Xue, Z., Welander, M., and Stummann, B.M. (2002). Characterization of two CTR-like protein kinases in Rosa hybrida and their expression during flower senescence and in response to ethylene. Journal of Experimental Botany 53, 1223–1225.

Muller, R., Owen, C.A., Xue, Z., Welander, M., and Stummann, B. (2003). The transcription factor EIN3 is constitutively expressed in miniature roses with differences in postharvest life. Journal of Horticultural Science and Biotechnology 78, 10–14.

Ogata, J., Kanno, Y., Itoh, Y., Tsugawa, H., and Suzuki, M. (2005). Plant biochemistry: Anthocyanin biosynthesis in roses. Nature 435, 757–758.

Sane, A.P., Tripathi, S.K., and Pravendra, N. (2007). Petal abscission in rose (Rosa bourboniana var Gruss an Teplitz) is associated with the enhanced expression of an alpha expansin gene, RbEXPA1. Plant Science 172, 481–487.

Scalliet, G., Journot, N., Jullien, F., Baudino, S., Magnard, J.L., Channeliere, S., Vergne, P., Dumas, C., Bendahmane, M., Cock, J.M., and Hugueney, P. (2002). Biosynthesis of the major scent components 3,5-dimethoxytoluene and 1,3,5-trimethoxybenzene by novel rose O-methyltransferases. FEBS Letters 523, 113–118.

Scalliet, G., Lionnet, C., Bechec, M.l., Dutron, L., Magnard, J.L., Baudino, S., Bergougnoux, V., Jullien, F., Chambrier, P., Vergne, P., Dumas, C., Cock, J.M., and Hugueney, P. (2006). Role of petal-specific orcinol O-methyltransferases in the evolution of rose scent. Plant Physiology 140, 18–29.

Shalit, M., Guterman, I., Volpin, H., Bar, E., Tamari, T., Menda, N., Adam, Z., Zamir, D., Vainstein, A., Weiss, D., Pichersky, E., and Lewinsohn, E. (2003). Volatile ester formation in roses. Identification of an acetyl-coenzyme A. Geraniol/Citronellol acetyltransferase in developing rose petals. Plant Physiology 131, 1868–1876.

Suwabe, K., Tsukazaki, H., Iketani, H., Hatakeyama, K., Kondo, M., Fujimura, M., Nunome, T., Fukuoka, H., Hirai, M., and Matsumoto, S. (2006). Simple sequence repeat-based comparative genomics between Brassica rapa and Arabidopsis thaliana: The genetic origin of clubroot resistance. Genetics 173, 309–319.

Tan, H., Liu, X., Ma, N., Xue, J., Lu, W., Bai, J.H., and Gao, J. (2006). Ethylene-influenced flower opening and expression of genes encoding Etrs, Ctrs, and Ein3s in two cut rose cultivars. Postharvest Biology and Technology 40, 97–105.

Tanaka, Y., Fukui, Y., Fukuchi-Mizutani, M., Holton, T.A., Higgins, E., and Kusumi, T. (1995). Molecular cloning and characterization of Rosa hybrida dihydroflavonol 4-reductase gene. Plant and Cell Physiology 36, 1023–1031.

Wang, D., Fan, J., and Ranu, R.S. (2004a). Cloning and expression of 1-aminocyclopropane-1-carboxylate synthase cDNA from rosa (Rosa × hybrida). Plant Cell Reports 22, 422–429.

Wang, K.L.C., Li, H., and Ecker, J.R. (2002). Ethylene biosynthesis and signaling networks. Plant Cell 14, S131–S151.

Weigel, D., and Meyerowitz, E.M. (1994). The ABCs of floral homeotic genes. Cell (Cambridge) 78, 203–209.

Wu, S., Watanabe, N., Mita, S., Ueda, Y., Shibuya, M., and Ebizuka, Y. (2003). Two O-methyltransferases isolated from flower petals of Rosa chinensis var. spontanea involved in scent biosynthesis. Journal of Bioscience and Bioengineering 96, 119–128.

Wu, S., Watanabe, N., Mita, S., Dohra, H., Ueda, Y., Shibuya, M., and Ebizuka, Y. (2004). The key role of phloroglucinol O-methyltransferase in the biosynthesis of Rosa chinensis volatile 1,3,5-trimethoxybenzene. Plant Physiology 135, 95–102.

Xu, Q., Wen, X., and Deng, X. (2007). Cloning of two classes of PR genes and the development of SNAP markers for powdery mildew resistance loci in chestnut rose (Rosa roxburghii Tratt). Molecular Breeding 19, 179–191.

# 19. Genetic Engineering and Tissue Culture of Roses

**Thomas Debener and Laurence Hibrand-Saint Oyant**

## 1 Introduction

The recent advances in rose genetics and in functional genetics described in the previous two chapters have improved our knowledge about interesting characteristics of the rose. Gene transfer technologies may facilitate the introgression of homologous or heterologous genes to improve major ornamental traits as e.g., scent, plant architecture and color as well as biotic and abiotic stress responses and yield.

Genetic transformation of roses requires the availability of reliable protocols for in vitro culture, for the transfer of genes and for selection and regeneration of transgenic plants.

As for many other ornamental crops, in vitro culture of roses can be used (i) for rapid multiplication of commercial cultivars, (ii) to produce healthy- and disease-free plants and finally (iii) as a source of explants for plant regeneration as a prerequisite for transformation techniques. In vitro propagation of roses is common practice, particularly for the propagation of pot roses. First reports of in vitro culture of rose (*Rosa multiflora*) were made by Elliot (1970). The methods of micro propagation of roses *in vitro* were reviewed several times (Skirvin et al., 1984; Rout et al., 1999; Borissova et al., 2000; Jabbarzadeh and Khosh-Khui, 2005; Pati et al., 2006).

An efficient in vitro plant regeneration protocol with high multiplication rates is the 'key' technology for various biotechnological techniques such as multiplication of clonal plants, mutation breeding programs by exposing to irradiation or by somaclonal variation, and biolistic or *Agrobacterium*-mediated transformation. However, the wide range of explants and experimental approaches that have been employed with different rose species and cultivars strongly suggest that a universal, cultivar-independent method for the production of regenerating tissues, coupled with efficient conversion and ex vitro acclimation of regenerants, will be difficult

T. Debener (✉)
Department of Molecular Breeding, Institute of Plant Genetics, Hannover, Germany

K.M. Folta, S.E. Gardiner (eds.), *Genetics and Genomics of Rosaceae*, Plant Genetics and Genomics: Crops and Models 6, DOI 10.1007/978-0-387-77491-6_19,

to achieve for rose (Ibrahim and Debergh, 2001). This is further demonstrated by various studies that have shown that diverse rose varieties respond differently to the various regeneration protocols (de Wit et al., 1990; Hsia and Korban, 1996; Kim et al., 2004a; Burrell et al., 2006).

## 2 Regeneration

Several protocols for whole plant regeneration have been published for roses mainly via somatic embryogenesis or organogenesis.

### *2.1 Somatic Embryogenesis (SE)*

In vitro multiplication via somatic embryogenesis offers a great potential for plant regeneration as in most cases embryos develop from single cells which subsequently may be germinated to completely clonal plants. However, reports of SE in roses did not emerge until early 1990s. The mechanisms to obtain SE, involve the induction of embryogenic callus, the differentiation of somatic embryos and finally the germination of embryos or the regeneration of plants via adventitious shoot formation.

Somatic embryos have been induced on a variety of culture media (Table 1) from in vitro explants of several rose genotypes (reviewed by Roberts et al. (1995)), but germination frequencies are often low reaching only a few percent.

The reports have described the induction of somatic embryogenesis on embryogenic callus derived from leaves (de Wit et al., 1990; Rout et al., 1991), immature leaves and stem segments (Rout et al., 1991), immature seeds (Kunitake et al., 1993), petioles and roots (Marchant et al., 1996; Sarasan et al., 2001), in vitro grown leaf explants (Kim et al., 2004a), filaments and petioles (Burrell et al., 2006), and leaflets and petioles (Estabrooks et al., 2007). Van der Salm et al. (1996a) have reported plant regeneration from embryogenic callus derived from adventitious roots produced by stem sections of the rootstock cv. Moneyway.

2,4-D is an auxin frequently used to induce callus and/or SE in rose (Roberts et al., 1990; Matthews et al., 1991; Noriega and Sondahl, 1991; Marchant et al., 1996; van der Salm et al., 1996b) although it is ineffective for several cultivars (Li et al., 2002a). Kim et al. (2004a) showed that 2,4-D was useful for callus induction but not for SE induction. Hsia and Korban (1996) showed, in a high 2,4-D concentration, an increased frequency of both organogenic and embryogenic callus from *R. hybrida* leaf explants. In addition to 2,4 D, other auxins as e.g. NAA were used in roses as well (Dohm et al., 2001). Recently, Estabrooks et al. (2007) compared 2,4,5-T, a synthetic auxin not previously tested in rose, with 2,4-D and concluded that 2,4,5-T can be used over a greater concentration range and resulted in significantly greater embryo yield. However, in *R. rugosa*, where immature seeds were used as explants, no growth regulators were used in the media for embryo induction and germination (Kunitake et al., 1993).

**Table 1** In vitro somatic embryogenesis of rose

| Species/cultivar | Explant source | Media | Others | References |
|---|---|---|---|---|
| Domingo and Vickey Brown | Leaf | ■ ½ MS- kinetin- NAA or NOA | | de Wit et al. (1990) |
| *Rosa hybrida* cv. Landora | Immature leaf<br>Stem segment | ■ ½ MS + BA + NAA + $GA_3$ | L-proline | Rout et al. (1991) |
| *Rosa hybrida* cv. Royalty, Soraya | Anther filament | ■ Modified MS and B5 + 2,4-D + Zeatin<br>■ Modified MS + NAA + Zeatin + $GA_3$ | Callus induction<br>Embryogenic callus | Noriega and Sondahl (1991) |
| *Rosa rugosa* | Immature seed | ■ MS | | Kunitake et al. (1993) |
| *Rosa hybrida* cv. Moneyway | Root->callus<br>Callus->SE | ■ SH + 2,4-D<br>■ SH | 4 weeks in the dark | van der Salm et al. (1996a) |
| *Rosa hybrida* cv. Trumpeter and Glads Tidings | Petiole<br>Root explant | ■ SH or MS + 2,4-D<br>■ SH or MS + 2,4-D + ABA + $GA_3$ | L-proline | Marchant et al. (1996) |
| *Rosa hybrida* cv. Carefree Beauty *Rosa chinensis* cv Red Sunblaze and Baby Katie | Leaf<br>Stem segment | ■ MS + NAA or 2,4-D | Glucose | Hsia and Korban (1996) |
| *Rosa hybrida* cv Soraya | Leaf | ■ MS + Kinetin<br>■ MS + BA + IAA | Vitamins | Kintzios et al. (2000) |
| Rosa Heritage × Allister stella gray | Roots | ■ ½ MS + 2,4-D | Mela | Sarasan et al. (2001) |

**Table 1** (continued)

| Species/cultivar | Explant source | Media | Others | References |
|---|---|---|---|---|
| *Rosa hybrida* cv Carefree Beauy and Grand Gala<br>*Rosa chinensis* cv Red Sunblaze | Leaf<br>Leaflet | ■ MS + 2,4-D<br>■ and ½ MS + BA + TDZ + $GA_3$ | | Li et al. (2002a) |
| *Rosa hybrida* cv 4th of July, Tournament of Roses, Graham Thomas, Sequoia Ruby<br>*R. multiflora* | In vitro grown leaf explants | ■ MS + 2,4 D, NAA or dicamba + zeatin<br>■ MS + zeatin | Callus induction<br>SE induction<br>Silver nitrate useful to increase SE germination | Kim et al. (2004a) |
| *Rosa hybrida*<br>15 cultivars | Petioles<br>Filaments | ■ B5 + 2,4 D + zeatin<br>■ MS + 2,4 D + zeatin | Large genotype effect<br>SE productivity inherited additively | Burrell et al. (2006) |
| 'Livin'Easy' | Leaflet<br>petiole | ■ 2,4,5-T or 2,4-D | 4 weeks in the dark | Estabrooks et al. (2007) |

**Media:** MS: Murashige and Skoog (1962), SH: Schenk and Hildebrandt (1972), B5: Noriega and Sondahl (1991)
**Auxins:** NAA : α-napthyl-acetic acid; NOA : 2,4,5-T : 2,4,5-trichlorophenooxyacetic acid; 2,4-D : 2,4,5-dichlorophenoxyacetic acid
**Cytokinins:** BA : 6-benzyladenine = 6-benzylaminopurine, 2,4-D: 2,4-dichlorophenoxyacetic acid; 2,4,5-T : 2,4,5-trichlorophenoxyacetic acid
**Others:** Mela : methyl laurate

Several investigations also established the crucial role of other growth regulators and supplements to enhance SE in roses. $GA_3$ was found to induce somatic embryogenesis in several rose cultivars (Rout et al., 1991; Marchant et al., 1996; Li et al., 2002a), L-proline (Rout et al., 1991; Marchant et al., 1996) nicotinic acid and cysteine (Kintzios et al., 2000) were used in embryo proliferation medium to enhance somatic embryogenesis. Sarasan et al. (2001) used a methyl ester (methyl laurate) to increase development and germination of SE.

## 2.2 Organogenesis

The first successful rose regeneration via in vitro shoot organogenesis was reported by Hill (1967). Direct regeneration of adventitious buds in *Rosa* has been obtained for a number of different genotypes including species roses using explants such as leaves, roots or leaflets (Table 2). Lloyd et al. (1988) observed adventitious buds with media containing BAP and NAA from non-organogenic callus on internodal segments and from leaves and roots without callus from tissue derived from *R. persica* × *R. xanthina* hybrid but not from tissue derived from *R. laevigata* or *R. wichurana*.

The latest references (Dubois and Vries, 1995; Ibrahim and Debergh, 2001; Pati et al., 2004) demonstrated the importance of two phases in this organogenesis process. In the first induction phase, the explants were cultured on media containing TDZ, NAA or IBA and silver nitrate (AgNO3). TDZ, a cytokinin-like phenyl urea derivative, is able to induce diverse morphogenic responses, ranging from tissue proliferation to adventitious shoot and embryo-formation in many woody plant species (Chevreau et al., 1989; Fiola et al., 1990; Korban et al., 1992; Escalettes and Dosba, 1993; Huetteman and Preece, 1993; Ledbetter and Preece, 2004; Song and Sink, 2005). The beneficial effects of $AgNO_3$ during the induction phase (Dubois and Vries, 1995; Ibrahim and Debergh, 2001; Pati et al., 2004) and for SE germination (Kim et al., 2004a) have been reported. This is presumably caused by silver nitrate's indirect inhibition of ethylene synthesis.

Other factors that improve organogenesis have been studied, for instance the age of donor explants (4 weeks of pruning, Ibrahim and Debergh, 2000; Pati et al., 2004), an incubation of 7 days in the dark for the first induction phase, a low PAR (photosynthetically active radiation: 15 $\mu$mol/m$\mu$/s) during the regeneration phase and a relative humidity of 85% in culture vessel of donor explants (Ibrahim and Debergh, 2001; Castillon and Kamo, 2002).

# 3 Genetic Transformation

The methods applied for rose transformation are based either on biological vectors, e.g. *Agrobacterium*-mediated transformation or non-biological vectors, e.g. particle-bombardment or biolistic transformation. *Agrobacterium* is a logical approach given

**Table 2** Direct organogenesis in rose

| Species/cultivar | Explant source | Induction medium | Regeneration medium | References |
|---|---|---|---|---|
| *Rosa persica* × *xanthina* | Internodal segments<br>Leaf<br>Root | – | MS + BAP | Lloyd et al. (1988) |
| *Rosa hybrida* cv. Melrutral | Leaf<br>root | – | MS + BAP | Arene et al. (1993) |
| 24 different genotypes from *Rosa hybrida* cvs *R. canina* and *R. multiflora* hybrids | Parts of leaflets<br>Parts of petiole | ½ MS + TDZ + NAA + $AgNO_3$ | MS + BAP + NAA + FeEDDHA | Dubois and Vries (1995) and Dubois et al. (2000) |
| *Rosa hybrida* L. | leaflets | MS + TDZ + IBA + $AgNO_3$ | MS + BAP + IBA + FeEDDHA | Ibrahim and Debergh (2001) |
| *Rosa damascena* | Petiole | ½ MS + TDZ + NAA + $AgNO_3$ | MS + BAP + NAA | Pati et al. (2004) |

only the ribosome inhibiting protein and not the chitinase or the glucanase gene showed a reduction of black spot susceptibility. Transformed rose plants with the T4-lysosyme gene inserted showed reduced powdery mildew disease symptoms (Li et al., 2003). As all theses genes encoded antifungal proteins, the generated transgenic plants may have resistance to a broad range of pathogens (Dohm et al., 2001). Consequently, the transgenic lines generated by Dohm et al. (2001) were evaluated for resistance to black spot, powdery mildew and downy mildew of roses. Preliminary results show that a number of transgenics show enhanced resistance to these pathogens (Debener unpublished results). Surprisingly, resistance for these three pathogens turns out to be dependent on the strain of the pathogen with some plants being resistant to one strain but susceptible to another and vice versa. In addition plants resistant to powdery mildew may be susceptible to black spot and vice versa indicating that this lytic enzyme gene insertion approach in roses does not necessarily lead to pathogen independent resistance.

Several reports discuss the problem of somaclonal variation associated with the regeneration process (Arene et al., 1993; Souq et al., 1996; Dohm et al., 2001; Condliffe et al., 2003; Kim et al., 2004a) but so far the amount of observed somaclonal variation in transgenic plants regenerated via somatic embryogenesis is low.

The time for the generation of a whole transformed rose is long, usually between 6 to 12 month to obtain independent shoots after callus induction on roots (van der Salm et al., 1996a; Dohm et al., 2001) and more than 12 months to promote shoot regeneration (Li et al., 2003).

## 4 Gene Function Studies by Expression in Model Organisms

Because of the disadvantages of roses as genetics systems (high heterozygosity, polyploidy) only a few genes have been analysed molecularly and biochemically as compared to other plant species. Furthermore, as roses are considered to be recalcitrant to plant transformation, no functional studies by expression of candidate genes in roses have been published to date. Instead, the few studies on rose genes published used model systems as e.g. *E. coli* and *Arabidopsis thaliana* to verify the function of cloned rose genes.

### *4.1 Genes Related to Flower Color*

Genes coding for proteins with functions in the flavonoid metabolism are among the best studied genes in ornamental plants. Therefore, they also were among the first genes cloned from cultivated roses. Tanaka et al., (1995) isolated and characterised a rose dihydroflavonol 4-reductase gene which catalyses the stereo specific reduction of (+) dihydroflavonols to 3,4-cis leucoanthocyanidins the immediate precursors of the anthocyanidins. As DFRs from different plant species have different substrate specificities leading to specific anthocyanidins, the specificity of the rose DFR was analysed in transgenic petunias. A full length cDNA was cloned into the binary vector pCGP293 under the control of the constitutive Mac promoter and transferred to petunia. All transgenic plants had a distinctive new salmon pink petal color and accumulated up to ten fold more pelargonidin compared to wild type plants indicating a predominant substrate specificity of rose DFR for dihydrocaempherol.

## *4.2 Genes Related to Flower Structure*

As previously described in the functional genomics part (this chapter, F. Foucher), putative organ identity genes presenting similarity with C type, *MASAKO C1* and *D1* (Kitahara and Matsumoto, 2000) or with B type, MASAKO P1, B3 and euB3 (Kitahara et al., 2001; Hibino et al., 2006) were isolated. To further characterise putative functional differences between the C function genes *Masako C1* and D1, Kitahara et al. (2004) introduced them under the constitutive 35 S promoter into *Arabidopsis* and *Torrenia* plants. For both *Masako C1* and *D1* transgenic *Arabidopsis* plants expressing these genes displayed homeotic transformations as to be expected for C function genes. Both genes induced the conversion from sepals to carpel like structures and from petals to stamens which in some cases also produced pollen. Both *Masako C1* and *D1* acted in a similar way indicating similar and maybe redundant functions in rose flower development.

The different functions of the class B genes were characterised by introducing all three genes with the constitutive 35S promoter separately into *Arabidopsis* and tobacco (Hibino et al., 2006). None of the plants expressing the transgenes displayed any homeotic transformation of flowers. However, when pair wise combinations of rose transgenes in *Arabidopsis* were generated by crossing lines transgenic for the individual constructs, the flower structures changed. Combinations of *Masako BP* and *Masako B3* showed a homeotic transition from sepals to petals and modified carpels into staminoid carpels corresponding to the observations on ectopic expression of the *Arabidopsis* genes *PI/AP3*. Combinations of *Masako BP* with *Masako euB3* plants only transformed the first whorl into petals. This indicates that in roses the combined action of the gene products of *Masako B3* and *Masako BP* is involved in petal and stamen development whereas the combined action of the gene products of *Masako B3* and *Masako euB3* is only involved in petal development. Further functions of *Masako BP* and *Masako euB3* in stamen development remain elusive as the number of transgenic lines investigated was not sufficient for further conclusions.

## *4.3 Scent Related Genes*

Several groups have isolated genes related to scent production in roses (see previous chapter on functional genomics). A candidate gene for Germacren D synthase was expressed in *E. coli* and feeding experiments with both *E. coli* extracts and cell free extracts from rose petals confirmed that this gene is responsible for the conversion of farnesyl diphosphate into Germacren D, a sesquiterpene and component of the floral scent of roses (Guterman et al., 2002).

Similarly, a gene for an alcohol acetyltransferase identified in an EST collection was analysed in feeding experiments of *E. coli* extracts which demonstrated that the preferred substrate of the *Rh AAT1* gene is the alcohol geraniol which is converted into geranylacetate (Shalit et al., 2003). In vitro experiments also showed that this enzyme accepts citronellol, nerol, octanol, hexanol, phenylethylalcohol and benzyl alcohol as substrates. The acetylation of geraniol and octanol was later confirmed by expression of the gene under the 35S promoter in petunia (Guterman et al., 2006). GC analyses of the head space of non transgenic petunias revealed the emission of benzaldehyde, benzyl alcohol and 2-phenyl ethyl alcohol as well as benzyl acetate and methyl benzoate whereas several transgenic lines emitted five to ten times higher levels of benzyl acetate and phenyl ethyl acetate. When transgenic plants were artificially fed either with geraniol or 1-octanol they emitted geranylacetate and octylacetate volatiles which are not detected in non transgenic petunia.

The largest group of scent related genes characterised from roses so far are the O-methyltransferases (OMT). Although all characterised O-methyltransferases from roses accept several substrates, there are distinct substrate patterns of different OMT enzymes. Apart from transcript analyses via northern blot and RT-PCR, all genes have been cloned into expression vectors expressed in *E. coli* and either purified proteins or protein extracts from *E. coli* cultures were analysed for substrate specificity in assays *in vitro*. To date several O-O-methyltransferases with orcinol as the major substrate have been cloned and analysed from a number of rose varieties which share DNA sequence similarities of around 95% in their coding regions (Lavid et al., 2002; Scalliet et al., 2002; Wu et al., 2003). Therefore, it has not been possible to determine which of these genes are allelic to each other and where in the rose genome these genes are located.

Apart from the orcinol-O-methyltransferases, phloroglucinol-O-methyltransferases (Wu et al., 2004) and C-O-methyltransferases utilising caffeic acid as a substrate have been analysed molecularly and biochemically by heterologous expression in *E. coli* (Scalliet et al., 2002; Wu et al., 2003)

## *4.4 Senescence Related Genes*

A number of genes with functions in rose flower senescence, mainly from the ethylene metabolism, have been characterised both on the DNA and expression level. However the only gene for which the gene product has been characterised functionally is a lipoxygenase induced by the senescence of rose petals (Fukuchi-Mizutani et al., 2000). Lipoxygenase activity was confirmed by heterologous expression of the full length cDNA in *E. coli* and a subsequent lipoxygenase enzyme assay. Both the low optimum pH at 4.8 and the expression pattern (increase during petal senescence) parallels findings for a carnation lipoxygenase also thought to be involved in petal senescence (Rouet-Mayer et al., 1992).

## 4.5 2.5 $Na^+/H^+$ Antiporter Genes

$Na^+/H^+$ antiporter proteins are responsible for the control of the Na influx, efflux and vacuolar compartmentation and therefore important in the adaptation of plants to increased salt concentrations in the environment. Based on degenerate primers designed on sequences of previously isolated $Na^+/H^+$ antiporter genes, Kagami and Suzuki (2005) isolated a cDNA clone from flower buds of *Rosa hybrida* cv. Watarase. The *RhNHX1* cDNA with an open reading frame of 1632 nucleotides displays a 74.1% amino acid identity to the *Arabidopsis thaliana* gene *AtNHX1* therefore not only suggesting a function as a $Na^+/H^+$ antiporter but also indicating that the protein is localised in the tonoplast. Southern blot analyses indicates that *RhNHX1* is a member of a multigene family similar to those plant species studied so far and RNA expression increases after treatment of plants with NaCl solutions suggesting a role in the salt stress response of roses. However, the key experiment for the assignment of a function to *RhNHX1* was the expression in the $Na^+/H^+$ antiporter defective yeast mutant *nhx1* which could be complemented by the rose gene. Apart from its function in salt stress, the authors speculate the utility of this gene in flower color modification in rose after ectopic expression.

## 4.6 Proteomic Studies

Only two studies have been published on analyses of whole organ proteomics of roses to date.

## 4.7 Defense Related Gene Products

Suo and Leung (2002) investigated the changes in protein expression in rose leaves after treatment with BTH (benzo-1,2,3-thiadiazole-7-carbothioic acid S-methyl ester) a synthetic resistance inducing chemical. As previous studies had shown an increase in resistance of BTH treated roses to black spot the authors proposed to identify proteins induced as candidates for resistance factors. Proteins from intercellular washing fluids were extracted 0 to 7 days after BTH treatment and were analysed via 2D gel electrophoresis and western blots. Western blots probed with antibodies against tobacco PR proteins (PR 1, 2, 3 and 5) revealed the induction of PR proteins between 2 days (PR1) and 4 days (PR 3) after treatment with BTH. In addition, 2D Gels revealed 7 protein spots that increased in intensity after BTH treatment and 7 spots that only occurred after the treatment. The protein expression patterns of BTH treated leaves were found to be very similar to black spot inoculated leaves, only the BTH treatment lead to a faster and stronger induction of some PR genes. However, no further analysis of the differentially expressed spots was conducted to identify the proteins. As for other pathosystems, the authors conclude that the PR proteins and the differentially expressed proteins analysed via 2D gels have a role in pathogen resistance.

## 4.8 Scent Related Gene Products

As an extension of their work on the transcriptome studies on the expression of scent related genes in rose petals Dafny-Yelin et al. (2005) published a report on proteome maps of developing rose petals. They extracted total proteins from petals of the strongly scented variety Fragrant Cloud at three developmental stages (1: buds closed, 4: buds open, and 6: flower fully opened) and analysed protein maps with 2D gel electrophoresis and comassie blue staining. A comparison of the maps from these time points revealed that from a total of 978 distinct protein spots, 421 were present at all stages and 575 were present only in one or two stages. 12.5 % of all proteins were stage specific. The subset of spots from stage 4 that were analysed and annotated fell into 11 functional classes among which genes for energy metabolism and general metabolism were the largest groups. Noteworthy is a large number of stress related proteins for which corresponding genes have already been identified during the transcriptome studies conducted previously (Guterman et al., 2002). Also surprising is a lack of correlation between protein expression and transcript expression as revealed by micro array analyses. This could be explained by the observation that multiple proteins are generated by posttranslational modifications and a precise determination of single gene expression is not possible due to the tetraploid nature of the target genome and a lack of strictly gene specific probes for genes from multigene families.

## References

Arene, L., Pellegrino, C., and Gudin, S. (1993). A comparison of the somaclonal variation level of Rosa hybrida L. cv. Meirutral plants regenerated from callus or direct induction from different vegetative and embryonic tissues. Euphytica 71, 83–90.

Borissova, A., Tsolova, V., Angeliev, C., and Atanassov, A. (2000). Somatic embryogenesis of *Rosa hybrida* L. Biotechnology & Biotechnological Equipment 14, 44–51.

Burrell, A.M., Lineberger, R.D., Rathore, K.S., and Byrne, D.H. (2006). Genetic variation in somatic embryogenesis of rose. HortScience 41, 1165–1168.

Castillon, J., and Kamo, K.K. (2002). Maturation and conversion of somatic embryos of three genetically diverse rose cultivars. HortScience 37, 973–977.

Chevreau, E., Skirvin, R.M., Abu-Qaoud, H.A., Korban, S.S., and Sullivan, J.G. (1989). Adventitious shoot regeneration from leaf tissue of three pear (Pyrus sp.) cultivars in vitro. Plant Cell Reports 7, 688–691.

Condliffe, P.C., Davey, M.R., Power, J.B., Koehorst-van Putten, H., and Visser, P.B. (2003). An optimised protocol for rose transformation applicable to different cultivars. Acta Horticulturae 612, 115–120.

Dafny-Yelin, M., Guterman, I., Menda, N., Ovadis, M., Shalit, M., Pichersky, E., Zamir, D., Lewinsohn, E., Adam, Z., Weiss, D., and Vainstein, A. (2005). Flower proteome: changes in protein spectrum during the advanced stages of rose petal development. Planta 222, 37–46.

de Wit, J.C., Esendam, H.F., Honkanen, J.J., and Tuominen, U. (1990). Somatic embryogenesis and regeneration of flowering plants in rose. Plant Cell Reports 9, 456–458.

Derks, F.H.M., Dijk, A.J.v., Hanisch ten Cate, C.H., Florack, D.E.A.,and (1995). Prolongation of vase life of cut roses via introduction of genes coding for antibacterial activity. Somatic embryogenesis and Agrobacterium-mediated transformation. Acta Horticulturae 405, 205–209.

Dohm, A., Ludwig, C., Schilling, D., and Debener, T. (2001). Transformation of roses with genes for antifungal proteins. Acta Horticulturae 547, 27–33.

Dohm, A., Ludwig, C., Schilling, D., and Debener, T. (2002). Transformation of roses with genes for antifungal proteins to reduce their susceptibility to fungal diseases. Acta Horticulturae 572, 105–111.

Dubois, L.A.M., and Vries, D.P.d. (1995). Preliminary report on the direct regeneration of adventitious buds on leaf explants of in vivo grown glasshouse rose cultivars. Gartenbauwissenschaft 60, 249–253.

Dubois, L.A.M., Vries, D.P.d., and Koot, A. (2000). Direct shoot regeneration in the rose: genetic variation of cultivars. Gartenbauwissenschaft 65, 45–49.

Elliot, R.F. (1970). Axenic culture of meristem tips of *Rosa multiflora*. Planta 95, 183–186.

Escalettes, V., and Dosba, F. (1993). In vitro adventitious shoot regeneration from leaves of Prunus spp. Plant Science (Limerick) 90, 201–209.

Estabrooks, T., Browne, R., and Zhongmin, D. (2007). 2,4,5-Trichlorophenoxyacetic acid promotes somatic embryogenesis in the rose cultivar 'Livin' Easy' (*Rosa* sp.). Plant Cell Reports 26, 153–160.

Fiola, J.A., Hassan, M.A., Swartz, H.J., Bors, R.H., and McNicols, R. (1990). Effect of thidiazuron, light fluence rates and kanamycin on in vitro shoot organogenesis from excised Rubus cotyledons and leaves. Plant Cell, Tissue and Organ Culture 20, 223–228.

Firoozabady, E., and Moy, Y. (2004). Regeneration of pineapple plants via somatic embryogenesis and organogenesis. In Vitro Cellular & Developmental Biology – Plant 40, 67–74.

Firoozabady, E., Moy, Y., Courtney-Gutterson, N., and Robinson, K. (1994). Regeneration of transgenic rose (Rosa hybrida) plants from embryogenic tissue. Bio/Technology 12, 609–613.

Fukuchi-Mizutani, M., Ishiguro, K., Nakayama, T., Utsunomiya, Y., Tanaka, Y., Kusumi, T., and Ueda, T. (2000). Molecular and functional characterization of a rose lipoxygenase cDNA related to flower senescence. Plant Science 160, 129–137.

Guterman, I., Masci, T., Chen, X.L., Negre, F., Pichersky, E., Dudareva, N., Weiss, D., and Vainstein, A. (2006). Generation of phenylpropanoid pathway-derived volatiles in transgenic plants: rose alcohol acetyltransferase produces phenylethyl acetate and benzyl acetate in petunia flowers. Plant Molecular Biology 60, 555–563.

Guterman, I., Shalit, M., Menda, N., Piestun, D., Dafny-Yelin, M., Shalev, G., Bar, E., Davydov, O., Ovadis, M., Emanuel, M., Wang, J.H., Adam, Z., Pichersky, E., Lewinsohn, E., Zamir, D., Vainstein, A., and Weiss, D. (2002). Rose scent: genomics approach to discovering novel floral fragrance-related genes. Plant Cell 14, 2325–2338.

Hibino, Y., Kitahara, K., Hirai, S., and Matsumoto, S. (2006). Structural and functional analysis of rose class B MADS-box genes MASAKO BP, euB3 and B3: paleo-type AP3 homologue MASAKO B3 association with petal development. Plant Science 170, 778–785.

Hill, G.P. (1967). Morphogenesis of shoot primordia in cultured stem tissue of a garden rose. Nature 216, 596–597.

Hsia, C., and Korban, S.S. (1996). Organogenesis and somatic embryogenesis in callus cultures of Rosa hybrida and Rosa chinensis minima. Plant Cell, Tissue and Organ Culture 44, 1–6.

Huetteman, C.A., and Preece, J.E. (1993). Thidiazuron: a potent cytokinin for woody plant tissue culture. Plant Cell, Tissue and Organ Culture 33, 105–119.

Ibrahim, R., and Debergh, P.C. (2000). Improvement of adventitious bud formation and plantlet regeneration from in vitro leaflet explants of roses (Rosa hybrida L.). Acta Horticulturae 520, 271–279.

Ibrahim, R., and Debergh, P.C. (2001). Factors controlling high efficiency adventitious bud formation and plant regeneration from in vitro leaf explants of roses (Rosa hybrida L.). Scientia Horticulturae 88, 41–57.

Jabbarzadeh, Z., and Khosh-Khui, M. (2005). Factors affecting tissue culture of Damask rose (*Rosa damascena* Mill.). Scientia Horticulturae 105, 475–482.

Kagami, T., and Suzuki, M. (2005). Molecular and functional analysis of a vacuolar Na+/H+ antiporter gene of Rosa hybrida. Genes and Genetic Systems 80, 121–128.

Kim, C.K., Oh, J.Y., Chung, J.D., Burrell, A.M., and Byrne, D.H. (2004a). Somatic embryogenesis and plant regeneration from in-vitro-grown leaf explants of rose. HortScience 39, 1378–1380.

Kim, C.K., Chung, J.D., Park, S.H., Burrell, A.M., Kamo, K.K., and Byrne, D.H. (2004b). Agrobacterium tumefaciens-mediated transformation of Rosa hybrida using the green fluorescent protein (GFP) gene. Plant Cell, Tissue and Organ Culture 78, 107–111.

Kintzios, S., Drossopoulos, J.B., and Lymperopoulos, C. (2000). Effect of vitamins and inorganic micronutrients on callus growth and somatic embryogenesis from young mature leaves of rose. Journal of Plant Nutrition 23, 1407–1420.

Kitahara, K., and Matsumoto, S. (2000). Rose MADS-box genes 'MASAKO C1 and D1' homologous to class C floral identity genes. Plant Science (Limerick) 151, 121–134.

Kitahara, K., Hirai, S., Fukui, H., and Matsumoto, S. (2001). Rose MADS-box genes 'MASAKO BP and B3' homologous to class B floral identity genes. Plant Science 161, 549–557.

Kitahara, K., Hibino, Y., Aida, R., and Matsumoto, S. (2004). Ectopic expression of the rose AGAMOUS-like MADS-box genes 'MASAKO C1 and D1' causes similar homeotic transformation of sepal and petal in Arabidopsis and sepal in Torenia. Plant Science 166, 1245–1252.

Korban, S.S., O'Connor, P.A., and Elobeidy, A. (1992). Effects of thidiazuron, naphthaleneacetic acid, dark incubation and genotype on shoot organogenesis from Malus leaves. Journal of Horticultural Science 67, 341–349.

Kunitake, H., Imamizo, H., and Mii, M. (1993). Somatic embryogenesis and plant regeneration from immature seed-derived calli of rugosa rose (Rosa rugosa Thunb.). Plant Science (Limerick) 90, 187–194.

Lavid, N., Wang, J.H., Shalit, M., Guterman, I., Bar, E., Beuerle, T., Menda, N., Shafir, S., Zamir, D., Adam, Z., Vainstein, A., Weiss, D., Pichersky, E., and Lewinsohn, E. (2002). O-methyltransferases involved in the biosynthesis of volatile phenolic derivatives in rose petals. Plant Physiology 129, 1899–1907.

Ledbetter, D.I., and Preece, J.E. (2004). Thidiazuron stimulates adventitious shoot production from Hydrangea quercifolia Bartr. leaf explants. Scientia Horticulturae 101, 121–126.

Li, X.Q., Krasnyanski, S.F., and Korban, S.S. (2002a). Somatic embryogenesis, secondary somatic embryogenesis, and shoot organogenesis in Rosa. Journal of Plant Physiology 159, 313–319.

Li, X.Q., Krasnyanski, S.F., and Korban, S.S. (2002b). Optimization of the uidA gene transfer into somatic embryos of rose via Agrobacterium tumefaciens. Plant Physiology and Biochemistry 40, 453–459.

Li, X.Q., Gasic, K., Cammue, B., Broekaert, W., and Korban, S.S. (2003). Transgenic rose lines harboring an antimicrobial protein gene, Ace-AMP1, demonstrate enhanced resistance to powdery mildew (Sphaerotheca pannosa). Planta 218, 226–232.

Lloyd, D., Roberts, A.V., and Short, K.C. (1988). The induction in vitro of adventitious shoots in Rosa. Euphytica 37, 31–36.

Marchant, R., Davey, M.R., Lucas, J.A., and Power, J.B. (1996). Somatic embryogenesis and plant regeneration in Floribunda rose (Rosa hybrida L.) cvs. Trumpeter and Glad Tidings. Plant Science (Limerick) 120, 95–105.

Marchant, R., Power, J.B., Lucas, J.A., and Davey, M.R. (1998a). Biolistic transformation of rose (Rosa hybrida L.). Annals of Botany 81, 109–114.

Marchant, R., Davey, M.R., Lucas, J.A., Lamb, C.J., Dixon, R.A., and Power, J.B. (1998b). Expression of a chitinase transgene in rose (Rosa hybrida L.) reduces development of blackspot disease (Diplocarpon rosae Wolf). Molecular Breeding 4, 187–194.

Matthews, D., Mottley, J., Horan, I., and Roberts, A.V. (1991). A protoplast to plant system in roses. Plant Cell, Tissue and Organ Culture 24, 173–180.

Murashige, T., and Skoog, F. (1962). A revised medium for rapid growth and bioassays with tobacco tissue cultures. Physiologia Plantarum 15, 473–479.

Noriega, C., and Sondahl, M.R. (1991). Somatic embryogenesis in hybrid tea roses. Biotechnology 9, 991–993.

Pati, P.K., Madhu, S., Anil, S., and Ahuja, P.S. (2004). Direct shoot regeneration from leaf explants of Rosa damascena Mill. In Vitro Cellular & Developmental Biology – Plant 40, 192–195.

Pati, P.K., Rath, S.P., Madhu, S., Anil, S., and Ahuja, P.S. (2006). In vitro propagation of rose – a review. Biotechnology Advances 24, 94–114.

Roberts, A.V., Horan, I., Matthews, D., and Mottley, J. (1990). Protoplast technology and somatic embryogenesis in Rosa. In Integration of in vitro techniques in ornamental plant breeding. Proceedings, Symposium, 10–14 November 1990., pp. 110–115.

Roberts, A.V., Yokoya, K., Walker, S., and Mottley, J. (1995). Somatic embryogenesis in *Rosa* ssp. In Somatic embryogenesis in woody plants, S.M. Jain, P. Gupta, and R. Newton, eds (Amsterdam: Kluwer Academic Publishers).

Rouet-Mayer, M.A., Bureau, J.M., and Lauriere, C. (1992). Identification and characterization of lipoxygenase isoforms in senescing carnation petals. Plant Physiology 98, 971–978.

Rout, G.R., Debata, B.K., and Das, P. (1991). Somatic embryogenesis in callus cultures of Rosa hybrida L. cv. Landora. Plant Cell, Tissue and Organ Culture 27, 65–69.

Rout, G.R., Samantaray, S., Mottley, J., and Das, P. (1999). Biotechnology of the rose: a review of recent progress. Scientia Horticulturae 81, 201–228.

Sarasan, V., Roberts, A.V., and Rout, G.R. (2001). Methyl laurate and 6-benzyladenine promote the germination of somatic embryos of a hybrid rose. Plant Cell Reports 20, 183–186.

Scalliet, G., Journot, N., Jullien, F., Baudino, S., Magnard, J.L., Channeliere, S., Vergne, P., Dumas, C., Bendahmane, M., Cock, J.M., and Hugueney, P. (2002). Biosynthesis of the major scent components 3,5-dimethoxytoluene and 1,3,5-trimethoxybenzene by novel rose O-methyltransferases. FEBS Letters 523, 113–118.

Schenk, R.U., and Hildebrandt, A.C. (1972). Medium and techniques for induction and growth of monocotyledonous and dicotyledonous plant cell cultures. Canadian Journal of Botany 50, 199–204.

Shalit, M., Guterman, I., Volpin, H., Bar, E., Tamari, T., Menda, N., Adam, Z., Zamir, D., Vainstein, A., Weiss, D., Pichersky, E., and Lewinsohn, E. (2003). Volatile ester formation in roses. Identification of an acetyl-coenzyme A. Geraniol/Citronellol acetyltransferase in developing rose petals. Plant Physiology 131, 1868–1876.

Skirvin, R.M., Chu, M.C., and Walker, J.C. (1984). Tissue culture of the rose. American Rose Annual 69, 91–97.

Song, G.Q., and Sink, K.C. (2005). Optimizing shoot regeneration and transient expression factors for Agrobacterium tumefaciens transformation of sour cherry (Prunus cerasus L.) cultivar Montmorency. Scientia Horticulturae 106, 60–69.

Souq, F., Coutos-Thevenot, P., Yean, H., Delbard, G., Maziere, Y., Barbe, J.P., and Boulay, M. (1996). Genetic transformation of roses, 2 examples: one on morphogenesis, the other on anthocyanin biosynthetic pathway. Acta Horticulturae 424, 381–388.

Suo, Y., and Leung, D.W.M. (2002). BTH-induced accumulation of extracellular proteins and blackspot disease in rose. Biologia Plantarum 45, 273–279.

Tanaka, Y., Fukui, Y., Fukuchi-Mizutani, M., Holton, T.A., Higgins, E., and Kusumi, T. (1995). Molecular cloning and characterization of Rosa hybrida dihydroflavonol 4-reductase gene. Plant and Cell Physiology 36, 1023–1031.

van der Salm, T.P.M., van der Toorn, C.J.G., Hanisch ten Cate, C.H., and Dons, H.J.M. (1996a). Somatic embryogenesis and shoot regeneration from excised adventitious roots of the rootstock Rosa hybrida L. 'Moneyway'. Plant Cell Reports 15, 522–526.

van der Salm, T.P.M., van der Toorn, C.J.G., Hanish ten Cate, C.H., van der Krieken, W.M., and Dons, H.J.M. (1996b). The effects of exogenous auxin and rol genes on root formation in Rosa hybrida L. 'Moneyway'. Plant Growth Regulation 19, 123–131.

van der Salm, T.P.M., van der Toorn, C.J.G., Bouwer, R., Hanisch ten Cate, C.H., and Dons, H.J.M. (1997). Production of ROL gene transformed plants of Rosa hybrida L. and characterization of their rooting ability. Molecular Breeding 3, 39–47.

van der Salm, T.P.M., Bouwer, R., Dijk, A.J.v., Keizer, L.C.P., Cate, C.H.H.t., van der Plas, L.H.W., and Dons, J.J.M. (1998). Stimulation of scion bud release by rol gene transformed rootstocks of Rosa hybrida L. Journal of Experimental Botany 49, 847–852.

Wu, S., Watanabe, N., Mita, S., Ueda, Y., Shibuya, M., and Ebizuka, Y. (2003). Two O-methyltransferases isolated from flower petals of Rosa chinensis var. spontanea involved in scent biosynthesis. Journal of Bioscience and Bioengineering 96, 119–128.

Wu, S., Watanabe, N., Mita, S., Dohra, H., Ueda, Y., Shibuya, M., and Ebizuka, Y. (2004). The key role of phloroglucinol O-methyltransferase in the biosynthesis of Rosa chinensis volatile 1,3,5-trimethoxybenzene. Plant Physiology 135, 95–102.

# Part IV
# Strawberry
# (Chapters 20 – 23)

On behalf of the *Fragaria* research community, we dedicate this section to the memory of Dr Günter Staudt. His death on May 22, 2008, at 81 years, deprives us of an inspirational and beloved colleague. His many works in the field of *Fragaria* systematics, taxonomy, and biogeography have guided our thinking and will illuminate the efforts of generations to come. Günter - we miss you.

# 20. Strawberry Genomics: Botanical History, Cultivation, Traditional Breeding, and New Technologies

**Kim E. Hummer and James Hancock**

## 1 Origin, Speciation, and Evolution

The Origin of Strawberries – Cherokee Nation

When first man was created, he lived with the mate Creator gave him. When they began to quarrel, first woman left her husband. The man followed, sad and crying, but first woman kept going and never looked behind. Unetlanv, the Creator, took pity on first man and asked him if he was still angry with his wife. He said he wasn't, so Unetlanv asked him if he would like to have her back. He answered, 'yes!'

Unetlanv put a patch of the ripest huckleberries in the path of first woman, but she passed right on by. A little further, he put a big clump of blackberries, but she didn't notice these, either. One by one, Unetlanv put fresh fruits in her path, but these she also refused to see. Suddenly, she saw a patch of large, ripe strawberries in front of her. She had never seen these before. She bent down to gather a few to eat, and as she picked them up, and she thought, 'My husband would love to eat these!' She gathered a bunch of the finest berries and started back along the path to give them to him. He met her with joy, and together they went home.

Today, strawberries are often kept in traditional homes. They remind us not to argue, and are a symbol of good luck.

(Cherokee Nation, 2007).

Strawberries, genus *Fragaria* L., are the most economically important soft fruit worldwide. This genus is a member of the Rosaceae, sub-Family Potentilloideae (formerly classified in Rosideae), has *Duchesnea* and *Potentilla* as close relatives. While Mabberley (2002) proposed reuniting, i.e., submerging, *Fragaria* under *Potentilla*, further research (Eriksson et al., 2003) has suggested that Mabberley was premature in this judgment, and in present majority botanical opinion the genus status has been retained.

The common dessert strawberry, *Fragaria* × *ananassa* Duchesne ex Rozier nothosubsp. *ananassa*, is a regular part of the diet of millions of people and is cultivated in the arable regions of the globe from the arctic to the tropics. More than

K.E. Hummer (✉)
USDA ARS National Clonal Germplasm Repository, 33447 Peoria Road, Corvallis, OR 97333-2521, USA

K.M. Folta, S.E. Gardiner (eds.), *Genetics and Genomics of Rosaceae*, Plant Genetics and Genomics: Crops and Models 6, DOI 10.1007/978-0-387-77491-6_20,

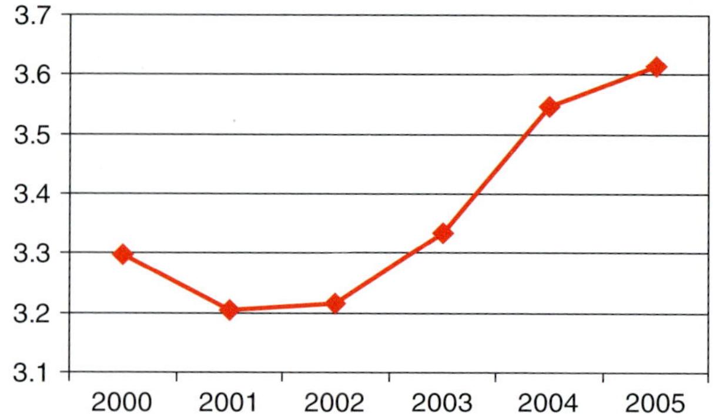


**Fig. 1** World strawberry production (FAO, 2007). MT per year

75 countries have significantly reportable amounts of strawberry production (FAO, 2007). Annual world production of strawberries has more than doubled in the last 20 years to over 3.6 million metric tons (Fig. 1). Most of the production occurs in the northern hemisphere (98%), though no genetic or climatic barriers prevent expansion to the south.

*Fragaria* includes 21 species (Table 1) distributed in the north temperate and holarctic zones (Staudt, 1989, 1999a, b; Rousseau-Gueutin et al., 2008). An accurate synonymy and phylogeny of the strawberry species is emerging. European and American species of *Fragaria* have been rigorously defined by Staudt (1989, 1999a), who, with colleagues, is proceeding to examine Asian species (Staudt 1999b, 2003, 2005; Staudt and Dickoré, 2001). Chinese species are under study (Dai et al., 2007; Lei et al., 2005) but require further collection and examination in light of global taxonomy. The distribution of specific ploidy levels within certain continents reflects the history and evolution of these species (Staudt, 1999a).

## 2 Diploid Species

Diploid strawberries can be distinguished by their plant habit, foliage, inflorescence structure, flowers and fruit (Table 2). *Fragaria viridis* has one of the smallest genomes of vascular plants (Hodgson, 2007). The *F. vesca* genome is also small. At 164 Mb (Akiyama et al., 2001) it is only slightly larger than that of *Arabidopsis thaliana*.

Although 12 diploid ($2n = 2x = 14$) species are native to Eurasia (Table 1), only *F. vesca* is indigenous in northern Eurasia and North America. It is also the only diploid species of North America. Along with other factors, the broader distribution of *F. vesca* suggests that it originated during the Cretaceous (Staudt, 1989). Its ancestor may be basal for the genus.

Although Darrow (1966) and subsequent authors (Hancock, 1999) describe *F. vesca* as native circumpolar boreal, and their distribution maps show *F. vesca* throughout Northern Europe, Asia and North America, this diploid species is not

**Table 1** World strawberry (*Fragaria* L.) species. After Hancock (1999) and Rousseau-Gueutin et al. (2008)

| Species | Ploidy | Geographic distribution |
|---|---|---|
| *F. bucarica* | 2*x* | Western Himalayas |
| *F. daltoniana* J. Gay | | Himalayas |
| *F. gracilis* A. Los. | | North China |
| *F. innumae* Makino | | Japan |
| *F. mandshurica* Staudt | | North China |
| *F. nilgerrensis* Schlect. | | Southeastern Asia |
| *F. nipponica* Lindl. | | Japan |
| *F. nubicola* Lindl. | | Himalayas |
| *F. pentaphylla* Lozinsk | | North China |
| *F. vesca* L. | | Europe, Asia west of the Urals, North America |
| *F. viridis* Duch. | | Europe and Asia |
| *F. yezoensis* | | Japan |
| *F. corymbosa* | 4*x* | Northern China |
| *F. gracilis* | | Northwestern China |
| *F. moupinensis* (French.) Card | | Northern China |
| *F. orientalis* Losinsk syn. = *F. corymbosa* Lozinsk | | Russian Far East/ China |
| *F. tibetica* spec. nov. Staudt | | China |
| *F.* × *bringhurstii* Staudt | 5*x* | California |
| *F. moschata* Duch. | 6*x* | Euro-Siberia |
| *F. chiloensis* (L.) Miller | 8*x* | Western N. America, Hawaii and Chile |
| *F. virginiana* Miller | | North America |
| *F.* × *ananassa* Duch. ex Lamarck | | Cultivated worldwide |
| *F. iturupensis* Staudt | 10*x* | Iturup Island, Kurile Islands |

native east of the Urals to Kamchatka (Hultén, 1927–1930), Hokkaido, Japan (Makino, 1979), western Alaska (Hultén, 1968) or Hawaii (Degener, 1975). This species has been recently introduced, i.e., since the time of European explorers, into these regions. *F. vesca* most likely initially arrived in North America from the east, i.e., Europe, and dispersed to the west developing into the North American subspecies (*F. v.* subsp. *americana* and *F. v.* subsp. *bracteata*).

While diploid strawberries have some barriers to inter-fertility, they can be crossed, and meiosis is regular even where interspecific hybrids are sterile (Hancock, 1999). At least three overlapping interfertile groups of diploid species have been suggested (Bors and Sullivan, 1998): (1) *F. vesca, F. viridis, F. nubicola* and *F. pentaphylla,* (2) *F. vesca*, *F. nilgerrensis*, *F. daltoniana* and *F. pentaphylla,* (3) *F. pentaphylla, F. gracilis and F. nipponica. Fragaria iinumae* may belong in group 3, as no fertile seeds have been recovered when it was crossed with either *F. vesca*, *F. viridis* or *F. nubicola*, but it has not been sufficiently artificially crossed with other species to accurately classify it. *Fragaria iinumae* does, however, have a glaucous leaf trait that is unique among the diploids, and its chloroplast RFLPs clusters it with *F. nilgerrensis* in a group that is isolated from the rest (Harrison et al., 1997).

**Table 2** Characteristics of diploid *Fragaria* species (modified from Sargent et al., 2003 and Folta and Davis, 2006)

| Species | Plant and stolon habit | Foliage | Inflorescence structure | Flowers | Fruit |
|---|---|---|---|---|---|
| *F. daltoniana* | Small; low-growing; short petioles; few hairs on all plant parts; deciduous habit | Small waxy leaflets; prominent petiolules | Single flowered; fully erect in flower and fruit; flowers above foliage; peduncles absent | Small; ovate, prominently veined and widely spaced petals; calyx clearly visible; round receptacle; large flat anthers; short stamens | Conic to cylindrical; bright pink skin; purple to black achenes; white flesh; woolly texture; no aroma |
| | Stolons deep red; sympodial | | | | |
| *F. gracilis* | Vigorous | Tri to penniform five foliate leaves with adpressed soft hairs | Self-incompatible, | Lanceolate sepals that reflex upon fruit ripening | Elongated ovoid |
| *F. iinumae* | Small; stout plants; thick leathery petioles and stolons; annual habit;leaves go dormant during the winter | Glabrous leaves with large serrations; short petiolules | Few flowers per inflorescence; few inflorescences per plant; short peduncles | Medium-sized; long slender petals, always more than five; small receptacles; large numbers of anthers; short stamens | Conic; bright red skin; soft watery flesh; bright yellow achenes; unpleasant, blackcurrant-like flavor; acidic |
| | Stolons pinky-red; thick and leathery; sympodial | | | | |
| F. *mandschurica* | Sympodial with short spreading hairs | Bright green leaves | Flowers bisexual, hermaphroditic, inflorescence mostly surpassing leaves, drooping with fruit | Petals very widely ovate to depressed ovate, almost orbicular, overlapping to not touching | Ovoid to broadly ovoid, sometimes globose. Achenes yellow-green to light brown when mature, in shallow pits or superficial |

**Table 2** (continued)

| Species | Plant and stolon habit | Foliage | Inflorescence structure | Flowers | Fruit |
|---|---|---|---|---|---|
| *F. nilgerrensis* | Robust; vigorous; long, thick leathery petioles; thick pubescence on all plant parts; evergreen habit<br><br>Stolons deep-red; thick and leathery; sparsely produced; sympodial | Thick, leathery leaves; leaflets almost round; deep, prominent veins; distinct petiolules | Complex inflorescence, supporting many flowers; thick peduncles, erect to semi-erect in fruit | Medium sized; prominent, large round receptacles; narrow petals do not overlap; calyx clearly visible; large flat anthers | Globose-conic fruit; white skin; deeply pitted; firm flesh; small brown achenes; highly aromatic, peach-like aroma; clasping to spreading calyx |
| *F. nipponica* | Small; compact deciduous habit. Stolons pinky-red; slender and filiform; support a large number of daughter plants; monopodial | Many leaves that fold upwards with a bluish hue; insignificant petiolules | Simple inflorescence; peduncles droop when in fruit | Medium-sized; very small receptacle; almost circular petals; long prominent outer filaments | Short wedge-shaped berries; pinky-red skin; pleasant aroma like that of cultivated strawberry; long prominent calyx |
| *F. nubicola* | Small; not vigorous; few leaves; short petioles; remontant flowering habit; deciduous habit<br><br>Stolons deep red; sympodial | Light-green leaflets, becoming darker around veins; short, almost absent petiolules | Simple inflorescence; few inflorescences per plant; peduncles clothed in thick pubescence; lie along ground in fruit; remontant flowering habit | Large; prominently veined petals small receptacle | Globose, necked berries; flattened on top; dark wine-red skin; tightly clasping calyx |

**Table 2** (continued)

| Species | Plant and stolon habit | Foliage | Inflorescence structure | Flowers | Fruit |
|---|---|---|---|---|---|
| *F. pentaphylla* | Compact; densely leafed; wavy petiolesStolons deep red; monopodial | Leaflets shiny, almost leathery; petioles support tertiary leaflets | Relatively simple inflorescence; slender peduncles; flowers above foliage | Large; large, contorted, wrinkled petals | Long conic berries, almost rectangular; knobbly appearance; pink to orangy-red skin; sunken achenes.White fruited form has larger, rounder berries |
| *F. vesca* | Robust and vigorous; long, gracile petioles; evergreen habit; *americana* subsp. is in general smaller, more gracile and less vigorous | Leaflets large; often curled under at edges | Complex inflorescence; peduncles stand erect in flower; *americana* subsp. has characteristically long pedicels and very short peduncles | Small; round, prominently veined petals; large receptacle; small anthers held on short filaments;Subsp. *americana* has smaller flowers | Long-conic to globose; usually bright red; distinctive, powerful aroma; large achenes raised or in shallow pits; clasping to reflex calyx |
| | Stolons green to brown; many slender, filiform stolons; sympodial | | | | |
| *F. viridis* | Slender and erect; long, slender petioles; deciduous habit | Yellowish leaflets; petiolules almost absent | Complex inflorescence; peduncles above leaves in flower but lie along ground in fruit; remontant flowering habit | Large; medium sized receptacles; large, overlapping petals | Oblate to globose berries; pale green skin with a red blush; firm flesh; acidic apple-like aroma; very large achenes |
| | Stolons pinky-red; monopodial | | | | |

## 3 Higher Ploidy Species

Polyploidy in *Fragaria* probably arose through unification of $2n$ gametes. Unreduced gametes are relatively common (Hancock, 1999; Hancock et al., 2007). Bringhurst and Senanayake (1966) found frequencies of giant pollen grains (a result of unreduced gametes) to be about 1% of the total. Over 10% of the natural hybrids generated of these two species resulted from unreduced gametes. Staudt (1989) observed restitution in microsporogenesis of a $F_1$ hybrid of *F. virginiana* × *F. chiloensis*.

From the biogeography of the genus, the pattern of occurrence of the polyploids, and the distribution of specific characteristics, Staudt (1999) speculated on *Fragaria* origin and evolution. He suggested that East Asia is a center of origin for diploid strawberries. The tetraploid species ($2n = 4x = 28$) are also East Asian natives. More recently Rousseau-Gueutin et al. (2008) examined phylogeny of *Fragaria* using *GBSSI-2* and *DHAR* nuclear genes sequences (Fig. 2). They concluded that their results provided 'evidence of the occurrence of multiple polyploidization events within *Fragaria* and of the allopolyploid origin of the hexaploid and octoploid species.' However, they could not discriminate between autopolyploid versus allopolyploid origins for the five tetraploid species.

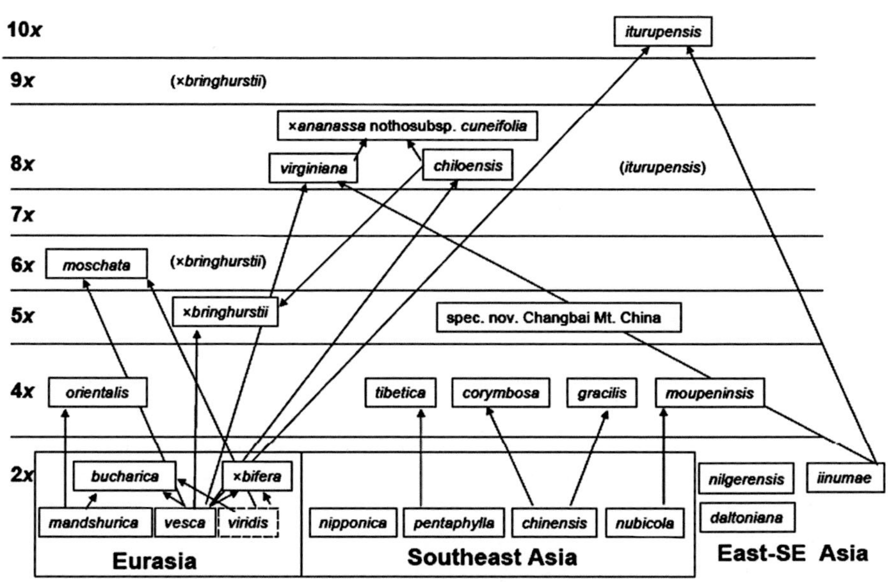


**Fig. 2** Phylogenetic hypothesis for intrageneric relationships in *Fragaria* based on *GBSSI-2* and *DHAR* nuclear genes sequences. Each *line* represents species from a specific ploidy level (diploid to octoploid). Bracketed species present congruent data between the two genes. Species in the same bracket belong to the same clade. *Dashed lines* indicate a soft incongruence between the two datasets.

They suggested that ;the analysis of additional variable nuclear genes might provide the needed information to better resolve the relationships between closely related diploid *Fragaria* species, to clarify the position of *F. bucharica* and *F. viridis*, and to better understand the origin of all *Fragaria* polyploid species.'

The tetraploids may have spread to the periphery of diploid *Fragaria's* ranges. *Fragaria orientalis* is probably an autopolyploid of *F. mandschurica* (Staudt, 1959, 2003). The tetraploid *F. corymbosa* Los. has been submerged under *F. orientalis* (Staudt, 2003). *F. nilgerrensis* is probably the diploid progenitor of the tetraploid *F. moupinensis* (Darrow, 1966; Staudt, 1999). *Fragaria tibetica* seems to be a tetraploid descendent of *F. pentaphylla*. Heteroecy occurs in strawberries in association with doubling of the chromosome number. Diploid *Fragaria* species are not dioecious, though tetraploids (Staudt and Dickoré, 2001), and octoploids are. Tetraploid *Fragaria* species are interfertile (Hancock, 1999).

Wild, naturally occurring pentaploid ($2n = 5x = 35$) strawberry species have been observed in California (*F.* × *brinhurstii* Staudt) and Jilin, China (Lei et al., 2005). These strawberries produce no fertile offspring.

The hexaploid ($2n = 6x = 42$), *F. moschata*, is solely European. The musk strawberry, as it is commonly known, is a dioecious, tall vigorous plant that produces few runners. Leaves are large, dark green, rugose, rhombic, prominently veined and pubescent. The flowers are large and the inflorescence is superior to the foliage but droops with ripe berries. The fruit is purplish red, soft, irregularly globose and has a strong flavor. The calyx reflexes. Red and white fruited forms are cultivated (Hancock, 1999).

Native octoploid strawberries are found primarily in North and South America, however, a small distribution of one octoploid species occurs on Iturup, one of the Kurile Islands (Staudt 1989). This pattern of distribution could be explained if the first hypothetical octoploid arose in East Asia and migrated via an Alaskan-Siberian land bridge to North America.

A limited distribution of a few colonies of an Asian decaploid *F. iturupensis,* (Hummer et al., 2009) grows on the rock skree on the eastern flank of Volcano Atsunupuri, on Iturup, in the Kurile Islands, now part of the Russian Federation. This location might have provided a refugia from the most recent glaciation, which is reported to have come only as far south as the northern part of Iturup Island. *F. iturupensis* has fruit similar in shape and flavor components to those of *F. vesca*, though its plant and leaves resemble *F. virginiana* subsp. *glauca* (Staudt, 1989, 2008).

After arriving in Northwestern America, the hypothetical octoploid may have differentiated into two ecologically distinct groups such as are present today (Staudt, 1999). *Fragaria chiloensis* and *F. virginiana* may be extreme forms of one species that separated during the Pleistocene, and subsequently evolved differential adaptations. One group, *F. chiloensis*, became adapted to coastal habitat; the second, *F. virginiana*, to montane continental conditions. While spreading along the coast, *F. chiloensis* developed the typical shiny, coriaceous, glabrous leaves of the species that we know today. The dispersal of *F. chiloensis* to Hawaii and Chile may have occurred via bird migrations from North America (Hancock, 1999). Potter et al. (2000) found that *F. virginiana* and *F. chiloensis* carry similar cpDNA restriction fragment mutations. Nucleotide sequences of the nuclear internal transcribed spacer (ITS) region and two contiguous non-coding regions of cpDNA also support a sister relationship that may indicate a monophyletic origin for these two species. Staudt (1999) proposed that an ancient *F. vesca* and a hypothetical octoploid *Fragaria* ancestor could have been members of the Arcto-Tertiary flora present from Alaska to Greenland and Siberia that occupied temperate upland areas at middle latitudes in North America during the Eocene. Arcto-Tertiary flora invaded lowlands as temperatures decreased in the Miocene. Towards the end of the Miocene, many species moved developed that are closely related to those of the present day (Wolfe, 1969; Ritchie, 1984).

Staudt (1999) postulated that *F. iturupensis* may be a primitive form related to *F. virginiana* subsp. *glauca*. Thus far molecular analyses have concurred (Njuguna et al., 2007). Strawberries are reported on multiple islands surrounding Hokkaido and in the greater and lesser Kuriles. Further exploration and study of strawberries of northern Pacific Islands is needed to determine where other higher ploidy strawberry colonies exist and what their phylogenetic role may have been.

## 4 Cultivated Octoploid Strawberries

The most economically important strawberry, *F. ×ananassa* nothosubsp. *ananassa,* has $2n = 8x = 56$ chromosomes. It is an accidental hybrid of two octoploids and arose in the mid-1700's when plants of *F. chiloensis* imported from Chile were planted in France near *F. virginiana* transplanted from the eastern North America.

Three hypothetical genome formulae have been suggested for the wild and cultivated octoploids: AAAABBCC (Federova, 1946), AAA'A'BBBB (Senanayake and Bringhurst, 1967), and AAA'A'BBB'B' (Bringhurst et al., 1990). Bringhurst et al.'s (1990) suggestion reflects the contention that octoploids are completely diploidized with strict disomic inheritance. *F. vesca* is likely to be the A genome donor; *F. innumae*, the B genome donor (Davis et al., 2006). Hybrids of *F. iinumae* that had been chromosome doubled and *F. ×ananassa* were highly fertile. Studies of the AdH gene also confirm that ancestors of *F. iinumae* could be linked to the 'B' genome (Davis et al., 2006). *F. viridis* may also be part of the background of octoploid strawberries. Like *F. vesca*, its chromosomes pair regularly with those of *F. chiloensis*, *F. virginiana* and *F. ×ananassa*. *F. viridis* could represent the A' genome. Davis and Yu (1997) demonstrated that *F. nubicola* or *F. pentaphylla* could also be represented in the A' group because each are interfertile with *F. viridis*. The chromosomes of these species share high levels of homology.

Inheritance patterns of the octoploids are in dispute. Lerceteau-Köhler et al. (2003) concluded that *F. ×ananassa* has mixed segregation ratios using amplified fragment length polymorphism (AFLP) markers, as they found that the ratio of coupling vs. repulsion markers fell between the fully disomic and polysomic expectations. However, three other studies evaluating isozyme, simple sequence repeat

(SSR) and restriction fragment polymorphism (RFLP) segregation observed predominantly disomic ratios, indicating that the octoploid strawberry is completely diploidized (Arulsekar and Bringhurst, 1981; Ashley et al., 2003).

The incorporation of traits from a number of lower ploid species has been accomplished through pollinations with native unreduced gametes or by artificially doubling chromosome numbers. The utility of this approach has been shown for a wide range of species in *Fragaria* and in the related genus *Potentilla* (Hancock, 1999). Particular success in incorporating lower ploidies into the background of *F.* × *ananassa* has come through combining lower ploidy species and then doubling to the octoploid level (Bors and Sullivan, 1998).

# 5 Intergeneric Hybrid

Recently *Potentilla palustris* (L.) Scop. was crossed with *Fragaria* ×*ananassa* Duchesne ex Rozier to produce inter-generic hybrid plants that resemble a strawberry but have pink flowers. These plants are everbearing, with strawberry like fruit. Hammer and Pistrick (2003) have proposed the name *F.* ×*rosea* (Mabb.) K. Hammer et Pistrik for this artificial hybrid. Cultivars of these hybrids have been named and patented (Ellis, 1989).

# 6 History of Cultivation

## *6.1 Classical*

'*Fraga*' is the Latin word for the strawberry. Linnaeus chose this as a derivative for the genus name. Roman poets Virgil and Ovid wrote of '*fraga*' in their poetry. Virgil (70 to 19 BCE), wrote '*humi nascentia fraga*' [child of the earth] in his third *Eclogue*. Virgil confirms that strawberries were not cultivated during his time when he writes only a warning to children picking wild strawberries to beware of serpents lurking in the grass. Ovid wrote of the '*arbuteos fructus mon-tanaque fraga'* [(They gathered) Arbutus berries and mountain strawberries] in his *Metamorphoses*, book I, v. 104, as furnishing a food of the golden age and again in the 13th book, '*mollia fraga*.' Pliny separates the, '*terrestribus fragis*,' [ground strawberry] from the arbutus tree in his lib. xv, c. 28. Cato, a Roman Senator (234-149 BCE) mentioned the medicinal uses of strawberries.

The Greeks, Theophrastus, Hippocrates, Dioscorides and Galen, however, did not mention strawberries; nor did other Latin writers on agriculture, Varro, Columella or Palladius. The strawberry is cited in Apuleius Platonicus for its medicinal value (Hedrick, 1919). Strawberries are not mentioned in the Bible, nor do they appear in any Egyptian or Greek art. This could be because of their northernly distribution of the species.

In the 12th century an abbess named Saint Hildegard von Binger declared strawberries unfit for consumption because they grew along the ground where snakes and toads most likely crawled upon them. Her words had such an effect on the local political figures that they, too, made similar declarations, discouraging the population from eating the berries. Among Europeans, this belief held for several years. In the mid 18th century, Charles Linnaeus, the Swedish botanist, put this superstition to rest by switching to a diet consisting only of strawberries to prove them edible (Darrow, 1966).

## 6.2 Old World

*Fragaria vesca,* the alpine strawberry or fraise de bois, was the first strawberry domesticated in the old world. The ancient Romans and Greeks originally cultivated it in gardens, and by the 1300's, this plant was being grown across Europe (Darrow, 1966). *F. vesca* had its widest popularity in the 1500's and 1600's in Europe before the introduction of strawberry species from the New World.

The musk-flavored *F. moschata* (hautbois or hautboy) was also planted in gardens by the late 15th century, along with the green strawberry, *F. viridis*. *F. viridis* was used solely as an ornamental all across Europe, while *F. moschata* was utilized for its fruit by the English, Germans and Russians.

## 6.3 New World

*Fragaria vesca* dominated strawberry cultivation in Europe, until *F. virginiana* from eastern Canada and Virginia began to replace it in the 1600's. Jacques Cartier, who discovered the St. Lawrence River in 1523, was most likely the first to bring *F. virginiana* to the Old World. Cartier mentioned strawberries numerous times in his diary (Hancock, 1999; Wilhelm and Sagen, 1974). The clones that arrived in Europe were wild because the aboriginal peoples of North America did not cultivate strawberries.

A Chilean clone of *F. chiloensis* was brought into Europe in the early 1700's by a French spy, Captain Amédée Frézier (Darrow, 1966; Wilhelm and Sagen, 1974). This strawberry had been domesticated in Chile for about 1,000 years by the indigenous Mapuches, and was spread widely by the Spanish during their colonization period (Hancock, 1999).

## 6.4 New World Species Brought to Europe

Unfortunately, after arriving in Europe, the Chilean strawberry did not bear fruit for several years and early reports on it were negative. The plants were barren because Frézier had brought back pistillate plants and the need for cross pollination was not recognized. The young French Botanist Antoine Nicholas Duschesne discovered that the 'Chili' would produce fruit when pollinized by *F. moschata* or *F. virginiana,* two other higher ploidy plants. The 'Chili' did not cross with the diploid *F. vesca*. The Chilean strawberry reached its highest acclaim in Brittany, and by the mid-1800's, probably more *F. chiloensis* was cultivated in France than in its native country.

Unusual seedlings with unique combinations of fruit and morphological characteristics began to appear in the gardens of Brittany after *F. chiloensis* was brought to France. While the origin of these seedlings was initially mysterious, Duchesne determined in 1766 that they were hybrids of *F. chiloensis* × *F. virginiana* and he named them *Fragaria* × *ananassa* to recognize the perfume of the fruit as smelling like pineapple (*Ananas*). The first hybrids of the 'Pineapple' or 'Pine' strawberry may have been selected early in the commercial fields of Brittany, and in botanical gardens across Europe.

The dessert strawberry, *F.* ×*ananassa*, now dominates strawberry cultivation and is grown in the arable regions of the world. *F. vesca* is generally restricted to home gardens where the small, aromatic fruit are considered a delicacy; most of the cultivars grown are everbearers. *F. chiloensis* is currently grown to a small extent in Chile, but has been largely replaced by *F.* ×*ananassa*. Neither *F. viridis nor F. moschata* is of current commercial importance.

## 6.5 Economic Importance

The USA is the leading producing nation with approximately 25% of the world's crop, followed by Spain, Japan, Poland, Italy and the Korean Republic. California dominates the strawberry industry in the USA with over 80% of the total production. The industries in Spain, the Korean Republic and the USA have grown steadily over the last two decades, while production in Japan, Italy and Poland have declined in the last decade, after dramatic increases in the 1970's and 1980's.

## 6.6 Nutritional Components

The strawberry is widely appreciated for its delicate flavor, aroma and nutritional value. Ripe strawberries are composed of approximately 90% water and 10% total soluble solids (Hemphill and Martin, 1992), and contain numerous important dietary components. They are extremely high in vitamin C and a standard serving of strawberries (10 fruit) supplies 95% of the recommended dietary requirements (Maas et al. 1996). The main soluble sugar components in strawberries are glucose and fructose, which are over 80% of the total sugars and 40% of the total dry weight (Wrolstad and Shallenberger, 1981). The primary organic acid is citric acid, which composed 88% of the total acids (Green, 1971). The strawberry also contains significant levels of ellagic acid, which is thought to be an anticarcinogenic (Maas et al., 1991).

Red color develops through the production of anthocyanins, primarily pelargonidin-3-glucosidase (Wrolstad et al., 1970; Kalt et al., 1993), although at least eight pelargonidin- and two cyanidin-based anthocyanins have been detected in strawberry juice (Bakker et al., 1994). Cyanidin 3-glucoside is the second most common anthocyanin. The total concentration of anthocyanins varies 16-fold across cultivars, and there is some variation in anthocyanin composition, although no clear associations between individual anthocyanins and color have been observed (Bakker et al., 1994).

Glucose, fructose and sucrose are the major soluble sugars found in the fruit of strawberries during all stages of ripening. Glucose and fructose are found in almost equal concentrations (Maas et al., 1996), and they rise continuously during fruit development from 5% in small green fruit to 6–9% in red berries (Kader, 1991). Sucrose levels are generally much lower, and show little accumulation until about the middle of fruit development (Forney and Breen, 1985). Invertases probably play an important role in regulating sweetness, by regulating hexose and sucrose levels (Ranwala et al., 1992; Manning, 1998).

Strawberry flavor is a complex combination of sweetness, acidity and aroma. The most intensely flavored fruits generally have high levels of both titratable acidity (TA) and soluble solids, while the blandest fruit are low in both these components (Kader, 1991). The primary components of flavor have not been completely elucidated, but strawberry aroma is thought to originate from a complex mixture of esters, alcohols, aldehydes and sulfur compounds (Dirinck et al., 1981; Pérez et al., 1996). Hundreds of volatile esters have been correlated with strawberry ripening and aroma development, with methyl- and ethyl-esters of butanoic and hexanoic acids being among the most prevalent (Larsen and Poll 1992; Pérez et al.; 1992, 1996). Other components in high concentration are trans-2-hexenyl acetate, trans-2-hexenal, trans-2-hexenol and 2,5-dimethyl-4-hydroxy-3(2H)-furanone (furaneol). Concentrations of these volatiles vary widely among cultivars and produce large variations in aroma quality (Hirvi, 1983; Shamaila et al., 1992; Pérez et al., 1996, 1997).

Aroma and fragrance content also varies across species. Several researchers consider 2,5-dimethyl-4-hydroxy-3(2H)-furanone (furaneol) and 2,5-dimethyl-4-methoxy-3(2H) furanone (mesifurane) as being particularly important aroma contributors (Pyysalo et al. 1979; Larson and Poll 1992; Sanz et al. 1994; Pérez et al., 1996). The wild species *F. vesca* and *F. virginiana* have much stronger aroma than the cultivated types (Hirvi and Honkanen, 1982). *F. vesca* contains high amounts of ethyl-acetate, but low amounts of methyl-butyrate, ethyl-butylate, and furanone. *Fragaria nilgerrensis* contains high levels of ethyl-acetate and furanone, but low levels of methyl-butyrate and ethyl-butyrate. Hybrids between *F. vesca* and *F.* ×*ananassa* have intermediate levels of fragrance and aroma, while crosses between *F. nilgerrensis* and *F.* ×*ananassa* more closely resemble *F. nilgerrensis*.

## 7 Traditional Breeding

The dessert strawberry is an outcrossing crop that is sensitive to inbreeding (Morrow and Darrow, 1952; Melville et al., 1980). It is asexually propagated by runners, so most breeding programs have been based on pedigree where elite parents are selected for intercrossing each generation. The strawberry germplasm base is relatively narrow (Sjulin and Dale, 1987), but if adequate population sizes are maintained, changes in levels of homozygosity across generations appear to be minimal (Shaw, 1995). Since highly heterozygous genotypes can be propagated as runners, few breeding programs have developed hybrid cultivars using inbred lines, although a few cultivars have been developed in this manner.

Selfing has been used in a number of instances to concentrate genes of interest (Hancock et al., 1996) and backcrossing can incorporate specific traits. Barritt and Shanks (1980) moved resistance to the strawberry aphid from native *F. chiloensis* to *F.* ×*ananassa*. Bringhurst and Voth (1978, 1984) transferred the day neutrality trait from native *F. virginiana* subsp. *glauca* to *F.* ×*ananassa*. About three generations were necessary to restore fruit size and yield to commercial levels.

In 1817, formal strawberry breeding was initiated in England by Thomas A. Knight (Darrow 1966; Wilhelm and Sagen 1974). He was one of the first systematic crop breeders. He used clones of both *F. virginiana* and *F. chiloensis* in his crosses. He produced ‘Downton’ and ‘Elton’ cultivars, noted for their large fruit, vigor and hardiness. Michael Keen, a market gardener near London, also became interested in strawberry improvement about this time and developed ‘Keen’s Imperial’ whose offspring, ‘Keen’s Seedling’ is in the background of many modern cultivars. This cultivar dominated strawberry acreage for about 100 years.

Thomas Laxton of England was the most active breeder during the later part of the 18th century. He released ‘Noble’ and ‘Royal Sovereign’. These two cultivars were grown on both sides of the Atlantic, and were popular until the middle of the 20th century. ‘Nobel’ was known for earliness, cold hardiness and disease resistance. ‘Royal Sovereign’ was popular because of earliness, productivity, flavor, attractiveness and hardiness.

In 1836, Charles Hovey, of Cambridge, Massachusetts, produced the first important North American strawberry, ‘Hovey’, by crossing the European pine strawberry, ‘Mulberry’ with a native clone of *F. virginiana*. This was the first American fruit cultivar produced from an artificial cross. For a while this strawberry was the major pomological product in the country (Hedrick 1925).

Albert Etter of California developed dozens of cultivars around the turn of the century with native *F. chiloensis* clones (Fishman 1987). His most successful cultivar was Ettersburg 80 (1910), which was widely grown in California, Europe, New Zealand and Australia. It was renamed ‘Huxley’ in England and was popular until 1953. Ettersburg 80 was extremely drought resistant, of high fresh and processing quality, because of the solid bright red color. Other outstanding Etter cultivars were ‘Ettersburg 121’, ‘Fendalcino’ and ‘Rose Ettersburg’. While his releases were very successful as cultivars, their greatest lasting impact was as breeding parents. Most California cultivars (and many others) have an Ettersburg cultivar in their background (Darrow 1966; Sjulin and Dale 1987).

In the middle of the 20th century, a number of particularly successful breeding programs emerged in Scotland, England, Germany and Holland. In Scotland, Robert Reid developed a series of red stele resistant cultivars utilizing American 'Aberdeen' as a source of resistance. His cultivar 'Auchincruive Climax' dominated acreage in Great Britain and northern Europe until its demise due to June yellows in the mid-1950's. In England, D. Boyle produced a large series of cultivars with the prefix 'Cambridge'. 'Cambridge Favorite' (1953) became the most important of the group and dominated the acreage in Great Britain by the 1960's. In Germany, R. von Sengbusch's produced a 'Senga' series, of which 'Senga Sengana' (1954) became paramount. 'Senga Sengana' was widely planted for its processing quality and is still important in Poland and other eastern European countries. In the Netherlands, H. Kronenberg and L. Wassenaar's released several cultivars, of which 'Gorella' (1960) made the greatest impact. It was noted for its size, bright red glossy skin and red flesh. B. Meelenbroek who followed in this program released 'Elsanta' (1981), considered the ideal fresh market cultivar for its bright color, flavor and regular size.

Many breeding advances in the eastern United States have come from the U. S. Department of Agriculture (Hancock, 1999). George Darrow at Beltsville, Maryland, developed 'Blakemore' which became the major southern US cultivar in the mid-1930's and 'Fairfax' was widely planted in the middle of this century from southern New England to Maryland and westward to Kansas. These two cultivars were used extensively in breeding, finding their way into the ancestry of a diverse array of cultivars grown in all parts of the US. Other important releases from Darrow were 'Pocahontas', 'Albritton', 'Surecrop' and 'Sunrise'. D. H. Scott, A. D. Draper and G. J. Galletta followed Darrow and released 'Redchief' (1968), 'Earliglow' (1975), 'Allstar' (1981), and 'Tribute' and 'Tristar' (1981). All of these cultivars are still grown today. Tribute and Tristar were the first day-neutrals widely grown in the eastern US and remain the leaders today. An active USDA breeding program has also been conducted at Corvallis, Oregon, initially by Darrow, G.F. Waldo and F.J. Lawrence, and now C. Finn. Some of the more important cultivars emerging from this program were 'Siletz' (1955) and 'Hood' (1965). 'Hood' is considered the premier berry for processing.

Several other state and federal supported programs have released important cultivars in the USA and Canada. Some of the most significant ones from the USA were 'Honeoye' and 'Jewel' (New York), 'Raritan' (New Jersey) and 'Sweet Charlie' (Florida). From Nova Scotia came 'Bounty', 'Glooscap' and 'Kent'.

H. Thomas and E. Goldsmith's of the University of California released the important cultivars 'Lassen' and 'Shasta' in 1945. 'Shasta' was widely grown in the central coast of California in the 1950's and 1960's because of its large size, firmness and long season. 'Lassen' was grown extensively in southern California about the same period, prized for its short rest period and high productivity. R. Bringhurst and V. Voth took over the California-Davis program in the 1950's and generated a succession of internationally important, Mediterranean adapted cultivars have including 'Tioga' (1964), 'Tufts' (1972), 'Aiko' (1975), 'Pajaro' (1979),

'Chandler' (1983), 'Selva' (1983), 'Camarosa' (1992) and 'Seascape' (1991). Most recently, Doug Shaw has released 'Diamonte' from this program.

The greatest concentration of breeding activity outside of Europe and the USA has been in Japan. Two very important cultivars were produced there: H. Fukuba's 'Fukuba' (1899), noted for its large size and high flavor (Darrow, 1966), and K. Tamari's 'Kogyoku' (1940), respected for its vigor, earliness and fruit size. 'Fukuba' was the most important cultivar in forcing culture until the early 1970's. 'Kogyoku' was one of the leading field grown cultivars after World War II, until it lost importance to the American import 'Donner' in the 1950's.

## 8 Structural Genomics

Marker systems have been developed in strawberry for genetic linkage mapping and QTL analysis (Sargent et al., 2004, 2007; Hadonou et al., 2004). These are broadly applicable across strawberry species, although SSRs developed from other Rosaceae species show only limited amplification. Davis and Yu (1997) provided the first diploid map of *F. vesca*, using RAPD markers and isozymes, plus some morphological traits. They crossed *F. vesca* f. *semperflorens* 'Baron Solemacher' and a wild clone of *F. vesca* ssp. *vesca* from New Hampshire, and developed an 80-marker map in the F2 population that represented the seven linkage groups and was 445 cM long. Unusually high levels of segregation distortion were noted (47%) that were skewed toward 'Baron Solemacher'. Davis and Yu speculated that the segregation distortion was caused by the maternal cytoplasm favoring maternal genes. Davis also used a candidate gene approach to determine the molecular basis of the yellow fruit color locus (c) in diploid strawberry. Using PCR they degenerated primer pairs to examine segregation patterns in intron length polymorphisms of genes involved in the anthocyanin biosynthetic pathway. They studied F2 progeny populations of a wild clone of northern California *F. vesca* × *F. vesca* 'Yellow Wonder' and 'Yellow Wonder' × *F. nubicola* from Pakistan, and were able to place five genes into their previously published map. They found F3H, the gene encoding flavanone 3-hydrolase, to be the likely candidate for the yellow fruit color locus.

Most recently, a diploid map of 78 markers was constructed from a hybrid population of *F. vesca* subsp. *vesca* f. *semperflorens* × *F. nubicola* (Sargent et al., 2004). The authors used a combination of SSRs, SCARs, gene specific markers and morphological markers that came from the GenBank data base and other studies. The seven linkage groups were identified in their 448 cM map. Segregation distortions were noted at 54% of the loci that were skewed toward the paternal parent F. nubicola. They speculated that the segregation distortions were due to meiotic irregularities or the self-incompatible nature of *F. nubicola*. Only one octoploid map has been published to date. Lerceteau-Köhler et al. (2003) used 727 AFLP markers and 119 individuals to build both a female and a male map from the cross of 'Capitola' × CF1116 [Pajaro × (Earliglow × Chandler)]. The female map was built with 235 markers and was 1604 cM long, while the male map was 1496 cM long with 280 markers. Only 3.2% of the markers displayed distorted segregation ratios. They detected 30 linkage groups on the female and 28 on the male side, but did not develop a consensus map of the two parents. The female genome was estimated at 2870 cM, while the male was 1861 cM. Two abstracts have described work on octoploid genetic maps, Viruel et al. (2002) and Weebadde et al. (2008). Viruel et al. used 300 SSR and RFLP markers and 86 progeny to build a consensus linkage map with 17 linkage groups and a total distance of 627 cM. 120 markers were unlinked or linked to only one marker, suggesting the need for more markers to build a complete map. Only 10% of the markers showed distorted segregation ratios. Weebadde et al. genotyped fifty-seven individuals of the cross 'Tribute' × 'Honeoye' with AFLP markers. Out of 611 polymorphic bands obtained using 52 primer combinations, 410 single dose fragments (SDRFs) were identified and 23 linkage groups. Most of the markers (255 out of 410) remained unlinked, indicating the need for more markers and larger population sizes to build a map with wide genome coverage.

## 9 Functional Genomics

Only a few strawberry QTL analyses have been conducted. In Weebadde et al. (2008), two AFLP markers were significantly associated with segregation of the day-neutrality trait at a 0.01% level and five at a 0.1% level. Several of these markers were not linked, indicating that day-neutrality is a quantitative trait in the octoploids. Haymes found AFLP markers linked to three red stele resistance genes (Hokanson and Maas, 2001). Lerceteau-Köhler et al. (2003) found fourteen QTL associated with seven characters (fruit height, ratio fruit height/diameter, fruit color, firmness, malate content, glucose content and ratio fructose to glucose). The percentages of phenotypic variance explained by the QTLs ranged from 12 – 20%.

## 10 Biotechnological Approaches to Genetic Improvement

Currently, two transgenic herbicide resistant cropping systems are common for soybean, maize, rapeseed, and cotton: RoundupReady® (active agent: glyphosate) and Liberty Link® (active agent: glufosinate). These systems may have application for strawberry. Weed infestation in fields is one of the major problems in all small fruit crops, particularly strawberry. To control weeds in many crops, non-selective and broad-spectrum herbicides, such as glyphosate and phosphinothricin, are used although they can only be applied as a directed spray under the bushes and avoiding any contact with the green tissues. Herbicide-resistant plants can be expected to broaden the application of non-selective herbicides and to provide a simple, inexpensive, potent, and environmentally friendly management for weed control.

With the impeding ban on methyl bromide (MB) fumigation in strawberry production, growers have lost the most effective control for weeds and soil pathogens.

Yields in today's cultivars are reduced by 50% in non-fumigated soil. Limited genetic variability for resistance has been found to the broad range of pathogens controlled by MB fumigation (Particka and Hancock, 2005). Herbicide resistant strawberries with the *CP4.EPSP* synthase gene or *PAT* would allow for the effective control of weeds, and a transgene like *pcht28* that provided resistance to a broad range of fungal pathogens could restore considerable yield potential. To minimize marketers concerns about the acceptance of transgenic fruit, targeted expression of these genes to just vegetative tissue would likely be required. The use of marker-free selection systems might also prove beneficial.

The incorporation of transgenes for fruit rot resistance and increased firmness in strawberries, such as the antisense of genes for pectate lyase and polygalacturonase-inhibiting protein, would be highly beneficial if the public will accept them. Strawberries are highly perishable, and even with controlled atmosphere storage and refrigeration, a high proportion of fruit are lost due to softening and fungal disease. In particular, *Botrytis* and *Phytopthora* are sources of substantial crop losses.

Transgenic breeding could also provide resistance to major virus diseases such as strawberry crinkle and yellows. In addition to expression of coat protein for protection against these diseases, application of RNA interference (RNAi) technology could be exploited to obtain virus-resistant plants (Tenllado et al. 2004; Hoffmann et al. 2006).

The distribution of small fruit cultivation is often restricted by low temperature stress. Frost tolerance during bloom and winter cold hardiness may reduce crops. While some genetic variability for these characteristics exists, very little improvement has been made by plant breeders. The incorporation of genes such as *CBF1* could provide the necessary genetic variability to improve this trait, if phenotypic effects could be limited by targeting expression to floral tissues or cold periods (Kasuga et al. 2004).

Cool temperatures during flowering also limits yields. When temperatures are low during bloom, pollinator activity is greatly reduced leading to poor seed set, reduced fruit size, irregularly shaped fruits and low yield. The incorporation of a gene that induces parthenocarpic fruiting like the *defH9-iaaM* auxin-synthesizing gene would be very beneficial in areas where cool conditions prevail during pollination.

Transgenic approaches might also be employed to modify the secondary metabolism of strawberries to improve their nutritional quality and flavor (Scalzo et al., 2005). Successful modification of the nutritional value of tomatoes through metabolic engineering and transformation has provided a novel example of the use of organ-specific gene silencing to enhance the nutritional value of fruits. Biochemical pathways might be altered through transgenic approaches to enhance anthocyanin production and strengthen aroma. Numerous genes have been identified in strawberry that are associated with aroma (Aharoni et al., 2000, 2004) and flavonoid metabolism (Manning, 1998).

Several obstacles are working against the acceptance of transgenic strawberries. The economic value of these fruit crops is limited compared to many of the agronomic crops and as a result there is only modest private stimulus to develop new biotechnological products. A second issue is that strawberries outcross and have widespread, native relatives in close proximity to cultivated fields. Most transgenic releases to date have been with species that do not have nearby congeners, greatly reducing the risk of the movement of the transgene into wild species populations. The release of transgenic strawberries will require more scrutiny and in-depth ecological studies than have been required of previous releases. A third issue is reluctance of the fruit industry to introduce products with a potential negative backlash from people leery of consuming transgenic crops.

A strong influx of federal and state funds, along with a careful analysis of what people's perceptions are regarding transgenic fruit is needed to stimulate strawberry biotechnology research. Until this happens, transgenic strawberries will remain as a research tool without commercialization. Using marker-free transformation systems and targeted expression of transgenes will minimize public concern, but the fear of the technology must be abated before transgenic strawberries will be commonly accepted.

## References


Aharoni A, Giri A., Verstappen F, Bertea C, Sevenier R, Sun Z, Jongsma M, Schwab W, and Bouwmeester H (2004). Gain and loss of fruit flavor compounds produced by wild and cultivated species. *Plant Cell* 16: 3110–3131

Aharoni A, Keizer L, Bouwmeester H, Sun Z, Alvarez-Huerta M, Verhoeven H, Blaas J, van Houweilingen A, De Vos C, van der Voet H, Jansen R, Guis M, Mol J, Davis R, Schena M, van Tunen A, and O'Connell A (2000). Identification of the SAAT gene involved in strawberry flavor biogenesis by use of DNA microarrays. *Plant Cell* 12: 647–661

Akiyama Y, Yamamoto Y, Ohmido N, Oshima M, and Fukui K (2001). Estimation of the nuclear DNA content of strawberries (*Fragaria* spp.) compared with *Arabidopsis thaliana* by using dual-step flow cytometry. *Cytologia* 66: 431–436

Arulsekar S, and Bringhurst RS, (1981). Genetic model for the enzyme marker PGI in diploid California *Fragaria vesca*. *J. Hered.* 73: 117–120

Ashley MV, Wilk JA, Styan SMN, Craft KJ, Jones KL, Feldheim KA, Lewers KS, and Ashman TL (2003). High variability and disomic segregation of microsatellites in the octoploid *Fragaria virginiana* Mill. (Rosaceae). *Theor. Appl. Genet.* 107: 1201–1207

Bakker J, Bridle P, and Bellworthy SJ (1994) Strawberry juice colour: A study of the quantitative and qualitative pigment composition of juices from 39 genotypes. J. Sci. Food Agric. 64: 31–37

Barrit BH, and Shanks Jr, CH (1980) Breeding strawberries for resistance to aphids Chaetosiphon fragaefolii and C. tomassi. HortScience 15:287–288

Bors B, and Sullivan JA (1998). Interspecific crossability of nine diploid *Fragaria* species. *HortScience* 32: 439

Bringhurst RS, and Senanayake YDA (1966). The evolutionary significance of natural *Fragaria chiloensis* × *F. vesca* hybrids resulting from unreduced gametes. *Amer. J. Bot.* 53: 1000–1006

Bringhurst RS, and Voth V, (1978) Origin and evolutionary potentiality of the day-neutral trait in octoploid Fragaria. Genetics 90:510

Bringhurst RS, and Voth V (1984) Breeding octoploid strawberries. Iowa State j. Res. 58:371–381

Bringhurst RS, Voth V, and Shaw D (1990). University of California strawberry breeding. *HortScience* 25:834

Cherokee Nation (2007). http://www.cherokee.org/home.aspx?section=culture&culture=culinfo&cat=PdWeE5zX1DE=&ID=drv5fzSkUvg= accessed 19 July 2007

Dai H, Lei J, and Deng M (2007). Investigation and studies on classification of wild *Fragaria* spp. distributed in the Changbai Mountains. *Acta Horticulturae Sinica* 34(1):63–66 [in Chinese with English summary]

Darrow GM (1966). *The Strawberry. History, Breeding and Physiology*. Holt, Rinehart and Winston, New York, 447 pp

Davis TM, DeMeglio LM, Yang R, Styan SMN, and Lewers KS (2006). Assessment of SSR Marker Transfer from the cultivated strawberry to diploid strawberry species:functionality, linkage group assignment, and use in diversity analysis. *J. Amer. Soc. Hort. Sci.* 131(4): 506–512

Davis TM, and Yu H (1997). A linkage map of the diploid strawberry, *Fragaria vesca*. *J. Hered.* 88:215–221

Degener O (1975). *Plants of Hawaii National Park Illustrative of Plants and Customs of the South Seas*. Braun-Brumfield, Inc., Ann Arbor, MI, 316 pp

Dirinck PJ, De Pooter HL, Willaert GA, and Schamp NM (1981). Flavor of cultivated strawberries: The role of the sulfur compounds. *J. Sci. Food Agric.* 29: 316–321

Ellis JR (1989). Frel. US Plant Patent http://patft.uspto.gov/netacgi/nph-Parser?Sect1=PTO1&Sect2=HITOFF&d=PALL&p=1&u=%2Fnetahtml%2FPTO%2Fsrchnum.htm&r=1&f=G&l=50&s1=PP07,598.PN.&OS=PN/PP07,598&RS=PN/PP07,598

Eriksson T, Hibbs MS, Yoder AD, Delwiche CF, and Donoghue MJ (2003). The phylogeny of Rosoideae (Rosaceae) based on sequences of the internal transcribed spacers (ITS) of nuclear ribosomal DNA and the trnL/F region of chloroplast DNA. *Int. J. Plant Sci.* 164: 197–211

FAO (2007). Food and Agriculture. http://faostat.fao.org/site/339/default.aspx Accessed 19 July 2007

Federova NJ (1946). Crossibility and phylogenetic relations in the main European species of *Fragaria Compt. Rend (Doklady) Acad. Sci. USSR*. 30:545–547

Fishman R (1987) Albert Etter: Fruit breeder. Fruit Varieties J. 41: 40–46

Folta KM, and Davis TM (2006). Strawberry genes and genomics. *Crit. Rev. Plant Sci.* 25:399–413

Forney CF, and Breen PJ (1985). Dry matter partitioning and assimilation in fruiting and deblossomed strawberry. *J. Amer. Soc. Hort. Sci.* 110:181–185

Green A (1971) Soft fruits. In: Hulme AC (ed) The Biochemistry of Fruits and Their Products, Vol.2 Academic Press, New York, pp 375–410

Hadonou A, Sargent MDJ, Wilson F, James CM, and Simpson DW (2004). Development of microsatellite markers in *Fragaria* their use in genetic diversity analysis and their potential for genetic linkage mapping. *Genome* 47:429–438

Hammer K, and Pistrick K (2003). New versus old scientific names in strawberries (*Fragaria* L.). *Genet. Resour. Crop Evol.* 50(7):789–791

Hancock JF (1999). *Strawberries*. CABI Pub. University Press, Cambridge, MA, 237 pp

Hancock JF, Scott DH, and Lawrence FJ (1996) Strawberries. In: Janick J and Moore JN (eds.). Fruit Breeding, Volume II. Vine and Small Fruit Crops. John Wiley and Sons, Inc., New York

Harrison RE, Luby JJ, and Furnier GR (1997). Chloroplast DNA restriction fragment variation among strawberry (*Fragaria* spp.) taxa. *J. Amer. Soc. Hort. Sci.* 122:63–68

Hedrick UP (1919). *Sturtevant's Edible Plants of the World*. J. B. Lyon, Albany, NY

Hedrick UP (1925). *The Small Fruits of New York*. J. B. Lyon, Albany, NY

Hemphill R, and Martin LH (1992). Microwave oven-drying method for determining soluble solids in strawberries. *HortScience* 27:1326

Hirvi T (1983). Mass fragmentographic and sensory analysis in the evaluation of the aroma of some strawberry cultivars. *Lebensn. Wiss. Technol.* 16:157–161

Hirvi T, and Honkanen E (1982). The volatiles of two new strawberry cultivars, "Annelie" and Alaska Pioneer obtained by backcrossing of cultivated strawberries with wild strawberries, *Fragaria vesca*, Rügen and *Fragaria virginiana. Zeitschrift fur Lebensmittel-Untersuchung und-Forschung* 175: 113–116

Hodgson J (2007). Evolution of sequencing technology. *Science*. May 11, 2007. 316:846a pullout-poster

Hokanson S, and Maas JL (2001). Strawberry biotechnology. In: Janick, J. (ed.) *Plant Breeding Reviews*, Vol. 21. John Wiley & Sons, Inc, New York

Hultén E (1927–1930) Flora of Kamchatka and the adjacent islands. *Kungl. Svenska Vet. Akad. Handl.*, (1927)5(1): 1–346; (1928). 8(3): 1–213; (1930). 8(4), 1–358 [in Russian]

Hultén E, (1968). *Flora of Alaska and Neighboring Territories. A Manual of the Vascular Plants.* Stanford University Press, Stanford, CA

Kader AA (1991). Quality and its maintenance in relation to the postharvest physiology of strawberry. In: Luby JJ and Dale A (eds.). *The Strawberry into the 21st Century*. Timber Press, Portland, OR. pp. 145–152

Kalt W, Prange RK, and Lidster PD (1993) Postharvest color development of strawberries: The influence of maturity, temperature and light. Can. J. Plant Sci. 73: 541–548

Kasuga M, Miura S, Shinozaki K, and Yamaguchi-Shinozaki K (2004). A combination of the *Arabidopsis* DREB1A gene and stress-inducible *rd29A* promoter improved drought- and low-temperature stress tolerance in tobacco by gene transfer. *Plant Cell Physiol.* 45: 346–350

Larsen M, and Poll L (1992). Odour thresholds of some important aroma compounds in strawberries. *Zeitschrift für Lebensmittel- Untersuchung und-Forschung* 195: 120–123

Lerceteau-Köhler E, Guérin G, Laigret F, and Denoyes-Rothan B (2003). Characterization of mixed disomic and polysomic inheritance in the octoploid strawberry (*Fragaria* × *ananassa*) using AFLP mapping. *Theor. Appl. Genet.* 17: 619–628

Lei J, Li Y, Du G, Dai H, and Deng, M (2005). A natural pentaploid strawberry genotype from the Changbai Mountains in Northeast China. *HortScience* 40(5): 1194–1195

Maas JL, Galletta GJ, and Stoner GD (1991). Ellagic acid, an anticarcinogen in fruits, especially strawberry: A review. *HortScience* 26: 10–14

Maas JL, Wang SY, and Galletta GJ (1996). Health enhancing properties of strawberry fruit. In: Pritts MP, Chandler CK, and Crocker TE. *Proceedings of the IV. North American Strawberry Conference,* Orlando, FL. pp. 11–18

Mabberley DJ (2002). *Potentilla* and *Fragaria* (Rosaceae) reunited. *Telopea* 9(4):793–801

Makino T (1979). *The New Illustrated Flora of Japan*. Hokuryukan Co. Ltd., Tokyo

Manning K (1998) Genes for fruit quality in strawberry. In: Cockshull KE, Gray D, Seymour GB and Thomas B (eds.) *Genetic and Environmental Manipulation of Horticultural Crops*. CAB International, Wallingford, UK. pp. 51–61

Melville AH, Galletta GJ, Draper AD, and Ng TJ (1980). Seed germination and early seedling vigor in progenies of inbred strawberry selections. *HortScience* 15: 49–750

Morrow EB, and Darrow GM. 1952. Effects of limited inbreeding in strawberries. *Proc. Amer. Soc. Hort. Sci.* 39:262–268

Njuguna W, Bassil N, Hummer K, and Davis T (2007). Genetic Diversity of *Fragaria iinumae* and *F. nipponica* based on microsatellite markers. *HortScience* 42(4):915

Particka C, and Hancock JF (2005). Field evaluation of strawberry genotypes for tolerance to black root rot in fumigated and nonfumigated soil. *J. Amer. Soc. Hort. Sci.* 130: 688–693

Pérez AG, Olías R, Sanz C, and Olías JM (1996). Furanones in strawberries: Evolution during ripening and postharvest shelf life. *J. Agric. Food Chem.* 44: 3620–3624

Pérez AG, Ríos JJ, Sanz C, and Olías JM (1992) Aroma components and free amino acids in strawberry cultivar Chandler during ripening. J. Agr. Food Chem. 40: 2232–2235

Pérez AG, Sanz C, Olías R, Ríos JJ, and Olías JM (1997). Aroma quality evaluation of strawberry cultivars in southern Spain. *Acta Hort.* 439: 337–340

Potter D, Luby J, and Harrison R (2000). Phylogenetic relationships in *Fragaria* L. (Rosaceae) inferred from non-coding nuclear and chloroplast sequences. *Syst. Bot.* 25:337–348

Pyysalo T, Honkanen E, and Hirvi T (1979). Volatiles of wild strawberries, *Fragaria vesca* L. compared to those of cultivated strawberries, *Fragaria* × *ananassa* 'Senga sengana.' *J. Agric. Food Chem.* 27: 19–22

Ranwala AP, Suematsu C, and Masuda H (1992). Soluble and wall-bound invertases in strawberry fruit. *Plant Sci.* 84: 59–64

Ritchie JC (1984). Past and present vegetation of the far northwest of Canada. University of Toronto Press. Toronto 251 pp

Rousseau-Gueutin M, Gaston A, Aïnouche A, Aïnouche M.L, Olbricht K, Staudt G, Richard L, and Denoyes-Rothan B. (2008). Origin and evolution of the polyploid *Fragaria* species through phylogenetical analyses of *GBSSI* and *DHAR* low-copy nuclear genes. http://www.adelaide.edu.au/acad/events/ accessed 07/07/2008.

Sanz C, Pérez AG, and Richardson DG (1994). Simultaneous HPLC determination of 2,5-dimethyl-4-hydroxy-3(2H)-furanon and related flavor components in strawberries. *J. Food Sci.* 59:139–141

Sargent D, Geibel JM, Hawins JA, Wilkinson MJ, Battey NH, and Simpson DW (2004) Quantitative and Qualitative Differences in Morphological Traits Revealed between Diploid *Fragaria* Species. *Ann. Bot.* 94(6):787–796

Sargent DJ, Hadonou AM, Simpson DW (2003) Development and characterization of polymorphic microsatellite markers from Fragaria viridis, a wild diploid strawberry. Mol. Ecol. Notes 3: 550–552

Sargent, DJ, Rys A, Nier S, and Simpson DW (2007). The development and mapping of functional markers in Fragaria and their transferability and potential for mapping in other genera. *Theor. Appl. Genet.* 114:373–384.

Senanayake YDA, and Bringhurst RS (1967). Improved techniques for the induction and isolation of polyploids in the genus *Fragaria*. Euphytica 25:725–734

Scalzo J, Battino M, Costantini E, and Mezzetti B (2005). Breeding and biotechnology for improving berry nutritional quality. *BioFactors* 23: 213–220

Shamaila M, Baumann TE, Eaton GW, Powrie WD, and Skura BJ (1992). Quality attributes of strawberry cultivars grown in British Columbia. *J. Food Sci.* 57: 696–699

Shaw DV (1995) Comparison of ancestral and current-generation inbreeding in an experimental strawberry breeding population. Theor. Appl. Genet. 90: 237–241

Sjulin T, and Dale A (1987) Genetic diversity of North American strawberry cultivars. J. Amer. Soc. Hort. Sci. 112:375–386

Staudt GS (1989). The species of *Fragaria*, their taxonomy and geographical distribution. *Acta Hort*. 265: 23–34

Staudt GS (1999a) *Systematics and Geographic Distribution of the American Strawberry Species*. Vol. 81, University of California Pub. Botany. 162 pp.

Staudt GS (1999b). Notes on Asiatic *Fragaria* species: *Fragaria nilgerrensis* Schltdl. ex *J. Gay. Bot. Jahrb. Syst*. 121:297–310.

Staudt GS (2003). Notes on Asiatic *Fragaria* species: III. *Fragaria orientalis* Losinsk. and *Fragaria mandshurica* spec. nov. *Bot. Jahrb. Syst*. 124:397–419

Staudt GS (2005). Notes on Asiatic *Fragaria* species: IV. *Fragaria iinumae* Makino. *Bot. Jahrb. Syst*. 126:163–175

Staudt, GS (2008). *Fragaria iturupensis*, a new source for strawberry inprovement? *Proceedings of the VI International Strawberry Symposium. Huelva, Spain*, abstract 887

Staudt GS, and Dickoré WB (2001). Notes on Asiatic Fragaria species: *Fragaria pentaphylla* Losinsk. and *Fragaria tibetica* spec. nov. *Bot. Jahrb. Syst*. 123:341–354

Tenllado F, Llave C, and Ramón Díaz-Ruíz J (2004). RNA interference as a new biotechnological tool for the control of virus diseases in plants. *Virus Res.* 102 (1):85–96

Hoffmann T, Kalinowski G, and Schwab W (2006). RNAi-induced silencing of gene expression in strawberry fruit (*Fragaria* × *ananassa*) by agroinfiltration: A rapid assay for gene function analysis. *Plant J* 48: 818–826

Viruel M, Sanchex AS, and Arus P (2002). An SSR and RFLP Linkage Map for the Octoploid Strawberry (Fragaria × ananassa). Plant and Animal Genome X Conference. San Diego, CA (Abstract)

Weebadde CK, Wang D, Finn, CE, Lewers KS, Luby JJ, Bushakra J, Sjulin TM, and Hancock JF (2008). Using a linkage mapping approach to identify QTL for day-neutrality in the octoploid strawberry. *Plant Breed.* 127(1): 94–101

Wilhelm S, and Sagen JA (1974). *A History of the Strawberry*. University of California Division of Agriculture Pub. 4031. Berkeley CA

Wolfe, JA (1969). Neogene floristic and vegetational history of the Pacific Northwest. *Madrono* 20:83–100

Wrolstad RE, Putnam TP, and Varseveld GW (1970) Color quality of frozen strawberries: Effect of anthocyanin, pH, total acidity and ascorbic acid variability. J. Food Sci. 35: 448–452

Wrolstad RE, and Shallenberger RS (1981) Free sugars and sorbitol in fruits-a compilation from the literature. J. Assoc. Off. Anal. Chem. 64:91–103

# 21. Strawberry (*Fragaria* spp.) Structural Genomics

**Daniel J. Sargent, Thomas M. Davis, and David W. Simpson**

## 1 Cytology and Interfertility

The genus *Fragaria* has a basic chromosome number of seven ($x$=7) (Ichijima, 1926), and four main fertility groups are recognized: the diploids (2$n$=2$x$=14) which include the model species for the genus, *F. vesca*, (Oosumi et al., 2006) amongst the 14 described species; the tetraploids (2$n$=4$x$=28) including *F. orientalis*; the single hexaploid species *F. moschata* (2$n$=6$x$=42); and four octoploid species (2$n$=8$x$=56): *F. chiloensis*, *F. iturupensis*, *F. virginiana* and the hybrid cultivated strawberry, *F.* ×*ananassa*. The octoploid species are thought to be allo-octoploids, with a proposed genome composition of AAA'A'BBB'B' implying contributions from at least four distinct ancestral diploid genomes (Bringhurst, 1990). However, the prior genome composition models (AAAABBCC and AAA'A'BBBB) of Fedorova (1946) and Senanayake and Bringhurst (1967), respectively, recognized the possibility of an autopolyploid component, and the genome compositions of the octoploid species have yet to be rigorously established.

Numerous diploid progenitors have been proposed for the octoploids, including the broadly distributed *F. vesca*, and the Japanese endemic species *F. iinumae* (Hancock, 1999). In addition to the intensive research attention accorded to *F. vesca,* other diploids are being utilized in *Fragaria* genomic research, including *F. iinumae* (Folta and Davis, 2006), *F. viridis* (Sargent et al., 2003; Nier et al., 2006), and *F. nubicola* (Sargent et al., 2004a; Sargent et al., 2006; Sargent et al., 2007; Vilanova et al., 2008).

Recently, Staudt (2006) addressed the taxonomy of three Himalayan diploid *Fragaria* species, *F. bucharica*, *F. daltoniana* and *F. nubicola*. In that monograph, species accessions that had previously been identified as *F. nubicola* were reassigned to *F. bucharica*. Accessions of what is now *F. bucharica*, including the IPK accession 94056-33.K (*F. nubicola* 601) = NCGR accession PI551851 (CFRA520),

D.J. Sargent (✉)
East Malling Research, East Malling, UK

K.M. Folta, S.E. Gardiner (eds.), *Genetics and Genomics of Rosaceae*, Plant Genetics and Genomics: Crops and Models 6, DOI 10.1007/978-0-387-77491-6_21,

have been cited by many researchers previously as *F. nubicola* (Lin and Davis, 2000; Potter et al., 2000; Deng and Davis, 2001; Sargent et al., 2004a, b, 2006, 2007; Davis et al., 2006a; Monfort et al., 2006; Vilanova et al., 2008), and in the light of the report of Staudt (2006) must now be reclassified as *F. bucharica*. The diploid *Fragaria* reference map FV×FN (see section on linkage mapping below) must therefore be renamed FV×FB.

In addition to the recognized species groups described above, rarely occurring pentaploids ($2n=5x=32$) have been found growing in the wild. The pentaploid *F.* ×*bringhurstii* Staudt. is thought to have arisen from cross-fertilization between the octoploid *F. chiloensis* and the diploid *F. vesca* (Bringhurst and Khan, 1963). Also, in a survey of 15 *F. orientalis* genotypes, a pentaploid plant was found that had been collected in the Julin Province of North Eastern China (Lei et al., 2005). In addition, various fertile polyploids have been produced artificially, such as the decaploid strawberries ($2n=10x=70$) produced from chromosome doubling of a hybrid between *F. chiloensis* and *F. nilgerrensis* ($2n=2x=14$) (Noguchi et al., 2002). The hybrid species *F.* ×*bifera*, consisting of both diploid (2n=2x=14) and triploid (2n=3x=21) members, was shown to have arisen from hybridization between *F. vesca* (as female) and *F. viridis* (Staudt et al., 2003).

The chromosomes of *Fragaria* are rather small, ranging in size from 0.9 – 1.7 microns in length (Yarnell, 1928) and display very little variation in morphology between species (Iwatsubo and Naruhashi, 1989, 1991). Ichijima (1926) studied the cytology of *Fragaria* chromosomes at various levels of ploidy. In the diploid species studied, no irregular chromosomal behavior was observed in the course of heterotypic division and the 14 somatic chromosomes were clearly seen and could be readily counted at metaphase. In the octoploid *Fragaria*, it was difficult to count the somatic chromosomes, which were closely and irregularly packed, and it was found that counts were best made during late diakinesis. Chromosome number in these experiments could not be determined definitely, but it was reported that it appeared to be 56, which would be expected for an octoploid species with a haploid chromosome number of $x=7$. Byrne and Jelenković (1976) observed only bivalents in meiotic spreads from nine cultivated genotypes. More recently, Lim (2004) applied fluorescent *in situ* hybridization (FISH) to *F. vesca* using the 45S and 5S rRNA genes, which revealed six 45S and two 5S sites within the 14 somatic chromosomes of *F. vesca*. The experiments of Lim (2004) enabled the construction of an *F. vesca* karyotype with three pairs of marker chromosomes, but the technique has yet to be extended to the genetically and cytologically more complex genomes of the octoploid *Fragaria* species.

High levels of interfertility have been reported between diploid *Fragaria* species (Senanayake and Bringhurst, 1967; Evans, 1964; Jones, 1955; Sargent et al., 2004b; Bors and Sullivan, 2005a) as well as between diploid species and those at other levels of ploidy (Bors and Sullivan, 2005b; Noguchi et al., 2002). Such interfertility suggests that species within the genus are not highly diverged.

## 2 Phylogenetic Analyses

The phylogenetic relationships between the different *Fragaria* species have been studied using chloroplast DNA (cpDNA) fingerprints (Harrison et al., 1997a) and sequence data from both cpDNA and the nuclear encoded ITS region (Potter et al., 2000). Both these studies reported low phylogenetic resolution with only a few monophyletic species groups such as *F. viridis* and *F. nilgerrensis*, while a major clade containing *F. vesca*, *F. bucharica* (formerly *F. nubicola*) and the polyploid *Fragaria* was recovered with strong bootstrap support, further supporting the hypothesis that all *Fragaria* species are closely-related.

Sargent (2005) studied the phylogenetic relationships between many of the diploid *Fragaria* in more depth, utilizing sequence data from the ITS and four cpDNA regions, and resolved a phylogeny that contained three major diploid clades, the first containing a single species, *F. iinumae*, another containing many of the Asiatic diploid species, and a third containing *F. vesca*, *F. bucharica* and *F. viridis*. However, since he did not include any of the polyploid species in his analyses, this study did not provide any further evidence towards the origins of the octoploid *Fragaria* species.

A soon-to-be submitted study based on an intron-containing region of the *Alcohol dehydrogenase* (*ADH*) gene (DiMeglio and Davis, unpublished data), and encompassing most of the *Fragaria* diploid and polyploid species, suggests diploids *F. vesca*, *F. bucharica*, *F. mandshurica*, and *F. iinumae* as possible genome donors to the octoploid species. Thus, this study provides enhanced illumination of the octoploid species' ancestry, and will help elucidate the pattern of reticulate evolution that is thought likely to have occurred in the evolution of the economically-important octoploid species.

## 3 Molecular Markers

The term 'molecular marker' can be applied to a large number of different techniques that detect variation at the DNA level, all of which have been used to a greater or lesser degree to assay variation between *Fragaria* species and varieties. Historically, arbitrary PCR-based marker systems, such as amplified fragment length polymorphisms (AFLP), random amplified polymorphic DNA (RAPD) and inter-simple sequence repeats (ISSR) have been used to detect variation between cultivated strawberry varieties for the purposes of cultivar identification and fingerprinting (Arnau et al., 2003; Degani et al., 2001; Gidoni et al., 1994; Graham et al., 1996; Kuras et al., 2004; Tyrka et al., 2002; Debnath et al., 2008.) and for population genetics studies (Harrison et al., 1997b; Porebski and Catling, 1998; Harrison et al., 2000; Carrasco et al., 2007), and many such applications have been summarized in detail by Hokanson and Maas (2001). However, when using arbitrary marker systems, there are difficulties in reproducing or interpreting the fingerprints produced between different laboratories. Consequently, standard reference fingerprints for cultivated strawberry varieties have not been established for the purposes of plant variety protection using these marker systems. Despite these difficulties, RAPD marker evaluation continues to find applications in diversity assessment and cultivar identification (Korbin et al., 2002; Kuras et al., 2004; Milella et al., 2006) and has been used in at least one forensic investigation to identify patented plant varieties (Congiu et al., 2000).

Arbitrary marker systems have also been used both directly and indirectly in the construction of genetic linkage maps in *Fragaria*. Building upon a strategy for the identification, and targeted development of codominant RAPD markers in *F. vesca* (Davis et al., 1995), RAPD markers were employed as the primary framework of the first *Fragaria* (*F. vesca*) linkage map (Davis and Yu, 1997), while Lerceteau-Köhler et al. (2003) used AFLP markers to produce an outline map of the cultivated strawberry *F.* ×*ananassa*. Such marker systems have the advantage of generating large numbers of segregating markers per primer combination used, and are thus useful for saturating molecular maps as they are rapid to use and relatively inexpensive. However, the suitability of such markers for the purposes of linkage mapping is hampered, because their interpopulational transferability is limited.

Therefore, for the purposes of developing marker-assisted selection in a wide range of germplasm and for comparative mapping, arbitrary markers must first be converted to sequence-characterized amplified regions (SCARs). A limited number of SCARs, including three that are linked to traits of economic importance, have been developed for *Fragaria*. These include markers for resistance to *Phytopthora fragariae* (Haymes et al., 2000) and for the *Rca2 Colletotrichum acutatum* resistance locus (Lerceteau-Köhler et al., 2005) in the cultivated strawberry, and markers tightly linked to the *Seasonal-flowering locus* (the mutant form of which has been assigned the gene name *semperflorens*) in *F. vesca* (Albani et al., 2004). Davis et al. (1995) also converted a codominant RAPD marker from *F. vesca* into a SCAR marker and this SCAR was subsequently mapped as marker B194 in the FV×FB (formerly FV×FN) map of Sargent et al. (2004a; 2006). SCAR markers closely linked to traits of interest are powerful tools for pre-selection of germplasm with superior characteristics for breeding, but have the disadvantage of being difficult to develop.

The advent of large numbers of molecular markers with a high degree of reproducibility for the genus *Fragaria*, such as microsatellites, gene-specific intron length polymorphisms, and cleaved-amplified polymorphic sequence (CAPS) markers, has largely superseded the use of arbitrary markers in recent years for both the construction of linkage maps (Deng and Davis, 2001; Sargent et al., 2006. 2007) and for the studies of relatedness and identification of strawberry cultivars (Davis et al., 2006a; Gil-Ariza et al., 2006; Govan et al., 2008; Hadonou et al., 2004; Kunihisa et al., 2003, 2005).

## *3.1 Microsatellites*

Microsatellites, or simple sequence repeats (SSRs), are short stretches of tandemly-repeated DNA motifs that are abundantly dispersed throughout eukaryotic genomes.

Recently, they have become the marker of choice for *Fragaria* due to their codominant, multiallelic, and highly polymorphic nature, the ease with which they can be rapidly and reliably genotyped, and their high degree of reproducibility and transferability between cultivars and species. Recent genetic investigations in *Fragaria* employing SSRs include linkage analyses (Sargent et al., 2004a, 2006) and cultivar identification (Gil-Ariza et al., 2006; Govan et al., 2008).

Microsatellite markers have been developed for *Fragaria* by a large number of different groups (Nourse et al., 2002; Ashley et al., 2003; James et al., 2003; Sargent et al., 2003, 2004a, 2006; Cipriani and Testolin, 2004; Hadonou et al., 2004; Bassil et al., 2006; Monfort et al., 2006; Lewers et al., 2005; Cipriani et al., 2006; Gil-Ariza et al., 2006;Shimomura and Hirashima, 2006) and to date, more than 200 distinct SSR primer pairs have been reported, isolated from *Fragaria* species at both the diploid (*F. bucharica*, *F. vesca* and *F. viridis*) and octoploid (*F.* ×*ananassa* and *F. virginiana*) level. Their inherent usefulness as genetic markers stems from the instability of the SSR repeat, which leads to their highly polymorphic nature, coupled with flanking DNA sequence that exhibits much greater stability and from which locus-specific primers are designed. The relative stability of the primer sequences between species within *Fragaria*, particularly when SSRs are designed from coding DNA sequence, means that SSRs have proved to be highly transferable between species within the genus (Davis et al., 2006a), and thus SSRs that amplify single loci in diploid species often amplify multiple, distinct loci in the octoploid species (Ashley et al., 2003; Govan et al., 2008)

Despite their high levels of transferability and thus utility for mapping and diversity studies within *Fragaria*, the transferability of *Fragaria* SSR markers to other closely-related genera such as *Rubus* and *Rosa* is much reduced (Lewers et al., 2005), and in more distantly-related genera within Rosaceae such as *Prunus* and *Malus*, amplification and polymorphism of *Fragaria* SSRs has been shown to be minimal (Sargent et al., unpublished data), with almost no markers proving useful for comparative mapping. Even EST-SSRs have not proved useful for comparative mapping studies between *Fragaria* and more distantly-related genera, due mainly to the fact that the majority are isolated from the 3' and 5' un-translated regions of genes which are generally less conserved than exon sequences. Thus, while SSRs remain the marker of choice for linkage mapping studies within *Fragaria*, their utility for synteny studies and comparative mapping within the Rosaceae is severely limited.

## *3.2 Gene-Specific Markers*

Gene-specific markers are useful for mapping because, being derived from coding sequence, they represent markers with biological function. In addition, since primer pairs are usually designed within coding regions of the gene, they are often highly conserved and thus, gene-specific markers can be transferred between species, and in many cases between distantly-related genera.

The first gene-specific sequence-tagged site (STS) marker in *Fragaria* was developed from the *ADH* sequence of Wolyn and Jelenković (1990) by Davis and Yu (1997). Primer pairs for the *ADH* marker were designed flanking introns II and III of the gene from exon sequence. These revealed intron length polymorphisms between the parents of the *F. vesca* mapping population of Davis and Yu (1997) that could be scored via agarose gel electrophoresis. Subsequently, using a similar approach to primer pair design to that employed by Davis and Yu (1997), Deng and Davis (2001) developed a set of degenerate primer pairs that amplified STS markers for genes of the anthocyanin biosynthesis pathway, utilizing DNA and protein sequences deposited into the EMBL database. Their primer pairs were able to amplify markers for five genes in the anthocyanin biosynthesis pathway and one associated transcription factor. By exploiting intron length polymorphisms, the authors were able to map these gene-specific markers in a number of interspecific diploid *Fragaria* progenies. The mapping of these gene-specific STS markers provided strong evidence, through co-segregation of the *yellow/white fruit color* locus *c* and the anthocyanin gene *Flavanone 3-hydroxylase* (*F3H*), that a mutation in this gene may be responsible for the yellow fruit color phenotype in *F. vesca* ssp. *vesca* f. *semperflorens* 'Yellow Wonder'.

More recently, Sargent et al. (2007) designed a set of primer pairs for 24 genes of known function, predominantly from *Fragaria* mRNA sequences deposited in the EMBL database. The primer pairs they designed flanked polymorphic introns and they were able to map all 24 genes to 29 loci on the diploid *Fragaria* genome, bringing the total number of markers for genes of known-function mapped in the genus to 35. In addition, the transferability of many of these markers to both *Malus* and *Prunus* was demonstrated, and they were able to locate five of these genes on maps of those species. Additionally, Brese and Davis (unpublished data) have mapped two ribosomal protein genes isolated from a *Fragaria* EST library to the FV×FB reference map, however their transferability to other rosaceous genera has yet to be demonstrated.

### *3.3 Cleaved Amplified Polymorphisms*

The cleaved amplified polymorphic sequence (CAPS) marker concept is finding useful application in strawberry. In the original CAPS application in *Arabidopsis* (Konieczny and Ausubel, 1993), and its initial use in strawberry (Kunihisa et al., 2003, 2005), restriction enzymes were used to digest PCR products generated by primer pairs located in conserved exon sequences flanking introns, and fragment length polymorphisms were visualized on agarose gels. Upon refining their CAPS primer design protocol to enhance locus-specificity in the octoploid strawberry, Kunihisa et al. (2005) were able to distinguish between 64 cultivars using a minimum of nine CAPS markers.

In a novel application of the CAPS concept, Davis et al. (2007a, 2008) have described a new molecular marker type called the "gene pair" marker. While the conventional CAPS approach specifically mines polymorphisms in introns, gene pair markers are aimed at exploiting polymorphisms in intergenic regions, the detection of which is facilitated by the comparatively small (1–5 kb) intergenic distances encountered in *Fragaria* (Davis et al., 2008). A gene pair marker is generated using a primer pair in which one primer is located in an exon of one gene, and the second primer is located in an exon of an immediately adjacent gene. Thus, the respective PCR product includes an entire intergenic interval, as well as any exon and intron sequence flanked by the primer pair. The initially described *CDPK1-BHLH1* strawberry gene pair locus was also amplifiable in *Prunus*, *Rubus*, and *Rosa* (Davis et al., 2007a), suggesting that gene pair markers will find application in intergeneric synteny studies in the Rosaceae.

## 3.4 Isoenzymes

The early history of isozyme marker utilization in *Fragaria* centered at the University of California, Davis (Arulsekar and Bringhurst 1983), where the Peroxidase (PX) and Phosphoglucoisomerase (PGI) isozyme systems were employed by Hancock and Bringhurst (1978, 1979) in studies of variation in natural populations of diploid *F. vesca* and the octoploids *F. chiloensis* and *F. virginiana*. Monogenic segregation ratios for the first molecular marker scored in *Fragaria*; the codominant isoenzyme Phosphoglucose isomerase (*Pgi*-2), were then reported in *F. vesca* (Arulsekar and Bringhurst, 1981). Subsequently, electrophoretically detectable variation among octoploid strawberry cultivars (*F.* ×*ananassa*) was studied using three isoenzyme systems: PGI, Leucine aminopeptidase (LAP), and Phosphoglucomutase (PGM) (Bringhurst et al., 1981). Genetic models were developed for the inheritance of both *Pgi* and *Lap* in *F.* ×*ananassa* (Arulsekar et al., 1981), on the basis of which the authors advanced the claims that all four detected *Pgi* loci were active and that the cultivated octoploid strawberry is diploidized. Subsequent studies of isozyme variation in the cultivated strawberry have been reviewed in detail by Hokanson and Maas (2001). Isozyme markers also played a key role in the first two reports of quantified genetic linkage in *Fragaria*, and these are discussed in the section below on linkage map development.

## 3.5 Restriction Fragment Length Polymorphisms (RFLPs)

Restriction fragment length polymorphism markers (RFLPs) have been characterized and used for mapping in a number of rosaceous genera, including *Malus* (Maliepaard et al., 1998) and *Prunus* (Dirlewanger et al., 2004), but to date, no nuclear RFLP probes have been directly developed from *Fragaria* species (although chloroplast restriction fragments were used to study the phylogeny of *Fragaria* by Harrison et al. (1997a). Viruel et al. (2002) successfully used a set of RFLPs previously developed from and mapped in *Prunus*, to construct a linkage map for the cultivated strawberry. Despite only 22% of the 123 probes tested revealing useable polymorphisms for mapping in the cultivated strawberry, they successfully demonstrated the transferability of these probes between distantly-related rosaceous genera, and thus the applicability of RFLPs for comparative mapping between *Prunus* and *Fragaria*. More recently, Vilanova et al. (2008) screened 65 *Prunus* RFLP probes in the grandparents of the diploid *Fragaria* reference map FV×FB, and were able to map 40 of these probes, allowing the first direct comparisons between the diploid *Prunus* and *Fragaria* genomes to be made.

## 4 Resistance Gene Analogues

Degenerate PCR primers targeting conserved regions of nucleotide binding site (NBS) domains (Leister et al., 1996) have been used to isolate resistance gene analogues (RGAs) from *F. vesca*, *F. chiloensis*, and *F.* ×*ananassa* (Martinez-Zamora et al., 2004). No RGA loci have been incorporated into any published *Fragaria* linkage map; however, four RGA-based gene pair loci (see below) have been located on the *F. vesca* and FV×FB linkage maps (Davis et al., in preparation) and Marchese et al. (unpublished data) have used the arbitrary RGA primers reported in Rajesh et al. (2002) to locate eight RGA loci to the FV×FB reference map.

## 5 Linkage and Physical Maps-Diploid

Due to its allo-octoploid, hybrid nature, the cultivated strawberry *F.* ×*ananassa*, is one of the most genetically complex fruit crop species. As a result, researchers studying genetics and genomics in the genus *Fragaria* have been concentrating the majority of their efforts to date on the study of a series of model diploid systems, utilizing the diploid *F. vesca* (Cipriani et al., 2006; Davis and Yu, 1997; Davis et al., 2007b) and its interfertile relatives such as *F. bucharica* (Sargent et al., 2004a, 2006, 2007) and *F. viridis* (Nier et al., 2006).

### *5.1 Morphological Markers*

Richardson (1914, 1918, 1920, 1923) was the first researcher to discuss the phenomenon of linkage between phenotypic traits under major-gene control in diploid *Fragaria*. Subsequently, Brown and Wareing (1965) presented a study of the mode of inheritance of morphological traits in diploid *Fragaria*. They reported the production of $F_2$ and backcross ($BC_1$) progenies to study the inheritance of four morphological traits and concluded that non-runnering (or *runnerless*) (*r*), perpetual flowering (or *semperflorens*) (*s*) and *yellow/white fruit color* (*c*) were controlled by three independently segregating major genes, and that the 'bushy' phenotype of one parental line was under the control of the *r* gene, or a second gene tightly linked to this locus. Since then, only two further morphological characters under single gene control have been described in diploid *Fragaria*, the long-stemmed phenotype *arborea* (*arb*) (Guttridge, 1973) and more recently, a pale-green leaf phenotype (*pg*) was reported by Sargent et al. (2004b) in *F. vesca*.

## 5.2 Molecular Markers

The report of a close (1.1 cM) linkage between the Shikimate dehydrogenase (*Sdh*) isozyme locus and the *c* locus governing the yellow (versus red) fruit color trait in *F. vesca* (Williamson et al., 1995) was followed quickly by the description of linkage (18.9 cM) between the Pgi-2 isozyme locus and the *r* locus governing the runnerless (versus runnering) trait in *F. vesca* (Yu and Davis, 1995). Based on the priority of these initial reports, the respective marker pairs were used to anchor *F. vesca* linkage groups I and II, respectively, in the first *Fragaria* linkage map (Davis and Yu, 1997), and this nomenclature has been followed on all subsequent diploid linkage maps in *Fragaria* (Cipriani et al., 2006; Nier et al., 2006; Sargent et al., 2004a, 2006, 2007). More recently, Albani et al. (2004) reported the development of three SCAR markers derived from ISSRs linked at 3.0, 1.7 and 0 cM from the *s* locus on linkage group VI (which they denoted SFL for *Seasonal flowering locus*), in a $BC_1$ *F. vesca* progeny of 1,049 individuals.

## 5.3 Linkage Map Construction

Davis and Yu (1997) were the first to report a genetic linkage map for diploid *Fragaria*. Their map was developed from an intraspecific $F_2$ population derived from the cross *F. vesca* ssp. *vesca* 'Baron Solemacher' × *F. vesca* ssp. *americana* 'WC6' (BS×WC6) and was constructed using 78 molecular loci, including 11 codominant and 64 dominant RAPD markers, the STS marker for the *ADH* gene and the isoenzymes *Pgi*-2 and *Sdh*-1. The BS×WC6 progeny also segregated for two morphological characters *s* and *r*, one of which (*r*) was mapped in their population. The resultant map covered seven discrete linkage groups and a total genetic distance of 445 cM. The transferability of some of the RAPD markers employed by Davis and Yu (1997) was sufficient to allow linkage groups and map positions of five genes and one transcription factor involved in the anthocyanin biosynthetic pathway in diploid *Fragaria* to be determined. The co-segregation of one of the genes in this pathway, *Flavanone 3-hydroxylase* (*F3H*), with the *yellow/white fruit color* locus (*c*) was identified in a progeny of 40 $F_2$ individuals through a candidate gene approach (Deng and Davis, 2001).

The growth of interest in the development of transferable genetic markers for *Fragaria*, primarily in the development of microsatellite markers from genomic DNA and EST libraries, prompted the production of a second generation linkage map for diploid *Fragaria* from a segregating interspecific $F_2$ population of the cross

*F. vesca* 815 ×*F. bucharica* 601 (formerly *F. nubicola* 601) (Sargent et al., 2004a, b, 2006, 2007), designated FV×FB. This population segregates for three morphological characters, *r*, *s* and *pg*. To date, 234 molecular loci (172 SSRs, 16 gene-specific STSs 40 RFLPs, five ESTs, and one SCAR) have been mapped through collaboration between laboratories in the UK, Spain, France and the USA using the full FV×FB mapping population. The map covers seven linkage groups with a total genetic distance of 606.8 cM (just 25% greater coverage than the first generation diploid *Fragaria* RAPD map of Davis and Yu (1997)) and has since been adopted as the international reference population for the genus (Fig. 1).

## 5.4 Selective, or Bin Mapping

Traditionally, adding further loci to a linkage map has meant analyzing co-segregations of markers in the entire mapping progeny from which the map was derived. However, using the approach of 'selective', or 'bin' mapping, it is possible to place new markers onto well-saturated framework maps using a minimum number of well-chosen seedling genotypes that maximize the recombination breakpoints observed in the entire population (Vision et al., 2000; Howad et al., 2005). Sargent et al. (2008) recently defined a set of bins for the FV×FB mapping population spanning the *Fragaria* genome, and collaboration between researchers in the UK, Spain and Italy has resulted in 99 SSRs and four gene-specific markers being added to the FV×FB map using a reduced set of just six seedlings. The use of a selective mapping strategy in *Fragaria* affords considerable savings in labor and laboratory costs over conventional mapping, and thus permits markers to be assigned map positions with the minimum of time and effort.

## 5.5 Comparative Mapping

Since the adoption of SSR and gene-specific markers for map construction in *Fragaria*, and the recent application of RFLPs to the *Fragaria* reference map (Vilanova et al., 2008), it has been possible to compare maps of different diploid *Fragaria* mapping progenies and also to align the reference maps of different genera within the Rosaceae. Nier et al. (2006) constructed a linkage map of a backcross progeny from a wide interspecific cross between *F. vesca* and *F. viridis*. Their map was composed of 33 molecular markers and covered 241.6 cM.

Despite being significantly shorter than the genetic distance covered by the same markers in the FV×FB reference map, marker order was conserved between the two maps. Likewise, Cipriani et al. (2006) developed a set of novel SSRs and mapped these markers in a small intraspecific *F. vesca* population. These markers were subsequently bin mapped in the FV×FB population of Sargent et al. (2008) and, once again, conservation of marker order between the two maps was observed, indicating

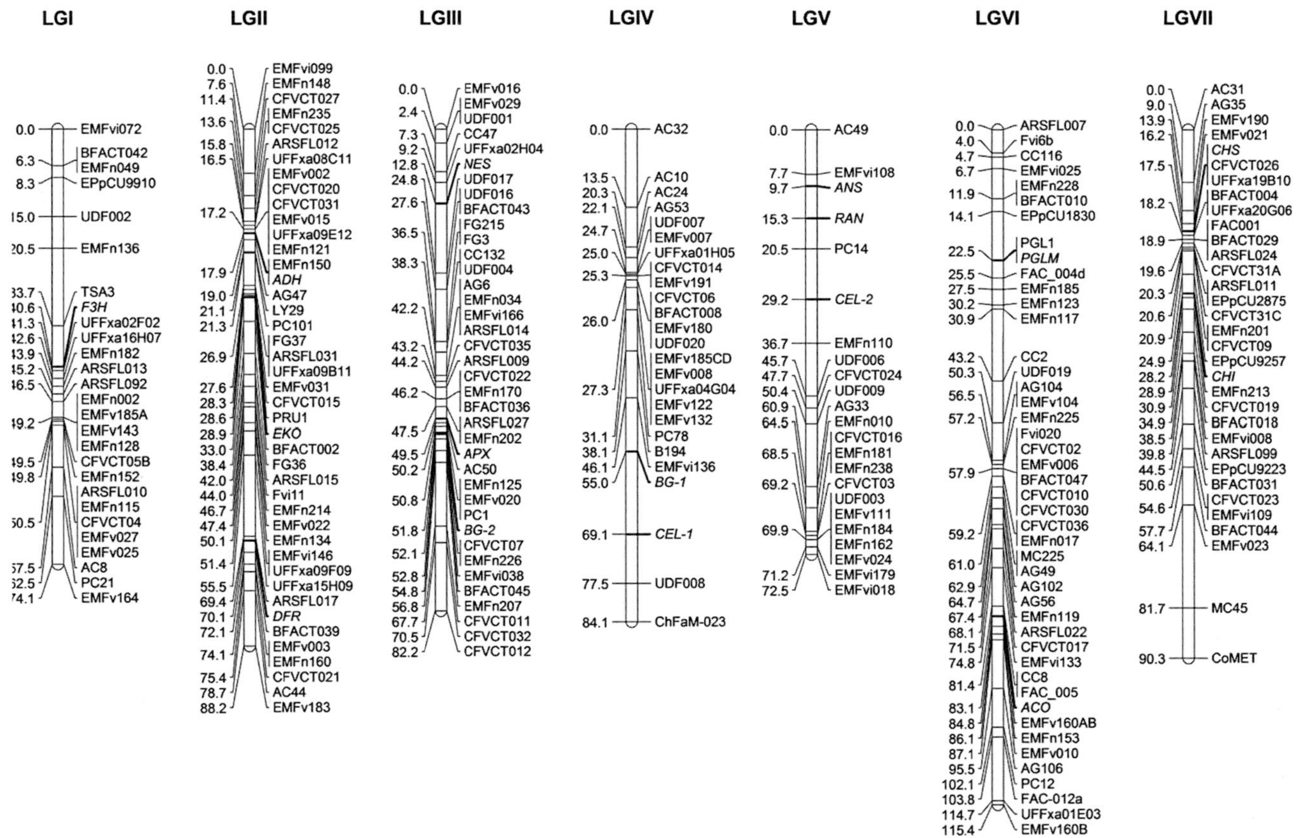


**Fig. 1** The diploid *Fragaria* $F_2$ reference linkage map (FV×FB) derived from the cross *F. vesca* 815 ×*F. bucharica* 601. The map is composed of 234 molecular loci (172 SSRs, 16 gene-specific STSs 40 RFLPs, five ESTs, and one SCAR) in seven linkage groups and covers a total genetic distance of 606.8 cM

that the chromosome structure of *F. bucharica*, *F. vesca* and *F. viridis* is essentially colinear at the macrosynteny level.

Through the mapping in the *Fragaria* and *Prunus* reference maps of 71 anchor loci (40 RFLPs, 23 ESTs and eight gene-specific STS markers) Vilanova et al. (2008) were able to compare, for the first time, the genomes of two of the most distinct taxa within Rosaceae. Their results showed a high degree of synteny conservation between the genomes of the two genera, with most of the loci of five of the eight linkage groups of *Prunus* showing correspondence with only one *Fragaria* linkage group homologue each. Their results suggested that there have been many chromosomal rearrangements, with at least 47 breakpoints occurring since divergence of the two genera. However, the conserved synteny displayed between the genomes of *Fragaria* and *Prunus* is sufficient to indicate that information gained on marker or gene position in one species within the Rosaceae can be used to inform studies in another. The development of structural genomics resources for *Prunus* is currently at a more advanced stage than for *Fragaria* and this knowledge will permit the use of *Prunus* resources for the benefit of research in *Fragaria*.

## 5.6 Octoploid Mapping

As previously stated, linkage mapping utilizing octoploid *Fragaria* species has been hampered by their highly complex genetic nature, and thus, mapping resources are less well developed than for the diploid species. However, the advent of the diploid *Fragaria* reference map (FV×FB) saturated with transferable markers (Sargent et al., 2004a, 2006, 2007; Vilanova et al., 2008), has facilitated the development of linkage maps of *F.* ×ananassa.

In the first report of marker-trait linkage in the cultivated strawberry, bulked segregant analysis (BSA) was used to identify a cluster of seven RAPD markers linked to the *Rpf1* locus, which confers race-specific resistance to red stele root rot (Haymes et al., 1997). The RAPD marker OPX-16C, which was linked in repulsion phase to *Rpf1*, was later converted to a SCAR marker that was linked to the resistance gene in coupling phase (Haymes et al., 2000). The occurrence of 1:1 segregation patterns for the RAPD markers scored in the mapping population (Md683 × Senga Sengana) of Haymes et al. (1997, 2000) stands in contrast to the claim of Zhang et al. (2003) that single copy RAPD loci are undetectable in octoploids.

The first linkage map of the cultivated strawberry (*F.* ×ananassa) was published by Lerceteau-Köhler et al. (2003). Their map was derived from the $F_1$ population 'Capitola' (CA75.121-101 × 'Parker') × CF1116 ['Pajaro' × ('Earliglow' × 'Chandler)] and consisted of 119 individuals mapped using a two-way pseudo test-cross strategy utilizing single-dose markers that segregated 1:1. Initially, all markers linked in coupling phase were mapped, after which, the data was inverted to allow markers linked in repulsion to be identified. Once all linkages were determined, separate male and female maps for the progeny were constructed. These initial linkage maps were composed of 515 single dose AFLP markers that segregated in 28 male and 30 female linkage groups with an average of ten and 7.8 markers per linkage group respectively. The female map spanned a total of 1604 cM, while the male map covered 1496 cM, equating to average linkage group lengths of 53.5 cM and 53.4 cM respectively.

Most of the linkage groups identified were shown to be in coupling/repulsion, indicating that the *F.* ×ananassa genome exhibits predominantly disomic behavior. However, the Lerceteau-Köhler et al. (2003) linkage map contained at least one large group of markers that all segregated in coupling phase, possibly indicative of polysomic segregation behavior. More recently, the map of Lerceteau-Köhler et al. (2003) was extended using a larger population of 213 individuals (Rousseau-Gueutin et al., 2008). This extended population was mapped using the original AFLP markers of Lerceteau-Köhler et al. (2003), along with the addition of further AFLP, SCAR and SSR markers. Based upon SSR marker commonality, four *F.* ×ananassa linkage groups with apparent homoeology to linkage group III of the FV×FB reference map were initially described (Davis et al., 2007b). The addition of codominant, transferable SSR markers and the larger progeny size employed in their study allowed Rousseau-Gueutin et al. (2008) to integrate the majority of the male and female linkage groups of Lerceteau-Köhler et al. (2003) and to further identify the four distinct homoeologous linkage groups for each of the seven diploid *Fragaria* linkage groups of the FV×FB reference map.

The mapping of additional codominant markers also allowed the authors to conclude that segregation in the cultivated strawberry is predominantly disomic, which is consistent with the findings of Ashley et al. (2003) who observed only disomic Mendelian segregations in the study of two controlled crosses of accessions of the species *F. virginiana* and concluded that the genome of this octoploid was highly diploidized. However, linkage groups that contained markers exclusively in coupling were again observed in the study of Rousseau-Gueutin et al. (2008), suggesting that polysomic segregation may still occur in isolated incidences within the *F.* ×ananassa genome.

A consensus map of 43 linkage groups encompassing 429 AFLP markers that could be mapped as single dose resitriction fragments (SDRFs) was constructed using 127 progeny from the cross ‘Tribute’ × ‘Honeoye’, and was 1541 cM in total length (Weebadde et al., 2008). The population was screened for flowering habit in four locations, and eight QTL were identified, none of which explained more than 36% of the variation, indicating that day-neutrality was inherited as a polygenic trait.

A less extensive linkage mapping study has been conducted in *F.* ×ananassa which attempted to isolate markers closely linked to the gene governing the everbearing trait (*Ev*) (Sugimoto et al., 2005). In their investigation, RAPD markers were identified closely linked to the *Ev* locus in a small subset (N=26) of the $F_1$ progeny from the cross ‘Everberry’ × ‘Toyonoka’. Five linked markers were then mapped in the full progeny of 199 seedlings and a map was produced spanning 39.7 cM, with the closest flanking markers located 11.8 cM and 15.8 cM on either side of the *Ev* locus, which segregated in a 1:1 Mendelian ratio expected for a single gene trait.

### 5.7 Association Mapping

SCAR markers derived from RAPD marker OPX-16C, which belonged to a cluster of RAPDs linked to red stele resistance/susceptibility, were assessed for association with resistance in a broad sampling of cultivars and breeding selections (Haymes et al., 2000). The presence of marker SCAR-R1A correctly identified 23 of 34 resistant genotypes. Eight of the 11 divergent genotypes belonged to a common pedigree, and their loss of linkage could be explained by a single crossover early in the pedigree. Only one of 53 susceptible genotypes possessed SCAR-R1A. Two AFLP-derived SCAR markers that were closely linked to the *Rca2* gene for *Colletotrichum acutatum* resistance in cultivar 'Capitola' were assessed in 43 *F.* ×*ananassa* cultivars (Lerceteau-Köhler et al., 2005). Of 28 resistant cultivars tested, 75 and 46% carried the STS-Rca2˙417 and STS-Rca2˙240 markers, respectively, while neither marker was detected in any of the 14 susceptible cultivars.

The Pedigree Genotyping concept (van de Weg et al., 2004) aims at integrating linkage and pedigree-based data as a means of exploiting breeding materials to bring higher resolution to the analysis of marker-trait associations. Pedigree Genotyping requires the use of multi-allelic markers such as SNPs or SSRs, and a dedicated software package to calculate the genetic values of different marker alleles. In this approach, all of the alleles of a given locus are traced through a multigenerational pedigree, and their associations with measured phenotypic values are assessed. By providing a basis for predicting phenotypic value based upon genotype, Pedigree Genotyping is envisioned as a potentially powerful tool in the service of marker assisted selection (van de Weg et al., 2004).

## 6 Whole Genome Sequencing

The small size of the basic *Fragaria* genome is favorable for whole genome sequencing. The cytologically determined C value for *Fragaria vesca* was reported to be 164 Mbp (Akiyama et al., 2001), from which a 206 Mbp C value was extrapolated (Folta and Davis, 2006; Davis et al., 2007b) to account for the discrepancy between the 125 Mbp genome size attributed to the *Arabidopsis thaliana* standard by Akiyama et al. (2001) and its 157 Mbp C value as cytologically determined by Bennett et al. (2003). Data suggesting that the subgenomes of the octoploid species may have diminished in size since these species arose from their diploid ancestors have been considered by Davis et al. (2007b).

An initial ~1% (1.75 Mbp) sampling of the *F. vesca* genome has been obtained by sequencing 50 genomic inserts (GenBank numbers EU024823-EU024872) from a fosmid library of *F. vesca* ssp. *americana* 'Pawtuckaway' (Davis et al., 2006b). Of these 50 sequenced clones, 30 were chosen at random and 20 were identified by using candidate gene sequences to probe library filters. Initial analysis indicates a protein-encoding gene density of about 1 gene per 6 kb, while SSR loci with five or more repeats occurred at a frequency of about 1 per 5 kb.

Porebski S, and Catling PM (1998) RAPD analysis of the relationship of North and South American subspecies of *Fragaria chiloensis*. Canadian Journal of Botany-Revue Canadienne de Botanique 76: 1812–1817

Potter D, Luby, JJ, and Harrison, RE (2000) Phylogenetic relationships among species of *Fragaria* (Rosaceae) inferred from non-coding nuclear and chloroplast DNA sequences. Systematic Botany 25: 337–348

Rajesh PN, Tekeoglu M, Gupta VS, Ranjekar PK, and Muehlbauer FJ (2002) Molecular mapping and characterization of an RGA locus RGAPtokin $1\text{–}2_{171}$ in chickpea. Euphytica 128: 427–433

Richardson CW (1914) A preliminary note on the genetics of *Fragaria*. Journal of Genetics 3: 171–177

Richardson CW (1918) A further note on the genetics of *Fragaria*. Journal of Genetics 7: 167–170

Richardson CW (1920). Some notes on *Fragaria*. Journal of Genetics 10: 39–46

Richardson CW (1923) Notes on *Fragaria*. Journal of Genetics 13: 147–152

Rousseau-Gueutin M, Barrot, L, Lerceteau-Köhler E, Sargent DJ, Monfort A, Simpson DW, Arús P, and Denoyes-Rothan B (2008) Comparative genetic mapping of diploid and octoploid *Fragaria* (strawberry) revealing essentially disomic behavior of the octoploid genome and complete synteny between species. Genetics 179: 2045–2060

Sargent DJ, Hadonou AM, and Simpson DW (2003) Development and characterisation of polymorphic microsatellite markers from *Fragaria viridis*, a wild diploid strawberry. Molecular Ecology Notes 3: 550–552

Sargent DJ, Geibel M, Hawkins JA, Wilkinson MJ, Battey NH, and Simpson DW (2004a) Quantitative and qualitative differences in morphological traits revealed between diploid *Fragaria* species. Annals of Botany 94: 787–796

Sargent DJ, Davis TM, Tobutt KR, Wilkinson MJ, Battey NH, and Simpson DW (2004b) A genetic linkage map of microsatellite, gene specific and morphological markers in diploid *Fragaria*. Theoretical and Applied Genetics 109: 1385–1391

Sargent DJ (2005) A genetic investigation of diploid *Fragaria*. PhD thesis, The University of Reading pp. 223

Sargent DJ, Clarke J, Simpson DW, Tobutt KR, Arús P, Monfort A, Vilanova S, Denoyes-Rothan B, Rousseau M, Folta KM, Bassil NV, and Battey NH (2006) An enhanced microsatellite map of diploid *Fragaria*. Theoretical and Applied Genetics 112: 1349–1359

Sargent DJ, Rys A, Nier S, Simpson DW, and Tobutt KR (2007) The development and mapping of functional markers in *Fragaria* and their transferability and potential for mapping in other genera. Theoretical and Applied Genetics 114: 373–384

Sargent DJ, Cipriani G, Vilanova S, Gil-Ariza D, Arús P, Simpson DW, Tobutt KR, and Monfort A (2008). The development of a bin mapping population and the selective mapping of 103 markers in the diploid *Fragaria* reference map. Genome 51: 120–127.

Senanayake YDA, and Bringhurst RS (1967) Origin of *Fragaria* polyploids I. Cytological analysis. American Journal of Botany 54: 221–228

Shimomura K, and Hirashima K (2006) Development and characterization of simple sequence repeats (SSR) as markers to identify strawberry cultivars (*Fragaria* × *ananassa* Duch.) Journal of the Japanese Society for Horticultural Science 75: 399–402

Staudt G (2006) Himalayan species of *Fragaria* (Rosaceae). Botanische Jahrbücher für Systematik, Pflanzengeschichte und Pflanzengeographie 126:483–508

Staudt G, DiMeglio LM, Davis TM, and Gerstberger P (2003) *Fragaria* ×*bifera* Duch.: Origin and taxonomy. Botanische Jahrbücher für Systematik, Pflanzengeschichte und Pflanzengeographie 125: 53–72

Sugimoto T, Tamaki K, Matsumoto J, Yamamoto Y, Shiwaku K, and Watanabe K (2005) Detection of RAPD markers linked to the everbearing gene in Japanese cultivated strawberry. Plant Breeding 124: 498–501

Tyrka M, Dziadczyk P, and Hortynski JA (2002) Simplified AFLP procedure as a tool for identification of strawberry cultivars and advanced breeding lines. Euphytica 125: 273–280 van de Weg WE, Voorips RE, Finkers R, Kodde LP, Jansen J, and Bink MCAM (2004) Pedigree genotyping: a new pedigree-based approach of QTL identification and allele mining. Acta Horticulturae 663: 45–50

Vilanova S, Arús P, Sargent DJ, and Monfort A (2008) Synteny conservation between two distantly-related Rosaceae genomes: *Prunus* (the stone fruits) and *Fragaria* (the strawberry). BMC Plant Biology 8:67

Viruel MA, Sánchez D, and Arús P (2002) An SSR and RFLP linkage map for the octoploid strawberry (*Fragaria* × *ananassa*). In: Plant, Animal and Microbe Genomes. 10th Conference, San Diego, California, USA. http://www.intl-pag.org/pag/10/abstracts/PAGX˙p660.html

Vision TJ, Brown DG, Shmoys DB, Durrett RT, and Tanksley SD (2000) Selective mapping: A strategy for optimizing the construction of high-density linkage maps. Genetics 155: 407–420

Weebadde CK, Wang D, Finn CE, Lewers KS, Luby JJ, Bushakra J, Sjulin TM, and Hancock JF (2008) Using a linkage mapping approach to identify QTL for day-neutrality in the octoploid strawberry. Plant Breeding 127: 94–101.

Williamson SC, Yu H, and Davis,TM (1995) Shikimate dehydrogenase allozymes: Inheritance and close linkage to fruit color in diploid strawberry. Journal of Heredity 86: 74–76

Wolyn DJ, and Jelenković G (1990) Nucleotide-Sequence of an Alcohol-Dehydrogenase Gene in Octoploid Strawberry (*Fragaria* × *ananassa* Duch). Plant Molecular Biology 14: 855–857

Yarnell SH (1928) Notes on the somatic chromosomes of the seven-chromosome group of *Fragaria*. Genetics 14: 78–83

Yu H, and Davis TM (1995) Genetic linkage between runnering and phosphoglucoisomerase allozymes, and systematic distortion of monogenic segregation ratios in diploid strawberry. Journal of the American Society for Horticultural Science 120: 687–690

Zhang Z, Fukino N, Mochizuki T, and Matsumoto S (2003) Single-copy RAPD marker loci undetectable in octoploid strawberry. Journal of Horticultural Science and Biotechnology 78: 689–694.

# 22. Functional Molecular Biology Research in *Fragaria*

**Wilfried Schwab, Jan G. Schaart, and Carlo Rosati**

## 1 Introduction

The most economically important strawberry species is the octoploid *Fragaria* × *ananassa*, grown worldwide over a wide range of latitudes for fresh market and processing industry. The small-fruited diploid *F. vesca* and the musky-flavored hexaploid *F. moschata* are grown on a much smaller scale only for local fresh markets and pastry industry. The largely unexploited genetic pool contain species and genotypes differing for important breeding traits such as: fruit size, aroma, firmness and chemical composition; dioecy/hermaphroditism; response to photoperiod; resistance to pathogens and pests; hardiness, etc. (Darrow, 1966; Faedi et al., 2002; Døving et al., 2005; Ulrich et al., 2007; and references therein).

From the biotechnology point of view, the relatively small genome of *Fragaria* species (<1.5-fold the *Arabidopsis* genome for diploid *Fragaria* genotypes) makes them an attractive target for future genomics initiatives. Concomitantly with other sequencing efforts in the last years, the number of *Fragaria* DNA sequences in public databases has exponentially increased to exceed 53,000 entries – including both core nucleotide sequences and expressed sequence tags (ESTs) of as many as fifteen *Fragaria* species – as of June 2008. Yet, the developed 'omics'-based tools and databases (Aharoni et al., 2000, 2002a, 2002b; Folta et al., 2005; Carbone et al., 2006; Gil-Ariza et al., 2006; Hjernø et al., 2006; Jung et al., 2008) have to be further implemented into comprehensive knowledge platforms. Therefore, one of the main challenges of the 'post-genomic era' remains the integration of high-throughput technologies coupled to efficient mathematical algorithms and modeling, aimed at unraveling metabolic networks and a better understanding of plant physiological processes.

This review focuses on progress in functional molecular and biotechnology studies, pinpointing research reports related to major breeding traits. Readers are also referred to previous reviews for more exhaustive coverage of existing literature

C. Rosati (✉)
Department of Genetics and Genomics, ENEA, Trisaia Research Centre, Rotondella (MT), Italy

K.M. Folta, S.E. Gardiner (eds.), *Genetics and Genomics of Rosaceae*, Plant Genetics and Genomics: Crops and Models 6, DOI 10.1007/978-0-387-77491-6_22,

(Hokanson and Maas, 2001; Folta and Davis, 2006) and to other chapters of this book for investigations on translational genomics and molecular markers.

## 2 Fruit-Related Traits

Strawberry fruits contain thousands of metabolites, many of which have a strong impact on consumer's senses and health. Analytical research has focused on the identification and quantification of such metabolites throughout fruit development and in the post-harvest chain of fresh and processed fruit. To date, most analytical biochemistry studies on strawberry have relied on specifically devised extraction/separation methods to identify/quantify specific compounds of interest (see below). The only report on comprehensive non-targeted metabolic analysis of strawberry identified 5,844 unique mass peaks by analyzing fruits at four developmental stages, with eight combinations of extraction/ionization methods and FTMS (Fourier transform mass spectrometry) technology (Aharoni et al., 2002a). Hundreds of compounds belonging to different biochemical classes (e.g., sugars, fatty and phenolic acids, amino acids, esters, terpenes, alkaloids, flavonoids) could be grouped according to differences in developmental accumulation profiles. Ultra high resolution and mass accuracy ($<1$ ppm) and throughput make FTMS approach luring, but high cost and cumbersome use of the instrument, difficulty in positive structural identification and inability to distinguish isomers are drawbacks that have limited its use in plant research. However, this work is a milestone for the development of high-throughput research in *Fragaria* and other fruit species.

## 3 Volatile Compounds

Studying the origin of volatiles emitted by strawberry is a worthwhile and challenging area of research because strawberry flavor is extremely popular worldwide (Bood and Zabetakis, 2002). More than 360 volatiles are produced by the strawberry fruit but not all are equally important for the overall aroma as it was demonstrated by application of an aroma extract dilution analysis. Among the volatiles only about fifteen odorants were shown to contribute to the smell of strawberry due to their high flavor dilution factors (Schieberle and Hofmann, 1997; Jetti et al., 2007). Furthermore, calculation of odor activity values (ratio of concentration to odor threshold) revealed six compounds (Z)-3-hexenal, 4-hydroxy-2,5-dimethyl-3(2H)-furanone (HDMF), methyl butanoate, ethyl butanoate, methyl 2-methylpropanoate, and 2,3-butanedione as the key flavor compounds in the typical strawberry-like odor of the juice. However, studies using five different strawberry cultivars showed that there are species-specific significant volatiles such as ethyl butanoate, methyl butanoate and $\gamma$-decalactone, and generally important strawberry aroma compounds such as HDMF, linalool and ethyl hexanoate (Larsen et al., 1992).

Volatile esters are produced by virtually all soft fruits. They probably serve a dual role as attractants for animals, insects and humans and as protectants against pests and pathogens. The last step in their biosynthesis is catalyzed by alcohol acyltransferases (AATs), which link alcohols to CoA-activated acyl moieties (Fig. 1A). The substrates are provided by the metabolism of either fatty acids or amino acids. A strawberry *AAT* cDNA (*SAAT*) was identified during profiling studies of ripening-induced genes utilizing cDNA microarrays (Aharoni et al., 2000). The recombinant SAAT enzyme catalyzed the formation of esters found in strawberry, using aliphatic, medium chain alcohols in combination with various chain length acyl-CoAs as substrates. The activity of SAAT with aromatic substrates and the tertiary monoterpenol linalool was negligible. The identification of *SAAT*-related genes in a number of fruit species led also to the isolation and heterologous expression of an *AAT* gene from wild strawberry (*Fragaria vesca*) *VAAT* (Beekwilder et al., 2004). Although the VAAT enzyme is closely related to SAAT, its activity is quite distinct. VAAT is much more active on short chain alcohol substrates which are not preferred by SAAT. These characteristics are in agreement with the substrate preference reported for enzyme extracts of wild and cultivated strawberry fruits (Olias et al., 2002). Thus, substrate preference could not be predicted on the basis of sequence information and was not necessarily reflected in the representation of esters in the corresponding fruit volatile profiles. The results suggested that the specific profile of a given fruit species is, to a significant extent, determined by the supply of precursors. A correlation between the expression of an *AAT* gene, total AAT activity and the presence of related esters in strawberry fruit headspace was demonstrated by molecular and biochemical studies (Carbone et al., 2006). Recently, overexpression of *SAAT* in lactic acid bacteria resulted in the production of octyl acetate, demonstrating the availability of suitable substrates in the host and offering the opportunity to use the plant gene to develop biotechnological process for the production of natural esters (Hernández et al., 2007). Ethanol, the precursor of ethyl butanoate and ethyl hexanoate is probably formed by an enzymatic reduction of acetaldehyde, which is formed from pyruvate by the action a pyruvate decarboxylase (PDC). Two strawberry *PDC* genes have been cloned but only *FaPDC1* is induced during fruit ripening (Moyano et al., 2004).

Terpenoids are derived from the mevalonate pathway, which is active in the cytosol, or from the plastidial 2-*C*-methyl-D-erythritol-4-phosphate pathway. Both pathways produce the C5 units that can be used by prenyl transferases in condensation reactions to produce the monoterpene precursor geranyl diphosphate (GDP), the sesquiterpene precursor farnesyl diphosphate (FDP), and diterpene and carotenoid precursors. Linalool, nerolidol, α-pinene, and limonene are the quantitatively predominant volatile terpenes in strawberry and can reach up to 20% of the total fruit volatiles (Loughrin and Kasperbauer, 2002). A substantial advance in the understanding of mono- and sesquiterpene formation in strawberry fruit and their functions during strawberry domestication has been made by cloning and functional expression of a nerolidol synthase 1 (*NES1*) gene and characterization of its gene product as a bifunctional mono- and sesquiterpene synthase (Fig. 1B, Aharoni et al., 2004). The recombinant FaNES1 enzyme generated (S)-linalool

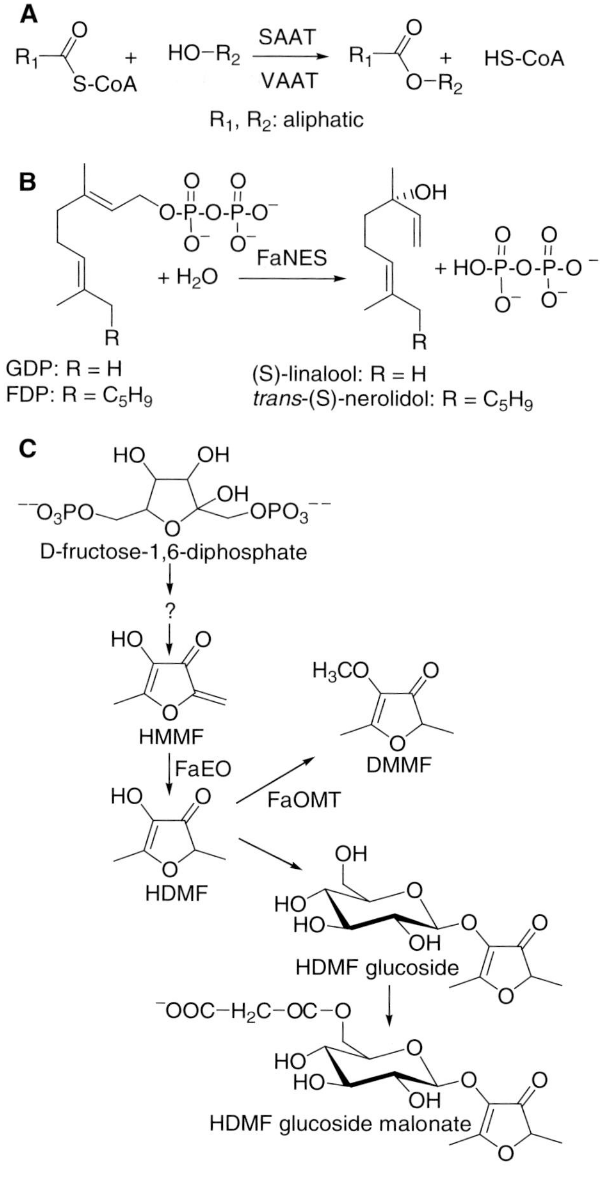


**Fig. 1** Proposed biosynthetic pathway and metabolism of: (**A**) fruit ester, (**B**) (S)-linalool and trans-(S)-nerolidol and (**C**) 4-hydroxy-2,5-dimethyl-3(2H)-furanone (HDMF). DMMF, 2,5-dimethyl-4-methoxy-3(2H)-furanone. FaEO, *Fragaria* × *ananassa* enone oxidoreductase. FaNES, *Fragaria* × *ananassa* nerolidol synthase. FaOMT, *Fragaria* × *ananassa O*-methyltransferase. FDP, Farnesyl diphosphate. GDP, Geranyl diphosphate. HMMF, 4-hydroxy-5-methyl-2-methylene-3(2H)-furanone. SAAT, Strawberry alcohol acyl-CoA transferase. VAAT, Vesca alcohol acyl-CoA transferase

and *trans*-(S)-nerolidol from GDP and FDP, respectively. Although similar genes were found in wild and cultivated strawberry species, only *FaNES1* is exclusively present and highly expressed during fruit ripening in cultivated octoploid varieties. Green fluorescent protein localization studies demonstrated that a change in subcellular localization – due to a truncation at the N-terminus – let the FaNES1 enzyme encounter both GDP and FDP, resulting in the production of (S)-linalool and *trans*-(S)-nerolidol. In contrast, a second terpene synthase (pinene synthase, *PINS*) gene is primarily expressed in wild strawberry species while an insertional mutation reduced its expression in cultivated species (Aharoni et al., 2004). In *Fragaria vesca*, FvPINS produced α-pinene and thus provided the substrate for a pinene hydroxylase which catalyzed the C10 hydroxylation of α-pinene to myrtenol. In *Fragaria* × *ananassa*, the loss of α-pinene further influenced the fruit flavor profile because the olefin is no longer available as a substrate for the downstream compound myrtenol and myrtenyl acetate (Aharoni et al., 2004). In vivo feeding experiments using [5,5-$^2H_2$]mevalonic acid lactone and [5,5-$^2H_2$]-1-deoxy-D-xylulose confirmed the biosynthesis of (S)-linalool, *trans*-(S)-nerolidol, and α-pinene in the cytosol via the mevalonic pathway (Hampel et al., 2006). These results nicely demonstrate how simple mutations can affect the overall flavor profiles that are eventually selected for during domestication.

The furanone HDMF is quantitatively a minor constituent in the headspace of strawberry fruit but, because of its low odor threshold value, it is considered as one of the most important volatiles of the overall flavor (Fig. 1C; Schieberle and Hofmann, 1997). The first indication for the enzymatic formation of HDMF was provided by a study demonstrating the correlation between fruit ripening stage and HDMF levels. Incorporation experiments with radiolabeled precursors and substances labeled with stable isotopes revealed D-fructose-1,6-diphosphate as an efficient biogenetic precursor of HDMF. In strawberry, D-fructose-1,6-diphosphate is transformed by an as yet unknown enzyme to 4-hydroxy-5-methyl-2-methylene-3(2H)-furanone (HMMF), which serves as substrate for an oxidoreductase recently isolated from ripe fruit (Raab et al., 2006). Sequence analysis of two peptide fragments showed identity with the protein sequence of a strongly ripening-induced, auxin-dependent putative quinone oxidoreductase (*Fa*QR). Because the recombinant protein reduced only the artificial substrate 9,10-phenanthrene quinone out of a number of quinones tested and formed HDMF from HMMF, the enzyme was renamed enone oxidoreductase (EO) (Klein et al., 2007). Highly similar sequences were identified in a tomato and pineapple expressed sequence tag collection. Both species also produce HDMF in their fruits. The *Solanum lycopersicun* EO was expressed and biochemical studies confirmed the involvement of SlEO in the biosynthesis of HDMF (Klein et al., 2007). In strawberry HDMF is further metabolized by an *O*-methyltransferase (OMT) to 2,5-dimethyl-4-methoxy-3(2H)-furanone (DMMF) (Fig. 1C). The corresponding cDNA was cloned from strawberry and functionally expressed (Wein et al., 2002). A common structural feature of the accepted substrates was an ortho-diphenolic structure also present in HDMF in its dienolic tautomer. *FaOMT* transcripts were absent in root petiole, leaf and flower but accumulated during ripening in strawberry fruit. Transformation of strawberry with the *FaOMT* sequence in antisense orientation, under the control of the CaMV 35S promoter, resulted in a near total loss of DMMF, demonstrating the in vivo methylation of HDMF by *Fa*OMT (Lunkenbein et al., 2006c). Due to the high significance of HDMF as flavor molecule, these studies provide a foundation for the improvement of strawberry flavor, for the biotechnological production of HDMF and for the development of markers to identify plants with high flavor potential from breeding populations.

## 4 Polyphenol Compounds

Ripe strawberry fruits accumulate large amounts of anthocyanin pigments and other polyphenols. These secondary metabolites have attracted much attention since they provide both visual and nutraceutical qualities to the fruit, as well as prevent/protect from degenerative and cardiovascular diseases (e.g., Hannum, 2004; Mink et al., 2007; Seeram, 2008). Levels and qualitative composition of flavonoids in strawberry fruits depend on genetic, developmental and environmental factors as well as post-harvest manipulations (e.g., Bakker et al., 1994; Kalt et al., 1999; Häkkinen and Törrönen, 2000; Wang and Lin, 2000; Anttonen et al., 2006; da Silva et al., 2007; Tulipani et al., 2008) also through transcription factors. The vast literature on phenylpropanoid and flavonoid pathways (e.g., Davies and Schwinn, 2006 and references therein) have boosted molecular research on berry species (e.g., Jaakola et al., 2002) and strawberry (see below). Because of the obvious change in color associated to anthocyanin biosynthesis that can be visually scored, the flavonoid pathway is also an excellent system for functional studies.

Strawberry fruit (i.e., receptacle) development lasts about 35 days from pollination to full ripening. Within 10–12 days after anthesis, a first peak of gene/enzyme activity is associated with the production of flavan-3-ol-derived compounds; 15–20 days later, a second peak boosts the synthesis of anthocyanins – mostly pelargonidin-3-glucoside (Perkins Veazie, 1995) – in fully developed fruits, which accumulate at levels higher than those of other polyphenols. (Poly)phenols are synthesized through a series of enzymatic reactions involving different types of enzymes (Fig. 2A). Hence, end products are produced through a series of reactions such as glycosylations, condensations, acylations, methylations, esterifications, etc. . A remarkable example of cross-talk between secondary metabolite and hormone pathways in strawberry is represented by the reversible inhibition of phenylpropanoid and flavonoid gene expression by auxins, produced by achenes (Moyano et al., 1998; Aharoni et al., 2002b; Lunkenbein et al., 2006a). Phenylpropanoids are synthesized from phenylalanine by the sequential action of phenylalanine ammonia-lyase (PAL), cinnamic acid 4-hydroxylase (C4H) and *p*-coumarate:CoA ligase (4CL). Such compounds can be directed towards the synthesis of flavonoids or lignin precursors. At early strawberry ripening stages, PAL activity has been associated with fruit astringency and the synthesis of proanthocyanidin (PA) compounds (Cheng and Breen, 1991).

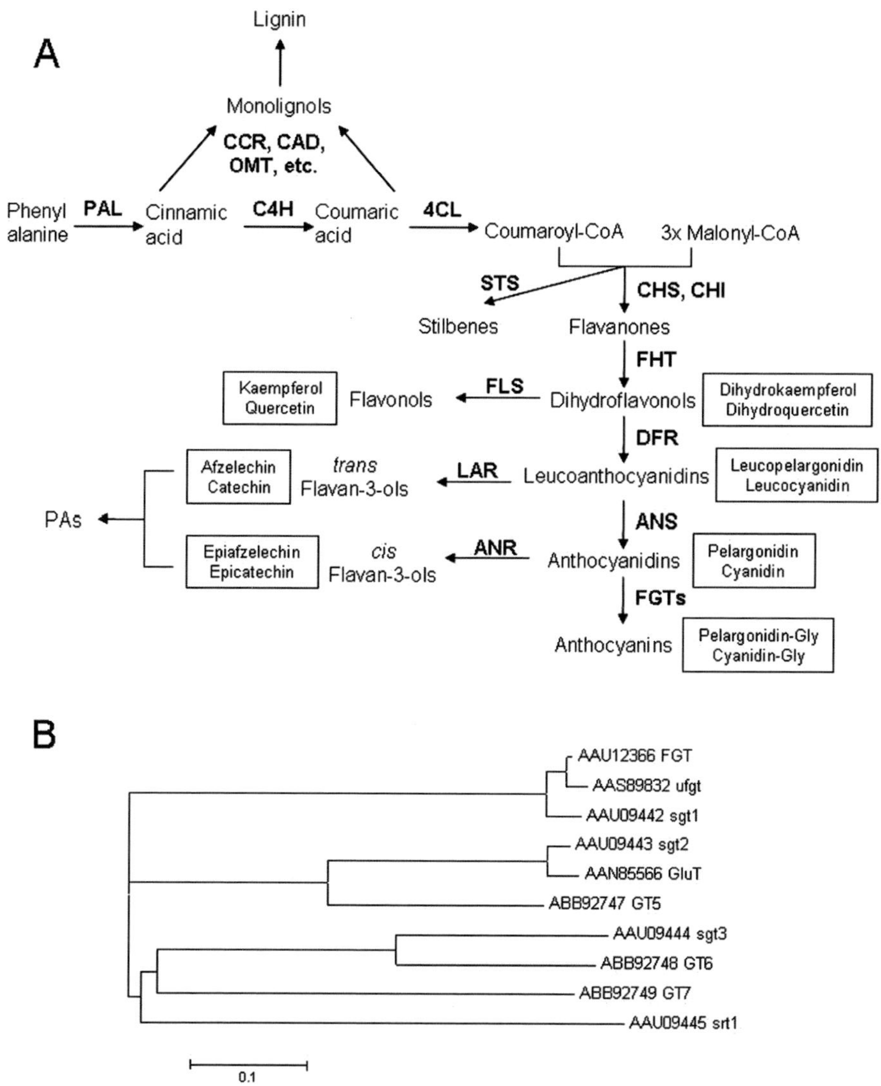


**Fig. 2** (**A**). Simplified representation of selected polyphenol metabolic routes most relevant to strawberry. For major flavonoid classes, 4'- (above) and 3',4'-hydroxylated (below) aglycones are indicated in each *box*, respectively. For enzyme acronyms, see text. (**B**). Phylogenetic analysis of strawberry glycosyltransferases derived from full length cDNAs present in Genbank as of July 31, 2007. Protein sequences were aligned using ClustalW (http://www.ebi.ac.uk/clustalw/); the dendrogram was obtained with Treeview (http://taxonomy.zoology.gla.ac.uk/rod/treeview.html)

A dual *O*-methyltransferase (FaOMT), implicated in HDMF biosynthesis, was shown to be able to methylate phenylpropanoids as well (Lunkenbein et al., 2006c). A further example of multifunctional enzymes is given by strawberry cinnamoyl aldehyde dehydrogenase (CAD) and coumaroyl CoA reductase (CCR), generally involved in the biosynthesis of monolignols and associated with vascular system (Anterola and Lewis, 2002; Blanco-Portales et al., 2002), but which play a role also in fruit firmness (see below). Wang et al. (2007) have recently reported the occurrence of *cis*- and *trans*-resveratrol (stilbenes) in strawberry fruits, which accumulate in the receptacle at higher levels than in achenes, with a decreasing trend over fruit development. The levels of resveratrol were also found to be affected by genotype, cultural practices, and environmental conditions. This report opens new perspectives for the characterization and improvement of stilbene in strawberry, whose biosynthesis is catalyzed by the enzyme stilbene synthase (STS; Fig. 2A).

The first steps in flavonoid biosynthesis are catalyzed by chalcone synthase (CHS) and chalcone isomerase (CHI), which condense one phenylpropanoid and three malonyl-CoA molecules and shape the C15 three-ring flavonoid structure, respectively (Fig. 2A). Because of its key position at the beginning of the flavonoid pathway, CHS has been a target of several studies also in strawberry. Molecular and biochemical investigations showed that the expression of four *FaCHS* genes follows a two-phase pattern, with two peaks at early and late fruit development stages; the single early peak of FaCHS enzyme activity might indicate cultivar-dependent variation or post-translational regulation of enzyme expression (Halbwirth et al., 2006; Almeida et al., 2007). Inhibition of *FaCHS* expression blocks the biosynthesis of downstream flavonoids: transgenic and Agroinfiltrated plants, respectively overexpressing antisense and RNA interference (RNAi) *FaCHS* constructs, had fruits with reduced anthocyanin and flavonoid content, and redirection of metabolite flux towards phenylpropanoid compounds, in correlation with reduced *FaCHS* transcript levels (Lunkenbein et al., 2006b; Hoffmann et al., 2006). Using *F. vesca* or *F. vesca* × *F. nubicola* segregating populations, Deng and Davis (2001) mapped *CHS* and *CHI* genes to linkage groups (LGs) III and VII, respectively; on the other hand, Sargent et al. (2004) found such genes tightly linked in LGVII using a larger *F. vesca* × *F. nubicola* $F_2$ population.

The sequential action of flavanone 3β-hydroxylase (FHT), dihydroflavonol 4-reductase (DFR) and anthocyanidin synthase (ANS) enzymes leads to the synthesis of anthocyanidin pigments. These enzymes in octoploid strawberry have a preference for 4'-hydroxylated flavonoids (Almeida et al., 2007), in line with the massive production of pelargonidins in ripe fruits. Expression of *FaFHT* and *FaANS* genes has a two-peak pattern, consistent with the synthesis of PAs and anthocyanins at early and final fruit development stages, respectively. Contrasting data on FaDFR gene and enzyme expression during fruit development exist in literature: a steady increase of *FaDFR* expression with a single peak at turning/red stages was scored by Real Time PCR (Almeida et al., 2007; Schaart et al., unpublished), while Northern analysis and enzymatic assays pointed out a two-phase expression, like for previous flavonoid genes (Moyano et al., 1998; Halbwirth et al., 2006). Genetic studies in diploid *Fragaria* populations mapped *FHT*, *DFR* and *ANS* genes respectively to LGs I, II and V, and associated yellow fruit color (absence of anthocyanins) to the *c* locus, encoded by *FHT* (Deng and Davis, 2001). Such results were confirmed in larger studies using SSR and functional markers (Sargent et al., 2004, 2006, 2007,

2008). In *F.* × *ananassa*, the only flavonoid gene mapped so far has been *FaDFR* (Lerceteau-Köhler et al., 2003).

Side-chain products flavonols and flavan-3-ols are produced by the action of flavonol synthase (FLS), leucoanthocyanidin reductase (LAR) and anthocyanidin reductase (ANR). Flavonols are compounds present in variable amounts in ripe fruits of different genotypes (Hernanz et al., 2007 and references therein), and their concentration was shown to decrease from turning to red stage consistently with FaFLS enzyme activity (Halbwirth et al., 2006). Catechin and epicatechin are the most common flavan-3-ols, although afzelechin and epiafzelechin accumulate in fruits of some genotypes. FaFLS, FaLAR and FaANR were shown to accept preferentially 3',4'-hydroxylated flavonoid substrates (Almeida et al., 2007), but this feature is likely to be cultivar-dependent. Flavan-3-ol monomers can form heterodimers with anthocyanins (Fossen et al., 2004), homodimers, and PA polymers; the enzymes responsible for the formation of such compounds are still unknown. Because of their large molecular weight, intact PAs are difficult to analyze, but recent progress in analysis of PA cleavage products, obtained e.g. by phloroglucinol treatment followed by LC, enabled to decompose PAs into their flavan-3-ol constituents and determine their levels (Kennedy and Jones, 2001). Indeed, a large PA pool, higher than that of anthocyanins, was found also at red ripe stage (Almeida et al., 2007), a finding opposite to the decreasing trend of flavan-3-ols during fruit development (Halbwirth et al., 2006).

The numerous flavonoid glycosyltransferase (FGT) enzymes add various sugar moieties prior to compartmentalization into the vacuole. Plant GTs accept a wide range of substrates: thus, their properties need to be functionally demonstrated (Vogt and Jones, 2000). In strawberry, a number of GTs have been isolated, which clusterize into different clades (Lunkenbein et al., 2006a; Fig. 2B). One FGT (AAU12366), very similar to two other entries (Fig. 2B), is a UDP-glucose:flavonoid 3-*O*-glucosyltransferase (F3GluT), showing a higher specificity for anthocyanidins than for flavonols; its gene and enzyme expression was induced at late ripening stages in different cultivars (Manning, 1998; Halbwirth et al., 2006; Almeida et al., 2007; Griesser et al., 2008). Downregulation of *FaFGT* in agroinfiltrated fruits by RNAi reduced levels of pelargonidin – but not of cyanidin – derivatives, and led to an increase of upstream flavan-3-ols. Most *FaFGT*-silenced fruits had a less intense red color and of a different hue. The absence of a white fruit phenotype, and steady levels of 3-*O*-glucosides of cyanidin, flavonols and flavan-3-ol: anthocyanin heterodimers respectively indicate the presence of other glycosyltransferases with redundant and specific F3GluT function in the genome of *F.* × *ananassa* (Griesser et al., 2008). FaGT2 (AAU09443, Fig. 2B) is a UDP-glucose:cinnamate glucosyltransferase, whose transcript levels correlated with the presence of relevant phenylpropanoid glucosides in fruits and other organs in wild type and transgenic plants (Lunkenbein et al., 2006a). Most interestingly, recombinant FaGT2 is active not only on cinnamic and *p*-coumaric acids, but also on other aromatic and aliphatic substrates, pointing its role of as detoxification enzyme (Landmann et al., 2007). The other strawberry glycosyltransferases still await molecular and biochemical characterization to correlate their properties with chemical composition of strawberry fruits.

Although an increasing amount of information on structural genes involved in flavonoid biosynthesis has become available, so far only one transcription factor (TF) controlling flavonoid biosynthesis has been described for strawberry. Aharoni et al. (2001) identified and characterized *FaMyb1*, a ripening regulated member of the Myb family of TFs. Overexpression of *FaMyb1* in tobacco led to the suppression of anthocyanin and flavonol synthesis in flowers of transgenic lines, suggesting that *FaMyb1* acts as a repressor of transcription rather than as a transcriptional activator. At present, additional strawberry genes homolog to flavonoid biosynthesis-related TFs – three of the Myb gene family, four of the Myc (bHLH) gene family and one WD 40 repeat transcription factor gene – are being individually characterized at PRI Wageningen. These genes are homolog to TFs controlling flavonoid biosynthesis in other species and show an expression pattern consistent with flavonoid accumulation during strawberry fruit development (Schaart et al. unpublished results).

Although a vast literature on flavonoids has been accumulated, some critical steps in the metabolism and ecology of polyphenol compounds still have to be unraveled – e.g., PA biosynthesis, the possible involvement of degradation/oxidation enzymes (e.g., polyphenol oxidases) in regulating metabolite levels, the role of polyphenol compounds in strawberry resistance to biotic and abiotic stress – possibly by the combination of reverse genetics- and 'omics'-based approaches.

## 5 Fruit Firmness

Fruit firmness is one of the most important quality traits in strawberry breeding programs. Fruit firmness is a genetically complex trait: therefore, breeding for optimal fruit firmness is a difficult task in the polyploid background of most *Fragaria* species. Moreover, the observed inverse correlation between firmness and flavor emission (Salentjn et al., 2003) makes the discovery of genetic key factors for fruit firmness and susceptibility to damage (vulnerability) a challenge.

Strawberry cultivars can differ significantly in textural aspects like internal fruit firmness, vulnerability and post-harvest shelf life. Texture of fruits is determined by cell turgor pressure and by the intrinsic characteristics of cell wall. Modification and turn-over of the primary cell wall is required for both growth and softening of fruits. The timing and extent of polysaccharide solubilization and depolymerization may vary among genotypes, causing different softening rates. Rosli et al. (2004) have analyzed the polysaccharide content of fruit cell walls during fruit development and ripening. Among three cultivars that differed in firmness of ripe fruits – 'Camarosa' the firmest, 'Pajaro' intermediate and 'Toyonaka' the softest – main differences in cell wall content were only detected at immature stages. During fruit development, levels of water-soluble polymers increased in all three cultivars until white fruit stage, after which they remained constant in 'Camarosa' and 'Pajaro', but decreased in 'Toyonaka'. On the contrary, the amount of HCl-soluble pectin fraction decreased during ripening. It was suggested that during fruit development the accumulation of water-soluble polymers is due to *de novo* synthesis rather than a change of solubility of the fraction. Then, during ripening, the water-soluble polymers in 'Camarosa' and 'Pajaro' are supposed to come from degradation of HCl-soluble pectins, as *de novo* synthesis was found to be absent at late ripening stages (Knee et al. 1977). In contrast, in 'Toyonaka' no net accumulation of water-soluble polymers was observed, suggesting that both pectin fractions are actively degraded.

The amount of hemicellulosic polysaccharides and cellulose also decreased during fruit ripening in all three cultivars tested, but no differences in hemicellulose or cellulose content among cultivars were detected at ripe stages. This suggests that this polysaccharide does not have a key-role in strawberry fruit softening. This conclusion was not supported by experiments describing the antisense down-regulation of *FaCel1* alone (Woolley et al., 2001) or in combination with *FaCel2* (Palomer et al., 2006), both genes encoding an endo-$\beta$-(1,4)-glucanase. Fruits of these antisense *FaCel* strawberry plants showed only a partial reduction of $\beta$-(1,4)-glucanase specific activity and no change in fruit softening, suggesting a more complex situation with possible redundant $\beta$-(1,4)-glucanase activities present in strawberry. In a recent paper, differential expression of *FaXyl1*, coding for fruit-specific $\beta$-xylosidase that participates in hemicellulose metabolism, was described (Bustamante et al., 2006). Expression of *FaXyl1* and $\beta$-xylosidase activity were higher in 'Toyonaka' than in 'Camarosa', suggesting a correlation between *FaXyl1* expression and fruit softness, thereby supporting a key role of hemicellulose degradation during fruit softening.

Several fruit-specific pectin-modifying genes have also been cloned from strawberry, like $\beta$-galactosidase (Trainotti et al., 2001), polygalacturonases (Redondo-Nevado et al., 2001; Figueroa et al., 2008) and pectin (methyl) esterases (Castillejo et al., 2004). Pectate lyase genes isolated from ripe fruits have been proposed as other candidates for pectin degradation in both *F.* × *ananassa* and *F. chiloensis* (Medina-Escobar et al., 1997; Figueroa et al., 2008). The role of this ripening-specific gene in fruit softening was assessed in antisense transgenic strawberry plants showing a suppression of *FaPel1* expression (Jiménéz-Bermúdez et al., 2002). Eleven out of twenty antisense *FaPel1*-transgenic lines showed higher external fruit firmness, ranging from 149 to 179% when compared with fruits of control plants. Interestingly, these plants showed no differences in general fruit traits like weight, color, shape and soluble acids. However, analysis of flavor and aroma was not reported. At PRI Wageningen, antisense transgenic strawberry plants have been made using a multigene antisense construct for the joint repression of five different cell wall modifying genes. This construct contained antisense sequences of *FasPG* (Redondo-Nevado et al., 2001), *FaPG-like* (Salentjn et al., 2003), *FaPel1* and *FaPel2* (Medina-Escobar et al., 1997) and *FaExp2* (Civello et al., 1999) under the regulation of a single CaMV 35S promoter. Only one transgenic line, CWas34, showed significant firmer fruits than fruits of wild type plants. In this line the expression of all five genes was stably repressed to levels of 4–11% for *FasPG*, 21–33% for *FaPG-like*, 45–79% for *FaPel1*, 3–29% for *FaPel2* and 10–11% for *FaE*xp2, as compared to wild type fruits and measured over two successive years. In order to study the relationship between firmness and aroma, the aroma of fruits from different transgenic lines was compared by a panel test: 11 out of 12 panelists marked fruits of two different transgenic lines (including CWas34) as having a recognizable different aroma (Salentjn EMJ and Schwab W, unpublished data). GC-MS analysis of these fruits has to be performed to characterize differences in volatile composition in transgenic fruits. The above described results indicate that cell wall modifications may have an effect on fruit flavor and aroma, supporting the above-mentioned negative association between texture and flavor.

Also expansins are intimately involved in the fruit softening process. These proteins help cell wall-modifying enzymes to gain access to the cell wall. Expansin activity was detected in ripening strawberry fruits and seven different expansin genes (*FaExp 1-7*) were isolated (Civello et al. 1999, Harrison et al., 2001). A direct correlation between mRNA expression levels and fruit firmness was found for three out of five *FaExp* genes. The transcript levels were higher in the soft 'Toyonaka', than in the two firmer 'Camarosa' and 'Selva' (Dotto et al., 2006). Moreover, accumulation of expansin proteins during ripening started earlier in the softest cultivar.

Salentjn et al. (2003) used DNA microarray technology to investigate differential gene expression during fruit ripening, with a focus on fruit firmness. Several candidate genes related to cell wall degradation were found to be differentially expressed in fruits of the soft 'Gorella' compared to the firm 'Holiday', further confirmed by Northern analysis. Two different pectin methyl esterase-like genes and a polygalacturonase-like gene were up-regulated in 'Gorella'. The cell wall loosening enzyme expansin (*FaExp2*) and another polygalacturonase showed a slight increase in transcript level. Remarkably, mRNAs coding for enzymes involved in lignin biosynthesis showed a clear differential expression profile among both cultivars. Most dramatic differences were observed for a cinnamoyl CoA reductase (*FaCCR*) homologue. Two different *FaCCR* clones showed a more than 20-fold higher level of expression in fruits of 'Gorella' (soft), whereas transcript levels of a cinnamyl alcohol dehydrogenase (*FaCAD*) gene were 2–3 times higher in 'Holiday' (firm). CCR and CAD catalyze successive steps in the lignin biosynthetic pathway, and differences in expression may lead to differences in lignin content and composition in both cultivars. Since lignin is found in vascular tissues (Aharoni et al., 2002b) and the amount of vascular tissue is associated with fruit texture (Jewell et al., 1973), the genetic background controlling differences in lignin content and composition may be important for fruit firmness. The finding that high fruit firmness was correlated with up-regulation of *FaCAD* and down-regulation of *FaCCR* in five additional genotypes (Carbone et al., 2006) allows to speculate a general involvement of CAD and CCR in the control of such an important trait in strawberry.

Altogether, it is clear that ripening-related fruit softening is a complex process involving the concerted action of different cell wall modifying enzymes. Future research will identify the functional allelic variation that plays a key role in fruit softening, to develop markers for breeding strawberry cultivars with improved firmness.

Alternatively, such information may be useful for improvement of fruit firmness in existing important strawberry cultivars using a transgenic approach.

## 6 Hormone Metabolism

Already by 1950, Nitsch demonstrated that the presence of auxins produced by achenes is essential for receptacle expansion during strawberry fruit development. Removal of achenes at early developmental stages stops fruit development, but this effect can be restored by applying synthetic auxins to de-achened fruits. Much later, the essential role of auxin during ripening was demonstrated by Given et al. (1988), who showed that the declining production of auxin in the achenes modulates the rate of ripening. This finding was supported by the observation that declining auxin levels trigger fruit ripening by inducing the expression of specific ripening-associated genes. This relates in particular to genes involved in flavonoid biosynthesis and softening (Medina-Escobar et al., 1997; Manning, 1998; Harpster et al., 1998; Moyano et al., 1998; Aharoni et al., 2002b; Castillejo et al., 2004; Palomer et al., 2004). Nevertheless, examples of ripening-related genes that are not affected by auxin, e.g. *FaExp2*, a cell wall-related expansin gene, have also been described (Civello et al., 1999; Aharoni et al., 2002b).

Free auxin levels peak in the receptacle and achenes before white fruit stage and then decline as fruit ripens. Auxin is also found as ester or amide conjugated forms (Archbold and Dennis, 1984) or as a protein conjugate (Park et al., 2006b). The interconversion between free and conjugated forms may be part of the mechanism that determines active auxin levels in the receptacle, which regulates fruit development and ripening.

Ethylene plays a prominent role in ripening of climacteric fruits, but the function of ethylene in the non-climacteric strawberry ripening process is still unclear. In strawberry, ethylene production is low, decreasing during fruit development and then increasing at red ripe stage (Knee et al., 1977; Abeles and Takeda, 1990). Administration of inhibitors of ethylene synthesis and ethylene action to large green strawberry fruits did not affect ripening, indicating that ethylene is not essential for regulation of ripening of strawberry fruits (Given et al., 1988). However, ethylene may have a role in strawberry fruit development and ripening, since application of 1-methylcyclopropane (1-MCP), an inhibitor of ethylene action, resulted in a modified expression of many genes associated with ripening such as pectate lyase and *ß*-galactosidase (Balogh et al., 2005). A similar observation was done by Castillejo et al. (2004) who described that a fruit-specific pectin esterase (*FaPE1*) gene was up-regulated in fruits treated with 1-MCP. On the other hand, in fruits treated for 24 hours with ethylene, as well as in overripe senescing fruits, *FaPE1* mRNA levels decreased. These results indicate a negative regulation of *FaPE1* expression exerted by ethylene. Three different strawberry ethylene receptor genes have been cloned (Trainotti et al., 2005). Their expression is induced by ethylene and the three genes show differential expression during fruit ripening, giving additional evidence for a possible role of ethylene in the ripening of strawberries.

## 7 Vitamins

Strawberries are rich in vitamin C (vitC or ascorbic acid), which has a primary role for human health. In nature, vitC is synthesized from D-glucose-6-phosphate (D-Glu-6-P) via different routes: in animals, D-Glu-6-P is channeled through the D-glucuronic acid pathway to form the precursor gulono-1,4-lactone; in plants, a more complex route, involving different sugar compounds (fructose, mannose), takes to the synthesis of precursor galactono-1,4-lactone (Wheeler et al., 1998). An alternative plant pathway to vitC synthesis, supposed to proceed from the degradation of the cell wall constituent pectin through galacturonic acid (GalUA), was validated by Agius et al. (2003). A strawberry GalUA reductase gene was cloned and found to be expressed in ripening fruits at levels directly correlated with vitC content. Its overexpression in *Arabidopsis* increased vitC levels in leaves, providing a proof-of-concept demonstration for engineering vitC in fruits through this alternative GalUA pathway. Following molecular genetics studies implemented in other fruit species (e.g., Davey et al., 2006), major QTLs involved in strawberry vitC biosynthesis could be determined and mapped for molecular breeding purposes.

## 8 Sugar and Acid Composition

Consumers' perception of fruit quality is based on hedonistic and sensorial attributes. Among the latter, levels of sugars and organic acids, and the relative balance/ratio of the two pools, play an important role in taste perception.

Glucose, fructose and sucrose –in order of abundance – are the main sugars in strawberry fruits; their concentration is affected by genetic and environmental factors (Shaw, 1988). Glucose provides also the building block for the synthesis of starch, the main carbohydrate reserve material in plants. The key step of glucose to starch conversion is catalyzed by ADP-glucose pyrophosphorylase (AGPase), mostly located in plastids. Park and Kim (2007) isolated the sequences of AGPase large (FagpL1 and FagpL2) and small (FagpS) subunits. The expression of *FagpL1* and *FagpS* was constantly high throughout fruit development, while *FagpL2* was weakly expressed only in leaves, suggesting a control of the expression of AGPase genes at both transcriptional and post-transcriptional levels. In order to modify sugar content in strawberry, Park et al. (2006a) inhibited the expression of *FagpS* by an antisense approach with a fruit-specific promoter: transgenic plants had a decreased starch content and concomitant increased soluble solid content, limited to fruits and most pronounced at red stage. This work opens the way to molecular breeding of sugar content in fruits. Another recent report analyzed the enzyme activities of five key enzymes (soluble acid invertase (sAIV), sucrose synthase (SS), sucrose phosphate synthase (SPS), hexokinase (HK) and fructokinase (FK)) in sugar biosynthesis during fruit ripening, highlighting different developmental patterns of enzyme activity and sugar accumulation (Xie et al., 2007). The critical study of this article, written in Chinese, was limited only to its English abstract. Fortunately, sequences of strawberry sAIV, SS, FK, two SPS and three HK are available in Genbank, allowing further molecular analyses. Duangsrisai et al. (2007) described the cloning and expression analysis of two genes involved in sorbitol metabolism in strawberry. $NAD^+$-dependent sorbitol dehydrogenase (*FaSDH*) and sorbitol-6-phosphate dehydrogenase (*FaS6PDH*) genes were cloned by RT-PCR strategies relying on sequence homology. Transcript levels of *FaSDH* and *FaS6PDH* quantified in fruits, leaves and shoot tips. However, enzyme activity in fruits and leaves was detected only for FaSDH.

To date, the limited information on molecular physiology of strawberry fruit's main organic acids (citric, malic and succinic, in order of concentration) hampers research on such compounds. Famiani et al. (2005) analyzed the information from other crops (grape, tomato), which showed an inverse correlation between the increase of phosphoenolpyruvate carboxykinase (*PEPCK*) enzyme activity and citrate and/or malate content in fruits. The hypothesis that *PEPCK* can have a role in the catabolism of such organic acids was indirectly confirmed in analyzing strawberry fruits, which have steady citric and malic acid content and are devoid of PEPCK activity throughout fruit development, a situation different from that of other berry species investigated. Thus, the characterization of strawberry *PEPCK* gene(s) is a first target for molecular breeding of fruit acid content.

## 9 Allergens

Similarly to other fruits, strawberries contain proteins which give rise to allergic reactions. The strawberry Fra a 1 protein family, homologous to the major birch pollen allergene Bet v 1, include several IgE-binding peptides with small intra- and inter-genotype sequence variability, though subjected to post-translational modifications accounting for different protein mobility in 2-D gel electrophoresis (Karlsson et al., 2004; Musidlowska-Persson et al., 2007). Proteomics studies compared white- and red-fruited varieties, pointing a number of Fra a 1 protein spots associated with low-allergen content in concert with reduced anthocyanin biosynthesis in the white genotype, and assessed the degree of biological variation due to genetic and environmental factors (Hjernø et al., 2006; Alm et al., 2007). *Ad hoc*-devised procedures for protein extraction from strawberry fruit samples were also reported (Zheng et al., 2007).

Other strawberry allergens include profilins and lipid transfer proteins (LTPs), represented also in Rosaceaous species. Strawberry LTPs were shown to be strongly expressed in all fruit tissues and accumulated upon abiotic stress (Yubero-Serrano et al., 2003). Other LTPs and profilins were isolated by screening a strawberry fruit cDNA library and expressed in yeast to test their properties, finding that a strawberry LPT had a lower allergenicity than its apple or peach homologs, thus being suitable for immunotherapy purposes (Zuidmeer et al., 2006).

## 10 Control of Flowering

Flower formation in strawberries is induced by low temperatures and shortened photoperiods. However, genotypes of all *Fragaria* species show, in their development of flower buds, a great range of response to temperature and day length (Heide and Sonsteby, 2007 and references therein). They exist as short-day, long-day or photoperiod insensitive (day-neutral or 'everbearing') cultivars. Phytohormones such as auxin and cytokinins have been shown to play a role in floral induction and differentiation in strawberry (Hou and Huang, 2005; Eshghi and Tafazoli, 2007). It is likely that auxin is produced in young leaves and transported through the vascular tissues to the shoot apical meristem. A gene encoding an auxin-binding protein (ABP1) has already been isolated from strawberry by screening a genomic library with an *ABP1* cDNA from maize (Lazarus and MacDonald, 1996). A cDNA with homology to a gene family known as *GAST* (gibberellic acid-stimulated transcript) was isolated during a search for genes differently expressed in strawberry fruits (de la Fuente et al., 2006). Members of this gene family have been associated with events like cell division, cell elongation or cell elongation arrest. *FaGAST1* was highly expressed in fruits at the white and red-ripe stages and in roots where the expression was confined to the cells at the end of the elongation zone. Transgenic *F. vesca* lines carrying *FaGAST* under the control of the CaMV 35S promoter showed delayed growth of the plant and fruits with reduced size and exhibited late flowering and low sensitivity to gibberellin. But photoperiod effects on petiole growth rate appear not to be mediated by changes in the levels of endogenous gibberellins (Wisemann and Turnbull, 1999).

A consequence of the transition to flowering is that meristem growth often becomes determinate. Once the initiation of the floral organs has been completed, stem cell division is inhibited. *AGAMOUS* (*AG*) imposes determinacy on the meristem and is required for stamen and carpel initiation (Battey and Tooke, 2002). Expression of a putative homolog of *AG* from strawberry (*FaAG1*) was confined to stamens, carpels, and developing fruit consistent with the role to control the patterning and timing of floral organ development (Rosin et al., 2003). Histone H4 is one of the 5 main histone proteins involved in the structure of chromatin in eukaryotic cells. It is a core element of the nucleosome and is conserved among eukaryotes. Transcription of the *H4* gene occurs in the 'S phase' of the cell cycle. In shoot apices, cell division can be detected by observing the expression of *H4* transcripts. When strawberry plants were placed under flower-inducing conditions, expression of the *H4* gene and thus cell division could be observed in the central zone before enlargement of the shoot apex occurred (Kurokura et al., 2006). Changes in the expression pattern of *H4* gene took place before morphological changes were visible.

Current research is interested in finding sources of genes for day-neutrality that are better adapted to continental climates (Weebadde et al., 2008). It is assumed that a dominant allele of a single gene causes seasonal instead of continuous flowering (everbearing) in *F. vesca* and some *F.* × *ananassa* species (Battey and Tooke, 2002; Albani et al., 2004; Sugimoto et al., 2005). When genetics of photoperiod sensitivity and flowering were investigated in progeny between sets of elite *F. virginiana* selections and *F.* × *ananassa* cultivars it was confirmed that day-neutrality is regulated by a single dominant gene (Hancock et al., 2002). However, only a few day-neutral *F. virginiana* produced ratios consistent with a single dominant gene model when crossed with short-day *F.* × *ananassa*. This result suggested that day-neutrality is either being regulated by different genes in the different *Fragaria* species or its regulation is polygenically inherited in both species but may be a threshold trait. This implies that several different sources of day-neutrality may exist in natural populations. However, the availability of markers and the small genome size of *F. vesca* (164 Mb) make the isolation and characterization of the dominant gene conferring restricted-flowering period feasible. Randomly amplified polymorphic DNA (RAPD) markers for the 'everbearing' gene in *F.* × *ananassa* were found using Japanese octoploid cultures (Sugimoto et al., 2005). The development of more closely linked inter-simple sequence repeats (ISSR)-derived sequence characterized amplified regions (SCAR) markers around the *SEASONAL FLOWERING LOCUS* (*SFL*) were reported in *F. vesca* (Albani et al., 2004). The markers can be used to assist large-scale screenings of the mapping population for future positional cloning efforts.

## 11 Resistance to Abiotic Stress

In their natural environment, strawberry plants are exposed to different abiotic stresses such as cold, water deficit, high temperature, salinity, heavy metals and mechanical wounding. It is estimated that such stress conditions can potentially reduce the yield of crop plants by more than 50% (Vij and Tyagi, 2007). Thus, studies of the physiological, biochemical and molecular aspects of stress tolerance have been conducted to unravel the intrinsic mechanisms developed during evolution to mitigate against stress by plants.

Freezing tolerance is induced in perennial and winter annual plants by low temperature and/or a short photoperiod. Screening of a differential cDNA library prepared from cold-acclimated strawberry plants yielded several cDNAs (*Fragaria* cold-regulated, *Fcor*) showing differential expression at low temperature (NDong et al., 1997). *Fcor1* and *Fcor2* are expressed in all tissues while *Fcor3* is specific to leaves. It was suggested that *Fcor1* may be useful as a molecular marker to select for this trait but the effect of differential gene expression on the freezing tolerance was not studied. A full-length cDNA encoding a calcium-dependent protein kinase (CDPK) has been isolated from a strawberry fruit library (Llop-Tous et al., 2002). *FaCDPK1* is expressed in roots, stolons, meristems, flowers, and leaves. Transcripts were not detected in young fruits, but accumulated as fruit turned to white. A significant increase in *FaCDPK1* mRNA level in fruit was observed after 10 h of cold treatment (4°C). The results suggested a role of *FaCDPK1* in fruit development as well as in the response of the fruit to low temperature. Similarly, an ortholog of a cold-induced transcription factor (*CBF1*) important in the cold acclimation response in *Arabidopsis thaliana* were cloned from strawberry (Owens et al., 2002). *FaCBF1* transcript levels were up-regulated in leaves following exposure to

4°C, but receptacles of two transgenic lines carrying a *CaMV35S-CBF1* construct showed no significant change in freezing tolerance, although they expressed the transgene at low level. But also foreign genes were utilized to confer cold resistance to strawberry plants. The wheat *Wcor410a* acidic dehydrin gene has been shown to be associated with the plasma membrane and its level has been correlated with the degree of freezing tolerance in different wheat genotypes (Houde et al., 2004). Transgenic strawberry lines containing *Wcor410a* expressed the protein at a level comparable with that in cold-acclimated wheat. However, improvement of freezing tolerance was only observed in cold-acclimated transgenic plants, suggesting that the WCOR410 protein needs to be activated by additional factors induced during cold acclimation. Another means of increasing frost resistance is the transfer of genes encoding antifreeze protein (AFP) from Antarctic fish. Strawberry plants were successfully transformed with AFP genes by using codons that were optimally expressed in the strawberry plant, but the effect on cold tolerance was not reported (Khammuang et al., 2005). Although a number of cold-induced genes have been cloned from strawberry and other crops, it appears that the knowledge of their function in cold resistance is still in its infancy.

Contrarily, glycine betaine has been known to accumulate during cold acclimation in different crops. In strawberry leaves a two-fold accumulation of glycine betaine increased the cold tolerance from –5.8°C to −17°C (Rajashekar et al., 1999). Similarly, exogenous application of either abscisic acid, which triggers betaine production, or glycine betaine improved freezing survival and re-growth in whole plants. But also osmotic and drought stress can cause significant accumulation of the amino acid derivative in many plants. This has led to propose that the accumulation of such compatible solute in plants is an adaptive mechanism against environmental stresses, making it an attractive target to confer tolerance in strawberry via genetic engineering of the biosynthetic pathways.

Osmotin was originally identified as the predominant protein accumulated in salt-adapted tobacco cell cultures and belongs to the pathogenesis-related (PR) proteins. Many studies have demonstrated that the expression of osmotin-like proteins (OLPs) can be activated by a variety of abiotic stress factors but also by microbial infections. An *OLP2* gene that is expressed at different levels in leaves, crowns, roots, green fruit, and red fruit has been recently cloned from strawberry (Zhang and Shih, 2007). The positive response of *FaOLP2* to three abiotic stimuli (abscisic acid, salicylic acid and mechanical wounding) suggested that it may help to protect against environmental stresses and pathogens. Transgenic strawberry plants carrying an osmotin gene from tobacco showed improved tolerance to salt stress in leaf disc senescence assay (Husaini and Abdin, 2008). Similarly, the expression of a strawberry non-specific lipid transfer (*Fxaltp*) gene is induced by abscisic acid, salicylic acid and mechanical wounding but repressed by cold stress (Yubero-Serrano et al., 2003). The investigation of gene expression responding to hydrodynamic stress in an agitation tank has allowed the cloning of an additional novel stress-response gene (*tuf,* turbulent flow) from strawberry (Takeda et al., 2003). The deduced amino acid sequence contained a nucleotide binding site domain of plant disease resistance genes, although some functional motifs showed differences. The study also

tern of a cynnamyl alcohol dehydrogenase gene from strawberry (*Fragaria* × *ananassa* cv. Chandler). J Exp Bot 53: 1723–1734

Bood KG, Zabetakis I (2002) The biosynthesis of strawberry flavor (II): Biosynthetic and molecular biology studies. J Food Sci 67: 2–8

Bustamante CA, Rosli HG, Añon MC, Civello PM, and Martínez GA (2006) *ß*-Xylosidase in strawberry fruit: isolation of a full length gene and analysis of its expression and enzymatic activity in cultivars with contrasting firmness. Plant Sci 171: 497–504

Carbone F, Mourgues F, Biasioli F, Gasperi F, Märk TD, Rosati C, and Perrotta G (2006) Comparative analysis of gene expression of strawberry fruit and correlation with profiles of volatile compounds and other quality traits in different genotypes. Mol Breed 18: 127–142

Castillejo C, de la Fuente JI, Iannetta P, Botella MA, and Valpuesta V (2004) Pectin esterase gene family in strawberry fruit: study of *FaPE1*, a ripening-specific isoform. J Exp Bot 55: 909–918

Chandler CK, Mertely JC, and Peres N (2006) Resistance of selected strawberry cultivars to anthracnose fruit rot and *Botrytis* fruit rot. Acta Hort 708: 123–126

Cheng GW, Breen PJ (1991) Activity of phenylalanine ammonialyase (PAL) and concentrations of anthocyanins and phenolics in developing strawberry fruit. J Am Soc Hortic Sci 116: 865–869

Cipriani G, Testolin R (2004) Isolation and characterization of microsatellite loci in *Fragaria*. Mol Ecol Notes 4: 366–368

Civello PM, Powell ALT, Sabehat A, and Bennett AB (1999) An expansin gene expressed in ripening strawberry fruit. Plant Physiol 121: 1273–1279

da Silva FL, Escribano-Bailn MT, Pérez Alonso JJ, Rivas-Gonzalo JC, and Santos-Buelga C (2007) Anthocyanin pigments in strawberry. LWT – Food Sci Technol 40: 374–382

Darrow GM (1966) The strawberry: history, breeding, and physiology. 1st ed. Holt, Rinehart and Winston, New York

Davey MW, Kenis K, and Keulemans J (2006) Genetic control of fruit vitamin C contents. Plant Physiol 142: 343–351

Davies KM, Schwinn KE (2006) Molecular biology and biotechnology of flavonoid biosynthesis. In: Flavonoids: chemistry, biochemistry and applications. CRC Press, Boca Raton FL, pp 143–218

de la Fuente J, Amaya I, Castillejo C, Sanchez-Sevilla JF, Quesada MA, Botella MA, and Valpuesta V (2006) The strawberry gene FaGAST affects plant growth through inhibition of cell elongation. J Exp Bot 57: 2401–2411

Deng C, Davis TM (2001) Molecular identification of the yellow fruit color (*c*) locus in diploid strawberry: a candidate gene approach. Theor Appl Genet 103: 316–322

Denoyes-Rothan B, Guérin G, Lerceteau-Köhler E, and Risser G (2005) Inheritance of a race-specific resistance to *Colletotrichum acutatum* in *Fragaria* × *ananassa*. Phytopathology 95: 405–412

Dotto MC, Martínez GA, and Civello PM (2006) Expression of expansin genes in strawberry varieties with contrasting fruit firmness. Plant Physiol Bioch 44: 301–307

Døving A, Måge F, and Vestrheim S (2005) Methods for testing strawberry fruit firmness: a review. Small Fruits Rev 4: 11–34

Duangsrisai S, Yamada K, Bantog NA, Shiratake K, Kanayama Y, and Yamaki S (2007) Presence and expression of $NAD^+$-dependent sorbitol dehydrogenase and sorbitol-6-phosphate dehydrogenase genes in strawberry. J Hort Sci Biotechnol 82: 191–198

Eshghi S, Tafazoli E (2007) Possible role of cytokinins in flower induction in strawberry. Am J Plant Physiol 2: 167–174

Faedi W, Mourgues F, and Rosati C (2002) Strawberry breeding and varieties: situation and perspectives. Acta Hort 567: 51–59

Famiani F, Cultrera NG, Battistelli A, Casulli V, Proietti P, Standardi A, Chen ZH, Leegood RC, and Walker RP (2005) Phosphoenolpyruvate carboxykinase and its potential role in the catabolism of organic acids in the flesh of soft fruit during ripening. J Exp Bot 56: 2959–2969

Figueroa CR, Pimentel P, Gaete-Eastman C, Moya M, Herrera R, Caligari PDS, Moya-León MA (2008) Softening rate of the Chilean strawberry (*Fragaria chiloensis*) fruit. Postharvest Biol Technol 49: 210–220

Filippone MP, Diaz Ricci J, Mamaní de Marchese A, Farías RN, and Castagnaro A (1999) Isolation and purification of a 316 Da preformed compound from strawberry (*Fragaria ananassa*) leaves active against plant pathogens. FEBS Lett 459: 115–118

Folta KM, Davis TM (2006) Strawberry genes and genomics. Crit Rev Plant Sci 25: 399–415

Folta KM, Staton M, Stewart PJ, Jung S, Bies DH, Jesdurai C, and Main D (2005) Expressed sequence tags (ESTs) and simple sequence repeat (SSR) markers from octoploid strawberry (*Fragaria* × *ananassa*). BMC Plant Biol 5: 12

Fossen T, Rayyan S, and Andersen ØM (2004) Dimeric anthocyanins from strawberry (*Fragaria ananassa*) consisting of pelargonidin 3-glucoside covalently linked to four flavan-3-ols. Phytochemistry 65: 1421–1428

Gil-Ariza DJ, I Amaya I, Botella MA, Muñoz-Blanco J, Caballero JL, López-Aranda JM, Valpuesta V, and Sánchez-Sevilla JF (2006) EST-derived polymorphic microsatellites from cultivated strawberry (*Fragaria* × *ananassa*) are useful for diversity studies and varietal identification among *Fragaria* species. Mol Ecol Notes 6: 1195–1197

Giménez G, Ballington JR (2002) Inheritance of resistance to *Colletotrichum acutatum* Simmonds on runners of garden strawberry and its backcrosses. HortScience 37: 686–690

Given NK, Venis MA, and Grierson D (1988) Hormonal regulation of ripening in the strawberry, a non-climateric fruit. Planta 174: 402–406

Griesser M, Hoffmann T, Bellido ML, Rosati C, Fink B, Kurtzer R, Aharoni A, Muñoz-Blanco J, and Schwab W (2008) Redirection of flavonoid biosynthesis through the down-regulation of an anthocyanidin glucosyltransferase in ripening strawberry fruit. Plant Physiol 146: 1528–1539

Gulen H, Eris A (2004) Effect of heat stress on peroxidase activity and total protein content in strawberry plants. Plant Sci 166: 739–744

Häkkinen SH, Törrönen AR (2000) Content of flavonols and selected phenolic acids in strawberries and *Vaccinium* species: influence of cultivar, cultivation site and technique. Food Res Int 33: 517–524

Halbwirth H, Puhl I, Haas U, Jesik K, Treutter D, and Stich K (2006) Two-phase flavonoid formation in developing strawberry (*Fragaria* × *ananassa*) fruit. J Agr Food Chem 54: 1479–1485

Hammerschlag F, Garcés S; Koch-Dean M, Ray S, Lewers K, Maas J, and Smith BJ (2006) *In vitro* response of strawberry cultivars and regenerants to *Colletotrichum acutatum*. Plant Cell Tiss Org 84: 255–261

Hampel D, Mosandl A, and Wüst M (2006) Biosynthesis of mono- and sesquiterpenes in strawberry fruits and foliage; $^2$H labeling studies. J Agric Food Chem 54: 1473–1478

Hancock JF, Luby JJ, Dale A, Callow PW, Serce S, and El-Shiek (2002) Utilizing wild *Fragaria virginiana* in strawberry cultivar development: inheritance of photoperiod sensitivity, fruit size, gender, female fertility and disease resistance. Euphytica 136: 177–184

Hannum SM (2004) Potential impact of strawberries on human health: a review of science. Crit Rev Food Sci 44: 1–17

Harpster MH, Brummell DA, and Dunsmuir P (1998) Expression analysis of a ripening-specific, auxin-repressed endo-1,4-betaglucanase gene in strawberry. Plant Physiol 118: 1307–1316

Harrison EP, McQueen Mason SJ, and Manning K (2001) Expression of six expansin genes in relation to extension activity in developing strawberry fruit. J Exp Bot 52: 1437–1446

Haymes KM, Henken B, Davis TM, and van de Weg WE (1997) Identification of RAPD markers linked to a *Phytophthora fragariae* resistance gene (*Rpf1*) in the cultivated strawberry. Theor Appl Genet 94: 1097–1101

Haymes KM, van de Weg WE, Arens P, Maas JL, Vosman B, and Den Nijs APM (2000) Development of SCAR markers linked to a *Phytophthora fragariae* resistance gene and their assessment in European and North American strawberry genotypes. J Am Soc Hortic Sci 125: 330–339

Heide OM, Sonsteby A (2007) Interactions of temperature and photoperiod in the control of flowering of latitudinal and altitudinal populations of wild strawberry (*Fragaria vesca*). Physiol Plant 130: 280–289

Hernández I, Molenaar D, Beekwilder J, Bouwmeester HJ, and van Hylckama Vlieg JET (2007) Expression of plant flavor genes in *Lactococcus lactis*. Appl Environ Microb 73: 1544–1552

Hernanz D, Recamales AF, Meléndez-Martínez AJ, González-Miret ML, and Heredia FJ (2007) Assessment of the differences in the phenolic composition of five strawberry cultivars (*Fragaria* × *ananassa* Duch.) grown in two different soilless systems. J Agric Food Chem 55: 1846–1852

Hjernø K, Alm R, Canbäck B, Matthiesen R, Trajkovski K, Björk L, Roespstorff P, and Emanuelsson C (2006) Down-regulation of the strawberry Bet v 1-homologous allergen in concert with the flavonoid biosynthesis pathway in colorless strawberry mutant. Proteomics 6: 1574–1587

Hoffmann T, Kalinowski G, and Schwab W (2006) RNAi-induced silencing of gene expression in strawberry fruit (*Fragaria* × *ananassa*) by agroinfiltration: a rapid assay for gene function analysis. Plant J 48: 818–826

Hokanson SC, Maas JL (2001) Strawberry biotechnology. In : Plant breeding reviews, Vol. 21. John Wiley and Sons Inc., New York, pp 139–180

Hou ZX, Huang WD (2005) Immunohistochemical localization of IAA and ABP1 in strawberry shoot apexes during floral induction. Planta 222: 678–687

Houde M, Dallaire S, N'Dong D, and Sarhan F (2004) Overexpression of the acidic dehydrin WCOR410 improves freezing tolerance in transgenic strawberry leaves. Plant Biotechnol J 2: 381–387

Husaini AM, Abdin MZ (2008) Development of transgenic strawberry (*Fragaria* × *ananassa* Duch.) plants tolerant to salt stress. Plant Sci 174: 446–455

Jaakola L, Määttä K, Pirttilä AM, Törrönen R, Kärenlampi S, and Hohtola A (2002) Expression of genes involved in anthocyanin biosynthesis in relation to anthocyanin, proanthocyanidin, and flavonol levels during bilberry fruit development. Plant Physiol 130: 729–739

Jetti RR, Yang E, Kurnianta A, Finn C, and Qian MC (2007) Quantification of selected aroma-active compounds in strawberries by headspace solid-phase microextraction gas chromatography and correlation with sensory descriptive analysis. J Food Sci 72: S487–496

Jewell GG, Rantsios A, and Scholey J (1973) Factors influencing the breakdown of fruit in strawberry jam. J Texture Stud 4: 363–370

Jiménéz-Bermúdez S, Redondo Nevado J, Muñoz-Blanco J, Caballero JL, Lopez Aranda JM, Valpuesta V, Pliego Alfaro F, Quesada MA, and Mercado JA (2002) Manipulation of strawberry fruit softening by antisense expression of a pectate lyase gene. Plant Physiol 128: 751–775

Jung S, Staton M, Lee T, Blenda A, Svancara R, Abbott A, and Main D (2008) GDR (Genome Database for Rosaceae): integrated web-database for Rosaceae genomics and genetics data. Nucl Acids Res 36: D1034–D1040

Kalt W, Forney CF, Martin A, and Prior RL (1999) Antioxidant capacity, vitamin C, phenolics, and anthocyanins after fresh storage of small fruits. J Agr Food Chem 47: 4638–4644

Kappers I, Aharoni A, van Herpen TWJM, Luckerhoff LLP, Dicke M, and Bouwmeester HJ (2005) Genetic engineering of terpenoid metabolism attracts bodyguards to *Arabidopsis*. Science 309: 2070–2072

Karlsson AL, Alm R, Ekstrand B, Fjelkner-Modig S, Schiött A, Bengtsson U, Björk L, Hjernø K, Roepstorff P, and Emanuelsson CS (2004) Bet v 1 homologues in strawberry identified as IgE-binding proteins and presumptive allergens. Allergy 59: 1277–1284

Keniry A, Hopkins CJ, Jewell E, Morrison B, Spangenberg GC, Edwards D, and Batley J (2006) Identification and characterization of simple sequence repeat (SSR) markers from *Fragaria* × *ananassa* expressed sequences. Mol Ecol Notes 6: 319–322

Kennedy JA, Jones GP (2001) Analysis of proanthocyanidin cleavage products following acid-catalysis in the presence of excess phloroglucinol. J Agric Food Chem 49: 1740–1746

Khammuang S, Dheeranupattana, Hanmuangja P, and Wongroung S (2005) Agrobacterium-mediated transformation of modified antifreeze protein gene in strawberry. Songklanakarin J Sci Technol 27: 693–703

Klein D, Fink B, Arold B, Eisenreich W, and Schwab W (2007) Functional characterization of enone oxidoreductases from strawberry and tomato fruit. J Agric Food Chem 55: 6705–6711

Knee M, Sargent JA, and Osborne DJ (1977) Cell wall metabolism in developing strawberry fruit. J Exp Bot 28: 377–396

Kurokura T, Inaba Y, and Sugiyama N (2006) Histone H4 gene expression and morphological changes on shoot apices of strawberry (*Fragaria* × *ananassa* Duch.) during floral induction. Sci Hortic 110: 192–197

Landmann C, Fink B, and Schwab W (2007) FaGT2: a multifunctional enzyme from strawberry (*Fragaria* × *ananassa*) fruits involved in the metabolism of natural and xenobiotic compounds. Planta 226: 417–428

Larsen M, Poll L, and Olsen CE (1992) Evaluation of the aroma composition of some strawberry (*Fragaria ananassa* Duch) cultivars by use of odour threshold values. Z Lebensm Unters Forsch 195: 536–539

Lazarus CM, MacDonald H (1996) Characterization of a strawberry gene for auxin-binding protein, and its expression in insect cells. Plant Mol Biol 31: 267–277

Lerceteau-Köhler E, Guérin G, Laigret F, and Denoyes-Rothan B (2003) Characterization of mixed disomic and polysomic inheritance in the octoploid strawberry (*Fragaria* × *ananassa*) using AFLP mapping. Theor Appl Genet 107: 619–628

Lerceteau-Köhler E, Guérin G, and Denoyes-Rothan B (2005) Identification of SCAR markers linked to Rca2 anthracnose resistance gene and their assessment in strawberry germplasm. Theor Appl Genet 111: 862–870

Lewers KS, Styan SMN, and Hokanson SC (2005) Strawberry GeneBank-derived and genomic simple sequence repeat (SSR) markers and their utility with strawberry, blackberry, and red and black raspberry. J Am Soc Hort Sci 130: 102–115

Llop-Tous I, Dominguez-Puigjaner E, and Vendrell M (2002) Characterization of a strawberry cDNA clone homologous to calcium-dependent protein kinases that is expressed during fruit ripening and affected by low temperature. J Exp Bot 53: 2283–2285

Lopez AP, Portales RB, Lopez-Raez JA, Medina-Escobar N, Muñoz-Blanco J, and Franco AR (2006) Characterization of a strawberry late-expressed and fruit-specific peptide methionine sulphoxide reductase. Physiol Plant 126: 129–139

Loughrin JH, Kasperbauer MJ (2002) Aroma of fresh strawberries is enhanced by ripening over red versus black mulch. J Agric Food Chem 50: 161–165

Lunkenbein S, Bellido M, Aharoni A, Salentijn EM, Kaldenhoff R, Coiner HA, Muñoz-Blanco J, and Schwab W (2006a) Cinnamate metabolism in ripening fruit. Characterization of a UDP-glucose: cinnamate glucosyltransferase from strawberry. Plant Physiol 140: 1047–1058

Lunkenbein S, Coiner H, de Vos CHR, Schaart JG, Boone MJ, Krens FA, Schwab W, and Salentijn EMJ (2006b) Molecular characterization of a stable antisense chalcone synthase phenotype in strawberry (*Fragaria* × *ananas*sa). J Agr Food Chem 54: 2145–2153

Lunkenbein S, Salentijn EM, Coiner HA, Boone MJ, Krens FA, and Schwab W (2006c) Up- and down-regulation of *Fragaria* × *ananassa O*-methyltransferase: impacts on furanone and phenylpropanoid metabolism. J Exp Bot 57: 2445–2453

Maas JL, Gouin-Behe C, Hartung JS, and Hokanson SC (2000) Sources of resistance for two differentially pathogenic strains of *Xanthomonas fragariae* in *Fragaria* genotypes. HortScience 35: 128–131

Manning K (1998) Isolation of a set of ripening-related genes from strawberry: their identification and possible relationship to fruit quality traits. Planta 205: 622–631

Martínez M, Abraham Z, Gambardella M, Echaide M, Carbonero P, and Diaz I (2005) The strawberry gene *Cyf1* encodes a phytocystatin with antifungal properties. J Exp Bot. 56: 1821–1829

Martínez-Zamora MG, Castagnaro AP, and Díaz Ricci JC (2004) Isolation and diversity analysis of resistance gene analogues (RGAs) from cultivated and wild strawberries. Mol Genet Genomics 272: 480–487

Medina-Escobar N, Cárdenas J, Moyano E, Caballero JL, and Muñoz-Blanco J (1997) Cloning, molecular characterization and expression pattern of a strawberry ripening-specific cDNA with sequence homology to pectate lyase from higher plants. Plant Mol Biol 34: 867–877

Medina-Escobar N, Cárdenas J, Muñoz-Blanco J, and Caballero JL (1998) Cloning and molecular characterization of a strawberry fruit ripening-related cDNA corresponding a mRNA for a low-molecular-weight heat-shock protein. Plant Mol Biol 36: 33–42

Mehli L, Kjellsen TD, Dewey FM, and Hietala AM. (2005) A case study from the interaction of strawberry and *Botrytis cinerea* highlights the benefits of comonitoring both partners at genomic and mRNA level. New Phytol 168: 465–474

Mink PJ, Scrafford CG, Barraj LM, Harnack L, Hong CP, Nettleton JA, and Jacobs Jr DR (2007) Flavonoid intake and cardiovascular disease mortality: a prospective study in postmenopausal women. Am J Clin Nutr 85: 895–909

Monfort A, Vilanova S, Davis TM, and Arús P (2006) A new set of polymorphic simple sequence repeat (SSR) markers from a wild strawberry (*Fragaria vesca*) are transferable to other diploid *Fragaria* species and to *Fragaria* × *ananassa*. Mol Ecol Notes 6: 197–200

Mori S, Kitamura H, and Kuroda K (2005) Varietal differences in *Fusarium* wilt-resistance in strawberry cultivars and the segregation of this trait in $F_1$ hybrids. J Jpn Soc Hortic Sci 74: 57–59

Moyano E, Portero-Robles I, Medina-Escobar N, Valpuesta V, Muñoz-Blanco J, and Caballero JL (1998) A fruit-specific putative dihydroflavonol 4-reductase gene is differentially expressed in strawberry during the ripening process. Plant Physiol 117: 711–716

Moyano E, Encinas-Villarejo S, Lopez-Raez JA, Redondo-Nivado J, Blanco-Portales R, Bellido ML, Sanz C, Caballero JL, and Muñoz-Blanco J (2004) Comparative study between two strawberry pyruvate decarboxylase genes along fruit development and ripening, post-harvest and stress condition. Plant Sci 166: 835–845

Musidlowska-Persson A, Alm R, and Emanuelsson C (2007) Cloning and sequencing of the Bet v 1-homologous allergen Fra a 1 in strawberry (*Fragaria ananassa*) shows the presence of an intron and little variability in amino acid sequence. Mol Immunol 44: 1245–1252

NDong C, Quellet F, Houde M, and Sarhan F (1997) Gene expression during cold acclimation in strawberry. Plant Cell Physiol 38: 863–870

Nitsch JP (1950) Growth and morphogenesis of the strawberry as related to auxin. Am J Bot 37: 211–215

Olias R, Perez AG, and Sanz C (2002) Catalytic properties of alcohol acyltransferase in different strawberry species and cultivars. J Agric Food Chem 50: 4031–4036

Orsomando G, Lorenzi M, Raffaelli N, Dalla Rizza M, Mezzetti B, and Ruggieri S (2001) Phytotoxic protein PcF, purification, characterization, and cDNA sequencing of a novel hydroxyproline-containing factor secreted by the strawberry pathogen *Phytophthora cactorum*. J Biol Chem 276: 21578–21584

Osorio S, Castillejo C, Quesada MA, Medina-Escobar N, Brownsey GJ, Suau R, Heredia A, Botella MA, and Valpuesta V (2008) Partial demethylation of oligogalacturonides by pectin methyl esterase 1 is required for eliciting defence responses in wild strawberry (*Fragaria vesca*). Plant J 54: 43–55

Owens CL, Thomashow MF, Hancock JF, and Iezzoni AF (2002) CPF1 orthologs in sour cherry and strawberry and the heterologous expression of CBF1 in strawberry. J Am Soc Hort Sci 127: 462–710

Palomer X, Domínguez-Puigjaner E, Vendrell M, and Llop-Tous I (2004) Study of the strawberry Cel1 endo-*β*-(1,4)-glucanase protein accumulation and characterization of its in vitro activity by heterologous expression in *Pichia pastoris*. Plant Sci 167: 509–518

Palomer X, Llop-Tous I, Vendrell M, Krens FA, Schaart JG, Boone MJ, van der Valk H, and Salentijn EMJ (2006) Antisense down- regulation of strawberry *endo-β*-(1,4)-glucanase genes does not prevent fruit softening during ripening. Plant Sci 171: 640–646

Park JI, Kim IJ (2007) Changes in the expression of ADP-glucose pyrophosphorylase genes during fruit ripening in strawberry. Food Sci Biotechnol 16: 343–348

Park JI, Lee YK, Chung WI, Lee IH, Choi JH, Lee WM, Ezura H, Lee SP, and Kim IJ (2006a) Modification of sugar composition in strawberry fruit by antisense suppression of an ADP-glucose pyrophosphorylase. Mol Breeding 17: 269–279

Park S, Cohen JD, and Slovin JP (2006b) Strawberry fruit with a novel indole-acyl modification. Planta 224: 1015–1022

Perkins Veazie, P (1995) Growth and ripening of strawberry fruit. Hort Rev 17: 267–297

Raab T, Lopez-Raez JA, Klein D, Caballero JL, Moyano E, Schwab W, and Muñoz-Blanco J (2006) *FaQR*, required for the biosynthesis of the strawberry flavor compound 4-hydroxy-2,5-dimethyl-3(2H)-furanone, encodes an enone oxidoreductase. Plant Cell 18: 1023–1037

Rajashekar CB, Zhou H, Marcum KB, and Prakash O (1999) Glycine betaine accumulation and induction of cold tolerance in strawberry (*Fragaria* × *ananassa* Duch.) plants. Plant Sci 148: 175–183

Redondo-Nevado J, Moyano E, Medina-Escobar N, Caballero JL, and Muñoz-Blanco J (2001) A fruit-specific and developmentally regulated endopolygalacturonase gene from strawberry (*Fragaria* × *ananassa* cv. Chandler). J Exp Bot 52: 1941–1945

Rosin FM, Aharoni A, Salentijn EMJ, Schaart JG, Boone MJ, and Hannapel DJ (2003) Expression patterns of a putative homolog of AGAMOUS, STAG1, from strawberry. Plant Sci 165: 959–968

Rosli HG, Civello PM, and Martínez GA (2004) Changes in cell wall composition of three *Fragaria* × *ananassa* cultivars with different softening rate during ripening. Plant Physiol Biochem 42: 823–831

Salentjin EMJ, Aharoni A, Schaart JG, Boone MJ, and Krens FA (2003) Differential gene expression analysis of strawberry cultivars that differ in fruit-firmness. Physiol Plant 118: 571–578

Sargent DJ, Davis TM, Tobutt KR, Wilkinson MJ, Battey NH, and Simpson DW (2004) A genetic linkage map of microsatellite, gene-specific and morphological markers in diploid *Fragaria*. Theor Appl Genet 109: 1385–1391

Sargent DJ, Clarke J, Simpson DW, Tobutt KR, Arús P, Monfort A, Vilanova S, Denoyes-Rothan B, Rousseau M, Folta KM, Bassil NV, and Battey NH (2006) An enhanced microsatellite map of diploid *Fragaria*. Theor Appl Genet 112: 1349–1359

Sargent DJ, Rys A, Nier S, Simpson DW, and Tobutt KR (2007) The development and mapping of functional markers in *Fragaria* and their transferability and potential for mapping in other genera. Theor Appl Genet 114: 373–384

Sargent DJ, Cipriani G, Vilanova S, Gil-Ariza D, Arús P, Simpson DW, Tobutt KR, and Monfort A (2008) The development of a bin mapping population and the selective mapping of 103 markers in the diploid *Fragaria* reference map. Genome 51: 120–127

Schaart JG, Mehli L, and Schouten HJ (2005) Quantification of allele-specific expression of a gene encoding strawberry polygalacturonase-inhibiting protein (PGIP) using Pyrosequencing. Plant J 41: 493–500

Schieberle P, Hofmann T (1997) Evaluation of the character impact odorants in fresh strawberry juice by quantitative measurements and sensory studies on model mixtures. J Agric Food Chem 45: 227–232

Seeram NP (2008) Berry fruits: compositional elements, biochemical activities, and the impact of their intake on human health, performance, and disease. J Agric Food Chem 56: 627–629

Shaw D (1988) Genotypic variation and genotypic correlations for sugars and organic acids of strawberries. J Am Soc Hortic Sci 113: 770–774

Shaw DV, Hansen J, Browne GT, and Shaw SM (2008) Components of genetic variation for resistance of strawberry to *Phytophthora cactorum* estimated using segregating seedling populations and their parent genotypes. Plant Pathol 57: 210–215

Smith BJ, Galletta GJ, and Gupton CL (1996) USDA-ARS strawberry resistance breeding, disease biology, and management progress. In: Proceedings of IV North American Strawberry Conference. University of Florida, Gainesville, FL, pp 253–258

Sugimoto T. Tamaki K, Matsumoto J, Yamamoto Y, Shiwaku K, and Watanabe K (2005) Detection of RAPD markers linked to the everbearing gene in Japanese cultivated strawberry. Plant Breed 124: 498–501

Takeda T, Tamura M, Ohtaki M, and Matsuoka H (2003) Gene expression in cultured strawberry cells subjected to hydrodynamic stress. Biochem Eng J 15: 211–215

Trainotti L, Spinello R, Piovan A, Spolaore S, Casadoro G, Schultz DJ, Craig R, Cox-Foster DL, Mumma RO, and Medford JI (2001) Beta-galactosidases with a lectin-like domain are expressed in strawberry. J Exp Bot 52: 1635–1645

Trainotti L, Pavanello A, and Casadoro G (2005) Different ethylene receptors show an increased expression during the ripening of strawberries: does such an increment imply a role for ethylene in the ripening of these non-climacteric fruits? J Exp Bot 56: 2037–2046

Tulipani S, Mezzetti B, Capocasa F, Bompadre S, Beekwilder J, de Vos CH, Capanoglu E, Bovy A, and Battino M (2008) Antioxidants, phenolic compounds, and nutritional quality of different strawberry genotypes. J Agric Food Chem 56: 696–704

Ulrich D, Komes D, Olbricht K, and Hoberg E (2007) Diversity of aroma patterns in wild and cultivated *Fragaria* accessions. Genet Resour Crop Evol 54: 1185–1196

van de Weg WE (1997) Gene-for-gene relationships between strawberry and the causal agent of red stele root rot, *Phytophthora fragariae* var. *fragariae*. PhD Thesis, University of Wageningen, The Netherlands, 93 p

Vij S, Tyagi AK (2007) Emerging trends in the functional genomics of the abiotic stress response in crop plants. Plant Biotechnol J 5: 361–380

Vogt T, Jones P (2000) Glycosyltransferases in plant natural product synthesis: characterization of a supergene family. Trends Plant Sci 5: 380–386

Wang SY, Lin HS (2000) Antioxidant activity in fruits and leaves of blackberry, raspberry, and strawberry varies with cultivar and developmental stage. J Agric Food Chem 48: 140–146

Wang J, Ge H, Peng S, Zhang H, Chen P, and Xu J (2004) Transformation of strawberry (*Fragaria ananassa* Duch.) with late embryogenesis abundant protein gene. J Hort Sci Biotech 79: 735–738

Wang SY, Chen CT, Wang CY, and Chen P (2007) Resveratrol content in strawberry fruit is affected by preharvest conditions. J Agric Food Chem 55: 8269–8274

Warabieda W, Miszczak A, and Olszak RW (2005) The influence of methyl jasmonate (JA-Me) and B-glucosidase on induction of resistance mechanisms of strawberry against two-spotted spider mite (*Tetranychus urticae* Koch.). Commun Agric Appl Biol Sci 70: 829–836

Weebadde CK, Wang D, Finn CE, Lewers KS, Luby JJ, Bushakra J, Sjulin TM, Hancock JF (2008) Using a linkage mapping approach to identify QTL for day-neutrality in the octoploid strawberry. Plant Breeding 127: 94–101

Wein M, Lavi N, Lunkenbein S, Lewinsohn E, Schwab W, and Kaldenhoff R (2002) Isolation, cloning and expression of a multifunctional *O*-methyltransferase capable of forming 2,5-dimethyl-4-methoxy-3(2H)-furanone, one of the key aroma compounds in strawberry fruits. Plant J 31: 755–765

Wheeler GL, Jones MA, and Smirnoff N (1998) The biosynthetic pathway of vitamin C in higher plants. Nature 393: 365–369

Winterbottom CQ (1991) Resistance of strawberry to *Colletotrichum acutatum*. Master of Science Thesis, University of Davis, California, USA, 69 p

Wisemann NJ, Turnbull CGN (1999) Endogenous gibberellin content does not correlate with photoperiod-induced growth changes in strawberry petioles. Aust J Plant Physiol 26: 359–366

Woolley L, James D, and Kenneth M (2001) Purification and properties of an endo-beta-1,4-glucanase from strawberry and downregulation of the corresponding gene, *cel1*. Planta 214: 11–21

Xie M, Chen JW, Qin QP, Jiang GH, Sun CB, Zhang HQ, and Xu HX (2007) (translated title from Chinese) The control of sugar accumulation within strawberry aggregate fruit by invertase and hexokinase. Zhi Wu Sheng Li Yu Fen Zi Sheng Wu Xue Xue Bao 33: 213–218

Xue S, Bors RH, and Strelkov SE (2006) Resistance sources to *Xanthomonas fragariae* in non-octoploid strawberry species. HortScience 40: 1653–1656

Yubero-Serrano EM, Moyano E, Medina-Escobar N, Muñoz-Blanco J, and Caballero JL (2003) Identification of a strawberry gene encoding a non-specific lipid transfer protein that responds to ABA, wounding and cold stress. J Exp Bot 54: 1865–1877

Zebrowska J, Hortyński J, Cholewa T, and Honcz K (2006) Resistance to *Verticillium dahliae* (Kleb.) in the strawberry breeding lines. Commun Agric Appl Biol Sci 71: 1031–1036

Zhang Y, Shih DS (2007) Isolation of an osmotin-like protein gene from strawberry and analysis of the response of this gene to abiotic stresses. J Plant Physiol 164: 68–77

Zheng Q, Song J, Doncaster K, Rowland E, and Byers DM (2007) Qualitative and quantitative evaluation of protein extraction protocols for apple and strawberry fruit suitable for two-dimensional electrophoresis and mass spectrometry analysis. J Agric Food Chem 55: 1663–1673

Zuidmeer L, Salentijn E, Rivas MF, Mancebo EG, Asero R, Matos CI, Pelgrom KTB, Gilissen LJWJ, and van Ree R (2006) The role of profilin and lipid transfer protein in strawberry allergy in the Mediterranean area. Clin Exp Allergy 36: 666–675.

# 23. GMO Strawberry: Methods, Risk and Benefits

**Bruno Mezzetti**

## 1 Introduction

The potential for profitable applications of biotechnology to many fruits and vegetables, tree fruits and nuts, can be limited by the high cost of research, development and regulatory approval combined with the small plantations and the diversity of varieties. Further, experimentation with perennials such as fruit trees, berry and nuts is comparatively expensive (because the experimental unit is larger and takes more time), and it is costly and not easy to bring into new plantations or replacing existing orchards with a new variety. For these reasons biotech application in horticultural crops are quite limited at research level and commercialization is almost completely absent. Horticultural crops use much less land but their production is of much higher value. Upon this situation, the development of biotechnology applications in horticultural crops and the achievement of market acceptance can be expected, in the near future, mainly for those crops with a broader importance in cultivation and market.

Among fruits, strawberry is perceived as an interesting crop for the development and application of most the advanced biotechnological application, including gene cloning and recombinant technology, because of its large-scale production, based on a quite limited variety diversity (almost the 60% of world production is covered by a variety), and with the interesting peculiarity to be grown as annual fruit (Mezzetti, 2003).

New tools and transgenic systems may aid in the dissection of molecular-genetic mechanisms that underlie traits of interest to strawberry researchers and the strawberry industry. However, the rapid, high-throughput transformation and regeneration system required to efficiently study functional and translational genomics in cultivated strawberry has not been developed. Transformation of octoploid strawberry has been well documented (James, 1987; Liu and Sanford, 1988; Nehra et al., 1990; Barcélo et al., 1998; Passey et al., 2003; Ricardo et al., 2003), but the efficiency

B. Mezzetti (✉)
Dipartimento di Scienze Ambientali e delle Produzioni Vegetali – Marche Polytechnic University, Via Brecce Bianche, Ancona, 60100-IT

K.M. Folta, S.E. Gardiner (eds.), *Genetics and Genomics of Rosaceae*, Plant Genetics and Genomics: Crops and Models 6, DOI 10.1007/978-0-387-77491-6_23,

varies among cultivars, requiring 40–112 days for the emergence of the first visible shoots. Furthermore, a given media formulation only works well with a subset of cultivars (Passey et al., 2003; Gruchala et al., 2004). The genetic lines and culture conditions that support efficient transformation of diploid strawberry have also been carefully defined (Alsheikh et al., 2002; Oosumi et al., 2006). These systems have great potential for supporting reverse-genetic analyses for genes of interest (Oosumi et al., 2006) as well as an advanced genetic linkage map (Davis and Yu, 1997; Sargent et al., 2006, 2007) but the diploid systems still require substantial average regeneration time and the results may not be directly applicable to crop species.

However, for the availability, at least for some genotypes, of efficient regeneration and transformation protocols this crop can be considered an interesting model for the development of basic genomics and recombinant DNA studies among rosaceous species.

## 2 Strawberry Regeneration and *Agrobacterium* Transformation

Several studies on strawberry have clearly demonstrated the importance of various factors such as genotype (Liu and Sanford, 1988; James et al., 1990, Nehra et al., 1990), type and source of explant (Nehra et al., 1989), hormonal balance and incubation conditions (Liu and Sanford, 1988; Nehra et al., 1989) on successful regeneration via adventitious organogenesis of transformed clones.

Recently, there have been important changes in the use of non-conventional approaches to breeding berries. A considerable number of studies have used the strawberry in biotechnological experiments of gene cloning, genetic transformation, fingerprinting and mapping for disease resistance, although detailed developmental and physiological characterization of the whole sequence of the organogenic processes in strawberry somatic tissues is still lacking (Mezzetti, 2003). Despite the sufficient regeneration levels achieved from leaf explants for some commercial strawberry genotypes, the regeneration of transformed strawberry plants remains difficult and appears to be strongly genotype dependent.

*Agrobacterium-tumefaciens*-mediated transformation of regenerated strawberry plants from calli or leaf discs has been demonstrated by several groups (James et al., 1990; Nehra et al., 1990; El Mansouri et al., 1996; Barcélo et al., 1998). However, it is clear that the production of a new transformed strawberry plant is strictly influenced by the virulence of the *Agrobacterium* strain (James et al., 1990; Uratsu et al., 1991) and by the availability of a highly efficient regeneration protocol.

To date, the most successful plant differentiation process for recombinant DNA technology in strawberry remains adventitious shoot organogenesis directly from somatic tissue or via previous callus formation. The types of explants used for inducing regeneration is also an important factor. For several *Fragaria* × *ananassa* genotypes, efficient regeneration protocols have been identified using different types of somatic tissues (Foucault and Letouze, 1987; Liu and Sanford, 1988; Rugini and Orlando, 1992; Graham et al., 1995; Passey et al., 2003), although leaf tissue has been the most studied (Liu and Sanford, 1988; Nehra et al., 1989; Sorvari et al., 1993) and in most cases has displayed the highest regeneration frequencies (Passey et al., 2003). Leaf explants have also been particularly useful for shoot regeneration and genetic transformation of wild strawberry (*F. vesca*; El Mansouri et al., 1996; Haymes and Davis, 1997; Balokhina et al., 2000; Jimenez-Bermudez et al., 2002; Agius et al., 2003; Mezzetti et al., 2004a).

The shoot regeneration response of leaf tissue has mainly been related to the genotype and the medium composition (plant growth regulators [PGRs] and the type of nutrient medium). The light and dark cycles are also known to affect tissue regeneration responses, although the effects seen can differ, depending on the genotype-PGR interactions (Liu and Sanford, 1988; Nehra et al., 1989). The major factor affecting regeneration efficiency, however, appears to be the genotype. Cultivated strawberry varieties have shown large variability in the differentiation competence of their somatic tissues, and the effects of PGR treatments in the induction of this process appear to be related to specific genetic factors (Passey et al., 2003).

The regeneration media that have generally been seen to produce the greatest shoot regeneration are Murashige and Skoog (1962) (MS) medium supplemented with 6-benzylaminopurine (BA) and indole-3-butyric acid (IBA) (Barcélo et al., 1998; Passey et al., 2003). The ability of 1-phenyl-3-(1,2,3-thiadiazol-5-yl) urea (thidiazuron; TDZ) to induce a high shoot regeneration efficiency in woody plant tissues, in particular, has also been reported (Huetteman and Preece, 1993; Bhagwat and Lane, 2004; Meng et al., 2004). In strawberry, the effect of TDZ has been explored recently in a restricted number of *Fragaria* × *ananassa* cultivars. This cytokinin-like PGR showed specific responses that were dependent on the genotype and the type of tissue (Passey et al., 2003). Among the auxins, 3-benzo[b]selenienyl acetic acid (BSAA) is a new molecule that is highly active and that has already been tested on some crops, where it was shown to be highly effective for the induction of somatic embryogenesis (Lamproye et al., 1990). However, it has not yet been tested for its ability to induce organogenesis in woody plants, with particular reference to the strawberry.

Among several strawberry genotypes tested by Landi and Mezzetti (2006) the regeneration and transformation response varied mainly depending to the genotype. The highest efficiency in genetic transformation (about 5%) was observed from the genotypes with the highest leaf tissue regeneration efficiency (100% of leaf tissue regeneration), such as for example Sveva, a newly released octoploid cultivar (*F.* × *ananassa*) and the diploid *F. vesca* cv. Alpina W.O. Genotypes such as cvs. Onda and Paros, performing with an intermediate leaf tissue regeneration efficiency (about 80% of leaves with adventitious shoots) showed a reduced *Agrobacterium* transformation efficiency (1–3%). Strawberry genotypes showing much lower percentages of leaf tissue regeneration (lower then 40%) can be quite more difficult to be transformed.

The *Agrobacterium tumefaciens* mediated transformation represents one of the most common techniques of recombinant DNA, and it is largely employed to obtain genetically modified plants. Efficient transformation and regeneration methods are a priority for successful application of recombinant DNA technology to vegetative propagated plants such as strawberry.

To date, the most resourceful plant differentiation process for recombinant DNA technology in strawberry remains adventitious shoot organogenesis directly from somatic tissue or a previous callus formation. The genotype and type of explants represent important factors affecting the regeneration process and consequently the genetic transformation efficiency.

The achievement of stable genetically transformed plants is also strictly related to the antibiotic (kanamycin) selection protocol, starting from the early stage of leaf tissue regeneration, immediately after the *Agrobacterium* infection and co-culture, and including the in vitro proliferation and rooting of newly selected shoots. Depending to the efficiency of the standard regeneration and transformation protocol, the first stable newly produced transgenic line can be available after 5–6 months from the first transformation experiment.

A new challenge for rapid regeneration and transformation system for cultivated strawberry based on discovery of a new genetic line derived from a productive Florida cultivar, 'Strawberry Festival', was attempted by Folta et al. (2006). Efforts to develop an efficient regeneration system from 'Strawberry Festival' directly did not yield results comparable to published benchmarks. A new approach was devised. Instead of attempting to identify a media formulation that supported high frequency regeneration in a recalcitrant cultivar, the segregation of the octoploid strawberry genome was explored to produce a rapid-cycling genotype that regenerates well on a given medium formulation. This was the first example of a strawberry breeding program specifically addressed to the selections of octoploid strawberry progeny for ease of genetic manipulation and vigorous growth in vitro. A comparable approach proved valuable in creating the rapid cycling 'Regen-S' alfalfa (Bingham, 1989), REL-1 and REL-2 sugar beet and 'Coker 310FR' cotton (Kumar et al., 1998) genetic lines. This genetic selection strategy resulted in a new strawberry genotype and a supporting regeneration protocol that allows progression from explant to a rooted plant in fewer than 60 days. The genetic line has been designated as 'Laboratory Festival #9' (LF9) to commemorate its lineage as well as its potential to serve as a valuable resource for studies of gene function in *Fragaria*. The regeneration capacity of various tissues was examined carefully to identify the tissue that best supports rapid regeneration. The results indicate that explant sources vary in the number of shoots produced and the time it takes to produce them. Although leaves and stolons readily regenerate, the best response was observed in petioles. The regeneration potential from all explants is higher in young tissue. Petiole segments favor polar production of callus and shoots, biasing production on the basipetal end. This observation suggests that a polar distribution of growth regulators, or sensitivity to growth regulators exists along the organ, leading to gradients within the explant. The differential accumulation results in more vigorous callus growth and eventual shoot generation from the basipetal tissues of the explant. Petiole explants originating from positions adjacent to the leaf perform better than those adjacent to the crown, indicating a possible difference in sensitivity to growth regulators between these tissues. Together, the data indicate that the optimal tissue for regeneration is the 1 cm petiole segment below an expanding young leaf. Regeneration and transformation is so efficient in this tissue that many transgenic shoots may be derived using only a few leaf-adjacent petiole segments as starting tissue.

The LF9 system represents a potentially valuable tool to the Rosaceae research community. The successful identification of the LF9 genetic line and derivation of protocols supporting efficient and rapid transformation and regeneration represents an important basis for high-throughput studies of gene function in cultivated strawberry, an under-studied yet economically important plant species (Folta et al., 2006).

## 3 Genes for Strawberry Improvement

The Rosaceae family is comprised of many species exhibiting great phenotypic diversity. Although the family members represent valuable tree crops, such as peach and apple, as well as brambles, roses, almonds, and strawberries, development of efficient transgenic systems to test interspecies gene function have not been widely reported. Currently, transformation of tree crops is only possible in a time scale of years. Recent efforts have identified genes involved in important processes relevant to dormancy, fruit quality and fruit production, yet transgenic assessment of gene function is slow. Members of the Rosaceae family share strong sequence similarity across many genes studied, suggesting that tests of gene function in a heterologous strawberry system will be informative.

Strawberry genotypes with a well characterized rapid-cycling transformation protocol can serve as the foundation for tests of gene function in cultivated strawberry, but also for other Rosaceae important specie. These resources could be used for strawberry improvement either through transgenic approaches or through development of markers linked to genes of relevant function that may be used in marker-assisted selection.

The regeneration and transformation protocols developed for several berry genotypes are now used for the insertion of new gene constructs known for controlling important agronomic characters.

### *3.1 Disease and Pest Resistance*

*Agrobacterium*-mediated transformation was used to introduce the CpTi gene in strawberry against attack by vine weevil (*Otiorhynchus sulcatus* F. Coleoptera: Curculionidae) (Graham et al., 1997a). The degree of protection against insect feeding conferred upon transgenic strawberry lines expressing the Cowpea trypsin inhibitor was evaluated under glasshouse conditions. Insect bioassays were carried out using *Otiorhynchus sulcatus* in two experiments and in both experiments there was a highly significant reduction in damage by weevil larvae on the transgenic lines. Strawberry cultivars (*Fragaria* × *ananassa*) Melody, Rhapsody and Symphony were used in transformation experiments carried out using the cowpea

[*Vigna unguiculata*] protease [proteinase] trypsin inhibitor (CpTi) gene construct (Graham et al., 1995). A simple visual assay technique was developed to identify those plants expressing CpTi. Trypsin acts on the substrate alpha-N-benzoyl-DL-arginine-p-nitroanilide (BAPNA) to produce p-nitroaniline (p-na).

Strawberry plants were transformed with the coat protein gene from strawberry mild yellow edge luteovirus to confer resistance to the associated disease (Finstad et al., 1995).

Both for raspberry and strawberry the gene controlling polygalacturonase (PG) enzyme was mostly studied both for his effect in increasing resistance to some important fungi (*Botrytis* and *Colletotrichum*) and also for controlling fruit maturation.

Grey mold, caused by *Botrytis cinerea*, is a common and destructive disease of strawberries in all growing regions throughout the world. Primary infection occurs on strawberry flowers but development of the typical grey mold symptoms occurs after the fruit has developed and begun to ripen. The most commonly infected floral organs are the petals and stamens. The receptacle, which will develop into the strawberry fruit, normally becomes infected by the fungus growing down the filament but infection via the petals is also possible, particularly when petal abscission is delayed.

Currently no cultivars are resistant to *B. cinerea* and true resistance has not been reported in any *Fragaria* × *ananassa* germplasm. There are differences in susceptibility but these can be attributed mainly to plant morphological characteristics that affect the microclimate around the flowers and fruit. The use of heterologous *PGIP* genes to control *B. cinerea* infection has proved successful in experiments carried out in California with transgenic tomatoes. The *PGIP* gene (from pear fruit) that had been used in these tomato experiments was introduced into strawberry cultivar Pegasus at HRI East Malling using *Agrobacterium*-mediated transformation (Simpson et al., 1999). The disease response of transgenic clones of Pegasus, confirmed positive for the *PGIP* gene by southern blot analysis, was tested challenging detached flowers by spraying each flower with 1 ml of a conidial suspension of *B. cinerea* ($1 \times 10^4$ / ml).

The results have also been encouraging for three other clones but for these the results have not been entirely consistent for all flower parts across all four experiments. There was thus no correlation between levels of *PGIP* and observed differences in susceptibility to *B. cinerea*. These results were not conclusive but it is encouraging that a small number of the transgenic clones showed reduced susceptibility to infection by *B. cinerea* on detached flowers.

Researchers have reported that enhanced chitinase levels in transgenic strawberry plants can indeed reduce the damage caused by powdery mildew fungi (Asao et al., 1997; Asao et al., 2003). Chalavi et al. (2003) isolated a chitinase gene (*pcht28*) from *Solanum chilense* and transferred it into 'Joliette' strawberry using *Agrobacterium*-mediated transformation. Introduction of the *pcht28* gene was verified by Southern blot analysis and its expression by Northern blot. In growth chamber studies, the transgenic strawberry plants that expressed *pcht28* had significantly higher resistance to *Verticillium dahliae*. Ricardo et al. (2006) obtained transgenic strawberry lines by expressing three defense genes: *ch5B* (encoding a chitinase protein from kidney bean (*Phaseolus vulgaris*)), *gln2* (encoding a glucanase protein from tobacco (*Nicotiana tabacum*)), and *ap24* (encoding a thaumatin-like protein from tobacco). The results showed that the expression of the *ch5B* gene in transgenic strawberry increased the resistance to gray mold while it had no significant effect on anthracnose disease resistance.What reported above are the few examples available on the GM technology applications for introducing genes for disease resistance in strawberry, when much more should be done on this topic by considering all pest and diseases affecting this fruit during cultivation and post-harvest.

## *3.2 Flower and Fruiting*

Among the major agronomic characters, flowering and fruiting are developmental processes of both heuristic and applied interest. In this regard, modification of flowering and fruiting can improve agricultural production in both a quantitative and qualitative manner.

Studies on the effect of changing of plant hormonal regulation on plant development and fruiting is seen with an important knowledge for the improvement of strawberry agronomic performances, including plant adaptability and resistance to diseases. These modifications can be easily achieved with the introduction of genes able to modify endogenous plant growth regulators (PGRs).

The *DefH9-iaaM* gene that is expressed specifically in placenta and ovules (Rotino et al., 1997), consistent with results obtained in other species, e.g. eggplant, tomato, tobacco (Spena and Rotino, 2001), also promotes parthenocarpy in *Rosaceae* (Mezzetti et al., 2004b), however, pathenocarpic fruits (i.e. from emasculated flowers) did not develop fully. The *DefH9-iaaM* auxin synthesizing gene when expressed and biologically active in strawberry can promote, under cultivation conditions that allow pollination and fertilization, the increase of fruit productivity by enhancing the number of inflorescences per plant and the number of flowers per inflorescence. So far auxin plays a role in plant fecundity in this perennial rosaceous species.

Another powerful tool for manipulation endogenous phyto-hormones is the transformation with oncogenes originating from *Agrobacterium tumefaciens* or *A. rhizogenes* an example from the latter being the *rol* genes which have proven highly useful for improving certain agronomic traits and even for producing novel plant morphogenesis. The *rol* genes are showing an increasing importance in floricultural crops (Zuker et al., 2001; Casanova et al., 2005), while few authors have studied the effect of the *rol* genes, and particularly of *rolC*, in fruit plants (Welander and Zhu, 2006).

In most plants, the *rolC* gene was driven by the constitutive promoter CaMV-35S to obtain more pronounced changes in phenotype, as described by Schmulling et al. (1988), Nilsson et al. (1996), Gardner et al. (2006), and in some cases producing the most dramatic changes on plant and flower architecture (Mitiouchkina and Dolgov, 2000; Winefield et al., 1999).

The potential use of the *rolC* gene in improving strawberry cultivation was assessed by the evaluation of the modifications induced by the expression of the gene on plant phenotype and as a consequence on their adaptability, production and quality. The effect of the *rol*C gene in strawberry was studied on 4 *rolC* clones of 'Calypso', transformed by using the *Agrobacterium*-mediated approach (Mazzara et al., 1998). *In vitro* root development of the *rol*C lines differed from the control with an increase in root number and a sharp decrease in root apparatus length. Furthermore, correlations were detected in the transgenic lines between the *rol*C expression level and some characters *in vivo*. Lines with the highest level of expression produced more leaves of smaller dimension, while showing a reduced height. This leads to the "bushy" phenotype determined by the modification induced with the *rol*C gene (Mezzetti et al., 2004b). Agronomic trials showed that the over-expression of *rolC* gene in strawberry has modified traits of economic interest such as increased plant adaptability, productivity and tolerance to a soil-borne/diseases, as well as fruit quality. These important agronomic improvements were associated to an increased symbiotic *rolC* – strawberry root/mycorrhize interaction (Landi and Mezzetti, 2006).

## *3.3 Quality – Maturation*

All major breeding and biotechnology programs on berries has as major priority the improvement of fruit quality. In general, a clear definition of quality is quite difficult and changes depending to the different destinations of the fruits. However, for berries the following components can be considered of major importance: flavor (a complex combination of sweetness, acidity, and aroma), firmness and self-life.

All these aspects are controlled by the complex developmental process related to fruit ripening, that involves specific changes in gene expression and cellular metabolism (Manning, 1994). In climateric fruits these events are coordinated by the gaseous hormone ethylene, which is synthesized autocatalytically in the early stages of ripening. Non-climacteric fruits do not synthesize or respond to ethylene in this manner, yet undergo many of the same physiological and biochemical changes associated with the production of a ripe fruit.

Wild strawberry is an attractive model system for studying ripening in non-climacteric fruit, because of its small diploid genome, its short reproductive cycle, and its capacity for transformation. Eight ripening-induced cDNAs were isolated from this species after differential screening of a cDNA library (Nam et al., 1999). The predicted polypeptides of seven clones exhibit similarity to database protein sequences, including acyl carrier protein, caffeoyl-CoA 3-O-methyltransferase, sesquiterpene cyclase, major latex protein, cystathionine gamma-synthase, dehydrin and an auxin-induced gene. None of these proteins appear to be directly related to events generally associated with ripening such as cell wall metabolism or the accumulation of sugars and pigments, rather, their putative functions are indicative of the wide range of processes upregulated during fruit ripening.

mRNA populations in ripening strawberry (*Fragaria* × *ananassa*) fruit were examined using polymerase chain reaction (PCR) differential display (Wilkinson et al., 1995). Five mRNAs with ripening-enhanced expression were identified using this approach. Three of the mRNAs appeared to be fruit-specific, with little or no expression detected in vegetative tissues. Sequence analysis of the cDNA clones revealed positive identities for three of the five mRNAs based on homology to known proteins. These results indicate that the differential display technique can be a useful tool to study fruit ripening and other developmental processes in plants at the RNA level.

### 3.4 Tissue Softening

Tissue softening accompanies the ripening of many fruits and initiates the processes of irreversible deterioration. Expansins are plant cell wall proteins proposed to disrupt hydrogen bonds within the cell wall polymer matrix. Expression of specific expansin genes has been observed in tomato (*Solanum lycopersicum*) meristems, expanding tissues and ripening fruit. It has been proposed that a tomato ripening-regulated expansin might contribute to cell wall polymer disassembly and fruit softening by increasing the accessibility of specific cell wall polymers to hydrolase action. Expansin gene expression was examined in strawberry (*F.* × *ananassa*) (Civello et al., 1999). Strawberry differs significantly from tomato in that the fruit is derived from receptacle rather than ovary tissue and strawberry is non-climacteric. A full-length cDNA encoding a ripening-regulated expansin, *FaExp2*, was isolated from strawberry fruit. The deduced amino acid sequence of *FaExp2* is most closely related to an expansin expressed in early tomato development and to expansins expressed in apricot fruit rather than the previously identified tomato ripening-regulated expansin, *LeExp1*. Nearly all previously identified ripening-regulated genes in strawberry are negatively regulated by auxin. Surprisingly, *FaExp2* expression was largely unaffected by auxin. Overall, the results suggest that expansins are a common component of ripening and that non-climacteric signals other than auxin may coordinate the onset of ripening in strawberry.

A novel E-type endo-beta-1,4-glucanase with a putative cellulose-binding domain is highly expressed in ripening strawberry fruits of the octoploid cultivar Chandler (Trainotti et al., 1999). Two full-length cDNA clones (*faEG1* and *faEG3*, respectively) have been isolated by screening a cDNA library representing transcripts from red fruits. Southern blot analysis of genomic DNA suggests that the strawberry endo-beta-1,4-glucanases (EGases) are encoded by a multigene family and are predominantly expressed during the ripening process. In agreement with other ripening-related genes in strawberry, the expression of *faEG1* and *faEG3* is also down-regulated by treatment with the auxin analogue NAA. Differences in temporal expression of the two EGase genes in fruits are not accompanied by differences in spatial expression. The pattern of expression and the sequence characteristics of the two polypeptides suggest that the two strawberry EGases operate in a synergistic and coordinate manner.

Antisense technology resulted an useful tool to prevent strawberry fruit from softening by suppressing particular genes involved in fruit softening without altering fruit quality (Mathews et al., 1995; Woolley et al., 2001; Jimenez-Bermudez et al., 2002; Palomer et al., 2006; Sesmero et al., 2007).

Llop-Tous et al. (1999) isolated two cDNA clones (*Cel1* and *Cel2*) from a cDNA library obtained from ripe strawberry (*Fragaria* × *ananassa*) fruit encoding divergent endo-beta-1,4-glucanases (EGases) [cellulases]. The EGases differ in their secondary and tertiary structures and in the presence of potential N-glycosylation sites. By *in vitro* translation it was shown that *Cel1* and *Cel2* bear a functional signal peptide, the cleavage of which yields mature proteins of 52 and 60 kDa, respectively. The *Cel2* EGase was expressed in green fruit, accumulating as the fruit turned from green to white and remaining at an elevated, constant level throughout fruit ripening. In contrast, the *Cel1* transcript was not detected in green fruit and only a low level of expression was observed in white fruit. The level of *Cel1* mRNA increased gradually during ripening, reaching a maximum in fully ripe fruit. The high levels of *Cel1* and *Cel2* mRNA in ripe fruit and their overlapping patterns of expression suggest that these EGases play an important role in softening during ripening.

### 3.5 Carbohydrate

Carbohydrate content and balance play an important role in determining the *flavor and processing quality* of the strawberry. As the fruit ripens, the sugar balance favors the hexoses glucose and fructose rather than sucrose. This will have consequences for the osmotic potential of the fruit cells and may result in an overall adjustment of water import, fruit growth and potential for processing. In many fruit species the enzyme invertase [beta-fructofuranosidase] is responsible for catalysing the breakdown of sucrose. An invertase that is located in the cell wall may regulate the unloading of sucrose from the phloem and control assimilate accumulation whereas an invertase located in the vacuole may regulate sucrose and hexose storage. Invertase genes cloned from potato, encoding cell wall and vacuolar forms, were integrated into two strawberry cultivars, Symphony and Senga Sengana, via *A. tumefaciens*-mediated transformation (Bachelier et al., 1997). Transgenic plants were be assessed for modified growth, sugar balance, flavor and processing quality (Graham et al., 1997b).

Recently, Park et al. (2006) generated transgenic plants that incorporated an antisense cDNA of ADP-glucose pyrophosphorylase (*AGPase*) small subunit (*FagpS*) driven by the strawberry fruit-dominant ascorbate peroxidase (APX) promoter, to evaluate the effects on carbohydrate contents during fruit development. The results showed that the levels of *AGPase* mRNA were drastically reduced in the red stage of fruits in all the transgenic plants. The suppression of the *AGPase* small subunit in transgenic plants resulted in a 16–37% increment of total soluble sugar content and a 27–47% decrease of the starch content in mature fruit without significantly affecting other fruit characteristics such as color, weight and hardness. Results from previous studies suggested that, through biotechnological alternation, the *AGPase* gene might be used for improving soluble sugar content and decreasing starch content in strawberry fruits.

### 3.6 Fruit Flavor

Fruit Flavor is a result of a complex mixture of numerous compounds. The formation of these compounds is closely correlated with the metabolic changes occurring during fruit maturation. DNA microarrays and appropriate statistical analyses were used to identified a novel strawberry alcohol acyltransferase (*SAAT*) gene that plays a crucial role in flavor biogenesis in ripening fruit (Aharoni et al., 2000). Volatile esters are quantitatively and qualitatively the most important compounds providing fruity odours. Biochemical evidence for involvement of the *SAAT* gene in formation of fruity esters is provided by characterizing the recombinant protein expressed in *Escherichia coli*. The *SAAT* enzyme showed maximum activity with aliphatic medium-chain alcohols, whose corresponding esters are major components of strawberry volatiles. The enzyme was capable of utilizing short- and medium-chain, branched, and aromatic acyl-CoA molecules as cosubstrates.

The results suggest that the formation of volatile esters in fruit is subject to the availability of acyl-CoA molecules and alcohol substrates and is dictated by the temporal expression pattern of the *SAAT* gene(s) and substrate specificity of the SAAT enzyme(s).

MADS-box genes encode putative transcription factors that are highly conserved among eukaryotes. The name arises from the four original members of this family, which are *MCM1* in yeast, *AG* in Arabidopsis, *DEFA* from *Antirrhinum*, and *SRF* in humans (Schwarz-Sommer et al., 1992). Genetic analyses have shown that plant MADS-box genes are homeotic and control the spatial and temporal locations of specific organs.

*AGAMOUS*, from Arabidopsis, is a MADS-box gene that is responsible for stamen and carpel identity. It is also involved in maintaining the floral meristem and preventing it from reverting from a determinate state to an indeterminate state (Mizukami and Ma, 1995). Homologs of this gene have been found in diverse plant species including tobacco (Kempin et al., 1993), tomato (Pnueli et al., 1994) and recently also in strawberry (Ahroni et al., 1999).

All these molecular studies on factors affecting strawberry fruit quality will open for the future really interesting opportunity to improve knowledge on the mechanisms controlling such complex but important agronomic character.

## 4 Biotechnology for Berry Nutritional Quality

Conventional breeding is one mean of achieving the development of berries rich in nutritional compounds; the genetic diversity available within sexual compatible species of any given crop will limit the extent of improvement. Transgenic approaches can provide an alternative, although there is currently public concern about their use in contemporary agriculture, particularly when genes derived from organisms other then plants are used.

Up to date, transgenic approaches were successfully used to increase the nutritional value of several worldwide important crops, for example rice (Paine et al., 2005) and tomato (Davuluri et al., 2005). However, these approaches have not increased both carotenoids and flavonoids simultaneously, except in some cases associated with several plant development defects (Gilberto et al., 2005).

Regarding berries, a first interesting result was the demonstration that the biosynthesis of L-ascorbic acid in ripe strawberry fruit can occur through D-galacturonic acid. Furthermore, it was demonstrated that vitamin C levels can be increased in *A. thaliana* plants by over-expressing the strawberry gene *GaiUR,* encoding a D-galacturonic acid reductase (Agius et al., 2003).

The role of *DefH9-iaaM* and *rolC* genes in improving fruit production and nutritional quality was studied for the first time in transgenic strawberries (Mezzetti et al., 2004a, b). The aim of this study was to better identify the potential use of both genes in improving strawberry agronomic performances and the antioxidant attributes of control and transgenic lines of two strawberry (*F.* × *ananassa*) genotypes (breeding selection AN93.231.53 and Calypso). Also for this type of study, differences in fruit Total Antioxidant Capacity (TAC) were determined by using TEAC (Scalzo et al., 2005) and in Total Polyphenols (TPH) by Folin-Ciocalteu. The increased productivity lead by *DefH9-iaaM* did not altered TAC, while the pleiotropic changes induced by *rolC* were also improving fruit TAC. The increase in plant strawberry productivity, caused by the *DefH9-iaaM*, gene is bound to be due, either directly or indirectly, to transgene expression, and consequently to increase auxin synthesis (Indole Acetic Acid – IAA) in the flower buds (Mezzetti et al., 2004b). The increased flowers IAA content induced by the expression of the *DefH9-iaaM* gene strongly affected plant morphogenic development, but without changing fruit components controlling quality and antioxidant capacity (Scalzo et al., 2005). The increased plant cytokinin metabolism induced by the expression of *rolC* gene had an effect on plant development (increased vigour and adaptability), but also lead to an improvement of fruit nutritional quality (mainly sugar content and TAC). The highest yield of *rolC* lines was mainly related to a larger fruit number with a reduced fruit weight but also associated to increased fruit total sugar content and TAC (Scalzo et al., 2005). This study represents the first evidence of *rolC* effect on fruit quality and antioxidant capacity. Further agronomical studies for the genetically modified strawberry, as well as the appropriate risk assessment, are now required.

Recently a great stride has been taken in identifying and characterizing, through biochemical and molecular means, the major enzymes and genes involved in flavonoid and proanthocyanidin biosynthesis during fruit development (Almeida et al., 2007).

The cloning and biochemical characterization of a glucosyltransferase involved in anthocyanin biosynthesis in strawberry fruit was reported by Griesser et al. (2008). Data were reported on the ripening-related and auxin-controlled expression of this gene. Whereas, by using other techniques to verify the function of glycosyltransferases in planta, was also reported the RNA interference (RNAi)-mediated down-regulation of an anthocyanidin-3-*O*-glycosyltransferase gene in a commercially important fruit crop, thus confirming its function in planta.

While an ongoing work has produced *AmDFR* and *MiANS* transgenic lines of Calypso and Sveva and preliminary transcriptomic studies on Sveva – DFR evidenced a deep perturbation of the whole pathway. These new transgenic lines represent unique new material for molecular and biochemical studies to elucidate the regulation of flavonoid pathway and improve the nutritional properties of strawberry (Montironi et al., 2009).

## 5 GM Strawberry Field Trials and Risk Assessment

GMO acceptability is linked to an appropriate evaluation for their potential risks for the environment (gene flow and gene impact on other micro-organism) and for their benefits in agricultural practices (reduce inputs, improvement of yield, quality and nutritional value).

The recent evolution of molecular biotechnology has promoted the development of new DNA recombinant technologies with important perspectives of application on agricultural systems and food industries, but at the same time opening a controversial discussion on their risks and benefits for the environment and the consumers.

These novelties for agriculture and food industries have induced the European and National institutions to identify specific and restricted rules able to guarantee the safeguard of human health and environment and useful to overcome the uncertainty showed by the EU consumers. With this aim the European Commission (EC) has adopted Directive 2001/18/EC (repealing Directive 90/220/EEC) to govern the deliberate release of GMOs into the environment. The new 2001/18/EC directive was adopted in different EU countries by supporting mainly the following two specific needs: (a) update the authorization procedures by facing more in deep also the environmental and health factors linked with the deliberate release of GM plants in the environment; (b) overcome the EU and National moratoria for GMO production and commercialization by introducing the traceability rules on GMO products. Beside this, the situation for EU research on GM crops still remain critical in comparison with rules adopted in all the other world areas.

Regulatory authorities and consumers are concerned about the environmental safety of GMO and are demanding for commercial transgenic plants to be free of unnecessary genes, such as marker genes (antibiotic resistance) or vector backbone sequences (Schaart et al., 2004, Mihály et al., 2006). Schaart et al. (2004), for example, used an inducible site-specific recombinase to obtain marker-free transgenic strawberry plants using a bifunctional selectable marker gene for the initial positive selection of transgenic tissue. When shoots had regenerated on kanamycin (Kan), leaf explants from $Kan^R$ shoots were treated with dexamethasone, which induces the recombinase gene, and subsequently subjected to a second round of regeneration in the presence of 5-fluorocytosine (5-FC). The *codA* gene converted the non-toxic

5-FC to cytotoxic fluorouracil (Stougaard, 1993), leading to the death of cells that still contain the marker genes. After negative selection, they successfully obtained 30 putative marker-free transgenic plants.

Field trials are a prerequisite step for assessing risk and benefit of a new GM plants and as consequence for market approval. The aim of the field trials is to test, in small-scale experiments, the stability of the inserted gene, the characteristics of the GM crop compared to the conventional one (e.g. growth characteristics), and most importantly to assess any potential risk to human health, animal heath and to the environment. The data obtained from field trials constitute a core part of the information submitted to regulators for safety assessment. The field trials often represent a long and expensive stage in the development process (i.e. from laboratory design to the submission for marketing) of a GMO.

Regarding gene flow two main lines of research can be envisaged:

(1) Hybridization potential of transgenic strawberries with their wild relatives. This issue can be addressed on the following levels:
   - Potential gene flow from transgenic to wild strawberries by honeybees, important if not main pollen vectors of strawberries, should be tested in the greenhouse with arrays of transgenic, wild and hybrid strawberries, arranged randomly or aggregated and exposed to honeybees, whose foraging behaviour will be monitored.
   - The potential for hybrid formation can be tested with hand-cross pollinations of emasculated flowers, and hybrid fertility will be tested by sowing hybrid seeds and examining seedling establishment.

(2) Potential ecological consequences of gene flow from transgenic to wild strawberries.

A prerequisite for addressing this issue is the study of the basic demographies of transgenic, wild and hybrid strawberries. Demographies of the variants can be examined using matrix models. Competition experiments between the three variants can allow to assess the invasive potential of transgenic and hybrid strawberries.

Both main approaches are instrumental for assessing the extinction risk of wild strawberries by hybridization with transgenic strawberries, or alternatively for providing arguments for a possible cultivation of transgenic strawberries.

To evaluate long term effects and alterations on non-target organisms the use *Drosophila melanogaster*, a classical non-target organism that can normally eat strawberries cultivated in open field, is seen as the most easily applicable model. Tests in drosophila seem to be the most desiderable (Wurgler et al., 1984), as the species has several useful qualities: a short generation time (10–12 days), numerous progeny, few chromosomes, many diverse markers, and a genome completely sequenced. These organisms are grown easily, research with this species is relatively inexpensive, and the organization of genes and chromosomes is comparable. It is a relatively long-lived organism to which substances to test can be applied in different way, and their effects on individual stages of the development of gametes can be studied. Furthermore, indirect mutagens are activated by the enzymatic system the same as they are in mammals.

A further point to study will be the question of genotoxicity, especially at high concentrations of single components, so *Drosophila* can be used to measure a wide range of genetic changes: recessive lethal, visible mutations, deletions, translocations, chromosome loss, dominant lethal, non disjunction. To assess short-term effects and appearance of unforeseen cellular damages, fractionated extracts should be tested on transformed and primary cell cultures and *in vivo* on model mice.

Finally, NMR should be applied to assess the metabolic profiles and evaluate possible alterations triggered by using different strawberry extracts on distant cellular or vertebrate model systems. Furthermore, mice biological fluids (especially urine) can be investigated to detect potential consequence for humans from a reliable animal model.

A possible final goal is to propose the NMR approach as a simplified, non-invasive and accurate method to evaluate metabolic alteration in organisms fed with different food components either GM or non-GM. In particular, such an NMR protocol should be oriented towards the use of mice urine as the fluid of election to track metabolic profiles upon GM food.

## 6 Conclusions

The past EU moratoria created a negative impact on the development of appropriate and accurate research programs on the assessment of GMOs risk and benefit, as a consequence of the dramatic reduction of the trial notification in most of the EU countries and particularly in Italy. This situation clearly created a lack of science based information useful to better identify the real perspectives of GMOs application for improving the productivity and quality of the major crops of interest particularly for EU countries.

Now we have the important need to implement research trial notifications in the EU countries and to create a European network for the larger and coordinated activities related to the assessment GM plants risk and benefits.

Essential knowledge is required for a sound risk assessment of planting transgenic strawberries commercially under outdoor conditions. In particular, to asses potential detrimental effects such as unwanted gene flow from transgenic to wild strawberries and potential ecological consequences on biodiversity, including the potential of extinction of wild strawberries by hybridization with transgenic strawberries. No systematic investigation of these issues is available so far, even though potential detrimental effects of growing transgenic strawberries are clearly and repeatedly stated in the literature.

As strawberries are a niche product, these findings will mainly be relevant for smaller farms. Under increasing global warming conditions, the importance of growing strawberries may actually increase, but also with the need of new high quality varieties well adapted to different growing conditions and/or cultivation systems.

Furthermore, strawberry is a charismatic crop and consumed directly without being processed, and hence the population will react particularly sensibly to a potential offer of transgenic strawberries. This requires a particularly sound ecological risk assessment for a commercial cultivation of transgenic strawberries.

In addition, strawberries may also serve as a model for other cultivated *Rosaceae* such as fruit trees of which transgenic varieties have already been developed (e. g. plum and cherry trees) or are under development (e. g. apple or peach trees). Strawberries have the same basic flower structure and the same main pollen vectors (honey bees), but much shorter life cycles, and hence are much more easily amenable to experimentation.

Among the different research carried out on strawberry genetic manipulation really few reached the field for a deep risk and benefits evaluation, the hope for the future is that a changed situation in rules and public acceptance produce a favorable climate for complete lab. to field GM strawberry risks and benefits evaluation.

## References


Agius F, Gonzalez-Lamothe R, Caballero JL, Munoz-Blanco J, Botella MA, and Valpuesta V (2003) Engineering increased vitamin C levels in plants by over-expression of a D- galacturonic acid reductase. Nat Biotechnol 21: 177–81.

Ahroni A, van Tunen AJ, Rosin FM, and Hannapel DJ (1999) Isolation of an AGAMOUS cDNA (STAG1) from Strawberry (*Fragaria* × *ananassa* cv. Elsanta) (Accession No. AF168468). (PGR99-153) Plant Physiol 121: 686.

Aharoni A, Keizer LCP, Bouwmeester HJ, Sun-Zhong K, Alvarez-Huerta M, Verhoeven HA, Blaas J, Houwelingen AMML van, Vos-RCH de, Voet H van-der, Jansen RC, Guis M, Mol J, Davis RW, Schena-M, Tunen AJ van, O'-Connell AP, and van-der-Voet-H (2000) Identification of the SAAT gene involved in strawberry flavor biogenesis by use of DNA microarrays. Plant Cell 12(5): 647–661.

Almeida JRM, D'Amico E, Preuss A, Carbone F, Ric de Vos CH, and Deiml B (2007) Characterization of major enzymes and genes involved in flavonoid and proanthocyanidin biosynthesis during fruit development in strawberry (*Fragaria*×*ananassa*). Arch Biochem Biophys 465: 61–71.

Alsheikh MK, Suso HP, Robson M, Battey NH, and Wetten A (2002) Appropriate choice of antibiotic and *Agrobacterium* strain improves transformation of antibiotic-sensitive *Fragaria vesca* and *F.v. sepmerflorens*. Plant Cell Rep 20:1173–1180.

Asao HG, Nishizawa Y, Arai S, Sato T, Hirai M, and Yoshida K (1997) Enhanced resistance against a fungal pathogen *Sphaerotheca fumuli* in transgenic strawberry expressing a rice chitinase gene. Plant Biotechnol 14:145–149.

Asao HG, Arai S, and Nishizawa Y (2003) Environmental risk evaluation of transgenic strawberry expressing a rice chitinase gene. *Seibutsu Kogakkaishi* 81:57–63 (in Japanese with English Abstract).

Bachelier C, Graham J, Machray G, Manoir J du, Roucou JF, McNicol RJ, Davies H, Du-Manoir J, Scheer HAT. vander, Lieten F, and Dijkstra J (Ed.) (1997) Integration of an invertase gene to control sucrose metabolism in strawberry cultivars. Proceedings of the third international strawberry symposium, Veldhoven, Netherlands, 29 April-4 May 1996. Volume I. Acta Horticulturae 439: 161–163.

Balokhina NV, Kaliayeva MA, and Buryanov Ya I (2000) The elaboration of a shoot regeneration system for the genetic transformation of the wild strawberry (*Fragaria vesca* L.). Biotekhnologiya 16: 46–51.

Barcélo M, El Mansouri I, Mercado JA, Quesada MA, and Alfaro FP (1998) Regeneration and transformation via Agrobacterium tumefaciens of the strawberry cultivar Chandler. Plant Cell Tiss Org Cult 54: 29–36.

Bhagwat B, and Lane WD (2004) In vitro shoot regenerations from leaves of sweet cherry (*Prunus avium*) 'Lapins' and 'Sweetheart'. Plant Cell Tiss Org Cult 78: 173–181.

Bingham ET (1989) Registration of regen-S Alfa germplasm useful in tissue culture and transformation research. Crop Sci 29:1095–1096.

Casanova E, Trillas MI, Moysset L, and Vainstein A (2005) Influence of rol genes in floriculture. Biotechnol Adv 23(1):3–39.

Chalavi V, Tabaeizadeh Z, and Thibodeau P (2003) Enhanced resistance to *Verticillium dahliae* in transgenic strawberry plants expressing a *Lycopersicon chilense* chitinase gene. J Am Soc Hortic Sci 128:747–753.

Civello PM, Powell ALT, Sabehat A, and Bennett AB (1999) An expansin gene expressed in ripening strawberry fruit. Plant Physiol 121(4): 1273–1279.

Davis TM, and Yu H (1997) A linkage map of the diploid strawberry, *Fragaria vesca*. J Hered 88:215–221.

Davuluri GR, Van Tuinen A, Fraser PD, Manfredonia A, Newman R, Burgess D, Brummell DA, King SR, Palys J, Uhlig J, Bramely PM, Pennings HJ, and Bowler C (2005) Fruit-specific RNAi-mediated suppression of *DET1* enhances carotenoid and flavonoids content in tomatoes. Nat Biotechnol 23: 890–895.

El Mansouri L, Mercadi JA, Valpuesta V, Lopez-Aranda JM, Pliegi-Alfaro F, and Quesada MA (1996) Shoot regeneration and *Agrobacterium*-mediated trasformation of *Fragaria vesca* L. Plant Cell Rep. 15: 642–646.

Finstad K, Martin RR, Ramsdell DC, and Barba M (1995) Transformation of strawberry for virus resistance. VII th International symposium on small fruit virus diseases. Acta Hort No. 385: 86–90.

Folta KM, Dhingra A, Howard L, Stewart PJ, and Chandler CK (2006) Characterization of LF9 an octoploid strawberry genotype selected for rapid regeneration and transformation. Planta 224:1058–1067.

Foucault C, and Letouze R (1987) In vitro: regeneration des plant de Fraisier a partir de fragmentes de petiole et de bourgeons floraux. Biol. Plantarum 29: 409–414.

Gardner N, Melberg T, George M, and Smith AG (2006) Differential expression of rolC results in unique plant phenotypes. J Am Soc Horticult Sci 131(1): 82–88.

Gilberto L, Perrotta G, Pallara P, Weller JL, Fraser PD, Bramley PM, Fiore A, Tavazza M, and Giuliano G (2005) Manipulation of the blue light photoreceptor cryptochrome 2 in tomato affects vegetative development, flowering time, and fruit antioxidant content. Plant Physiol 137:199–208.

Graham J, McNicol RJ, and Greig K (1995) Transgenic apples and strawberries: advances in transformation, introduction of genes for insect resistance and field studies of tissue cultured plants. Ann Appl Biol 127: 163–173.

Graham J, Gordon SC, and McNicol RJ (1997a) The effect of the CpTi gene in strawberry against attack by vine weevil (Otiorhynchus sulcatus F. Coleoptera: Curculionidae). Ann Appl Biol 131(1): 133–139.

Graham J, Machray G, Manoir J du, Roucou JF, McNicol RJ, Davies H, and Du Manoir J (1997b) Integration of an invertase gene to control sucrose metabolism in strawberry cultivars. Acta Hort 439: 161–163.

Griesser M, Hoffmann T, Bellido ML, Rosati C, Fink B, Kurtzer R, Aharoni A, Muñoz-Blanco J, Schwab W (2008) Redirection of flavonoid biosynthesis through the down-regulation of an anthocyanidin glucosyltransferase in ripening strawberry fruit. Plant Physiol 146(4):1528–1539.

Gruchala A, Korbin M, and Zurawicz E (2004) Conditions of transformation and regeneration of Iduka and Elista strawberry plants. Plant Cell Tiss Organ Cult 79: 153–160.

Haymes KM, and Davis TM (1997) *Agrobacterium* – mediated transformation of 'Alpine' *Fragaria vesca* and transmission of transgenes to R1 progeny. Plant Cell Rep. 17: 279–283

Huetteman CA, and Preece JE (1993) Thidiazuron: a potent cytokinin for woody plant tissue culture. Plant Cell Tiss Org Cult 33: 105–119.

James DJ (1987) Cell and tissue culture technology for the genetic manipulation of temperate fruit trees. In: Biotechnol Genet Eng Rev. 5: 33–79. G.E. Russell (Ed.) Intercept, Newcastle-upon Tyne

James DJ, Passey AJ, and Barbara DJ (1990) *Agrobacterium*-mediated transformation of the cultivated strawberry (*Fragaria* × *ananassa* Duch.) using disarmed binary vectors. Plant Sci 69: 79–94.

Jimenez-Bermudez S, Redondo-Nevado J, Munoz-Blanco J, Caballero JL, Lopez-Aranda JM, Valpuesta V, Pliego-Alfaro F, Quesada MA, and Mercado JA (2002) Manipulation of strawberry fruit softening by antisense expression of a pectate lyase gene. Plant Physiol 128: 751–759.

Kempin SA, Mandel MA, and Yanofsky MF (1993) Conversion of perianth into reproductive organs by ectopic expression oh the tobacco floral homeotic gene NAG1. Plant Physiol, 103:4 1041–1046.

Kumar S, Sharma P, and Pental D (1998) A Gentic approach to in vitro regeneration of non-regenerating cotton (*Gossypium hirsutum* L.) cultivars. Plant Cell Rep 15:59–63.

Lamproye A, Hofinger M, Berthon JY, and Gaspar T (1990) 3-(Benzo(b)selenienyl)acetic acid, a potent synthetic auxin in somatic embryogenesis. Comptes rendus Acad Sci Paris 311 (série III): 127–132.

Landi L, and Mezzetti B (2006) TDZ, auxin and genotype effects on leaf organogenesis in Fragaria. Plant Cell Rep 25(4):281–8.

Liu ZR, and Sanford JC (1988) Plant regeneration by organogenesis from strawberry leaf and runner tissue. HortScience 23: 1057–1059.

Llop-Tous I, Dominguez-Puigjaner E, Palomer X, and Vendrell M (1999) Characterization of two divergent endo-beta-1,4-glucanase cDNA clones highly expressed in the nonclimacteric strawberry fruit. Plant Physiol 119:(4): 1415–1421.

Manning K (1994) Changes in gene expression during strawberry fruit ripening and their regulation by auxin. Planta 194:(1): 62–68.

Mathews H, Wagoner W, Kellog J, and Bestwick R (1995). Genetic transformation of strawberry: stable integration of a gene to control biosynthesis of ethylene. In Vitro Cell Dev Biol Plant 31:36–43.

Mazzara M, Mezzetti B, James JD, and Negri P (1998) Il gene rolC in fragola. L'informatore Agrario 29:46–49.

Meng RG, Chen THN, Finn CE, and Li YH (2004) Improving in vitro plant regeneration from leaf and petiole explants of 'Marion' blackberry. Hort Science 39: 316–320.

Mezzetti B (2003) Genetic transformation in strawberry and raspberry. In: Plant Genetic Engineering, Vol. 6 Improvement of Fruit Crops. Eds. Pawan K Jaiwal and Rana P Singh. SCI Tech Publishing LLC, Houston, TX

Mezzetti B, Costantini E, Cionchetti F, Landi L, Pandolfini T, and Spena A (2004a) Genetic transformation in strawberry and raspberry for improving plant productivity and fruit quality. Euroberry Symposium, Acta Hort (ISHS) 649:107–110. http://www.actahort.org/books/649/649˙19.htm.

Mezzetti B, Landi L, Pandolfini T, and Spena A (2004b) The defH9-iaaM auxin-synthesizing gene increases plant fecundity and fruit production in strawberry and raspberry. BMC Biotechnol 4:4 http://www.biomedcentral.com/1472–6750/4/4.

Mihály K, Ingrid MM, and Bánfalvi Z (2006) Generation of marker- and backbone-free transgenic potatoes by site-specific recombination and a bi-functional marker gene in a non-regular one-border *Agrobacterium* transformation vector. Transgenic Res 15:729–737.

Mitiouchkina TY, and Dolgov SV (2000) Modification of chrysanthemum plant and flower architecture by rolC gene from Agrobacterium rhizogenes introduction. Acta Hort 508: 163–169.

Mizukami Y, and Ma H (1995) Separation of Ag function in floral meristem determinacy from that in reproductive organ identity by expressing antisense AG RNA. Plant Mol Biol 28(5): 767–784.

Montironi E, Costantini E, Mourgues F, Rosati C, and Mezzetti B (2009) Engineering strawberry anthocyanin levels by transformation with late flavonoid pathway genes. Int. Strawberry Symposium Huelva (SP). Acta Hortic., in press

Murashige T, and Skoog F (1962) A revised medium for rapid growth and bioassays with tobacco tissue cultures. Physiol Plant 15: 473–497.

Nam YW, Tichit L, Leperlier M, Cuerq B, Marty I, and Lelievre JM (1999) Isolation and characterization of mRNAs differentially expressed during ripening of wild strawberry (Fragaria vesca L.) fruits. Plant Mol Biol 39: 629–636.

Nehra NS, Stushnoff C, and Kartha KK (1989) Direct shoot regeneration from leaf discs. J Am Soc Horti Sci 114:1014–1018.

Nehra NS, Chibbar RN, Kartha KK, Datla RSS, Crosby WL, and Stushnoff C (1990) Genetic transformation of strawberry by *Agrobacterium tumefaciens* using a leaf disk regeneration system. Plant Cell Rep 9: 293–298.

Nilsson O, Moritz T, Sundberg B, Sandberg G, and Olsson O (1996) Expression of the *Agrobacterium rhizogenes* rolC gene in a deciduous forest tree alters growth and development and leads to stem fasciation. Plant Physiol 112(2):493–502.

Oosumi T, Gruszewski HA, Blischak LA, Baxter AJ, Wadl PA, Shuman JL, Veilleux RE, and Shulaev V (2006) High-efficiency transformation of the diploid strawberry (*Fragaria vesca*) for functional genomics. Planta 223(6):1219–30.

Paine JA, Shipton CA, Chaggar S, Howells RM, Kennedy MJ, Vernon G, Wright SY, Hinchliffe E, Adams JL, Silverstone AL, and Drake R (2005) Improving the nutritional value of Golden Rice through increased pro-vitamin A content. Nat Biotechnol 23: 482–487.

Palomer X, Llop-Tous I, Vendrell M, Krens FA, Schaart JG, and Boone MJ (2006) Antisense down-regulation of strawberry endo-β-(1,4)-glucanase genes does not prevent fruit softening during ripening. Plant Sci 171: 640–646.

Park JI, Lee YK, Chung WI, Lee IH, Choi JH, and Lee WM (2006) Modification of sugar composition in strawberry fruit by antisense suppression of an ADP-glucose pyrophosphorylase. Mol Breed 17:269–279.

Passey AJ, Barrett KJ, and James DJ (2003) Adventitious shoot regeneration from seven commercial strawberry cultivars (*Fragaria* × *ananassa* Duch.) using a range of explant types. Plant Cell Rep 21: 397–401.

Pnueli L, Hareven D, Rounsley SD, Yanofsky MF, and Lifschitz E (1994) Isolation of the tomato AGAMOUS gene TAg1 and analysis of its homeotic role in transgenic plants. Plant Cell 6(2): 163–173.

Ricardo VG, Coll Y, Castagnaro A, and Ricci JCD (2003) Transformation of a strawberry cultivars using a modified regeneration medium. Hort Science 38:277–280.

Ricardo VG, Ricci JCD, Hernández L, and Castagnaro AP (2006) Enhanced resistance to *Botrytis cinerea* mediated by the transgenic expression of the chitinase gene *ch5B* in strawberry. Transgenic Res 15:57–68.

Rotino GL, Perri E, Zottini M, Sommer H, and Spena A (1997). Genetic engineering of parhenocarpic plants. Nat Biotechnol 15: 1398–1320.

Rugini E, and Orlando R (1992) High efficiency shoot regeneration from calluses of strawberry (*Fragaria* × *ananassa* Duch.) stipules of *in vitro* shoot cultures. J Hortic Sci 67: 577–582.

Sargent DJ, Clarke J, Simpson DW, Tobutt KR, Arús P, Monfort A, Vilanova S, Denoyes-Rothan B, Rousseau M, Folta KM, Bassil NV, and Battey NH (2006) An enhanced microsatellite map of diploid Fragaria. Theor Appl Genet 112(7):1349–59.

Sargent DJ, Rys A, Nier S, Simpson DW, and Tobutt KR (2007) The development and mapping of functional markers in Fragaria and their transferability and potential for mapping in other genera. Theor Appl Genet 114(2):373–384.

Scalzo J, Battino M, Costantini E, and Mezzetti B (2005) Breeding and biotechnology for improving berry nutritional quality. Biofactors 23(4):213–20.

Schaart JG, Krens FA, Pelgrom KTB, Mendes O, and Rouwendal GJA (2004) Effective production of marker-free transgenic strawberry plants using inducible site-specific recombination and a bifunctional selectable marker gene. Plant Biotechnol J 2:233–240.

Schmulling T, Schell J, and Spena A (1988) Single genes from *Agrobacterium rhizogenes* influence plant development. EMBO J 7: 2621–2629.

Schwarz-Sommer Z, Saedler H, Sommer H, Russo VEA, Brody S, Cove D, and Ottolenghi S (Ed) (1992) Homeotic genes in the genetic control of flower morphogenesis in *Antirrhinum majus*. Development: the molecular genetic approach. Springer Verlagm, Heidelberg, pp. 242–256.

Sesmero R, Quesada MA, and Mercado JA (2007) Antisense inhibition of pectate lyase gene expression in strawberry fruit: characteristics of fruits processed into jam. J Food Eng 79: 194–199.

Simpson DW, James DJ, Passey AJ, Massiah A, and Sargent DJ (1999) Genetic modification of strawberry for enhanced resistance to *Botrytis cinerea*. Abstract meeting COST836, Integrated Research in Berries, INRA – Paris.

Sorvari S, Ulvinen S, Hietarante T, and Hiirsalmi H (1993) Pre-culture medium promotes direct shoot regeneration from micropropagated strawberry leaf disks. HortScience 28: 55–57.

Spena A, and Rotino GL (2001) Parthenocarpy: state of the art. In: Current trends in the embryology of Angiosperm. Eds. Bhojwani SS and Soh WY) Kluwers Academic Publishers, Dordrecht, pp. 435–450.

Stougaard J (1993) Substrate-dependent negative selection in plants using a bacterial cytosine deaminase gene. Plant J 3:755–761.

Trainotti L, Spolaore S, Ravanello A, Baldan B, and Casadoro G (1999) A novel E-type endo-beta-1,4-glucanase with a putative cellulose-binding domain is highly expressed in ripening strawberry fruits. Plant Mol Biol 40(2): 323–332.

Uratsu SL, Ahmadu H, Bringust RS, and Dandekar AM, (1991) Relative virulence of *Agrobacterium tumefaciens* strains on strawberry. HortScience 26: 196–199.

Welander M, and Zhu Li H (2006) Rol genes: molecular biology, physiology, morphology, breeding uses. Plant Bree Rev 26: 79–103.

Wilkinson JQ, Lanahan MB, Conner TW, and Klee HJ (1995) Identification of mRNAs with enhanced expression in ripening strawberry fruit using polymerase chain reaction differential display. Plant Mol Biol 27(6): 1097–1108.

Winefield C, Lewis D, Arathoon S, and Deroles S (1999) Alteration of Petunia plant form through the introduction of the rolC gene from Agrobacterium rhizogenes. Mol Breed 5(6): 543–551.

Woolley LC, James DJ, and Manning K, 2001. Purification and properties of an endo-β-1,4-glucanase from strawberry and down-regulation of the corresponding gene, cel1. Planta 214:11–21.

Wurgler FE, Sobels FH, and Vogel E (1984) *Drosophila* as an assay system for detecting genetic changes. In: Hanbook of Mutagenicity Test Procedures, 2nd edn, Kildey BJ, Legator M, Nichols W, and Ramel C (Eds). Elsevier Science, Amsterdam, pp. 551–602.

Zuker A, Tzfira T, Scovel G, Ovadis M, Shklarman E, Itzhaki H, and Vainstein A (2001) *RolC*-transgenic carnation with improved horticultural traits: quantitative and qualitative analysis of greenhouse-grown plants. J Am Soc Hort Sci 126:13–18.

# 24. Raspberries and Blackberries: The Genomics of *Rubus*

**Julie Graham and Mary Woodhead**

## 1 Taxonomy

The genus Rubus is comprised of a highly heterozygous series of some 500 species, with ploidy levels ranging from diploid to dodecaploid (Jennings, 1988; Meng and Finn, 2002). Members of the genus can be difficult to classify into distinct species for a number of reasons, including hybridization between species and apomixes (Robertson, 1974; Dickinson et al., 2007; Evans et al., 2007). For a description of the genus and the species contained within see Skirvin et al. (2005). The domesticated subgenera *Idaeobatus* and *Eubatus* contain the raspberries, blackberries, arctic fruits and flowering raspberries, all of which have been utilized in breeding programs. The most economically important *Rubus* species are the raspberries, the European red raspberry, R. *idaeus* L. subsp. *idaeus*, the North American red raspberry *R. idaeus* subsp. *strigosus* Michx and the black raspberry (*R. occidentalis* L.). The red raspberry being diploid has been the focus of most of the developments in molecular genetics.

*Rubus* subgenus *Idaeobatus* is distributed principally in Asia but also East and South Africa, Europe and North America. In contrast, subgenus *Eubatus* is mainly distributed in South America, Europe and North America (Jennings, 1988). The members of subgenus *Idaeobatus* are distinguished by the ability of their mature fruits to separate from the receptacle.

## 2 Diversity

Roach (1985) and Jennings (1988) gave accounts of the early domestication of red raspberry (*Rubus idaeus* L). During the 19th Century, the North American red raspberry (*R. idaeus* subsp. strigosus Michx) was introduced into Europe and

J. Graham (✉)
SCRI, Invergowrie, Dundee, UK

K.M. Folta, S.E. Gardiner (eds.), *Genetics and Genomics of Rosaceae*, Plant Genetics and Genomics: Crops and Models 6, DOI 10.1007/978-0-387-77491-6_24,

subsequently crossed with the European sub-species (R. *idaeus* subsp. vulgatus Arrhen.). Five parent cultivars dominate the ancestry of red raspberry; 'Lloyd George' and 'Pynes Royal' entirely derived from R. *idaeus* var. vulgatus and 'Preussen', 'Cuthbert' and 'Newburgh' derived from both sub-species. Controlled crossing began slightly earlier in the US than the UK with the introduction of 'Latham' in 1914 (McNicol and Graham, 1992). Domestication has resulted in a reduction of both morphological and genetic diversity in red raspberry (Graham et al., 1996; Haskell, 1960; Jennings, 1988) with modern cultivars being genetically similar (Dale et al. 1993; Graham and McNicol, 1995). Similar work on the genetic relatedness of black raspberries using RAPD markers was carried out and raised similar concerns of a narrow genetic base with the need for more incorporation of more diverse germplasm into black raspberry breeding (Weber, 2003). Relatedness in blackberries has also been examined using pedigree analysis with similar findings recommending the diversification of the gene pool (Stafne and Clark, 2004). This restricted genetic diversity is of serious concern for the future of *Rubus* breeding, especially when seeking durable host resistance to intractable pests and diseases for which the repeated use of pesticides in some regions is ineffective, unsustainable or unacceptable for certain selected markets, such as 'organic production'. The gene base can and is being increased by the introduction of unselected raspberry clones and species material (Knight et al., 1989). However, the time required to produce finished cultivars from this material can be considerable, particularly if several generations of backcrossing are required to remove undesirable traits. Efforts are being made to conserve the biodiversity of berries across Europe (Bartha-Pichler, 2006) and an interest in the conservation of genetic resources has led to studies on wild raspberry populations in various countries. Studies in Scotland have show wild raspberries to be genetically and physiologically differentiated from each other and from cultivars (Marshall et al., 2001; Graham et al., 1997, 2003). Similar studies using phenotypic characteristics have been carried out on twelve wild raspberry populations in Russia (Ryabova, 2007) where wild populations were examined for characteristics which may be useful in cultivated raspberries. Romanian (Rusu et al., 2006a) and Bulgarian red raspberries (Badjakov, personal communication) have been studied to determine their similarity with European and American germplasm using SSR markers. Genetic diversity has been examined in natural populations of black raspberry (*R. coreanus*) in Korea using ISSR markers (Hong et al., 2003) and overall genetic relationships among populations were associated with geographic location. Black raspberry (*R. leucodermis*) populations have also been evaluated for traits of importance for use in red and black raspberry breeding (Finn et al., 2003). With concern over climate change coupled with the desire for limiting 'flown food' to reduce the carbon footprint of agriculture and horticulture, availability of suitable varieties will be crucial for the future success of raspberry commercial cultivation. This conservation of genetic resources may prove to be invaluable in securing germplasm for future breeding programs.

## 3 Breeding Objectives

*Rubus* breeding is a long slow process hampered by several genetic problems, which depending on species, include polyploidy, apomixes, pollen incompatability and poor seed germination. As far as we are aware there are 30 *Rubus* breeding programs in 19 countries, almost all of which are in Europe or North America. Breeding programs sponsored by end-users or government aim to develop appropriate germplasm enabling their particular industry to realize its potential and thus goals vary from program to program. As new challenges arise and new production systems are developed, breeding programs are faced with meeting these demands with new cultivars. The core primary objectives in raspberry breeding include: high quality fruit, good yield, shelf life and suitability for shipping, if for the fresh market, suitability for mechanical harvesting for the processing market, adaptation to the local environment and improved pathogen resistance (Graham and Jennings, 2008).

### *3.1 Fruit Characters*

While many characteristics are important in the successful acceptance of new cultivars, fruit quality must be considered the premier factor. Flavor, appearance and shelf life are the main attributes of fresh market quality and are essential for repeat purchase of fruit by consumers. Flavor is a highly subjective trait but can be broken down into multiple descriptors for taste, texture and other sensory characteristics. Good, acceptable flavor in raspberry tends to be fruity, sweet and floral with a desire for some acidity but no bitterness (Harrison et al., 1999). Color, brightness, size and shape contribute to the appearance and success of a variety and are crucial for initial purchase of fruit by consumers. A naturally dark color can be perceived as overripe by fresh market retailers, whereas a darker color is desirable for processing. Large fruit size is an attractive characteristic to both consumers and producers as it is more cost effective to pick.

### *3.2 Plant Characters*

Plant habit is important for plantation management and has a major effect on yield potential. In summer fruiting types, the most important characteristics include the number and height of young canes, consistency of bud break, internode length, lateral length and position of laterals (Dalman, 1991). In primocane fruiting types (where fruit is produced on first year canes) the amount of branching and extent of lateral development on the primocanes are major yield components. In both types erect, spineless canes are desirable (Jennings, 1988).

### 3.3 Phytochemicals

Raspberries were first used in Europe for medicinal purposes (Jennings, 1988), and once again there is heightened interest focused on these fruit as major sources of antioxidants, such as anthocyanins, catechins, flavonols, flavones and ascorbic acid; compounds that protect against a wide variety of human diseases, particularly cardiovascular disease and epithelial (but not hormone-related) cancers (Deighton et al., 2000; Moyer et al., 2002). There has been a recent explosion of data on berry fruit and their impact on human disease (e.g. Marinova and Ribarova, 2007; Rupasinghe and Clegg, 2007) As a result, the consumption of these berries is expected to increase substantially in the near future as their value in the daily diet is publicized. A concerted effort by the public health authorities in Finland, for example, has promoted the consumption of small berry fruits to their populations (Puska et al., 1990). Recent epidemiological studies have suggested the efficacy of compounds found at high concentrations in berries for the prevention of a number of chronic diseases and studies are now aimed at understanding the mechanisms of action of specific groups of phytochemicals. For a review of the literature on the potential health benefits of berry fruits see Hancock et al. (2007).

### 3.4 Production

While strawberries remain the best selling soft fruit, other fruits such as raspberry, are gaining popularity because of the increasing all year round availability. Raspberries have always been attractive as fresh dessert fruits or for processing from frozen berries into conserves, purees and juices.

The Scottish bred 'Glen Ample', released in the mid-1990's (www.fruitgateway.co.uk), supercedes older varieties such as 'Malling Jewel', 'Glen Clova', 'Glen Prosen' and 'Glen Moy'. 'Glen Ample', along with 'Tulameen' and recently the new cultivar 'Octavia' dominate the UK market and acreage due to their desirable fresh market characteristics. Serbia is a major world producer and exporter of raspberries, producing one quarter of the world tonnage. 90% of the acreage is dominated by the North American 'Willamette'. Protected cropping and out of season production in European countries is expanding so that in areas of southern Spain nearly 100% of fresh market raspberries are being grown under polytunnels. The early season 'Glen Lyon' has a low chilling requirement which makes it suitable for re-propagation and manipulation of canes and is currently the ideal variety for this production system. These protected cropping systems have been adopted by the UK industry to improve fruit quality and extend the season. Since the majority of fresh market production goes to large supermarket chains, the demand for good fruit quality, flavor and shelf life is high. In other European countries, Pacific Northwest-bred cultivars have led the industry, such as 'Meeker', 'Willamette' and 'Tulameen'. The primocane-fruiting (plants fruit in their first year) 'Heritage' has led the industry in many countries. In Scandinavia, the hardy Norwegian variety

'Veten' has been the mainstay for many years, now 'Glen Ample' has taken the lead. In the US, 'Meeker' and 'Willamette' developed in the mid-1900s are the primary cultivars although recent publicly developed 'Cowichan' and 'Coho' are being widely planted. Black raspberry (*R. occidentalis* L.) production has traditionally been concentrated almost completely in Oregon with 'Munger' and 'Jewel' being the leading varieties; however, a strong South Korean industry has developed over the past five years. A recent study carried out on the costs associated with blackberries (*Rubus* subgenus *Rubus*) as a crop for the southeastern United States for both the pick your own and wholesale markets (Safley et al., 2006) has shown this could be a profitable venture. Blackberry as a commercial crop is also gaining popularity in Europe, showing one of the biggest rises in cultivation aside from blueberries (Jennings, personal communication.).

Machine harvesting of processing raspberries is the standard practice for most major raspberry production regions around the world and is essential where picking labor is expensive or unavailable. Despite advances in machine technology, it appears that the major improvements in harvesting will come from plant breeding (Cormack, 1989). No single attribute has been found to determine successful machine harvest-ability but a range of interacting traits such as uniform, strong vigor and good cane density with an upright habit governs harvest performance. Medium length laterals with good fruit presentation is also desirable. Maturity, physical shape of the berry and receptacle all contribute to ease of pick. This will help ensure that a high percentage of uniform, ripe fruit with acceptable process quality and minimal green fruit are harvested throughout the season (Hall et al., 2002).

In the UK and Europe, a move from outside field plantations to protected cropping systems has taken place in an effort to extend the cropping season and to improve fruit quality. Protected cropping and out-of-season production in European countries is expanding, so that in areas of southern Spain nearly 100% of fresh market dessert raspberries for early, main and late season are being grown under tunnels (Brennan et al., 1999) mainly 'Glen Lyon'. Such changes in agronomic practices affect plant growth, seasonality and fruit quality and have implications for a shift in pest and pathogen pressures. Recently, breeding programs in the UK have responded to this change in production by trialing and selecting germplasm under protected cropping systems (FruitGateway.Co.UK) www.fruitgateway.co.uk.

## 4 Limitations

Concern over environmental impact and sustainability of agricultural and horticultural practices is leading to a greater emphasis on pest and disease resistance, as well as the ability of plants to withstand local environmental stresses. The changes in environmental, cultural and agronomic practices within the industry will impact strongly on the nature of the germplasm required for the future. Greater conservation of genetic resources and utilization of diverse locally adapted germplasm will be required for future viability of raspberry production.

The incorporation of novel resistance/tolerance to pests and diseases is regarded as essential for the development of cultivars suitable for culture under integrated pest management (IPM) systems. Sources of resistance in diverse *Rubus* sp. to many pests and diseases have been identified and exploited in conventional cross-breeding (Keep et al., 1977; Jones et al., 1984; Jennings, 1988; Knight, 1991; Williamson and Jennings, 1992; Briggs, 1965; Briggs et al., 1982; Jones, 1986; Martin, 2002; Woodford et al., 2002). However, germplasm bearing single resistance genes, when planted over extensive areas, can eventually be overcome by the rapid evolution of new biotypes of pests, so that new types of host resistance are required to sustain plant protection (Birch et al., 2002; Jones et al., 2002). Pest and diseases of raspberry in Europe have been extensively reviewed in Gordon et al., (2006).

## 5 Structural Genomics

Breeding methods used in raspberry have changed very little over the last 40 years or so, and little novel germplasm has made its way into commercial cultivars. However, with the narrowing genetic base coupled with the increasing demands from consumers, new breeding methods are required to meet demands. The speed and precision of breeding can be improved by the deployment of molecular tools for germplasm assessment, management and the development of genetic and physical maps. Red raspberry (*Rubus idaeus*) is a good species for the application of molecular and genomics techniques, being diploid ($2n = 2x = 14$) with a very small genome (275 Mbp) making it highly amenable to complete physical map construction and map-based gene cloning.

The availability of abundant genetic variation in natural and experimental populations and adaptation to a range of diverse habitats (Keep, 1972; Graham et al. 1997, 2003; Marshall et al. 2001; Balciuniene et al., 2005; Ryabova, 2007) offers researchers a rich source of variation in morphology, anatomy, physiology, phenology and response to a range of biotic and abiotic stressors. The ability to vegetatively propagate individual plants provides opportunities to capture genetic variation over generations and replicate individual genotypes to partition and quantify the environmental and genetic components of variation of genetic linkage maps. These are necessary to develop diagnostic markers for polygenic traits and, in the future, possibly identify the genes behind the traits. Understanding the genetic control of commercially and nutritionally important traits and the linkage of these characteristics to molecular markers on chromosomes is the future of plant breeding. This facilitates the development of diagnostic markers for polygenic traits and the identification of genes controlling complex phenotypes.

## 6 DNA Markers

The development and application of molecular markers has been reviewed by Antonius-Klemola (1999), Hokanson (2001) and Skirvin et al. (2005). As well as the deployment of anonymous DNA markers such as RAPDs (Graham et al.,

1994, 1997; Pattison and Weber, 2003) and AFLPs (Graham et al., 2006), SSR and EST-SSR markers have been developed (Graham et al., 2002, 2004, 2006; Stafne et al., 2005; Lewers et al., 2005; Lopes et al., 2006; Woodhead et al., 2008) which allow the development of genetic linkage mapping, fingerprinting and assessments of diversity to be undertaken in raspberry. Work is now underway at SCRI to develop single nucleotide polymorphism (SNP) markers for specific genes and transcription factors of interest.

## 7 Linkage Mapping

To date there are several genetic linkage maps available for raspberry (Pattison and Weber, 2003; Pattison et al., 2007; Graham et al., 2004, 2006; Sargent et al., 2007) which have largely been constructed in order to identify markers for particular pest and disease resistances, although they can of course be used to identify markers for other traits. Graham et al. (2004) utilised a cross between the phenotypically diverse European red raspberry 'Glen Moy' and the North American 'Latham'. This wide cross segregates for a large number of important traits, thus the same population and this reference map can be used for the major breeding objectives (Table 1). SSR markers from both genomic and cDNA libraries from 'Glen Moy' and 'Autumn Bliss' were used together with AFLP markers to create a linkage map. An enhanced map with further SSR and EST-SSR and gene markers has recently been completed (Graham et al., 2006). This work has highlighted the importance of maps and markers for raspberry breeding with demonstration of a tight association between Gene H and resistance to two fungal diseases, cane botrytis (*Botry-*

**Table 1** Important plant characters segregating in the 'Glen Moy' and 'Latham' cross utilized for generating a raspberry linkage map (Graham et al., 2004, 2006)

| Character | Glen Moy | Latham |
|---|---|---|
| Canes | Hairy (Hh) | Not hairy (hh) |
| | Spine free | Spiny |
| | Green/brown | Brown/purple |
| Fruit | Large | Small |
| | Sweet | Sour |
| | Thimble shape | Round |
| | Early ripening | Late ripening |
| | Pale red | Dark red |
| Pest and Disease | | |
| | Root rot susceptible | Root rot resistant |
| | Rust Susceptibility | Rust resistance |
| | Resistance to spur blight | Susceptible to spur blight |
| | Resistance to cane botrytis | Susceptible to cane botrytis |
| Other | Not hardy | Hardy |

*tis cinerea*) and spur blight (*Didymella applanata* (Niessl) Sacc.) on linkage group 2 (Graham et al., 2006). This gene is a valuable marker as raspberry breeders in general have limited resources and rarely include a primary screen for fungal diseases. It had been reported previously that some disease resistances were associated with distinctive morphological traits, most notably that of cane pubescence (fine hairs) determined by Gene H (genotype HH or Hh), the recessive allele of which gives glabrous canes (genotype hh). Raspberry cultivars and selections with fine hairs (pubescent canes) were reported to be more resistant to cane botrytis (*Botrytis cinerea*), cane blight (*Leptosphaeria coniothyrium*) and spur blight than non-hairy ones (Knight and Keep, 1958; Jennings and Brydon, 1989), but more susceptible to cane spot (*Elsinoe veneta*), powdery mildew (*Sphaerotheca macularis*) and yellow rust (*Phragmidium rubi-idaei*) (Keep, 1968; Anthony et al., 1986; Jennings and McGregor, 1988; Williamson and Jennings, 1992). No association between Gene H and rust or cane spot susceptibility was detected in this recent study. How Gene H contributes to the disease resistance has not been determined, and work to saturate the map region around Gene H is underway using AFLPs, as well as identifying candidate genes in the region (Woodhead, Graham and Smith, personal communication.).

Raspberry root rot caused by *Phytophthora fragariae* var. *rubi* is probably the most destructive disease in raspberry plantations (Wilcox et al., 1993; Wilcox and Latorre, 2002; Seemüller et al., 1986) and it has been the focus of several studies (Graham and Smith, 2002; Pattison and Weber, 2003; Pattison et al., 2007). Resistance to *Phytophthora* root rot (PRR) is found in cultivars derived from *Rubus idaeus subsp strigosus*, the native North American red raspberry but less so in those derived from *Rubus idaeus vulgatus,* the European red raspberry (Pattison and Weber, 2005). Thus, generating crosses between the two can facilitate the identification of the genes underpinning this resistance. Using a 'Glen Moy' (*R. idaeus vulgatus*) × 'Latham' (*R. idaeus strigosus*) mapping population two regions, one on each of two linkage groups have been identified and further research aimed at confirming these in a second population through glasshouse and field trials has been completed (Graham et al., unpublished data). BAC clones have been mapped into the resistance regions and will be sequenced to identify any genes in the region (Graham and Smith personal data). Using a RAPD-based linkage map from a cross between 'Latham' and 'Titan', Pattison and Weber (2003) also identified markers clustered on 2 linkage groups which were associated with disease score QTL for PRR using bulk segregant analysis. Recently this work has been extended to include AFLP, RAPD and resistance gene analog polymorphism (RGAP) markers in other 'Latham' and 'Titan' populations (Pattison et al., 2007). Considerable progress towards identifying markers and ultimately the gene(s) responsible for resistance to this disease is being made and this can be incorporated into raspberry breeding programs, allowing the rapid identification and selection of durable resistant genotypes.

Aphids (particularly *Amphorophora idaei (Borner)* and *Aphis idaei van der Goot*) are one of the most damaging arthropod pests in raspberry (Gordon et al., 1997), due to direct feeding damage to susceptible cultivars and because they act as vectors for virus transmission (Gordon et al., 2006). Breeding for host plant resistance to raspberry aphids, over the past 40 years has reduced the need for pesticides and controlled the spread of aphid borne viruses (Birch et al., 2005; Birch and Jones, 1988). However, over the 10–15 years it takes to produce a new variety by conventional breeding methods, insect pests are constantly adapting and overcoming plant resistance genes. Several types of aphid resistance genes, minor/multi-gene and single major genes e.g. A1 and A10 with different mechanisms have been used against *A. idaei* in sequence by raspberry breeders, however in the UK each type of major gene resistance has been broken. To date, minor gene-based aphid resistance remains durable in raspberry but it provides only partial resistance (Birch et al., 2005). Efforts to identify new sources of aphid resistance from wild species and other cultivars is underway, as is the development of molecular markers to speed up the selection of promising genotypes (Birch et al., 2005). In addition, efforts to map aphid resistance genes by anchoring marker data from appropriate segregating populations to the published raspberry maps are underway (Sargent et al., 2007) with the determination of the linkage group for A1. Determining the map location of a number of aphid resistance genes from various sources will allow the discrimination of different genes and gene pyramiding in new raspberry cultivars.

Recent surveys of commercial raspberry crops in the UK have found that many virus diseases are becoming increasingly prevalent (Jones and McGavin, 2004). Another virus Raspberry bushy dwarf virus (RBDV) is also causing serious concern. RBDV is a pollen borne virus which has been re-emerging in the past 15 years throughout raspberry growing regions of the world (Martin et al., 2004), mainly because RBDV-immune plants (carrying the Bu gene) are being replaced by RBDV-susceptible plants with superior agronomic traits (Jones et al., 1998; Jones and McGavin, 2004). The disease causes crumbly fruit and yield reduction (Martin et al., 2004), and is therefore an important target for crop improvement. Several isolates of RBDV have been sequenced (MacLeod et al., 2004) and genetic transformations have been undertaken to produce RBDV resistant plants (MacLeod et al., 2004; Martin et al., 2004). Attempts to develop markers for other viral resistance genes have been carried out for raspberry leaf spot and raspberry vein chlorosis utilizing the ‘Glen Moy’ × ‘Latham’ cross of Graham et al. (2004). Field screening was carried out to measure symptom production of leaf spot and vein chlorosis in 2 different environments. These traits were analyzed for significant linkages to mapped markers and resistance loci were found on linkage groups 2 and 8 (Rusu et al., 2006b).

Work towards the genetic mapping of health-related compounds has been initiated in *Rubus* (Stewart et al., 2007). Berries are extremely high in antioxidants, exhibiting up to 4 times more antioxidant capacity than non-berry fruits, 10 times more than vegetables and 40 times more than cereals (Halvorsen et al., 2002). They contain high levels of the antioxidant vitamins A, C and E and very high levels of non-essential but strongly antioxidant phenolic compounds. Phenolics can account for 90% or more of the overall antioxidant capacity found in berry fruit (Deighton et al., 2000), the most readily visible of which are the anthocyanin pigments. These pigments impart the deep, vibrant colors of berries and can be found at concentrations of up to 500 mg 100 g $FW^{-1}$. Berries represent a significant dietary source of anthocyanins, as only 24 out of 100 common foods contain anthocyanins and non-berry anthocyanin containing foods typically contain less than 100 mg 100 g $FW^{-1}$ (Wu et al., 2006). Progress in mapping anthocyanins has been made by Kassim et al., 2009). Here high performance liquid chromatography (HPLC) was used to quantify eight major anthocyanins cyanidin and pelargonidin glycosides: -3-sophoroside, -3-glucoside, -3-rutinoside and -3-glucosylrutinoside across two seasons and two environments in progeny from a cross between two *Rubus* subspecies, *Rubus idaeus* (cv. Glen Moy) × *Rubus strigosus* (cv. Latham). The eight antioxidants mapped to the same chromosome region on linkage group (LG) 1 of the map of Graham et al., (2006), across both years and from fruits grown in the field and under protected cultivation. Seven antioxidants also mapped to a region on LG 4 across years and for both field and protected sites. A chalcone synthase (PKS 1) gene sequence (Zheng et al., 2001; Zheng and Hrazdina, 2008) mapped to LG 7 but did not underlie the anthocyanin QTLs identified. However other candidate genes including bHLH (Espley et al., 2007), NAM/CUC2 (Ooka et al., 2003) like protein and bZIP transcription factor (Holm et al., 2006; Mallappa et al., 2006) underlying the mapped anthocyanins were identified (Kassim et al., 2009). The shift in focus from vitamin C and micronutrients towards the polyphenolics causes something of a challenge for any breeding effort, since the polyphenolics are chemically diverse and the content of individual health-promoting compounds varies in raspberry fruit due to both developmental and genetic factors (Beekwilder et al. 2005). However, with the emergence of metabolomics the simultaneous analysis of multiple metabolites at specific time points is now feasible. In *Rubus* a metabolomic approach has been used to identify bioactive compounds in a segregating mapping population planted under two different environments (Stewart et al., 2007). As a greater understanding of the relative importance and bioavailability of the different antioxidant compounds is achieved, it may become possible develop and identify those raspberry genotypes with enhanced health-promoting properties from breeding programs (Beekwilder et al. 2005).

## 8 Physical Mapping

Large insert genomic libraries (BACS) are both invaluable tools and a source of genomic DNA for physical mapping, positional cloning and as a scaffold for whole genome sequencing. *Rubus idaeus* is an ideal candidate for BAC library construction, since it is diploid ($2n = 2x = 14$) and has a very small genome (275 Mbp). The small genome size of raspberry makes it highly amenable to complete physical map construction, and thereby provides a platform for map-based gene cloning and comparative mapping with other members of the Rosaceae (Dirlewanger et al., 2004).

One of the most challenging steps required for the construction of plant large-insert genomic libraries is the isolation of high molecular weight DNA (HMW-DNA), either in the form of embedded protoplasts or nuclei. Raspberry and other soft-fruit species have, however, proven recalcitrant to standard genomic DNA extractions as they contain very high levels of carbohydrates, particularly polysaccharides, and polyphenolic compounds. They require heavily modified methods for ordinary genomic DNA isolations (Woodhead et al., 1998; Millan-Mendoza and Graham, 1999) and in order to prepare HMW-DNA suitable for the construction of BAC libraries a novel nuclei isolation procedure was developed (Hein et al., 2005). This DNA is of high quality and has been used for the construction of the first publicly available red raspberry BAC library from the European red raspberry, 'Glen Moy'. Currently, the library comprises over 15,000 clones with an average insert size of approximately 130 kb (6–7 genome equivalents). Hybridization screening of the BAC library with chloroplast (*rbcL*) and mitochondrial (*nad1*) coded genes revealed that contamination of the genomic library with chloroplast and mitochondrial clones was very low (>1%) (Hein et al., 2004).

Initial screening of the BAC library employed probes for chalcone synthase, phenylalanine ammonia lyase and a MADS-box gene involved in bud dormancy (Hein and Williamson, personal communication). More recently, the library has been probed with genes involved in epidermal cell fate (Woodhead, Graham and Williamson, personal communication), fruit quality genes (Woodhead and McCallum, personal communication) and a peach ever-growing gene (Abbott, personal communication).

Future work will focus on anchoring the physical map to the genetic map, which will enable alignment of the maps and the identification of genomic regions harbouring genes controlling important phenotypes. Some progress has been made here specifically for linkage groups 2, 3 and 6 of the map of Graham et al., (2006) (Graham et al. unreported data). An integrated physical/genetic map will also allow the extent of synteny or colinearity of the *Rubus* genome with other members of the Rosaceae to be determined.

The availability of a detailed genetic linkage map, together with a deep coverage bacterial artificial chromosome library, will be of great value in the identification of the genetic factors that underpin a wide range of commercially important characteristics such as the appearance, texture and the sensory attributes (taste and aroma) of raspberry fruit and the genetic resistance to pests and diseases. The establishment of gene-phenotype relationships will allow gene-based selection in breeding and the functional assignment of genes for commercially important traits.

## 9 Functional Genomics

Understanding the function of genes and other parts of the genome is known as functional genomics. In raspberry, as with many other species, functional genomics is at an early stage.

The advances in genomics technologies have lead to a massive increase in the numbers of DNA sequences held in public databases. However, a search of the NCBI nucleotide database for *Rubus* retrieved only 2239 sequences, a large number of which are actually viral sequences. In comparison, a similar search for the peach yielded 100,021 hits. The number of raspberry sequences is, however, very likely to increase rapidly as efforts are under way to sequence EST libraries generated from different tissues and developmental stages. At the Scottish Crop Research

Institute, cDNA libraries have been generated from leaves (approximately 6500 clones), canes (approximately 8000 clones) and roots (approximately 7300 clones) and further libraries are being constructed from fruit and shoots (Graham, Smith, Woodhead and McCallum unpublished data). As well as providing sequence information on genes expressed in these tissues, these resources are being used to identify DNA markers (EST-SSRs and SNPs) for use in the genetic mapping programs.

Fruit quality characteristics in raspberry are key drivers for breeding programs and a greater understanding of the processes and genes involved in quality will enable more targeted breeding in the future. Genes up-regulated during fruit ripening have been identified using classical plus/minus screening of cDNA libraries (Jones et al., 1999) and RNA fingerprinting techniques (Jones et al., 2000). Amongst the genes identified were cell wall hydrolases involved in fruit softening and ACC oxidase (Jones et al., 2000) involved in the ethylene biosynthetic pathway. Raspberry fruit ripen in response to ethylene (Burdon and Sexton, 1990a, b) but the ripening-related increase in ethylene biosynthesis is not associated with an increase in respiration rate (Jennings, 1988; Perkins-Veazie and Nonnecke, 1992).

The phenylpropanoid pathway is important in raspberry, as end products contribute to the color and aroma of the fruit and are involved in other processes such as lignin production. Aroma and color in raspberry fruit are partly derived from the polyketide derivatives benzalacetone and dihydrochalcone which are formed during fruit ripening as a result of the action of several enzymes, polyketide synthases (PKS), benzalcetone synthase and chalcone synthase (CHS) during fruit development. A number of PKS genes have been characterised from raspberry (Zheng et al., 2001; Kumar and Ellis, 2003). Kumar and Ellis (2003) reported the PKS gene family in *Rubus* consists of at least 11 members and expression analysis of 3 cDNAs showed they exhibited tissue-specific and developmental patterns of expression, with two cDNAs up-regulated during fruit ripening. Work is underway to identify and map these genes in the 'Glen Moy' × 'Latham' mapping population and raspberry BACs containing these PKS genes have been identified and are being characterized (Kassim, Paterson, Graham and Woodhead, personal communication).

Genes encoding 4-coumarate:CoA ligase, an enzyme that activates cinnamic acid and its derivatives to thioesters which then serve as intermediates for the production of phenylpropanoid-derived compounds that influence fruit quality have also been studied. Kumar and Ellis (2003) have characterized the 4-coumarate:CoA ligase (4CL) genes in raspberry found there are three genes which are differentially expressed in various organs and during fruit development and ripening. Based on the expression patterns and substrate utilization profiles of the recombinant proteins, they suggest that 4CL1 is involved in the biosynthesis of phenolics in leaves, 4CL2 in cane lignification and 4CL3 in the flavonoid and/or flavor pathway in fruit. These genes are also being targeted in the *Rubus* mapping program (Woodhead, Graham and Smith, personal communication).

A further project to characterise bud dormancy phase transition in woody perennial plants at a molecular level generated a total of 5300 ESTs from endodormant (true dormancy) and paradormant (apical dominance) raspberry meristematic bud tissue (Mazzitelli et al., 2007). PCR-products from these cloned cDNA fragments have been spotted onto glass slides and have been used in microarray experiments to identify genes that show differential expression. At present, approximately 380 clones exhibit up or down regulation during the endodormancy – paradormancy transition. Some of these ESTs, including one encoding a MADs-box gene, a MYB gene and several containing SSRs have been identified and mapped in the 'Glen Moy' × 'Latham' mapping population and these underlie mapped ripening stage QTL and will form the basis of future studies (Woodhead et al (2008) Graham et al (2009)).

## 10 Future Strategies

Through the creation of genetic and physical map resources in raspberry, there is now the real possibility of linking phenotype with genotype in raspberry. Initial research will focus on mapping key commercial and environmentally important traits. Closer links with other members of Rosaceae community to facilitate sharing of marker data/co-linearity/synteny etc. should allow progress to be made in a timely manner.

Ultimately, the information derived from *Rubus* genomic programs will be adapted and used to drive more efficient, targeted breeding to provide superior cultivars with high fruit quality (flavor, nutrition etc) and durable resistance to pests and diseases as well as a greater understanding of the key processes driving plant and fruit quality characteristics.

## References


Anthony VM, Williamson B, Jennings DL and Shattock RC (1986) Inheritance of resistance to yellow rust (*Phragmidium rubi-idaei*) in red raspberry. Ann App Biol 109: 365–374

Antonius-Klemola K (1999) Molecular markers in *Rubus* (Rosaceae) research and breeding. J Hort Sci Biotech 74: 149–160

Balciuniene L, Labokas J and Radaitiene D (2005) Primary evaluation of field collections of wild Rubus idaeus. Botanica Lithuanica 11: 3–15

Bartha-Pichler B (2006) Endangered biodiversity of berries in Switzerland – a National Berry Collection in Riehen (Basel-Stadt, Switzerland) and strategies for conservation of ProSpecieRara facing the increasing gene erosion. Mitteilungen der Naturforschenden Gesellschaften beider Basel, **9**: 33–45

Beekwilder J, Jonker H, Meesters P, Hall RD, van der Meer IM and de Vos CHR (2005) Antioxidants in raspberry: on-line analysis links antioxidant activity to a diversity of individual metabolites. J Agri and Food Chem 53: 3313–3320

Birch ANE and Jones AT (1988) Levels and components of resistance to *Amphorophora idaei* in raspberry cultivars containing different resistance genes. Ann of Appl Biol 113: 567–578

Birch ANE, Jones AT, Fenton B, Malloch G, Geoghegan I, Gordon SC, Hillier J and Begg G (2002) Resistance-breaking raspberry aphid biotypes: constraints to sustainable control through plant breeding. Acta Hort 587: 315–317

Birch ANE, Gordon SC, Brennan R and Jones AT (2005) Breeding for resistance to the large raspberry aphid: an update on durability of current genes and future prospects. Bulletin OILB/SROP 28: 21

Brennan RM, McNicol R, Gillespie T and Raffle S (1999) Factors affecting out-of-season Rubus production. Acta Horticulturae 505: 115–120

Briggs JB (1965) The distribution, abundance and genetic relationships of four strains of *Rubus* aphid (*Amphorophora rubi*) in relation to raspberry breeding. J Hort Sci 40: 109–117

Briggs JB, Danek J, Lyth M and Keep E (1982) Resistance to the raspberry beetle, *Byturus tomentosus*, in *Rubus* species and their hybrid derivatives with *R. idaeus.* J Hort sci **57**: 73–78

Burdon JN and Sexton R (1990a) Fruit abscission and ethylene production of red raspberry cultivars. Scientia Horticulturae 43: 95–102

Burdon JN and Sexton R (1990b) The role of ethylene in the shedding of red raspberry fruit. Ann Bot 66: 111–120

Cormack MR (1989) Growers and processors set to meet the challenge. Grower. November 19

Dalman P (1991) The effect of new cultivation practices on the yield, cane growth and health status of red raspberry (*Rubus idaeus* L.) in Finland. Ann Agric Fenn 30: 421–439

Dale A, Moore PP, McNicol RJ, Sjulin TM and Burmistrov LA (1993) Genetic diversity of red raspberry varieties throughout the world. J Amer Soc Hort Sci 118: 119–129

Deighton N, Brennan R, Finn C and Davies H V (2000) Antioxidant properties of domesticated and wild *Rubus* species. J Sci Food Agric 80: 1307–1313

Dickinson TA, Lo E and Talent N (2007) Polyploidy, reproductive biology, and Rosaceae: understanding evolution and making classifications. Plant Syst Evol 266: 59–78

Dirlewanger E, Graziano E, Joobeur T, Garriga-Caldere F, Cosson P, Howad W and Arus P (2004) Comparative mapping and marker-assisted selection in Rosaceae fruit crops. Proc Natl Acad Sci USA 101: 9891–9896

Espley RV, Hellens RP, Putterill J, and Stevenson DE (2007) Red colouration in apple fruit is due to the activity of the MYB transcription factor, MdMYB10. Plant J 49: 414–427

Evans KJ, Symon DE, Whalen MA, Hosking JR, Barker RM and Oliver JA (2007) Systematics of the *Rubus fruticosus* aggregate (Rosaceae) and other exotic taxa in Australia. Aust Syst Bot 20: 187–251

Finn C, Wennstrom K, Link J and Ridout J (2003) Evaluation of *Rubus leucodermis* populations from the Pacific Northwest. HortScience 38: 1169–1172

Gordon SC, Woodford JAT and Birch ANE (1997) Arthropod pests of Rubus in Europe: pest status, current and future control strategies. J Hort Sci 72: 831–862

Gordon SC, Williamson B and Graham J (2006) Current and future control strategies for major arthropod pests and fungal diseases of red raspberry (*Rubus idaeus*) in Europe. In: CROPS: Growth, Quality and Biotechnology ed. Dris WFL Publisher, Finland 925–950

Gordon SC, Williamson B and Graham J (2006) Current and future control strategies for major arthropod pests and fungal diseases of red raspberry (*Rubus idaeus*) in Europe. In: CROPS: Growth, Quality and Biotechnology, WFL Publisher, Helsinki, pp 925–950

Graham J, McNicol RJ, Greig K, Van de Ven WTG (1994) Identification of red raspberry cultivars and an assessment of their relatedness using fingerprints produced by random primers. J Hortic Sci 69: 123–130

Graham J and McNicol RJ (1995) An examination of the ability of RAPD markers to determine the relationships within and between *Rubus* species. Theor Appl Genet 90:1128–1132

Graham J, Gordon SC and Williamson B (1996) Progress towards the use of transgenic plants as an aid to control soft fruit pests and diseases. In: Proceedings of Brighton Crop Protection Conference-Pests & Diseases 1996. 3, pp. 777–781

Graham J, Squire GR, Marshall B and Harrison RE (1997) Spatially-dependent genetic diversity within and between colonies of wild raspberry *Rubus idaeus* detected using RAPD markers. Mol Ecol 6: 272–281

Graham J, Smith K, Woodhead M and Russell J (2002) Development and use of SSR markers in *Rubus* species. Mol Ecol Notes 2: 250–252

Graham J and Smith, K (2002). DNA markers for use in raspberry breeding. ActaHort 585:51–56

Graham J, Marshall B and Squire G (2003) Genetic differentiation over a spatial environmental gradient in wild *Rubus idaeus* populations. New Phyt 157: 667–675

Graham J, Smith K, MacKenzie K, Jorgenson L, Hackett C and Powell W (2004) The construction of a genetic linkage map of red raspberry (*Rubus idaeus* subsp. *idaeus*) based on AFLPs, genomic-SSR and EST-SSR markers. Theor Appl Genet 109:740–749.

Graham J, Smith K, Tierney I, MacKenzie K and Hackett CA (2006) Mapping gene *H* controlling cane pubescence in raspberry and its association with resistance to cane botrytis and spur blight, rust and cane spot. Theor Appl Genet 112: 818–831

Graham J and Jennings SN (2008) Raspberry breeding. In : Breeding Tree Crops. IBH & Science Publication Inc, Oxford

Graham J, Hackett CA, Smith K, Woodhead M, Hein I and McCallum S (2009) Mapping QTLs for developmental traits in raspberry from bud break to ripe fruit. Theor Appl Genet DOI 10/1007s00122-009-0969-6

Hall HK, Stephens MJ, Alspach P, Stanley CJ and Kempler C (2002) Traits of importance for machine harvest of raspberries. Acta Hort 585: 607–610

Halvorsen BL, Holte K, Myhrstad MCW, Barikmo I, Hvattum E, Remberg SF, Wold AB, Haffner K, Baugerod H, Andersen LF, Moskaug O, Jacobs DR and Blomboff R (2002) A systematic screening of total antioxidants in dietary plants. J Nutr 132: 461–471

Hancock RD, McDougall GJ and Stewart D (2007) Berry fruit as 'superfood': hope or hype? Biologist 54: 73–79

Harrison RE, Brennan RM, Hunter EA, Morel S and Muir, DD 1999 Genotypic, environmental and processing effects on the sensory character of *Rubus* and *Ribes* Acta Hort 505: 25–32

Haskell G (1960) The raspberry wild in Britain. Watsonia 4: 238–255

Hein I, Williamson S, Russell J, Graham J, Brennan R and Powell W (2004) Development of genomic resources for red raspberry: BAC library construction, analysis and screening. Proceedings of Plant and Animal Genome XII Conference, San Diego, CA, USA, p 148

Hein I, Williamson S, Russell J and Powell W (2005) Isolation of high molecular weight DNA suitable for BAC library construction from woody perennial soft-fruit species. BioTechniques 38: 69–71

Holm M, Ma LG, Qu LJ and Deng XW (2006) Two interacting bZIP proteins are direct targets of COP1-mediated control of light-dependent gene expression in Arabidopsis. Genes Dev 16: 1247–1259

Hokanson SC (2001) SNiPs, Chips, BACs and YACs: are small fruits part of the party mix? HortScience 36: 859–871

Hong YP, Kim MJ and Hong KN (2003) Genetic diversity in natural populations of two geographic isolates of Korean black raspberry. J Hort Sci Biotech 78: 350–354

Jennings DL (1988) Raspberries and Blackberries: Their Breeding, Diseases and Growth. Academic Press London, 230 pp

Jennings DL and McGregor GR (1988) Resistance to cane spot (*Elsinoe veneta*) in red raspberry and its relationship to resistance to yellow rust (*Phragmidium rubi-idaei*). Euphytica 37: 173–180

Jennings DL and Brydon E (1989) Further studies on breeding for resistance to *Leptosphaeria coniothyrium* in red raspberry and related species. Ann Appl Biol 115: 499–506

Jones AT, Gordon SC and Jennings DL (1984) A leaf-blotch disorder of Tayberry associated with the leaf and bud mite (*Phyllocoptes gracilis*) and some effects of three aphid-borne viruses. J Hort Sci 59: 523–528

Jones AT (1986) Advances in the study, detection and control of viruses and virus diseases of *Rubus*, with particular reference to the United Kingdom. Crop Res 26: 127–171

Jones AT, McGavin WJ, Mayo MA and Graham J (1998) Natural infection with raspberry bushy dwarf virus (RBDV) of the putatively RBDV-resistant red raspberry cultivar Glen Moy, and the demonstration that it does not contain the RBDV resistance gene, *Bu*. Ann Appl Biol 133: 403–414

Jones CS, Davies HV, Taylor MA and McNicol RJ (1999) Analysis of genes differentially expressed during fruit ripening in raspberry. Acta Hort 505: 401–402

Jones CS, Davies HV and Taylor MA (2000) Profiling of changes in gene expression during raspberry (Rubus idaeus) fruit ripening by application of RNA fingerprinting techniques. Planta 211: 708–714

Jones AT, McGavin WJ and Birch ANE (2002) Effectiveness of resistance genes to the large raspberry aphid, *Amphorophora idaei* Börner, in different raspberry (*Rubus idaeus* L.) genotypes and under different environmental conditions. Ann Appl Biol 136: 147–152

Jones AT and McGavin WJ (2004) Resurgence of virus problems in Rubus in the United Kingdom: possible effects on crop production and certification. Bulletin OILB/SROP 27: 73–78

Kassim A, Poette J, Paterson A, Zait D, McCallum S, Woodhead M, Smith K, Hackett C, and Graham J. (2009) Environmental and seasonal influences on red raspberry anthocyanin antioxidant contents and identification of QTL. Mol Nutr Food Res 53.

Keep E (1968) Incompatibility in Rubus with special reference to *R. idaeus* L. Can J Genet Cytol 10: 253–262

Keep E. (1972) Variability in the wild raspberry. New Phyt 71: 915–924

Keep E, Knight VH and Parker JH (1977) *Rubus coreanus* as donor of resistance to cane diseases and mildew in red raspberry breeding. Euphytica 26: 505–510

Knight RL and Keep E (1958) Developments in soft fruit breeding at East Malling. Rept East Malling Res. Stn. for 1957, 62–67

Knight VH, Jennings DL and McNicol RJ (1989) Progress in the UK raspberry programme. Acta Hort 262: 93–103

Knight VH (1991) Use of salmonberry, *Rubus spectabilis* Pursh., in red raspberry breeding. J Hort Sci 66: 575–581

Kumar A and Ellis BE (2003) A family of polyketide synthase genes expressed in ripening *Rubus* fruits. Phytochemistry 62: 513–526

Lewers KS, Styan SMN, Hokanson SC and Bassil NV (2005) Strawberry GenBank-derived and genomic simple sequence repeat (SSR) markers and their utility with strawberry, blackberry and red and black raspberry. J. Amer Soc Hort Sci 130: 102–115

Lopes MS, Maciel G, Mendonca D, Gil FS and Machado ADC (2006) Isolation and characterization of simple sequence repeat loci in *Rubus hochstetterorum* and their use in other species from the Rosaceae family. Mol Ecol Notes 6: 750–752

MacLeod MR Wood NT, Sinclair JA, Mayo MA, McGavin WJ, Jorgensen L and Jones AT (2004) Further studies on the molecular biology of raspberry bushy dwarf virus and the development of resistance to it in plants. Proceedings of the Tenth International Symposium on Small Fruit Virus Diseases 656: 159–163

Mallappa C, Yadav V, Negi P and Chattophayay SA (2006) Basic leucine zipper transcription factor, G-box-binding factor 1, regulates blue light-mediated photomorphogenic growth in *Arabidopsis*. J Biol Chem 281: 22190–22199

Marinova D and Ribarova F (2007) HPLC determination of carotenoids in Bulgarian berries. J Food Comp Anal 20: 370–374

Marshall B, Harrison RE, Graham J, McNicol JW, Wright G and Squire GR (2001) Spatial trends of phenotypic diversity between colonies of wild raspberry *Rubus idaeus*. New Phyt 151: 671–682

Martin RR (2002) Virus diseases of *Rubus* and strategies for their control. Acta Hort 585: 265–270

Martin RR, Keller KE and Matthews H (2004) Development of resistance to raspberry bushy dwarf virus in Meeker red raspberry. Proceedings of the 10th International Symposium on Small Fruit Virus Diseases 656: 165–169

Mazzitelli L, Hancock RD, Haupt S, Walker PG, Pont SDA, McNicol J, Cardle L, Morris J, Viola R, Brennan R, Hedley PE and Taylor M (2007) Co-ordinated gene expression during phases of dormancy release in raspberry (*Rubus idaeus L.*) buds. J Exp Bot 58: 1035–1045

McNicol RJ and Graham J (1992) Temperate small fruit. In: Biotechnology of Perennial Fruit Crops. CAB International, Wallingford, pp 303–321

Meng R and Finn CE (2002) Determining ploidy level and nuclear DNA content in *Rubus* by flow cytometry. J Amer Soc Hort Sci 127: 223–227

Millan-Mendoza B and Graham J (1999) Organogenesis and micropropagation in red raspberry using forchlorfenuron (CPPU). J Hort Sci Biotechnol 74: 219–223

Moyer RA, Hummer KE, Finn CE, Frei B and Wrolstad RE (2002) Anthocyanins, phenolics, and antioxidant capacity in diverse small fruits: *Vaccinium*, *Rubus*, and *Ribes*. J Agri Food Chem 50: 519–525

Ooka H, Satoh K, Doi K, Nagata T, Otomo Y, Murakami K, Matsubara K, Osato N, Kawai J, Carninci P, Hayashizaki Y, Suzuki K, Kojima K, Takahara Y, Yamamoto K and Kikuchi (2003) Comprehensive analysis of NAC family genes in *Oryza sativa* and *Arabidopsis thaliana*. DNA Res 10: 239–247

Pattison JA and Weber CA (2003) Molecular genetic analysis of phytophthora root rot resistance in red raspberry (*Rubus idaeus* L.). Hortscience 38: 711

Pattison JA and Weber CA (2005) Evaluation of red raspberry cultivars for resistance to *Phytophthora* root rot. J Am Pomolog Soc 59: 50–56

Pattison JA, Samuelian SK and Weber CA (2007) Inheritance of Phytopthora root rot resistance in red raspberry determined by generation means and molecular linkage analysis. Theor Appl Genet 115: 225–236

Perkins-Veazie P and Nonnecke G (1992) Physiological changes during ripening of raspberry fruit. HortSci 27: 331–333

Puska P, Vartiainen E, Korhonen HJ, Kartovaara L, Berg M-A, Pietinen A, Nissinen P and Tuomilehto J (1990) The North Karelia Project: results of a major national demonstration project on CHD prevention in Finland since 1972. Atheroscler Rev 21: 109–117

Rupasinghe HPV and Clegg S (2007) Total antioxidant capacity and total phenolic content mineral elements and histamine concentrations in wine from different fruit sources. J Food Comp Anal 20: 133–137

Rusu AR, Pamfil D, Graham J, Smith K, Balteanu VA, Groza Gh, Bondrea I and Patrascu, B (2006a). Simple sequence repeat SSR markers used in Rubus species from Romanian flora and north-European and north-American rubus cultivars (*Rubus ideaeus*). USAMV-CN 63

Rusu AR, Pamfil D and Graham J (2006b) Mapping resistance of red raspberry (*Rubus idaeus* subsp. *idaeus*) to viral diseases – Leaf spot (RLSV) and Vein chlorosis (RVCV) on the genetic linkage map. Seria Zootehnie si Biotehnologii 62: 318

Roach FA (1985) Cultivated Fruits of Britain: Their Origin and History. Blackwell, Oxford

Robertson KR (1974) The genera Rosaceae in the southeastern United States. J Arnold Arboretum 55: 352–360

Ryabova D (2007) Population evaluation in crop wild relatives for in situ conservation: a case study for raspberry *Rubus idaeus* L in the Leningrad region of Russia. Gen Res Crop Evol 54: 973–986

Safley CD, Boldea O and Fernandez GE (2006) Estimated costs PF producing, harvesting and marketing blackberries in the southeastern United States. HortTechnology 16: 109–117

Sargent DJ, Fernandez F, Rys A, Knight VH, Simpson DW and Tobutt KR (2007) Mapping A1 conferring resistance to the aphid *Amphorophora idaei* and dw (dwarfing habit) in red raspberry using AFLP and microsatellite markers. BMC Plant Biol 7(15)

Seemüller E, Duncan JM, Kennedy DM and Riedel M (1986). *Phytophthora* sp. als Ursache einer Wurzelfäule an Himbeere. Nachrichenblatt des Deutschen Pflanzenschutzdienstes 38: 17–21.

Skirvin RM, Motoike S, Coyner M and Norton MA (2005) *Rubus* spp. cane fruit. In: Biotechnology of Fruit and Nut Crops. CABI Publishing, Wallingford

Stafne ET and Clark JR (2004) Genetic relatedness among eastern North American blackberry cultivars based on pedigree analysis. Euphytica 139: 95–104

Stafne ET, Clark JR, Weber CA, Graham J and Lewers KS (2005) Simple sequence repeat markers for genetic mapping of raspberry and blackberry. J Am Soc Hort Sci 130: 722–728

Stewart D, McDougall GJ, Sungurtas J, Verrall S, Graham J and Martinussen I (2007) Metabolomic approaches to identifying bioactive compounds in berries: advances towards fruit nutritional enhancement. Mol Nutr Food Res 51: 645–651

Weber CA (2003) Genetic diversity in black raspberry detected by RAPD markers. HortScience 38 : 269–272

Wilcox WF, Scott P H, Hamm PB, Kennedy DM, Duncan JM, Brasier CM and Hansen, EM (1993) Identity of a *Phytophthora* species attacking raspberry in Europe and North America. Mycol Res 97: 817–831

Wilcox WF and Latorre BA (2002) Identities and geographic distributions of *Phytophthora* spp. causing root rot of red raspberry in Chile. Plant Dis 86: 1357–1362

Williamson B and Jennings DL (1992) Resistance to cane and foliar diseases in red raspberry (*Rubus idaeus*) and related species. Euphytica 63: 59–70

Woodford JAT, Williamson B and Gordon SC (2002) Raspberry beetle damage decreases shelf-life of raspberries also infected with *Botrytis cinerea*. Acta Hort 585: 423–427

Woodhead M, Davies HV, Brennan RM and Taylor, MA (1998) The isolation of genomic DNA from blackcurrant (*Ribes nigrum* L.). Mol Biotechnol 9: 243–246

Woodhead M, McCallum S, Smith S, Cardle L Mazzitelli L and Graham J (2008) Identification, characterisation and mapping of simple sequence repeat (SSR) markers from raspberry root and bud ESTs. Mol Breed 22: 555–563

Wu X, Beecher GR, Holden JM, Haytowitz DB, Gebhardt SE and Prior RL (2006) Concentrations of anthocyanins in common foods in the United States and estimation of normal consumption. J Agric Food Chem 54: 4069–4075

Zheng D, Schroder G, Schroder J and Hrazdina G (2001) Molecular and biochemical characterization of three aromatic polyketide synthase genes from *Rubus idaeus*. Plant Mol Biol 46: 1–15

Zheng D and Hrazdina G (2008) Molecular and biochemical characterization of benzalacetone synthase and chalcone synthase genes and their proteins from raspberry (*Rubus idaeus* L). Arch Biochem Biophys 470: 139–145

# 25. Loquat (*Eriobotrya* Lindl.)

**Maria L. Badenes, Shunquan Lin, Xianghui Yang, Chengming Liu, and Xuming Huang**

## 1 Origin and Botany

Loquat (*Eriobotrya japonica* Lindl., Rosaceae, Maloideae) is a subtropical evergreen fruit tree that blooms in fall and early winter. Records on loquat in China span over 2,000 years. Loquat species are native to the South East of China (Fig. 1). The medium and low region of Dahube river and South East of the Gongga Montains

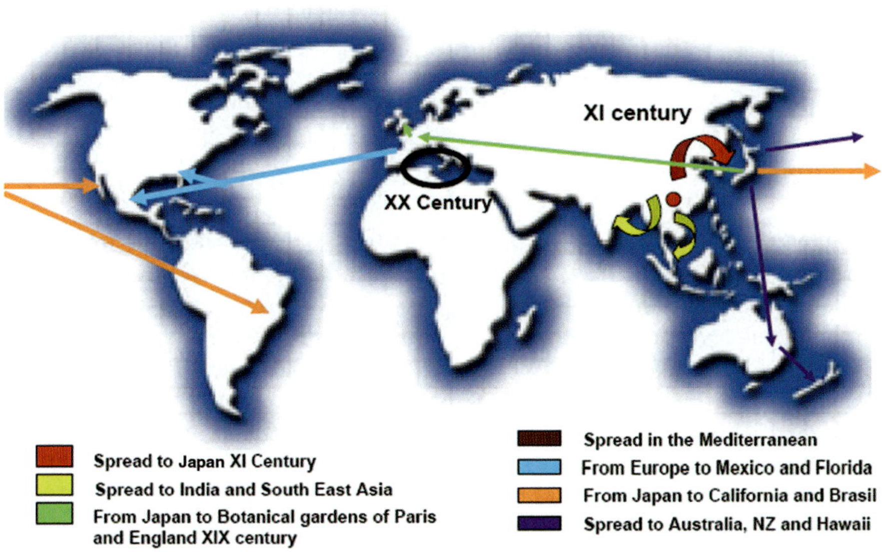


**Fig. 1** Origin and spread of loquat species

M.L. Badenes (✉)
Fruit Breeding Department, Instituto Valenciano de Investigaciones Agrarias, Apartado Oficial 46113, Moncada, Valencia, Spain

K.M. Folta, S.E. Gardiner (eds.), *Genetics and Genomics of Rosaceae*, Plant Genetics and Genomics: Crops and Models 6, DOI 10.1007/978-0-387-77491-6_25,

are considered the primary center of origin for *Eriobotrya japonica* (Zhang et al., 1990), with the Yunnan region as a secondary center (Yang et al., 2005).

The loquat cultivated in Japan was introduced from China in ancient times and its cultivation in Japan was described as early as 1180 (Ichinose, 1995). In 1784, the loquat was introduced from Guangdong into the National Garden at Paris, and in 1787 it was introduced into the Royal Botanical Gardens at Kew, England. Thereafter, loquat was distributed around the Mediterranean countries, including Algeria, Cyprus, Egypt, Greece, Israel, Italy, Spain, Tunisia and Turkey. Sometime between 1867 and 1870, loquat was introduced to Florida from Europe, and to California from Japan. Chinese immigrants are assumed to have carried the loquat to Hawaii (Morton, 1987). Cultivation spread to India and Southeastern Asia, the East Indies, Australia, New Zealand, Madagascar, and South Africa. Loquats are now distributed in many Asian countries, for example, Laos, Nepal, Pakistan, South Korea, and Vietnam; in Armenia, Azerbaiian and Georgia; and in the Americas, including Argentina, Brazil, Chile, the mountains of Ecuador, Guatemala, Mexico, and Venezuela. Generally, loquats are found between latitudes 20 and 35° North or South. but can be cultivated up to latitude 45° under marine climates (Lin et al., 1999). Currently, the crop is grown in temperate and subtropical areas, sharing the same environmental areas as citrus.

# 2 Genetic Resources and Diversity

## *2.1 Studies According to Pomological Characters*

Loquat was described first time as a species in 1784 by Thunberg, who included it in the *Mespilus* genus. In 1822, the English botanist, John Lindley, named the genus *Eriobotrya* from Greek, *erio*-, wool, and *botrys*, a cluster, referring to the woolly, clustered panicles (Huxley, 1992). *Eriobotrya* is often confused with *Mespilus*, and sometimes with *Crataegus* and *Photinia*. The number of loquat species is disputed and opinions vary among authors. However, it is known that there are at least 22 species and 10 varieties or *forma*. Most of them originated in Southern China, and the remainder in Southeastern Asia (Lin, 2007 ). The taxonomy of genus *Eriobotrya* is also still under dispute, one system is based on the lower surface of young leaves having hair or not, another is based on blooming from late autumn to early winter, or spring, unfortunately neither is accepted widely.

Only *E. japonica* is cultivated for its fruit. *Eriobotyra deflexa* and *E. prinoides* have been used as rootstocks, but they are less widely used than *Photinia serrulata* Lindl. in China and *Cydonia, Malus, Pyrus,* and *Pyracantha* in Mediterranean regions. Because more than 95% of loquats are grafted on *E. japonica* seedlings, it is clear that other species of *Eriobotrya* could be used as rootstocks and this approach is under evaluation with 10 species at South China Agricultural University. *Eriobotrya cavaleriei* Rehd. has been used to make wine. *E. prinoides* Rehd. & Wils.

has been used for Chinese medicinals and other species are under evaluation for this purpose.

A summary of the main cultivars grown in each country can be found in Badenes et al. (2006) and Llácer et al. (2004).The self-incompatibility of many cultivars (Carrera et al., 2007)) and propagation from seed has provided a range of cultivars optimal for different cropping areas. In all areas established, loquat species exist as diverse genetic lines, and germplasm collections have been established in China, Japan and Spain (Zheng, 2007; Lin, 2004; Llácer et al., 2004; Terai, 2002). Species from China possess great diversity; more than any other location. *E. japonica* has been grown in 10 different regions on China, and there are more than 1000 accessions described in the various Chinese germplasm collections (Zheng, 2007). For example, there are 83 cultivars (or selections) of *E. japonica* in Zhejiang, 78 in Fujian, 57 in Jiangsu, 31 in Anhui, 18 in Guangdong, and 9 in Sichuan (Zhang et al., 1990). The largest collection of germplasm, more than 250 cultivars, is located in Fuzhou, however less than 50 cultivars are widely cultivated in China.

A number of these cultivars have been classified as either "whitish flesh" or "orange" in China, the former make up 30 percent of the number of total cultivars in China (Ding et al., 1995), and some whitish flesh cultivars, such as 'Zhaozhong' and 'Baiyu', are the leading cultivars in Jiangsu province. Among orange cultivars, loquats have formed different ecological types in various zones during the long course of their cultivation and acclimatization. Ecotypes in China can be divided into two cultivar groups: the north subtropical cultivar group (NSCG) and the south subtropical cultivar group (SSCG) (Ding et al., 1995). Hence, cultivars in China are divided into three groups, namely, whitish group, north subtropical group, and south subtropical group.

Most cultivars cultivated in Japan belong to the north subtropical group. Several cultivars, such as 'ShiroMogi', could be placed in the whitish-flesh group. Three cultivars ('Mogi', 'Tanaka', and 'Nakasakiwase) account for 95% of the total crop area. Spanish commercial production depends on only four cultivars: 'Algerie', 'Magdal', 'Golden Nugget' (from USA), and 'Tanaka' (from Japan) with 'Algerie' being the main cultivar.

Assembly of germplasm and diversity studies based on pomological traits have also been carried out in the Spanish collection (Badenes et al., 2000; Leguizamón and Badenes, 2003; Martínez-Calvo et al., 2006). Separate genetic resources reside in Japan, where the crop was introduced more than 1000 years ago the collections are maintained at the National Institute of Fruit Tree Science (http://www.fruit.affrc.go.jp). Most cultivars come from local selection, as only a few cultivars grown in Japan or introduced from Japan in other countries are derived from breeding. Ranking in third place in terms of diversity are the European Mediterranean countries. The European germplasm collection is held at Instituto Valenciano de Investigaciones Agrarias (IVIA), in Spain. This collection was built from surveys made in the Mediterranean countries under the framework program GENRES, section of underutilized fruits (Badenes et al., 2004b).

However, the usage of loquat germplasm is very low. In Mediterranean areas, loquat is still referred to as an under-utilized fruit, and this may be due to under-development of germplasm resources. Several countries such as Spain, Japan, and China are now attaching importance to cultivar improvement. It is anticipated that genetic diversity of genus *Eriobotrya* will be more widely exploited in the near future.

Crop improvement is being carried out both by breeding programs and selection of accessions from germplasm resources. Breeding programs are located in China (Zheng, 2007; Shih, 2007), Japan (Terai, 2002) and recently one has been started in Spain. However, the varieties grown mostly come from selections made by growers in the cropping areas. Surveys and selection of accessions are being carried out in China (Zheng, 2007; Mediterranean countries (Llácer et al., 2004), Turkey (Polat, 2007; Karadeniz and Şenyurt, 2007), and Pakistan (Hussain et al., 2007).

## *2.2 Use of Molecular Markers for Diversity Studies*

### 2.2.1 Genetic Diversity in Cultivars and Allied Species

Genetic diversity is an important parameter utilized for fruit improvement, either by selection or application of various breeding methodologies. Information on genetic diversity is also valued for the management of germplasm and for evolving conservation strategies. Although DNA markers are generally considered the best tools for determining genetic diversity, as they are unlimited in number and independent of environmental interaction, and show high polymorphism, valuable insights have been gained from isoenzyme studies.

#### Isoenzymes

Degani and Cai (1987) were the first to use isoenzyme markers in loquat genetic studies, characterizing the major cultivars of *E. japonica* 'AKK01', 'AKK013' and 'Zikm', all cultivated in Israel.

Since 1990, when *E. prinoides* Rehd. & Wils. *var.dadunensis* H.Z. Zhang was identified and named, isozyme analysis has been used in studies concerning the genetic relationships among *E. japonica* (used for its fruit), *E. prinoides* Rehd. & Wils. *var. dadunensis* H.Z. Zhang and *E. prinoides* Rehd. & Wils. A combination of morphological observation of pollen grains under the scanning electric microscope and peroxidase isozyme analysis enabled the conclusion that *E. prinoides* Rehd. & Wils. *var.dadunensis* H.Z. Zhang, was a relatively independent population which was located between *E. prinoides* Rehd. & Wils. and *E. japonica* in evolutionary position and might be one of the ancestors of *E. japonica* (Zhang et al., 1990) .

In contrast, analyses of peroxidase isozyme profiles led Li et al. (1992) to draw different conclusions: that *E. prinoides* Rehd. & Wils., *E. prinoides* Rehd. & Wils. *var. dadunensis* H.Z. Zhang was an independent taxon, and *E. japonica* had originated from a hybrid of *E. prinoides* Rehd. & Wils. *var. dadunensis* H.Z.Zhang and *E. prinoides* Rehd. & Wils. Tang (1997) investigated the chromosomal karyotype and peroxidase isozymes and found that *E. prinoides* Rehd. & Wils. *var. dadunensis*

H.Z. Zhang have two chromosomal karyotypes, which were similar to *E. japonica* (3A) and *E. prinoides* Rehd. & Wils.(2A), and also concluded that *E. prinoides* Rehd. & Wils. *var. dadunensis* H.Z. Zhang was a hybrid of *E. japonica* and *E. prinoides* Rehd. & Wils.

However, Cai et al. (2004) in their initial study of 120 accessions from 4 species of *Eriobotrya* and a later study of 4 species and 1 variant of *Eriobotrya* with 12 allozymes, confirmed that *E. prinoides* Rehd. & Wils. *var. dadunensis* H.Z. Zhang was a relatively independent population located between *E. prinoides* Rehd. & Wils. in evolutionary position, serving as a link between *E. japonica* and the other members in *Eriobotyra* (Cai et al. (2005). The same authors extended this study with more accessions (Cai et al., 2007) and the final results supported the classification made by Zhang et al., (1990), agreeing with former classifications of the genus based on morphology and biological characters,

RAPDs

Vilanova et al. (2001), used RAPDs (Random Amplified polymorphic DNA) (Williams et al., 1990) to analyze the genetic diversity of loquat with accessions that included 33 loquat cultivars from Italy, Spain, Japan and Portugal The cluster analysis results were consistent with the geographical and genetic origins of the accessions. For example, 'Algerie', 'Cardona', 'Peluche', 'Buenet' and 'Marc' were grouped in the same cluster. They also pointed out that RAPD markers could be used to identify seedling cultivars that were derived from the same ancestor, but not sports. A larger group of accessions from surveys made in Spain, Italy and Greece was studied by Badenes et al. (2002), who concluded that the diversity was low due to the relatively recent introduction of loquat into Europe, and suggested the broadening the collection with more accessions from Asia. In China, RAPD markers have been used to study both taxonomic and phylogenetic relationships. Chen et al. (2003) analyzed the genetic relationships of *Eriobotyra* by UPGMA, and divided 11 loquat genetic resource accessions into 2 groups. Eight cultivars were grouped together and 3 wild accessions made up another group. Later, in 2007, they studied the genetic relationships and taxonomy of 65 representative loquat germplasm accessions. According to UPGMA cluster analysis of RAPD bands, 65 loquat plants were clustered two main groups. *E. prinoides* Rehd. & Wils., 'Guizhouyesheng', 'Taiwangpipa' and 'Hainanyesheng' belonged to the wild group and the other accessions to the cultivated group. This result was consistent with that of the traditional classification in the previous studies, however the classification of the cultivated group based on cluster analysis was somewhat different from the traditional classification (Chen et al., 2007).

### 2.2.2 Genetic Diversity in Wild Species

Most of the above mentioned studies on loquat have been performed on different cultivars of *E. japonica,* with few involving wild species. Since South China Agricultural University started to collect wild loquat resources in 2000, 19 wild loquats (including species, variants and *forma*) have been collected and conserved. Analysis of genetic diversity of species in genus *Eriobotrya* is being undertaken using molecular markers such as RAPDs, AFLPs and SSRs.

### RAPDs/AFLPs

Yang et al. (2007) studied the genetic relationships among *E. japonica*, *E. prinoides* Rehd. & Wils. *var. dadunensis* H.Z. Zhang. and *E. prinoides* Rehd. & Wils. using RAPD and AFLP marker analyses. They showed that *E. prinoides* Rehd. & Wils. *var .dadunensis* H.Z. Zhang might have originated from the hybridization of *E. japonica* and *E. prinoides* Rehd. & Wils., confirming the opinion of Tang (1997) based on isoenzyme markers (see above).

### ISSR

The phylogenetic relationships among 17 accessions of genus *Eriobotrya* from China analyzed by ISSR (Xie et al., 2007), indicated that the accessions could be clustered into 4 groups and that *E. kwangsiensis* Chun. was a new species.

### rDNA ITS Sequences

Phylogenetic relationships within *Eriobotrya* were established based on the rDNA ITS (ribosomal DNA internal transcribed spacers sequences). A phylogenetic tree of 15 loquat species was generated using *Photinieae* serrulaia L. and *Rhaphiolepis* indica (L.) Lindl. as outgroups. The result showed that *Eriobotrya* Lindl. species formed a monophyletic group, closely related to *Rhaphiolepis indica* rather than *Photinieae* serrulaia (Li et al., 2007b) (Fig. 2). In the consensus tree, the loquat (*Eriobotrya* Lindl.) was divided into two main clades. Seguin Loquat (*E. seguinii* Card) and Henry Loquat (E. *henryi* Nakai) formed a clade which may be the most primitive taxon in loquat (*Eriobotrya* Lindl.). The second clade was composed of the other taxa, which were divided into four groups: Tibet Loquat (*E.elliptica* Lindl.), Bengal Loquat (*E.bengalensis* Hook. f.), Narrowleaf Bengal Loquat (*forma angustifolia* Vidal) formed one group; Malipo Loquat (*E. malipoensis* Kuan), Daduhe Loquat (*var. dadunensis* H.Z.Zhang), Loquat (*E. japonica* Lindl.) another; Bigflow Loquat (*E. cavaleriei* Rehd) , Fragrant Loquat *(E. fragrans* Champ); Taiwan Loquat (E. *deflexa* Nakai ) Kokshun Loquat (*var. koshunensis* Nakai) and Salwin Loquat (*E. salwinensis* Hand-Mazz the third group; and the final group included Taiwan Loquat (*E. deflexa* Nakai) and Kokshun Loquat (*var.koshunensis* Nakai), which may be the most recent species of loquat ( *Eriobotrya* Lindl.).

### SSRs

SSR (Simple Sequence Repeat) markers (Rafalski et al., 1996) have also been used for diversity studies. Badenes et al. (2004a) used sequences of SSRs cloned from *Malus* species to study a set of loquat accessions and found a high degree

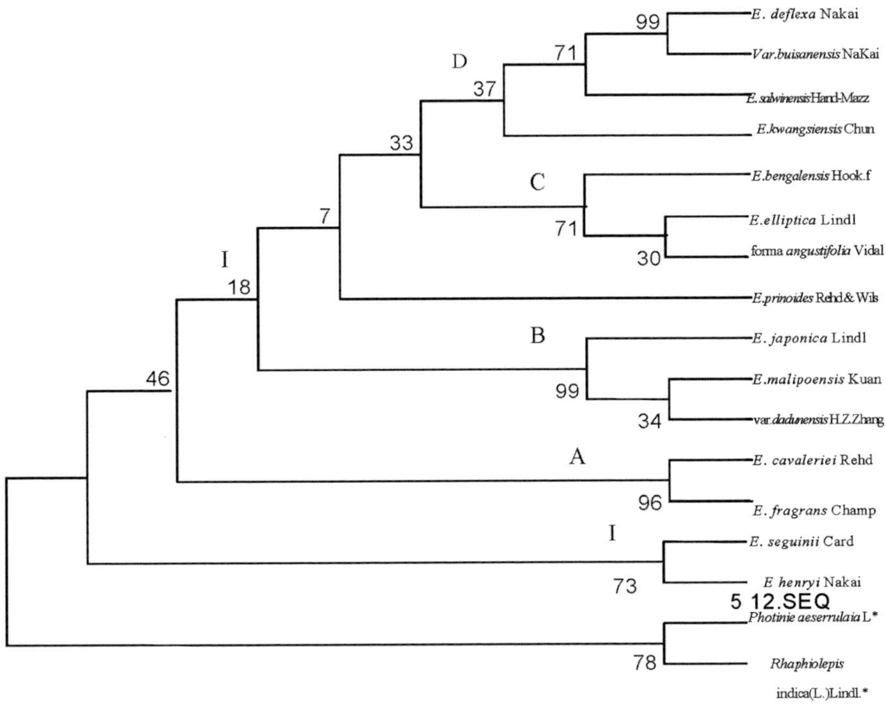


**Fig. 2** Majority-rule consensus trees based on ITS sequences. Numbers above branches are bootstrap percentages after 1000 replications. Asterisk indicates outgroups. (Figure from Li et al., 2007b)

of transferability between the *Malus* and *Eriobotrya* genera. This transferability was confirmed by Soriano et al. (2005), who used 30 SSR sequences from *Malus* (Guilford et al., 1997; Gianfranceschi et al., 1998) to study a group of 40 loquat accessions and found that 43% of sequences from *Malus* amplified in *Eriobotrya.* Thirty-nine polymorphic alleles allowed the identification of origin and genetic distance among accessions.

In general, application of molecular markers in studies of loquat genetic diversity and genetic relationships is still at its initial stage. Compared with other fruit crops, relevant studies in loquat lag behind, and should be strengthened so that molecular markers play an more important role in loquat breeding.

### 2.2.3 Cultivar Identification

The traditional methods of identifying fruit cultivars depend on botanical, morphological, palynological and biochemical traits including isoenzymes etc. In recent years, DNA based molecular markers have been widely used in identification of fruit varieties. Pan et al. (2002) identified loquat cultivars using the RAPD technique, in which two primers screened (S255, S232) distinguished 16 cultivars. Fukuda (2002) also reported that 108 polymorphic bands generated by 28 random primers could identify 69 cultivars from Japan, China, Israel, Greece, Mexico and America. RAPDs have also been used to elucidate the identity of cultivars in Chinese collections (He et al., 2007).

Soriano et al. (2005) investigated the genetic relationships among 40 loquat accessions that originated from different countries, and were part of the germplasm collection at IVIA. Thirty primer pairs of flanking microsatellites previously identified in *Malus* × *domestica* (Borkh.) were tested for amplification in loquat. Thirteen of them amplified polymorphic products and unambiguously distinguished 34 genotypes within the 40 accessions analyzed. Six accessions showing identical marker patterns were Spanish local varieties thought to have been derived from 'Algerie' by a mutational process very common in loquat species. Using SSRs and AFLPs, Feng et al. (2007) confirmed that 'Ninghaibai' was a new white flesh cultivar.

### 2.2.4 Pedigree Analysis

In fruit crops, there are many natural hybrids. Some cultivars are derived from seedling selection, some from deliberate crosses, and others provide useful techniques for pedigree analysis of fruit crops.

'Zaozhong 6' is the first hybrid loquat released in China and is cultivated widely across China. Chen et al. (2004) used RAPD markers to confirm that 'Zaozhong 6' was a hybrid of 'Jiefangzhong' (a Chinese cultivar) with 'Moriowase' (a Japanese cultivar). 'Chuannong No.1' is a new variety developed recently, but has an unknown pedigree. Its pedigree analysis was carried out by Luo et al. (2007) using the RAPD technique. The results showed that 'Longquan No. 5 (77-1)' was likely to be the parent of 'Chuannong No. 1'.

## 3 Genetic Mapping

Gisbert et al. (2009) developed microsatellite sequences from *Eriobotrya japonica* cultivar 'Algerie' and screened them across a set of 20 accessions that included plant material from Asian countries. These results confirmed that SSRs from loquat, along with sequences from *Malus* and *Pyrus* provide enough markers for genetic studies of loquat. Currently, 27 SSRs from loquat and 163 SSRs from *Malus* have been used to construct the first genetic linkage map of loquat shown in Fig. 3 (Gisbert et al., 2009). The markers from *Malus* have been used as anchor loci for establishing synteny with the maps of other members of the Rosaceae.

In addition, construction of a molecular genetic map of loquat using SRAP (sequence related amplified polymorphism) markers is being carried out at South China Agricultural University.

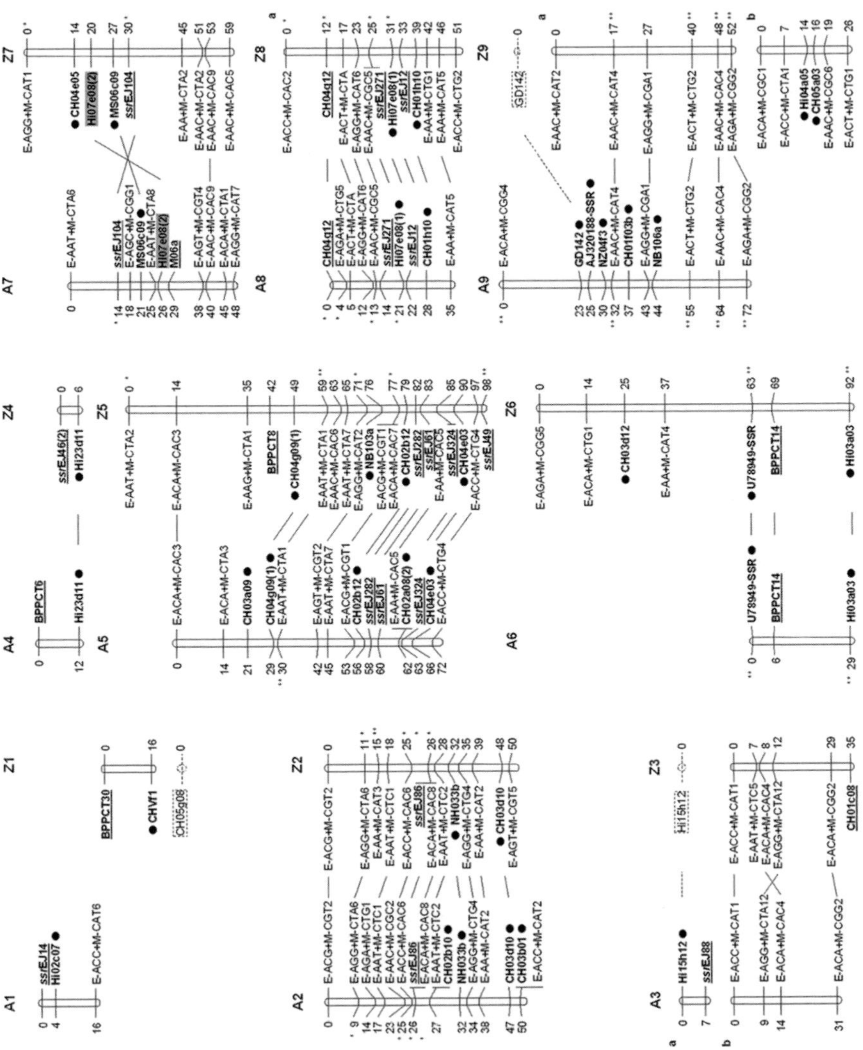


**Fig. 3** (continued)

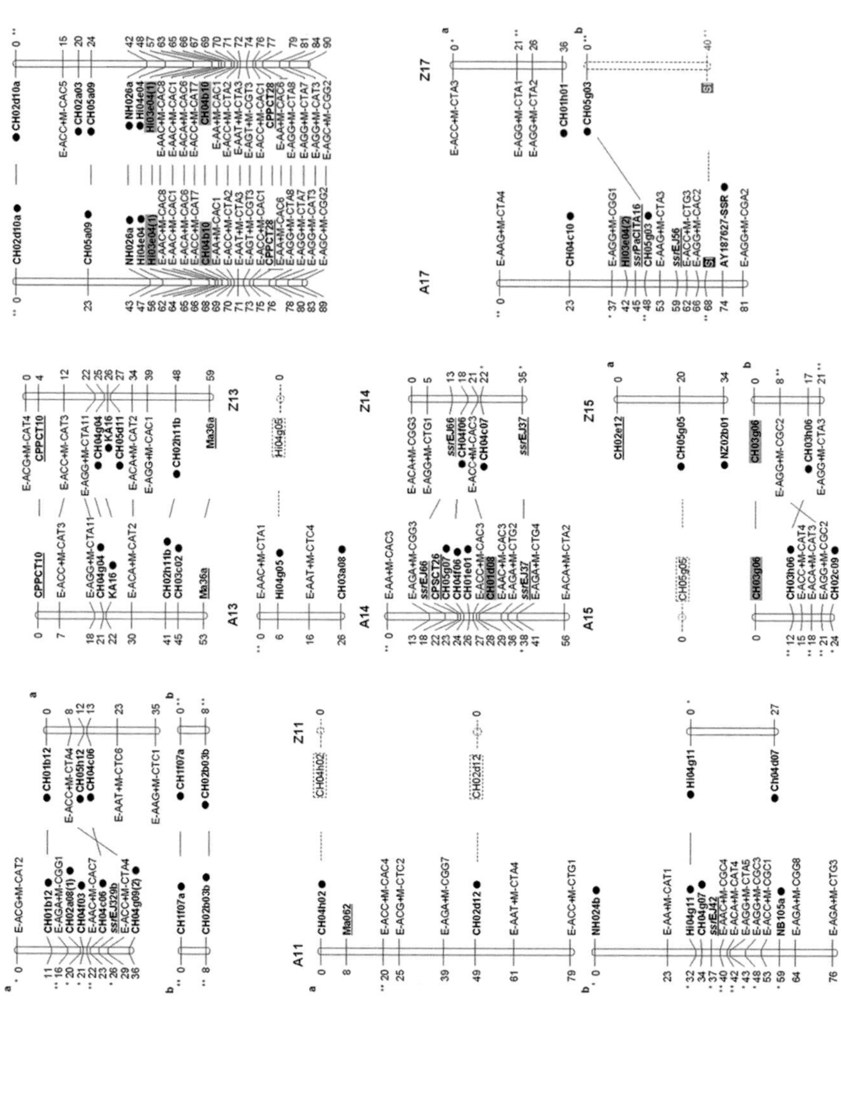


**Fig. 3** (continued) Molecular linkage maps obtained from the 'Algerie' and 'Zaozhong-6' Linkage group numbers are according to Liebhard et al. (2002). Figure from Gisbert et al., (2009)

# 4 Biotechnology Techniques Applied in Loquat Breeding

## 4.1 Genetic Transformation

Experiments are under way to explore shortening of the loquat juvenile period (Song et al., 2007) using the *LEAFY* gene. Such efforts may speed breeding efforts and possibly affect the physiology of dormancy and fruit yield. *LEAFY* is one of the key genes expressed early in plant floral meristem differentiation and it has been shown that the transcriptional factor expressed by *LEAFY* gene can activate its downstream gene *AP1* in model-plants such as *Arabidopsis thaliana* (Weigel and Nilsson, 1995), rice (He, et al., 2000), tobacco (Kelly et al., 1995). The gene *ejLFY* related to floral formation was isolated from loquat and used to transform leaf discs of tobacco plants mediated by *Agrobacterium tumefaciens*. Transgenic plants of loquat carrying this gene are not available yet. Efforts in this direction are aimed at speeding breeding of loquat and studies of a possible effect on physiology of dormancy and fruit yield.

## 4.2 Use of Polyploidy in Breeding

The production of seedless fruits is an important breeding objective in loquat. Several techniques that use the sterility produced by polyploidy are under study. Direct embryogenesis from anther culture of loquat (*Eriobotrya japonica* L.) has been investigated for its application in obtaining polyploid genotypes with different levels of ploidy (Li et al., 2007a). Occurrence of natural polyploidy has been described as another interesting source of variability for breeding (Guo et al., 2007)

## 4.3 Research on Loquat Breeding by Mutagenesis

Jiang et al., (2007) studied the effect of different doses of Co $\gamma$ ray irradiation on loquat shoots. The plant material showed a strong sensitivity to $\gamma$ ray irradiation, and the mutagenesis effect was significant. The mutants followed a normal distribution, suggesting irradiation is a useful method to increase variability and that selection of favorable mutants could be a good breeding strategy.

**Acknowledgments** The loquat research at IVIA is funded by the grant INIA RF2004-017. Support for Lin et al. was provided by...

# References

Badenes, M.L., Canyamás, T., Romero, C., Soriano, J.M. and Llácer, G. 2004a. Genetic diversity of an European collection of loquat based on RAPD and SSR molecular markers. Options Méditerranéennes, 58: 53–56.

Badenes, M.L., Canyamás, T.P., Romero, C., Giordani, E., Martínez-Calvo, J. and Llácer, G. 2004b. Characterization of underutilized fruits by molecular markers, loquat a case of study. Genetic Resources and Crop Evolution, 51: 335–341.

Badenes, M.L., Canyamás, T.P., Romero, C., Soriano, J.M., Martínez-Calvo J. and Yacer, G. 2002. Genetic diversity of an European collection of loquat. Acta Horticulturae, 620: 169–174.

Badenes, M.L., Castro, M., Gisbert, A.D., Martínez-Calvo, J. and Yacer, G. 2006. Clasificación botánica, variedades y patrones. (27–48). In: El Cultivo del Níspero Japonés. Agustí, M., Reig, C. and Undurraga P. (eds). Universidad Politécnica de Valencia (España) y Pontificia Universidad Católica de Valparaiso (Chile). 302 pp.

Badenes, M.L., Martínez-Calvo, J. and Llácer, G. 2000. Analysis of a germplasm collection of loquat (Eriobotrya japonica Lindl.) Euphytica, 114, 187–194.

Cai, L.H., Huang, H.W., Deng, X.X. and Zhang, W.C. 2004. Allozyme analysis of genetic diversity, genetic structure and interspecific relationship in genus *Eriobotrya*. Options Méditerranéennes, 58: 31–40.

Cai, L.H., Li, Z.Z., Huang, H.W. and Zhang, W.C. 2005. Genetic changes of isozyme in Eriobotrya and DNA fingerprinting of different cultivars. Journal of Wuhan Botanical Research, 23:. 406–416

Cai, L.H., Yan, F.H., Luo, Z.R. and Huang, H.W. 2007. Allozyme analysis of interspecific relationships and cultivar identification in Eriobotrya. Acta Horticulturae, (ISHS) 750:215–220.

Carrera, L., Santol, J., Herrero, M. and Hormaza, J.I. 2007. Characterization of S alleles and sestablishment of incompatibility groups in Eryobotrya japonica Lindl. Proceedings of the XII Eucarpia Symposium of Fruit Breeding Section. Zaragoza September 16–19. Acta Horticulturae in press

Chen, Y.T., Lai, Z.X. and Chen, J.Y. 2003. RAPD analysis of main species and cultivars in Eriobotrya. Acta Agriculturae Universitis Jiangxiensis, 25(2): 258–261

Chen, Y.T., Lai, Z.X. and Chen, J.Y. 2004. RAPD analysis of genetic relationship among loquat cultivars Zaozhong 6, Jiefangzhong and Moriowase. Journal of Fujian Agricultural University (Natural Science), 33(1):46–50

Chen, Y.T., Lai, Z.X. and Chen, J.Y. 2007. RAPD analysis of 65 accessions of loquat germplasm (Eriobotrya spp.). Chinese Journal of Tropical Crops, 28 (1):35–71

Degani, C. and Cai, W.J. 1987. The application for isozyme analysis for identification of varieties in Loquat. Subtropical Plant Science, 2:7–11

Ding, C.K, Chen, Q.F., Sun, T.L. and Xia, Q.Z. 1995. Gemplasm resources and breeding of Eriobotrya japonca Lindl. in China, Acta Horticulturae, 403:121–126.

Feng, J.J., Liu, Q. and Wang, X.D. 2007. Characterization of a new loquat cultivar 'Ninghaibai'. Acta Horticulturae, 750: 117–124

Fukuda, S. 2002. Cultivar identification in loquat assessed by RAPD analysis. Journal of the Japanese Society for Horticultural Science 71:826–828

Gianfranceschi, L., Seglias, N. and Tarchini, R. 1998. Simple sequence repeats for the genetic analysis of apple. Theoretical and Applied Genetics, 96:1069–1076

Gisbert, A.D., Martinez-Calvo, J., Llácer, G., Badenes, M.L. and Romero, C. 2009. Development of two loquat [*Eriobotrya japonica* (Thunb.) Lindl.] linkage maps based on AFLPs and SSR markers from different Rosaceae species. Molecular Breeding, DOI 10.1007/ s11032-008-9253-8

Guilford, P., Prakash, S., Zhu, J.M., Rikkerink, E., Gardiner, S., Bassett, H. and Forster, R. 1997. Microsatellites in Malus x domestica (apple): abundance, polymorphism and cultivar identification. Theoretical and Applied Genetics, 94:249–254.

Guo, Q.G., Li, X.L., Xing, W.W., He, Q. and Liang, G.L. 2007. Occurrence of natural triploids in loquat. Acta Horticulturae. (ISHS) 750:125–128.

He, Q., Zhao, X.C., Guo, Q.G., Li, X.L. and Liang, G.L. 2007. Molecular identification of 5 loquat cultivars. Acta Horticulturae. (ISHS), 750:155–158.

He, Z.H, Zhu, Q. and Dabi, T. 2000.Transformation of rice with the Arobidopsis floral regulator *LEAFY* cause early heading. Transgenic Research, 9:223–227

Hussain, A., Abbasi, N.A. and Akhtar, A. 2007. Fruit characteristics of different loquat genotypes cultivated in Pakistan. Acta Horticulturae. (ISHS), 750:287–292.

Huxley, A. (ed.) 1992. The New Royal Horticultual Society Dictionary of Gardening. Macmillan, London, p. 196.

Ichinose, L. 1995. The origin and development of loquat (in Japanese), Series of Agricultural Technology, 4 (Supp):l–5.

Jiang, J.M., Chen, X.P., Xu, J.H., Liu, Y.J., Liu, H.Y. and Zheng, S.Q. 2007. Irradiation mutagenesis in loquat: fruit trait mutations. Acta Horticulturae. (ISHS), 750:249–252.

Karadeniz, T. and Şenyurt, M. 2007. Pomological characterization of loquat selections of the black sea region of Turkey. Acta Horticulturae. (ISHS), 750:113–116.

Kelly, A.J., Bonnlander, M.B. and Meeks-Wagner, D.R. 1995.The tobacco homolog of *FLO-RICAULA* and *LEAFY* is transcriptionally expressed in both vegetative and floral meristems. Plant Cell, 7:225–234.

Leguizamón J. and Badenes, M.L. 2003. Multivariate analysis as a tool for germplasm studies, example of analysis of germplasm loquat data. International Workshop on Characterization of Genetic Resources of Temperate Zone Fruits for the Tropics and Subtropics. Acta Hor., 606: 29–35.

Li, J.Q., Hou, C.X., Luo, N., Deng, Q.X. and Wang, Y.Q. 2007a. Direct embyogenesis from anther culture of loquat. Acta Horticulturae. (ISHS) 750:209–214.

Li, P., Lin, S.Q., Hu, G.B. and He, Y.H. 2007b. A Preliminarily Phylogeny Study of the Eriobotrya based on the ITS sequences. Acta Horticulturae, 750:241–244.

Li, X.L. Jiang, L.G. and Jiang, J.G. 1992. Studies on the relationship of the genus Eriobotrya in Sichuan. Journal of Southwest Agricultural University, 14: 539–542

Liebhard, R., Gianfranceschi, L., Koller, B., Ryder, C.D., Tarchini, R., Van De Weg, E. and Gessler, C. 2002. Development and characterisation of 140 new microsatellites in apple (*Malus* × *domestica* Borkh.). Molecular Breeding 10:217–241.

Lin, S. 2004. Plant material of loquat in Asian countries. First International Symposium on Loquat, Valencia, Spain, April 2002. Options Méditerranéennes, 58:41–44.

Lin, S.Q. 2007. World loquat production and research with special reference to China. Acta Horticulturae 750:37–44.

Lin, S.Q., Sharpe, R.H. and Janick, J. 1999. Loquat: botany and horticulture. Horticultural Reviews, 23:233–276.

Llácer, G., Badenes, M.L. and Martínez-Calvo, J. 2004. Plant material of loquat in Mediterranean countries. First International Symposium on Loquat, Valencia, Spain, April 2002. Options Méditerranéennes, 58: 45–52

Luo, N., Fan, J.X., Li, J.Q., Deng, Q.X. and Wang, Y.Q. 2007. Pedigree analysis of 'Chunannong No. 1' loquat using RAPD molecular markers. Acta Horticulturae. 750: 143–147

Martínez-Calvo, J., Badenes, M.L. and Llácer, G. 2006. Nuevas variedades de níspero japonés del banco de germoplasma del IVIA. Monografías del INIA. Serie Agrícola.

Morton, J.F. (ed.). 1987. LoquaL. In: Fruits of Warm Climates, Creative Resource Systems, Winterville. FL, p, 103–108

Pan, X.F., Meng, X.X. and Cao, G.L. 2002. Application of RAPD-PCR technique for identification of varieties in Loquat. Journal of Fruit Science, 19:136–138

Polat, A.A. 2007. Selection studies on loquat growing in Bakras (Hatay), Turkey. Acta Horticulturae. (ISHS), 750:169–174.

Rafalski, J.A., Vogel, J.M., Morgante, M., Powell, W., Andre, C. and Tingey, S.V. 1996. Generating and using DNA markers in plants. In: Birren, B. and Lai, E. (eds) Non-mammalian genomic analysis. A practical guide. Academic Press, San Diego, pp 75–134

Shih, J.-.C. 2007. Loquat production in Taiwan. Acta Horticulturae. (ISHS), 750:55–60.

Song, H.W., Liu, Y.X., Lin, S.Q. and Hu, G.B. 2007. Transformation of tobacco with ejlfy from loquat and molecular identification of transgenic plants. Acta Horticulturae. (ISHS), 750: 233–240.

Soriano, J.M., Romero, C., Vilanova, S., Llácer, G. and Badenes M.L. 2005. Genetic diversity of loquat [Eriobotrya japonica (Thunb Lind)] assessed by SSR markers. Genome 48: 108–114.

Tang, B. 1997. A study of the sibship among Eriobotrya japonica Lindl. E. prinoides Rehd. & Wils. var. daduheensis H. Z. Zhang and E. prinoides Rehd. & wils.. Journal of Chongqing Teachers college (Natural Science Edition). 14: 18–25

Terai, O. 2002. Breeding loquat in Japan. First International Symposium on Loquat, Valencia, Spain, April 2002. Abstracts

Vilanova S., Badenes, M.L., Marinez-Calvo, J. and Placer, G. 2001. Analysis of loquat (Eriobotrya japonica Lindl) germplasm by RAPD molecular markers. Euphytica 121:25–29

Weigel, D. and Nilsson, O. 1995. A developmental switch sufficient for flower initiation in diverse plants. Nature, 377:495–500.

Williams, J.G.K., Kubelik, A.R., Livak, K.J., Rafalski, J.A. and Tingey, S.V. 1990. DNA polymorphism amplified by arbitrary primers are useful as genetic markers. Nucleic Acids Research, 18:6531–6535

Xie, J.H., Yang, X.H. and Lin, S.Q. 2007. Analysis of genetic relationships among Eriobotrya germplasm in China using ISSR markers. Acta Horticulturae. 750: 203–208

Yang, X.H., Glarpe, K., Lin, S.Q., Hu, Y.L., Ruan, S. and Liu, Y.X. 2005. Taxa of genus Eriobotrya plants around the world and specializing Southeastern Asia. Journal of Fruit Science 22: 55–59.

Yang, X.H., Liu, C.M. and Lin, S.Q. 2007. The genetic relationship among common loquat, Daduhe loquat and Oakleaf loquat, According to RAPD and AFLP analysis. Subtropical Plant Science, 36(2):9–12

Zhang, H.Z., Peng, S.A. and Cai, L.H. 1990. The germplasm resources of the genus Eriobotrya with special reference on the origin of E. japonica Lindl. Acta Horticulturae Sinica, 17:5–11

Zheng, S.Q. 2007. Achievement and prospect of loquat breeding in china. Acta Horticulturae. (ISHS) 750:85–92.

# 26. Genomics Tools Across Rosaceae Species

**Elisabeth Dirlewanger, Béatrice Denoyes-Rothan, Toshiya Yamamoto, and David Chagné**

## 1 Introduction – Toward a Rosaceae Genome as a Unique System?

Rosaceae geneticists and molecular biologists, whether focusing on ornamental, fruit crops or trees grown for their timber, have very similar challenges, which are to create varieties that can compete in a globalized market. Rosaceae breeders' objectives are to improve the product quality (e.g. improved fruit quality) and productivity (e.g. improved phenology). There is also a pressure from the growers in response to the consumer demand to develop pest and disease resistant cultivars, production of which will have a lesser impact on the environment and produce fruits with high health benefits. One could wonder if the processes underlying comparable characters found in several species throughout the Rosaceae family are also similar at the molecular level. The assumption of comparative genomics is, 'yes they are'. However such a hypothesis can be a risky statement because of the great diversity found within the Rosaceae family. For instance, the type of fruit varies among peach, strawberry and apple, varying from drupe, false fruit (the true fruit being the akene) and pome, respectively. Consequently, one could wonder how the genetic and molecular factors controlling a trait like fruit texture, for instance, could be conserved between all fruit types. Hence, one great challenge of the Rosaceae genomics community is to integrate their work and validate their findings across species and genera, in order to verify whether the comparative genomics hypothesis of existence of conserved candidate genes, Quantitative Trait Loci (QTLs) and major loci is true or not. Comparative genome mapping is the part of comparative genomics that aims at transferring genetic information, such as the position of major genes, QTL or candidate genes between species, in order to validate their position or assist analysis of less-studied species. Practically, comparative genome mapping determines the linkage between homologous genes of related

E. Dirlewanger (✉)
Institut National de la Recherche Agronomique, Centre de Bordeaux, Unité de Recherches sur les Espèces Fruitières, BP 81, 33883, Villenave d'Ornon Cedex, France

K.M. Folta, S.E. Gardiner (eds.), *Genetics and Genomics of Rosaceae*, Plant Genetics and Genomics: Crops and Models 6, DOI 10.1007/978-0-387-77491-6_26,

species and is often measured by the degree of synteny, that is the conservation of gene content, and colinearity, that is the conservation of the gene order between conserved genomic regions. Comparative genome mapping has recently been identified by the international Rosaceae research community as a key area which needs some substantial efforts in the near future (http://en.wikisource.org/wiki/International_Rosaceae_Whitepaper#FUNCTIONAL_GENOMICS). In addition to assessing the conservation of genes and the traits they control traits between species, comparative genome mapping is also a powerful tool for studying the genome evolution within the family, e.g. detection of chromosomal rearrangements (Schmidt, 2002).

However, compared to other groups of organisms, the understanding of the Rosaceae family genome synteny, in term of chromosomal rearrangements and identification of conserved regions still relies on few examples (Dirlewanger et al., 2004b; Yamamoto et al., 2004a; Arús et al., 2005; Denoyes-Rothan et al., 2006). Cereal crops are a good example of plant species for which efforts have been made in the 1990s to integrate all the research carried out in separate species (Causse et al., 1994; Da Silva et al., 1995). Cereal researchers have done so by using common tools between species, such as high density maps enriched in orthologous transferable markers and by developing a public online database (Gramene http://www.gramene.org/plant_ontology). This made it possible to identify conserved syntenic blocks between a wide range of species, and then strategies for the choice of the first species for whole genome sequencing (rice) (Kilian et al., 1995; Asnaghi et al., 2000). At the first glance, the grass model seems be attractive for the Rosaceae researchers and a few years ago, peach was chosen as the model species for Rosaceae (Horn et al., 2005; Shulaev et al., 2008). This choice is, at present, not so clear and arguments have been made for apple and also strawberry (Folta and Dhingra, 2006; Folta and Davis, 2006). The recent announcement of the whole genome sequencing of apple and peach (both due in 2008) makes these two species good candidates as reference species. However, these species lack the optimized reverse genetics tools that exist in strawberry (Oosumi et al., 2006; Folta et al., 2006). A consensus was recently reached by the Rosaceae community that there is no one specific model species for Rosaceae. Hence the Rosaceae model looks more like a complex system without a reference species that other species that gravitate around, and consisting of a more multifaced model with each species bringing some original information, thanks to specific resources developed in each (Shulaev et al., 2008). This underlines the requirement for comparative genomics, that will make this information useable for all Rosaceae, including those for which no genome sequence has been initiated. The identification of homologous chromosomes or chromosomal fragments between species of the whole family and the development of a large set of common orthologous markers transferable to a wide range of Rosaceae are desired resources. We will discuss the development of such tools in the following chapter.

# 2 The Genome Structure of Rosaceae

## *2.1 Chromosome Numbers and Genome Size*

Most of Rosaceae species have a chromosome number ranging from $x = 7$ to $x = 9$, except for species of the Maloideae subfamily which have a chromosome number of $x = 17$ (Table 1).

**Table 1** Mean 2C values and ploidy levels of Rosaceae species

| | DNA content (pg/2C) | Ploidy | Source |
|---|---|---|---|
| Spiraeoideae $x = 9$ | | | |
| *Physocarpus opulifolius (L.)* Maxim. | 0.42 | 2x | 1 |
| *Spiraea chinensis* Maxim. | 0.42 | | 1 |
| *Physocarpus bracteatus* (Rydb.) | | | |
| Rehd. | 0.43 | | 1 |
| *Spiraea crenata* L. | 0.46 | | 1 |
| *Stephanandra incisa* (Thunb.) Zabel | 0.53 | | 1 |
| *Neillia simensis* D. Oliver | 0.54 | 2x | 1 |
| *Spiraea pubescens* Turcz. | 0.94 | 2x | 1 |
| *Exochorda giraldii* Hesse $x = 8$ | 1.11 | 2x | 1 |
| *Spiraea wilsonii* Duthie | 1.59 | | 1 |
| *Spiraea nipponica* Maxim. | 1.75 | 2x | 1 |
| *Spiraea sargentiana* Rehd. | 1.84 | | 1 |
| Amygdaloideae $x = 8$ | | | |
| *Prunus persica* (L.) Batch. 'Madison' | 0.54 | 2x | 3 |
| *Prunus persica* (L.) Batch. 'Red Haven' | 0.55 | 2x | 1 |
| *Prunus armenaica* L. 'Sundrop' | 0.60 | 2x | 3 |
| *Prunus subhirtella* Miq. | 0.60 | 2x | 1 |
| *Prunus avium* (L.) L. 'Van' | 0.67 | 2x | 3 |
| *Osmaronia cerasiformis* $x = 6$ | 0.98 | 2x | 1 |
| *Prunus serotina* J.E. Ehrh. | 1.00 | 4x; 5x; 6x | 1 |
| *Prunus x* spp. '4X' | 1.36 | 4x | 3 |
| *Prunus cerasus* L. 'Montmorency' | 1.42 | 4x | 3 |
| *Prunus* x spp. 'Standley' | 1.83 | 6x | 3 |
| *Prinsepia uniflora* Batal. | 3.09 | 4x | 1 |
| Rosodeae | | | |
| *Rosa wichuraiana* Crepin $x = 7$ | 0.20 | 2x | 2 |
| *Rubus idaeus* L. $x = 7$ | 0.58 | 2x | 3 |
| *Rosa blanda* Aiton $x = 7$ | 0.60 | 3x | 2 |
| *Acaena magellanica* (Lam.) Vahl. $x = 7$ | 0.60 | 6x | 4 |
| *Potentilla fructicosa* L. | 0.80 | 2x; 4x; 6x | 1 |
| *Rhodotypus scandens* (Thunb.) Mak. $x = 9$ | 0.74 | 2x | 1 |
| *Rubus odoratus* L. $x = 7$ | 0.76 | 2x | 1 |
| *Neviusia alabamensis* A. Gray $x = 9$ | 1.02 | 2x | 1 |
| *Sanguisorba minor* Scop. $x = 7$ | 1.10 | 4x | 2 |
| *Aphanes arvensis* L. $x = 8$ | 1.10 | 6x | 5 |

**Table 1** (continued)

| | DNA content (pg/2C) | Ploidy | Source |
|---|---|---|---|
| *Dryas octopetala* L. $x = 9$ | 1.16 | 2x | 1 |
| *Rosa acicularis* Lindley $x = 7$ | 1.30 | 6x | 2 |
| *Rosa multiflora* Thunb. ex J. Murr. $x = 7$ | 1.65 | 2x; 4x | 1 |
| *Duchesnea indica* (Andr.) Focke $x = 7$ | 3.00 | 6x; 12x | 1 |
| *Fragaria vesca* $x = 7$ | 0.55 | 2x | 6 |
| *Fragaria moschata* $x = 7$ | 1.53 | 6x | 6 |
| *Fragaria ananassa* $x = 7$ | 1.85 | 8x | 6 |
| Maloideae x = 17 | | | |
| *Pyracantha* 'Royal' | 0.99 | | 1 |
| *Pyrus communis* L. 'Bartlett' | 1.11 | 2x | 3 |
| *Chaenomeles speciosa* (Sweet) Nakai | 1.20 | 2x | 1 |
| *Pyrus calleryana* Decne. | 1.26 | 2x | 1 |
| *Amelanchier* sp. | 1.31 | | 1 |
| *Sorbus alnifolia* (Siebold & Zucc.) Koch | 1.36 | 2x | 1 |
| *Sorbus americana* Marsh | 1.30 | 2x | 1 |
| *Pyracantha coccinea* E. J. Roem. | 1.41 | 2x | 1 |
| *Cydonia oblonga* Mill. | 1.45 | 2x | 1 |
| *Mespilus germanica* L. | 1.48 | 2x | 1 |
| *Eriobotrya japonica* (Thunb.) Lindl. | 1.54 | 2x | 1 |
| *Cotoneaster melanocarpa* Lodd. | 2.24 | 4x | 1 |
| *Photinea parvifolia* (E. Pritz.) Schneid. | 2.29 | | 1 |
| *Aronia arbutifolia* (L.) Pers. | 2.57 | 2x; 4x | 1 |
| *Crataegus crus-galli* L. | 2.71 | 3x; 4x | 1 |
| *Malus × domestica* (Borkh.) 'Gala' | 1.57 | 2x | 7 |
| *Malus ionensis* | 2.57 | 3x | 7 |
| *Malus floribunda* | 1.46 | 2x | 7 |
| *Arabidopsis thaliana* ($x = 5$) | 0.33–0.36 | 2x | 8 |

Table reproduced and updated from Dickson et al. (1992). References used: (1) Dickson et al. (1992); (2) Bennett and Smith (1991); (3) Arumuganathan and Earle (1991); (4) Bennett et al. (1982); (5) Bennett and Smith (1976); (6) Denoyes-Rothan, unpublished results; (7) Tatum et al. (2005); (8) Bennett et al. (2003).

Overall, Rosaceae species have small genomes compared to other families of angiosperms (Bennett and Leitch, 2004). Some species such as *Rosa wichuraiana* (0.2 pg/2C) maintain a DNA content (2C values, evaluated by flow cytometry) comparable to *Arabidopsis thaliana* (0.33–36 pg/2C) (Bennett et al., 2003) (Table 1). When values of subfamily are compared between diploid species, the Spiraeoideae subfamily has the smallest average genome size, closely followed by Amygdaloideae. In general, the range of variation of DNA content of Rosaceae is rather small compared to other taxa (Dickson et al., 1992). Not surprisingly, the range of variation within a genus and among species of similar level of ploidy is very small.

Nuclear DNA content in diploid Maloideae ranges from 1.11 pg/2C to 1.57 pg/2C, 2–3 times larger than species of the other subfamilies. This observation, in combination with the fact that the basic chromosome number for all of the genera is $x = 17$ (Sax, 1931, 1932; Kovanda, 1965), indicates that a duplication event occurred during the Maloideae genome evolution. A debate exists about the allo- or auto-polyploid origin of Maloideae. Recent molecular genetic studies denied the wide-hybridization hypothesis that the subfamily Maloideae originated from allopolyploidization between Amygdaloideae ($x = 8$) and Spiraeoideae ($x = 9$), and supported the autopolyploid origin of hybridization between closely related members of Spiraeoideae (Morgan et al., 1994; Evans and Campbell, 2002). Morgan et al. (1994) reported that phylogenetic analysis of nucleotide sequences of ribulose-1,5-bisphosphate carboxylase/oxygenase large subunit (*rbcL*) gene do not provide evidence for close relationships between Maloideae and Amygdaloideae, however do indicate a close relationship between Maloideae and some genera of Spiraeoideae. Sequence analysis of GBSSI (granule-bound starch synthase, waxy) genes has demonstrated that the subfamily Maloideae originated from an autopolyploidisation event involving only members of a lineage that contained the ancestors of *Gillenia*, in the subfamily Spiraeoideae (Evans and Campbell, 2002).

## *2.2 Ploidy Levels*

Polyploidy is widely spread in plants, where at least 30–80% of angiosperms have experienced one or more polyploidization events in their evolutionary history (Masterson, 1994; Wendel, 2000; Bennett, 2004; Keith and Wendel, 2005). A recent comparison of the whole genome sequences generated for four plant species (*Arabidopsis*, rice, poplar and grapevine) suggested the three dicotyledonous species, *Arabidopsis*, poplar and grapevine, had a common hexaploid ancestor (The French-Italian Public Consortium for Grapevine Genome Characterization, 2007). The evolution of genome structure and of associated rearrangements that are due to polyploidization can be studied through comparative genetic mapping between polyploid species and their diploid relatives. When high degrees of macrosynteny and colinearity are revealed, genetic information can then be transferred from diploids to polyploids (Sorrells, 1992). In the genus *Fragaria*, which displays species showing a range of ploidy level from diploid ($2x = 14$) to octoploid ($8x = 56$), high levels of macrosynteny and colinearity between diploid and octoploid species have been shown (Denoyes-Rothan et al., 2006). These result suggested that major chromosomal rearrangements have not been frequent in the evolution of *Fragaria* after polyploidisation events. However, the evident reduction of the genome size of the octoploid genome compared to the size expected if the octoploid genome was four times the size of the diploid one) (16% see values above) could be due to events which followed the origin of a polyploid, such as loss of DNA segments which were not detected by comparative mapping (reviewed in Osborn et al., 2003). Alternatively, this discrepancy prompts speculation that the diploid genomes originally contributed to the octoploid species were not of uniform size.

The Rosaceae family also includes genera that are considered as allopolyploid such as *Malus* and *Pyrus*, where triploid and tetraploid plants have been found (Sax, 1932; Kovanda, 1965).

# 3 Tools for Genomics Analyses in Rosaceae

Comparative genetic studies have demonstrated that molecular marker tools such as Simple Sequence Repeats (SSRs), developed in *Prunus* are applicable to other species in the genus (Lambert et al., 2004; Dirlewanger et al., 2004b). These results suggest that development of structural and functional genomics tools in one family member might greatly enhance genomics efforts for many, if not most of the other species in the Rosaceae.

## *3.1 Rosaceae Genomics Database Resources*

Centralized database and bioinformatics tools are critical for comparative genomics research in Rosaceae. The Genomics Database for the Rosaceae (GDR; http://www.rosaceae.org) serves as a communication hub within the scientific community and incorporates most of the family's publicly available structural and functional genomics knowledge. The GDR is the outcome of a US-based initiative for Rosaceae genomics (Jung et al., 2004, 2008). The database gives access to comprehensive and frequently updated data on genetic and physical maps, molecular markers and gene annotation, and facilitates communication among researchers (e.g. by posting conference announcements). The genetic linkage maps of various Rosaceae species including apple and pear (Maloideae), peach, almond, apricot and cherry (Prunoideae), strawberry, raspberry and rose (Rosoideae) are accessible and can be aligned using common markers.

The other public database for Rosaceae comparative genomics is ESTree, an Inter-university Centre for several research units in Italy devoted to the implementation of genomics and functional genomics in peach (Lazzari et al., 2005). The ESTree database (http://www.itb.cnr.it/estree) comprises a collection of *Prunus persica* and *Prunus amygdalus* expressed sequenced tags (EST) and is intended as a resource for functional genomics. Sequenced clones are derived from twelve peach libraries and three almond libraries produced in nine different labs. Libraries were prepared from various tissues (fruit mesocarp at different developmental stages, skin fruit, and shoot). Both databases are interconnected and constitute useful tools for genomic studies.

## *3.2 Structural Genomic Tools*

### 3.2.1 Physical Maps

Genome-wide physical maps have already proven to be powerful tools and infrastructures for advance genomics research and especially for map-based cloning of

Mendelian loci (Arondel et al., 1992) and QTL (Frary et al., 2003). In plants, physical maps have been constructed for a wide range of species, including *Arabidopsis thaliana* (Mozo et al., 1999), sorghum (Klein et al., 2000), rice (Chen et al., 2002), soybean (Wu et al., 2004) and poplar (Kelleher et al., 2007). In Rosaceae, progress has been made mainly on peach and apple. Efforts toward constructing a physical map and integrating the physical map with the linkage map in *Prunus*, almond, and peach, are well under way (Horn et al., 2005; Wang et al., 2002; Joobeur et al., 1998; Georgi et al., 2002; Jung et al., 2004; see also GDR website). More recently, an apple physical map was developed (Han et al., 2007). Both physical maps will play a critical role in advanced genomics research, including marker development in targeted chromosome regions, fine-mapping and isolation of genes and QTL, conducting comparative genomics analyses of plant chromosome and large-scale genomic sequencing.

In peach, the use of marker hybridization data from the TxE *Prunus* genetic map to the physical map has provided evidence of duplication of chromosomal regions within the peach genome (Abbott et al., 2007).

### 3.2.2 Molecular Markers

A large number of molecular markers of varying types have been developed for Rosaceae genetic map construction and represent an invaluable resource for comparative genome mapping. However, not all these markers are potential orthologous markers.

#### gDNA SSRs

Markers developed from genomic SSRs have been shown to be transferable between species belonging to the same genus and to the same subfamily (Dirlewanger et al., 2004b; Bao et al., 2007). Within the genus *Prunus,* the level of transportability is extremely high. For example, among 277 *Prunus* SSRs (141, 58, 31, 9, 4 and 6 from peach, apricot, almond, sweet cherry, sour cherry and Myrobolan plum, respectively), 95.3 % showed a PCR amplification in Myrobolan plum (Dirlewanger et al., 2004a). For this reason, gDNA SSR markers were intensively used to compare *Prunus* linkage maps (Dirlewanger et al., 2004b). Detailed comparisons were made using SSR markers in common SSR markers between the reference *Prunus* map TxE (Joobeur et al., 2000) and the genetic maps of *P. armeniaca* (Lambert et al., 2004), *P. davidiana* (Foulongne et al., 2003), *P. avium* (Dirlewanger et al., 2003) and *P. cerasifera* (Dirlewanger et al., 2004a).

In the Maloideae, genomic SSRs developed in Japanese pear (*Pyrus pyrifolia*) could be transferred into other species of the genus *Pyrus*, e.g. European pear (*P. communis*), Chinese pears (*P. ussuriensis, P. bretshneideri*) and *P. calleryana* (Kimura et al., 2002; Yamamoto et al., 2002; Bao et al., 2007). It was reported that SSR markers developed in apple produced discrete amplified fragments in several *Pyrus* genotypes, indicating that apple SSRs could also be applicable to pear (Yamamoto et al., 2001). Liebhard et al. (2002) reported that apple

SSRs successfully amplified in species of other Maloideae genera (*Amelanchier*, *Cotoneaster*, *Crataegus*, *Cydonia*, *Mespilus*, *Pyrus* and *Sorbus*). Cross-genus application of SSR markers developed in pear and apple was examined for quince (*Cydonia oblonga*). It was revealed that 77 out of 118 SSR markers producing one or more reproducible amplified bands could be used in quince, including 20 SSRs from pear and 57 SSRs from apple (Yamamoto et al., 2004b).

In Rosoideae, microsatellites have been mainly developed in *Fragaria* from species of different ploidy levels: from the cultivated octoploid strawberry, *F.* × *ananassa* (Davis et al., 2006; Rousseau-Gueutin et al., 2008), from the octoploid *F. virginiana* (Ashley et al., 2003), from three diploid species, *F. vesca* (James et al., 2003; Cipriani and Testolin, 2004; Hadonou et al., 2004), *F. viridis* (Sargent et al., 2003) and *F. nubicola* (Sargent et al., 2004). Genomic SSRs have also been developed in other Rosoideae genera, such as *Rubus* (Graham et al., 2002; Stafne et al., 2005; Lopes et al., 2006) and *Rosa* (Zhang et al., 2006; Hibrand-Saint Oyant et al., 2008). The transferability of genomic SSRs was reported to be high within the genus *Fragaria*, with 70–100% being transferable (Ashley et al., 2003; Davis et al., 2006; Rousseau-Gueutin et al., 2008), but only 20% could be transferred to other genera in the Rosoideae, such as *Rosa* (Rousseau-Gueutin, unpublished results) and *Rubus* (Stafne et al., 2005).

At the inter-sub-family level, gDNA SSRs transferability is very low, as been shown between *Prunus* and *Malus*. Cipriani et al. (1999) found that only 18% of peach SSRs could be amplified in apple. Similarly, Yamamoto et al. (2004b) observed that only about one-tenth of the *Prunus* SSRs could be transferred in to the genetic maps of 'Bartlett' and 'Housui' the genus *Pyrus*. Only one out of 15 apple SSR markers was transportable to Amygdaloideae (Liebhard et al., 2002).

All those results suggest that genomic SSRs are limited to comparison between closely related species, most likely due to the fact that they arise from non-coding regions with great variability between species. Therefore they are inadequate for map comparisons between species of two subfamilies.

### EST-Based Markers

The increase in availability of ESTs for Rosaceae has provided a valuable resource of molecular markers. For instance, SSR motifs were found in approximately 17% of apple ESTs (Newcomb et al., 2006). EST-based SSRs are supposed to be more transferable than the genomic ones as they are linked to more conserved expressed sequences. They have been obtained for several genera of Rosaceae: peach (www.rosaceae.org; strawberry (Gil-Ariza et al., 2006; Bassil et al., 2006a and 2006b; Lewers et al., 2005; Folta et al., 2005), *Rubus* (Lewers et al., 2008) and apple (Celton et al., 2008).

However, very little is known about their transferability within Rosaceae. Unpublished data showed that the SSR motif can be lost from one species to another one as observed between *Fragaria* and *Rosa* (Denoyes-Rothan, pers comm.) or between apple and peach (E. Rikkerink, personal communication). However some cross-species transferability has been observed between apple and pear (Celton et al., 2008).

Because of their abundance in all regions of the genome (i.e. to the order of one SNP every 100 bp), markers based on single nucleotide polymorphisms (SNPs) can be developed readily from ESTs. Bioinformatics tools make it possible to identify them in EST databases, as has been shown in apple (Newcomb et al., 2006). Although most SNPs are not conserved between species, their high abundance makes them attractive for the development of orthologous markers. Because of the high sequence homology found in ESTs developed from different plant species (Van der Hoeven et al., 2002), orthologous markers can be developed for virtually any gene. A set of EST-based SNPs detected in candidate genes involved in the production of anthocyanin pigments was mapped in apple (Chagné et al., 2007). This strategy was also chosen for the mapping of candidate genes involved in fruit quality in the ISAFRUIT Integrated European project (http://www.isafruit.org/Portal/index.php). In this project, 150 candidates genes selected from the *Prunus* and *Malus* ESTs will be mapped (Howad et al., 2005) in the TxE reference map and those located in QTL regions will be mapped in peach and apricot maps. Some of them are already mapped in apple linkage maps (Hidras European project).

Several COS (Conserved Ortholog Set) markers, developed from Solanaceae (Fulton et al., 2002), have been tested as a source of transportable markers (Dirlewanger unpublished results). Among the 24 primer pairs already available in the SOL Genomics Network (SGN) database (http://www.sgn.cornell.edu), eight give amplification on the five Rosaceae species analysed: peach, sour cherry, almond × peach, sweet cherry and strawberry. However, the polymorphism observed after sequencing the PCR product was very low. COS markers constitute a source of transposable markers, even between different families, however but the polymorphism may be lower than observed with other EST markers.

### Gene Pair Markers

Gene pair markers are an innovative tool for comparative linkage mapping recently developed in *Fragaria vesca*, where genes are closely spaced, i.e. approximately one gene per 6 kb (Davis et al., 2008). As a consequence of the small intergenic distances, PCR primers targeted to conserved exon sequences in two adjacent genes (referred to here as a 'gene pair') can be used to amplify a product that contains an entire intergenic region flanked by gene-defining exon sequences. Such intergenic regions are rich sites of useful polymorphisms, including SNPs, insertion/deletions and SSRs.

Comparison of strawberry exon sequences with homologous EST sequences from other rosaceous species (e.g. cherry, rose, raspberry and apple) species-specific will enable amplification of gene pair loci in multiple rosaceous species. Gene pair markers constitute a useful new tool for comparative linkage mapping in the Rosaceae family, and this approach may be applied in other taxa featuring small genomes with commensurately short intergenic distances. The use of this marker will probably increase with the availability of the upcoming whole genome sequence for apple and peach, which will make it possible to access intergenic sequence information readily.

## *3.3 Functional Genomics Tools*

### 3.3.1 EST and Candidate Gene Sequencing

In the last five years, a considerable effort has gone into developing more advanced genomics resources in Rosaceae. Extensive EST collections developed from strawberry (Folta et al., 2005), apricot (Grimplet et al., 2005), apple (Korban et al. 2005; Newcomb et al., 2006; Park et al., 2006) and peach are available (Horn et al., 2005). All these ESTs constitute a powerful source of candidate genes. These sequences were used to identify candidate genes for pest and disease resistance (Rikkerink et al., 2003; Casado-Diaz et al., 2006), flavor (Souleyre et al., 2005) and color (Espley et al., 2007; Chagné et al., 2007) in apple and are currently used in the ISAFRUIT European Integrated project.

### 3.3.2 Microarray

Microarrays or DNA/RNA chips are a powerful tool used for the characterization of gene expression in plants. The expression of a large number of genes can be simultaneously assayed by hybridizing oligonucleotide probes with labelled RNA prepared from plants subjected to different treatment or from different tissue types, and looking for hybridization patterns that suggest a significant change in expression. Microarrays have been developed for several members of the Rosaceae, such as apple (Crowhurst et al. 2005), peach (Trainotti et al. 2005) and strawberry (Aharoni et al., 2000; Casado-Diaz et al., 2006). Microarrays made it possible to study the expression pattern of genes involved in fruit ripening and flavor in strawberry (Aharoni et al., 2000 and 2004) and apple (Schaffer et al., 2007), and in resistance to diseases (Casado-Diaz et al., 2006). In peach, microarrays were used to analyse transcriptome changes during the transition from pre-climacteric to climacteric phase in the fruit (Trainotti et al., 2005).

Microarrays can also be used for comparative genomics, and preliminary results have shown that a microarray developed from apple ESTs could be used to hybridize other Rosaceae's RNA such as pear and peach (R. Schaffer et al., Apple Functional Genomics, this volume).

### 3.3.3 Transformation

Initially transformation was used to introduce various heterologous (trans)genes largely aimed at providing pathogen or pest protection. More recently, it has been used to deliver endogenous genes in order to identify their function. This strategy was used for the *Vf* gene candidate HcrVf2 introducing an apple homologue of the

*Arabidopsis* regulatory gene NPR1 as well as apple proteins that are known to interact with the *Erwinia amylovora* secreted type III effector protein DspE (Belfanti et al., 2004; Aldwinckle et al., 2003). Another strategy for functional analysis attempts to turn off the genes in order to either identify function (Dandekar et al., 2004), or modify plant development to create a novel phenotype such as been shown with an apple knock-out for 1-aminocyclopropane-1-carboxylic acid (ACC) oxidase and/or ACC synthase (Schaffer et al., 2007). Methods for gene knock-down using RNA interference technology have also begun in apple (Gilissen et al., 2005). In strawberry, several cultivars have been successfully transformed, making studies of *Fragaria* gene function possible in an homologous system. Two systems, using a diploid 'Hawaii-4' and an octoploid 'LF9' have been defined and present a high level of efficient transformation and a rapid cycling (Folta et al., 2006). In peach, several reviews have been published (Scorza and Hammerschlag, 1992; Srinivasan and Scorza, 1999, 2005), however but the transformation efficiency remains very low and is affected by many factors such as the environment and the antibiotic selection pressure. However, plum has been repeatedly used to develop transgenic trees and this system can be used routinely (Padilla et al., 2003).

## *3.4 Comparative Genome Mapping – Colinearity in Rosaceae*

One contribution of genetic mapping to fundamental studies in plant biology is the ability to study genome organization and evolution in closely related species by studying the degree of conservation at the level of the whole genome, chromosome or chromosomal segments (macrocolinearity) and at the level of gene sequences (microcolinearity) (Schmidt, 2002; Paterson et al., 2004). Comparative genome mapping aims at identifying conserved genome blocks, and regions of lesser conservation. In addition to gaining information gained on genome evolution, comparative genome mapping can serve as a starting point for initial mapping and gene cloning investigations in less characterized species.

### 3.4.1 Colinearity Within Genera and Within Subfamilies

Prunus v Prunus

The position of common markers from the reference TxE map was compared to the map position in several *Prunus* species and interspecific hybrids. This made it possible to align the maps of seven species: peach, almond, apricot, sweet cherry, Myrobalan plum, *P. ferganensis* and *P. davidiana* (Dirlewanger et al., 2004b; Arús et al., 2005). The distribution and order of markers in all *Prunus* species resulted in a general pattern of complete synteny, except for a reciprocal translocation between peach and almond on linkage groups 6 and 8 (Jáuregui et al., 2001; Dirlewanger et al., 2004a). This suggests that the *Prunus* genome could be treated as a single entity.

### *Fragaria* Diploid *v* Octoploid

By developing genetic linkage maps of the complex octoploid cultivated strawberry *F.* × *ananassa* using single dose AFLP and microsatellite (SSR) markers, two parental maps comprising 28 and 26 linkage groups were obtained for the female and the male parents, respectively (Rousseau-Gueutin et al., 2008). Almost all of these linkage groups were in coupling/repulsion phase, however a small proportion were observed to be in coupling phase only. These results suggest that although disomic behavior is prevalent in the cultivated strawberry *F.* × *ananassa,* that residual levels of polysomic segregation may still occur.

Two maps, one developed from the octoploid cultivated strawberry *F.* × *ananassa* (Lerceteau-Kohler et al., 2003) and another developed from the diploid inter-specific cross *F. vesca* × *F. bucharica* (Sargent et al., 2006), were compared using 51 common SSR markers (Denoyes-Rothan et al., 2006; Rousseau-Gueutin et al., 2008). The octoploid genetic parental maps were assembled into seven homoeologous groups (HGs) and each HG was found to be homologous to one linkage group of the diploid map. The observed macrosynteny and colinearity were very high, except for two putative inversions, identified by single SSR markers. These results indicate that polyploidisation in *Fragaria* was not accompanied by important structural rearrangements and that the diploid and octoploid genomes are highly conserved.

### Malus v Pyrus

Colinearity between *Malus* and *Pyrus* was examined. When the maps of the European pears 'Bartlett' and 'La France', and the Japanese pear 'Housui' were compared with those of apple cultivars 'Fiesta' and 'Discovery' (Liebhard et al., 2003) using apple SSRs, pear linkage groups could be successfully aligned to the apple ones (Yamamoto et al., 2004a, 2007). These findings showed that positions and linkages of SSR loci are well conserved between pear and apple, suggesting that partial synteny exists between both genera. Pierantoni et al. (2004) also reported that more than 100 apple SSR markers could be positioned on pear genetic linkage maps. Silfverberg-Dilworth et al. (2006) described the mapping of 6 pear SSRs in the linkage maps of apples 'Fiesta' and 'Discovery', whose positions were identified in the same homologous linkage groups. This was confirmed by a later study based on a larger set of pear SSRs mapped in a 'Malling 9' × 'Robusta 5' apple rootstock map (Celton et al. 2008).

### Fragaria v Rosa

A comparative mapping study has been initiated among two closely related genera in the *Rosoideae* tribe, *Fragaria* and *Rosa.* A total of 125 primer pairs developed from gDNA SSRs and EST-SSRs were tested. Only twelve (10%) were polymorphic in both populations (Rousseau et al., 2006). In addition to the SSRs, markers derived from floral initiation candidate genes were tested. Preliminary results using 20 markers provide anchor points between the homologous linkage groups and important structural rearrangements have been found. These results suggest *Fragaria* linkage groups are composed of at least two *Rosa* linkage groups.

### 3.4.2 Colinearity Between Genera Belonging to Different Subfamilies

Malus and Prunus

Only a small amount of information is available concerning the comparison of the *Malus* and *Prunus* genome and this is based on studies using only restriction fragment length polymorphism (RFLP) and isoenzyme markers (Dirlewanger et al., 2004b). Based on the few common markers (24 RFLPs and 6 isozymes) located on the *Prunus* TxE and on the apple Prima × Fiesta maps, the *Prunus* linkage group (called G) G3 and G7 are homologous to two apple linkage groups each as expected if they correspond to two homeologous chromosomes. The G1, corresponding to the longest Prunus chromosome corresponds to two homeologous apple linkage groups (called L) (L13) and L16) on its upper part, and to the apple L8 on its lower part.

Concerning microsynteny, no information is available until now but the complete sequence of the two genomes in a near future will clarify relationship among them.

Fragaria and Prunus

The diploid genomes of *Prunus* ($2n = 2x = 16$), that include all the stone fruits (peach, apricot, cherry and plum) and almond, and *Fragaria* ($2n = 2x = 14$), the wild and cultivated strawberries, were compared using 71 orthologous markers, including 40 restriction fragment length polymorphisms (RFLPs), 29 indels or single nucleotide polymorphisms (SNPs) derived from expressed sequence tags (ESTs) in both genera and two simple-sequence repeats (SSRs) (Vilanova et al., 2008). These markers were mapped in the reference maps of both species, provided good coverage of the *Prunus* (75%) and *Fragaria* (77%) genomes, and indicate a clear pattern of synteny, with most markers on each chromosome of one of these species mapping to one or two chromosomes of the other. A large number of rearrangements (36) were identified, most of which produced by inversions (27) and the rest (9) by translocations or fission/fusion events. A simple example are the markers of two highly syntenic groups (*Fragaria* linkage group V and Peach linkage group 5), which would require a single inversion to be placed in the same order. These results provide the first framework for the comparison of the position of genes or DNA sequences for these two economically valuable and yet distantly-related genera of the Rosaceae.

### 3.4.3 Beyond the Family: Comparative Mapping Among Peach or Apple, *Arabidopsis* and *Populus*

The comparison of genome structure among plants that diverged over a significant time allows an insight into plant genome evolution. Such comparisons have been made at the whole genome sequence level between Arabidopsis, rice, poplar and grapevine, making it possible to identify a possible paleo-hexaploid ancestral genome for dicots (The French-Italian Public Consortium for Grapevine Genome Characterization 2007). In addition, identification of significantly conserved regions can identify functional chromosomal units. As no genome sequence is available yet for Rosaceae, the macrosyteny between peach and *Arabidopsis* was studied by comparing the *Prunus* map with the *Arabidopsis thaliana* genome sequence (Jung et al., 2006). Using a set of RFLP markers mapped in the TxE map (Dominguez et al., 2003), 37 syntenic regions were identified, covering 23 and 17% of the *Prunus* and *Arabidopsis* genomes, respectively indicating a high level of macrosynteny between the species. The longest of these regions included 13 markers for a distance of 25 cM in linkage group 2 of *Prunus* and 16 homologous sequences spanning 5.4 Mb in chromosome 5 of *Arabidopsis*. At the microsyteny level, BACs and BAC end sequences spread along the peach genome were compared to *Arabidopsis* sequences (Georgi et al., 2003). Predicted genes in these sequences were homologous to genes scattered along the five chromosomes of *Arabidopsis*. These two studies concur in detecting a fragmentary preservation between these two genomes putatively separated for more than 90 million years.

Very recently, homologies between peach and poplar genomes were analysed (Jung, personal communication). A peach BAC sequence was reported to be highly homologous to two poplar fragments located on two linkage groups in poplar. These new results suggest that the whole genome sequence already available from poplar (Tuskan et al., 2006) will be useful for Rosaceae studies. The same type of study was performed with apple by mapping apple BAC end pairs to the *Populus*, *Arabidopsis* and rice genomes (Xu et al., 2001, 2002; Han and Korban, 2008). Results indicate that apple has a higher degree of synteny with the closely related *Populus* than with the distantly related *Arabidopsis*.

# 4 Challenges for the Future

## *4.1 Structural and Comparative Genomics; Whole Genome Sequencing*

Comparative genomics is just beginning in the Rosaceae, with the genomes of only a few species having been compared in depth. One of the key elements that have facilitated this progress has been the existence of a reference high-density genetic map for *Prunus* (TxE), which was constructed with transferable markers. An in-depth knowledge of the synteny among and between different subfamilies will help in designing strategies at the whole family level to localise homologous regions containing genes/QTLs of interest, describing allelic variation, identifying candidate genes located at a target region and finding markers tightly linked to a gene/QTL.

The future priorities would be:

- Continuing to develop highly polymorphic orthologous markers that are transportable to a wider range of species. Such markers could be based on gene pair markers and EST-based SNPs.
- Developing genetic maps or whole genome sequence for Rosaceae species that are not yet available such as Spiroideae.
- Exploring the three Rosaceae genome sequences that will soon be available (peach, apple and strawberry): the alignment of these genomes will contribute a wealth of information on genome evolution in the Rosaceae family.
- Completing physical maps/genetic maps of key species. Is it predictable that with the ongoing reducing cost for sequencing and expected advances in bioinformatics, more genome sequences will be coming in the near future. Considering the small genome size of some species belonging to the Spiroideae or Rosoideae subfamilies, it would not be surprising to see them sequenced soon.

### *4.2 Association Mapping and Other Ways to Link Genotype to Phenotype*

A major goal of research in the future will consist of developing faster and better methods to link genotype information to desirable phenotype information. Association genetics, i.e. genetic analysis based on linkage disequilibrium (LD) using collections of unrelated genotypes, is a strategy complementary to other approaches such as gene/QTL mapping and gene cloning. The extent of LD conservation depends on the history of the species, its reproduction mode (i.e. self-pollinated or outcrossing) and whether the population used is structured or not. Association studies attract an especially high interest in fruit tree genetic studies, because they do not require the creation of large segregating populations, that takes a long time and is expensive. The phenotype information, sometimes already available, on genotypes of interest can be found in germplasm repositories or breeders' collections. Knowledge of the syntenic relationships between different species would allow the choice of the best genotype collection and species for these applications. For example, in the highly syntenic genus *Prunus* where it is likely that species with high (peach) and low (apricot, almond, cherry) LD conservation exists (Aranzana et al., 2008), peach collections may be adequate for whole genome scans to find gene or QTL positions and apricot, almond or cherry populations may be more useful for the validation of candidate genes.

The development of methodologies that can generate and analyse large numbers of data points, in the order of hundreds of thousands SNPs per individual assay, will be required to improve the chances of identifying candidate genes for any given phenotype by whole genome scans. Such a density of markers would be required to cover the genome of Roseceae species, in particular species with a slow decay of

LD and a relatively small genome such as peach. For the moment such technologies have only been applied to human genetics. For example, the use of a 500 K SNPs assay (GeneChip 500 K Mapping Array Set, Affymetrix) made it possible to identify genes associated with common diseases in human (WTCCC 2007). Even though these technologies are not developed yet for Rosaceae, it is certain that Rosaceae geneticists will benefit from the technical and statistical advances made by model systems.

## References

Abbott AG, Arús P, Scorza R (2007) Peach. In Kole C (ed) Genome Mapping & Molecular Breeding in Plants. Vol. 4: Fruits and Nuts, Berlin: Springer, 137–156

Aharoni A, Giri AP, Verstappen FWA, Bertea CM, Sevenier R, Sun Z, Jongsma MA, Schwab W, Bouwmeester HJ (2004) Gain and loss of fruit flavor compounds produced by wild and cultivated strawberry species. Plant Cell 16:3110–3131

Aharoni A, Keizer LCP, Bouwmeester HJ, Sun Z, Alvarez-Huerta M, Verhoeven HA, Blaas J, van Houwelingen AMML, De Vos RCH, van der Voet H, Jansen RC, Guis M, Mol J, Davis RW, Schena M, van Tunen AJ, O'Connell A (2000) Identification of the *SAAT* gene involved in strawberry flavor biogenesis by use of DNA microarrays. Plant Cell 12:647–662

Aldwinckle HS, Borejsza-Wysocka EE, Malnoy M, Brown SK, Norelli JL, Beer SV, Meng X, He SY, Jin QT (2003) Development of fire blight resistant apple cultivars by genetic engineering. Acta Hort 622:105–111

Aranzana MJ, Abbassi EK, Howad W, Pere Arus P (2008) Estimating the extent of linkage disequilibrium in peach (*Prunus persica*) American and European cultivars. Plant & Animal Genomes XVI Conference, San Diego, CA, January 12–16

Arondel V, Lemieux B, Hwang I, Gibson S, Goodman HM, Somerville CR (1992) Map-based cloning of a gene controlling omega-3 fatty acid desaturation in *Arabidopsis*. Science 992:1353–1355

Arumuganathan K and Earle ED (1991) Nuclear DNA content of some important plant species. Plant Mol Biol Rep 9: 208–218

Arús P, Yamamoto T, Dirlewanger E, Abbott AG (2005) Synteny in the Rosaceae. Plant Breeding Reviews, Vol 27, Ed Janick J. London: John Wiley and Sons Inc. 175–211

Ashley MV, Wilk JA, Styan SMN,. Craft KJ, Jones KL, Feldheim KA, Lewers KS, Ashman TL (2003) High variability and disomic segregation of microsatellites in the octoploid *Fragaria virginiana* Mill. (Rosaceae). Theor App Genet 107:1201–1207

Asnaghi C, Paulet F, Kaye C, Grivet L, Deu M, Glaszmann JC, D'Hont A (2000) Application of synteny across Poaceae to determine the map location of a sugarcane rust resistance gene Theor Appl Genet 101:962–969

Bao L, Chen K, Zhang D, Cao Y, Yamamoto T, Teng Y (2007) Genetic diversity and similarity of pear (*Pyrus* L.) cultivars native to East Asia revealed by SSR (simple sequence repeat) markers. Genet Res Crop Evol 54: 959–971

Bassil NV, Gunn M, Folta K, Lewers KS (2006a) Microsatellite markers for *Fragaria* from 'Strawberry Festival' expressed sequence tags. Mol Ecol Notes 6(2): 473–476

Bassil NV, Njuguna W, Slovin JP (2006b). EST-SSR markers from *Fragaria vesca* L. cv. yellow wonder. Mol Ecol Notes 6(3):806–809

Belfanti E, Silfverberg-Dilworth E, Tartarini S, Patocchi A, Barbieri M, Zhu J, Vinatzer BA, Gianfranceschi L, Gessler C, Sansavini S (2004) The HcrVf2 gene from a wild apple confers scab resistance to a transgenic cultivated variety. Proc Natl Acad Sci 101:886–890

Bennett, MD (2004) Perspectives on polyploidy in plants – ancient and neo. Biol. J. Linn Soc. 82: 411–423

Bennett MD, Leitch IJ (2004) Plant DNA C-values database (release 5.0, Dec. 2004, http://www.rbgkew.org.uk/cval/homepage.html)

Bennett MD, Leitch IJ, Price HJ, Johnston JS (2003) Comparisons with *Caenorhabditis* (~100 Mb) and Drosophila (~175 Mb) using flow cytometry show genome size in arabidopsis to be ~157 Mb and thus ~25 % larger than the *Arabidopsis* genome initiative estimate of ~125 Mb. Ann Bot 91: 547–557

Bennett MD, Smith JB (1976) Nuclear DNA amounts in angiosperms. Philos Trans R Soc Lond 274:227–274

Bennett MD, Smith JB (1991) Nuclear DNA amounts in angiosperms. Philos Trans R Soc Lond 334:309–345

Bennett MD, Smith JB, Heslop-Harrison JS (1982) Nuclear DNA amounts in angiosperms. Philos Trans R Soc Lond 216:179–199

Casado-Diaz A, Encinas-Villarejo S, de los Santos B, Schiliro E, Yubero-Serrano EM, Amil-Ruiz F, Pocovi MI, Pliego-Alfaro F, Dorado G, Rey M, Romero F, Munoz-Blanco J, Caballero JL (2006) Analysis of strawberry genes differentially expressed in response to *Colletotrichum* infection. Physiologia Plantarum 128: 633–650

Causse MA, Fulton TM, Cho YG, Ahn SN, Chunwongse J, Wu K, Xiao J, Yu Z, Ronald PC, Harrington SE, Second G, McCouch SR, Tanksley SD (1994) Saturated molecular map of the rice genome based on an interspecific backcross population. Genetics 138 1251–1274

Celton J-M, Tustin DS, Chagné D, Gardiner SE (2008) Construction of a dense genetic linkage map for apple rootstocks using SSRs developed from *Malus* ESTs and *Pyrus* genomic sequences, Tree Genet 6 Genomes, 5:93–107

Chagné D, Carlisle CM, Blond C, Volz RK, Whitworth CJ, Oraguzie NC, Crowhurst RN, Allan AC, Espley RV, Hellens RP, Gardiner SE (2007) Mapping a candidate gene (MdMYB10) for red flesh and foliage colour in apple. BMC Genomics 8:212

Chen M, Presting G, Barbazuk WB, Goicoechea JL, Blackmon B, Fang G, Kim H, Frisch D, Yu Y, Sun S, Higingbottom S, Phimphilai J, Phimphilai D, Thurmond S, Gaudette B, Li P, Liu J, Hatfield J, Main D, Farrar K, Henderson C, Barnett L, Costa R, Williams B, Walser S, Atkins M, Hall C, Budiman MA, Tomkins JP, Luo M, Bancroft I, Salse J, Regad F, Mohapatra T, Singh NK, Tyagi AK, Soderlund C, Dean RA, Wing RA (2002) An integrated physical and genetic map of the rice genome. Plant Cell 14:537–545

Cipriani G, Lot G, Huang WG, Marrazzo MT, Peterlunger E, Testolin R (1999) AC/GT and AG/CT microsatellite repeats in peach (*Prunus persica* (L) Batsch): Isolation, characterisation and cross-species amplification in *Prunus*. Theor App Genet 99:65–72

Cipriani G, Testolin R (2004) Isolation and characterization of microsatellite loci in *Fragaria*. Mol Ecol Notes 4(3): 366–368

Crowhurst RN, Allan AC, Atkinson RG, Beuning LL, Davey M, Friel E, Gardiner SE Gleave AP, Greenwood DR, Hellens RP, Janssen BJ, Kutty-Amma S, Laing WA, MacRae EA, Newcomb RD, Plummer KM, Schaffer R, Simpson RM, Snowden KC, Templeton MD, Walton EF, Rikkerink EHA (2005) The HortResearch apple EST database – a resource for apple genetics and functional genomics. In: Plant & Animal Genome XIII Conference, San Diego, CA, USA

Dandekar AM, Teo G, Defilippi BG, Uratsu SL, Passey AJ, Kader AA, Stow JR, Colgan RJ, James DJ (2004) Effect of down-regulation of ethylene biosynthesis on fruit flavor complex in apple fruit. Transgenic Res 13:373–384

Da Silva J, Honeycutt R, Burnquist W, Al-Janabi SM, Sorells ME, Tanksley SD, Sobral WS (1995) *Saccharum spontaneum* L. ''SES 208'' genetic linkage map combining RFLP- and PCRbased markers. Mol Breed 1: 165–179

Davis TM, DiMeglio LM, Yang RH, Styan SMN, Lewers KS (2006) Assessment of SSR marker transfer from the cultivated strawberry to diploid strawberry species: Functionality, linkage group assignment, and use in diversity analysis. J Am Soc Hort Sci 131:506–512

Davis TM, Folta KM, Shields M, and Zhang Q (2008) Gene pair markers: An innovative tool for comparative linkage mapping. *In* Proceedings of 6th North American Strawberry Symposium. North American Strawberry Growers Association, Kemptville, Canada (in press)

Denoyes-Rothan B, Rousseau M, Barrot L, Lerceteau-Köhler E, Sargent D, Simpson D, Monfort A, Arus P, Guerin G (2006). Comparative linkage maps of two levels of Ploidy in *Fragaria*. Plant and Animal Genome XIV (San Diego 15–20 January 2006)

Dickson EE, Arumuganathan K, Kresovich S, Doyle JJ (1992) Nuclear DNA content variation within the Rosaceae. Am J Bot 79: 1081–1086

Dirlewanger E, Cosson P, Howad W, Capdevill G, Bosselu N, Claverie M, Voisin R, Poizat C, Lafargue B, Baron O, Laigret F, Kleinhentz M, Arús P, Esmenjaud D (2004a) Microsatellite genetic linkage maps of myrobalan plum and an almond-peach hybrid – location of root-knot nematode resistance genes. Theor Appl Genet 109:827–838

Dirlewanger E, Cosson P, Poizat C, Laigret F, Aranzana MJ, Arus P, Dettori MT, Verde I, Quarta R (2003) Synteny within the *Prunus* genomes detected by molecular markers. 26th International Horticultural Congress & Exhibition (IHC2002), Toronto, USA, 11–17 août 2002. Acta Hort 622:177–187

Dirlewanger E, Graziano E, Joobeur T, Garriga-Calderé F, Cosson P, Howad W, Arús P (2004b) Comparative mapping and marker assisted selection in Rosaceae fruit crops. Proc Natl Acad Sci 101:9891–9896

Dominguez I, Graziano E, Gebhardt C, Barakat A, Berry S, Arús P, Delseny M, Barnes S (2003) Plant genome archaeology: Evidence for conserved ancestral chromosome segments in dicotyledonous plant species. Plant Biotechnol J 1:91–99

Espley RV, Hellens RP, Puterill J, Kutty-Amma S, Allan AC (2007) Red colouration in apple fruit is due to the activity of a MYB transcription factor, MdMYB10. Plant J 49:414–427

Evans RC, Campbell CS (2002) The origin of the apple subfamily (Maloideae; Posaceae) is clarified by DNA sequence data from duplicated GBSSI genes. Am J Bot 89:1478–1484

Folta KM, Davis T (2006) Strawberry genes and genetics. Crit Rev Plant Sci 25: 399–415

Folta KM, Dhingra A (2006) Transformation of strawberry: The basis for translational genomics in Rosaceae. In Vitro Cell Dev Biol Plant 42:482–490

Folta K, Dhingra A, Howard L, Stewart PJ, Chandler CK (2006) Characterization of LF9, an octoploid strawberry genotype selected for rapid regeneration and transformation. Planta 224: 1058–1067

Folta K, Staton M, Stewart PJ, Jung S, Bies DH, Jesdurai C, Main D (2005) Expressed sequence tags (ESTs) and simple sequence repeat (SSR) markers from octoploid strawberry *(Fragaria* × *ananassa).*BMC Plant Biol 5:12

Foulongne M, Pascal T, Pfeiffer F, Kervella J (2003) QTLs for powdery mildew resistance in peach × *Prunus davidiana* crosses: Consistency across generations and environments. Mol Breed 12:33–50

Frary A, Doganlar S, Frampton A, Fulton T, Uhlig J, Yates H, Tanksley S (2003) Fine mapping of quantitative trait loci for improved fruit characteristics from *Lycopersicon chmielewskii* chromosome 1. Genome 46:235–243

The French-Italian Public Consortium for Grapevine Genome Characterization (2007) The grapevine genome sequence suggests ancestral hexaploidization in major angiosperm phyla. Nature 449: 463–468

Fulton TM, Van der Hoeven R, Eannetta NT, Tanksley SD (2002) Identification, analysis, and utilization of conserved ortholog set markers for comparative genomics in higher plants. Plant Cell 14:1457–1467

Georgi LL, Wang Y, Reighard GL, Mao L, Wing RA, Abbott AG (2003) Comparison of peach and *Arabidopsis* genomic sequences: Fragmentary conservation of gene neighborhoods. Genome 46:268–276

Georgi LL, Wang Y, Yvergniaux D, Ormsbee T, Iñigo M, Reighard G, Abbott AG (2002) Construction of a BAC library and its application to the identification of simple sequence repeats in peach [*Prunus persica* (L.) Batsch]. Theor Appl Genet 105:1151–1158

Gil-Ariza DJ, Amaya I, Botella MA, Blanco JM, Caballero JL, Lopez-Aranda JM, Valpuesta V, Sanchez-Sevilla JF (2006) EST-derived polymorphic microsatellites from cultivated strawberry

(*Fragaria* × *ananassa*) are useful for diversity studies and varietal identification among Fragaria species. Mol Ecol Notes 6(4):1195–1197

Gilissen LJ, Bolhaar ST, Matos CI, Rouwendal GJ, Boone MJ, Krens FA, Zuidmeer L, van Leeuwen A, Akkerdaas J, Hoffmann-Sommergruber K, Knulst AC, Bosch D, van de Weg WE, van Ree R (2005) Silencing the major apple allergen Mal d 1 by using the RNA interference approach. J Allergy Clin Immunol 115:364–369

Graham J, Smith K, Woodhead M, Russell J (2002) Development and use of simple sequence repeat SSR markers in *Rubus* species. Mol Ecol Notes 2(3):250–252

Grimplet J, Romieu C, Audergon JM, Marty I, Albagnac G, Lambert P, Bouchet JP, Terrier N (2005) Transcriptomic study of apricot fruit (*Prunus armeniaca*) ripening among 13 006 expressed sequence tags. Physiologia Plantarum 125:281–292

Hadonou AM, Sargent DJ, Wilson F, James CM, Simpson DW (2004) Development of microsatellite markers in Fragaria, their use in genetic diversity analysis, and their potential for genetic linkage mapping. Genome 47:429–438

Han Y, Gasic K, Marron B, Beever JE, Korban SS (2007) A BAC-based physical map of the apple genome. Genomics 89:630–637

Han Y, Korban SS (2008) An overview of the apple genome through BAC end sequence analysis. Plant Mol Biol 67:581–588

Hibrand-Saint Oyant L, Crespel L, Rajapakse S, Zhang L, Foucher F (2008) Genetic linkage maps of rose constructed with new microsatellite markers and locating QTL flowering traits. Tree Gene Genomes 4:11–23

Horn R, Lecouls AC, Callahan A, Dandekar A, Garay L, McCord P, Howad W, Cha H, Verde I, Main D, Jung S, Georgi L, Forrest S, Mook J, Zhebentyayeva T, Yu Y, Kim HR, Jesudurai C, Sosinski B, Arús P, Baird V, Parfitt D, Reighard G, Scorza R, Tomkins J, Wing R, Abbott AG (2005) Candidate gene database and transcript map for peach, a model species for fruit trees. Theor Appl Genet 110:1419–1428

Howad W, Yamamoto T, Dirlewanger E, Testolin R, Cosson P, Cipriani G, Monforte AJ, Georgi L, Abbott AG, Arús P (2005) Mapping with a few plants: Using selective mapping for microsatellite saturation of the *Prunus* reference map. Genetics 171:1305–1309

James CM, Wilson F, Hadonou AM, Tobutt KR (2003). Isolation and characterization of polymorphic microsatellites in diploid strawberry (*Fragaria vesca* L.) for mapping, diversity studies and clone identification. Mol Ecol Notes 3(2):171–173

Jáuregui B, de Vicente MC, Messeguer R, Felipe A, Bonnet A, Salesses G, Arús P (2001) A reciprocal translocation between 'Garfi' almond and 'Nemared' peach. Theor Appl Genet 102: 1169–1176

Joobeur T, Periam N, de Vicente MC, King GJ, Arús P (2000) Development of a second generation linkage map for almond using RAPD and SSR markers. Genome 43:649–655

Joobeur T, Viruel MA, de Vicente MC, Jáuregui B, Ballester J, Dettori MT, Verde I, Truco MJ, Messeguer R, Batlle I, Quarta R, Dirlewanger E, Arús P (1998) Construction of a saturated linkage map for *Prunus* using an almond × peach $F_2$ progeny. Theor Appl Genet 97: 1034–1041

Jung S, Jesudurai C, Staton M, Du Z, Ficklin S, Cho I, Abbott A, Tomkins J, Main D (2004) GDR (Genome Database for Rosaceae): Integrated web resources for Rosaceae genomics and genetics research, BMC Bioinform. 5:130

Jung S, Main D, Staton M, Cho I, Zhebentyayeva T, Arús P, Abbott A (2006) Synteny conservation between the Prunus genome and both the present and ancestral *Arabidopsis* genomes BMC Genomics 7:81

Jung S, Staton M, Lee M, Blenda A, Svancara R, Abbott A and Main D (2008) GDR (Genome Database for Rosaceae): Integrated web-database for Rosaceae genomics and genetics data. Nucleic Acids Res 36 (Database issue):D1034–D1040

Keith AL, Wendel JF (2005) Polyploidy and genome evolution in plants. Curr Opin Plant Biol 8:135–141

Kelleher CT, Chiu R, Shin H, Bosdet IE, Krzywinski MI, Fjell CD, Wilkin J, Yin T, DiFazio SP, Ali J, Asano JK, Chan S, Cloutier A, Girn N, Leach S, Lee D, Mathewson CA, Olson T, O׳connor

K, Prabhu AL, Smailus DE, Stott JM, Tsai M, Wye NH, Yang GS, Zhuang J, Holt RA, Putnam NH, Vrebalov J, Giovannoni JJ, Grimwood J, Schmutz J, Rokhsar D, Jones SJ, Marra MA, Tuskan GA, Bohlmann J, Ellis BE, Ritland K, Douglas CJ, Schein JE. (2007) A physical map of the highly heterozygous Populus genome: Integration with the genome sequence and genetic map and analysis of haplotype variation. Plant J 50:1063–1078

Kilian A, Kudrna DA, Kleinhofs A, Yano M, Kurata N, Steffenson B, Sasaki T (1995) Rice-barley synteny and its application to saturation mapping of the barley *Rpg1* region. Nucleic Acids Res 23: 2729–2733

Kimura T, Shi YZ, Shoda M, Kotobuki K, Matsuta N, Hayashi T, Ban Y, Yamamoto T (2002) Identification of Asian pear varieties by SSR analysis. Breed Sci 52:115–121

Klein PE, Klein RR, Cartinhour SW, Ulanch PE, Dong J, Obert JA, Morishige DT, Schlueter SD, Childs KL, Ale M, Mullet JE (2000) A high-throughput AFLP-based method for constructing integrated genetic and physical maps: Progress toward a sorghum genome map. Genome Res. 10:789–807

Korban SS, Vokin LO, Liu L, Aldwinckle HS, Ksenija GS, Gasic K, Orlando-Gonzales D, Malnoy M, Thimmapuram J, Caroll NJ, Golsborough P, Orvis K, Clifton S, Pape D, Martin M, Meyer R (2005) Large-scale analyses of EST sequences in apple genome. In: Plant & Animal Genome XIII Conference, January, San Diego, CA, USA: http://www.intl-pag.org/pag/13/abstract/GAG13_W130.html

Kovanda M (1965) On the generic concepts in the Maloideae. Preslia (Praha) 37:27–34

Lambert P, Hagen LS, Arús P, Audergon JM (2004) Genetic linkage maps of two apricot cultivars (*Prunus armeniaca* L.) compared with the almond 'Texas' × peach 'Earlygold' reference map for *Prunus*. Theor Appl Genet 108:1120–1130

Lazzari B, Caprera A, Vecchietti A, Stella A, Milanesi J, Pozzi C (2005) ESTree db: A tool for peach functional genomics. BMC Bioinformatics 6: S16

Lerceteau-Kohler E, Guerin G, Laigret F, Denoyes-Rothan B (2003) Characterization of mixed disomic and polysomic inheritance in the octoploid strawberry (*Fragaria* × *ananassa*) using AFLP mapping. Theor App Genet 107:619–628

Lewers KS, Saski CA, Cuthbertson BJ, Henry DC, Staton ME, Main DS, Dhanaraj AL, Rowland LJ, Tomkins JP (2008) A blackberry (*Rubus* L.) expressed sequence tag library for the development of simple sequence repeat markers. BMC Plant Biol, 8:69 doi:10.1186/1471-2229-8-69

Lewers KS, Styan SMN, Hokanson SC, Bassil NV (2005). Strawberry GenBank-derived and genomic simple sequence repeat (SSR) markers and their utility with strawberry, blackberry, and red and black raspberry. J Am Soc Hort Sci 130:102–115

Liebhard R, Gianfranceschi L, Koller B, Ryder CD, Tarchini R, Van de Weg E, Gessler C (2002) Development and characterisation of 140 new microsatellites in apple (*Malus* × *domestica* Borkh.). Mol Breed 10: 217–241

Liebhard R, Koller B, Gianfranceschi L, Gessler C (2003) Creating a saturated reference map for the apple (*Malus* × *domestica* Borkh.) genome. Theor Appl Genet 106:1497–1508

Lopes MS, Maciel G, Mendonca D, Gil FS, Machado ADC (2006) Isolation and characterization of simple sequence repeat loci in *Rubus hochstetterorum* and their use in other species from the Rosaceae family. Mol Ecol Notes 6:750–752

Masterson J (1994) Stomatal size in fossil plants – evidence for polyploidy in majority of angiosperms. Science 264: 421–424

Morgan DR, Soltis DE, Robertson KR (1994) Systematic and evolutionary implications of rbcL sequence variation in Rosaceae. Am J Bot 81:890–903

Mozo T, Dewar K, Dunn P, Ecker JR, Fischer S, Kloska S, Lehrach H, Marra M, Martienssen R, Meier-Ewert S, Altmann T (1999) A complete BAC-based physical map of the Arabidopsis thaliana genome. Nat Genet 22:271–275

Newcomb RD, Crowhurst RN, Gleave AP, Rikkerink EHA, Allan AC, Beuning LL, Bowen JH, Gera E, Jamieson KR, Janssen BJ, Laing WA, McArtney S, Nain B, Ross, GS, Snowden KC, Souleyre EJF, Walton EF, Yauk YK (2006) Analyses of expressed sequence tags from apple. Plant Physiol 141:147–166

Osborn TC, Butrulle DV, Sharpe AG, Pikering KJ, Parkin AIP, Parker JS, Lydiate DJ (2003) Detection and effects of a homeologous reciprocal transposition in *Brassica napus*. Genetics 165:1569–1577

Oosumi T, Gruszewski HA, Blischak LA, Baxter AJ, Wadl PA, Shuman JL, Veilleux RE, Shulaev V (2006) High-efficiency transformation of the diploid strawberry (*Fragaria vesca)* for functional genomics. Planta 223:1219–1230

Padilla IMG, Webb KR, Scorza R (2003) Early antibiotic selection and efficient rooting and acclimatization improve the production of transgenic plum plants (*Prunus domestica* L.) Plant Cell Rep 22:38–45

Paterson, AH, Bowers JE, Chapman BA, Peterson DG, Rong J, Wicker TM (2004) Comparative genome analysis of monocots and dicots, toward characterization of angiosperm diversity. Curr Opin Biotechnol 15(2):120–125

Park S, Sugimoto N, Larson MD, Beaudry R, van Nocker S (2006) Identification of genes with potential roles in apple fruit development and biochemistry through large-scale statistical analysis of expressed sequence tags. Plant Physiol 141(3): 811–824

Pierantoni L, Cho KH, Shin IS, Chiodini R, Tartarini S, Dondini L, Kang SJ, Sansavini S (2004) Characterisation and transferability of apple SSRs to two European pear F1 populations. Theor Appl Genet 109: 1519–1524

Rikkerink E, Hilario E, Rusholme R, Gardiner S, Bus V, Gleave A, Crowhurst R (2003) Mining the HortResearch apple EST database- in silico tissue expression analyses of resistance gene candidates and resistance gene classes. Plant & Animal Genomes XI Conference. January 11–15, 2003

Rousseau M, Hibrand Saint Oyant L, Foucher F, Barrot L, Lalanne D, Sargent D, Simpson D, Laigret Fand, Denoyes-Rothan B (2006) Comparative mapping in the Rosoideae tribe: *Rosa* and *Fragaria*. 3rd Internation Rosaceae Genomic Conference. Napier, New Zealand March 2006

Rousseau-Gueutin M, Lerceteau-Köhler E, Barrot L, Sargent D, Monfort A, Simpson D, Arùs P, Guérin G, Denoyes-Rothan B (2008) Comparative genetic mapping between octoploid and diploid *Fragaria* species reveals a high level of colinearity between their genomes and the essentially disomic behavior of the cultivated octoploid strawberry. Genetics 179:2045–2060

Sargent DJ, Clarke J, Simpson DW, Tobutt KR, Arus P, Monfort A, Vilanova S, Denoyes-Rothan B, Rousseau M, Folta KM, Bassil NV, Battey NH (2006) An enhanced microsatellite map of diploid Fragaria. Theor App Genet 112:1349–1359

Sargent DJ, Davis TM, Tobutt KR, Wilkinson MJ, Battey NH, Simpson DW (2004) A genetic linkage map of microsatellite, gene-specific and morphological markers in diploid *Fragaria*. Theor App Genet 109:1385–1391

Sargent DJ, Hadonou AM, Simpson DW (2003) Development and characterization of polymorphic microsatellite markers from *Fragaria viridis*, a wild diploid strawberry. Mol Ecol Notes 3(4):550–552

Sax K (1931) The origins and relationships of the Pomoideae. J. Arnold Arbor. 12: 3–22

Sax K (1932) Chromosome relationships in the Pomoideae. J. Arnold Arbor. 13: 363–367

Schmidt R. (2002) Plant genome evolution: Pessons from comparative genomics at the DNA level. Plant Mol Biol 48:21–37

Scorza R, Hammerschlag F (1992) Emerging technologies for the genetic improvement of stone fruit. In Hammerschlag FA, Litz RE (eds) Biotechnology of perennial Fruit Crops. CABI, Wallingford, Oxon, pp 277–302

Schaffer RJ, Friel EN, Souleyre EJF, Bolitho K, Thodey K, Ledger S, Bowen JH, Ma JH, Nain B, Cohen D, Gleave AP, Crowhurst RN, Janssen BJ, Yao JL, Newcomb RD (2007) A Genomics approach reveals that aroma production in apple is controlled by ethylene predominantly at the final step in each biosynthetic pathway([w]). Plant Physiol.144: 1899–1912

Shulaev V, Korban SS, Sosinski B, Abbott AG, Aldwinckle HS, Folta KM, Iezzoni A, Main D, Arús P, Dandekar AM, Lewers KS, Brown SK, Davis TM, Gardiner SE, Potter D, Veilleux RE (2008) Multiple models for Rosaceae genomics. Plant Phy Rev DOI:10.1104/107.115618

Silfverberg-Dilworth E, Matasci CL, Van de We E, Van Kaauwen MPW, Walser M, Kodde LP, Soglio V, Gianfranceschi L, Durel CE, Costa F, Yamamoto T, Koller B, Gessler C, Patocchi A (2006) Microsatellite markers spanning the apple (*Malus* × *domestica* Borkh.) genome. Tree Genet Genomics 2: 202–224

Sorrells, ME (1992) Development and application of RFLPs in polyploids. Crop Sci 32: 1086–1091

Souleyre EJF, Greenwood DR, Friel EN, Karunairetnam S, Newcomb RD (2005) An alcohol acyl transferase from apple (cv. Royal Gala), MpAAT1, produces esters involved in apple fruit flavour. FEBS J 272(12): 3132–3144

Srinivasan C, Scorza R (1999) Transformation of somatic embryos of trees and pravine. In Jain SM, Gupta PK, Newton NJ (eds) Somatic Embryogenesis in Woody Plants. Kluwer Acad Press, London, Vol. 5, pp 313–330

Srinivasan C, Scorza R (2005) The influence of genotype on the induction of somatic embryos from in vitro cultured zygotic embryos and adventitious shoot regeneration from cotyledons of peach and nectarine. Acta Hort 738: 512–542

Stafne ET, Clark JR, Weber CA, Graham J, Lewers KS (2005). Simple sequence repeat (SSR) markers for genetic mapping of raspberry and blackberry. J Am Soc Hor Sci 130:722–728

Tatum TC, Stepanovic S, Biradar DP, Lane Rayburn A, Korban SS (2005) variation in nuclear DNA content in *Malus* species and cultivated apples. Genome 48:924–930

Trainotti L, Bonghi C, Ziliotto F, Zanin D, Rasori A, Casadoro G, Ramina A, Tonutti P (2005) The use of microarray mPEACH1.0 to investigate transcriptome changes during transition from pre-climacteric to climacteric phase in peach fruit. Plant Sci 170:606–613

Tuskan GA, DiFazio S, Jansson S, Bohlmann J, Grigoriev I, Hellsten U, Putnam N, Ralph S, Rombauts S, Salamov A, ScheinJ, SterckL, Aerts A, Bhalerao RR, Bhalerao, Blaudez D, Boerjan W, Brun A, Brunner A, Busov V, Campbell M, Carlson JRP, Chalot M, Chapman J, Chen G-L, Cooper D, Coutinho PM, Couturier J, Covert S, Cronk Q, Cunningham R, Davis J, Degroeve S, Déjardin A, dePamphilis C, Detter J, Dirks B, Dubchak I, Duplessis S, Ehlting J, Ellis B, Gendler K, Goodstein D, Gribskov M, Grimwood J, Groover A, Gunter L, Hamberger B, Heinze B, Helariutta Y, Henrissat B, Holligan D, Holt R, Huang W, Islam-Faridi N, Jones S, Jones-Rhoades M, Jorgensen R, Joshi C, Kangasjärvi J, Karlsson J, Kelleher C, Kirkpatrick R, Kirst M, Kohler A, Kalluri U, Larimer F, Leebens-Mack J, Leplé J-C, Locascio P, Lou Y, Lucas S, Martin F, Montanini B, Napoli C, Nelson DR, Nelson C, Nieminen K, Nilsson O, Pereda V, Peter G, Philippe R, Pilate G, Poliakov A, Razumovskaya J, Richardson P, Rinaldi C, Ritland K, Rouzé P, Ryaboy D, Schmutz J, Schrader J, Segerman B, Shin H, Siddiqui A, Sterky F, Terry A, Tsai C-J, Uberbacher E, Unneberg P, Vahala J, Wall K, Wessler S, Yang G, Yin T, Douglas C, Marra M, Sandberg G, Van de Peer Y, Rokhsar D (2006) The genome of black cottonwood, *Populus trichocarpa* (Torr. & Gray). Science 313: 1596–1604

Van der Hoeven R, Ronning C, Giovannoni J, Martin G, Tanksley S (2002) Deductions about the number, organization, and evolution of genes in the tomato genome based on analysis of a large expressed sequence tag collection and selective genomic sequencing. Plant Cell 14(7): 1441–1456

Vilanova S, Sargent DJ, Arús P, Monfort A (2008) Synteny conservation between two distantly-related Rosaceae genomes: *Prunus* (the stone fruits) and *Fragaria* (the strawberry) BMC Plant Biol 8:67

Wang Y, Georgi LL, Zhebentyayeva TN, Reighart GL, Scorza R, Abbott AG (2002) High-throughput targeted SSR marker development in peach (*Prunus persica*), Genome 45: 319–328

Wendel JF (2000) Genome evolution in polyploids. Plant Mol Biol 42:225–249

The Welcome Trust Case Control Consortium (2007) Genome-wide association study of 14,000 cases of seven common diseases and 3,000 shared controls. Nature 447:661–678

Wu C, Sun S, Nimmakayala P, Santos FA, Meksem K, Springman R, Ding K, Lightfoot DA, Zhang H-B (2004) A BAC- and BIBAC-based physical map of the soybean genome, Genome 14:319–326

Xu M, Korban SS, Song J, Jiang J (2002) Constructing a bacterial artificial chromosome library of the apple cultivar GoldRush, Acta Hort 595:103–112

Xu M, Song J, Cheng Z, Jiang J, Korban SS (2001) A bacterial artificial chromosome (BAC) library of Malus floribunda 821 and contig construction for positional cloning of the apple scab resistance gene Vf. Genome 44:1104–1113

Yamamoto T, Kimura T, Sawamura Y, Kotobuki K, Ban Y, Hayashi T, Matsuta N (2001) SSRs isolated from apple can identify polymorphism and genetic diversity in pear. Theor Appl Genet 102: 865–870

Yamamoto T, Kimura T, Shoda M, Ban Y, Hayashi T, Matsuta N (2002) Development of microsatellite markers in Japanese pear (*Pyrus pyrifolia* Nakai). Mol Ecol Notes 2: 14–16

Yamamoto T, Kimura T, Saito T, Kotobuki K, Matsuta N, Liebhard R, Gessler C, van de Weg WE, Hayashi T (2004a). Genetic linkage maps of Japanese and European pears aligned to the apple consensus map. Acta Hort 663: 51–56

Yamamoto T, Kimura T, Soejima J, Sanada T, Ban Y, Hayashi T (2004b) Identification of quince varieties using SSR markers developed from pear and apple. Breed Sci.54: 239–244

Yamamoto T, Kimura T, Terakami S, Nishitani C, Sawamura Y, Saito T, Kotobuki K, Hayashi T (2007) Integrated reference genetic linkage maps of pear based on SSRs and AFLPs. Breed. Sci 57:321–329

Zhang LH, Byrne DH, Ballard RE, Rajapakse S (2006). Microsatellite marker development in rose and its application in tetraploid mapping. J Am Soc Hortic Sci 131:380–387

# 27. Application of Genetic Markers in Rosaceous Crops

Vincent G.M. Bus, Daniel Esmenjaud, Emily Buck, and François Laurens

## 1 Introduction

Genetic markers can assist plant breeders to improve their breeding outcomes in several ways, from assessing genetic diversity of the germplasm used in breeding to marker assisted selection (MAS) to variety protection (Charcosset and Moreau, 2004). Major investments in genomic research over the last 15 years have made a wealth of markers available to breeders for MAS, sometimes called marker assisted breeding (MAB), for traits of interest in the different rosaceous crops. Research on marker identification and mapping is most advanced in apple (Gardiner et al., 2007), followed by peach/nectarine (Abbott et al., 2007), and has been reviewed in previous chapters of this book. Breeders could see the advantages of MAS when the technology was still in its infancy in the 1980s, when the potential of MAS as an integrated tool in breeding was demonstrated through modelling (Lande and Thompson, 1990; Lande, 1992). However, while markers have been developed for many traits in nearly as many crops (Young, 1999), their application in MAS has been relatively low even in the breeding of cereals (Dekkers and Hospital, 2002), and is even lower in perennial crops, including the rosaceous crops (Byrne, 2007). In a recent survey of over 100 fruit and ornamental breeding programs of perennial crops from around the world, cost was identified as one of the main reasons why less than 50% of breeders used molecular markers in their programs (Byrne, 2007). Also, the use of genetic markers may be underestimated, as some commercial breeding companies may prefer to keep the genetic markers as a trade secret (Davis et al., 2007).

Markers are increasingly used for MAS in breeding particularly in apple (Gardiner et al., 2006; Kellerhals et al., 2009). MAS typically involves targeted

V.G.M. Bus (✉)
The New Zealand Institute for Plant and Food Research Limited, Hawke's Bay Research Centre, 265 Lawn Rd, Hastings 4172, New Zealand

K.M. Folta, S.E. Gardiner (eds.), *Genetics and Genomics of Rosaceae*, Plant Genetics and Genomics: Crops and Models 6, DOI 10.1007/978-0-387-77491-6_27,

selection with a few markers in large breeding families (Dayteg et al., 2007). The reason for the low uptake of marker technology in the Rosaceae may initially have been as much about having access to the technology as well as the cost of development and implementation, which is relatively high for small crops if the resources are not shared. Also, MAS development is not always straightforward (Young, 1999) because of: (1) a limited understanding of gene networks, epistatic and pleiotropic effects, and genotype by environment effects (Ribaut and Ragot, 2007); (2) traits that are mostly quantitatively inherited; and (3) a limited knowledge on allelic diversity of quantitative trait loci (QTL). Furthermore, a degree of complexity is added in the case of the Rosaceae, as most of them are outcrossing species with levels of heterozygosity that are higher than the phenotypic levels of variation. This, however, has not hampered the major efforts going into the mapping of QTL for complex traits, such as fruit quality in apple (King et al., 2001; Liebhard et al., 2003a; Costa et al., 2009; Kenis et al., 2008; Picañol et al., 2009; Tartarini et al., 2008) and disease resistance in several crops (Liebhard et al., 2003b; Calenge et al., 2004, 2005a; Dondini et al., 2004; Linde et al., 2006). MAS for QTL has advanced to the validation and application stages in apple following the identification of QTL for scab and fire blight resistance in apple (Liebhard et al., 2003b; Calenge et al., 2004; Khan et al., 2007; Tartarini et al., 2008).

Assuming that polygenic traits are controlled by a number of QTL that are inherited in a Mendelian fashion (Hospital and Charcosset, 1997), the principles of MAS as applied to major genes apply to QTL, too. The issue with QTL, however, is that because of their often minor effects on the traits of interest, their map positions are less precise than for major genes, and are defined by an interval with a high likelihood to contain the QTL. As a consequence, it is recommended to use several well-chosen markers spanning the interval to ensure the QTL is introgressed (Gimelfarb and Lande, 1995). However, when comparing MAS with phenotypic selection, MAS was found to be very efficient for fixing QTL alleles, but that genetic gain was not necessarily larger than with phenotypic selection because of genetic and/or environmental interactions (Moreau et al., 2004). The application of MAS for QTL has been shown to be effective for the introgression of desired traits through backcrossing in self-pollinating crops where application of this technology is more advanced. In tomato, five chromosomal regions with 11 QTL for fruit quality traits were successfully introgressed from cherry tomato into large-fruited cultivars with the aid of MAS (Lecomte et al., 2004), although some QTL lacked consistency even in related backgrounds because of gene interactions (Chaïb et al., 2006). One to three markers were used for each chromosomal region, where two markers were sometimes more effective than three. To reduce the cost through lower marker data points (MDP), stepwise selection was advised, as the selection involved MAS for five putative chromosomal regions for acylsugar-mediated multiple pest (Lawson et al., 1997) and black mold (*Alternaria alternata*) (Robert et al., 2001) resistance in tomato. Breeding families were screened first with polymerase chain reaction (PCR)-based markers before restriction fragment length polymorphism (RFLP) markers were applied on sub-sets of the families. Whereas in the case of black mold MAS was successful, as the heritability was increased compared with phenotypic selection, with only a few of the selected plants not carrying all the resistance QTL (Robert et al., 2001), the introgression of the pest resistance was less effective, as the genetics of the resistance were not completely understood and additional phenotypic evaluation was required (Lawson et al., 1997). In maize, the marker-assisted backcross approach (MABC) targeting five chromosomal regions involved in drought-related yield and flower traits was regarded as very effective, even though large populations requiring many MDP for genotyping were generated for each breeding cycle (Ribaut and Ragot, 2007). Although most rosaceous crops involve outbreeding species, the expectation is that MAS will also be successful in these crops.

In this section, we review the progress in the application of genetic markers in rosaceous crops. We will discuss: (1) the cost-effectiveness of MAS; (2) advances in marker technology; and (3) the success of marker application. For the last, we will evaluate their implementation in the Rosaceae as a selection tool in foreground selection (Hospital and Charcosset, 1997), based on some examples, and touch on their use in background selection approaches (Hillel et al., 1990). In the absence of information in the Rosaceae, examples on selection strategies for genetic gain based on genotypic and phenotypic selection will be drawn from experience gained in other, mostly self-pollinating, crops (Charcosset and Moreau, 2004). A major contrast between these and the rosaceous crops is that in the latter, MAS is mostly still at the validation stage, whereas for some of the major food crops, complete systems covering several generations have been designed. This is a reflection of MAS having primarily been applied in the Rosaceae for the pyramiding of resistance genes that have been introgressed separately in the final stages of backcrossing, a strategy that has proved effective in e.g. rice (Hittalmani et al., 2000; Singh et al., 2001; Gopalakrishnan et al., 2008) and wheat (Nocente et al., 2007).

## 2 The Cost/Benefit of Marker Assisted Selection

### *2.1 Genotypic Versus Phenotypic Selection*

As trees of many of the rosaceous fruit crops have long juvenile periods, early genotypic selection promises major advantages over phenotypic selection. In resistance breeding, it is the only route to early selection for pyramided resistance genes, by-passing a cumbersome test-cross generation otherwise needed because epistatic effects hinder phenotypic selection (Luby and Shaw, 2001). The general premise for successful MAS is that markers should be closely linked to the gene of interest, efficient for the screening of large populations, highly reproducible, economical to use, and user-friendly (Mohan et al., 1997), in order to improve the probability of selecting superior genotypes (Knapp, 1998). MAS provides the opportunity for early selection for traits expressed in mature plants and increased accuracy of difficult-to-phenotype traits, particularly when they have a low heritability (Xie and Xu, 1998; Kumar, 1999; Oraguzie et al., 2004b; Collard et al., 2005), and much smaller populations are required to ensure the selection of progeny with the desired combination of traits (Knapp, 1998). Once the genetics of a trait are fully understood, MAS could be reduced to 'genotype building' to create individuals with the desired combination of alleles at each locus (Dekkers and Hospital, 2002). Modelling, to date mostly done for self-pollinating crops, of the foreground selection process in which a limited number of loci are targeted, has shown that fewer generations are required than with phenotypic selection (Servin et al., 2004). The effectiveness of MAS can be improved further by extending it to background selection to select ideotypes for a range of desired traits (Hospital and Charcosset, 1997). As this requires a large number of genetic markers, optimal breeding and selection strategies need to be designed to reduce the otherwise high costs.

Not only the cost of genotypic *versus* phenotypic selections needs to be weighed up (Luby and Shaw, 2001), but also the return of investment on getting a new product to market. The cost of MAS is determined by the number of MDP as a function of the number of progeny times the number of generations required to achieve breeding goals (Frisch and Melchinger, 2001). Generally, minimal MDP are required when several selection steps are applied, the population size is increased over the generations, and target genes are combined in early backcross stages. While there are some reports of MAS being more cost-effective than phenotypic screening when targeting a single gene, particularly when it is recessive (e.g. Iyer-Pascuzzi and McCouch, 2007), or QTL (e.g. Yu et al., 2000), the choice between MAS and conventional selection involves a trade-off between time and money, which usually reflects the contrast between public and private breeding programs (Morris et al., 2003). A plant breeder must have a detailed understanding of the cost structure for both genotypic and phenotypic screening to determine the economic benefits of precision selection of MAS (Dreher et al., 2003). This includes the time savings achieved for early foreground selection of mature traits, as well as for background selection for the donor genome to reduce populations. As these costs are highly specific to each breeding program, cost savings will only be discussed in general terms in this chapter, except to say that labour usually is the highest cost factor in both phenotypic and genotypic selection. A major reduction in labour costs in genotypic selection can be achieved by automating the most labour-intensive step of DNA extraction (Dilworth and Frey, 2000; Mace et al., 2003) once leaves have been harvested from the plants (Dayteg et al., 2007). Leaf harvesting not only is a labour-intensive step, but also is one prone to errors related to sampling the correct plants and placing the samples in the correct wells of the plates, as they generally are the carrier of choice. While the financial gains in staple arable crops (e.g. Wang et al., 2005) mostly involve the improvement of production-related traits (Lande and Thompson, 1990), in horticultural crops they more often are based on value-added traits that differentiate new products in the marketplace, such as colored skin and flesh, and novel flavor and texture traits in rosaceous crops. The value of new products is greatly enhanced by being first to market (Morris et al., 2003) and the chances of this are increased through the application of MAS to accelerate development (Dekkers and Hospital, 2002). The economic advantage of a gene should be at least equal to the duration of its introgression multiplied by the average economic progress of the crop per unit of time (Charcosset and Moreau, 2004).

## 2.2 Marker Assisted Selection Technology

The current high costs of the technology from DNA extraction and genotyping to data analysis (Dekkers and Hospital, 2002), will see MAS implemented in major crops first e.g. rice (Hittalmani et al. 2000; Singh et al. 2001) and wheat (Nocente et al. 2007), however, the expectation is that the development of high-throughput systems will reduce costs. Polymerase chain reaction (PCR)-based markers have a considerable advantage over other types of markers (Batley and Edwards, 2007). For example, PCR markers have replaced isozyme markers and genomic DNA blot analysis in genotyping *Prunus* accessions for their incompatibility alleles because of their simplicity, reproducibility, and ease of use (Tao et al., 1999; Barckley et al., 2006). Removing the electrophoresis step will considerably improve the efficiency of PCR-based MAS (Gu et al., 1995) and the high resolution melting technique (Liew et al., 2004; Montgomery et al., 2007) is a major enabling step in the direction of high-throughput MAS. The high specificity of simple sequence repeat (SSR) or microsatellite markers combined with the possibility of multiplexing (Frey et al., 2004) makes marker-intensive selection, such as background selection, more readily achievable. SSR analysis requires DNA of high quality involving more purification steps (Kasem et al., 2008), which the latest generation of automated DNA extraction systems are able to deliver (Mace et al., 2003), but is compensated for by the relatively minor cost increase of multiplexing compared with single assays (Dayteg et al., 2007). Another major strength of SSR markers is that they are often highly polymorphic and hence very informative.

A major cost of MAS is upfront capital investment in equipment, which often is ignored in the cost estimates (Morris et al., 2003). The use of a single core system is advisable as it not only minimises the investment required, but also provides more flexibility and improves the reliability (Dayteg et al., 2007). Another significant cost is the process of identifying genetic markers linked to traits of interest (Young, 1999), which becomes more expensive as the genetic complexity of the trait of interest increases and the contribution of the individual alleles to the trait decreases. Strategies developed to increase the efficiency of marker identification, such as the candidate gene approach (Gardiner et al., 2003; Baldi et al., 2004; Calenge et al., 2005b; Peace et al., 2005; Naik et al., 2006; Chagné et al., 2007), genome scanning approach (GSA) (Patocchi et al., 2005; Erdin et al., 2006; Silfverberg-Dilworth et al., 2006), bin mapping (Howad et al., 2005; Celton et al., 2008), and the use of genomics technologies to screen expressed sequence tags (ESTs) (Newcomb et al., 2006; Chagné et al., 2007), have been applied to the rosaceous crops. A more extensive technique of QTL identification is the pedigree genotyping approach (Van de Weg et al., 2004, 2009), which takes advantage of the several generations of backcrossing through which many breeding programs have gone. Genomic regions and allelic variants of genes are linked with phenotypic traits across generations and families, i.e. identity by descent, rather than using a single large family to map these traits, as is commonly done in QTL mapping. The FlexQTL™ software (Biometris, Wageningen UR, NL) was developed for multiple QTL mapping (Bink et al., 2009) and was successfully validated for acidity and fruit firmness QTL (Stankiewicz-Kosyl et al., 2009; Tartarini et al., 2009).

Once markers have been identified, MAS can increase economic returns, as the larger selection gains compensate for the higher costs of MAS because higher selection gains compared with phenotypic selection (Moreau et al., 2000) will accelerate the breeding process (Yousef and Juvik, 2001).

## 2.3 Advances in Marker Technology

As a result of the large numbers of breeding lines that need to be genotyped for MAS, molecular breeders of rosaceous crops face a technical challenge to increase throughput. Laboratory automation is becoming a 'must do' if one wants to reduce the genotyping cost. Robotic liquid handling platforms and fully automated DNA extraction systems are being utilized in rosaceous crops to remove bottlenecks such as DNA purification, PCR reaction set up, electrophoresis (Frey et al., 2004; Cook and Gardiner, 2004) and data collection. MAS on tens or even hundreds of thousands seedlings per year (Dayteg et al., 2007) is now achievable for genotyping by a single laboratory equipped with up-to-date automation devices. Automated systems are less prone to human errors and enable better traceability of the developed breeding lines and their genotypes, which is important for breeders who want to achieve high quality standards for commercial outcomes.

As genotyping a large number of seedlings becomes more feasible, the next challenge for the breeders is to select for a larger number of traits/markers. Some types of molecular markers are suitable for multiplexing, but may be limited to a maximum number per reaction. PCR-free large parallel assays are the future for high-throughput genotyping technologies based on a high density of informative markers covering the whole genome. Because of their abundance in the genome, SNPs, single feature polymorphisms (SFPs) and restriction site-associated DNA (RAD) markers will be some of the markers of choice in array-based MAS (Edward et al., 2008; Gupta et al., 2008). A high-throughput technique of up to 500,000 scores per reaction using the Affymetrics GeneChip 500 K array has been developed for the human genetics project (Consortium TWTCC, 2007). Finding a causative mutation using the candidate gene approach seemed like a daunting task a few years ago, but has since been applied to rosaceous crops. The decreasing price of sequencing and fast-paced improvement in the sequencing technologies makes it conceivable for parents of each breeding line, and even each progeny, to be sequenced in the near future. The relevant polymorphisms could readily be screened in their progeny using high-throughput assays, such as the diversity array technology (DArT) (Jaccoud et al., 2001; Kilian et al., 2005) and MassARRAY system (Irwin, 2008).

## 2.4 Breeding Strategies and Marker Assisted Selection

In some cases, a combination of phenotypic and genotypic selection will be most effective in achieving the highest genetic gain. An example is where QTL have remained undetected in mapping studies (Wilde et al., 2007) or where QTL are too small to justify the investment in genotyping a large population with the many MDP required (Lande and Thompson, 1990). Stepwise or tandem selection based on both genotype and phenotype within each generation or across generations, starting with marker only selection in the first generation(s), will be most efficient in identifying progeny carrying all the genetic factors that contribute to the trait being expressed at the desired level (Charcosset and Moreau, 2004; Liu et al., 2004) and was demonstrated for *Fusarium* head blight resistance in wheat (Wilde et al., 2007, 2008).

Closely linked markers, preferably within 1 cM of the gene (Gupta et al., 1999), will ensure their long-term usefulness for MAS over many generations of breeding (Hospital et al., 2000, 2001). The availability of several more distant markers that flank the gene on either side will also result in sufficient precision of MAS in the short-term (Spelman and Bovenhuis, 1998; Gardiner et al., 2007), but not in the long-term, as linkages are expected to be broken. Higher transportability is expected of markers identified through linkage disequilibrium (LD) mapping, as linkage is preserved following numerous recombination events in diverse germplasm (Oraguzie et al., 2007b). Genes themselves functioning as markers or functional markers derived from the gene (Andersen and Lübberstedt, 2003; Forster et al., 2008) enable gene assisted selection (Byrne, 2003; Wilcox et al., 2007) and the very highest selection efficiency. A very good example of this are the endopolygalacturonase markers endoPG-4 and endoPG-5 that can discriminate and perfectly predict the three phenotypes based on the allelic combinations at the *Freestone-Melting flesh* locus in peach (Peace et al., 2005).

# 3 Marker Applications in the Rosaceae

## *3.1 Genetic Diversity and Breeding Parent Selection*

A set of diverse germplasm is required for the long-term success of any breeding program with evolving breeding objectives. The maintenance of germplasm often is more about maintaining genes, rather than accessions with specific phenotypes, hence an understanding of the diversity and phylogeny of species at the molecular level is required for the acquisition of germplasm (Rajapakse, 2003). Genetic markers can help breeders to maintain and exploit gene pools efficiently by reducing the duplication of genes and establishing a ‘core collection’ that represents most of the diversity of a genus (Kumar, 1999). Such a core collection has been established for apple (Hokanson et al., 1998, 2001, 2001) and has been studied for example health compounds (Stushnoff et al., 2003) and winter hardiness (Luby et al., 2004). A specific core collection has been developed for *Malus sieversii* (Volk et al., 2005), which is regarded as a major progenitor of the domesticated apple (Vavilov, 1951; Hokanson et al., 1997; Forsline et al., 2003). The discriminatory power of markers is relied upon for the fingerprinting of cultivars for variety protection (Smith and Helentjaris, 1996). Fingerprinting has been applied specifically to rosaceous crops for this purpose (Koller et al., 1993; Gerlach and Stosser, 1998; Testolin et al., 2000); however, it is more extensively used to understand the genetic relationships within and/or between species (e.g. Dunemann et al., 1994; Oliveira et al., 1999; Oraguzie et al., 2001, 2005; Teng et al., 2001, 2002; Harris et al., 2002; Kimura et al., 2003; Nybom, 2003; Coart et al., 2006; Guarino et al., 2006; Volk et al., 2006, 2008; Wiedow, 2006; Ramos-Cabrer et al., 2007; Sorkheh et al., 2007; Zhebentyayeva et al., 2008a). SSR markers are also very useful for studying synteny between distant species within a genus as well as across genera within the Rosaceae (Dirlewanger et al., 2004b; Yamamoto et al., 2004; Howad et al., 2005; Arús and Gardiner, 2007; Dondini et al., 2007; Celton et al., 2008) and breeders can take advantage of these findings to identify markers for traits of interest in their specialist crop.

Markers can help to identify putative parents of cultivars derived from chance seedlings (Nybom and Schaal, 1990; Harada et al., 1993; Gardiner et al., 1996a), and help to rectify 'breeder errors'. For example, accessions 'Nova Easygro' (Crowe, 1975) and PRI 1640–100 (Giongo et al., 2001) were thought to carry the *Vr* scab resistance gene from Russian apple R12740-7A (Dayton and Williams, 1968), but the marker evidence for 'Nova Easygro' actually carrying *Rvi6* (previously *Vf* (Bus et al., 2009)) is particularly strong (Cheng et al., 1995; Gianfranceschi et al., 1996). Another example of breeder error is accession A722-7, which was expected to carry the *Rvi12* (*Vb*) gene from Hansen's baccata #2 (Alston, pers. comm.), was shown to carry the *Rvi11* (*Vbj*) gene from *M. baccata jackii* (Gygax et al., 2004). Also, two accessions of OR18T26 ('Starking' × *M. baccata jackii*) were shown to be genetically different, with only one of them actually having derived from *M. baccata jackii*. A third example are the $F_2$ selections 26829-2-2 and 26830-2 of *M. floribunda* 821 (Hough et al., 1953), which both carried a CHVf-1 marker allele that was present in neither parent of the 'Rome Beauty' × *M. floribunda* 821 cross to commence the introgression of this gene into apple cultivars (Vinatzer et al., 2004). Markers can also help to clarify the genetics of a trait. Although the scab resistance of Russian apple has long been known to be complex (Dayton et al., 1953), it has mostly been referred to simply as *Vr*, suggesting it to be a single gene resistance. Later, the name *Vr* was associated with a gene conditioning stellate necrotic (SN) resistance reactions (Aldwinckle et al., 1976); however, following the mapping of this gene (Hemmat et al., 2002) and the *Rvi2* (*Vh2*) gene from Russian apple R12740-7A, which also conditions SN reactions, to the same region on linkage group 2 (LG2), it has recently been argued that they are the same gene, and that *Vr* conditions chlorotic (Chl) reactions similar to those induced by *Rvi6* (Bus et al., 2005b). As resistance genes often reside in gene clusters that may carry several functional alleles, the ability of pathogen strains to overcome multiple resistance genes clouds the interpretation of gene-for-gene relationships and hence the mapping of resistance genes. This was demonstrated by the *Rvi2* and *Rvi8* (*Vh8*) scab resistance genes, for which differential races of *Venturia inaequalis* were required to discern the genes where most of the markers were unable to do so (Bus et al., 2005a). Also, markers for resistance genes can sometimes be present in susceptible accessions, e.g. randomly amplified polymorphic DNA (RAPD) markers for the scab resistance genes *Rvi2*, *Rvi5* (*Vm*), *Rvi6* and *Rvi12* were present in *M. sieversii* germplasm that showed scab symptoms (Luby et al., 2001). The 433 bp allele of a sequence characterised amplified region (SCAR) marker, L19SCAR, for the latter gene (Bus et al., 2005a) is present in several high quality susceptible apple cultivars, and emphasizes the need for breeders to ensure they select informative markers for use in MAS (Gardiner et al., 2006).

# 4 Marker Assisted Selection

## *4.1 Foreground Selection*

The introduction of a specific trait into a new cultivar generally involves a two-stage selection process, where foreground selection for one or a few specific traits is performed, followed by a second selection step for the combination of all the other traits of interest within the subpopulation carrying the main trait or traits. This process is commonly applied to several generations of (pseudo)backcrossing to introgress specific traits from inferior genetic backgrounds into high quality cultivars (Dekkers and Hospital, 2002). Early selection for individuals carrying the favorable alleles for the main trait is essential to increase the efficiency of the breeding process and MAS is an effective tool to achieve this, as it removes the need for test crosses in the case of pyramided resistance genes, and enables selection for traits that are not expressed until the plants are mature.

MAS is a useful way to select for resistance to pests and diseases that are not present in the geographical location of the breeding program (Collard et al., 2005). For example, the availability of genetic markers for the *Sd-1* (Roche et al., 1997; Cevik and King, 2002), *Vn-k* (Terakami et al., 2006) and PPV (Sicard et al., 2008; Soriano et al., 2008; Zhebentyayeva et al., 2008a) genes allows New Zealand breeders to select for apple leaf-curling aphid (*Dysaphis devecta*), pear scab (*Venturia nashicola*) and plum pox virus resistance, respectively, as none is present in this country.

## *4.2 Resistance Gene Pyramiding*

In general terms, gene pyramiding is defined as the accumulation of a number of genes to create an ideotype (Servin et al., 2004). However, most reports on this topic have focused on the pyramiding of resistance genes to develop durable resistances to pests and diseases. MAS is mostly applied in resistance breeding because of its potential value in the selection of breeding lines with durable resistance, and genetic marker development was mainly initiated with this in mind. While there are some examples of single resistance genes having proved to be durable (Johnson, 2000), there are more examples of single genes, or sometimes combinations of resistance genes, that have been overcome by pathogens. Validation of the durability of resistance gene combinations is difficult as it is determined *a posteriori* (Alexander and

Bramel-Cox, 1991); however, the validity of the strategy can be deduced from the fact that many natural pyramids, which in the case of apple often consist of major effect genes (Bus, 2006), have shown to be durable. In the absence of information as to which gene combinations may provide the most durable resistance, it seems prudent to breed for pyramids of major genes. The gene pyramiding itself may, however, be delayed until the genes are in a genetic background suitable for cultivar breeding, as this approach will require fewer MDP across the whole process of gene introgression, provided the initial maintenance of separate breeding lines in the field is cost-effective.

As many of these major genes condition distinct phenotypic resistance reactions, they lend themselves well to studying the effectiveness of MAS for gene pyramiding. In most cases the pyramiding is applied to resistance genes that are inherited independently, but sometimes it involves the selection of linked genes. This requires a well-planned breeding strategy (Bus et al., 2004) as well as markers that are either gene-specific or produce gene-specific alleles to identify the individual genes. For example, L19SCAR, as mentioned above, does not distinguish the linked *Rvi2* and *Rvi8* genes (Bus et al., 2005a). However, several multi-allelic SSR markers with specific alleles linked to each of the genes are able to discriminate between them (Gardiner et al., 2006).

In the following we provide several examples of breeders working towards MAS application in the Rosaceae (most crops) and their reasons to do so, and actually applying it (mainly for resistance genes in apple and self-incompatibility in *Prunus*). The examples show the usefulness of genetic markers and MAS covering the principles of e.g. epistatic effects and synteny between species.

## *4.3 Malus*

There are many breeding programs for apple, which all have improving fruit quality and disease resistance as their main objectives (Laurens, 1999). Markers increasingly become available for traits related to fruit quality and are ready for use in MAS (Chagné et al., 2007; Gardiner et al., 2007; Oraguzie et al., 2004; Costa et al., 2008, 2009; Kenis et al., 2008; Zhu and Barritt, 2008). However, most MAS validation and applications concern disease resistance, where the focus to date has been on the pyramiding of resistance genes to apple scab (Bus et al., 2000, 2002; Kellerhals et al., 2008) as most markers have been developed for this disease (Gardiner et al., 2007).

## *4.4 Pyramiding Apple Scab Resistance Genes 1: Rvi6 × Rvi2*

Genetic markers that have been developed from the resistance genes themselves are very powerful, as they are expected to provide a 100% success rate. The Vfa2SCAR is derived from one of the four paralogs in the *Rvi6* scab resistance gene cluster

(Vinatzer et al., 2001; Xu and Korban, 2002). In the sub-set of plants showing the typical Chl resistance reactions with limited sporulation typical of the *Rvi6* gene of a A157R08T149 (*Rvi6*) × A068R03T057 (*Rvi2*) family, the marker did indeed not show any recombination (Table 1). The Chl progeny comprised both the Class 3A and 3B (Chevalier et al., 1991) seedlings, which confirmed that seedlings of both these classes indeed carry the gene (Gardiner et al., 1996b). With *Rvi6* mapping to LG1 (Maliepaard et al., 1998) and *Rvi2* to LG2 (Bus et al., 2005b), the genes were expected to segregate independently. However, the phenotypes did not meet the segregation ratio of SN:Chl:S = 2:1:1 ($P(\chi^2>8.33; df=2) = 0.02$) as there was a distortion towards susceptible (S) plants (Table 1). Such distortions are common for both genes (Bus et al., 2005b; Gao and van de Weg, 2006). Out of the 354 seedlings that showed either Chl resistance reactions or S symptoms, 9.0% were recombinants, all of them susceptible seedlings with the Vfa2 marker (Table 1). The issue of glasshouse screening conditions inducing sporulation on seedlings carrying resistances applies to other resistance genes that condition hypersensitive response (HR) reactions, such as *Rvi2*, too, which would have compounded the error rate of MAS in this example. Considering all 648 seedlings, 15.3% were recombinants for the L19SCAR linked to the *Rvi2* gene, with 9.9% of the SN seedlings not showing the marker, but 19.8% of the non-SN plants showing the marker. The overall recombination rate for this gene was much higher than expected based on earlier research, where the marker was mapped much closer to the gene (Bus et al., 2002, 2005b; Gardiner et al., 2007). Genotypic selection in the A157R08T149 × A068R03T057 family for pyramided scab resistance genes based on the L19SCAR and Vfa2SCAR markers required 1,296 MDP and reduced the population to 158 seedlings, of which, however, only 107 (68%) actually carried the *Rvi2* gene based on the number of progeny showing the SN phenotype typical of this gene (Table 1). We attribute such a high level of sporulating seedlings to the high disease pressure of the glasshouse screening, which will make some of the seedlings carrying the gene appear susceptible to varying degrees, depending on the gene (Bus et al., unpublished data).

**Table 1** Validation of marker assisted selection for progeny with the pyramided *Rvi2(Vh2)* and *Rvi6(Vf)* scab resistance genes with the L19SCAR and Vfa2SCAR markers in a A157R08T149 × A068R03T057 family. L19SCAR is linked to *Rvi2*, which conditions stellate necrotic (SN) resistance reactions, and Vfa2SCAR is linked to *Rvi6*, which conditions chlorotic (Chl) resistance reactions that show some sporulation under glasshouse screening conditions. Susceptible (S) seedlings show abundant sporulation

| Marker | | Phenotype | | | |
|---|---|---|---|---|---|
| L19 | Vfa2 | SN | Chl | S | Total |
| + | + | 107 | 43 | 8 | 158 |
| + | − | 158 | 0 | 19 | 177 |
| − | + | 24 | 119 | 24 | 167 |
| − | − | 5 | 0 | 141 | 146 |
| Total | | 294 | 162 | 192 | 648 |

Additional field observations on the resistance status of apple plants will enable us to determine the accuracy of both the genotypic and the phenotypic glasshouse screens (King et al., 1998; Tartarini et al., 2000). Underperformance of the latter tip the balance in favor of MAS as a more cost-effective strategy in breeding programs even for major resistance genes, particularly if functional markers derived from the genes of interest are available, as a single marker will be sufficient to achieve a 100% success rate.

The variability in the resistance reactions conditioned by the *Rvi6* resistance and its genetics have long been a point of discussion (Gessler, 1989). The variability in resistance expression (Tartarini et al., 2000) is perhaps influenced by the size of the chromosomal region around the gene introgressed from the donor parent *M. floribunda* 821 (King et al., 1999; Huaracha et al., 2004) through linkage drag. No accessions have been reported to date where the introgressions have been reduced to the *Rvi6* gene itself. Such accessions will be superior breeding parents if the extended introgression contained undesirable genes (Hospital, 2001), as has been demonstrated in several cases with tomato (Robert et al., 2001; Lecomte et al., 2004), but they may at the same time have lost possible scab resistance or modifier genes or QTL that enhance the *Rvi6* resistance (Durel et al., 2003). MAS is the primary tool to remove introgressions on either side of the gene of interest, but separate generations will be required to achieve this, as double recombinations generally are very low within a distance of 15–20 cM (Kearsey, 1997). The consistent phenotypes (3A and 3B, rarely 2 or 1) of the *Rvi6* seedlings in the glasshouse scab screening at Plant and Food Research indeed suggest that the resistance-enhancing factors are absent in the breeding parents and that MAS for this gene specifically targets *Rvi6*.

## 4.5 Pyramiding Apple Scab Resistance Genes 2: Rvi2 × Rvi4

The A176R02T281 × A176R03T191 cross was designed to re-combine the *Rvi2* and Rvi4 (*Vh4*) scab resistance genes, which both originated from Russian apple and both map to LG2 at a distance of about 40 cM (Bus et al., 2005b). As the genes were combined through intermediate hosts, they segregated independently without segregation distortion, as confirmed by the phenotypic segregation ratios with the HR resistance (*Rvi4* gene) reactions masking the expressing the SN reactions (*Rvi2*) (HR:SN:S = 2:1:1 (P($\chi^2$>1.53; df=2) = 0.47) (Table 2). The recombination rate of 12.7% between the 8283SNP marker and *Rvi4* was more favorable than expected based on a mapping distance of 17 cM (Gardiner et al., unpublished data), while a recombination rate of 7.0% between the CH05e03SSR marker and *Rvi2* was considerably higher than expected based on earlier mapping distance of about 4 cM or less (Bus et al., 2005b; Gardiner et al., 2007), but less than half compared with L19SCAR in the *Rvi6* × *Rvi2* family in the example above. Similarly to the latter family, the population would have been reduced by more than 75% to 90 plants (Table 2), which is 22.8% of the population, when relying on MAS alone. However, based on the recombination rate of 7% between CH05e03SSR and *Vh2*, only an

**Table 2** Validation of marker assisted selection for progeny with the pyramided *Rvi2* and *Rvi4* scab resistance genes with the 8283SNP and CH05e03SSR markers in a A176R02T281 × A176R03T191 family. 8283SNP (= M4) is linked to *Rvi4*, which conditions hypersensitive response (HR) resistance reactions, and CH05e03SSR (= M2) is linked to *Rvi2*, which conditions stellate necrotic (SN) resistance reactions under glasshouse screening conditions. Susceptible (S) seedlings show abundant sporulation. The genotypes (all assumed in heterozygous state; only one allele is shown for each gene and marker locus) are presented for each marker-phenotype combination, with the four expected genotypes printed in bold

| Marker | | Phenotype | | | |
|---|---|---|---|---|---|
| 8283 | CH05e03 | HR | SN | S | Total |
| + | + | 81<br>***Rvi4* M4 *Rvi2* M2**<br>*Rvi4*M4*rvi2*M2 | 9<br>*rvi4*M4*Rvi2*M2 | 0<br>*rvi4*M4*rvi2*M2 | 90 |
| + | – | 97<br>***Rvi* M4 *rvi2* m2**<br>*Rvi4*M4*Rvi2*m2 | 0<br>*rvi4*M4*Rvi2*m2 | 11<br>*rvi4*M4*rvi2*m2 | 108 |
| – | + | 20<br>*Rvi4*m4*Rvi2*M2<br>*Rvi4*m4*rvi2*M2 | 79<br>***rvi4* m4 *Rvi2* M2** | 3<br>*rvi4*m4*rvi2*M2 | 104 |
| – | – | 10<br>*Rvi4*m4*Rvi2*m2<br>*Rvi4*m4*rvi2*m2 | 10<br>*rvi4*m4*Rvi2*m2 | 75<br>***rvi4* m4 *rvi2* m2** | 95 |
| Total | | 208 | 98 | 89 | 395 |

estimated 75 progeny, i.e. those with genotype *Rvi4*M4*Rvi2*M2, of the 81 HR plants with both markers actually carry both scab resistance genes, which is 83% out of the 90 seedlings showing both markers, or 19.0% of the whole population. Preceding MAS with a phenotypic selection for the HR seedlings reduces the progeny, and therefore the number of MDP with CH05e03SSR, to about half, which in this case will result in the selection of 101 seedlings, of which 94 seedlings (23.8 % of the population) carry both genes (Table 2). Nevertheless, given the relatively long genetic distances of the markers to their respective genes, this is a respectable outcome of MAS, as the findings were in general agreement with the expected outcome for two independently segregating genes.

### 4.6 Pyramiding Woolly Apple Aphid Resistance Genes: *Er1* × *Er3*

In contrast to the scab resistance genes, genes conferring resistance to other pests and diseases usually do not show distinct phenotypes, while phenotyping often is much more cumbersome. For example, apple trees carrying woolly aphid resistance genes either show no symptoms or low numbers of galls that do not appear to vary in size or form among the different genes conditioning them. Recently, the first three *Er* genes were mapped, and *Er1* from 'Northern Spy' and *Er3* from 'Aotea 1' were

**Table 3** Validation of marker assisted selection with the NZsn_O05 marker for pyramided woolly apple aphid resistance genes *Er1* and *Er3* in a 'Northern Spy' ($a_1b$) × S26R01T053 ($a_3c$) family. While the *a* alleles of NZsn_O05 were identical for both resistance genes, the subscripts (1 for *Er1* and 3 for *Er3*) were added to show that the low numbers of $a_1a_3$ and $a_3b$ genotypes are the result of segregation distortions associated with the *Er3* gene in S26R01T053

| | Phenotype | | | | | | | | Segregation | |
|---|---|---|---|---|---|---|---|---|---|---|
| Genotype | 0 | 1 | 2 | 3 | 4 | 5 | R[a] | Total | S[a] | |
| $a_1a_3$ | | 34 | 1 | 0 | 0 | 1 | 2 | 38 | 35 | 3 |
| $a_3b$ | 29 | 2 | 0 | 1 | 1 | 9 | 42 | 31 | 11 | |
| $a_1c$ | | 102 | 25 | 2 | 2 | 3 | 12 | 146 | 127 | 19 |
| $bc$ | 23 | 6 | 4 | 1 | 5 | 109 | 148 | 29 | 119 | |
| Total | | 188 | 34 | 6 | 4 | 10 | 132 | 374 | 222 | 152 |

[a] R = resistant (classes 0–1) ; S = susceptible (classes 2–5).
Adapted from (Bus et al., 2009)

found to be linked, as they both mapped near the top of LG8 (Bus et al., 2009). The genes are used in both apple rootstock and scion cultivar breeding and have been pyramided to develop cultivars with durable resistances. Validation of MAS for progeny carrying both genes in a 'Northern Spy' (*Er1*) × S26R01T053 (*Er3*) family showed that a large segregation distortion associated with the *Er3* gene had occurred (Table 3). As was demonstrated previously (Bus et al., 2000), this did not affect the effectiveness of MAS, as 35 out of 38 plants that were homozygous for the *a* allele of the NZsn˙O05 marker, were indeed resistant to the pest. However, with the marker mapping about 8 cM (p = 0.08) from *Er1* and 4 cM ($q = 0.04$) from *Er3*, an estimated $n(1–p)(1–q) = 31$ progeny, or 81% of all 38 homozygous progeny (Table 3), would indeed carry both genes, with the other resistant plants carrying either *Er1* or *Er3* only.

In the first generation, all plants with both resistance genes will carry them in repulsion phase, but in the next generation it is possible to select progeny carrying them in coupling phase (Bus et al., 2004). This will facilitate the efficient backcrossing of the combined genes into new cultivars, as well as the addition of other woolly aphid resistance genes into the pyramid.

# 5 Prunus

## 5.1 Genetic Mapping of Traits

In *Prunus*, genetic maps have been developed for all the economically important crops of this genus and MAS for horticultural traits is progressing towards the implementation phase. With 22 major genes and 14 QTL mapped in peach/nectarine (Abbott et al., 2007), breeders can commence applying MAS to a number of traits of interest in this crop, but there are no reports of the success of MAS to date. In plum, the focus has been on resistance to root-knot nematodes (RKN) (see below) and plum pox virus (Esmenjaud and Dirlewanger, 2007), with the latter resistance also being an important trait in apricot (Lalli et al., 2008; Sicard et al., 2008; Soriano et al., 2008). In almond, seven major genes (Martínez-Gómez et al., 2007) and five candidate genes for drought resistance (Campalans et al., 2001) have been mapped. MAS is routinely applied in almond breeding to assign breeding parents to cross-incompatibility groups (Ortega and Dicenta, 2003; Barckley et al., 2006). MAS for self-incompatibility is also successfully applied to apricot (Hormaza et al., 2007) and cherry (Dirlewanger et al., 2007), where markers derived from S-ribonucleases (Tao et al., 1999) are used to genotype sweet cherry (reviewed by Iezzoni et al., 2005), Japanese apricot (Tao et al., 2000; Yaegaki et al., 2001) and Chinese apricot (Qi et al., 2005). A high-throughput system based on microarray technology has now been developed for the rapid and accurate genotyping of *Prunus* germplasm for self-incompatibility (Pasquer et al., 2008).

Since many traits, such as self-incompatibility, fruit texture, and pest and disease resistance are of common interest to the different *Prunus* crops, comparative mapping (e.g. Joobeur et al., 1998; Cipriani et al., 1999; Dirlewanger et al., 2004a; Clarke et al., 2008; Olmstead et al., 2008) is well advanced in this genus. This research, together with the development of physical maps in the major rosaceous genera *Prunus* (Zhebentyayeva et al., 2008b) and *Malus* (Han et al., 2007; Han and Korban, 2008), will serve as a springboard to the Rosaceae at large (e.g. Dirlewanger et al., 2004b).

An example of this is the investigation into the evolutionary relationships among self-incompatibility alleles, which have been studied extensively in apple (e.g. Nybom et al., 2008) and pear (e.g. Sanzol and Robbins, 2008) as well as in *Prunus*, showing that the S-ribonucleases from *Malus*/*Pyrus* are distinctly different from those of *Prunus*, since the *Maloidae* diverged more recently (Ma and Oliveira, 2002; Sassa et al., 2007).

## *5.2 Root-Knot Nematode Resistance in Prunus Rootstocks*

*Prunus* species have long been severely affected by various nematodes (Sharpe et al., 1969; Nyczepir, 1991; Nyczepir and Halbrendt, 1993). The most damaging pests among them are the highly polyphagous root knot nematode (RKN), *Meloidogyne* spp., which form characteristic gall symptoms on the roots. As the highly toxic nematicides are progressively removed from the market, resistant rootstocks are the most economic and environmentally sound method for nematode control (Layne, 1987; Cook and Evans, 1987). The predominant species occurring world-wide (Sasser and Freckman, 1987) are *M. arenaria*, *M. incognita* and *M. javanica*. A fourth species, the 'peach RKN', *M. floridensis*, restricted to Southern USA where it is highly destructive (Handoo et al., 2004), overcomes the resistance of the commonly used peach and almond rootstocks. The primary target for gene pyramiding is to combine major gene resistances to multiple nematode species rather than several genes to a single species, since the risk of a nematode species overcoming resistance with extensive economic consequences is limited. This risk is low because the three predominant RKN species reproduce by obligate parthenogenetic mitosis (Triantaphyllou, 1985) and thus have less adaptive capability than meiotic species, and have poor self-dispersal abilities in the soil for virulent strains to spread, unless aided by sub-standard sanitation practices by growers. Despite this, an evaluation test applying a high and continuous pressure of nematode inoculum has been developed to be used at INRA (Esmenjaud et al., 1992) for the screening of high-level resistance genes.

Sources of resistance characterized in *Prunus* spp. exhibit different spectra to the RKN and have either an isolate-, species-, or a genus-specific range of resistance (Esmenjaud et al., 1994). For example, differential resistances to several RKN isolates have been identified in the peach rootstock GF.305. Species-specific resistances have been characterized in the peach cultivar 'Shalil', which is resistant to *M. arenaria* and *M. incognita*, but not to *M. javanica*. A genus-specific resistance has been identified in the 'Myrobalan' plum accessions P.2175 and P.2980 (alleles *Ma1* and *Ma3*, respectively; Table 4) (Esmenjaud et al., 1996; Rubio-Cabetas et al., 1998). They are highly resistant to the four RKN mentioned above (Lecouls et al., 1997; Rubio-Cabetas et al., 1998) and to other tested *Meloidogyne* species, such as *M. mayaguensis* (Esmenjaud et al., 1994, 1997; Rammah and Hirschmann, 1988; Rubio-Cabetas et al., 1999) as no virulent populations are known to date. The *Ma* gene has been mapped to LG7 of the *Prunus* reference map (Claverie et al., 2004a; Dirlewanger et al., 2004a, 2004b), where it forms part of a cluster of three TIR-NBS-LRR candidate genes (Claverie et al., 2006; Esmenjaud and Dirlewanger, 2007).The *Mi* gene conferring *M. incognita* resistance was mapped in the middle of LG2 of 'Nemared', a rootstock derived from 'Nemaguard' (Ramming and Tanner, 1983). STS markers for resistance to *M. incognita* have been obtained from the Japanese peach 'Juseitou' (Yamamoto and Hayashi, 2002). Lastly, a single gene designated *RMia*, with two alleles from 'Nemaguard' ($RMia_{Nem}$) and GF.557 ($RMia_{557}$) controlling resistance to both *M. arenaria* and *M. incognita*, has been demonstrated (Table 4). Since the *RMia* gene shares markers with *Mi* (Yamamoto and Hayashi, 2002), and maps to the same region on LG2 (Claverie et al., 2004a; Dirlewanger et al., 2004a, 2004b), both genes are presumably closely linked or allelic, if not the same. In almond, resistance to *M. javanica* has been identified from the 'Alnem' seedlings and shown to be controlled by a single major gene (Kochba and Spiegel-Roy, 1975, 1976). This source is also resistant to *M. arenaria* (Scotto La Massèse et al., 1984). The gene to *M. javanica* has been preliminarily located on LG 7 in the vicinity of *Ma* and has been designated *RMja* (Esmenjaud, unpublished), but it is not yet known whether *RMja* also controls *M. arenaria* (Table 4).

In *Prunus* rootstock material, the three major RKN resistance genes with a complete dominance, from plum (*Ma*), peach (*RMia*) and almond (*RMja*), are being pyramided (Esmenjaud et al., 2009) in interspecific hybrids of the type 'Myrobalan' × peach (*Ma+Rmia*), 'Myrobalan' × almond (*Ma+RMja*), or 'Myrobalan' × almond-peach (*Ma* + *RMia*), created by INRA fruit breeding units (UREF Bordeaux and UGAFL Avignon). Breeding is particularly stressed for the tri-species hybrid progeny P.2175 × ('Garfi' × 'Nemared')$_{22}$ which is being selected for pyramiding *Ma* and *RMia* genes. In 'Myrobalan' plum, several reliable SCAR markers

**Table 4** Spectra and genetics for root-knot nematode resistance in *Prunus* sources. P.2646 and 'Garfi' are susceptible control rootstocks

| Sub-genus and source material | Resistance spectrum | | | | Genetics of resistance (linkage group) | References |
|---|---|---|---|---|---|---|
| | MA[a] | MI | MJ | MF | | |
| Prunophora | | | | | | |
| *P. cerasifera* (Myrobalan plum) | | | | | | |
| P.2175 (*Ma1/ma*) | R[b] | R | R | R | Single gene (*Ma*) to MA, MI, MJ and MF (LG7) | Esmenjaud et al. (1996) |
| P.2980 (*Ma3/ma*) | R | R | R | R | | Lecouls et al. (1997) |
| P.2646 (*ma/ma*) | S | S | S | S | | |
| Amygdalus | | | | | | |
| *P. persica* (peach) | | | | | | |
| Nemared ($RMia_{Nem}/RMia_{Nem}$) | R | R | R/S | S | Single gene (*RMia*) to MI and MA (LG2) | Lu et al. (2000) |
| Shalil ($RMia_{557}/RMia_{557}$) | R | R | S | S | Single gene (*RMia*) to MI and MA (LG2) | Claverie et al. (2004b) |
| GF.557[c] ($RMia_{557}/rmia_{557}$) | R | R | S | S | | Claverie et al. (2004b) |
| *P. dulcis* (almond) | | | | | | |
| Alnem 1 (*RMja/RMja*) | R | S | R | S | Single gene (*RMja*) to MJ (LG7) | Kochba and Spiegel-Roy (1975); Esmenjaud (unpublished) |
| Alnem 88 (*RMja/RMja*) | R | S | R | S | | |
| Garfi (*rmja/rmja*) | S | S | S | S | | |

[a] MA = *Meloidogyne arenaria*, MI = *M. incognita*, MJ = *M. javanica*, MF = *M. floridensis*
[b] R = resistant, S = susceptible, R/S resistance behaviour depending on the RKN isolate
[c] GF.557 = almond (H) × 'Shalil' (R).
Adapted from Esmenjaud and Dirlewanger (2007).

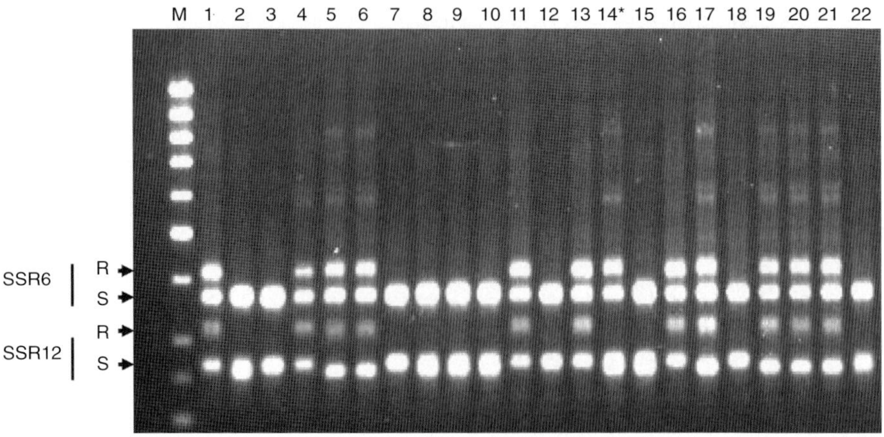


**Fig. 1** Multiplex amplification of the two co-dominant single sequence repeat (SSR) markers SSR6 and SSR12 flanking the *Ma1* allele among 22 segregating *Prunus* individuals of the cross P.2175 × P.2646 (*Ma1/ma* × *ma/ma*). R = resistant marker allele, S = susceptible marker allele. M = 100 bp molecular weight marker ladder. 14∗ = recombinant individual within the window of the two markers

flanking (SCAL19 and SCAFLP4, both <1 cM) or co-segregating (SCAFLP2) with the *Ma1* allele in P.2175 (Lecouls et al., 1999, 2004; Claverie et al., 2004b), have been identified by bulked segregant analysis (BSA) (Michelmore et al., 1991) and are available for MAS of all interspecific crosses. SSR markers for *Ma* have also been identified through the cloning strategy and will be useful, particularly when they flank the gene and can be multiplexed for the direct genotyping of the individuals (Fig. 1). For the *RMia* peach gene, STS markers developed by Yamamoto and Hayashi (2002) can be used, but more reliable markers are being investigated (Esmenjaud, unpublished). Selection of this tri-species material is also in progress under conventional methods for complementary adaptive characteristics, such as tolerance to waterlogging and drought (Esmenjaud, 2004; Dirlewanger et al., 2004c).

# 6 Fragaria

## *6.1 Anthracnose Resistance*

Anthracnose in strawberry is attributed to three species: *Colletotrichum acutatum* Simmonds, *C. fragariae* Brooks and *C. gloeosporioides* (Penz.) Penz. and Saac., which can infect most parts of the plants (Smith, 2008; Denoyes and Baudry, 1995). However, in terms of economic losses that accumulate to millions of dollars worldwide each year, the most severe effects have been attributed to *C. acutatum* (Smith, 2008). The European Economic Community now have strict quarantine controls in place, with destruction of affected plants, as various spraying and management schemes have resulted in mixed success (Freeman, 2008; Sjulin, 2008). Resistant cultivars are seen as the long-term solution to sustainable disease control and MAS plays an important role in developing new cultivars with durable resistance.

Plant accessions and cultivars of *Fragaria* species resistant to *C. acutatum* have been identified and developed through intensive breeding programs (Simpson et al., 1994; Winterbottom, 1991; Denoyes et al., 2005). Two major modes of resistance have been determined: a polygenic quantitative resistance to strains of pathogenicity group 1, and a single dominant gene, *Rca2*, controlling resistance to strains of pathogenicity group 2, although some minor genes may contribute to this resistance in certain cultivars (Denoyes et al., 2005). Two SCAR markers derived from amplified fragment length polymorphism (AFLP) markers were mapped to within 2.8 and 0.6 cM onto one side of *Rca2* in an octoploid cultivated strawberry (*F.* × *ananassa*) 'Capitola' × 'Pajaro' $F_1$-population (Lerceteau-Köhler et al., 2002, 2005). The SCAR markers were then validated within other populations and germplasm to ensure they could be used effectively over several sources of resistance and genetic backgrounds, by screening them against 43 genotypes or cultivars of *F.* × *ananassa* (Lerceteau-Köhler et al., 2005). Their success rate was high, with 81.4% of resistant/susceptible genotypes correctly identified using the closer STS-Rca2˙417 marker. Research is now in progress on the polygenic resistance to pathogenicity group 1. Initial investigations have resulted in the detection of five QTL, none of which correlates to the linkage group containing *Rca2* (Denoyes-Rothan et al., 2004). It is anticipated that the use of STS-Rca2-417 for *Rca2* and markers for major QTL will enable the pyramiding of the resistances to provide a wider spectrum of control of *C. acutatum* in strawberry (Denoyes-Rothan et al., 2004, 2005; Lerceteau-Köhler et al., 2005).

The development and application of MAS technologies within the 'smaller' Rosaceae crops, such as *Rubus* spp. and strawberry, has focused on foreground selection. The main emphasis has been for the selection of disease and pest resistance; however, there also are a few examples of markers for physiological/horticultural traits (Albani et al., 2004; Folta and Davies, 2006; Graham and Smith, 2002; Graham et al., 2006; Weebadde et al., 2008). More interestingly, research into the transferability of molecular markers between rosaceous crops could open up new doors for selection for shared/related traits. This new era of synteny holds great interest for these smaller crops.

## 7 Rubus

### *7.1 Phytophthora Root Rot Resistance in Raspberry*

One of the main diseases in raspberry is root rot caused by the oomycete *Phytophthora fragariae* var. *rubi* (Harrison et al., 1998) and it is regarded as one of the main reasons for raspberry (*Rubus idaeus*) decline within Europe and the Americas

(Wilcox et al., 1993). Fungicides provide marginal disease control in the field and nursery stocks (Wilcox et al., 1999; Duncan and Cooke, 2002); hence attention largely shifted to sourcing resistances that now are extensively used in both American and European programs (Knight, 1991; Jennings and Brennan, 2002; Spiegler and Thoss, 1993; Barritt et al., 1979; Daubeny et al., 1992; Kempler et al., 2002). In 2005, the need for further breeding for *Phytophthora* resistance in raspberry was raised as one of two main areas breeding programs need to focus on (Finn et al., 2005) and several institutes are investigating the use of MAS to increase selection efficiency to develop improved cultivars.

Resistance to *Phytophthora* root rot appears to be under quantitative control (Pattison et al., 2007) with additive (Barritt et al., 1979) or non-additive (Nestby and Heiberg, 1995) gene effects. Using the resistant cultivar 'Latham' and $F_1$, $F_2$, $B_1$ and $B_2$ populations led to the conclusion that a dominant two gene-model was the most likely form of inheritance in these populations, with other minor genes involved. More recent work by Weber et al. (2008) has resulted in the development of two molecular markers being mapped to two QTL involved in disease resistance, plant disease index, petiole lesion and root regeneration score. These markers, SCAR618 and CAP464, were then tested for efficiency on 18 different genotypes, varying in resistance levels, and various populations known to segregate for resistance. Overall, SCAR618 was more efficient than CAP464, with accuracies of 62 and 56%, respectively, but their combined use resulted in the correct identification of 85% of resistant individuals (Weber et al., 2008). The development of these markers and their validation in various breeding material has now made the application of MAS for *Phytophthora* resistance possible. Further marker development with improved efficiencies across a wider range of material facilitated by the identification of resistance gene analogues (RGAs) (Samuelian et al., 2008) is planned for the future, to enable better utility of these markers within breeding programs (Weber, personal communication).

# 8 Background Selection

## *8.1 Whole Genome Selection*

Where foreground selection focuses on the identification of individuals with a targeted trait, either monogenic or polygenic, followed by a second stage selection for general traits, background selection aims to combine both steps into one. Many resistance genes, but also some fruit quality traits have been introgressed from wild species that have little in common with the cultivated species. When the species are crossed, the genome of the resultant progeny will consist of a combination of both parental genomes at variable levels, i.e. the genome of the high quality backcross parent has been diluted. With the aid of whole-genome selection (WGS) (Pradhan et al., 2003; Varshney et al., 2005), the genotype of the cultivated parent can be restored by using MAS to select against the chromosomal regions from the donor parent (Paterson et al., 1991). Applying WGS without targeting specific genes in a recurrent selection program is more suitable for population improvement based on random mating of individuals selected for their genotype (Ribaut and Ragot, 2007). The targeted introduction of traits, however, is more efficient for the development of the ideotype for a crop because of *a priori* knowledge of marker linkage to a trait (Charcosset and Moreau, 2004). It can be achieved by foreground selection for the trait of interest, through e.g. MABC (Ribaut and Ragot, 2007) or pedigree breeding (Christopher et al., 2007), while selecting against the genome of the donor parent with three neutral markers per chromosome (Hospital and Charcosset, 1997). The expectation is that fewer generations would be required for the introgression of the desired trait(s), particularly when a combination of targeted traits are introgressed from several donor parents (Servin et al., 2004). Simulation studies showed that, under the assumption of tight marker-trait linkage, the targeted selection of 50 QTL in a markers-only recurrent selection approach was very effective in increasing the frequencies and fixation of favorable alleles (Hospital et al., 2000).

## 8.2 Reducing the Juvenile Period

In most tree fruit crops, the juvenile period lasts several years, hence the 'fast-breeding' approach is being evaluated, where the breeding process is accelerated by reducing both the number of generations through background selection and the generation times required. The Plant and Food Research approach is based on applying the principles of tree physiology and involves the use of a climate chamber, which greatly limits the number of seedlings to which fast-breeding can be applied. Hence, background selection is applied to reduce progenies to a few plants that carry the target gene(s) in a genetic background most similar to that of the high quality parent. The technology is being validated on a 'Royal Gala' × A689-24 family segregating for the *Pl2* powdery mildew resistance gene (Bus et al., 2000) derived from the crabapple ancestor *M. zumi* and alleles from higher quality ancestors. Since marker-based WGS can be applied to select for differing contributions of the grand-parental genomes, the theoretical limits of such selection pressure were assessed by estimating the contribution of each grandparent to each progeny. A genetic map has been developed for this population and a selected set of markers covering the whole genome is being screened across the progeny, to identify seedlings with the desired new trait as well as high proportions of genome inherited from the high quality grandparent(s)/low proportions of the low quality ancestor(s). The 'Royal Gala' × A689-24 progeny carrying the gene indeed carried a higher proportion of the genome of A368-12, the *Pl2* parent of A689-24 (Fig. 2) (Volz et al., 2009). The elite seedlings are grown on their own roots in controlled environment rooms to advance flowering within two years of crossing, from which a further introgression cycle can be repeated.

Alternative methods of bypassing the juvenile period involve the use of transgenic plants. Overexpression of the *BpMADS4* gene from silver birch (*Betula*

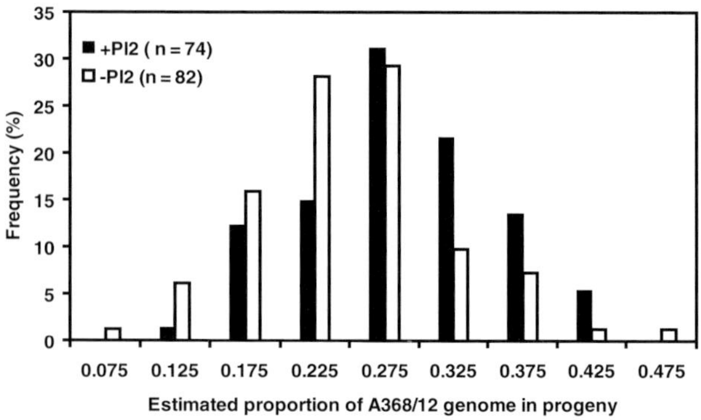


**Fig. 2** Frequency distribution of the estimated proportion of grandparental genome from the *Pl2*-donor A368-12 in 'Royal Gala' × A689-24 apple progeny for powdery mildew resistant (+*Pl2*) and susceptible (–*Pl2*) plants

*pendula*) in apple resulted in some transformants flowering after about four months since the introduction of the gene while still in the *in vitro* phase (Flachowsky et al., 2007). Most flowers were morphologically normal and developed normally into fruit following successful pollination by non-transformed accessions, which confirmed the proof of concept for fast-breeding. An alternative approach to reducing the juvenile period has been the silencing of the *PcTFL*1-1 gene, which was isolated from the pear cultivar 'Spadona' (Flaishman et al., 2009). The transgenic 35S:*PcTFL*1–1 'Spadona' plants produced flowers after five to eight months, which developed into fruit after pollination and produced fertile seed. The seedlings grown from these seed produced flowers five months after sowing, while grafted trees flowered continuously and produced fruit with normal shape, color and taste (Flaishman et al., 2009). The pollen and resultant seed when applied to non-transformed cultivars were also viable. The high and consistent flowering rates give the transgenic plants an advantage over the physiological approach, which has a lower and more unpredictable success rate.

## 9 Conclusion and Perspectives

In this chapter we have presented, without aiming to be complete, an overview of the 'state of the art' of genetic marker application in the rosaceous crops. It is clear that genetic markers greatly assist breeders in their ability to identify the best performing potential cultivars in what ultimately is a phenotypic selection that integrates all traits (Kumar, 1999). However, the level at which genotypic selection will replace phenotypic selection is different for each breeding programme, depending on the available resources. The diverse range of relatively small crops, together with their often disadvantageous genetic and physiologic characteristics, such as outcrossing and therefore heterozygous nature, long juvenile period and large plant size means that genetic and genomic studies in the Rosaceae are more time-consuming and expensive compared with those in the major crops. The continued effort into the mapping of horticultural traits reflects the confidence breeders have in MAS playing an important role in cultivar development as it progresses to the implementation phase in the larger rosaceous crops. These first experiences have shown that some practical limitations will have to be overcome before MAS becomes a selection tool routinely applied for many traits on thousands of seedlings. The current levels at which technologies, such as automated DNA extraction, multiplexing and the creation of dense genetic maps operate are well below the requirements for the future. A major hurdle before cheap high-throughput systems become the norm is the requirement for fast and error-free leaf sampling techniques. The collation and analysis of the large amount of data generated with new technologies, such as SNP discovery involving the miniaturization of high-throughput systems, create an additional level of complexity and need adequate analytical capacity to process them (Gupta et al., 2008; Pattemore et al., 2008). High resolution mapping of a large number of QTL through e.g. association mapping (Oraguzie et al., 2007a) will allow the association of complex phenotypes with alleles of genes, themselves used as functional markers, and their interactions (Paran and Zamir, 2003; Kroymann and Mitchell-Olds, 2005). Knowledge of the allelic diversity can then be converted into a selection index involving the weighting of the different loci as a practical and efficient tool for the breeders.

With genotyping technology advancing rapidly to the stage where genome sequencing of each progeny will be commonplace, accurate phenotyping forms the bottleneck to breeders taking full advantage of MAS. The identification of QTL in general and small QTL in particular needs a considerable investment in time, and development of the appropriate phenotyping skills needs to be weighed against the economic benefits of MAS for the traits of interest.

## References


Abbott AG, Arús P, and Scorza R (2007) Peach. In: Kole C (ed) Genome Mapping and Molecular Breeding in Plants, Fuits and Nuts. Springer, Berlin, pp 137–156

Albani MC, Battey NH, and Wilkinson MJ (2004) The development of ISSR-derived SCAR markers around the *SEASONAL FLOWERING LOCUS* (*SFL*) in *Fragaria vesca*. Theor Appl Genet 109: 571–579

Aldwinckle HS, Gustafson HL, and Lamb RC (1976) Early determination of genotypes for apple scab resistance by forced flowering of test cross progenies. Euphytica 25: 185–191

Alexander HM, and Bramel-Cox PJ (1991) Sustainability of genetic resistance. In: Sleper DA, Barker TC, Bramel-Cox PJ (eds) Plant Breeding and Sustainable Agriculture: Considerations for Objectives and Methods. Crop Science Society of America, Madison, WI, pp 11–27

Andersen JR, and Lübberstedt T (2003) Functional markers in plants. Trends Plant Sci 8: 554–560

Arús P, and Gardiner S (2007) Genomics for improvement of rosaceae temperate tree fruit. In: Varshney R, Tuberosa R (eds) Genomics-Assisted Crop Improvement. Springer, New York, pp 307–357

Baldi P, Patocchi A, Zini E, Toller C, Velasco R, and Komjanc M (2004) Cloning and linkage mapping of resistance gene homologues in apple. Theor Appl Genet 109: 231–239

Barckley KK, Uratsu SL, Gradziel TM, and Dandekar AM (2006) Multidimensional analysis of S-alleles from cross-incompatible groups of California almond cultivars. J Amer Soc Hort Sci 131: 632–636

Barritt BH, Crandall PC, and Bristow PR (1979) Breeding for root-rot resistance in red raspberry. J Am Soc Hort Sci 104: 92–94

Batley J, and Edwards D (2007) SNP applications in plants. In: Oraguzie NC, Rikkerink EHA, Gardiner SE, De Silva HN (eds) Association Mapping in Plants. Springer, New York, pp 95–102

Bink M, Voorrips R, van de Weg E, and Jansen H (2009) Statistical tools for QTL mapping in multiple, pedigreed populations. Acta Hort (in press)

Bus VGM, Chagné D, Bassett HCM, Bowatte D, Calenge F, Celton J-M, Durel C-E, Malone MT, Patocchi A, Ranatunga AC, Rikkerink EHA, Tustin DS, Zhou J, and Gardiner SE (2008) Genome mapping of three major resistance genes to woolly apple aphid (*Eriosoma lanigerum* Hausm.). Tree Genet Genomes 4: 233–236

Bus V, Ranatunga C, Gardiner S, Bassett H, and Rikkerink E (2000) Marker assisted selection for pest and disease resistance in the New Zealand apple breeding program. Acta Hort 538: 541–547

Bus V, van de Weg WE, Durel C-E, Gessler C, Calenge F, Parisi L, Rikkerink E, Gardiner S, Patocchi A, Meulenbroek M, Schouten H, and Laurens F (2004) Delineation of a scab resistance gene cluster on linkage group 2 of apple. Acta Hort 663: 57–62

Bus V, White A, Gardiner S, Weskett R, Ranatunga C, Samy A, Cook M, and Rikkerink E (2002) An update on apple scab resistance breeding in New Zealand. Acta Hort 595: 43–47

Bus VGM (2006) Gene-for-gene relationships and durable resistance to apple scab. In: Mercer CF (ed) Proceedings of 13th Australasian Plant Breeding Conference. Breeding for Success: Diversity in Action. Christchurch, NZ, pp 1159–1169

Bus VGM, Laurens FND, van de Weg WE, Rusholme RL, Rikkerink EHA, Gardiner SE, Bassett HCM, Kodde LP, and Plummer KM (2005a) The *Vh8* locus of a new gene-for-gene interaction between *Venturia inaequalis* and the wild apple *Malus sieversii* is closely linked to the *Vh2* locus in *Malus pumila* R12740-7A. New Phytol 166: 1035–1049

Bus VGM, Rikkerink EHA, van de Weg WE, Rusholme RL, Gardiner SE, Bassett HCM, Kodde LP, Parisi L, Laurens F, Meulenbroek EJ, Plummer K (2005b) The *Vh2* and *Vh4* scab resistance genes in two differential hosts derived from Russian apple R12740-7A map to the same linkage group of apple. Molec Breed 15: 103–116

Bus V, Rikkerink E, Aldwinckle HS, Caffier V, Durel C-E, Gardiner S, Gessler C, Groenwold R, Laurens F, Le Cam B, Luby J, Meulenbroek B, Kellerhals M, Parisi L, Patocchi A, Plummer K, Schouten HJ, Tartarini S, and van de Weg E. (2009) A proposal for the nomenclature of *Venturia inaequalis* races. Acta Hort (in press)

Byrne DH (2003) Marker-assisted selection. In: Roberts AV, Debener T, Gudin S (eds) Encyclopedia of Rose Science. Elsevier, Oxford, UK, pp 350–357

Byrne DH (2007) Molecular marker use in perennial plant breeding. Acta Hort 751: 163–167

Calenge F, Drouet D, Denancé C, van de Weg WE, Brisset MN, Paulin J-P, and Durel C-E (2005a) Identification of a major QTL together with several minor additive or epistatic QTLs for resistance to fire blight in apple in two related progenies. Theor Appl Genet 111: 128–135

Calenge F, Faure A, Goerre M, Gebhardt C, van de Weg WE, Parisi L, and Durel C-E (2004) Quantitative trait loci (QTL) analysis reveals both broad-spectrum and isolate-specific QTL for scab resistance in an apple progeny challenged with eight ioslates of *Venturia inaequalis*. Phytopathology 94: 370–379

Calenge F, van der Linden CG, van de Weg E, Schouten HJ, van Arkel G, Denancé C, and Durel C-E (2005b) Resistance gene analogues identified through the NBS-profiling method map close to major genes and QTL for disease resistance in apple. Theor Appl Genet 110: 660–668

Campalans A, Pages M, and Messeguer R (2001) Identification of differentially expressed genes by the cDNA-AFLP technique during dehydration of almond (*Prunus amygdalus*). Tree Physiol 21: 633–643

Celton J-M, Tustin DS, Chagné D, and Gardiner SE (2008) Construction of a dense genetic linkage map for apple rootstocks using SSRs developed from *Malus* ESTs and *Pyrus* genomic sequences. Tree Genet Genomes 5: 93–107

Cevik V, and King GJ (2002) High-resolution genetic analysis of the *Sd-1* aphid resistance locus in *Malus* spp. Theor Appl Genet 105: 346–354

Chagné D, Carlisle CM, Blond C, Volz RK, Whitworth CJ, Oraguzie NC, Crowhurst RN, Allan AC, Espley RV, Hellens RP, and Gardiner SE (2007) Mapping a candidate gene (*MdMYB10*) for red flesh and foliage colour in apple. BMC Genomics 8: Article 212

Chaïb J, Lecomte L, Buret M, and Causse M (2006) Stability over genetic background, generations and years of quantitative trait locus (QTLs) for organoleptic quality in tomato. Theor Appl Genet 112: 934–944

Charcosset A, and Moreau L (2004) Use of molecular markers for the development of new cultivars and the evaluation of genetic diversity. Euphytica 137: 81–94

Cheng FS, Brown SK, Weeden NF, and Aldwinckle HS (1995) Molecular markers for scab resistance from 'Nova Easygro' apple. Plant Genome III Conference, San Diego, CA. http://www.intl-pag.org/3/abstracts/81pg3.html

Chevalier M, Lespinasse Y, and Renaudin S (1991) A microscopic study of the different classes of symptoms coded by the *Vf* gene in apple for resistance to scab (*Venturia inaequalis*). Plant Pathol 40: 249–256

Christopher M, Mace E, Jordan D, Rodgers D, McGowan P, Delacy I, Banks P, Sheppard J, Butler D, and Poulsen D (2007) Applications of pedigree-based genome mapping in wheat and barley breeding programs. Euphytica 154: 307–316

Cipriani G, Lot G, Huang WG, Mazzarro MT, Peterlunger E, and Testolin R (1999) AC/GT and AG/CT microsatellite repeats in peach [*Prunus persica* (L) Batsch]: isolation, characterization and cross-species application in *Prunus*. Theor Appl Genet 99: 65–72

Clarke JB, Sargent DJ, Bošković RI, Belaj A, and Tobutt KR (2008) A cherry map from the interspecific cross *Prunus avium* 'Napolean' × *P. nipponica* based on microsatellite, gene-specific and isozyme markers. Tree Genet Genomes DOI 10.1007/s11295-008-0166-9

Claverie M, Bosselut N, Lecouls AC, Voisin R, Poizat C, Dirlewanger E, Kleinhentz M, Lafargue B, Laigret F, and Esmenjaud D (2004a) Location of independent root-knot nematode resistance genes in plum and peach. Theor Appl Genet 108: 765–773

Claverie M, Dirlewanger E, Bosselut N, Lecouls AC, Voisin R, Kleinhentz M, Lafargue B, Caboche M, Chalhoub B, and Esmenjaud D (2004b) High resolution mapping and chromosome landing at the root-knot nematode resistance locus *Ma* from Myrobalan plum using a large-insert BAC DNA library. Theor Appl Genet 109: 1318–1327

Claverie M, Dirlewanger E, Cosson P, Bosselut N, Lecouls AC, Voisin R., Kleinhentz M, Lafargue B, Rosso MN, Abad P, Chalhoub C, and Esmenjaud D (2006) The *Ma* gene from Myrobalan plum (*Prunus cerasifera* Ehr.) conferring a complete-spectrum resistance to root-knot nematodes (*Meloidogyne* spp.) is a member of a TIR-NBS-LRR gene cluster. 28th Symposium of the European Society of Nematology, Blagoevgrad, Bulgaria, 5–9 June 2006

Coart E, Van Glabeke S, De Loose M, Larsen AS, and Roldán-Ruiz I (2006) Chloroplast diversity in the genus *Malus*: new insights into the relationship between the European wild apple (*Malus sylvestris* (L.) Mill.) and the domesticated apple (*Malus domestica* Borkh.). Molec Ecol 15: 2171–2182

Collard BCY, Jahufer MZZ, Brouwer JB, and Pang ECK (2005) An introduction to markers, quantitative trait loci (QTL) mapping and marker-assisted selection for crop improvement: the basic concepts. Euphytica 142: 169–196

Consortium TWTCC (2007) Genome-wide association study of 14,000 cases of seven common diseases and 3,000 shared controls. Nature 447: 661–678

Cook M, and Gardiner S (2004) Development of a fully automated system to extract DNA from difficult plant tissues for genomics research. Plant & Animal Genome XII, San Diego, CA. http://www.intl-pag.org/12/wwwsubmit/P2c˙Poster˙182.html

Cook R, and Evans K (1987) Resistance and tolerance. In: Brown RH, Kerry BR (eds) Principles and Practice of Nematode Control in Crops. Academic Press, Sydney, pp 179–231

Costa F, Alba R, Soglio V, Schouten HJ, Gianfranceschi L, Costa G, Sansavini S, and Giovannoni J (2009) Comparative translational genomics to target candidate genes impacting on fruit quality in apple (*Malus* × *domestica* Borkh.). Acta Hort (in press)

Costa F, van de Weg WE, Stella S, Dondini L, Pratesi D, Musacchi, and Sansavini S (2008) Map position and functional allelic diversity of *Md-Exp7* , a new putative expansin gene associated with fruit softening in apple (*Malus* × *domestica* Borkh.) and pear (*Pyrus communis*). Tree Genet Genomes 4:575–586

Crowe A (1975) 'Nova Easgro' apple. Fruit Varieties J 29: 26

Daubeny H, Pepin HS, and Levesque CA (1992) Breeding for resistance to aphids and root rot in red raspberry. Acta Hort 317: 187–190

Davis TM, Denoyes-Rothan B, and Lerceteau-Köhler E (2007) Strawberry. In: Kole C (ed) Genome Mapping and Molecular Breeding in Plants, Fuits and Nuts. Springer, Berlin, pp 189–205

Dayteg C, Tuveson S, Merker A, Jahoor A, and Kolodinska-Brantestam A (2007) Automation of DNA marker analysis for molecular breeding in crops: practical experience of a plant breeding company. Plant Breed 126: 410–415

Dayton DF, Shay JR, and Hough LF (1953) Apple scab resistance from R12740-7A, a Russian apple. Proc Am Soc Hort Sci 62: 334–340

Dayton DF, and Williams EB (1968) Independent genes in *Malus* for resistance to *Venturia inaequalis*. Proc Am Soc Hort Sci 92: 89–94

Dekkers JC, and Hospital F (2002) The use of molecular genetics in the improvement of agricultural populations. Nat Rev Genet 3: 22–32

Denoyes B, and Baudry A (1995) Species identification and pathogenicity study of French *Collentotrichum* strains isolated from strawberry using morphological, and cultural characteristics. Phytopathology 85: 53–57

Denoyes-Rothan B, Guérin G, Lerceteau-Köhler E, and Risser G (2005) Inheritance of a race-specific resistance to *Colletotrichum acutatum* in *Fragaria* × *ananassa*. Phytopathology 95: 405–412

Denoyes-Rothan B, Lerceteau-Köhler E, Guérin G, Bosseur S, Bariac J, Martin E, and Roudeillac P (2004) QTL Analysis for resistances to *Colletotrichum acutatum* and *Phytophthora cactorum* in octoploid strawberry (*Fragaria* × *ananassa*). Acta Hort 663: 147–151

Dilworth E, and Frey JG (2000) A rapid method for high throughput DNA extraction from plant material for PCR amplification. Plant Molec Biol Rep 18: 61–64

Dirlewanger E, Cosson P, Howad W, Capdeville G, Bosselut N, Claverie M, Voisin R, Poizat C, Lafargue B, Baron O, Laigret F, Kleinhentz M, Arús P, and Esmenjaud D (2004a) Microsatellite genetic linkage maps of Myrobalan plum and an almond-peach hybrid – location of root-knot nematode resistance genes. Theor Appl Genet 109: 827–838

Dirlewanger E, Graziano E, Joobeur T, Garriga-Caldere F, Cosson P, Howad W, Arús P (2004b) Comparative mapping and marker-assisted selection in Rosaceae fruit crops. PNAS 101: 9891–9896

Dirlewanger E, Kleinhentz M, Xiloyannis C, Dichio B, Claverie M, Bosselut N, Howad W, Voisin R, Gomez-Aparisi J, Rubio-Cabetas MJ, Poessel JL, Di Vito M, Arús P, Laigret F, and Esmenjaud D (2004c) Breeding for a new generation of *Prunus* rootstocks based on marker-assisted selection: a European initiative. Acta Hort 663: 829–833

Dirlewanger E, Claverie J, Wünsch A, and Iezzoni AE (2007) Cherry. In: Kole C (ed) Genome Mapping and Molecular Breeding in Plants, Fuits and Nuts. Springer, Berlin, pp 103–118

Dondini L, Lain O, Geuna F, Banfi R, Gaiotti F, Tartarini S, Bassi D, and Testolin R (2007) Development of a new SSR-based linkage map in apricot and analysis of synteny with existing *Prunus* maps. Tree Genet Genomes 3: 239–249

Dondini L, Pierantoni L, Gaiotti F, Chiodini R, Tartarini S, Bazzi C, and Sansavini S (2004) Identifying QTLs for fire-blight resistance via a European pear (*Pyrus communis* L.) genetic linkage map. Molec Breed 14: 407–418

Dreher K, Khairallah M, Ribaut J-M, and Morris M (2003) Money matters (I): costs of field and laboratory procedures associated with conventional and marker-assisted maize breeding at CIMMYT. Molec Breed 11: 221–234

Duncan JM, and Cooke DEL (2002) Work on raspberry root rot at the Scottish Crop Research Institute. Acta Hort 585: 271–277

Dunemann F, Kahnau R, and Schmidt H (1994) Genetic relationships in *Malus* evaluated by RAPD 'fingerprinting' of cultivars and wild species. Plant Breed 113: 150–159

Durel C-E, Parisi L, Laurens F, van de Weg WE, Liebhard R, and Jourjon MF (2003) Genetic dissection of partial resistance to race 6 of *Venturia inaequalis* in apple. Genome 46: 224–234

Edward KJ, Poole RL, and Barker GL (2008) SNP discovery in plants. In: Henry RJ (ed) Plant Genotyping II, SNP Technology. CAB International, Wallingford, UK, pp 1–29

Erdin N, Tartarini S, Broggini GAL, Gennari F, Sansavini S, Gessler C, and Patocchi A (2006) Mapping of the apple scab-resistance gene *Vb*. Genome 49: 1238–1245

Esmenjaud D (2004) Breeding for durable resistance to RKN in perennials: a European initiative for *Prunus* rootstocks. In: Cook R, Hunt DJ (eds) Nematology Monographs and Perspectives 2. Proceedings International Congress of Nematology IV, Tenerife, Spain, 2002. Brill Press, Leiden, Netherlands, pp 279–287

Esmenjaud D, and Dirlewanger E (2007) Plum. In: Kole C (ed) Genome Mapping and Molecular Breeding in Plants, Fruits and Nuts. Springer, Berlin, pp 119–136

Esmenjaud D, Minot JC, Voisin R, Bonnet A, Salesses G (1996) Inheritance of resistance to the root-knot nematode *Meloidogyne arenaria* in Myrobalan plum. Theor Appl Genet 92: 873–879

Esmenjaud D, Minot JC, Voisin R, Pinochet J, Salesses G (1994) Inter- and intraspecific resistance variability in Myrobalan plum, peach, and peach-almond rootstocks using 22 root-knot nematode populations. J Am Soc Hortic Sci 119: 94–100

Esmenjaud D, Minot JC, Voisin R, Pinochet J, Simard MH, and Salesses G (1997) Differential response to root-knot nematodes in *Prunus* species and correlative genetic implications. J Nematol 29: 370–380

Esmenjaud D, Scotto La Massese C, Salesses G. Minot JC, and Voisin R (1992) Method and criteria to evaluate resistance to *Meloidogyne arenaria* in *Prunus cerasifera* Ehr. Fund Appl Nematol 15: 385–389

Esmenjaud D, Voisin R, Van Ghelder C, Bosselut N, Lafargue B, Di Vito M, Dirlewanger E, Poëssel JL, and Kleinhentz M (2009) Genetic dissection of resistance to root-knot nematodes *Meloidogyne* spp. in plum, peach, almond and apricot, from various segregating interspecific *Prunus* progenies. Tree Genet Genomes Online First DOI 10.1007/s11295.008-0173-x

Finn CE, Moore PP, and Kempler C (2005) Rasberry cultivars: what's new? What's succeeding? Where are breeding programs heading? Acta Hort 777: 33–40

Flachowsky H, Peil A, Sopanen T, Elo A, and Hanke V (2007) Overexpression of *BpMADS4* from silver birch (*Betula pendula* Roth.) induces early-flowering in apple (*Malus* × *domestica* Borkh.). Plant Breed 126: 137–145

Flaishman MA, Cohen Y, Freiman A, Golubowicz S, Korchinsky R, Shlizerman L, and Yablowicz Z (2009) Development of a fast and compact breeding system in pear by the use of juvenile-free transgenic plants. Act Hort (in press)

Folta KM, and Davies TM (2006) Strawberry genes and genomics Critic Rev Plant Sci 25: 399–415

Forsline PL, Aldwinckle HS, Dickson EE, Luby JJ, and Hokanson SC (2003) Collection, maintenance, characterization, and utilization of wild apples of Central Asia. Hort Rev 29: 1–61

Forster JW, Cogan NOI, Dobrowolski MP, Francki MG, Spangenberg GC, and Smith KF (2008) Functionally associated molecular genetic markers for temperate pasture plant improvement. In: Henry RJ (ed) Plant Genotyping II, SNP Technology. CAB International, Wallingford, UK, pp 154–186

Freeman S (2008) Management, survival strategies, and host range of *Colletotrichum acutatum* on strawberry. Hort Sci 43: 66–68

Frey JE, Frey B, Sauer C, and Kellerhals M (2004) Efficient low-cost DNA extraction and multiplex fluorescent PCR method for marker-assisted selection in breeding. Plant Breed 123: 554–557

Frisch M, and Melchinger AE (2001) Marker-assisted backcrossing for simultaneous introgression of two genes. Crop Sci 41: 1716–1725

Gao Z, van de Weg W (2006) The *Vf* gene for scab resistance in apple is linked to sub-lethal genes. Euphytica 151: 123–133

Gardiner S, Bus V, Volz R, and Bassett H (2006) Marker assisted selection in apple breeding internationally. In: Mercer CF (ed) Proc 13th Australasian Plant Breeding Conference. Breeding for Success: Diversity in Action. Christchurch, NZ, pp 681–686

Gardiner S, Murdoch J, Meech S, Rusholme R, Bassett H, Cook M, Bus V, Rikkerink E, Gleave A, Crowhurst R, Ross G, and Warrington I (2003) Candidate resistance genes from an EST database prove a rich source of markers for major genes conferring resistance to important apple pests and diseases. Acta Hort 622: 141–151

Gardiner SE, Bassett HCM, Madie C, and Noiton DAM (1996a) Isozyme, randomly amplified polymorphic DNA (RAPD), and restriction fragment-length polymorphism (RFLP) markers used to deduce a putative parent for the 'Braeburn' apple. J Am Soc Hort Sci 121: 996–1001

Gardiner SE, Bassett HCM, Noiton DAM, Bus VG, Hofstee ME, White AG, Ball RD, Forster RLS, and Rikkerink EHA (1996b) A detailed linkage map around an apple scab resistance gene demonstrates that two disease resistance classes both carry the *Vf* gene. Theor Appl Genet 93: 485–493

Gardiner SE, Bus VGM, Rusholme RL, Chagné D, and Rikkerink EHA (2007) Apple. In: Kole C (ed) Genome Mapping and Molecular Breeding in Plants, Fruits and Nuts. Springer, Berlin, pp 1–62

Gerlach HK, and Stosser R (1998) Kettenreaktionen im Obstbau: Sortenidentifizierung mit Hilfe des DNA-Fingerprinting. Erwerbsobstbau 40: 103–106

Gessler C (1989) Genetics of the interaction *Venturia inaequalis – Malus*: the conflict between theory and reality. IOBC wprs Bulletin 12(6): 168–190

Gianfranceschi L, Koller B, Seglias N, Kellerhals M, and Gessler C (1996) Molecular selection in apple for resistance to scab caused by *Venturia inaequalis*. Theor Appl Genet 93: 199–204

Gimelfarb A, and Lande R (1995) Marker-assisted selection and marker-QTL associations in hybrid populations. Theor Appl Genet 91: 522–528

Giongo L, Bergamini A, Rigatti R, and Komjanc M (2001) Marker assisted selection for scab resistant apple. Acta Hort 546: 581–589

Gopalakrishnan S, Sharma RK, Rajkumar KA, Joseph M, Singh VP, Singh AK, Bhat KV, Singh NK, and Mohaptra TM (2008) Integrating marker assisted background analysis with foreground selection for identification of superior bacterial blight resistant recombinants in Basmati rice. Plant Breed 127: 131–139

Graham J, and Smith K (2002) DNA markers for use in raspberry breeding. Acta Hort 585: 51–56

Graham J, Smith K, Tierney I, Mackenzie K, and Hackett CA (2006) Mapping gene *H* controlling cane pubescence in raspberry and its association with resistance to cane botrytis and spur blight, rust and cane spot. Theor Appl Genet 112: 818–831

Gu WK, Weeden NF, Yu J, and Wallace DH (1995) Large-scale, cost-effective screening of PCR products in marker-assisted selection applications. Theor Appl Genet 91: 465–470

Guarino C, Santoro S, De Simone L, Lain O, Cipriani G, and Testolin R (2006) Genetic diversity in a collection of ancient cultivars of apple (*Malus × domestica* Borkh.) as revealed by SSR-based fingerprinting. J Hort Sci Biotech 81 39–44

Gupta PK, Varshney RK, Sharma PC, and Ramesh B (1999) Molecular markers and their applications in wheat breeding. Plant Breed 118: 369–390

Gupta PK, Rustgi S, and Mir RR (2008) Array-based high-throughput DNA markers for crop improvement. Heredity 101: 5–18

Gygax M, Gianfranceschi L, Liebhard R, Kellerhals M, Gessler C, and Patocchi A (2004) Molecular markers linked to the apple scab resistance gene *Vbj* derived from *Malus baccata jackii*. Theor Appl Genet 109: 1702–1709

Han Y, Gasic K, Marron B, Beever JE, and Korban SS (2007) A BAC-based physical map of the apple genome. Genomics 89: 630–637

Han YP, and Korban SS (2008) An overview of the apple genome through BAC end sequence ananlysis. Plant Molec Biol 67: 581–588

Handoo ZA, Nyczepir AP, Esmenjaud D, van der Beek JG, Castagnone-Sereno P, Carta LK, Skantar AM, and Higgins JA (2004) Morphological, molecular and differential-host characterization of *Meloidogyne floridensis* n. sp. (Nematoda: Meloidogynidae), a root-knot nematode parasitizing peach in Florida. J Nematol 36: 20–35

Harada T, Matsukawa K, Sato T, Ishikawa R, Niizeki M, and Saito K (1993) DNA-RAPDs detect genetic variation and paternity in *Malus*. Euphytica 65: 87–91

Harris SA, Robinson JP, and Juniper BE (2002) Genetic clues to the origin of the apple. Trends in Genet 18: 426–430

Harrison RE, McNicol RJ, Cooke DEL, and Duncan JM (1998) Recent developments in *Phytopthora fragariae* var. *rubi* research at the Scottish Crop Research Institute. Acta Hort 505: 327–240

Hemmat M, Brown SK, and Weeden NF (2002) Tagging and mapping scab resistance genes from R12740-7A apple. J Am Soc Hort Sci 127: 365–370

Hillel J, Schaap T, Haberfield A, Jeffreys A, Plotzky Y, Cahaner A, and Lavi U (1990) DNA fingerprints applied to gene introgression in breeding programs. Genetics 124: 783–789

Hittalmani S, Parco A, Mew TV, Zeigler RS, and Huang N (2000) Fine mapping and DNA marker-assisted pyramiding of the three major genes for blast resistance in rice. Theor Appl Genet 100: 1121–1128

Hokanson SC, Lamboy WF, Szewc-McFadden AK, and McFerson JR (2001) Microsatellite (SSR) variation in a collection of *Malus* (apple) species and hybrids. Euphytica 118: 281–294

Hokanson SC, McFerson JR, Forsline PL, Lamboy WF, Luby JJ, Djangaliev A, and Aldwinckle HS (1997) Collecting and managing wild *Malus* germplasm in its center of diversity. Hort Sci 32: 173–176

Hokanson SC, Szewc-McFadden AK, Lamboy WF, and McFerson JR (1998) Microsatellite (SSR) markers reveal genetic identities, genetic diversity and relationships in a *Malus* × *domestica* Borkh. core subset selection. Theor Appl Genet 97: 671–683

Hormaza JI, Yamane H, and Rodrigo J (2007) Apricot. In: Kole C (ed) Genome Mapping and Molecular Breeding in Plants, Fuits and Nuts. Springer, Berlin, pp 171–187

Hospital F (2001) Size of donor chromosome segments around introgressed loci and reduction of linkage drag in marker-assisted backcross programs. Genetics 158: 1363–1379

Hospital F, and Charcosset A (1997) Marker-assisted introgression of quantitative trait loci. Genetics 147: 1469–1485

Hospital F, Goldringer I, and Openshaw S (2000) Efficient marker-based recurrent selection for multiple quantitative trait loci. Genetic Research 75: 357–368

Hough LF, Shay JR, and Dayton DF (1953) Apple scab resistance from *Malus floribunda* Sieb. J Am Soc Hort Sci 62: 341–347

Howad W, Yamamoto T, Dirlewanger E, Testolin R, Cosson P, Cipriani G, Monforte AJ, Georgi L, Abbott AG, and Arús P (2005) Mapping with a few plants: using selective mapping for microsatellite saturation of the *Prunus* reference map. Genetics 171: 1305–1309

Huaracha EM, Xu ML, Gasic K, Pauwels E, Putte AVd, Keulemans JW, and Korban SS (2004) Phenotypic reaction and genetic analysis using AFLP-derived SCARs for resistance to apple scab. J Phytopathol 152: 260–266

Iezzoni AF, Andersen RL, Schmidt H, Tao R, Tobutt KR, and Wiersma PA (2005) Proceedings of the s-allele workshop at the 2001 International Cherry Symposium. Acta Hort 667: 25–35

Irwin D (2008) The MassARRAY system for plant genomics. In: Henry RJ (ed) Plant Genotyping II, SNP Technology. CAB International, Wallingford, UK, pp 98–113

Iyer-Pascuzzi AS, and McCouch SR (2007) Functional markers for *xa5*-mediated resistance in rice (*Oryza sativa*, L.). Molec Breed 19: 291–296

Jaccoud D, Peng K, Feinstein D, and Kilian A (2001) Diversity Arrays: a solid state technology for sequence information independent genotyping. Nucleic Acids Res 29: e25

Jennings SN, and Brennan RM (2002) Improvement of raspberry cultivars in Scotland. Acta Hort 585: 179–183

Johnson R (2000) Classical plant breeding for durable resistance to diseases. J Plant Pathol 82: 3–7

Joobeur T, Viruel MA, de Vicente MC, Jáuregui B, Ballester J, Dettori MT, Verde I, Truco MJ, Messeguer Battle I, Quarta R, Dirlewanger E, and Arús P (1998) Construction of a saturated linkage map for *Prunus* using an almond × peach F2 progeny. Theor Appl Genet 97: 1034–1041

Kasem S, Rice N, and Henry RJ (2008) DNA extraction from plant tissue. In: Henry RJ (ed) Plant Genotyping II, SNP Technology. CAB International, Wallingford, UK, pp 2 19–271

Kearsey MJ (1997) Genetic resources and plant breeding. In: Callow JA, Ford-Lloyd BV, Newbury HJ (eds) Biotechnology and Plant Genetic Resources. Conservation and Use. CAB International, Wallingford, UK, pp 175–202

Kellerhals M, Spuhler M, Patocchi A, and Frey J (2009) Selection efficiency in apple breeding. Acta Hort (in press)

Kempler C, Daubeny H, and Harding B (2002) Recent progress in breeding red raspberries in British Columbia, Canada. Acta Hort 585: 47–50

Kenis K, Keulemans J, and Davey M (2008) Identification and stability of QTLs for fruit quality traits in apple. Tree Genet Genomes DOI 10.1007/s11295-008-0140-6.

Khan MA, Durel CE, Duffy B, Drouet D, Kellerhals M, Gessler C, and Patocchi A (2007) Development of molecular markers linked to the 'Fiesta' linkage group 7 major QTL for fire blight resistance and their application for marker-assisted selection. Genome 50: 568–577

Kilian A, Huttner E, Wenzl P, Jaccoud D, Carling J, Caig V, Evers M, Heller-Uszynska K, Cayla C, Patarapuwadol S, Xia L, Yang S, and Thomson B (2005) The fast and the cheap: SNP and DArT-based whole genome profiling for crop improvement. In: Tuberosa R, Phillips RL, Gale M (eds) Proceedings of the International Congress In the Wake of the Double Helix: From the Green Revolution to the Gene Revolution, May 27–31, 2003, Bologna, Italy. Avenue Media, pp 443–461

Kimura T, Sawamura Y, Kotobuki K, Matsuta N, Hayashi T, Ban Y, and Yamamoto T (2003) Parentage analysis in pear cultivars characterized by SSR markers. J Jpn Soc Hort Sci 72: 182–189

King GJ, Alston FH, Brown LM, Chevreau E, Evans KM, Dunemann F, Janse J, Laurens F, Lynn JR, Maliepaard C, Manganaris AG, Roche P, Schmidt H, Tartarini S, Verhaegh J, and Vrielink R (1998) Multiple field and glasshouse assessments increase the reliability of linkage mapping of the *Vf* source of scab resistance in apple. Theor Appl Genet 96: 699–708

King GJ, Lynn JR, Dover CJ, Evans KM, and Seymour GB (2001) Resolution of quantitative trait loci for mechanical measures accounting for genetic variation in fruit texture of apple (*Malus pumila* Mill.). Theor Appl Genet 102: 1227–1235

King GJ, Tartarini S, Brown L, Gennari F, and Sansavini S (1999) Introgression of the *Vf* source of scab resistance and distribution of linked marker alleles within the *Malus* gene pool. Theor Appl Genet 99: 1039–1046

Knapp SJ (1998) Marker-assisted selection as a strategy for increasing the probability of selecting superior genotypes. Crop Sci 38: 1164–1174

Knight VH (1991) Use of the Salmonberry, *Rubus spectbilis* Pursh, in red raspberry breeding. J Hort Sci 66: 575–581.

Kochba J, and Spiegel-Roy P (1975) Inheritance to the root-knot nematode (*Meloidogyne javanica* Chitwood) in bitter almond progenies. Euphytica 24: 453–457

Kochba J, and Spiegel-Roy P (1976) 'Alnem 1', 'Alnem 88', 'Alnem 201' almonds: nematode-resistant rootstock seed source. Hort Science 11: 270

Koller B, Lehmann A, McDermott JM, and Gessler C (1993) Identification of apple cultivars using RAPD markers. Theor Appl Genet 85: 901–904

Kroymann J, and Mitchell-Olds T (2005) Epistasis and balanced polymorphism influencing complex trait variation. Nature 435: 95–98

Kumar LS (1999) DNA markers in plant improvement: an overview. Biotech Adv 17: 143–182

Lalli DA, Abbott AG, Zhebentyayeva TN, Badenes ML, Damsteegt V, Polák J, Krška B, and Salava J (2008) A genetic linkage map for an apricot (*Prunus armeniaca* L.) $BC_1$ population mapping *plum pox virus* resistance. Tree Genet Genomes 4: 481–493

Lande R (1992) Marker-assisted selection in relation to traditional methods of plant breeding. In: Stalker HT, Murphy JP (eds) Plant Breeding in the 1990s. Proceedings of the Symposium on Plant Breeding in the 1990s, CAB International, Wallingford, pp 437–451

Lande R, and Thompson R (1990) Efficiency of marker-assisted selection in the improvement of quantitative traits. Genetics 124: 743–756

Laurens F (1999) Review of the current apple breeding programmes in the world: objectives for scion cultivar improvement. Acta Hort 484: 163–170

Lawson DM, Lunde CF, and Mutschler MA (1997) Marker-assisted transfer of acylsugar-mediated pest resistance from the wild tomato, *Lycopersicon pennellii*, to the cultivated tomato, *Lycopersicon esculentum*. Molec Breed 3: 307–317

Layne REC (1987) Peach rootstocks. In: Rom RC, Carlson RF (eds) Rootstocks for Fruit Crops. John Wiley and Sons, New York, pp 185–216

Lecomte L, Duffe P, Buret M, Servin B, Hospital F, and Causse M (2004) Marker-assisted introgression of five QTLs controlling fruit quality traits into three tomato lines revealed interactions between QTLs and genetic backgrounds. Theor Appl Genet 109: 658–668

Lecouls AC, Rubio-Cabetas MJ, Minot JC, Voisin R, Bonnet A, Salesses G, Dirlewanger E, and Esmenjaud D (1999) RAPD and SCAR markers linked to the *Ma1* root-knot nematode resistance gene in Myrobalan plum (*Prunus cerasifera* Ehr.). Theor Appl Genet 99: 328–336

Lecouls AC, Rubio-Cabetas MJ, Voisin R, Bonnet A, Duval H, Salesses G, Dirlewanger E and Esmenjaud D (2004) Marker-assisted selection for the wide-spectrum resistance to root-knot nematodes conferred by the *Ma* gene from Myrobalan plum (*Prunus cerasifera*) in interspecific *Prunus* material. Molec Breed 13: 113–124

Lecouls AC, Salesses G, Minot JC, Voisin R, Bonnet A, and Esmenjaud D (1997) Spectrum of the *Ma* genes for resistance to *Meloidogyne* spp. in Myrobalan plum. Theor Appl Genet 85: 1325–2334

Lerceteau-Köhler E, Guérin G, and Denoyes-Rothan B (2005) Identification of SCAR markers linked to *Rca2* anthracnose resistance gene and their assessment in strawberry germplasm. Theor Appl Genet 111: 862–870

Lerceteau-Köhler E, Roudeillac P, Markocic M, Guérin G, Praud K, Denoyes-Rothan B (2002) The use of molecular markers for durable resistance breeding in the cultivated strawberry (*Fragaria* × *ananassa*). Acta Hort 567: 615–618

Liebhard R, Kellerhals M, Pfammatter W, Jertmini M, and Gessler C (2003a) Mapping quantitative physiological traits in apple (*Malus* × *domestica* Borkh.). Plant Molec Biol 52: 511–526

Liebhard R, Koller B, Patocchi A, Kellerhals M, Pfammatter W, Jermini M, and Gessler C (2003b) Mapping quantitative field resistance against apple scab in a 'Fiesta' × 'Discovery' progeny. Phytopathol 93: 493–501

Liew M, Pryor R, Palais R, Meadows C, Erali M, Lyon E, and Wittwer C (2004) Genotyping of single-nucleotide polymorphisms by high-resolution melting of small amplicons. Clinic Chem 50:1156–1164

Linde M, Hattendorf A, Kaufmann H, and Debener T (2006) Powdery mildew resistance in roses: QTL mapping in different environments using selective genotyping. Theor Appl Genet 113: 1081–1092

Liu P, Zhu J, and Lu Y (2004) Marker-assisted selection in segregating generations of self-fertilizing crops. Theor Appl Genet 109: 370–376

Lu ZX, Reighard GL, Nyczepir AP, Beckman TG, Ramming DW (2000) Inheritance of resistance to root-knot nematodes in *Prunus* rootstocks. Hort Sci 35: 1344–1346

Luby J, Forsline P, Aldwinckle H, Bus V, and Geibel M (2001) Silk road apples – collection, evaluation, and utilization of *Malus sieversii* from Central Asia. Hort Science 36: 225–231

Luby JJ, Bedford DS, and Forsline PL (2004) Winter hardiness in the U.S. Department of Agriculture *Malus* core collection. Acta Hort 663: 605–608

Luby JJ, and Shaw DV (2001) Does marker-assisted selection make dollars and sense in a fruit breeding program? Hort Science 36: 872–879

Ma R-C, and Oliveira MM (2002) Evolutionary analysis of S-RNase genes from Rosaceae species. Molec Genet Genomics 267: 71–78

Mace ES, Buhariwalla HK, and Crouch JH (2003) A high-throughput DNA extraction protocol for tropical molecular breeding programs. Plant Molec Biol Rep 21: 459a–459 h.

Maliepaard C, Alston FH, van Arkel G, Brown LM, Chevreau E, Dunemann F, Evans KM, Gardiner S, Guilford P, van Heusden AW, Janse J, Laurens F, Lynn JR, Manganaris AG, den Nijs APM, Periam N, Rikkerink E, Roche P, Ryder C, Sansavini S, Schnidt H, Tartarini S, Verhaegh JJ, Vrielink-van Ginkel M, and King GJ (1998) Aligning male and female linkage maps of apple (*Malus pumila* Mill.) using multi-allelic markers. Theor Appl Genet 97: 60–73

Martínez-Gómez P, Sánchez-Pérez R, Dicenta F, Howad W, Arús P, and Gradziel TM (2007) Almond. In: Kole C (ed) Genome Mapping and Molecular Breeding in Plants, Fuits and Nuts. Springer, Berlin, pp 229–242

Michelmore RW, Paran I, and Kesseli RV (1991) Identification of markers linked to disease resistance genes by bulked segregant analysis: a rapid method to detect markers in specific genomic regions by using segregating populations. PNAS 88: 9828–9832

Mohan M, Nair S, Bhagwat A, Krishna TG, Yano M, Bhadia CR, and Sasaki T (1997) Genome mapping, molecular markers and marker-assisted selection in crop plants. Molec Breed 3: 87–103

Montgomery J, Wittwer CT, Palais R, and Zhou L (2007) Simultaneous mutation scanning and genotyping by high-resolution DNA melting analysis. Nat Protoc 2: 59–66

Moreau L, Charcosset A, and Gallais A (2004) Experimental evaluation of several cycles of marker-assisted selection in maize. Euphytica 137: 111–118

Moreau L, Lemarie S, Charcosset A, and Gallais A (2000) Economic efficiency of one cycle of marker-assisted selection. Crop Sci 40: 329–337

Morris M, Dreher K, Ribaut J-M, and Khairallah M (2003) Money matters (II): costs of maize inbred line conversion schemes at CIMMYT using conventional and marker-assisted selection. Molec Breed 11: 235–247

Naik S, Hampson C, Gasic K, Bakkeren G, and Korban SS (2006) Development and linkage mapping of E-STS and RGA markers for functional gene homologues in apple. Genome 49: 959–968

Nestby R, and Heiberg N (1995) Genetic variation for resistance to *Phytophthora fragariae* var. *rubi* in red raspberries. Euphytica 81: 143–149

Newcomb RD, Crowhurst RN, Gleave AP, Rikkerink EHA, Allan AC, Beuning LL, Bowen JH, Gera E, Jamieson KR, Janssen BJ, Laing WA, McArtney S, Bhawana N, Ross GS, Snowden KC, Souleyre EJF, Walton EF, and Yauk Y (2006) Analyses of expressed sequence tags from apple. Plant Physiol 141: 147–166

Nocente F, Gazza L, and Pasquini M (2007) Evaluation of leaf rust resistance genes *Lr1*, *Lr9*, Lr24, *Lr47* and their introgression into common wheat cultivars by marker-assisted selection. Euphytica 155: 329–336

Nybom H (2003) DNA fingerprinting. In: Roberts AV, Debener T, Gudin S (eds) Encyclopedia of Rose Science. Elsevier, Oxford, UK, pp 318–325

Nybom H, and Schaal BA (1990) DNA "fingerprints" applied to paternity analysis in apples (*Malus* × *domestica*). Theor Appl Genet 79: 763–768

Nybom H, Sehic J, and Garkava-Gustavsson L (2008) Self-incompatibility alleles of 104 apple cultivars grown in Northern Europe. J Hort Sci Biotech 83: 339–344

Nyczepir AP (1991) Nematode management strategies in stone fruits in the United States. J Nematol 23: 334–341

Nyczepir AP, and Halbrendt JM (1993) Nematode pests of deciduous fruit and nut trees. In: Evans K, Trudgill DL, Webster JM (eds) Plant Parasitic Nematodes in Temperate Agriculture. CAB International, Oxon, pp 381–425

Oliveira CM, Mota M, Monte-Corvo L, Goulao L, and Silva DM (1999) Molecular typing of *Pyrus* based on RAPD Markers. Scientia Hort 79: 163–174

Olmstead JW, Sebolt AM, Cabrera A, Sooriyapathirana SS, Hammar S, Iriarte G, Wang D, Chen CY, van der Knaap E, and Iezzoni A (2008) Construction of an intra-specific sweet cherry

(*Prunus avium* L.) genetic linkage map and synteny analysis with the *Prunus* reference. Tree Genet Genomes DOI 10.1007/s11295-008-0161-1

Oraguzie NC, Gardiner SE, Bassett HCM, Stefanati M, Ball RD, Bus VGM, and White AG (2001) Genetic diversity and relationships in *Malus* sp. germplasm collections as determined by randomly amplified polymorphic DNA. J Am Soc Hort Sci 126: 318–328

Oraguzie NC, Iwanami H, Soejima J, Harada T, and Hall A (2004a) Inheritance of the *Md-ACS1* gene and its relationship to fruit softening in apple (*Malus* × *domestica* Borkh.). Theor Appl Genet 108: 1526–1533

Oraguzie NC, Rikkerink E, Gardiner S, Bus V, Currie A, Rusholme R, and Volz R (2004b) A review of developments in breeding techniques and gene technology tools for new cultivar development in apple. Recent Res Dev Genet Breed 1: 223–257

Oraguzie NC, Rikkerink EHA, Gardiner SE, and de Silva HN (2007a) Association Mapping in Plants. Springer, New York

Oraguzie NC, Wilcox PL, Rikkerink EHA, and de Silva HN (2007b) Linkage disequilibrium. In: Oraguzie NC, Rikkerink EHA, Gardiner SE, de Silva HN (eds) Association Mapping in Plants. Springer, New York, pp 11–39

Oraguzie NC, Yamamoto T, Soejima J, Suzuki T, and de Silva HN (2005) DNA fingerprinting of apple (*Malus* spp.) rootstocks using simple sequence repeats. Plant Breed 124: 197–202

Ortega E, and Dicenta F (2003) Inheritance of self-compatibility in almond: breeding strategies to assure self-compatibility in the progeny. Theor Appl Genet 106: 904–911

Paran I, and Zamir D (2003) Quantitative traits in plants: beyond the QTL. Trends Genet 19: 303–306

Pasquer F, Frey B, and Frey JE (2008) Identification of cherry incompatibility alleles by microarray. Plant Breed Doi:10.1111/j.1439-0523.2007.01476.x

Paterson AH, Tanksley SD, and Sorrells ME (1991) DNA markers in plant improvement. Adv in Agron 46: 39–90

Patocchi A, Walser M, Tartarini S, Broggini GAL, Gennari F, Sansavini S, and Gessler C (2005) Identification by genome scanning approach (GSA) of a microsatellite tightly associated with the apple scab resistance gene *Vm*. Genome 48: 630–636

Pattemore JA, Trau M, and Henry RJ (2008) Nanotechnology: the future of cost-effectiive plant genotyping. In: Henry RJ (ed) Plant Genotyping II, SNP Technology. CAB International, Wallingford, UK, pp 133–153

Pattison JA, Samuelian SK, and Weber CA (2007) Inheritance of *Phytophthora* root rot resistance in red raspberry by generation means and molecular linkage analysis. Theor Appl Genet 115: 225–236

Peace CP, Crisosto CH, and Gradziel TM (2005) Endopolygalacturonase: a candidate gene for *Freestone* and *Melting flesh* in peach. Molec Breed 16: 21–31

Picañol R, Lozano L, Alegre S, Bonany J, Arús P, and Howad W (2009) QTL analysis of peach fruit quality: map construction in three breeding progenies with SSR markers. Acta Hort (in press)

Pradhan AK, Gupta V, Mukhopadhyay A, Arumugam A, Sodhi YS, and Pental D (2003) A high-density linkage map in *Brassica juncea* (Indian mustard) using AFLP and RFLP markers. Theor Appl Genet 106: 607–614

Qi J, Gai S, Zhang J, Gu M, and H Shu (2005) Identification of self-incompatibility genotypes of apricot (*Prunus armeniaca* L.) by S-allele-specific PCR analysis. Biotech Lett 27: 1205–1209

Rajapakse S (2003) Molecular markers. In: Roberts AV, Debener T, Gudin S (eds) Encyclopedia of Rose Science. Elsevier, Oxford, UK, pp 334–341

Rammah A, and Hirschmann H (1988) *Meloidogyne mayaguensis* n. sp. (*Meloidogynidae*), a root-knot nematode from Puerto Rico. J Nematol 20: 58–69

Ramming DW, and Tanner O (1983) Nemared peach rootstock. Hort Science 18: 376

Ramos-Cabrer AM, Díaz-Hernández MB, and Pereira-Lorenzo S (2007) Morphology and microsatellites in Spanish apple collections. J Hort Sci Biotech 82: 257–265

Ribaut J-M, and Ragot M (2007) Marker-assisted selection to improve drought adaptation in maize: the backcross approach, perspectives, limitations, and alternatives. J Experim Bot 58: 351–360

Robert VJM, West MAL, Inai S, Caines A, Arntzen L, Smith JK, and StClair DA (2001) Marker-assisted introgression of blackmold resistance QTL alleles from wild *Lycopersicon cheesmanii* to cultivated (*L. esculentum*) and evaluation of QTL phenotypic effects. Molec Breed 8: 217–233

Roche P, Alston FH, and Maliepaard C (1997) RFLP and RAPD linked to the rosy leaf curling aphid resistance gene (*Sd1*) in apple. Theor Appl Genet 94: 528–533

Rubio-Cabetas MJ, Lecouls AC, Salesses G, Bonnet A, Minot JC, Voisin R, and Esmenjaud D (1998) Evidence of a new gene for high resistance to *Meloidogyne* spp. in Myrobalan plum (*Prunus cerasifera*). Plant Breed 117: 567–571

Rubio-Cabetas MJ, Minot JC, Voisin R, Esmenjaud D, Salesses G, and Bonnet A (1999) Response of the *Ma* genes from Myrobalan plum to *Meloidogyne hapla* and *M. mayaguensis*. Hort Science 34: 1266–1268

Samuelian SK, Baldo AM, Pattison JA, and Weber CA (2008) Isolation and linkage mapping of NBS-Lrr resistance gene analogs in red raspberry (*Rubus idaeus* L.) and classification among 270 Rosaceae NBS-LRR genes. Tree Genet Genomes DOI 10.1007/s11295-008-0160-2

Sanzol J, and Robbins TP (2008) Combined analysis of S-alleles in European pear by pollinations and PCR-based S-genotyping; correlation between S-phenotypes and S-RNase genotypes. J Amer Soc Hort Sci 133: 213–224

Sassa H, Kakui H, Miyamoto M, Suzuki Y, Hanada T, Ushijima K, Kusaba M, Hirano H, and Koba T (2007) S locus F-box brothers: multiple and pollen-specific F-box genes with S haplotype-specific polymorphisms in apple and Japanese pear. Genetics 175: 1869–1881

Sasser JN, Freckman DW (1987) A world perspective in nematology: the role of the society. In: Veech JA, Dickson DW (eds) Vistas on Nematology. Society of Nematologists, Hyattsville, MD, pp 7–14

Scotto La Massèse C, Grasselly C, Minot JC, and Voisin R (1984) Différence de comportement de 23 clones et hybrides de *Prunus* à l'égard de quatre espèces de *Meloidogyne*. Rev Nematol 7: 265–270

Servin B, Martin OC, Mézard M, and Hospital F (2004) Toward a theory of marker-assisted gene pyramiding. Genetics 168: 513–523

Sharpe RH, Hesse CO, Lownsbery BA, Perry VG, and Hansen CJ (1969) Breeding peaches for root-knot nematode resistance. J Am Soc Hortic Sci 94: 209–212

Sicard O, Marandel G, Soriano JM, Lalli DA, Lambert P, Salava J, Badenes ML, Abbott A, and Decroocq V (2008) Flanking the major *Plum pox virus* resistance locus in apricot with co-dominant markers (SSRs) derived from candidate resistance genes. Tree Genet Genomes 4: 359–365

Silfverberg-Dilworth E, Matasci CL, van de Weg WE, van Kaauwen MPW, Walser M, Kodde LP, Soglio V, Gianfranceschi L, Durel C-E, Costa F, Yamamoto T, Koller B, Gessler C, and Patocchi A (2006) Microsatellite markers spanning the apple (*Malus* × *domestica* Borkh.) genome. Tree Genet Genomes 2: 202–224

Simpson DW, Winterbottom CQ, Bell JA, and Maltoni ML (1994) Resistance to a single UK isolate of *Colletotrichum acutatum* in strawberry germplasm from Northern Europe. Euphytica 77: 161–164

Singh S, Sidhu JS, Huang N, Vikal Y, Li Z, Brar DS, Dhaliwal HS, Khush GS (2001) Pyramiding three bacterial blight genes (*xa5*, *xa13* and Xa21) using marker-assisted selection into indica rice cultivar PR106. Theor Appl Genet 102: 1011–1015

Sjulin TM (2008) Special problems in nursery propagation of day-neutral strawberry cultivars susceptible to *Colletotrichum acutatum*. Hort Science 43: 78–80

Smith BJ (2008) Epidemiology and pathology of strawberry anthraconose: a North American perspective. Hort Science 43: 69–73

Smith S, and Helentjaris T (1996) DNA fingerprinting and plant variety protection. In: Paterson AH (ed) Genome Mapping in Plants. R.G. Landes Company, Austin, TX, pp 95–110

Soriano JM, Vera-Ruiz EM, Vilanova S, Martínez-Calvo J, Llácer G, Badenes ML, and Romero C (2008) Identification and mapping of a locus conferring plum pox virus resistance in two apricot-improved linkage maps. Tree Genet Genomes 4: 391–402

Sorkheh K, Shiran B, Gradziel TM, Epperson BK, Martínez-Goméz P, and Asadi E (2007) Amplified fragment length polymorphism as a tool for molecular characterization of almond germplasm: genetic diversity among cultivated genotypes and related wild species of almond, and its relationships with agronomic traits. Euphytica 156: 327–344

Spelman R, and Bovenhuis H (1998) Genetic response from marker assisted selection in an outbred population for differing marker bracket sizes and with two identified quantitative trait loci. Genetics 148: 1389–1396

Spiegler G, and Thoss H (1993) Breeding for resistance to *Phytophthora* root rot in red raspberries. Acta Hort 352: 477–484

Stankiewicz-Kosyl M, Nowicka A, Krajewski P, Tomala K, Soska A, Laurens F, Govan C, Lateaur M, Costa F, Tartarini S, Guerra W, Lewandowski M, Rutkovski K, Zurawicz E, Gianfranceschi L, Durel C-E, Mathis F, Barbaro E, Mott D, Patocchi A, Gobbin D, Fernandez F, Evans K, Dunemann F, Boudichevskaja A, Jansen J, and van de Weg E (2009) QTL analysis of acidity in apple using pedigree-based approach. Acta Hort (in press)

Stushnoff C, McSay AE, Forsline PL, and Luby J (2003) Diversity of phenolic antioxidant content and radical scavenging capacity in the USDA Apple Germplasm core collection. Acta Hort 623: 305–311

Tao R, Yamane H, Sugiura A, Murayama H, Sassa H, and Mor H (1999) Molecular typing of S-alleles through identification, characterization and cDNA cloning for S-RNases in sweet cherry. J Amer Soc Hort Sci 124: 224–233

Tao R, Habu T, Yamane H, Sugiura A, and Iwamoto K (2000) Molecular markers for self-compatibility in Japanese apricot (*Prunus mume*). Hort Science 35: 1121–1123

Tartarini S, Costa F, Guerra W, Hoeller I, Evans K, Fernandez F, Kellerhals M, Eigenmann C, Korbin M, Keller S, and van de Weg E (2009) Marker-assisted breeding (MAB) for fruit firmness and acidity. Acta Hort (in press)

Tartarini S, Sansavini S, Vinatzer B, Gennari F, and Domizi C (2000) Efficiency of marker assisted selection (MAS) for the *Vf* scab resistance gene. Acta Hort 538: 549–552

Teng Y, Tanabe K, Tamura F, and Itai A (2001) Genetic relationships of pear cultivars in Xinjiang, China, as measured by RAPD markers. J Hort Sci Biotech 76: 771–779

Teng Y, Tanabe K, Tamura F, and Itai A (2002) Genetic relationships of *Pyrus* species and cultivars native to East Asia revealed by randomly amplified polymorphic DNA markers. J Am Soc Hort Sci 127: 262–270

Terakami S, Shoda M, Adachi Y, Gonai T, Kasumi M, Sawamura Y, Iketani H, Kotobuki K, Patocchi A, Gessler C, Hayashi T, and Yamamoto T (2006) Genetic mapping of the pear scab resistance gene *Vnk* of Japanese pear cultivar Kinchaku. Theor Appl Genet 113: 743–752

Testolin R, Marrazzo MT, Cipriani G, Quarta R, Verde I, Dettori MT, Pancaldi M, and Sansavini S (2000) Microsatellite DNA in peach (*Prunus persica* (L) Batsch) and its use in fingerprinting and testing the genetic origin of cultivars. Genome 43: 512–520

Triantaphyllou AC (1985) Cytogenetics, cytotaxonomy and phylogeny of root-knot nematodes. In: Sasser JN, Carter CC (eds) An Advanced treatise on *Meloidogyne*, Vol I. North Carolina State University Graphics, Raleigh, NC, pp 113–126

van de Weg WE, Voorrips RE, Finkers R, Kodde LP, Jansen J, and Bink MCAM (2004) Pedigree genotyping: a new pedigree-based approach of QTL identification and allele mining. Acta Hort 663: 45–50

van de Weg WE, Jansen H, Bink M, Voorrips RE, Durel C-E, Laurens F, Dunemann F, Evans K, Patocchi A, Guerra W, Komjanc M, Lateur M, Kellerhals M, Ryder C, Sansavini S, Tomala K, Zurawicz E, and Gianfranceschi L (2009) QTL mapping in multiple, pedigreed populations: the concept and the framework of statistical procedures. Acta Hort (in press)

Varshney RK, Graner A, and Sorrells ME (2005) Genomics-assisted breeding for crop improvement. Trends Plant Sci 10: 621–630

Vavilov NI (1951) The origin, variation, immunity, and breeding of cultivated plants. Chron Bot 13: 1–364

Vinatzer BA, Patocchi A, Gianfranceschi L, Tartarini S, Zhang HB, Gessler C, and Sansavini S (2001) Apple contrains receptor-like genes homologous to the *Cladosporium fulvum* resistance gene family of tomato with a cluster of genes cosegregating with *Vf* apple scab resistance. MPMI 14: 508–515

Vinatzer BA, Patocchi A, Tartarini S, Gianfranceschi L, Sansavini S, and Gessler C (2004) Isolation of two microsatellite markers from BAC clones of the *Vf* scab resistance region and molecular characterization of scab-resistant accessions in *Malus* germplasm. Molec Breed 123: 1–6

Volk GM, Richards CM, Henk AD, Reilley AA, Bassil NV, and Postman JD (2006) Diversity of wild *Pyrus communis* based on microsatellite analyses. J Am Soc Hort Sci 131: 408–417

Volk GM, Richards CM, Reilley AA, Henk AD, Forsline PL, and Aldwinckle HS (2005) Ex situ conservation of vegetatively propagated species: development of a seed-based core collection for *Malus sieversii*. J Am Soc Hort Sci 130: 203–210

Volk GM, Richards CM, Reilley AA, Henk AD, Reeves PA, Forsline PL, and Aldwinckle HS (2008) Genetic diversity and disease resistance of wild *Malus orientalis* from Turkey and southern Russia. J Amer Soc Hort Sci 133: 383–389

Volz RK, Rikkerink E, Austin P, Lawrence T, and Bus VGM (2009) "Fast-breeding" in apple: a strategy to accelerate introgression of new traits into elite germplasm. Acta Hort (in press)

Wang Y, Xue Y, Li J (2005) Towards molecular breeding and improvement of rice in China. Trends Plant Sci 10: 610–614.

Weber CA, Pattison J, and Samuelin S (2008) Marker assisted selection for resistance to root rot in red raspberry caused by *Phytophthora fragariae* var. *rubi*. Acta Hort 777: 311–315

Weebadde CK, Wang D, Finn CE, Lewers KS, Luby JJ, Buchakra J, Sjulin TM, and Hancock JF (2008) Using a linage mapping approach to identify QTL for day-neutrality in the octoploid strawberry. Plant Breed 127: 94–101

Wiedow C (2006) Characterization of phenotypic and molecular diversity in offsprings of *Malus sieversii* (Lebed.) Roem. as basis for a core collection of apple genetic resources. PhD thesis, Martin-Luther-Universität Halle-Wittenberg, Halle, Germany

Wilcox PL, Echt CE, and Burdon RD (2007) Gene-assisted selection: applications of association genetics for forest tree breeding. In: Oraguzie NC, Rikkerink EHA, Gardiner SE, De Silva HN (eds) Association Mapping in Plants. Springer, New York, pp 211–247

Wilcox WF, Scott PH, Hamm PB, Kennedy DM, Duncan JM, Brasier CM, and Hansen EM (1993) Identity of a *Phytophthora* species attacking raspberry in Europe and North America. Mycol Res 97: 817–831

Wilcox WF, Pritts MP, and Kelly MJ (1999) Integrated control of *Phytophthora* root rot of red raspberry. Plant Disease 83: 1149–1154

Wilde F, Korzun V, Ebmeyer E, Geiger HH, and Miedaner T (2007) Comparison of phenotypic and marker-based selection for Fusarium head blight resistance and DON content in spring wheat. Molec Breed 19: 357–370

Wilde F, Schon CC, Korzun V, Ebmeyer E, Schmolke M, Hartl L, and Miedaner T (2008) Marker-based introduction of three quantitative-trait loci conferring resistance to Fusarium head blight into an independent elite winter wheat breeding population. Theor Appl Genet 117: 29–35

Winterbottom CQ (1991). Resistance of strawberry to *Colletotrichum acutatum*. Master Thesis of Science, University of California.

Xie C, and Xu S (1998) Efficiency of multistage marker-assisted selection in the improvement of multiple quantitative traits. Heredity: 489–498

Xu M, and Korban SS (2002) A cluster of four receptor-like genes resides in the *Vf* locus that confers resistance to apple scab disease. Genetics 162: 1995–2006

Yamamoto T, Hayashi T (2002) New root-knot nematode resistance genes and their STS markers in peach. Scientia Hortic 96:81–90

Yamamoto T, Kimura T, Saito T, Kotobuki K, Matsuta N, Liebhard R, Gessler C, van de Weg WE, and Hayashi T (2004) Genetic linkage maps of Japanese and European pears aligned to the apple consensus map. Acta Hort 663: 51–56

Yaegaki H, Shimada T, Moriguchi T, Hayama H, Haji T, and Yamaguchi M (2001) Molecular characterization of S-RNase genes and S-genotypes in the Japanese apricot (*Prunus mume* Sieb. et Zucc.). Sex Plant Reprod 13: 251–257

Young ND (1999) A cautiously optimistic vision for marker-assisted breeding. Molec Breed 5: 505–510

Yousef GG, and Juvik JA (2001) Comparison of phenotypic and marker-assisted selection for quantitative traits in sweet corn. Crop Sci 41: 645–655

Yu K, Park SJ, and Poysa V (2000) Marker-assisted selection of common beans for resistance to common bacterial blight: efficacy and economics. Plant Breed 119: 411–415

Zhebentyayeva TN, Reighard GL, Lalli D, Gorina VM, Krška B, and Abbott AG (2008a) Origin of resistance to *plum pox virus* in apricot: what new AFLP and targeted SSR analyses tell. Tree Genet Genomes 4: 403–417

Zhebentyayeva TN, Swire-Clark G, Georgi LL, Garay L, Jung S, Forrest S, Blenda AV, Blackmon B, Mook J, Horn R, Howad W, Arús P, Main D, Tomkins JP, Sosinski B, Baird WV, Reighard GL, and Abbott AG (2008b). A framework physical map for peach, a model Rosaceae species. Tree Genet Genomes, DOI 10.1007/s11295-008-0147-z

Zhu Y, and Barritt BH (2008) Md-ACS1 and Md-ACO1 genotyping of apple (*Malus* × *domestica* Borkh.) breeding parents and suitability for marker-assisted selection. Tree Genet Genomes 4: 555–562

# 28. Rosaceaous Genome Sequencing: Perspectives and Progress

**Bryon Sosinski, Vladimir Shulaev, Amit Dhingra, Ananth Kalyanaraman, Roger Bumgarner, Daniel Rokhsar, Ignazio Verde, Riccardo Velasco, and Albert G. Abbott**

## 1 Plant Genome Sequencing Overview and History

The long-term goal of plant genomics is to identify, isolate and determine the function of plant genes that are associated with both vegetative and reproductive phenotypes. Most phenotypes require the coordinated activity and regulatory control of suites of genes over time and in precise positions within the plant. Until recently, the idea of establishing a comprehensive approach to isolate and characterize all the genes involved in any complex phenotype was a daunting one, and oftentimes it has been necessary to perform whole genome sequencing to obtain all of the gene sequences. The sequence of several plant genomes have been generated including the *Arabidopsis*, poplar and rice genomes, with the model legume *Medicago* and sorghum well underway. In addition, large amounts of expressed sequence tag (EST) information are being obtained for many other plants, including rosaceaous plants. The advances in genomics, informatics, and phylogenetics have been developed and refined by these reference projects, to the point that it is now thought that, in many cases, the vast majority of genes of a plant can be identified without the complete genome sequence, however the EST approach to gene identification does not provide valuable information regarding promoters and other non-coding regulatory elements.

One of the first eukaryotic genomes to be completely sequenced was that of the small mustard species *Arabidopsis thaliana*. During the past decade, *Arabidopsis* has emerged as one of the most widely used model organisms for studying the biology of higher plants. Its genome was chosen for sequencing because it is highly compact, about 125 Mb, with little interspersed repetitive DNA (Arabidopsis Genome Initiative, 2000). However, since *Arabidopsis* is rather distantly related to the cereal crops that provide the bulk of the world food supply, the genome of rice has also been sequenced. Rice was chosen because, in addition to its importance as a food source for about one quarter of the human population, it has one of

B. Sosinski (✉)
Department of Horticultural Science, North Carolina State University, Raleigh, NC, USA

K.M. Folta, S.E. Gardiner (eds.), *Genetics and Genomics of Rosaceae*, Plant Genetics and Genomics: Crops and Models 6, DOI 10.1007/978-0-387-77491-6_28,

the most compact genomes among the cereals. It contains about 3.5 times as much DNA as *Arabidopsis* but only about 20% as much DNA as maize and about 3% as much DNA as wheat (Bennett and Smith, 1991; Goff et al., 2002; Yu et al., 2002). However, the genome organization of the cereals appears to be very highly conserved; rice, wheat, maize, sorghum, millet and other cereals exhibit a high degree of synteny (Gale and Devos, 1998). The differences in genome size primarily reflect the amplification of interspersed repetitive sequences (Bennetzen et al., 1998); there is no evidence that angiosperms with large amounts of DNA per cell have substantially greater numbers of functional genes than angiosperms with relatively small DNA contents. This is particularly significant for our various approaches to sequence rosaceaous genomes. Because of extensive synteny among the cereal genomes, knowledge of gene order and organization in rice may be used to isolate and characterize the corresponding genes in other cereals (McCouch, 1998). Thus, for instance, if a genetic locus encoding a useful trait is mapped between a pair of closely linked molecular markers in wheat, it may be possible to identify candidate genes for the rice ortholog by analyzing the rice genome sequence located between the rice orthologs of the molecular markers.

The sequences of *Arabidopsis,* poplar and rice provide us with three reference genomes from which the genome contents of other angiosperms can be extrapolated. It seems likely that as the costs of DNA sequencing continue to drop and technologies improve, additional genomes of other important plants will ultimately be sequenced. However, for the foreseeable future, additional complete (euchromatin and heterochromatin) plant genome sequences will probably not be available because of the high costs for whole genome sequencing of any of the major crops. Extensive, partial cDNA and BAC sequence information may be publicly available for a majority of the genes from many important plant species, both crop and non-crop (Pennisi, 1998). There are currently more than 3,000,000 EST sequences from numerous plant species in public databases, and this number is expected to continue to grow rapidly during the next several years. Thus, as genes associated with functions or traits in one plant are cloned, it will usually be possible to identify the orthologs responsible for the trait in other plant species, including those in the more basal lineages, by a database search or by using the sequence information to clone the corresponding gene from the species of interest.

Although flowering plants have evolved during the past 130 million years or so, and might therefore be expected to be very similar at the genomic level, substantial morphological, developmental and metabolic diversity exists. A major challenge to understanding the genetic basis of interspecific diversity is that, in at least some cases, minor changes in the structure or expression of a gene can lead to major changes in phenotype. Understanding the basis of this diversity is key to understanding how to effect rational improvements in the utility of crop species. Knowledge of the genetic basis for intraspecies variation in specific traits should be useful in selecting or creating useful variation within a species. Because of the relatively recent radiation of the angiosperms, we consider it likely that there will be very few protein-encoding angiosperm genes that do not have orthologs or paralogs in the model genomes. Therefore, understanding the genetic basis for diversity may devolve to identifying the relevant differences in the control of expression or the function of essentially the same set of genes. Indeed, it has been hypothesized that the developmental diversity of angiosperms may largely result from changes in the cis-regulatory sequences of transcriptional regulators (Doebley and Lukens, 1998). As can be seen from an examination of the peach EST data shown elsewhere in this book, this is not entirely true. Indeed, a large proportion (approx. 20%) of the peach and almond ESTs remain 'orphans' in the databases, showing no significant homology to other ESTs present. These are likely to be genes that are Rosaceae, peach, tree or fruit specific transcripts that cannot appear in the existing models. Therefore it would appear that we need genome sequence from additional model species before we can truly realize the promise of plant genomics.

## *1.1 Genome Sequencing Strategies*

The majority of Eukaryotic genomes that have been sequenced (the rice and *Arabidopsis* plant genomes in particular) have employed a BAC by BAC approach (Arabidopsis Genome Initiative, 2000; Goff et al., 2002). This strategy requires that a complete, ordered physical map constructed from large insert (BAC, PAC, cosmid) genomic clones must first be available for the species under study. A minimal tile of contigs is extracted from the map, and the individual clones are then subjected to shotgun sequencing. The primary advantage of this approach over whole genome shotgun (WGS) is that reassembly of the sequences is much easier due the smaller number of sequences being assembled in each step of the project. In addition, the physical map provides positional context for each sequence. Moreover, problematic sequences are readily isolated to a defined region of the genome, facilitating resolution. The downside to this approach is the cost; physical maps are costly, require high levels of specialized expertise and are time-consuming to generate. That is in addition to the raw sequencing costs.

One of the most recent plant genome sequencing strategies that has been developed is known as gene enrichment. The first example of this was a method based on methylation filtration (Rabinowicz, et al., 1999). This strategy does not attempt to obtain the entire genome sequence for a plant, but instead allows for focusing upon gene rich regions (gene space) of the genome for sequencing. In this procedure, total plant DNA is isolated, and the highly methylated DNA is excluded from cloning. As the majority of non-coding, repetitive plant DNA is methylated and therefore excluded from cloning, the resulting genomic libraries are highly enriched for genes (Rabinowicz, 2003). Another approach to isolate the gene space of a genome employs renaturation kinetics (Cot) of DNA to clone just the low copy fraction of the DNA (Peterson et al., 2002). Several projects are underway to test this and other methods of genespace enrichment on the maize genome (2,500 Mb) (Palmer et al., 2003; Whitelaw, et al., 2003), tobacco (Opperman, personal communication), and *Medicago* (Young et al., 2005).

The whole genome shotgun (WGS) sequencing strategy has emerged as a useful method to decode genomes of species that have relatively small genomes, particularly bacterial. The list of prokaryotic genomes that have been sequenced using WGS is vast, but notably includes *Haemophilus influenzae* (Fleischmann et al., 1995) and several *Bacilli* (Kunst et al., 1997: Takami et al., 2000). For the most part, these genomes are simple having one or a few chromosomes and limited amounts of repetitive DNA. The WGS approach has also been successfully employed for the generation of the poplar genome sequence (Tuskan et al., 2006). Briefly, the WGS technique employs total genomic DNA that is sheared either mechanically or enzymatically, is then size selected and cloned. These clones are then randomly sequenced until as much as a twelve-fold genome coverage is achieved. These random sequence reads are then assembled, and the resulting sequence gaps are closed by primer walking, PCR etc. In the case of complex genomes, including plants, WGS alone can be a problematic sequencing solution due to genome size and complexity, as has been reported with the human genome project (Waterston et al., 2003) where difficulties arose with the assembly of repetitive regions. This consideration is often balanced with the fact that WGS is the most affordable way to rapidly make progress with genome sequencing.

## 2 Sequencing Technologies

Sequencing based on Sanger chemistry remains the most widely utilized technology to capture genomic information from small microbial genomes to the large human genome. Currently, the automated sequencers employing Sanger chemistry can generate 1–2 million base pairs (bp) in 24 hrs with an average read length of 550–800 bp and with great accuracy (Hall, 2007). However, the demand for sequence generation in resolving biological questions has far exceeded the output possible with Sanger chemistry-based projects and this approach is increasingly becoming prohibitive due to the time and costs involved. Recent years have seen the advent of novel high-throughput sequencing technologies that are aimed at generating large amount of sequence data at a fraction of the cost (Margulies et al., 2005; Shendure et al., 2005; Braslavsky et al., 2003). The sequence read length from the new sequencing technologies ranges between 35 bp to 400, which is far below the threshold required to resolve the repeat structure of larger complex genomes. However, there are other complementary methodologies that offshoot this disadvantage. A brief description of three new technologies that have been successfully commercialized is provided here.

### *2.1 Pyrosequencing*

Pyrosequencing, sequencing-by-synthesis, was first described over a decade ago (Ronaghi et al., 1999). However, it has been commercialized very recently by 454

Life Sciences (Roche) and is based on a novel and highly parallel system based on pyrosequencing chemistry (Nyren et al., 1993). As the name suggests, sequence is generated from a single DNA molecule template as the complementary DNA strand is synthesized in a complex reaction. The single DNA templates are coated on beads and preloaded into a microfabricated glass plate containing 1.6 million pico liter reaction wells. The DNA sequence is determined when four nucleotides are sequentially delivered through the plate. Incorporation of a new nucleotide results in a stoichiometric release of pyrophosphate that further generates ATP. The ATP thus generated catalyzes the conversion of luciferin resulting in emission of photons. The visible signal generated in each reaction well is captured by a CCD camera and the entire continuum of these signals is converted to corresponding sequence via signal processing (Margulies et al., 2005).

### *2.2 Illumina's Solexa Sequencing Technology*

The Solexa technology is also based on sequencing-by-synthesis principle and is similar to 454 platform till the library preparation stage. Thereafter, it relies on the solid phase bridge PCR technique (Chetverina and Chetverin, 1993) for amplification of the targeted DNA templates arranged as random array of clusters on a flow cell (Bentley, 2006). Sequence is generated from the DNA templates using four-color fluorescent chemistry via repeated primer extension with reversible dye-terminators.

### *2.3 SOLiD Technology*

The Supported Oligonucleotide Ligation and Detection Platform (SOLiD) technology has been developed by Applied Biosystems (ABI). It is based on the polymerase colony approach and employs hybridization-ligation chemistry (Shendure et al., 2005; Mitra et al., 2003). The sequencing is performed by iterative cycles of hybridization of a mixture of sequencing primers and fluorescent probes.

## 3 Next Generation Technologies

In addition to the above-mentioned technologies, research is ongoing to develop even better sequencing platforms to cut down the time and costs drastically. Some of these methods include a nanofluidic-based sequencing system, single molecule based approaches and nanopore sequencing. One of the new technologies to watch out for is microchip electrophoresis based sequencing. The technology allows for sequencing 600 bp in 6.5 minutes (Fredlake et al., 2008). This is the only technology that is near implementation amongst the new ones that are being developed currently.

# 4 Peach Genome Sequencing

## *4.1 Rationale*

Peach (*Prunus persica*) is considered one of the genetically best characterized species in the Rosaceae, peach has distinct advantages that make it suitable as a model genome species for the Rosaceae. In contrast to some *Prunus* species, such as cultivated plums and sour cherries, that are polyploid (Moore and Janick, 1975), peach is a diploid with $n = 8$ (Jelenkovic and Harrington, 1972) and has a comparatively small genome: $5.9 \times 10^8$ bp or 0.61 pg/diploid nucleus (Baird et al., 1994). This is only about twice the value for *Arabidopsis thaliana*, and is currently thought to be approximately 300 MB (Arumuganathan and Earle, 1991). Peach has a relatively short juvenility period of 2–3 years compared to most other fruit tree species that require 6–10 years. In addition, a number of genes for fundamentally important traits have been genetically described in peach, including genes controlling flower and fruit development, tree growth habit, dormancy, cold hardiness, and disease and pest resistance.

### 4.1.1 Sequencing Methodology

Given the relatively small genome size of peach (300 MB) combined with the fact that robust EST and physical map resources are available, an argument for each of the sequencing approaches can be made. Budgetary considerations make the WGS approach quite attractive, but sequence assembly issues may persist. The fact that peach has become a model species for *Prunus* and also for the family Rosaceae to some extent, compels us to consider the robust BAC by BAC approach in order to obtain high quality reference sequence, but it may not be necessary or desirable to obtain the sequence of the repetitive intergenic regions of the peach genome. On the other hand, targeting the gene space of peach for sequencing is quite attractive as it offers an affordable way to obtain the sequence of the most critical genomic regions – the genes. Given the small size of the peach genome and resulting likely small fraction of repetitive DNA; it is unlikely that the benefit of gene space enrichment will outweigh the technical complexities of its application.

A number of comparative studies have been conducted within Rosaceae and *Prunus* that indicate that the genomes within the family and species are highly collinear. The transferable markers (RFLPs, SSRs and isozymes) mapped in the *Prunus* reference (TxE) population have been used for the construction of linkage maps in other *Prunus* species. Detailed comparisons can be established between this map and those of almond (Joobeur et al., 2000), apricot (Lambert et al., 2004), *P. davidiana* (Foulongne et al., 2003), cherry (Dirlewanger et al., 2003) and *P. cerasifera* (Dirlewanger et al., 2004). The distribution of the markers into the eight linkage groups and their order within each linkage group was generally identical between species, suggesting a high degree of synteny. Excepting one published example of a translocation difference between linkage group 6 and 8 in peach and almond (Jauregui personal communication), these results strongly indicate that the group of *Prunus* species studied shares a nearly identical genome, and that the information on gene sequence and position obtained in one of them would be generally useful for the rest. Recently, significant genome synteny has been identified between peach and the model plant genomes excluding *Arabidopsis* and rice through comparisons of sequenced *Prunus* BACs with the model genome sequences (Jung et al., 2009).

At the Plant and Animal Genome XV Meeting on 01/16/07, Jerry Tuskan from the Joint Genome Institute (JGI) announced the new roster of plants that will be targeted for genome sequencing at JGI. Peach has been selected for an 8X whole genome shotgun (WGS) that will be completed by the end of 2009. This selection was made in part due to the fact that peach is a tree that is closely related to poplar, and there is a belief that the availability of the peach genome will assist in closing the remaining gaps in the poplar genome (Tuskan and Rokhsar, personal communication; Jung et al., 2009). Since the original genome sequencing announcement, genomic researchers in Italy have committed to providing 4X of the total 8X that will be generated.

As peach is primarily an outcrossing species, the majority of peach accessions are quite heterozygous, which can be problematic for genomic sequence assembly. To overcome this issue, both the US and Italian groups have agreed to employ a doubled haploid ‘Lovell’ accession for genome sequencing. In addition to being effectively homozygous across the genome (fixed), this doubled haploid was also employed as the substrate for physical mapping in peach (Zhebentyayeva et al. 2008). We anticipate that the doubled haploid nature of ‘Lovell’ will simplify the assembly of the WGS sequence in the near term, and will be directly applicable for sequence gap closing and finishing in the long term.

Our approach for sequencing the peach genome will combine both WGS and ultimately BAC by BAC approaches to provide a cost-effective, but effective means of obtaining this midsize genome. Initially we will obtain an 8X WGS of the peach genome using the robust standard Sanger sequencing. Eventually, BACs from the minimal tiling path of the peach physical map will be pooled, and shotgun sublibraries made for sequencing (Cai et al., 2001). The pooling approach will lower sheared library expenses by 10-fold over the course of the project, and the complexity of the genome strongly suggests that reconstruction will be achievable. Any pools of BACs that fail to assemble can be deconstructed for individual sequencing. While this approach will cost slightly more than a WGS approach alone, the data from other genome sequencing projects indicates that we will obtain a much more robust and accurately assembled genome. Moreover, this approach will let us selectively target certain areas of the genome for subsequent additional sequencing due to increased biological interest. It will likewise provide a more robust template for subsequent finishing if that becomes needed or feasible. Our goal is to obtain high-quality reference sequence of the peach genome. This resource will be of immediate utility to the Rosaceae community, as well as to those interested in comparative genomics and evolution.

## 4.2 *Annotation and Assembly*

For the initial annotation of the peach genome, we will employ the tools already in place at JGI that were employed for the poplar genome. Briefly, once the draft sequence is available we will identify a first-draft reference set of putative protein-coding gene loci using homology-based, and expressed sequence tag (EST)–based methods. As mentioned above, to aid the annotation process, additional full-length sequences, from full-length enriched cDNA libraries from peach, will be generated and used in training the gene-calling algorithms. In addition to gene prediction, repetitive sequences, including SSRs will be characterized (Tuskan et al., 2006). We anticipate that once this annotated draft of the peach genome is available, further community-based efforts will continue to refine the annotation.

The current (2007/2009) timeline for peach sequencing is as follows. We anticipate completion of an initial 2X of Sanger sequencing by the end of the summer '08 that will additionally include a 15X coverage of fosmid ends. An additional 4X contribution from the Italian group should follow in the early months of 2009. Depending upon the quality of assembly at that time, additional (up to 2X) WGS reads will be generated at that time. To enhance the annotation of the peach genome, up to 1 million 454 EST reads will be generated during the project.

# 5 Apple Genome Sequencing

## 5.1 *Rationale*

Apple (*Malus sps*) is an ancient tetraploid that behaves as a diploid and is comprised of 55 species (Phipps et al., 1990). The apple genome represents one of the larger genomes in Rosaceae; 1.54 pg DNA/2C nucleus or 750 Mb per haploid genome complement (Tatum et al., 2005). The International Rosaceae community recognizes apple as a model for functional genomics (Rosaceae White Paper – http://www.bioinfo.wsu.edu/gdr/community/rosexec/; Folta et al., 2006). This status has been bestowed on apple due to the availability of over 300,000 apple ESTs in the public database (Newcomb et al., 2006), genomic libraries, microarray expression data and physical maps (Han et al., 2007). Apple is highly amenable to genetic manipulation *in vitro* (Malnoy et al., 2007; Teo et al., 2006) enabling functional genomics studies (Norelli et al., 2007). Owing to long juvenility, phenotypic consequences of genetic perturbations manifest themselves after an extended duration of time. However, development of non-juvenile transgenic apple lines is expected to abbreviate this timeline to a few months (Flachowsky et al., 2007) enabling large scale testing of gene function for developing functionally-validated molecular markers.

## 5.2 *Sequencing Methodology*

Sequencing of the apple genome has benefitted from two independent initiatives, which have recently coalesced into an international consortium described below.

In the first project, the cultivated variety 'Golden Delicious' was chosen for sequencing by the Istituto Agrario San Michele all'Adige (IASMA) as this variety accounts for over 80% of this region's production and for its importance to the rest of the world (second most cultivated variety). The availability of a 'Golden Delicious'-derived double haploid (DH) genotype provided an additional advantage for undertaking an integrated approach to sequence the apple genome. The IASMA initiative concentrated on the cultivated heterozygous 'Golden Delicious' material and was funded by the Province of Trento, Italy. The choice to sequencing a heterozygous apple was based on our recent successful experience of sequence a largely heterozygous genotype of grape, i.e. the Pinot Noir (Velasco et al., 2007). The grape genome sequencing initiative resulted in collaboration between IASMA and Myriad Genetics Inc. that resulted in the development of several methodologies to assemble and anchor largely heterozygous genomes (Cartwright et al., 2007; Velasco et al., 2007; Zahrkikh et al., 2008). These methods are being applied to the current 'Golden Delicious' apple genome sequencing project that is being carried out jointly between IASMA, Myriad Genetics Inc. and 454/Roche. Based on our prior experience with grape genome sequencing and assembly, the strategy for obtaining and assembling the heterozygous apple genome sequence has been modified to increase the efficiency of this process. The amount of Sanger sequencing has been reduced to 4X coverage (7X coverage in grape) and that of 454 pyrosequencing has been increased to 10 X coverage (4X in grape; Velasco et al., 2007). This adjustment in the strategy has been catalyzed by new advances in pyrosequencing that generate much longer reads ($\sim$ 400 bp) and the possibility to perform paired-end sequencing on short and long insert (3–16 kb) libraries. These new developments obviate the need, at least partially, for obtaining fosmid and BAC-end sequences for genome scaffolding.

The Sanger approach consisted of obtaining sequence from 27 different libraries of various insert sizes (1.5–3 kb; 7–15 kb: fosmid and BACs) and was completed in the summer of 2007. The pyrosequencing approach was initially based on obtaining sequence from the existing chemistry that produces an average of 260 bases per read. During the current year that has been substituted with the new chemistry that increases the average read length to almost 400 bases.

The completion of the project is based on the collaboration between IASMA, INRA at Angers, France (PI. Charles-Eric Durel), and Plant and Food Research, New Zealand (PI: Sue Gardiner). These groups are sharing 5 crossing populations in apple to develop a reliable and an extremely dense integrated molecular map to anchor the genomic scaffolds to chromosomes. These maps are based on internationally shared SSRs and SNPs developed on the 'Golden Delicious' assemblies produced at various stages of sequencing progress (5x; 7x; 9.5x; 11x). The project data will be made publicly available via the Genomic Database for Rosaceae hosted at Washington State University (PI: Dorrie Main) and GenBank in the next few months.

A complementary public initiative to sequence the Double Haploid (DH) 'Golden Delicious' apple is underway at Washington State University, USA. This initiative was established with seed funding from WSU, the Washington Tree Fruit Research Commission and has now found additional support through USDA-NRI (PIs: Dhingra and Kalyanaraman). A collaboration between WSU

(Dhingra), University of Washington (Roger Bumgarner) and INRA at Angers, France (Lespinasse, Chevreau and Durel) served as the foundation of this initiative. This project aims at capturing the apple genome sequence circumventing the traditional BAC by BAC approach. The methodology comprises of assembling 4X whole genome shotgun sequence on a scaffold of ~37,000 BAC end sequences and paired-end reads generated by 454 sequencing. A 6X BAC library from the DH 'Golden Delicious' genotype is being constructed and will be publicly available to the community after completion of the project. A relatively simpler genetic organization of the DH material is expected to simplify the *in silico* genome assembly. The available data has been successfully assembled using a parallel assembler (Bumgarner lab). However, the repeat structures remain to be resolved. The development of higher-end computational methods continues to integrate the repeat elements in genome assemblies (Kalyanaraman and Dhingra Labs). The apple genome is several-fold larger than strawberry and peach genomes. It would be interesting to study the repeat structure and gene organization. We expect to observe a large degree of synteny between the genomes of the three members of Rosaceae.

As of this date, ~ 1.5X coverage of the DH apple has been obtained using pyrosequencing and this data will soon be available via GenBank upon quality validation. In addition, the data will be accessible to the general public via the apple genome project portal hosted at www.genomics.wsu.edu. The final dataset will be submitted to Genomic Database for Rosaceae hosted at WSU.

The IASMA project and the WSU-based public initiative have now metamorphosed into an International Consortium for Apple Genome Sequencing (iCAGeS). The consortium comprises of INRA at Angers (PIs – Lespinasse, Chevereau and Durel) and the University of Western Cape, South Africa (PI: J. Rees). The South African group is utilizing the DH material and generating sequences using Illumina's Solexa technology. The iCAGeS provides a platform to integrate various datasets being generated by the participants and provide a comprehensive understanding of apple genome structure to the community.

## 6 *Fragaria* Genome Sequencing

The woodland strawberry, *Fragaria vesca,* has the smallest genome of any economically important plant species, with relatively small numbers of repetitive sequences and, therefore, sequencing of its genome can fully utilize the advantages of the 454 technology and provide an invaluable resource to the plant research community.

The Strawberry Genome Sequencing Consortium, comprised of experts in a wide array of research areas, was created in the spring of 2008 with the aim to sequence the full genome of the *F. vesca* (http://strawberry.vbi.vt.edu).

At present, the largest publicly available genomic sequence resource in Rosaceae is that of *F. vesca*, from which approximately 1.75 Mb of genomic sequence derived from 50 genomic sites (as fosmid clones), has been deposited in GenBank under accession numbers EU024823-EU024872. This work was performed by Dr. Thomas

Davis, one of the Consortium participants. Preliminary analysis of these sequences has revealed a gene density of about 1 gene per 6 kb and an SSR density of about one SSR locus per 5 kb. An extensive collection of diploid and octoploid strawberry ESTs derived from different tissues and developmental stages that have also been deposited to GenBank by Consortium members. The available *Fragaria* genomic sequence database will provide a valuable resource for the assembly and annotation of other rosaceous genomes.

The genomic sequencing is being performed at the VBI Core Laboratory Facility (CLF; http://www.vbi.vt.edu/core˙laboratory˙facility/). In January 2007 the first Roche GS-FLX was installed in the VBI CLF. Since installation several full bacterial genome sequencing project have been completed at the VBI CLF. Recently we started a pilot project to sequence the *F. vesca* genome and have already generated ~8X full genome coverage. It is anticipated that we will need ~15X (3 Gb) coverage of the genome. In addition, the Consortium will perform sequencing of cosmids and BACs.

To complete the genome sequencing we will utilize a *F. vesca* genomic BAC library in the pECBAC1 vector that is currently available at Dr. Shulaev's lab at VBI. The average insert size is about 130 Kb (Fig. 1). Sequencing of the individual BAC clones will help with anchoring contigs derived from the GS-FLX sequencing and with completing the genome sequencing.

The genome will be assembled and annotated as a joint community project. Members of the Consortium have significant experience in all aspects of full genome sequencing, assembly and annotation and have pledged to fund and support this joint Rosaceae Community effort. The end result of this collaboration, the complete genome sequence of *F. vesca*, will be distributed freely to the research community and the annotated genome will be hosted by the Genome Database for Rosaceae (GDR; http://www.bioinfo.wsu.edu/gdr/).

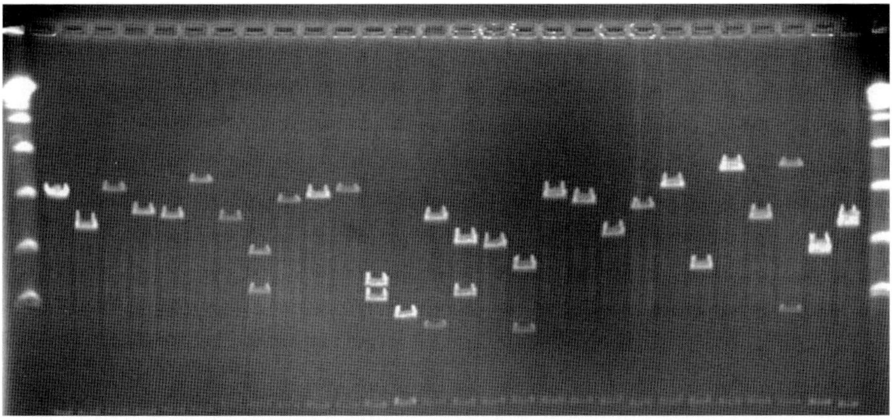

**Fig. 1** Agarose gel electrophoresis of DNA isolated from 28 randomly selected *F. vesca* BAC clones. DNA was digested with NotI enzyme. The marker is Lambda Ladder (with bands every 50 Kb). The cloning enzyme is MboI and the vector is pECBAC1 (BamHI site)

## References


Arabidopsis Genome Initiative. 2000. Analysis of the genome sequence of the flowering plant Arabidopsis thaliana. Nature 408(6814): 796–815.

Arumuganathan K and Earle E. 1991. Nuclear DNA content of some important plant species. Plant Mol Biol Rep 9: 208–218.

Baird WV, Estager AS, and Wells J. 1994. Estimating nuclear DNA content in peach and related diploid species using laser flow cytometry and DNA hybridization. J Amer Soc Hort Sci 119:1312–1316.

Bennett MD and Smith JB. 1991. Nuclear DNA amounts in angiosperms. Philos Trans R Soc Lond B 334: 309–345.

Bennetzen JL, SanMiguel P, Chen M, Tikhonov A, Francki M, Avramova Z. 1998. Grass genomes. Proc Natl Acad Sci USA 95(5):1975–1978.

Bentley DR. 2006. Whole-genome re-sequencing. Curr Opin Genet Dev 16: 545–552.

Braslavsky I, Hebert B, Kartalov E and Quake SR. 2003. Sequence information can be obtained from single DNA molecules. Proc Natl Acad Sci U S A 100: 3960–3964.

Cai WW, Chen R, Gibbs RA and A Bradley. 2001. A clone-array pooled shotgun strategy for sequencing large genomes. Genome Res. 11(10):1619–1623.

Cartwright DA, Troggio M, Velasco R and Gutin A. 2007. Genetic mapping in the presence of genotyping errors. Genetics 176: 2521–2527.

Chetverina HV and Chetverin AB. 1993. Cloning of RNA molecules in vitro. Nucleic acids Res 21: 2349–2353.

Dirlewanger E, Cosson P, Poizat C, Laigret F, Aranzana MJ, Arús P, Dettori MT, Verde I and Quarta R. 2003. Synteny within the *Prunus* genomes detected by molecular markers. Acta Hort. 622:177–187.

Dirlewanger E, Graziano E, Joobeur T, Garriga-Caldere F, Cosson P Howad W, Arús P. 2004. Comparative mapping and marker-assisted selection in Rosaceae fruit crops. Proc Natl Acad Sci U S A 101(26):9891–98916.

Flachowsky H, Peil A, Sopanen T, Elo A and Hanke V. 2007 Overexpression of *BpMADS4* from silver birch (*Betula pendula* Roth.) induces early-flowering in apple (*Malus* × *domestica* Borkh.). Plant Breed 126: 137–145.

Folta KM, Dhingra A, Howard L, Stewart PJ and Chandler CK. 2006. Characterization of LF9, an octoploid strawberry genotype selected for rapid regeneration and transformation. Planta 224: 1058–1067.

Foulongne M, Pascal T, Arús P and Kervella J. 2003. The potential of *Prunus davidiana* for introgression into peach [*Prunus persica* (L.) Batsch] assessed by comparative mapping Theor Appl Genet 107:227–238.

Doebley J and Lukens L. 1998. Transcriptional regulators and the evolution of plant form. Plant Cell 10:1075–1082.

Fleischmann RD, Adams MD, White O, Clayton RA, Kirkness EF, Kerlavage AR, Bult CJ, Tomb JF, Dougherty BA, Merrick JM, et al. 1995. Whole-genome random sequencing and assembly of *Haemophilus influenzae.* Rd. Science 269:496–512.

Fredlake CP, Hert DG, Kan CW, Chiesl TN, Root BE, Forster RE and Barron AE. 2008. Ultrafast DNA sequencing on a microchip by a hybrid separation mechanism that gives 600 bases in 6.5 minutes. Proc Natl Acad Sci U S A 105: 476–481.

Gale MD and Devos KM. 1998. Plant comparative genetics after 10 years. Science 282:656–659.

Goff SA, Ricke D, Lan TH, Presting G, Wang R, Dunn M, Glazebrook J, Sessions A, Oeller P, Varma H, Hadley D, Hutchison D, Martin C, Katagiri F, Lange BM, Moughamer T, Xia Y, Budworth P, Zhong J, Miguel T, Paszkowski U, Zhang S, Colbert M, Sun WL, Chen L, Cooper B, Park S, Wood TC, Mao L, Quail P, Wing R, Dean R, Yu Y, Zharkikh A, Shen R, Sahasrabudhe S, Thomas A, Cannings R, Gutin A, Pruss D, Reid J, Tavtigian S, Mitchell J, Eldredge G, Scholl T, Miller RM, Bhatnagar S, Adey N, Rubano T, Tusneem N, Robinson

R, Feldhaus J, Macalma T, Oliphant A, Briggs S. 2002. A draft sequence of the rice genome (Oryza sativa L. ssp. japonica). Science 296(5565): 92–100.

Hall N. 2007. Advanced sequencing technologies and their wider impact in microbiology. J Exp Biol 210: 1518–1525.

Han Y, Gasic K, Marron B, Beever JE and Korban SS. 2007. A BAC-based physical map of the apple genome. Genomics 89: 630–637

Jelenkovic G and Harrington E. 1972. Morphology of the pachytene chromosomes in *Prunus persica*. Can J Genet Cytol 14: 317–324.

Joobeur T, Periam N, de Vicente MC, King GJ and Arús P. 2000. Development of a second generation linkage map for almond using RAPD and SSR markers. Genome Aug;43(4):649–655.

Jung S, Jiwan D, Cho I, Lee T, Abbott A, Sosinski B and Main D. 2009. Synteny of *Prunus* and other model plant species. BMC Genomics 10: 76.

Kunst F, Ogasawara N, Moszer I, Albert AM, Alloni G, Azevedo V, Bertero MG, Bessieres P, Bolotin A, Borchert S, Borriss R, Boursier L, Brans A, Braun M, Brignell SC, Bron S, Brouillet S, Bruschi CV, Caldwell B, Capuano V, Carter NM, Choi S.K, Codani JJ, Connerton IF, Danchin A, et al. 1997. The complete genome sequence of the gram-positive bacterium *Bacillus subtilis*. Nature. 390:249–256.

Lambert P, Hagen LS, Arús P and JM Audergon. 2004. Genetic linkage maps of two apricot cultivars (*Prunus armeniaca* L.) compared with the almond Texas × peach Earlygold reference map for *Prunus*. Theor Appl Genet 108:1120–1130.

Malnoy M, Jin Q, Borejsza-Wysocka EE, He SY and Aldwinckle HS. 2007. Overexpression of the apple MpNPR1 gene confers increased disease resistance in malus × domestica. Mol Plant Microbe Interact 20: 1568–1580.

Margulies M, Egholm M, Altman WE, Attiya S, Bader JS, Bemben LA, Berka J, Braverman MS, Chen YJ, Chen Z, Dewell SB, Du L, Fierro JM, Gomes XV, Godwin BC, He W, Helgesen S, Ho CH, Irzyk GP, Jando SC, Alenquer ML, Jarvie TP, Jirage KB, Kim JB, Knight JR, et al. 2005 Genome sequencing in microfabricated high-density picolitre reactors. Nature 437: 376–380.

McCouch S. 1998. Toward a plant genomics initiative: thoughts on the value of cross-species and cross-genera comparisons in the grasses. Proc Natl Acad Sci USA. 95(5):1983–1985.

Mitra RD, Shendure J, Olejnik J, Edyta Krzymanska O and Church GM. 2003. Fluorescent in situ sequencing on polymerase colonies. Anal Biochem 320: 55–65.

Moore J and Janick J. 1975. (eds.) Advances in Fruit Breeding. Purdue University Press, West Lafayette, IN, 623 p.

Newcomb RD, Crowhurst RN, Gleave AP, Rikkerink EHA, Allan AC, Beuning LL, Bowen JH, Gera E, Jamieson KR, Janssen BJ, Laing WA, McArtney S, Nain B, Ross GS, Snowden KC, Souleyre EJF, Walton EF and Yauk YK. 2006. Analyses of expressed sequence tags from apple (Malus × domestica). Plant Physiol 141: 147–166.

Norelli JL, Borejsza-Wysocka E, Baldo AM, Aldwinckle HS, Bassett CL, Farrell RE, Malnoy M, Lalli DA, Korban SS, Gasic K and Wisniewski ME. 2007. Functional genomic analysis of apple (Malus) ESTs associated with fire blight (Erwinia amylovora). Phytopathology 97, S185–S185.

Nyren P, Pettersson B and Uhlen M. 1993. Solid phase DNA minisequencing by an enzymatic luminometric inorganic pyrophosphate detection assay. Anal Biochem 208: 171–175.

Palmer LE, Rabinowicz PD, O'Shaughnessy AL, Balija VS, Nascimento LU, Dike S, de la Bastide M, Martienssen RA and McCombie WR. 2003. Maize genome sequencing by methylation filtration. Science 302(5653): 2115–2117.

Pennisi, E. 1998. A bonanza for plant genomics. Science. 282(5389):652–654.

Peterson, DG, Schulze SR, Sciara EB, Lee SA, Bowers JE, Nagel A, Jiang N, Tibbitts DC, Wessler SR, and AH Paterson. 2002. Integration of cot analysis, DNA cloning, and high-throughput sequencing facilitates genome characterization and gene discovery. Genome Res 12(5): 795–807.

Phipps JB, Robertson KR, Smith PG and Rohrer JR. 1990. A checklist of the subfamily maloideae (rosaceae). Can J Bot Rev (Canadienne De Botanique) 68: 2209–2269.

Rabinowicz PD, Schutz K, Dedhia N, Yordan C, Parnell LD, Stein L, McCombie WR and Martienssen RA. 1999. Differential methylation of genes and retrotransposons facilitates shotgun sequencing of the maize genome. Nat Genet 23: 305–308.

Rabinowicz PD. 2003 Constructing gene-enriched plant genomic libraries using methylation filtration technology. Methods Mol Biol 236: 21–36.

Ronaghi M, Nygren M, Lundeberg J and Nyren P. 1999. Analyses of secondary structures in DNA by pyrosequencing. Anal Biochem 267: 65–71.

Shendure J, Porreca GJ, Reppas NB, Lin X, McCutcheon JP, Rosenbaum AM, Wang MD, Zhang K, Mitra RD and Church GM. 2005. Accurate multiplex polony sequencing of an evolved bacterial genome. Science (New York) 309: 1728–1732.

Takami H, Nakasone K, Takaki Y, Maeno G, Sasaki R, Masui N, Fuji F, Hirama C, Nakamura Y, Ogasawara N, Kuhara S., Horikoshi K. 2000. Complete genome sequence of the alkaliphilic bacterium *Bacillus halodurans* and genomic sequence comparison with *Bacillus subtilis*. Nucleic Acids Res 28: 4317–4331.

Tatum TC, Stepanovic S, Biradar DP, Rayburn AL and Korban SS. 2005. Variation in nuclear DNA content in Malus species and cultivated apples. Genome 48: 924–930.

Teo G, Suziki Y, Uratsu S L, Lampinen B, Ormonde N, Hu WK, DeJong TM and Dandekar AM. 2006. Silencing leaf sorbitol synthesis alters long-distance partitioning and apple fruit quality. Proc Natl Acad Sci U S A 103: 18842–18847.

Tuskan GA, Difazio S, Jansson S, Bohlmann J, Grigoriev I, Hellsten U, Putnam N, Ralph S, Rombauts S, Salamov A et al. 2006. The genome of black cottonwood, *Populus* trichocarpa (Torr. & Gray). Science 313:1596–1604.

Velasco R, Zharkikh A, Troggio M, Cartwright DA, Cestaro A, Pruss D, Pindo M, FitzGerald LM, Vezzulli S, Reid J, Malacarne G, Iliev D, Coppola G, Wardell B, Micheletti D, Macalma TM, Facci M, Mitchell JT, Perazzolli M, Eldredge G, Gatto P, Oyzerski R, Moretto M, Gutin N, Stefanini M, Chen Y, Segala C, Davenport C, Demattč L, Mraz A, Battilana J, Stormo K, Costa F, Tao Q, Si-Ammour A, Harkins T, Lackey A, Perbost C, Taillon B, Stella A, Solovyev V, Fawcett JA, Sterck L, Vandepoele K, Grando MS, Toppo S, Moser C, Lanchbury J, Bogden R, Skolnick M, Sgaramella V, Bhatnagar SK, Fontana P, Gutin A, Van de Peer Y, Salamini F, and Viola R. High quality draft consensus sequence of the genome of a heterozygous grapevine variety. PLoS ONE. 2007 2(12):e1326

Young ND, Cannon SB, Sato S, Kim D, Cook DR, Town CD, Roe BA and Tabata S. 2005. Sequencing the genespaces of Medicago truncatula and Lotus japonicus. Plant Physiol 137(4): 1174–1181.

Yu J, Hu S, Wang J, Wong GK, Li S, Liu B, Deng Y, Dai L, Zhou Y, Zhang X, Cao M, Liu J, Sun J, Tang J, Chen Y, Huang X, Lin W, Ye C, Tong W, Cong L, Geng J, Han Y, Li L, Li W, Hu G, Huang X, Li W, Li J, Liu Z, Li L, Liu J, Qi Q, Liu J, Li L, Li T, Wang X, Lu H, Wu T, Zhu M, Ni P, Han H, Dong W, Ren X, Feng X, Cui P, Li X, Wang H, Xu X, Zhai W, Xu Z, Zhang J, He S, Zhang J, Xu J, Zhang K, Zheng X, Dong J, Zeng W, Tao L, Ye J, Tan J, Ren X, Chen X, He J, Liu D, Tian W, Tian C, Xia H, Bao Q, Li G, Gao H, Cao T, Wang J, Zhao W, Li P, Chen W, Wang X, Zhang Y, Hu J, Wang J, Liu S, Yang J, Zhang G, Xiong Y, Li Z, Mao L, Zhou C, Zhu Z, Chen R, Hao B, Zheng W, Chen S, Guo W, Li G, Liu S, Tao M, Wang J, Zhu L, Yuan L, Yang H. 2002. A draft sequence of the rice genome (Oryza sativa L. ssp. indica). Science 296(5565): 79–92.

Waterston RH, Lander ES, and Sulston JE. 2003. More on the sequencing of the human genome. PNAS 100(6): 3022–3024.

Whitelaw CA, Barbazuk WB, Pertea G, Chan AP, Cheung F, Lee Y, Zheng L, van Heeringen S, Karamycheva S, Bennetzen JL, SanMiguel P, Lakey N, Bedell J, Yuan Y, Budiman MA, Resnick A, Van Aken S, Utterback T, Riedmuller S, Williams M, Feldblyum T, Schubert K, Beachy R, Fraser CM and Quackenbush J. 2003. Enrichment of gene-coding sequences in maize by genome filtration. Science 302(5653): 2118–2120.

Zharkikh A, Troggio M, Pruss D, Pindo M, Eldrdge G, Cestaro A, Mitchell JT, Vezzulli S, Bhatnagar S, Fontana P, Viola R, Gutin A, Salamini F, Skolnick M and Velasco R. 2008. Sequencing and assembly of highly heterozygous genome of Vitis vinifera L. cv. Pinot Noir: problems and solutions. J Biotech 136(1–2): 38–43.

Zhebentyayeva TN, Swire-Clark G, Georgi LL, Garay L, Jung S, Forrest S, Blenda AV, Blackmon B, Mook J, Horn R, Howad W, Arús P, Main D, Tomkins JP, Sosinski B, Baird WV, Reighard GL, and Abbott AG. 2008. A framework physical map for peach, a model Rosaceae species. Tree Genet Genomes (online first 1614–2950).

# Index

## E



## F

## G

Printed in the United States of America